The Medicinal and Poisonous Plants of India

REV. FR. JEAN FERDINAND CAIUS

SCIENTIFIC
PUBLISHERS (INDIA)

Scientific Publishers (India)
5-A, New Pali Road, P.O. Box 91
Jodhpur - 342 001 (India)
Tel.: +91-291-2433323, 2624154
Fax.: +91-291-2613449
E-mail: info@scientificpub.com
www.scientificpub.com

Print : 2021
Reprint in India, 2012

The work has been published in the
Journal of the Bombay Natural History Society
in Vol. 37(4) to 45 (3) (1935-1944)

ISBN: 978-81-7233-750-6

Printed in India

The Medicinal and Poisonous Plants of India

REV. FR. JEAN FERDINAND CAIUS

Acknowledgement

Publisher is grateful for kind permission accorded by the Principal, St. Xavier's College Bombay for and on behalf of the Superior, Society of Jesus, Bombay, who are the copyright holders of this work

The Medicinal and Poisonous Plants of India

Rev. Fr. Jean Ferdinand Caius

The book is a compilation of 24 papers published in the Journal Bombay Natural History Society during 1935-1944 on medicinal and Poisonous plants of India by Rev. Fr. J.F. Caius, S.J. a leading authority on the subject. This is the first attempt to produce J.F. Caius, work in a book form. The book will be indispensable for those working on medicinal and poisonous plants, not only in India but throughout the world.

The work contains medicinal and poisonous plants of important families, keys for identification of genera and species, their chemical substances, medicinal and poisonous importance, distribution, ethnobotanical notes and also vernacular names of plants, almost in all the languages of the World. An index to botanical names of plants has been appended by the publisher.

FR. JEAN FERDINAND CAIUS S.J
(1877-1944)

Born in France 17 January 1877 : died at Bombay 27 July 1944."One of that distinguished company of Jesuits, men of great learning and science, who devoted their lives to service in India and who passing, have left this country richer for their labours and poorer for their loss". On coming out to India in 1895 as a Scholastic in the Society of Jesus, he joined the Madura Mission and was subsequently attached to St. Joseph's College, Trichinopoly, where among other things, he started a Natural History Museum. Fr. Caius was Professor of Chemistry at St. Joseph's College, Trichy from 1911 to 1922 when he was transferred to Bombay as director of the Chemistry Dept. of St. Xavier's College. By 1924 he had established his reputation as a biochemist and was appointed by the Government of Bombay to take charge of the Pharmacological Laboratory of the Haffkine Institute where he worked until his retirement in 1932 at the age of 55. He was Chairman of the Committee of Trustees of the Natural History Section, Prince of Wales Museum of Western India, and Honorary Secretary of Bombay Natural History Society and one of the Editors of its journal between 1938 and 1944. His obituarist, who was the Curator of the Society at the time and in close and constant touch with him, records 'It was good to work with him. He brought to his work the acumen, the clarity of thought and the critical mind of the true man of science that he was. But he also brought to it that human sympathy and understanding which endeared him to all who worked with him. In the field of science he was unable to suffer fools glady. But in the field of charity his great heart got the better of him. For this he will be loved and remembered by us'.

One important contribution by Fr. Caius was his extensive studies of the poison apparatus of snakes, and of the remedies employed against snake poisons. In India he turned his attention to the investigation of the value of various remedies in the Ayurvedic and Unani systems of medicine supposed to be effective against snake bite. After long and laborious research he and his collaborators rendered great public service by demonstrating that these widely employed so - called remedies were completely ineffective against cobra and viper poison. Fr. Caius was one of the leading authorities on the medicinal and poisonous plants of India, and students of indigenous drugs will be ever indebted to him for his revision of Kirtilkar and Basu's out-of-print work on *Indian Medicinal Plants.* Another subject that interested Fr. Caius was " Earth-eating and Salt-licks." His chemical analyses of the salt-lick earth sent to him from various parts of India, published in the Society's Journal, should be of great relevance in the scientific management of wildlife reserves.

S.H. PRATER

(from the *J. Bombay Natural History Society* 80 : 321. 1983)

CONTENTS

Palms/Palmae	1
Sedges/Cyperaceae	26
Ferns/Filices	34
Grasses/Gramineae	55
Orchids/Orchidaceae	100
Aroids/Araceae	109
Spider worts/Commelinaceae	124
Campions/Caryophyllaceae	129
Crowfoots/Ranunculaceae	137
Magnoliads/Magnoliaceae	155
Dilleniads/Dilleniaceae	158
Anonads/Anonaceae	159
Menispermads/Menispermaceae	166
Berberids/Berberidaceae	177
Spurges/Euphorbiaceae	182
Waterlilies/Nymphaeaceae	232
Poppyworts/Papaveraceae	237
Fumitories/Fumariaceae	244
Crucifers/Cruciferae	247
Capparids/Capparidaceae	267
Mignonettes/Resedaceae	277
Violets/Violaceae	278
Rockroses/Cochlospermaceae	284
Bixads/Bixaceae	285
Flacourtiads/Flacourtiaceae	287
Pittosporads/Pittosporaceae	292
Milkworts/Polygalaceae	293
Seaheaths/Frankeniaceae	295
Purslanes/Portulacaceae	295
Tamarisks/Tamaricaceae	299
Composites/Compositae	302
Flaxworts/Linaceae	377
Labiates/Labiatae	381
Dammers/Dipterocarpaceae	422
Guttifers/Guttiferae	426
Silk-cottons/Bombacaceae	433
Teas/Ternstroemiaceae	438
Tutsans/Hypericaceae	441
Water-peppers/Elatinaceae	444
Mallowworts/Malvaceae	445
Lindenblooms/Tiliaceae	473
Sterculiads/Sterculiaceae	483
Index	495

THE MEDICINAL AND POISONOUS PALMS OF INDIA.

BY

J. F. CAIUS, S.J., F.L.S.

The **Palmae** include about 200 genera with about 1,500 species, natives of tropical and subtropical regions, a few being found in the warmer temperate countries.

The medicinal and poisonous Palms of the world belong to 34 different genera:—ACANTHOPHOENIX (Mascarene Islands); ACROCOMIA (Tropical America; Brazil); ARECA (Tropical Asia; Malay-Archipelago; New Guinea; Australia); ARENGA (Tropical Asia; Malay-Archipelago; New Guinea; Australia); ASTROCARYUM (Tropical America); ATTALEA (Tropical America); BORASSUS (Tropical Africa; India, from Ceylon to the Sunda Islands); CALAMUS (Tropical and Sub-tropical Asia; Tropical West Africa; Australia); CARYOTA (Tropical Asia; Malay-Archipelago; Australia); CHAMAEROPS (Mediterranean region); COCOS (Tropical and Sub-tropical America); COPERNICIA (America); CORYPHA (Tropical Asia; Malay-Archipelago); DIPLOTHEMIUM (Brazil); ELAEIS (Tropical Africa and America); EUTERPE (Tropical America; West Indies); HYPHAENE (Africa; India); LICUALA (Indo-Malayan region; New Guinea; North Australia); LODOICEA (Seychelles Islands); LOXOCOCCUS (Ceylon); MAURITIA (Northern Brazil; Guiana; West Indies); MAXIMILIANA (Tropical Brazil; Guiana; Trinidad); METROXYLON (Malay-Archipelago; New Guinea); NANNORHOPS (India; Afghanistan; Southern Persia); NEODYPSIS (Madagascar); NIPA (Tropical Asia; New Guinea; Australia); OENOCARPUS (Tropical America); ORANIA (Malay-Archipelago; Papua); PHOENIX (Sub-tropical Africa; Tropical Asia); PINANGA (Malay-Archipelago); RAPHIA (Tropical Africa; Madagascar; Tropical America); SERENOA (Florida, Southern California); TRACHYCARPUS (Northern India; Burma; Northern China; Japan); WALLICHIA (East India).

The medicinal and poisonous Palms of India belong to 24 genera, either indigenous or introduced:—ACANTHOPHOENIX, ARECA, ARENGA, ATTALEA, BORASSUS, CALAMUS, CARYOTA, CHAMAEROPS, COCOS, COPERNICIA, CORYPHA, ELAEIS, HYPHAENE, LICUALA, LODOICEA, LOXOCOCCUS, METROXYLON, NANNORHOPS, NIPA, PHOENIX, PINANGA, RAPHIA, TRACHYCARPUS, WALLICHIA.

A. Leaves pinnatisect, leaflets free with reduplicate sides or confluent as a plaited limb. Flowers monoecious or dioecious:—
   1. Male flowers minute, solitary or binate towards the tips of the branches, 3- or 6-androus; female larger, solitary ... ... ARECA
   2. Male flower one on each side of a female, stamens 6 or more:—
      a. Leaves unarmed ... ... PINANGA.
      b. Leaves with long spines ... ... ACANTHOPHOENIX.
   3. Male flowers 9-12-androus, female petals with valvate tips ... ... LOXOCOCCUS.

4. Male calyx tubular. Stamens 6. Albumen equable ... WALLICHIA.
5. Male sepals 3. Stamens many. Albumen equable ... ARENGA.
6. Male sepals 3. Stamens many. Albumen ruminate ... CARYOTA.
7. Male flower in catkin-like branches of a drooping androgynous spadix; female capitate at the apex of the spadix ... ... NIPA.

B. Leaves pinnatisect; leaflets with implicate sides. Spadices interfoliar; spathe solitary. Flowers dioecious ... ... PHOENIX.

C. Leaves flabelliform, orbicular or cuneiform. Spadices interfoliar; spathes many. Flowers usually bisexual:

1. Fruit with a smooth pericarp. 3 free carpels; each carpel with its own style or stigma:—
   - *a.* Seed curved-reniform; albumen with a single excavation near the raphe ... ... TRACHYCARPUS.
   - *b.* Seed erect elliptic; albumen ruminate ... ... CHAMAEROPS.
2. Fruit with a smooth pericarp. 3 appressed carpels, or carpels united where they meet; stigma one:
   - *a.* Spadix terminal; remains of stigma at the base of the fruit:
     - ¶ Style short. Embryo apical ... ... CORYPHA.
     - ¶¶ Style subulate. Embryo dorsal ... ... NANNORHOPS.
   - *b.* Spadix axillary lateral; remains of stigma apical:
     - ¶ Berry drupaceous; endocarp hard brittle. Albumen equable ... ... LICUALA.
     - ¶¶ Berry with a soft membranous endocarp. Albumen deeply ruminate ... ... COPERNICIA.

D. Leaves pinnatisect; leaflets with reduplicate sides. Spathes usually many. Fruit clothed with reflexed shining closely imbricating scales:—

1. Stem short, stout, annulated. Leaflets linear with the midrib and edges spinulose. No common spathe, but many small incomplete sheaths ... ... RAPHIA.
2. Stem elongate. Leaflets acuminate, quite entire, nerves parallel. Spathes tubular, persistent:—
   - *a.* Polycarpic spadices axillary ... ... CALAMUS.
   - *b.* Monocarpic spadix very large, terminal ... METROXYLON.

E. Leaves flabelliform. Spadices interfoliar; spathes numerous, sheathing. Flowers dioecious; males minute, sunk in the cavities of the catkin-like branches:

1. Stamens 6. Fruit mostly with 3 stones; pericarp thinly fleshy, seeds sinuate ... ... BORASSUS.
2. Stamens 6. Fruit by the abortion of the carpels, unilocular with one central, woody, ovate stone; pericarp fibrous with a shining epiderm ... ... HYPHAENE.
3. Stamens 8. Fruit generally with 1 bilobed kidney-shaped stone. Seeds bilobed ... ... LODOICEA.
4. Fruit large, ovate or oblong, with a dry fibrous outer covering, red or greenish-brown. Seeds usually 3, sometimes 4 or 5 ... ... ATTALEA.
5. Fruit obovoid, terete or trigonous, 1-seeded ... COCOS.
6. Fruit ovoid or obovoid, 1-3-seeded. Pericarp spongy and oily ... ... ELAEIS.

ACANTHOPHOENIX.

This genus numbers only four species, natives of the Mascarene Islands where they grow abundantly. Two of them, *A. rubra* with a globose fruit and *A. nobilis* with an oblong-cylindric fruit, are cultivated in some of our Indian gardens.

Acanthophoenix rubra Wendl. (= *Areca rubra* Bory) is a very elegant palm. Its root is a popular diuretic at La Reunion, and the light red decoction is a favourite cooling drink.

French: Palmiste rouge—; *German*: Rote Dornen-Areka.

ARECA.

The genus comprises about 40 species. They are very ornamental and graceful palms inhabiting Tropical Asia, the Malay Archipelago, New Guinea, and Australia.

Only one species, *A. catechu*, is known to be medicinal. Its seed, the betel-nut, is a masticatory of great antiquity with most of the Asiatic races. It is used as an astringent, stimulant and anthelmintic in China, Indo-China, the Malay Archipelago, Australia, Melanesia, Polynesia, the West Indies, Madagascar, La Reunion, and Guiana. It was officinal in the pharmacopoeias of Germany, Sweden, and Switzerland.

Mention must, however, be made of two other species, *A. concinna* and *A. nagensis*, the fruits of which are used as substitutes for the betel-nut.

1. Stamens 6. Fruit 1½-2 in., smooth, orange or scarlet ... *A. catechu*.
2. Stamens 6. Fruit 1½ in., umbonate, reddish-yellow . *A. concinna*.
3. Stamens 3. Fruit 1 in., narrowed at both ends ... *A. nagensis*.

1. **Areca catechu** Linn. The Betel Nut Palm is cultivated exclusively within the moist tropical tracts that fringe the coast of India, and practically within a belt of land that, with a few exceptions, does not extend inland for more than 200 miles. It rarely ascends to altitudes of 3,000 ft. and gradually disappears, even from the littoral area, as localities are entered where the duration of the dry hot months equals or exceeds the monsoons. It flourishes in the dry plateau of Mysore, Kanara and Malabar. It is usually seen as a garden plant, but occasionally and in certain localities, especially of Western and Southern India, of Ceylon and of Burma, it is grown in special gardens. In Eastern and Northern Bengal, in some portions of Assam and in Ceylon, its cultivation has assumed still greater dimensions.

In some parts of India a decoction of the ROOT is a reputed cure for sore lips.

The root is used in Cambodia for the treatment of liver complaints. It is mixed with numerous other drugs, and made into a decoction to be administered in jaundice.

Malay women use the young green SHOOTS as an abortifacient in early pregnancy.

The dry expanded PETIOLES serve as excellent ready-made splints for fractures.

In some parts of India the juice of the young tender LEAVES mixed with oil is applied as an embrocation in cases of lumbago.

In Cambodia the leaves are considered a good remedy for cough. Mixed with other drugs they are given in bronchitis after maceration in alcohol or vinegar. For lumbago they are reduced to pulp in a mortar and applied hot to the loins.

In Malabar an inebriating lozenge is prepared from the SAP of the tree.

In Malaya the green FRUIT in its unripe state is sometimes used as a poison in combination with opium.

The fact that the use of fresh betel-nuts gives rise to a sensation of strangling and giddiness is well known in the East, and it has also been observed that the nuts of certain trees in most betel plantations retain their poisonous properties when dried. These trees cannot be distinguished from the others, so that not unfrequently accidents happen from their nuts becoming mixed with the produce of the plantation before their presence has been detected. The poisonous properties are destroyed by heat, and consequently many people use the cooked or red betel-nuts of commerce to avoid the possibility of accident. When such nuts have been eaten by mistake, salt or lime juice, or acid pickles are the best remedies.

Yunani practitioners consider the NUT digestive, astringent, diuretic and emmenagogue. They recommend it as a cardiac and nervine tonic; and they use it as an astringent lotion for the eyes.

Hindu medical writers describe the unripe nuts as laxative and carminative; the fresh nuts as intoxicating, productive of giddiness, and harmful to the eyesight; when dried, they are said to sweeten the breath, strengthen the gums, remove bad taste from the mouth, and produce a stimulant and exhilarant effect on the system. Their use is recommended in calculous affections and other urinary disorders, and as an aphrodisiac; for the latter purpose a confection is made by boiling the nuts in milk and adding a number of aromatic and stimulant substances.

The nut is commonly used as an astringent for bleeding gums; in the Punjab, women employ it both internally and locally for stopping discharges from the vagina. In Ceylon it is scraped and applied externally to ulcers.

The seed, reduced to charcoal and powdered, is said to be an excellent dentifrice.

The water in which unripe nuts have been boiled is often given to women along with powdered red betel and other spices after confinement as a gentle stimulant.

The nut is useful in checking the heartburn of pregnancy. Control experiments made with tincture of catechu have shown the superiority of the nut, and would seem to demonstrate that this is not merely due to astringent action; it is quite possible that its property as a nervine stimulant enhances its utility.

In Cambodia the unripe fruit is considered a good remedy for diarrhoea. Five green nuts are heated over a slow fire until they become burning hot inside; they are then cut into slices and, while still hot, made into about a half-pint of infusion to be taken in the course of the day. In cases of dysentery the nut is combined with opium; a pellet of opium about the size of a large grain of maize for every nut. Seven incisions are made obliquely on the surface of the nut and they are filled with opium. The nut is then warmed over a very slow fire and whenever the opium swells it is forced back into the cuts until the whole mass has sunk into the flesh of the fruit. The cooking operation over, the nut is cut into slices and thrown in a pint of water. The whole is then boiled down to one third, and this is to be taken in the course of the day.

Concerning the immature fruit Koman writes: 'On the suggestion of a patient who informed me that he was cured of his consumption by a preliminary administration of the fresh juice of tender arecanuts followed by a prolonged use of pellitory, this drug was given to eight patients. The dose suggested was five ounces of the juice to be gradually sipped in the morning in the course of an hour or two. None of the patients could stand this dose, as it produced violent vomiting and purging. The dose was consequently reduced to half an ounce. Even half an ounce had a nauseating effect and generally produced two or three evacuations. I cannot understand why this preliminary treatment by this special method of purgation is considered necessary. As stated above the juice of the tender nuts acts in small doses as a laxative.'

In Travancore the nuts are variously prepared for use. Those that are used by families of rank are collected while the fruit is tender; the husks or the outer pod is removed; the kernel, a round fleshy mass, is boiled in water; in the first boiling of the nut, when properly done, the water becomes red, thick and starch-like, and this is afterwards evaporated into a substance like catechu. The boiled nuts being now removed, sliced and dried, the catechu-like substance is rubbed to the same and dried again in the sun, when they become of a shining black, ready for use. Whole nuts, without being sliced, are also prepared in the same form for use among the higher classes, while ripe nuts as well as young nuts in a raw state are used by all classes of people generally; and ripe nuts preserved in water with the pod are also used.

In Mysore the nuts are taken as they come from the tree, and boiled for some hours in an iron vessel. They are then taken out, and the remaining water is inspissated by continual boiling. This process furnishes *Kossa*, or most astringent *terra japonica*, which is black, and mixed with paddy-husks and other impurities. After the nuts are dried they are put into a fresh quantity of water and boiled again, and this water being inspissated like the former, yields the best or dearest kind of catechu, called *coony*. It is yellowish brown, has an earthy fracture, and is free from the admixture of foreign bodies.

In the north of China, small pieces of the nut are boiled and the decoction is taken as a domestic remedy in various visceral affections.

The powdered nut, in doses of ten or fifteen grains every three or four hours, has been used with satisfactory results in the relaxed condition of the bowels which sometimes occurs in tropical climates. Large doses, six drachms to one ounce, however, produce griping and irritation; and loose motions may start as a result of such irritation.

In some parts of China the nuts, bruised and powdered, are mixed with the green food given to horses, and they are thus considered a preventive against diarrhoea.

In Europe betel-nuts have been used as an anthelmintic for

tapeworm and as an astringent; and in veterinary practice their reputation as a vermifuge is well established.

In India and China the powdered nut is generally given for the expulsion of tapeworm in the dose of 4 to 6 drachms, taken in milk. It is said to be efficacious against round as well as tapeworms. Powell thought so highly of its anthelmintic properties that he expressed the opinion that the habit of chewing arecanut among the inhabitants of certain countries where intestinal parasites are common, is a protective habit instinctively acquired on account of its prophylactic value against these parasites. Waring, however, declared that the nut 'can hardly have any claim to this character, as amongst the Hindu and Burmese, who use it habitually as a masticatory, intestinal worms (lumbrici) are almost universally met with'.

Bentley (1904) and Schueffner (1912) treated a number of cases of hookworm disease with half to one ounce doses of the powdered nuts with little effect. Caius and Mhaskar (1924) gave four drachms of the recently-dried seeds in the form of a powder without any preliminary preparation and without any after purgative in cases of roundworm and hookworm infections. The patients passed one to three semi-solid stools, but no worms were expelled. The powdered fresh nut produced a stronger irritant effect on the intestine, but no worms were expelled. Chopra and Chandler (1928) believe that the chewing of betel-nut and betel-leaf does influence the number of hookworms harboured. This result is not, however, attributable to any anthelmintic power of the juice, which is not swallowed; but to the constant spitting which tends to eliminate the immature hookworms while making their way from the trachea to the oesophagus. The chewing of tobacco has a similar effect, and in some places is credited with anthelmintic power.

Sushruta recommends arecanut as an antidote to snake and scorpion venoms. By injecting cobra venom into dogs and scorpion venom into mice Mhaskar and Caius (1931, 1932) have shown experimentally that the nut has no preventive, antidotal, or therapeutic value.

The seeds were first examined by Bombelon (1886) and later by Jhans (1888-1891), who isolated, in addition to choline, the four alkaloids arecoline, arecaidine, arecaine, and guvacine, of which arecaine is now known to be identical with arecaidine. Emde (1915) added to these arecolidine, and K. Hess (1918) guvacoline.

Of these alkaloids arecoline alone exhibits markedly toxic properties. It belongs to the nicotine-pilocarpine group and acts on the central and peripheral nervous system producing paralysis, which may be preceded by convulsions. With nicotine the central and with pilocarpine the peripheral action is more marked, whilst both are about equal in arecoline. Arecoline hydrobromide is recognised in the pharmacopoeias of France, Germany, Hungary, Russia, Spain, Sweden, Switzerland and Turkey; it is used internally in small doses as a sialogogue and diaphoretic, and as an anthelmintic. It has also been employed like physostigmine to produce contraction of the pupil.

Amboina: Buah, Puah—; *Andamans*: Ahbuddah, Ahpurruddah—; *Annam*: Cay cau, Cay cau gia—; *Arabic*: Fofal, Fufal—; *Assam*: Tambul—; *Banda*: Pua—; *Bengal*: Gua, Supari—; *Burma*: Kun, Kunsi, Kuntheebin, Kwamtheebeng—; *Cagayan*: Bua—; *Cambodia*: Dom sla, Sla—; *Canarese*: Adake, Adaki, Adike, Betta, Bettadike, Bette, Chikaniyadike, Chikke, Cotadike, Kaungu, Khhapura, Poga, Puga, Pugiphala, Tambula—; *Caniarines*: Banga—; *Cantonese*: Pan Long—; *Chinese*: Ping Lang—; *Deccan*: Supari, Supyari—; *Dutch*: Arecapalmboom, Pinang—; *English*: Areca Nut Palm, Areca Palm, Betel Nut Palm, Betel Nut Tree, Betel Palm, Cashoo Nut Tree, Catechu Palm, Catechu Tree, Drunken Date Tree, Fasel Nut, Fauselnut Tree, Medicinal Cabbage Tree, Indian Nut Tree, Pinang Palm, Supari Palm—; *French*: Arec, Arec bétel Arec cachou, Arec de l'Inde, Arèque, Aréquier, Pinangue—; *French Guiana*: Arec—; *German*: Arecapalme, Arikapalme, Betelnusspalme, Betelpalme, Catechupalme, Katechupalme, Kaupalme, Pinangpalme—; *Guam*: Pugua—; *Gujerati*: Hopari, Phophal, Sopari—; *Hindi*: Supari, Suppari, Supyari—; *Ilocano*: Boa—; *Italian*: Areca—; *Java*: Bhunghanapenang, Jambe, Jebug—; *Konkani*: Fufal, Maddi, Supari—; *Laos*: Kok mak—; *La Reunion*: Arec, Pak—; *Malay*: Pinang—; *Malaya*: Chiniping—; *Malayalam*: Atekka, Chempalukka, Chhonta, Kalunnu, Kamuka, Kavunnu, Kazhangu, Khhapuram, Kramukam, Pakka, Pugam—; *Marathi*: Pophali, Pung, Supari—; *Mundari*: Kasailidaru—; *New Britain*: Bue—; *Pampangan*: Luyos—; *Pelew Islands*: Bua—; *Persian*: Girdchob, Popal, Pupal—; *Philippines*: Bongapalo, Bongasantol, Lugos, Mangupod—; *Porebunder*: Hopari—; *Portuguese*: Areca, Arequeira—; *Russian*: Areka, Kapustnaya palma—; *Sanskrit*: Akota, Chhataphala, Chikkana, Dirghapadapa, Dridhavalkala, Ghonta, Gopadala, Guvaka, Kapitana, Karamatta, Khapura, Kramuka, Puga, Pugi, Rajatala, Suranjana, Tambula, Tantusara, Valkataru—; *Santali*: Gua—; *Sinhalese*: Puvakka, Puwak—; *Solomon Islands*: Boa—; *Spanish*: Arequiero—; *Swedish*: Areka—; *Tagalog*: Bonga, Bongangmatulis, Bunga, Mangipod, Sacsic—; *Tamil*: Kamugu, Kandi, Kiramugam, Kugagam, Pakku, Pakkuppanai, Pugam, Tuvarkkay—; *Telugu*: Chikinamu, Chikini, Gautupoka, Khapuramu, Kolapoka, Kramukamu, Oppulu, Oppuvakkulu, Poka, Prakka, Pugamu, Vakka—; *Tulu*: Kangu—; *Urdu*: Supari—; *Uriya*: Gua, Pugo, Supari Trinodrumo—; *Visayan*: Bonga—.

2. **Areca concinna** Thw. is endemic in Ceylon and is to be found in the forests of the moist low country; Subarayamuwa, Reigam Korale, Pasdun Korale. It is known as *Len-teri*. It is occasionally planted; the fruits, however, which are also chewed with betel, like those of *A. catechu*, are generally obtained from wild trees.

3. **Areca nagensis** Griff. is found in the Naga Hills, up to 800 ft., very scarce, usually on high situations on river sides. The Nagas and Abors use its fruit as a substitute for the betel-nut.

Naga: Talpat—; *Singpho*: Tongtau—.

ARENGA.

This genus comprises about 10 species of tall, stout palms, distributed throughout Tropical Asia, the Malay Archipelago, New Guinea, and Australia.

A. Engleri Becc. and *A. saccharifera* Labill. are used medicinally by the Chinese.

A. obtusifolia Mart. and *A. saccharifera* Labill. are used as poisons by the Malays. They are both cultivated in India.

1. Leaflets bifarious ... ... *A. obtusifolia.*
2. Leaflets 4-5-fariously fascicled ... ... *A. saccharifera.*

1. **Arenga obtusifolia** Mart. (=*A. Westerhoutii* Griff.) is a native of the Malay Peninsula—Naning. Penang—cultivated in India.

The ripe fruit is an irritant to the mucous membranes. It causes an acute swelling of the mouth and fauces when taken internally, and this is probably due to the mechanical irritation of needle crystals. The juice obtained from it is used by the Malays to poison their enemies. The pulpy part of the fruit is boiled and crushed; and the juice, after straining, is administered in coffee. It is said to cause dyspnoea and restlessness. In the Philippine Islands it is used by the Tagals for poisoning fish.

Malacca: Anooee Kutaree—; *Penang*: Langkab, Langcap—.

2. **Arenga saccharifera** Labill. Sago Palm (=*Saguerus rumphii* Roxb.=*Borassus gomutus* Lour.) is a beautiful and magnificent palm found in Assam, Martaban and Tenasserim, occasionally on the Pegu Yoma; commonly cultivated in India. It is distributed throughout the Malay Peninsula and Archipelago, and ascends in Java up to 4,000 ft.

In Cambodia the root is considered stomachic and pectoral. It is said to be very serviceable in bronchitis marked with profuse expectoration.

The petioles are credited in Cambodia with diuretic and antithermic properties. They are prescribed in chronic paludism with enlargement of the spleen.

The juice of the fleshy outer covering of the fruit is highly stimulating and corrosive. If applied to the skin it causes great pain and inflammation. It is said that the inhabitants of the Moluccas were in the habit of using in the defence of posts during the war a liquor obtained by the maceration of this fruit, which the Dutch appropriately called 'hell-water'. It is used by the Malays to poison their enemies. The pulpy part of the fruit is boiled and crushed, and the juice, after straining, is given in coffee.

Burma: Taungong, Toungong—; *Cambodia*: Thnotnhi—; *Chinese*: So Mu Mien—; *Dutch*: Aren, Arenboom, Sreng, Arengboom, Arengpalm, Arenpalm, Gamoetoeboom, Gomoetipalm, Gemoetoepalm, Sagoeweerpalm, Sagueerboom, Sagueerpalm, Saguweerpalm, Suikerboom—; *English*: Areng Palm, Gomuti Palm, Malay Sago Palm, Sago Palm, Sugar Palm—; *French*: Anou, Arbre au sagou, Aren à sucre, Areng, Gomonto, Gomuti, Lantar, Lontar, Palmier areng, Palmier rondier, Palmier à sucre, Rondier—; *German*: Echte Zuckerpalme, Gomutipalme, Sagwirepalme, Zuckerpalme—; *Ilocano*: Ratipan—; *Java*: Aren, Buwah atap, Duk, Kolang Kaling, Lirang—; *Madura Island*: Dhuk, Edhuk—; *Malacca*: Gumuti—; *Malay*: Anan, Berkat, Kabong—; *Pangasinan*: Anibung—; *Philippines*: Baru—; *Sunda Islands*: Kawung—; *Tagalog*: Cauon, Iroc, Pugahan—; *Tamil*: Kichilippanai, Kumudippanai, Segovarisi—; *Visayan*: Bahi, Hibioc, Hidioc, Ibioc, Idioc—.

ATTALEA.

The genus is Tropical American. It consists of about 23 species distributed in various parts of South America, especially in the vicinity of the Amazon and its tributaries, from the level of the sea to a height of 4,000 ft. above it.

Attalea speciosa Mart. is the *Uanassu* Palm of Brazil and Guiana, cultivated in India. The oil from the fruit is much used in Guiana in the treatment of rheumatic pains.

BORASSUS.

Martelli (1913) has proved conclusively that there are at least 7 species belonging to this genus. They inhabit Tropical Africa, India, from Ceylon to the Sunda Islands, and Northern Australia.

B. flabellifer var. *aethiopum* Warb. is used medicinally in the Gold Coast.

Borassus flabellifer Linn. The Palmyra Palm is a native of India. It extends in North-West India as far north as Aligarh and Shahjehanpur. Isolated trees in gardens are found in Rohilkhand and the Upper Ganges Doab, as far as Saharanpur. Also on both sides of the Persian Gulf, attaining there about the same latitude as in North-West India. Immense groves of it are found on the Malabar Coast, extending from Cape Comorin through Travancore, Calicut, Goa, and the Bombay Presidency, on through Gujerat and, some distance up, on the banks of the Indus. But what are emphatically called the Palmyra regions may be included in a line extending along the Coromandel coast from Cape Comorin to Madras, including the northern portion of Ceylon, and from Madras all along, taking in a considerable belt of the coast between that and Point Palmyras, and then passing up to Gaya. After that the line should be carried on about due east until it reaches Ava, below which, on the banks of the Irrawaddy, there are immense groves of this palm.

The Palmyra Palm is found in various parts of the mountain district of Ceylon, including the vicinity of Kandy and Badulla at elevations of 1,680 and 2,450 ft., respectively, having a mean annual temperature of 74° at the former and 72° at the latter.

Although immense groves are found on the banks of the Irrawaddy, from the sea coast up to nearly as far as Ava or Amarapura in Burma, and as far inland as Gaya in Bengal, and in isolated patches all round and through Ceylon, still the most congenial places for the favourable development of the Palmyra Palm will be found in low sandy plains scarcely elevated above the level of the sea, and where they are exposed to the burning sun, and the force of at least one of the monsoons. Such are Jaffna, with the surrounding islands, and other portions of the Northern Province of Ceylon, the District of Tinnevelly, with portions of the Madura Collectorate, and portions of the Madras and Bombay Presidencies.

The therapeutic properties of the various parts of the Palmyra are described in detail in Sanskrit and Persian medical works. They may easily be gathered from the following records of popular therapy.

The ROOT is considered cooling and restorative. The expressed juice of the young root is used in cases of gastric catarrh and to check hiccup.

In Cambodia the root is said to be a good diuretic and anthelmintic; it is much employed as a cure for gonorrhoea. In Guinea the young roots are boiled in water and the decoction is given in diseases of the respiratory tract.

The decoction of the BARK, with a little salt added to it, is a good astringent gargle for strengthening the gums and the teeth.

Reduced to charcoal and pulverised, the bark makes a good dentifrice.

In some parts of Southern India an extract of the green LEAVES is used internally in secondary syphilis.

The light-brown cotton-like substance from the outside of the base of the leaves is employed in Ceylon as a styptic for arresting haemorrhage from superficial wounds.

The ash of the SPATHE is an antacid in heartburn. It acts as a powerful blister, and is applied on enlarged liver and spleen in combination with some other demulcents.

In Cambodia the SPADIX is considered anthelmintic, diuretic, and antithermic. It is useful in abdominal troubles complicated with swelling and fever, more especially in chronic malarial poisoning with enlargement of the spleen.

The ash of the spadix is given internally in bilious affections, and is largely used as an antiperiodic in many parts of India.

The fresh juice of the plant or TODDY is cooling, sweet and useful in inflammatory dropsy, and gastric catarrh. It acts as a laxative when taken regularly for several mornings. It is helpful in checking hiccup. It is a good diuretic in gonorrhoea, and a good tonic in emaciation or phthisis. It is used as an adjunct to stimulating drugs in the low stages of intermittent and remittent fevers.

The juice is used as a stimulant and antiphlegmatic. A useful stimulant application, called toddy-poultice, is prepared by adding fresh-drawn toddy to rice-flour till it has the consistence of a soft poultice, and this being subjected to a gentle fire, fermentation takes place. This, spread on a cloth and applied to the parts affected, acts as a valuable application to gangrenous ulcerations, carbuncles, and indolent ulcers.

The fermented juice is intoxicating. Slightly fermented it is prescribed in diabetes. The Ayurvedists of Southern India prescribe it also in consumptive cases.

The pulp of the ripe FRUIT is applied externally in skin diseases. The kernel is bruised and applied to the sting of scorpions; but, Caius and Mhaskar have shown experimentally that this has no preventive, antidotal or therapeutic value.

The gelatinous contents of the unripe SEEDS is considered cooling and restorative. It is used as a remedy for nausea and vomiting.

The embryo of the germinating seed is said to be cooling, nutritive and diuretic.

In Cambodia the YOUNG PLANT is given in gonorrhoea and is, moreover, considered antibilious and antidysenteric.

The TERMINAL BUD of the tree is credited with the same therapeutic properties as the embryo of the germinating seed, and is considered cooling, nutritive, and diuretic.

Palm SUGAR is antibilious and alterative; it is used in hepatic disorders and gleet. Candy is given in coughs and pulmonary affections.

The sugar from the Palmyra Palm is considered a valuable antidote in Cambodia. The Khmers say it is a specific for *Strychnos* poisoning.

Bengal: Tal, Talgachh—; *Burma*: Tan—; *Cambodia*: Thnot chhmoul—; *Canarese*: Karitale, Ole, Oleya, Pane, Tala, Tale, Tali, Trinaraja—; *Ceylon*: Pannamaram—; *Deccan*: Taarkdizaar—; *Dutch*: Jagerboom, Weingeevende palmboom—; *English*: Brab Tree, Char Palm, Desert Palm, Fan Palm. Palmyra Palm—; *French*: Cocotier de mer, Rondier, Rondier éventail—; *Fulah*: Doubé—; *German*: Palmyrapalme—; *Gujerati*: Tad—; *Hindi*: Tal, Tar, Tarkajhar—; *Konkani*: Tadmadd—; *Malayalam*: Ampana, Eta, Karimpana, Pana, Talam, Trinarajan—; *Malinke*: Cébé—; *Marathi*: Tad, Tamar—; *Porebunder*: Mototad—; *Portuguese*: Palmeira macha brava—; *Sanskrit*: Asavadru, Bhumipishacha, Chirayu, Dhvajadruma, Dirghadru, Dirghapadapa, Dirghaskanda, Dirghataru, Drumashreshtha, Drumeshvara, Guchhapatra, Karapatravan, Lekhyapatra, Madadhya, Madhurasa, Mahonnata, Patri, Shataparva, Tala, Taladruma, Tamsi, Tantugarbha, Tantuniryasa, Taruraja—; *Sinhalese*: Tal, Talgaha—; *Soussou*: Kanké—; *Tamil*: Anbanai, Edagam, Karadalam, Karambanai, Nilam, Nungu, Panai, Pondai, Pondu, Pul, Purbadi, Puttrani, Sattruppanai, Talai, Talam, Tali—; *Telugu*: Karatalamu, Namatadu, Pentitadu, Potutadu, Tadu, Trinarajamu—; *Tulu*: Ole, Tari—; *Urdu*: Tad—; *Uriya*: Talo, Tanlo, Trinorajo—.

CALAMUS.

This genus numbers over 200 species inhabiting tropical and sub-tropical Asia, Malaya, the Philippine Islands, New Guinea, and Australia; a few are met with in tropical Africa.

C. margaritae Hce. is used medicinally in China; *C. rotang* Linn. and *C. verus* Lour. in Annam and Cambodia.

C. verus also enters into the composition of poisons prepared by the Malays for arrows and blowpipe darts.

The resin from *C. draco* Willd. is officinal in Portugal.

Three out of the 52 Indian species are credited with medicinal properties.

1. Leaflets many, equidistant, linear uniform ... ... *C. rotang*.
2. Leaflets in distant opposite groups of 3-5 ... ... *C. travancoricus*.
3. Leaflets in very distant groups of 3 ... ... *C. rheedei*.

1. **Calamus rotang** Linn. The Rattan is distributed over the Central Provinces, the Deccan, the Carnatic, and Ceylon; not in Bengal.

The ROOT in combination with other drugs is prescribed for snake-bite in Bhavaprakasha, Rasaratnakara, Yogaratnakara, and Sharangdhar-samhita. It is also prescribed for scorpion-sting in Rasaratnakara, and Vaidyavinoda. Mhaskar and Caius have, however, proved conclusively that the drug either single or in combination does not in any way counteract the poisoning effects of either snake or scorpion venom.

The root is much used medicinally in Cambodia where it is credited with antidysenteric, antibilious, hypotensive, tonic, febrifuge, and depurative properties. In veterinary practice it is employed as a tonic and aperient.

The WOOD is given as a vermifuge in Annam.

Annam: Cay may—; *Bengal*: Bet, Beta, Chachibet—; *Bombay*: Bet—; *Cambodia*: Lompeak, Rompeak—; *Canarese*: Betta, Habbe, Nagabetta—; *Central Provinces*: Prabba, Pepa—; *Dutch*: Rotting, Rottinggewas, Rottingriet runt, Runtstek, Spaansch riet—; *English*: Cane, Chair-bottom Cane, Common Rattan, Rattan, Rotang, Slender Rattan, Water Rattan—; *French*: Canne de Bengale, Canne épineuse, Canne à main, Jonc de l' Inde, Ratin, Rotang, Rotin, Rottain—; *German*: Palmriet, Rattangpalme, Rotang, Rcttang, Rotting, Spanisches Rohr, Steinrottang, Stuhlrohr—; *Gujerati*: Natar—; *Hindi*: Bet, Beta, Chachibet—; *Malay*: Rottamm, Rotan—; *Malayalam*: Chural, Nirvanni, Purampu—;

Marathi: Beta, Veta—; *Persian*: Bed—; *Sanskrit*: Abhrapushpa, Dirghapatraka, Dirghavalli, Gandhapushpa, Kalana, Latavamsa, Manjari namra, Nichula, Ratha, Rathabhra, Shita, Sushena, Vanira, Vanjula, Vetasa, Vetasi, Vetra, Vidula—; *Sinhalese*: Wewel—; *Spanish*: Caña de Bengala, Caña de Indias—; *Tamil*: Arini, Mellisuppirambu, Nirvanji, Pirambu, Pisin, Sadi, Sural, Suvedagandam, Suvedam, Vaniram, Vanjikkodi, Vedasam, Vettiram—; *Telugu*: Bethama, Bettamu, Niruprabba, Pemu, Prabba, Prabili, Sannabettamu—; *Tulu*: Suralbetta—.

2. **Calamus travancoricus** Bedd. is found in the Deccan peninsula: from Malabar to Travancore. The tender leaves are used in dyspepsia, biliousness, and ear troubles; they are considered anthelmintic.

Canarese: Nayibetta—; *Malayalam*: Cheruchural, Kattuchural—.

3. **Calamus rheedei** Griff. grows in Malabar where its powdered dried seed is applied to ulcers.

Malayalam: Kattuchural—.

CARYOTA.

The 12 species which compose this genus are distributed throughout Tropical Asia, Malaya and Australia.

Of the three indigenous species two are used medicinally

1. Leaves 18-20 ft. long ... ... *C. urens*.
2. Leaves 4-9 ft. long ... ... *C. mitis*.

1. **Caryota urens** Linn. The Fish-tail Palm is found all over India in the forests of the western and eastern moist zones. It is common in Burma, Bengal and Orissa, ascending in Sikkim to 5,000 ft. On the Western Ghats it extends to near Mahableshwar. It is reported to abound in the south-eastern corner of Aheree in the Chanda District.

According to Ayurvedists the NUT is acrid and cooling; it allays thirst and fatigue.

The nut is used as an application to the head in cases of hemicrania.

A glass of the freshly drawn TODDY, taken early in the morning, acts as a laxative; it is thus much used by the people of Malabar.

Assam: Baraflawar—; *Bombay*: Birlimhad, Birlimhar—; *Burma*: Kimbo, Minbaw, Minbo—; *Canarese*: Bagani, Baini, Bayne—; *Deccan*: Marikajhar—; *Dutch*: Jagerieboom, Nieboom, Sagueerboom, Wilde Sagueerboom—; *English*: Bastard Sago Palm, East Indian Wine Palm, Elephant's Palm, Fish-tail Palm, Ghaut Palm, Hill Palm, Indian Sago-Palm, Jaggery Palm, Kittul Tree, Malabar Sago Palm, Mhar Palm, Toddy Palm, Wine Palm—; *French*: Caryote brûlant, Faux sagonier de l' Inde, Palmier céleri—; *German*: Bastardsagopalme, Ostindische Brennpalme, Sagopalme—; *Gujerati*: Shankrajata, Shivajata—; *Hindi*: Mari, Marikajhad—; *Khond*: Sarta—; *Konkani*: Birlamadd, Birlimad—; *Lepcha*: Runbong, Simong—; *Magahi*: Hlyamban—; *Malayalam*: Anappana, Chundapana, Chuntappana, Irampana, Kalapana, Vainavu—; *Marathi*: Ardhimpari, Ardhisupari, Berli, Berlimad, Berlimada, Berlimhar, Bherawa, Bherlamuda, Bherlimad, Bhirlimahad, Birli, Mad—; *Philippines*: Cabonegro, Sagu, Taguipan; *Porebunder*: Mervajata—; *Portuguese*: Palmeira brava—; *Sanskrit*: Dhoajavriksha, Dirgha, Mada, Madadruma, Madyadru, Madyadruma, Mohakari, Rajju, Vitanaka—; *Saora*: Jivalaggu—; *Sinhalese*: Kittul, Nepora—; *Tagalog*: Pugahan, Taquipan—; *Tamil*: Adam, Irambanai, Kondapauni, Kundarbanai, Pugam, Talam, Thippali, Tippilippanai, Udalarbanai—; *Telugu*: Bakini, Jivalaggu, Kondajivalaggu, Jiluga, Marc, Yatrakatari—; *Tulu*: Indu, Kannida—; *Uriya*: Modhura, Salapo, Solopo—.

2. **Caryota mitis** Lour. This very elegant palm is found in Burma—from Arakan southwards, Martaban, the Malay Peninsula, Penang, the Andaman Islands, and the Malay Archipelago.

In Malaya the FRUIT is put into wells with intent to cause annoyance. Bathing with well-water that has been treated in this way gives rise to an intense itching of the skin, and may cause an acute inflammation of the eyes. The fresh juice of the fruit when applied directly to the skin is extremely irritating.

The juice of the fruit, mixed with bamboo hairs and an extract of toad, is considered a very potent poison in Kelantan.

In Cambodia the soft fibres at the base of the LEAF-SHEATH are used in the cauterisation of wounds.

Cambodia: Ansc—; *Malay*: Beredin, Meredin, Tukkus—; *Penang*: Dudur—.

CHAMAEROPS.

The two species of this genus inhabit the western part of the Mediterranean region.

Chamaerops humilis Linn., Dwarf Fan Palm, is the only palm indigenous to Europe. It is found as far up as Nizza in Italy; but thence southwards it has a great range, being met with in all the countries bordering the Mediterranean Sea. It is very common in Algeria.

The fruit is astringent.

Andalusia: Palma—; *Arabic*: Doum—; *Catalan*: Margallo, Margallonera—; *English*: Dwarf Fan-Palm, European Palm—; *French*: Palmier nain, Palmiste—; *Italian*: Cefaglioni, Palma nana, Palmito—; *Maltese*: Giummar—; *Portuguese*: Palma das Vassouras, Palmeira—; *Sicily*: Curina, Giummara, Palmetta, Piumara—; *Spanish*: Palma enana, Palmito—.

COCOS.

The genus consists of about 30 species, all American, one of them cosmopolitan in the tropics.

C. coronata Mart., *C. flexuosa* Mart., *C. nucifera* Linn., *C. schizophylla* Mart., *C. yatai* Mart. are used medicinally in Brazil; *C. nucifera* Linn. is similarly used in India, Indo-China and China.

The fat from the kernel of *C. nucifera* Linn. is officinal in Holland and in Portugal.

Three of the therapeutically active species are to be met with in India.

1. Leaves 1.8-4.5 m. long ... ... *C. nucifera.*
2. Leaves 1.8-2.4 m. long ... ... *C. schizophylla.*
3. Leaves surrect, arcuate; leaflets concinnous ... *C. yati.*

1. **Cocos nucifera** Linn. The Coconut Palm is found at present in every part of the tropics, where it flourishes in the greatest luxuriance in the vicinity of the sea, especially a few feet above high-water mark. Although that is its chief habitat, it cannot be termed an exclusively littoral plant, for it has been met with far inland—at Merida in Yucatan, at Patna in Bihar, at Concepcion del Pao—. Whenever the Coconut Palm ventures beyond the limits of the tropics, it loses in elegance of aspect and power of productiveness. In the Sandwich Islands, just at the edge of the torrid

zone, it has a mean look, and yields fruit in comparatively small quantities.

The young ROOT is employed as an astringent gargle in sore throat. Boiled with ginger and salt it is efficacious in fevers.

The Ayurvedists credit it with anthelmintic value; but it is rarely, if ever, used as a worm remedy in India. It is popularly known as a diuretic and commonly used as such. It is also prescribed in uterine diseases and in dysentery.

In Cambodia an infusion of the root is considered as a valuable diuretic which is administered in gleet, bronchitis, and liver troubles with or without jaundice.

In Ceylon the root is said to strengthen the gums, and in Tahiti it is occasionally used instead of arecanut by betel-chewers.

In the Gold Coast the BARK is used for curing toothache and earache. Yunani practitioners in India recommend the ash of the bark as a dentifrice and an antiseptic; they prescribe it in scabies.

The TOMENTUM—a soft, downy, light-brown coloured substance found on the outside of the lower part of the leaves, where they spring from the stem—is used for stopping blood in cases of wounds, bruises, leech-bites, etc. It is said to be a good styptic.

The FLOWERS, when used medicinally, are given as an astringent. Sanskrit authors recommended them as useful in the treatment of diabetes, dysentery, leprosy, and urinary discharges, though liable to induce constipation.

If TODDY is drawn early in the morning and drunk at once, it forms a pleasant drink having a slightly stimulating effect and acting as a gentle laxative.

Toddy is refrigerant and diuretic. Its immoderate use causes great harm, resulting in sleepiness, loss of appetite, premature old age, extraordinary obesity, and diseases resembling dropsy and scurvy. Some of those who are addicted to it lose their intellectual faculties, are seized with trembling, or become stupid, absent-minded, or even insane.

Toddy is considered by Europeans as highly unwholesome during the rainy season.

A toddy-poultice is prepared by adding fresh-drawn toddy to rice-flour till the consistence of a soft poultice, subjecting to a gentle fire, and spreading on a cloth. It forms a valuable application to gangrenous ulcerations, indolent ulcers and carbuncles.

The outside scrapings of the HUSK and LEAVES applied to ulcers will cleanse and heal them rapidly if soaked in proof rum.

The husk of the fruit is used in the treatment of tapeworm.

The green FRUIT is given as a refrigerant, and especially employed as an astrigent in the sore-throats of children.

The MILK of the green fruit is a cooling drink, antiemetic and soothing, and useful in allaying urinary irritation. It is a good drink for cholera cases, and it is said to check vomiting when other means fail. If drunk in immoderate quantity the milk of young nuts may produce a slight degree of strangury and a predisposition to dropsical complaints. It is commonly admitted in Bengal that too much coconut milk induces a hydrocele swelling of the scrotum. Among the Tahitians the milk is considered as one of the exciting causes of elephantiasis.

In Brazil the milk of immature nuts is applied as a lotion to ulcers of the penis.

The expressed juice of the fresh kernel has been successfully employed in general ill-health, malnutrition, debility, incipient phthisis, thirst, fever and urinary disorders. In large doses it proves aperient, and in some cases actively purgative, on which account it has been suggested as a substitute for castor-oil.

The water of the ripe nut is unwholesome and can be drunk only sparingly, as it is strongly diuretic and is apt to produce an irritation of the bladder and urethra.

In French India the milk is given internally as an antidote for snake venom; but Mhaskar and Caius have proved satisfactorily that it has no such action. Bapat had already noted that two parts of cocoanut milk and 20 parts of sprouts of *Albizzia lebbek* given internally for three consecutive days cure all snake-bites 'with the exception of those of the cobra and the Russell's viper'.

Fresh cocoanut water is the antidote used by Malays to counteract *Sarcolobus* poison.

The fresh kernel or the tender pulp is nourishing, cooling and diuretic. In Ceylon it is given in sunstroke. The ash obtained from its calcination is said to be a valuable antacid and digestive.

The milk and meat of one nut eaten early in the morning, fasting, will expel the tapeworm if present. Caius and Mhaskar found this treatment ineffective in the case of hookworms.

The fresh OIL obtained by boiling the milk of the ripe fruit is used as an application for burns and in baldness. It is thought to have special properties in strengthening the hair, hence it is much in use as a local application in alopecia, and in loss of hair after fevers and debilitating diseases.

In the Konkan the oil which separates from the freshly-rasped kernel, alone or mixed with tamarind-seed oil, is used as an application to burns and rheumatic swellings.

An inunction of the oil to the whole body is used in fevers, and to the chest in lung diseases. In Ceylon the oil is applied to the head for cooling.

The oil is given in plethora and as a vermifuge in Jamaica. An emulsion of the oil and kernel is prescribed in coughs and pulmonary diseases generally.

In the Maldives cocoanut oil is esteemed a powerful antidote against the bite of poisonous reptiles.

The purified and prepared oil is used as a substitute for cod-liver oil in debility and phthisis. Its prolonged use, however, is apt to disturb the digestive organs and induce diarrhoea.

In Germany it is used in pharmacy, to a considerable extent, as a substitute for lard.

The TAR, obtained by covering the thoroughly ignited shell with a stone cup, is a rubefacient and an effectual domestic remedy for ringworm, itch and other parasitic affections.

The old and dried KERNEL is cut into thin slices and used as an aphrodisiac ingredient in confection; also as an anthelmintic, it is given to remove tapeworms.

The terminal bud or CABBAGE is nourishing and digestive. Its fresh juice is cooling and diuretic.

In Cambodia the WOOD enters into the composition of a fumigant prescribed in the treatment of piles.

Ahanta: Kukui—; *Annam*: Cay dua—; *Aowin*: Agye—; *Arabic*: Jadhirdah, Jouzehindi, Narjil, Shajratuljouzehindi, Shajratunnarjil—; *Banziri*: Diko—; *Bengal*: Dab, Narakel, Narikel, Nairyal—; *Betsimisaraka*: Voaniho—; *Bicol*: Niog—; *Bombay*: Maar, Mahad, Mar, Naralchajhada, Narel, Naril, Naural—; *Brazil*: Coco da Bahia, Coqueiro da India, Inajaguasuiba—; *Burma*: On, Ondi, Ong, Onsi, Onti, Ung, Ungbin—; *Cagayan*: Niog—; *Cambodia*: Daung—; *Canarese*: Gitaka, Gitaku, Kobari, Kobbari, Kurube, Matte, Narikela, Narikera, Siyala, Tare, Tenginakayi, Tengu, Trinaraja—; *Chinese*: Yeh Tzu—; *Deccan*: Narel, Narelkajhar—; *Dutch*: Calappusboom, Cocos, Cocosboom, Cocospalm, Kalappus, Klapper, Klapperboom, Klapperpalm, Koko, Kokosboom, Kokosnootenboom, Kokosnootpalm, Kokospalm—; *English*: Cocoa Nut, Cocoanut Palm, Coconut Palm, Cocoanut Tree—; *Ewe*: Yevunai—; *Fanti*: Kube—; *French*: Cocotier, Roi des végétaux, Cocotier commun, Cocotier des Indes, Cocotier nucifère, Cocotier ordinaire, Cocotier porte-noix, Palmier—; *Ga*: Akokoshicho—; *Gabon*: Omangata—; *German*: Calappabaum, Calappusbaum, Cocosnussbaum, Cocospalme, Echte Kokospalme, Indianischer Nussbaum, Kokosbaum, Kokosgalen, Kokospalme, Kokospalmenbaum, Wandernde Seeuferpalme—; *Guam*: Niyog—; *Gujerati*: Naliyer, Nariel, Nariera, Nariyela, Naryal—; *Hausa*: Kwakwar Attagara—; *Hindi*: Narel, Nariel, Nariyal, Nariyalkaper, Nariyel—; *Ilocano*: Niog—; *Italian*: Cocco—; *Java*: Bhungkana, Bhungkananjijor, Enjor, Ijor, Kalapa, Kerambil, Klapa, Klendah, Njejor, Njijor, Njor, Tangkalkalapa, Wiiklapa, Witkrambil—; *Konkani*: Madd—; *Krepi*: Agorne, Naiti, Yevone, Yevune—; *Krobo*: Kokosi, Ngmaicho—; *Malayalam*: Chentennu, Chocham, Karikku, Kulittai, Langalam, Nalikeram, Narikelam, Tenna, Tennu—; *Marathi*: Mad, Mada, Mahad, Mar, Naral, Naralchajhada, Naralmad, Narel, Narela, Narula, Tenginmar, Varala—; *Mundari*: Burkadaru, Narieldaru—; *Mysore*: Nur—; *New Caledonia*: Nou, Nou boibate, Nou bouangae, Nou do, Nou goine, Nou jomalate, Nou kigoute, Nou mia, Nou pougne, Nou tamen, Nou tiguit—; *Nzima*: Kukwe—; *Pahuin*: M'ban n'tang—; *Pampangan*: Ngon got—; *Persian*: Badinj, Darakhtebandinj, Darakhtenargil, Nargil—; *Philippines*: Coco, Lubi, Pangosin, Tapiasin—; *Polynesia*: Niu—; *Portuguese*: Coqueiro, Palmeira—; *Quittah*: Ene—; *Roumanian*: Cocotier—; *Russian*: Kokosovoe dyerevo—; *Sanskrit*: Dakshinatriya, Dridhanira, Dridhaphala, Duraruha, Garikera, Jataphala, Junga, Karakambha, Kaushikaphale, Kurchashekhara, Kurchashirashaka, Langali, Mahaphala, Mangalya, Mriduphala, Mutkuna, Nadikeli, Narikari, Narikela, Nilataru, Payodhara, Phalakeshara, Phalamunda, Putodaka, Rasaphala, Sadaphala, Sadapushpa, Shiraphala, Skandhapal, Skandhataru, Subhanga, Sutanga, Toyagarbha, Trinaraja, Tryakshaphala, Tryambakaphala, Uchhataru, Varaphala, Vishvamitrapriya—; *Sinhalese*: Pol, Polgaha, Polgass, Polnawasi, Tambili—; *Spanish*: Cocotero, Rey de los vegetales—; *Tagalog*: Adiavan, Niog, Pamocol—; *Tamil*: Edagam, Ilangali, Keli, Muppudaikay, Nadigelam, Naligelam, Naligeram, Narigelam, Papparettennai, Talai, Ten, Tengay, Tengu, Tennai—; *Telugu*: Ettabondalakobbari Gujjunarikadamu, Kobbari, Kobbera, Nalikeramu, Narikadamu, Narikelamu, Narikeramu, Langali, Mukkantipandu, Te, Temranu, Tenkaya, Trinarajamu—; *Tulu*: Tare—; *Twi*: Kokosi—; *Urdu*: Nariyel—; *Uriya*: Gotoma, Langoli, Nodia, Paido, Poiu, Trinodrumo—; *Visayan*: Anibong, Bonotan, Botong, Cayomanis, Dahili, Lobi, Lubacan, Niogngapoti, Lobingahinbaon, Lobingapilipog, Limbaon, Niog, Pangonn, Pilipog, Potot, Tamis, Tamisan, Tayomanis—; *Zambales*: Ongot—.

2. **Cocos schizophylla** Mart., Aracuri Palm, is a native of Brazil found in the Province of Bahia at Camamu and Bahia, San Jorge dos Lheos. It is cultivated in Indian gardens.

In the Province of Bahia the juice of the unripe fruit is used for inflammation of the eyes.

Brazil: Alicuri, Aracui, Aricuri, Ariri—; *English*: Aracuri Palm—.

3. **Cocos yatai** Mart. inhabits Argentine—provinces of Concordia, Corrientes and Entre Rios—and Brazil. It is cultivated in Indian gardens.

In Argentine, where it is known as *yatai*, the fleshy parts of the fruit are considered anthelmintic.

COPERNICIA.

The genus numbers 9 species spread over Brazil, Venezuela, Argentine, San Domingo, Cuba and New Granada.

Copernicia cerifera Mart., Brazilian Wax Palm, inhabits Brazil and is found in the Provinces of Bahia, Pernambuco and Piauhy. It is sometimes grown in Indian gardens.

The root is alterative and diuretic, and is in high popular esteem as a tonic and blood purifier.

Brazil: Carnauba—; *English*: Brazilian Wax Palm—.

CORYPHA.

The 6 species of this genus inhabit Tropical Asia.

Corypha umbraculifera Linn. The Talipot Palm is rather common in Ceylon in the moist low region below 2,000 ft. It is also met with on the Malabar Coast and in Kanara—in the moist forests of the Kumpta and Honavar talukas of Northern Kanara, covering extensive areas near the Gersoppa and Yena rivers, also on the Yellapur Ghats; sometimes planted in gardens near the coast. It extends to the South Andaman Islands, the Little Coco and the Great Coco. It is cultivated in tropical India and Burma.

The fruit is a fish poison.

Bengal: Bajarbattuler, Tali, Tallier, Tara, Tarit ; *Burma*: Pebin—; *Canarese*: Baini, Indu, Sritale, Sritali, Tali—; *English*: Fan Palm, Great Fan Palm, Holy Palm, Java Fan Palm, Malabar Coast Fan Palm, South Indian Talipot Palm, Talipot Palm, Umbrella Palm—; *Malayalam*: Kutappana, Sitalam, Talippana—; *Marathi*: Bajarbattu, Tali—; *Sanskrit*: Alpayushi, Karalika, Katakali, Pakti, Sritala, Tali—; *Sinhalese*: Tala—; *Tamil*: Kudaippanai Sidalam, Talappam, Talippanai—; *Telugu*: Dridhatalamu, Sritalamu—; *Tulu*: Panoli—.

ELAEIS.

The genus numbers 4 species, inhabitants of Tropical Africa and Eastern Tropical South America.

E. guineensis Jacq. is used medicinally in Equatorial West Africa and in Brazil; *E. melanococca* Gaertn. too is used in Brazil.

Elaeis guineensis Jack. The African Oil Palm is a native of Africa, cultivated in India.

In Guinea the oil from the sarcocarp is applied to wounds as a vulnerary. It is used as a liniment for rheumatism and courbature.

The Bubis of the Island of Fernando Po make an excellent poultice of the oil which they apply to wounds.

In Equatorial West Africa the roots are used as a diuretic, and the fresh sap as a laxative.

Angola: Dihoho—; *Bacongo*: Matebbe—; *Baffuru*: M'bila—; *Baga*: M'Bia—; *Banziri*: Bete—; *Brazil*: Coqueiro de Denté—; *Congo*: Leba—; *Dutch*: Afrikaansche awarra, Afrikaansche oliepalm, Obepalm, Oliepalm van Guinea, Oliepalm van West Africa, Oliepalm van de Kust van Guinea, Palmietboom—; *English*: African Oil Palm, Oil Palm, True Oil Palm—; *French*: Aouara d'Afrique, Aouara des Caraïbes, Aoura de Guinée, Aovora, Arouara des Caraïbes, Avoira de Guinée, Elaïs de Guinée, Eléide, Eléide de Guinée, Noix de palme, Noix de palmier, Palmier crocro, Palmier épineux, Palmier à huile, Palmiste épineux—; *Ga*: Ngmetsho—; *Gaboon*: Oila—; *German*: Afrikanische Oelpalme, Guineische Oelpalme, Oelpalme—; *Guinea*: Toehntis—; *Hausa*: Kwakwa—; *Malinke*: Tintulu—; *Nzima*: Arairlair—; *Pahuin*: Aline—; *St. Thomas Island*: Denden, Palmeira andim—; *Surinam*: Aaavora, Avoora, Avouara, Avuara, Maba, Obe—; *Susu*: Tugi—; *Twi*: Abair—; *Yakoma*: Zamba—.

HYPHAENE.

This genus numbers 27 species distributed all over tropical and sub-tropical Africa, Arabia, and Western India.

Hyphaene thebaica Mart. The Egyptian Doum Palm grows along the valley of the Nile in Middle and Upper Egypt, and is to be met with at Shaik Othman near Aden. Old specimens may be seen in many a garden of India and Ceylon, and, as a rule, they are much better developed than the tree growing in Egypt.

The fruit is considered anthelmintic.

Diu Island: Okamundel—; *Egypt*: Mama—; *English*: Egyptian Doum Palm, Gingerbread Tree—; *Ewe*: Sokuti—; *Gold Coast*: Dum Palm, Gingerbread Palm—.

LICUALA.

This genus comprises about 50 species inhabiting Asia, Australia and the Pacific Islands.

Licuala spinosa Wurmb. (=*Corypha pilearia* Lour.) is found in Malacca, common in wet places, particularly in hedges. It is also met with in the tidal forests of the Andamans, in the Nicobars and the Malay Islands.

In Cambodia the bark is used in combination with other drugs for the treatment of tuberculosis with spitting of blood.

Cambodia: Chak—; *Malay*: Plass—.

LODOICEA.

The genus is limited to 1 species endemic in two of the Seychelles Islands: Praslin and Curieuse.

Lodoicea seychellarum Labill., Double Cocoanut Palm.—The marvellous medicinal properties which are ascribed to the NUTS by ancient physicians, both European and Asiatic, have now been recognized as fanciful and due solely to the rarity of the fruit.

The water of the green fruit or its soft kernel is said to be antibilious and antacid when taken after meals. The ripe fruit is also used similarly, but is at the same time purgative.

The kernel is a tonic and stomachic. With rose water it is given in indigestion, diarrhoea, and for the relief of colic. In combination with other drugs it is made into a paste and applied to swollen glands.

In Southern India the kernel is given in woman's milk for the treatment of typhus fever. It is credited with antiscorbutic and antivenereal properties.

In Bombay it is prescribed as a tonic and febrifuge, and is used to check diarrhoea and vomiting, especially in cholera. It is also commonly given to children, mixed with the root of Nux-vomica, for colic.

The nut is in great repute among the Arabs as a tonic, preservative and alexipharmic.

The fibrous HUSK of the nut is used in the form of a decoction in the Tamil country. It is considered useful in reducing the quantity of sugar in the urine in cases of diabetes. Koman (1919) gave the decoction to two patients in the hospital with diabetes mellitus. The sugar eliminated decreased considerably, but did not disappear altogether. There was relapse as soon as the decoction was stopped.

Arabic: Narjilebahri—; *Bombay*: Jaharinaral, Jeharinaryal—; *Burma*: Pénle-on-si—; *Deccan*: Daryakanarel—; *Dutch*: Dubbele cocosnoot van de Seychelles, Dubbele Klapper, Maledivische noot. Seychellennoot, Zeeklapper—; *English*: Cocoanut of the Maldives, Double Cocoanut Palm, Sea Cocoanut Palm—; *French*: Coco de l'île Praslin, Coco des Echelles, Coco de mer, Coco de Salomon, Coco des Seychelles, Cocotier des îles Seychelles, Cocotier des Maldives, Cocotier des Seychelles, Cul de négresse, Double Coco, Lodoïce des Maldives, Lodoïcée, Lodoïcée des Seychelles, Lontar domestique, Rendier éventail, Rendier lontar, Tobel—; *German*: Doppelte Cocosnuss, Kokosartige Lodoicée, Maldivische Nuss, Meercocos, Seecocos, Wundernuss Salomons—; *Gujerati*: Daryanunariyal—; *Hindi*: Daryakanaryal—; *Indian Archipelago*: Calappalaut—; *Java*: Djenggi, Djenggli, Kelapalaut, Kepodjenggi, Pelokdjenggi—; *Malayalam*: Akraritennu, Kataltenna—; *Persian*: Nargilebahri—; *Portuguese*: Coco das Maldivas, Coco do Mar—; *Seychelles*: Coquinko, Tavacarre—; *Sinhalese*: Mudupol—; *Tamil*: Kadattengai—; *Telugu*: Samudraputenkaya—.

LOXOCOCCUS.

The genus is limited to 1 species endemic in Ceylon.

Loxococcus rupicola Wendl. and Drude is known in the vernacular as *dotalu*. Its seed is used for mastication with betel, like arecanut.

METROXYLON.

This genus comprises about 9 species, natives of the Malay Archipelago, New Guinea and Fiji.

1. Spikes 4-5 in. long, ½ in. in diameter ... ... *M. sagus*
2. Spikes 2½ in. long of the size of the little finger ... *M. rumphii*.

1. **Metroxylon sagus** Rottb. (—*Sagus laevis* Rumph.). The common Sago Palm is to be found in the entire group of the Moluccas, and in Borneo, Sumatra, Java, the Philippine Islands, the Malay Peninsula; chiefly cultivated.

The fruit is used as a poison in Malaya.

Dutch: Meeboom, Meelgevende palmboom, Molueksche palmboom, Sagoboom, Sagoeboom, Sagopalm—; *English*: Common Sago Palm, Rumphius's Sago Palm, Sago Palm, Sago Tree—; *French*: Arbre au sagon, Palmier du Japon, Palmier sagon, Sagonier, Sagonier farinifère, Sagoutier—; *German*: Echte Sagopalme, Molukkische Sapfenpalme, Rumph's Sagopalme, Sagobaum, Sagopalme—.

2. **Metroxylon rumphii** Mart. (=*Sagus rumphii* Willd.) is distributed throughout the Malay Archipelago.

The fruit is used as a poison by the Malays

English: Rumph's Sago Palm—; *Malay*: Sagu—.

NANNORHOPS.

The genus is limited to 1 species distributed through Sind, Baluchistan, Punjab and Afghanistan. It is abundant in the Peshawar Valley, in Kohat and in the trans-Indus territory along the eastern skirts of the Suliman range, ascending up to 3,000 ft. Also very common on the hills which form the western boundary of Sind. It is common locally on a limited area in the central Salt Range, between 2,000 and 5,000 ft. and on Mount Sakesar. In one place in the Siwalik tract east of the Jhelum, near Sumani above Bhimbur. It is common in the Khaiber Pass, and generally in the low arid mountains of eastern Afghanistan. It is found everywhere up to 5,000 ft. in Baluchistan and Makran, except near the coast.

Nannorhops ritchieana H. Wendl. is a low gregarious shrub, the leaves usually tufted from an underground, much-branched rhizome.

The young leaves are given in diarrhoea and dysentery. They are also purgative, but chiefly used in veterinary medicine.

Bangash Hills: Fiesch—; *Barkhan*: Dhora, Mazari—; *Duki*: Dhora, Mazari—; *Hindi*: Mazari, Mazri—; *Jhalawan*: Pish—; *Kohlu*: Dhora Mazari, Pish—; *Musa Khel*: Dhora, Mazari—; *Pushtu*: Maizurrye—; *Salt Range*: Kaliun, Kilu—; *Shahrig*: Dhora, Mazari, Pish—; *Sibi*: Dhora, Mazari, Pish—; *Sindh*: Dhora, Fease, Pease, Pesh, Pfarra, Phana, Pfis—; *Trans-Indus*: Mazari, Mazrai, Mzarai—.

NIPA.

The genus is limited to 1 species which covers thousands and thousands of acres of the salt marshes of the islands and coasts of the Indian Ocean. It inhabits the river estuaries and tidal forests of the Sunderbans, Chittagong, Burma and the Andamans, and extends throughout Malaya to Queensland. Its Western limit is Ceylon where it is to be found at the mouth of rivers on the south-west coast, Kalutara, Gindura River near Galle.

Nipa fruticans Wurmb. The Nipa Palm is used medicinally in the Philippine Islands.

The fresh leaves are much used in the treatment of ulcers in the form of cataplasm or lotion. They are considered a valuable remedy for the bites of centipedes.

The alcohol obtained from the toddy by distillation is mixed with water and used as an eye wash in inflammations of the eyelids and of the conjunctiva.

Andamans: Poothada—; *Bengal*: Gabna, Gulga—; *Burma*: Dane—; *Cagayan*: Tata—; *English*: Nipa Palm, Water Cocoanut—; *Guam*: Nipa, Sasa—; *Gujerat*: Paradeshitadio—; *Hindi*: Gulga—; *Philippines*: Lasa—; *Ponape*: Parran—; *Sinhalese*: Gimpol—; *Sulu Archipelago*: Ballang—; *Tagalog*: Nipa, Sasa—; *Telugu*: Kotitikaya, Nipamu—; *Zambales*: Saga—.

PHOENIX.

The genus consists of about 12 species which inhabit subtropical Africa and Tropical Asia.

Four of the species are credited with medicinal properties

1. Leaves 2-6 ft. long	...	...	...	...	*P. pusilla.*
2. Leaves 8-10 ft. long	...	...	...	...	*P. paludosa.*
3. Leaves 10-15 ft. long	...	...	...	...	*P. sylvestris.*
4. Leaves longer	...	...	...	...	*P. dactylifera.*

1. **Phoenix pusilla** Gaertn. is to be found on the Coromandel Coast, not far from the sea, and in the dry forests of the northern part of Ceylon.

The fresh SAP is cooling and laxative.

The GUM is used in diarrhoea and in genito-urinary diseases.

The FRUIT is prescribed in cough, asthma, fetid breath, also in fever and gonorrhoea.

The SEEDS are made into a paste by trituration with water and applied over the eye in inflammations and for opacity of the cornea.

Canarese: Hullichala, Ichalu, Sannayichalu—; *Ceylon*: Inchu—; *Hindi*: Palawat—; *Malayalam*: Chittintal, Inta—; *Tamil*: Ichu, Indu, Inju, Kalangu, Kurinji, Sagi, Siruyinju, Sittinju—.

2. **Phoenix paludosa** Roxb. The Wild Date Palm is tolerably common throughout India, wild or more often cultivated. It forms siderable portion of the impenetrable woods which cover the Sunderbans. It is met with along the Selwin, between Amherst and Moulmein, and extends to Penang, Siam and Cochin China.

The fruit is acidulous, sweet and cooling. It relieves flatulence and assuages pain. It is given in fever and inflammation.

Bengal: Golpatta, Hintal, Hital—; *Burma*: Thinboung—; *Hindi*: Hintal—; *Penang*: Dangsa—; *Sanskrit*: Hintala—; *Telugu*: Giruka tati, Hintalamu—.

3. **Phoenix sylvestris** Roxb. The Wild Date Palm is tolerably common throughout India, wild or more often cultivated. It forms extensive forests in Rohilkhand, on the low ground along the Ramganga River, and on the plateau of Mysore, between Shimoga and Tumkur, in the moister stretches of low ground which intersect and drain the rocky undulating granite hills. It is most abundant in Bengal, Behar, on the Coromandel Coast and in Gujerat. In the Bombay Presidency it is common in moist ground throughout the dry districts, usually along banks and in the beds of streams and watercourses. It is not uncommon in the Siwalik tract and the outer Himalaya, reaching up to 5,000 ft. in Kumaon. It is met with at Ghiaunla in Garhwal at 3,500 ft., along the banks of the Bias above Mandi, in the Jamu hills at 2,000 ft. elevation, and in the Salt Range

The ROOT is given for toothache.

The fresh TODDY is considered a cooling drink.

The FRUIT is prescribed in gonorrhoea and gleet.

The fruit, pounded and mixed with almonds, quince seeds, pistachio nuts, spices and sugar forms a very popular restorative remedy.

A paste, formed of the KERNELS and the root of *Achyranthes aspera*, is eaten with betel-leaves as a remedy for ague.

Bengal: Kajar, Kejur—; *Berar*: Seindi—; *Bombay*: Khajur, Khajura, Khajuri, Sendi—; *Canarese*: Andadayichalu, Ichala, Ichale, Ichalu, Ichela, Ichil, Kallichalu, Kallu, Siyindu—; *Deccan*: Sandolekanar—; *English*: Date-sugar Palm, Indian Wine Palm, Sugar Palm, Wild Date Palm—; *Gond*: Sindi—; *Gujerati*: Kajuri, Kharak, Tadi—; *Hindi*: Kejur, Khaji, Khajur, Khajuri, Salma, Sendhi, Thakil, Thalma—; *Kolami*: Khajur—; *Konkani*: Kajuri—; *Malayalam*: Inta, Intappana, Kattinta—; *Marathi*: Boichand, Sendri, Shindi, Sindikajuri—; *Mundari*: Darukita, Kitadaru—; *Porebunder*: Khalelananjhad, Tadi—; *Punjab*: Khaji, Khajur—; *Sanskrit*: Bhumikharjurika, Duraroha, Duraruha, Dushpradarsha, Haluka, Haripriya, Kakakarkati, Kapila, Kashayi, Kharju, Kharjuri, Mriduchhada, Nishreni, Skandhappala, Svadi, Svadumastaka, Yavaneshta—; *Santal*: Khijur—; *Tamil*: Ichambanai, Inju, Karavam, Kattinju, Madal, Periyayinju—; *Telugu*: Ita, Peddayita—; *Uriya*: Khorjuri, Khorjuro—.

4. **Phoenix dactylifera** Linn. The Edible Date Palm is cultivated and self-sown in Sind and in the Southern Punjab, particularly near Multan and Muzaffargarh, also in the Sind Sagar Doab and trans-Indus territory. Near Dehra Ghazi Khan, date-palms are very numerous on a strip 10-12 miles long from north to south. A few trees are found planted at many places in the Eastern Punjab, also at Saharanpur, and here and there in the Ganges Doab and Bandelkhand. It is also grown in the Deccan and Gujerat, but does not thrive in Bengal. It thrives luxuriantly in the arid rainless regions of North Africa and West Asia, where it is exposed to extreme heat in the daytime, and not uncommonly to frost at night, but it requires a certain amount of moisture in the soil. In Europe it is cultivated in Spain, on the Hyères Islands, the Riviera near Nice, San Remo and Genoa, where it attains its northernmost point. There is a wood of date-palms at Bordighera near San Remo, said to contain over 4,000 stems cultivated mainly to yield palms for Palm Sunday at Rome. In South Italy, Sicily and Greece, the tree is not uncommon, but the fruit is small and poor.

The tree yields a GUM which is considered a demulcent, diuretic and refrigerant. It is esteemed in the Punjab as a valuable remedy in diarrhoea and diseases of the genito-urinary system.

When the heart of the LEAVES is cut out a sweet thickish fluid collects in the cavity, which is very refreshing and slightly purgative. Yunani practitioners use the leaf as an aphrodisiac and prescribe it in liver complaints.

The fresh SAP is cooling and laxative.

The FLOWER is bitter. The Mahommedans say it is purgative, expectorant and tonic to the liver; they recommend its use in fever and blood diseases.

The fresh FRUIT is considered demulcent, expectorant, laxative, tonic and diuretic. But a long continued use is said to produce soreness of the gums. Water in which fresh dates have been steeped for a while is given to relieve alcoholic intoxication. Milk in which clean and fresh dates have been infused is a very nourishing and restorative drink especially during convalescence from fevers and small-pox.

Dates are said to promote expectoration, soothe the chest and also prevent constipation. They are prescribed in cases of cough, asthma, consumption and generally for all kinds of chest troubles. They are also given in fever, gonorrhoea, liver complaints and abdominal troubles; and they are recommended for the treatment of dysentery.

The dried fruit, pounded and mixed with almonds, quince seeds, pistachio nuts, spices and sugar, forms a much esteemed restorative. It is also used as an ingredient in various tonic and aphrodisiac preparations.

In Southern India the ground SEEDS are made into a paste by trituration with water and applied over the eyelids for opacity of the cornea. The paste is also said to be efficacious in relieving headaches and hemicrania.

A fine paste made of the seed of the date-fruit and the root of *Achyranthes aspera* is applied to betel-leaves and made into little packets together with clove, cardamom, catechu and arecanut. Three packets are administered at intervals of one hour before an expected attack of ague.

The smoke from the burning powdered seeds affords great relief when used as a fumigant for piles.

Arabic: Nakhleh—; *Badaga*: Gajjira—; *Bengal*: Khajur—; *Bhote*: Khasser—; *Bombay*: Khajur—; *Burma*: Sunbalun, Swonpalwon—; *Canarese*: Kajura, Karika, Karjura, Kharjura—; *Chinese*: Wu Lou Tzu—; *English*: Arabian Date Palm, Cultivated Date Palm, Edible Date, Large Date, Persian Date—; *French*: Dattier, Dattier cultivé—; *German*: Dattelbaum, Dattelpalme—; *Greek*: Phoinix—; *Gujerati*: Karek, Khajur—; *Hebrew*: Chhomer, Tomer—; *Hindi*: Khaji, Khajur—; *Italian*: Dattero, Palma—; *Kachhi*: Khaji—; *Kej*: Abdandan, Abraughan, Amiri, Anguro, Arrasht, Bambai, Banduk, Barni, Barral, Bingu, Bulediraughani, Burshakri, Buzband, Chafshak, Charpan, Dashtari, Gognai, Gozti, Gundgoreg, Gurbagu, Gwazo, Haleni, Husaini, Jafash, Jafshakjalgi, Jamsaki, Jowanabushmas, Johanajaski, Jozo, Kalerak, Kalunti, Karpaso, Khargi, Khurmazard, Kohijalgi, Kuleri, Kuzanabat, Mairisurkh, Makli, Mataftaza, Mulkijalgi, Nazamdazi, Nazantabaqi, Nazbibilanguk, Naznin, Pandi, Peshnai, Pingu, Pull, Ragom, Baughani, Santgwaragh, Shagashkhand, Shipga, Siahkanok, Sohri, Suhrebegamjangi, Tigal, Ushtarkor, Wakhshi, Washkalunt, Washkung, Washnao, Zard—; *Kohhaja*: Mach—; *Malayalam*: Itta, Ittappalam, Tenitta—; *Malta*: Palma, Palma da datteri, Palma tattamar—; *Marathi*: Kharjur—; *Nasirabad*: Khajji, Khurma—; *Nushki*: Mach—; *Panjgur*: Begamjangi, Dandari, Dashtlafashsabzo, Fard, Fofo, Haraksabzo, Hussainizard, Joshandakalut, Joshandasabzo, Khurmakarmachi, Khurmaikalut, Kungo, Mozawati, Rabi, Radag, Zardan, Zardpanjguri—; *Portuguese*: Palmeira, Tamareira—; *Punjab*: Khaji, Khajur—; *Pushtu*: Kajura, Shewa—; *Roumanian*: Curmal—; *Russian*: Phinikovoe dyerevo—; *Sanskrit*: Dipya, Hayabhaksha, Madhurasraoa, Mudarika, Phalapushpa, Pindakharjura, Pindakharjurika, Pindiphala, Rajajambu, Sapinda, Svadupinda—; *Sharig*: Khajur—; *Sibi*: Khajji, Khurma—; *Sind*: Kaji, Kurma, Pindchirdi, Tar—; *Sinhalese*: Ind—; *Spanish*: Datilera, Palma datilera, Palmera—; *Swedish*: Palmtrae—; *Tamil*: Ichu, Inju, Karchuram, Kuravam, Perindu, Perinju, Titti—; *Telugu*: Gajjuramu, Ita, Kharjuramu, Manjiyita, Muddakharjuramu, Peridu, Perita, Simakharjuramu—; *Turki*: Karmah—; *Urdu*: Khurma—; *Uriya*: Khorjjuri—; *Waziri*: Khajurra—.

5. According to the authors of the Encyclopaedia Mundarica *Kindad* is a stemless palm different from *Kita* (=*Phoenix acaulis* Buch.). The pinnae of its pinnatisect leaves are shorter and recurved at the end, they spread at a much greater angle to the axis than the pinnae in an *otekila* leaf.

The root-stock is pounded and thrown into pools and rivers to poison fish.

6. *Kitamuli* is said to be a stemless palm, different from *Phoenix acaulis* Buch. It has only two plicately multifid leaves like those of the male *acaulis*, and is said to die after two or three years. Its corm is slender having the size and form of a carrot, and containing nevertheless a farinaceous substance.

The mealy substance of the corm, edible in the *Kita*, is deadly poisonous in the *Kitamuli*. It acts as a vesicant and is, therefore, applied to the chest in pneumonia. It is also rubbed on the herpetic eruptions called *dinai*.

PINANGA.

This genus numbers about 50 species distributed over the Indo-Malayan region.

Pinanga dicksonii Bl. (=*Areca dicksonii* Roxb.) is found in great abundance on the mountains of Travancore and Malabar. It is gregarious and locally abundant in the evergreen forests near the Gersoppa and Nilkund Ghats of Northern Kanara.

The poorer classes eat the nut as a substitute for the common arecanut.

The dried husk of the fruit is used medicinally in China, Japan and Annam. It is in much request in flatulent, dropsical or obstructive diseases of the stomach. It is also given in choleraic affections.

China: Ta Fu Tzu—; *Malayalam*: Kanakamuka—; *Telugu*: Kondapoka—.

RAPHIA.

This genus numbers about 20 species which inhabit Tropical Africa and America.

Two of them are credited with medicinal properties.

1. Leaves rising straight up ... ... *R. ruffia.*
2. Leaves rising nearly vertically from the stem, bending out on every side in curves, forming a plume *R. vinifera.*

1. **Raphia ruffia** Mart. (=*R. pedunculata* Beauv.=*Sagus ruffia* Jacq.) is indigenous in Madagascar. It is extensively cultivated on the Mascarene Islands and naturalized in America.

The ROOT is given for toothache.

The fresh JUICE is a valuable laxative.

Dutch: Madagascarsche sagoboom, Sagodragende palm—; *English*: Raffia, Raffia Palm, Rafia Palm, Raphia Palm, Roffia, Roffia Palm—; *French*: Mouffia, Palmier de Mayotte, Raffia, Rafia, Raphia—; *German*: Bambuspalme, Madagaskarische Sagopalme, Raffiabastpalme, Raffiaweinpalme—; *Hova*: Rafia, Rofia, Voarafy—; *Madagascar*: Fomby, Maivanaty, Raofia.

2. **Raphia vinifera** Beauv. The Wine Palm grows in communities in the freshwater swamps of Lower Nigeria.

The SEEDS contain a bitter oil which in Guinea is much used for courbature and rheumatic pains.

In the Gold Coast the pounded seeds are thrown in pools and rivers to stupefy the fish.

Aowin: Dongkang—; *Ashanti*: Adobair—; *Awuna*: Alati—; *Dutch*: Raphiavezelpalm—; *English*: Bamboo Palm, Jupatic Palm, Pharaoh's Date-palm, Wine Palm—; *Ewe*: Ala—; *Fanti*: Adobair—; *French*: Bourdon, Palmier à vin,

Raphier—; *Ga*: Adobair—; *German*: Bambuspalme, Echte Weinpalme, Weingebende Sagopalme, Weinpalme—; *Krepi*: Adoka, Altai—; *Krobo*: Sowaircho—; *Malinke*: Ban—; *Nizima*: Dorka, Eduiri—; *Quittah*: Adoka—; *Twi*: Adobair—; *Wassaw*: Tombo—.

TRACHYCARPUS.

The genus consists of three species inhabiting the Himalayas, China and Japan.

Trachycarpus excelsa H. Wendl. .(=*Chamaerops excelsa* Thunb.) Fortune's Chusan Palm is found in Upper Burma and extends through Yunan to China and Japan.

The FRUIT is astringent.

Chinese: Tsung Lu—; *English*: Fortune's Chusan Palm—.

WALLICHIA.

The genus numbers three species which inhabit India and Malaya.

Wallichia disticha T. Anders. occurs in the Gonda Hills of Oudh, the valleys of Sikkim Himalaya to 2,000 ft., the Makum forest in Assam. It is found in Upper Burma ascending to 4,000 ft. in the hills east of Bhamo, and on the Pegu Yoma, chiefly on the eastern slopes.

The BERRIES irritate the skin.

Lepcha: Katong—; *Lower Burma*: Letme, Zanaung—; *Upper Burma*: Minbaw—.

THE MEDICINAL AND POISONOUS SEDGES OF INDIA

BY

J. F. CAIUS, S.J., F.L.S.

The CYPERACEAE are grass-like herbs broadly distributed all over the world, and especially in the cold regions of the northern hemisphere; they are gregarious in marshy plains, damp meadows, and the dry slopes of high mountains; they are less frequent in maritime estuaries. The family includes 85 genera, with about 2,600 species.

The medicinal and poisonous sedges of the world belong to 11 genera:—CAREX (cold and temperate regions); CYPERUS (Tropics and Subtropics); ERIOPHORUM (Arctic or Northern temperate regions); FIMBRISTYLIS (Tropics); GAHNIA (Australasia; from Singapore to the Sandwich Islands); JUNCELLUS (warm regions); KYLLINGA (Tropics and Subtropics); PYCREUS (warm and temperate regions); REMIREA (Tropics); SCIRPUS (cosmopolitan; as far as the Polar regions); SCLERIA (moist warm countries).

The medicinal and poisonous sedges of India belong to 7 genera:—CAREX, CYPERUS, FIMBRISTYLIS, JUNCELLUS, KYLLINGA, REMIREA, SCIRPUS.

A. Spikelets of few or many glumes; lower 1-2 glumes always empty, uppermost male empty or imperfect,

I. Flowering glumes usually many, all distichous no hypogynous bristles :

1. Style bifid; flowers 1 or 2 in spikelet; rhachilla of spikelet deciduous ... ... *Kyllinga*

2. Style bifid; rhachilla of spikelet persistent; nut dorsally compressed ... ... *Juncellus.*

3. Style trifid; rhachilla of spikelet persistent after fall of glumes ... ... *Cyperus.*

II. Flowering glumes usually many in a spiral; hypogynous scales or bristles often present :

1. Style-base persistent or completely deciduous; no hypogynous bristles; leafy plants ... *Fimbristylis.*

2. Style-base not articulate passing into nut gradually; hypogynous bristles 0-6, undivided, linear, rarely obloug ... ... *Scimirea*

B. Spikelets sessile in dense digitate cylindric spikes; glumes 4, 3 empty, uppermost with a perfect flower *Remirea.*

C. Spikelets of male or female flowars, or both; nuts enclosed in an utricle entire or 2-fid at the tip only *Carex.*

CAREX.

This genus includes over 900 species broadly dispersed in cold and temperate regions, a few being found in mountainous tropical regions.

C. arenaria Linn., *C. disticha* Huds., and *C. hirta* Linn, are used medicinally in Europe; *C. macrocephala* Willd. and *C. siderosticta* Hance are similarly used in China.

Carex cernua Boott is found in Assam and extends to Tongking and Japan.

The plant is said to be toxic to cattle producing lack of appetite, loss of milk, and definite nervous symptoms.

CYPERUS.

This genus includes 300 species, natives of tropical and subtropical regions, a few being found in temperate regions. It abounds on the shores of tropical rivers, and in the clearings of virgin forests.

The following species are used medicinally in Europe: *C. esculentus* Linn.. *C. longus* Linn., *C. rolundus* Linn.—; in China, Indo-China, Malaya, and the Philippine Islands: *C. rotundus* Linn.—; in West Africa: *C. articulatus* Linn., *C. esculentus* Linn.—; in South Africa: *C. esculentus* Linn., *C. fastigiatus* Rottb., *C. longus* Linn., *C. sexangularis* Nees—; in Madagasear: *C. aequalis* Vahl, *C. alboviridis* C. B. Clarke, *C. esculentus* Linn.—; in Guiana: *C. elegans* Linn.

A. Rhacilla of spikelets not much winged; spikelets spicate or subracemose, linear, 6-20 flowered. Leaves and bracts long ... ... 1. *C. iria.*
B. Rhachilla of spikelets (usually conspicuously) winged:
  1. Tall plants. Stolons long, hardening into creeping rhizomes. Leaves short. Glumes approximate, closely imbricate:
    a. Spikelets shortly spicate, 12-50 flowered ... 2. *C. articulatus.*
    b. Spikelets linear, pale straw-coloured ... ... 3. *C. scariosus.*
  2. Tall or medium plants, stoloniferous or rhizome woody. Leaves and bracts long. Rhachilla of spikelets winged. Glumes closely imbricate:
    a. Stem at base nodosely thickened:
      i. Leaves flaccid, flat; umbels expanded, usually compound ... ... 5. *C. rotundus.*
      ii. Leaves rigid, filiform, or very narrow; umbel simple, compact ... ... 7. *C. stoloniferus.*
    b. Stem at base not nodose:
      i. Stem 6-12 in high; leaves as long or nearly so 6. *C. esculentus.*
      ii. Stem 12-36 in. high; leaves $\frac{1}{2}$-$\frac{3}{4}$ stem ... ... 4. *C. longus.*
  3. Tall stout plants. Stolons absent. Leaves and bracts long. Umbel large, compound. Spikes and spikelets many. Rhachilla of spikelets conspicuously winged. Anthers apiculate, not rarely crested. Style with its three branches small. Nut small ashy black.
    Spikelets linear subterete suberect dirty straw-calour ... ... 8. *C. platyphyllus*

1. Cyperus iria Linn. is found throughout India and Ceylon in rice fields. It extends to Malaya, Indo-China, China, Japan, Australia, Afghanistan, Persia, the Mediterranean, and tropical Africa.

The plants is tonic, stimulant, stomachic, and astringent.

Among the Mundas of Chota Nagpur the tuber is ground together with that of *C. rotundus*, and drunk in fever.

Bengal : Burachucha—; *Hasada* : Jimtu—; *Hindi* : Burachucha, Nagarmotha—; *Lao* : Yak dan din—; *Malay* : Rumput tuloh belalang—; *Mundari* : Huringloeongjimtu, Loeongjimtu, Loeongjintu—; *Naguri* : Jindu—; *Rajputana* Moth—; *Sinhalese* : Welhiri—.

2. **Cyperus articulatus** Linn. occurs from Bengal to Ceylon on the banks of rivers and lakes. It inhabits the tropical and warm regions of both hemispheres.

The tuber is a good tonic and stimulant. It is used medicinally in Hausaland.

Arabic : Chasegj—; *Betsimisaraka* : Mita—; *Egypt* Goreyb—; *Guinea* : Gorhe, Madia—; *Kanuri* : Kajiji—; *La Reunion* : Jambelon—; *West Indies* : Adrue—

3. **Cyperus scariosus** Br. is fonnd in Bengal and Pegu, and extends thence to Australia.

The root is recognized by both Ayurveda and Yunani. Hindu writers describe it as pungent, acrid and coolin; promoting the flow of milk; regulating the bodily temperature by its antipyretic and sedative action; useful in the treatment of biliousness, fever, dysentery; relieving thirst, fatigue, burning sensations, and bad taste in the mouth.

Mohammedan authors qualify the taste of the root as bitter, hat, and unpleasant As a drug it relieves flatulence, stimulates the menstrual discharge, enriches the blood, and improves the appetite; it also checks abnormally profuse menstruation and urination; it is useful in the treatment of stuttering, offensive discharge from the nose, eye sore, brain and chest troubles, piles, swellings, lumbago, and scorpion sting.

In practice the root is considered cordial, stomachic, and desiccant ; also diaphoretic and diuretic. A decoction is used in gonorrhoea and in syphilitic affections; and it has been found useful as an astringent in diarrhoea.

In cases of epilepsy the root is given in combination with valerian.

Caius and Mhaskar have shown experimentally that the drug is not an antidote to scorpion venom.

Arabic: Soad, Soadekufi—; *Bengal*: Nagarmutha—; *Burma*: Vomonniu—; *Canarese* : Konnarigadda, Nagarmusthe—; *Deccan* Nagarmotah—; *Gujerati* Nagaramothya—; *Hindi*: Nagarmotha—; *Malayalam* : Korakizhanna—; *Marathi*: Lawala—; *Persian*: Mushkezamin—; *Sanskrit* : Chakranksha, Charukesara, Chudalapindamusta, Kachharuha, Kalapini, Nadeyi, Nagarmusta, Nagarotha, Shishire, Vrishadhmankshi, Uchchta—; *Tamil* : Koraikilangu—; *Telugu* : Kolatungamuste, Tungagaddalaveru—; *Urdu* : Nagarmotha—.

4. **Cyperus longus** Linn. occurs at Quetta and Mount Abu. It extends westwards to the Atlantic.

The bitter aromatic tuber is used in Spain as a stimulant, stomachic, and emmenagogue.

The Zulus prepare an enema from the tuber for children with stomach troubles. They also blow the powdered tuber into the nose and ears for colds and other troubles in these regions, and the tuber may be chewed for the same purpose.

At Filabusi in Southern Rhodesia the juice of the plant is regarded as being very poisonous and is said to burn the skin when applied to it.

Catalan : Castanyola—; *Chinese* : Hiang Fon Tzeu—; *Egypt* : Sa'ad—; *Filabusi*: Mlabie—; *French*: Souchet long, Souchet odorant—; *Italian*: Giunco odorato—; *Spanish* : Juncia olorosa—.

5. **Cyperus rotundus** Linn. is common in waste grounds, gardens, and roadsides in open spots. It is found throughout India and Ceylon, and in all warm parts of the world.

Ayurvedists describe the root as pungent, acrid, cooling, astringent, bitter, stomachic, and anthelmintic. It restores lost appetite, and is useful in leprosy, thirst, fever, diseases of the blood, biliousness, dysentery, intense itching, pain, vomiting, epilepsy, ophthalmia, and erysipelas.

According to Arabian and Persian writers the root is attenuant, diuretic, emmenagogue, diaphoretic, anthelmintic, ane vulnerary. They recommend it for the treatment of ulcers and sores, fevers, dyspepsic, and urinary calculus. They prescribe it in large doses as an anthelmintic.

The roots are commonly used as a diaphoretic and astringent. They are also credited with stimulant and diuretic properties. They are held in great esteem as a cure for disorders of the stomach and irritation of the bowels. They are scraped and pounded with green ginger and, in this form, mixed with honey, they are given in cases of dysentery in doses of about a scruple, and are said to have been given with benefit in cholera. They are used too as a worm remedy. In the Philippines the bruised root is applied to the face for toothache.

In the Konkan the fresh tubers are applied to the breast to stimulate the flow of milk. In, Chota Nagpur a decoction is used in fever.

In Malaya the tubers are smoked in cases of pain in the nose. In Cambodia they are a common remedy for fever and retention of urine.

According to the Chinese the small tubers act on the lungs and liver. Their general action is tonic, stimulating and stomachic.

In Ceylon a decoction is given in fever, diarrhoea, dyspepsia and stomach complaints.

The tuber is an Ayurvedic medicine for snake bites and scorpion stings; but Mhaskar and Caius have demonstrated that it is useless in the treatment of either snake bite or scorpon sting.

Sanjiva Rao, Panicker and Sudborough have studied the chemical composition of the essential oil from the tubers, but have not been able to isolate any pure substances from it (1925). Further work (1928) by Kimura and Ohtani does not throw any light on the therapeutic properties of the drug.

Annam : Cu co, Huong phu—; *Arabic* : Suad—; *Australia* : Yelka—; *Bengal* : Ghun, Motha, Mutha, Nagurmotha—; *Bicol* : Botobotones—; *Bombay* : Barikmoth, Musta—; *Cambodia* : Kravalchruk, Kravanhchruk—; *Canarese* : Tungegadde—; *Chinese* : Houi Teon Ts' in, Hsiang Fu, So Ts'ao—: *Deccan* : Korekijhar—; *Egypt* Burbeyt, Dis, Negil, Sa'ad, Sa'ad-el-homar, Sibl-el-ma'iz—; *English* : Nut Grass—; *Guam* : Chaguan, Humatag—; *Gujerati* : Motha—; *Hamadan* : So-ad—; *Hasada* : Sunumsangga—; *Hausa* : Ayaaya—; *Hindi* : Ghun, Motha, Mutha, Nagurmotha—; *Ho* : Ratesila—; *Iraq* : Ghusainiyah, Sa'ad, su'aid, Sijil—; *Japan* : Hamasage, Kobushi—; *Kut* : Ghusainiyah—; *Lao* : Ya m'niu mu—: *La Reunion* : Oumine—; *Malaya* : Heong foo, Rumput haliva hitam—; *Marathi* : Bimbal, Motha—; *Mundari* : Bathabijir—; *Naguri* : Mothatasad, Pirjimtu—; *Nasirabad*: Kabb—; *New Caledonia* : Jilio—*Pampangan* : Cusung, Galonalpas, Malaapolid, Mota, Omading, Omadiung, Onoran, Sursur—; *Persian* : Suad, Su'd, Muschk-i-zemin—; *Sadani* : Mothaghas—; *Sanskrit*: Abda, Arnoda, Bhadrakshi, Bhadramusta, Gangeya, Granthi

Gundra, Hima, Kachhola, Kakshottha, Kasheru, Krodeshtha, Kuru, Kurubilva, Kutannata, Musta, Mustaka, Sugandhigranthila, Valya, Varahi, Varida, Vindakhya—; *Santali* : Tandisura—; *Sinhalese* : Kalanduru—; *Sokoto* : Girigiri—; *Spanish* : Juncia redunda—; *Sudan* : Seid—; *Tagalog* : Mutha—; *Tamil*: Kora, Korai—; *Telugu*: Bhadramuste, Gandala, Kaivartakamuste, Mustakamu Shakhatungaveru, Tungamuste—; *United States of America* : Nut Grass—; *Uraon* : Utrubanda—.

6. **Cyperus esculentus** Linn. inhabits the Upper Gangetic Plain and is found scattered from the Punjab to the Nilgiris and Anamalais. It is distributed over Southern Europe, Africa and America.

According to Ayurveda the tuber is cooling, sweet, acrid; galactagogue, astringent to the bowels, aphrodisiac; sharpens the appetite and improves the taste; useful in eye troubles, burning sensations, and leprosy; disturbs both heat production and heat regulation in the body.

The juice expressed from the tubers is used as an aphrodisiac in Sierra Leone and on the Gold Coast. In Guinea the tubers are given as a cooling drink; the leaves are applied topically for headache.

The root is an Egyptian remedy for colic and hypochondriasis.

The Zulus chew portions of the root for the relief of indigestion, especially, when this condition is accompanied by foul breath. Zulu girls, with a view to hastening the inception of menstruation, eat porridge in which a handful of the boiled roots has been mashed.

In Madagascar the tuber is used as a stimulant and aphrodisiac.

The tubers are used as a vegetable in Southern Europe, and Northern and Southern Africa, and, after roasting and grainding, as a substitute for coffee and cocoa. They contain fatty acids, sucrose and starch; but no alkaloid, caffeine or asparagin.

Ada: Fie—; *Afrikaans* : Euntjie, Hoenderuintjie, Uintjie—: *Catalan* : Chufa—; *Egypt* : El-aswad, Es-sogheyyer, Habb-el-'aziz, Habb-el-'azizza. Sa'ad—; *English* : Chufa, Earth Almond, Ground-almond, Nut Sedge, Rush Nut, Tiger-nuts. Zulu Nut—; *Ewe* : Fie, Fio—; *Fanti* : Atadwe—; *French* : Souchet comestible, Souchet sultan, Souchet tubereux, Trasi—; *French West Africa* : N'ton—; *Fufulde* : Watujp—; *Ga* : Atangwe—; *Greek* : Kiperos - ; *Hausa* : Aya—; *Hindi* : Chichada—; *Hova* : Karepoka—; *Kreni* : Fio—; *Krobo*: Fai—; *Madagascar*: Karekika—; *Malinke*: Toki—; *Malta* : Babbagiggi, Chufa, Dolcichini, Edible Rushnut. Habbghaziz—; *Punjab* : Dila, Kaseru—; *Sanskrit* : Kaseruka, Kshudramusta. Sugandhi, Sukanda. Sukareshta—; *Spanish*: Chufa, Juncia avellanada—; *Transvad* : Einches—; *Twi* : Atadwe—; *United States* : Chufu—; *West Africa* : Rush Nut. Tiger Nut—; *Zulu* : in Dawo—.

7. **Cyperus stoloniferus** Retz is found on the shores of India. especially in sea sand, from Sind to Ceylon, Coromandel and the Malay Peninsula. It has also been found in the Nilgiri Mountains. It is common in Malaya where it is found in the sandy sea shores, and in tidal mud: in the former short and wiry, in the latter taller and longer-leaved. It is distributed to China, Australia and Mauritius.

The scented tuber is reputed stomachic, and considered a good stimulant for the heart.

Persian : Mutransialian—; *South India* : Jatamansi—.

8. **Cyperus platyphyllus** Roem. and Schult. occurs in the Decean Peninsula and in Ceylon.

The tuber is said to be tonic and stimulant.

FIMBRISTYLIS.

This genus consists of 225 species, chiefly tropical.

Fimbristylis junciformis Kunth occurs throughout India from Kashmir to Madras and Pegu, up to altitudes of 1, 500,-5,000 ft. It is also found in the islands of Ceylon and Madagascar and the Philippines.

The Santals use the root in dysentery.

Santali : Bindimuthi—.

JUNCELLUS.

The genus numbers about 20 species inhabiting all warm regions.

Juncellus inundatus C. B. Clarke is frequent in swamps, from Sylhet to the sea. It extends to China.

The tubers are used as a tonic and stimulating medicine.

Bengal : Pati—; *Hindi* : Pati—.

KYLLINGA.

This genus includes 50 species inhabiting tropical and subtropical regions.

K. odorata Vahl is used medicinally in Brazil.

A. Nut-bearing glume winged in upper half of keel ... 1. *K. monocephala*
B. Nut-bearing glume not winged in the upper half of its keel:
1. Rhizome practically absent ... ... 2. *K. triceps*
2. Rhizome elongate ... ... 3. *K. brevifolia*

1. **Kyllinga monocephala** Rottb. is found throughout India and Ceylon, generally growing in the shade. It is very common in the waste grounds and along the roadsides of Singapore and other parts of the Malay Peninsula. It occurs in the hot and warm regions of the Old World, except the Mediterranean.

The root is a popular cooling medicine much used in fevers. It is often used in the same way and for the same purposes as that of *K. triceps.*

The herb is used as an antidote in many parts of India; but Mhaskar and Cajus have shown that it is not an antidote to either snake venom or scorpion venom.

Bengal: Nirbishi, Svetagothubi—; *English*: Button Sedge—; *Guam*: Boton cillo, Changuan lemae—; *Hawaii* : Kaluja—; *Hindi* : Nirbishi, Swetgothubi—; *Malay* : Rumput butong—; *Malayalam* : Mottenga, Pimottenga—; *Marathi* : Mustu—; *Pampangan* : Malaapolid—; *Sanskrit* : Musta, Nirvisha—; *Tagalog* : Anuang—; *Visayan* : Barubatones, Barubotones, Bolobotones, Borobotones, Bosicad, Botonsilo, Malabotones, Mutha, Sudsud. Tobotobolangit—.

2. **Kyllinga triceps** Rottb. is to be found in Sind, Gujerat, Deccan, Konkan, the Southern Mahratta Country, the North-West Provinces, Rajputana, Burma, Ceylon. It extends to China, North Australia and Africa.

The herb is an Ayurvedic drug described as bitter and cooling, good against infection or poison, useful in healing wounds and regulating the heat of the body, valuable in the treatment of nervous troubles and of diseases of the blood.

In Malabar a decoction of the roots is used to relieve thirst in fevers and diabetes. Oil boiled with the roots is used to relieve intensive itching of the skin.

The roots yield an essential oil which is used to promote the action of the liver and to relieve thirst in fevers and diabetes.

Bengal: Nirbishi, Svethgothubi—; *Hindi:* Nirbisi, Shwetgothubi—; *Malayalam:* Mottenga, Pimottenga—; *Marathi:* Mustu—; *Sanskrit:* Apavisha, Avisha, Nirvisha, Vishabhava, Vishaha, Vishahantri, Vishavairini, Vivisha—.

3. **Kyllinga brevifolia** Rottb. is common throughout India, Ceylon and Malacca. It is common on roadsides and waste ground in the Malay Peninsula. It thrives in all warm regions, except the Mediterranean.

The Malays use the rhizome for poulticing sore legs.

Malay: Rumput tuki—.

REMIREA.

The only species of this genus. **R. maritima** Aubl., is found on all tropical sea-shores It is common in sea-sand Canara, Tenasserim, the Malay Peninsula, the Nicobars, Ceylon. It is very common in tropical America.

The stock is astringent and diuretic. An infusion of the root is given as a sudorific and diuretic in Brazil and Guiana.

Betsimisaraka: Bararatandrinka—; *Brazil:* Paratura—; *French Guiana:* Remire—.

SCIRPUS.

The genus includes about 200 species, dispersed all over the world, as far as the Polar regions.

S. lacustris Linn. is used medicinally in Europe. *S. maritimus* Linn. in China, *S. cernuus* Vahl and *S. paludicola* Kunth in South Africa.

A. Nut marked with transverse wavy lines; hypogynous bristles absent :
1. Spikelets in a single lateral dense head ... 1. *S. articulatus.*
2. Spikelets clustered in rays of a lateral umbel ... 2. *S. corymbosus*

B. Nut not marked with transverse wavy lines; hypogynous bristles present:
1. Glumes 2-fid at the apex:
a. Stems cylindrical. Rhizome thick ... ... 3. *S. lacustris.*
b. Stems trigonous. Rhizome with tuber-like round swellings ... ... 4. *S. maritimus*
2. Glumes not 2fid at the apex ... ... 5. *S. grossus.*

1. **Scirpus articulatus** Linn. is found throughout India from the Himalaya to Ceylon and Moulmein, ascending up to 3,000 ft. It extends to the Philippine Islands, Australia and Africa.

The plant is used as a purgative.

Bengal : Putputichechka—; *Hindi* : Chichora, Patpatichechka—: *Telugu* : Luttipittaalli—.

2. **Scirpus corymbosus** Heyne occurs in Western India, Mount Abu, Goonah, Jubbulpore. Hyderabad and Bangalore. It is also found in Africa and Madagascar.

In South Africa it is suspected to be poisonous to cattle.

3. **Scirpus lacustris** Linn. is to be found in Kashmir and Ladak reaching altitudes of 4,500-5,000 ft., also in Muneypore. It is distributed over Europe, Africa, Australia and North America.

The stock is astringent and diuretic.

French : Jone des chaisiers, Jone d'ean, d'etang, Scirpe des marais—; *Madagascar* : Sabotraka, Savotraka—; *Malta* ; Bass, Common Bulrush, Giunco di palude—.

4. **Scirpus maritimus** Linn. is found from Kashmir, Kashgar —10,000 ft., and Moradabad to Mysore and Malabar. It is distributed all over the Old World and is represented by varieties in Australia and America.

The root is used in China as an astringent and diuretic.

Chinese: Ching San Ling—; *Egypt*: Debshe—; *Malta*: Sea-side Clubrush Spurt Grass, Mosca, Erba nocca—; *Nasirabad* :Kab—; *Punjab*: Dila, Murak--; *Sibi* : Kab —.

5. **Scirpus grossus** Linn. fil. occurs throughout India with the exception of the North-Western area. It is found from Sind and Assam to Ceylon and Malacca, extending to the Malay Islands, Cochinchina and Indo-China.

According to Ayurveda its properties are the same as those of *Cyperus esculentus*.

Yunanists describe the root as slightly sweet, cooling, laxative, tonic to the liver, good against infection and poison, diuretic; useful in burning sensations, vomiting, diarrhoea, fever, and gonorrhoea.

The root has astringent properties and is given to check diarrhoea and vomiting.

Bengal : Kasura, Kesari, Kesor, Kesur—; *Bombay* : Kachera—: *Hindi* : Kasurn, Kesur—; *Malay* : Rumput mendarong, Rumput murong. Rumput musiang—; *Marathi* : Kasara—; *Mundari* : Jomekesari, Kesari, Khesari, Marangkesari—; *Sanskrit*: Gundakanda, Kaseru. Kaseruka—; *Telugu*; Gunda tungagaddi—; *Uraon* : Kesari—; *Urdu* : Kaseru—.

THE MEDICINAL AND POISONOUS FERNS OF INDIA.

BY

J. F. CAIUS, S.J., F.L.S.

The FILICES are a very large and important family of plants consisting of about 170 genera and upwards of 3,500 species. They are distributed all over the globe, although they find their headquarters in tropical America and tropical Asia. Even the Arctic zone is not excluded. They are found at all elevations, from 10,000 to 12,000 feet in the Tropics, down to the sea level.

The medicinal and poisonous ferns of the world belong to 44 genera:—ACROSTICHUM (Tropical America); ACTINIOPTERIS (North Africa, Mascarene Islands, Persia, Afghanistan, India, Ceylon); ADIANTUM (Cosmopolitan, especially tropical America); ALSOPHILA (South Australia); ASPIDIUM (Tropical regions); ASPLENIUM (Cosmopolitan); ATHYRIUM (Cosmopolitan); BALANTIUM (Cape Verde, Canary Islands, Madeira, Azores); BLECHNUM (Tropical America); BOTRYCHIUM (Cosmopolitan); CERATOPTERIS (Tropical countries); CHEILANTHES (Tropical and temperate regions, xerophytic); CIBOTIUM (Tropical America, Polynesia, Asia); CYCLOPHORUS (Tropical countries); CYSTOPTERIS (Temperate regions); DAVALLIA (South Spain, Portugal, Azores, Madeira, Canary Islands, Cape Verde); DICKSONIA (Australia); DRYMOGLOSSUM (Tropical regions); DRYNARIA (Palaeo-tropics); DRYOPTERIS (Northern temperate regions); GLEICHENIA (Tropical and subtropical regions); GYMNOGRAMME (Central America; xerophytic); HELMINTHOSTACHYS (Ceylon, Himalaya to Queensland); HEMIDICTYUM (India, Western Asia, Europe, North-West Africa); LOMARIA (India, Ceylon, Philippine Islands, Fiji, New Zealand, South Australia, Tasmania); LYGODIUM (Tropical Asia, Eastern Asia); MARSILIA (Europe and subtropical regions); MERTENSIA (Tropical and subtropical countries); MOHRIA (East Africa, Madagascar); NEPHRODIUM (Northern temperate regions); NEPHROLEPIS (Tropical and subtropical regions); NOTHOCLAENA (Subtropical regions, Mediterranean); ONOCLEA (East Asia, North America); OPHIOGLOSSUM (Tropical and temperate regions); OSMUNDA (Temperate and tropical countries); PELLAEA (Subtropical regions); PLEOPELTIS (Tropical and subtropical regions); POLYPODIUM (Tropical and temperate regions); POLYSTICHUM (Moist temperate regions); PTERIDIUM (Cosmopolitan); PTERIS (Cosmopolitan); SCOLOPENDRIUM (Europe, Asia); STENOLOMA (India, Ceylon, Malaya, China, Japan, Polynesia, East African Islands); WOODWARDIA (Tropical and subtropical regions).

The medicinal and poisonous ferns of India belong to the following 24 genera:—ACTINIOPTERIS, ADIANTUM, ASPIDIUM, ASPLENIUM, ATHYRIUM, BLECHNUM, BOTRYCHIUM, CERATOPTERIS, CHEILANTHES, CIBOTIUM, CYSTOPTERIS, DRYMOGLOSSUM, DRYNARIA, DRYOPTERIS, GLEICHENIA, HELMINTHOSTACHYS, HEMIDICTYUM, LYGODIUM,

8

OPHIOGLOSSUM, OSMUNDA, PELLAEA, PLEOPELTIS, PTERIS, STENOLOMA.

I. *Gleicheniaceae*—Sori dorsal, without indusium, composed of a few sporangia having a transverse or obliquely transverse complete ring and opening vertically.
Caudex creeping; stipes forked; segments small, almost round or pectinate ... ... GLEICHENIA.

II. *Dicksonieae*—Sori globose; indusium inferior subglobose, free, closed, at length bursting irregularly, more frequently cup-shaped, entire or with 2 lips.
Indusium apical on a vein 2-valved ... ... CIBOTIUM.

III. *Davallieae*—Indusium squamiform, suborbicular or tubular, open at the apex.
a. Indusium apical, compound, sub-orbicular, only open at the top ... ... STENOLOMA.
b. Indusium medial on a vein membranaceous, hood-like ... ... CYSTOPTERIS.

IV. *Pterideae*—Indusium oblong or linear, formed of the more or less changed and reflexed margin of the frond, opening inwardly.
a. Indusium globose to linear, usually many and distinct, sometimes confluent and continuous bearing the capsules on its under side; veins free ... ... ADIANTUM
b. Indusium rounded and distinct, or more or less confluent but not continuous; capsules on the frond ... ... CHEILANTHES.
c. Indusium quite continuous, sori at first dot-like, but soon running into a line ... ... PELLAEA.
d. Indusium quite continuous; sori linear continuous, occupying a slender filiform receptacle in the axis of the indusium; veins free ... ... PTERIS.
e. As in *Pteris*, but veins all anastomosing without free included veinlets ... ... DRYOPTERIS.
f. Anomalous ... ... CERATOPTERIS.

V. *Blechneae*—Indusium linear or oblong parallel with the midrib and opening towards it, not near the margin.
Indusium membranaceous, distinct from the margin of the frond, parallel with and usually contiguous to the midrib, veins free ... ... BLECHNUM.

VI. *Asplenieae*—Indusium linear or oblong or horseshoe-shaped, opening towards the midrib, sometimes double; sori attached to the veins.
a. Indusium linear or oblong, single; vein free ... ASPLENIUM.
b. Indusium linear or oblong, more or less curved. ATHYRIUM.
c. Indusium linear or oblong; veins anastomosing obliquely only towards margin of frond ... HEMIDICTYUM.
d. Indusium linear elongated submarginal, fronds fan-like ... ... ACTINIOPTERIS.

VII. *Aspidieae*—Indusium superior, elliptical, subglobose or reniform, fixed either by the centre or a sinus.
Indusium peltate, orbicular or reniform; veins copiously anastomosing with free included veinlets ... ASPIDIUM.

VIII. *Polypodieae*—Sori on the back of the lobes, round or rarely somewhat oblong.
a. Fronds with the base oak-leaf-like or with separate sterile oak-leaf-like small fronds ... ... DRYNARIA.

	b. Fronds various; veins copiously anastomosing with free included veinlets	PLEOPELTIS.
IX.	*Grammitideae*—Sori on the back of the lobes more than twice as long as broad, usually linear. Sori in a continuous linear, or interrupted central or submarginal line, veins reticulated, fronds simple and dimorphous	DRYMOGLOSSUM.
X.	*Osmundaceae*—Capsules 2-valved, opening across the apex, furnished with a short horizontal ring ...	OSMUNDA.
XI.	*Schizaeaceae*—Capsules 2-valved, opening down the side crowned by a complete operculiform ring. Capsule solitary in the axils of large imbricating clasping involucres. Scandent	LYGODIUM.
XII.	*Ophioglossaceae*—Capsules deeply 2-valved, opening down the side nearly to the base without a ring.	
	a. Capsules sessile in 2 rows on a narrow close spike	OPHIOGLOSSUM.
	b. Capsules in small crested clusters forming a loose spike	HELMINTHOSTACHYS.
	c. Capsules in 2 rows on the face of spikes which form a compound panicle	BOTRYCHIUM.

The parts used medicinally are the fronds and the rhizomes. The former are often aromatic, and they are commonly given in infusion for their mucilaginous, pectoral, and astringent properties. On the other hand, the rhizomes usually contain a bitter astringent principle together with a fixed oil which is poisonous to worms.

The common Male Fern, or Shield Fern, which has been known from the times of Theophrastus and Dioscorides as a specific remedy for intestinal worms, particularly the tape worm, is official in all pharmacopoeias. The true Maidenhair Fern is official in Belgium, Portugal, Switzerland, and Turkey. The Canadian Maidenhair and the common Hart's Tongue are recognized by the French Codex, and several species of Polypody Ferns are still retained in the Austrian pharmacopoeia.

ACTINIOPTERIS.

This genus consists of a single species, ***A. dichotoma*** Bedd., which resembles a miniature palm. It is found throughout India, epecially the Peninsula, in dry rocky places, below 4,000 ft. From Ceylon to the Mascarene Islands, Maarlisberg, Zambesiland, Angola, Upper Egypt, Abyssinia, Socotra, Mount Sinai, Arabia, Persia, Afghanistan.

It is used medicinally as a worm remedy, and as an astringent to arrest hemorrhage.

Bombay: Bhuitad, Mapursika, Mayursikha—; *English*: Peacock's Tail—; *North-Western Provinces*: Morpach, Morpankhi—; *Sanskrit*: Mayurshikha—.

ADIANTUM.

This cosmopolitan genus numbers about 190 species, mostly tropical American.

Tradition has attributed to various species of this genus of ferns valuable properties in chronic pulmonic catarrhs

The following species are used medicinally in Europe, Indo-China, Mexico—*A. capillus-veneris* Linn.—; in China—*A. capillus-veneris* Linn., *A. flabellulatum* Linn., *A. monochlamys* Eat.—; in North America—*A. capillus-veneris* Linn., *A. pedatum* Linn.—; in Brazil—*A. cuneatum* Langsd. and Fisch., *A. radiatum* Linn., *A. subcordatum* Sw., *A. tenerum* Sw.—; in South Africa and La Reunion—*A. capillus-veneris* Linn., *A. aethiopicum* Linn.—.

The fronds of *A. pedatum* Linn. are officinal in France; those of *A. capillus-veneris* Linn. in Belgium, Portugal, Switzerland, Turkey.

Nine therapeutically active species may be found growing in India.

I. Frond simply pinnate, rachis rooting at the apex.
  a. Pinnae half-moon shaped, distinctly stalked, $\frac{3}{4}$-$\frac{1}{2}$ inch by $1\frac{1}{2}$-1 inch. Texture herbaceous — 1. *A. lunulatum.*
  b. Pinnae wedge-shaped at the base, nearly sessile, $\frac{1}{2}$-$\frac{3}{4}$ inch by $\frac{1}{4}$ inch. Texture leathery ... — 2. *A. caudatum.*

II. Frond at least bipinnate; segments fan-shaped with the stalk near the centre; sori oblong or obversely kidney-shaped.
  a. Pinnules firm, membranaceous-chartaceous, glabrous, scarcely ever or but slightly 2-3 lobed, fertile lobes with 2, rarely 3 notches, each notch bearing a rather large sorus at the bottom ... ... — 3. *A. venustum.*
  b. Sori roundish or transversely oblong.
    i. Pinnules distinctly wedge-shaped at the base. Sori in shallow depressions of the lobes.
      † Pinnules deciduous ... ... — 4. *A. tenerum.*
      †† Pinnules not deciduous ... .. — 5. *A. capillus-veneris.*
    ii. Pinnules roundish, being straight, almost wedge-shaped or rounded at the base. Sori in deep hollows of the lobes ... ... — 6. *A. aethiopicum.*
  c. Sori obversely kidney-shaped in deep round hollows of the lobes ... ... — 7. *A. cuneatum.*

III. Frond dichotomous ... ... — 8. *A. pedatum.*

IV. Frond flabellate ... ... — 9. *A. flabellulatum.*

1. **Adiantum lunulatum** Burm. This very graceful fern occurs throughout Northern India in moist places; in South India it is very general on the western side in the plains and lower slopes of the hills. It is found in Burma, Ceylon, the Malay Peninsula, Perak, Cochin-China, China, Hongkong, Polynesia, Tropical Australia. It extends to Cape Verde Isles, Angola, Guinea, Zambesi-land, Madagascar. In Tropical America it grows from Mexico southward to the Organ Mountains in Brazil.

Ayurvedists describe the plant as pungent, cooling, alterative, alexiteric, and indigestible. They consider it useful in dysentery, diseases of the blood, ulcers, erysipelas, burning sensations, and epileptic fits. They recommend the rhizome for strangury and for fever due to elephantiasis.

In general the fronds are considered to be deobstruent, diuretic, emmenagogue, resolvent and pectoral. As a pectoral they are used in pulmonary catarrh. They are also demulcent and mildly stimulant. As a discutient their plaster is a useful application on chronic gouty and other swellings. Burnt with oil they are used as an application for itch.

In Gujerat the plant is extensively used in the treatment of children for febrile affections. The leaves are rubbed with water and given with sugar. The herbaceous parts pounded with ochre are applied topically for erysipelatous inflammations.

Arabic: Shirulajibala—; *Bengal*: Goyalelata, Kalijhant—; *Bombay*: Hansraj, Hansaraj, Kombada, Mubarak, Ratakombada, Rajahans—; *Canarese*: Navalad—; *Gujerati*: Hanspadi, Hansraj, Mubarkha, Mubarkhinipalo—; *Hindi*: Hansapadi, Hansapagi, Kalijhamp, Kalijhant, Paresiyavasan—; *Ilocano*: Dalipaco—; *Marathi*: Ghodkhuri, Hansari, Kamsaraj, Rajhans—; *Persian*: Parasiyavashana—; *Philippines*: Culantrillo—; *Porebunder*: Hansraj, Kalohansraj—; *Sanskrit*: Brahmadani, Chitrapada, Dharttarashtrapadi, Ghritamandalika, Godhangri, Godhapadika, Hansaghri, Hansapadi, Hansavati, Karnati, Kiramata, Kirapadika, Kitamari, Madhusrava, Padangi, Raktapadi, Sancharini, Shitangi, Sutapadika, Suvuka, Tamrapadi, Tridala, Tripadi, Triphala, Vikranta, Vishvagranthi—; *Tagalog*: Caicai, Gayomanmanoc, Lamotlamotan, Lomotlomotan—.

2. **Adiantum caudatum** Linn. occurs throughout India, Ceylon, and the Malay Peninsula in the plains and lower slopes of the hills. It extends to Java, Borneo, the Philippine Islands, China, and Arabia Felix. It is also found in the Cape Verde Islands, the banks of the Niger, Angola, Zambesiland, the Cape Colony, and Mauritius.

The fronds are used as a cure for cough and fever. They are also employed externally as a remedy for skin diseases.

Cutch: Mayurshika—; *Punjab*: Adhsaritakajhari, Gunkiri—; *Sanskrit*: Mayurashikha—.

3. **Adiantum venustum** Don. is a common Himalayan fern distributed from Afghanistan, through Kashmir and the Punjab, to Nepal. Blanford says this is 'one of the commonest and most abundant ferns of Simla, covering banks and sloping ground in shady places, and ranging from 4,500 ft. up to the top of Hattu at 10,500 ft.' It grows in the soil in forest, often forming for miles the most characteristic under-vegetation.

This maidenhair is a famous Yunani drug, and at one time it used to be exported in cartloads from Jaunsar, along the cart-road to Saharanpur, for some medicinal purpose. Mahommedan writers describe the fronds of the plant as slightly bitter, resolvent, deobstruent, expectorant, diuretic, emmenagogue, purgative, aphrodisiac, and alexipharmic. They consider them useful in biliousness, phlegmatic humours, inflammations, diseases of the chest, colds, headache, tumours, ophthalmia, and hydrophobia. They prepare an oil which they apply to piles, tuberculous glands, and wounds, and also 'to bring out a thorn which has penetrated into the body'. The leaves made into a plaster are applied topically to chronic tumours of various kinds.

There is no doubt as to this fern possessing astringent and aromatic properties, and being emetic in large doses. It is a tonic, febrifuge, and expectorant. The plant is very useful as a mild tonic, especially during convalescence from fevers. A vapour bath medicated by a decoction from the leaves of this fern is regarded as useful in fever.

In Chamba the plant is pounded and applied to bruises.

It is this fern which supplies in the Punjab most of the *hansraj* which is commonly administered as an anodyne in bronchitis, and is considered diuretic and emmenagogue.

As an ointment it is used for the prevention of hair from falling. The ashes of the plant mixed with olive oil and vinegar are used to make the hair grow upon the bald patches produced by ringworm of the scalp.

The plant is one of the ingredients in Sushruta's Vidaryadi-gana recommended for the treatment of scorpion sting. On the other hand Yogaratnakara, Nighantaratnakara, and Brihannighantaratnakara advise macerating the rhizome and putting the juice and pulp into the ear of the patient. But from their experimental work with mice Caius and Mhaskar have concluded that no part of the plant is an antidote to scorpion venom.

Arabic: Kuzburatelbir, Masifelaswad, Sakelasward, Shaerelfual, Shaereljibal, Shiruljibal, Shiruljinn—; *Bombay*: Mubarak—; *Hindi*: Hansraj, Kalijhanp, Kalijhant—; *Persian*: Hansraj, Paresiyawashan, Parsiawashan—; *Sanskrit*: Hansapadi—; *Tamil*: Mayirsikki—; *Urdu*: Mobarkha, Parsia ushan—.

4. **Adiantum tenerum** Sw. is found in Mexico and the West Indian Islands, southwards to Juan Fernandez and Peru. One solitary specimen was collected in Sion Wood, Bombay Island, by Blatter and d'Almeida, and regarded as an escape from cultivation.

The plant is demulcent, expectorant, and sudorific. It is much used in Brazil in bronchial and catarrhal affections.

5. **Adiantum capillus-veneris** Linn. appears to be found all over India, where circumstances are favourable; shade and permanent moisture being essentials. It occurs chiefly in the Western Himalaya, ascending to an altitude of 8,000 ft.; but found also far to the east in the valley of Manipur, extending to the mountains of the Burma-Manipur frontier and to Chittagong. It is common in the Punjab, descending even to the plains, where it is found in wells and damp places. It is also quite common in South India, in the Bombay and Madras Presidencies, particularly on the west side up to 5,000 ft. It is rare in Ceylon, but extends to Polynesia, Japan, and South-Eastern China. Through Afghanistan and Baluchistan it finds its way to Arabia, Syria, Siberia, the Caucasus, Central and Southern Europe, South-West England, Isle of Man, and Ireland. It is also met with in the Canary Islands and in many parts of the African Continent, both North and South. It extends from Florida southward to Venezuela and the Amazon Valley.

In the Punjab, the leaves along with pepper, are administered as a febrifuge; and in South India, when prepared with honey, they are used in catarrhal affections.

The herb is mucilaginous, pectoral, and expectorant; and is used as a popular cough medicine throughout most parts of

Europe. It has also been used to stimulate menstrual discharges. It is given in the form of infusion, sweetened with sugar or honey.

In France large quantities are employed in the preparation of *Sirop de Capillaire*, which is given as a favourite medicine in all coughs, throat affections, and bronchial disorders. It is flavoured with orange flowers, and acts as a demulcent with slightly stimulating effects. One part of the plant is gently boiled with ten parts of water, and with nineteen parts of white sugar. Dr. Johnson, says Boswell, used to put *Capillaire* into his port wine. Sir John Hill used to say that the fine syrup made in France from that Fern is not by any means to be thought a trifle, because barley water, sweetened with this, is one of the very best remedies for a violent cold.

'The True Maidenhair,' says Gerard, 'maketh the hair of the head and beard to grow that is fallen and pulled off.'

The Basutos smoke the leaf for head and chest colds.

In Southern California the plant is much used as an astringent, expectorant, and emmenagogue.

At Colomas, in Mexico, the leaf is used as a tea to relieve colic; but at Colothan it is taken as a tea for the absence or abnormal stoppage of the menses.

The rhizome is the part used in Persian medicine; it is credited with expectorant properties and is given for relieving difficult respiration.

Arabic: Shairuljin, Shiruljin—; *Catalan*: Capillera, Falsia—; *Colombia*: Cilantrillo, Culantrillo—; *Dutch*: Venushaar, Vrouhenhaar—; *Egypt*: Kuzbaret-el-bir—; *English*: Maidenhair Fern, Maria's Fern, Our Lady's Hair—; *French*: Adiante, Adianthe, Capillaire, Capillaire commun, Capillaire d'Italie, Capillaire de Montpellier, Capillaire vrai, Cheveux de Vénus—; *German*: Frauenhaar, Venushaar—; *Greek*: Adianton—; *Gujerati*: Hanspadi—; *Hindi*: Hansraj, Mubaraka, Pursha—; *Indo-China*: Duoi chon, Thiet tuyen thao—; *Italian*: Adianto, Capelvenere, Capillare, Capilvenere—; *Kashmir*: Dumtuli—; *Kumaon*: Mubaraka—; *La Reunion*: Capillaire—; *Lepcha*: Luk-sip, Ruk-sip—; *Malta*: Maidenhair, Capelvenere, Tursin il Bir—; *Pacific Coast*: Black Maiden's-hair, Lady's hair, Maidenhair, True Maiden's-hair, Venus's-hair Fern—; *Persian*: Barr-i-sija waschan, Kasnburat-el-bir, Parsia washan, Pusia wechame, Sirsiapeshane—; *Portuguese*: Avenca, Cabellos de Venus, Capillaria, Herva capillar—; *Roumanian*: Chica-voinicului, Perul fetei, Perul sfantei Marii, Vergura invelita—; *Russian*: Adiant, Krasnyi jenskiy volos—; *Salt Range*: Parasigavashan, Parshavarsha—; *Spanish*: Capilera, Capilera de Mompeller, Culantrillo de pozo—; *Suto*: Pata-lewana, Pata-mawa—; *Teheran*: Kashburat—; *Trans-Indus*: Bisfaif, Kirwatzei—; *Turkish*: Baldirikara—.

6. **Adiantum aethiopicum** Linn. (=*A. emarginatum* Bedd.) is found in North Kanara, and at the higher elevations on the Nilgiris, and the Pulney Hills. It extends to Ceylon, Australia, New Zealand, La Reunion, Madagascar, Cape Colony, Natal, Zambesiland, the Cameroon Mounts up to 7,000 ft., Abyssinia, Spain, and Afghanistan. It also occurs from Texas and California southward to Valparaiso and Montevideo

An infusion of the leaves is commonly used as an emollient in coughs and diseases of the chest.

At La Reunion the plant is a popular cough medicine, and much used as a sudorific.

In Basutoland a decoction of the caudex is used to promote

parturition. The natives smoke the leaf for colds in the head and the chest.

Afrikaans: Vrouehaar—; *South Africa*: Large Maidenhair—; *Suto*: Maorumetsoo, Pata-lewana, Pata-mawa—.

7. **Adiantum cuneatum** Langs. and Fisch. is a native of Brazil, very common in cultivation. It runs into many varieties and is difficult to identify.

In Brazil the leaves are a popular cough medicine. The plant is considered a good sudorific.

8. **Adiantum pedatum** Linn. grows all along the Himalayas from Kashmir to Sikkim; scattered, and not plentiful. It extends to China, Manchuria, Japan. It is found abundantly from Unalashka and Canada, southwards to Virginia and California, in wet lands and woods.

The plant is employed in France and North America as a pectoral in chronic pulmonic catarrhs. It is also said to be endowed with astringent and emmenagogue properties.

English: Canadian Maidenhair—; *French*: Capillaire du Canada—; *Pacific Coast*: American Maidenhair, Hair Fern, Rock Fern—; *Spanish*: Capilera del Canada, Culantrillo del Canada—.

9. **Adiantum flabellulatum** Linn. is very common in some parts of India: Nepal, Assam, Khasia, Sylhet. It is found in the Ouvah District of Ceylon, in the Malay Peninsula and the Malay Archipelago, Southern China, Japan.

The rhizome is used as an anthelmintic in the Manipur State and in Annam.

The herb is used in China as a cough medicine.

Annam: Thiet tien thao—; *Chinese*: T'ieh Hsien Ts'ao—.

Aspidium.

The genus consists of 50 species distributed throughout the Tropics.

The following species are used medicinally in Europe—*A. fragile* Sw., *A. roeticum* Linn.—; in China— *A. falcatum* Sw.—; in North America—*A. marginale* Sw., *A. spinulosum* (Mill.) Sw., *A. trifoliolatum* Sw.—; in South Africa—*A. aculeatum* Sw. var. *pungens* Klf., *A. athamanticum* (Hook.) Kuntze—.

Of the numerous species found growing in India, two are therapeutically active:—

I. Indusium orbicular, veins generally anastomosing acutely with free veinlets from their junction; fronds 1-2 ft. long by 6-9 in. broad; sori small copious, scattered ... 1. *A. falcatum*.

II. Indusium often quite absent, peltate orbicular, or reniform, veins copiously anastomosing with free included veinlets; fronds 1-4 ft. long by 1 ft. or more broad; sori on the netted veins, small and scattered in the uncontracted fronds, large and more or less in two rows between the main veins in the contracted ones. 2. *A. polymorphum*.

1. **Aspidium falcatum** Sw. grows in North-Western India and extends from Kashmir and the Punjab to Sikkim and Bhutan, Assam and the Khasia Mountains without being common any-

where; also in the Nilgiris at the higher elevations. It is found in China, Japan, the Sandwich Islands, Caffraria, Natal, and Madagascar.

The Chinese use the rhizome as an anthelmintic, chiefly for the expulsion of the tapeworm.

Chinese: Kuan Chung—.

2. **Aspidium polymorphum** (Wall.) Bedd. grows in the western forests of the Madras Presidency up to 4,000 ft.; in Northern India from Garhwal to Mishmee and Chittagong; in Burma, Ceylon, the Malay Archipelago, the Philippine Islands, and Fernando Po.

The rhizome is used as an anthelmintic.

Asplenium.

This cosmopolitan genus numbers 540 species.

The following species are used medicinally in Europe—*A. adiantum-nigrum* Linn., *A. ruta-muraria* Linn., *A. trichomanes* Linn.—; in La Reunion—*A. adiantum-nigrum* Linn.—; in South Africa—*A. adiantum-nigrum* Linn., *A. cuneatum* Linn., *A. furcatum* Thunb., *A. monanthemum* Linn., *A. trichomanes* Linn.—; in Brazil—*A. regulare* Sw., *A. sulcatum* Lam.—.

Five therapeutically active species are met with in India:—

I. Fronds pinnate.
- a. Fronds 6-12 in. long, about ½ in. broad; sori linear oblong, 3-6 on each side of the midrib — 1. *A. trichomanes.*
- b. Fronds 6 in. long to 2 ft. long or more, 4-8 in. broad; sori in long irregular lines reaching nearly to the margin ... ... 2. *A. falcatum.*

II. Fronds bi-quadripinnate.
- a. Fronds 1-2 in. long, about 1 in. broad; sori copious ... ... 3. *A ruta-muraria.*
- b. Fronds 6-12 in. long, 4-6 in. broad at the base; veins obscure, oblique; sori copious ... 4. *A. adiantum-nigrum.*
- c. Fronds coriaceous 6-18 in. long, 4-6 in. broad; veins flabellate, deeply channelled; sori linear 5 *A. furcatum.*

1. **Asplenium trichomanes** Linn. is a common Himalayan fern occurring from Afghanistan to Kumaon at altitudes of 5,000-11,000 ft. It is very common from 5,000 to 9,000 ft. in the Simla Region. It has been found in South India at Kulhatty on the Nilgiris. It occurs in Japan, Java, Australia, Tasmania, and New Zealand, sub-arctic Greenland, Canada and westward to the Rocky Mountains, British Columbia, the United States, Bermuda and southward to Panama, the West Indies Islands, and from Guiana and Venezuela to Bolivia.

This is the common Maidenhair of the British Isles, distributed over the continent of Europe 'throughout its length and breadth, from Iceland and Lapland to the Rock of Gibraltar; throughout the Mediterranean Region; and from the extreme west of Ireland to the extreme east of Europe', Siberia and Persia. It thrives equally well in the Azores and the Macaronesian Isles,

Algeria, Morocco, Abyssinia, Somaliland, the Ruwenzori Mountain of Central Africa, and the Cape of Good Hope.

Its leaves are sweet, mucilaginous, and expectorant, being therefore highly useful in many pulmonary disorders. A tea brewed from them is one of the common English remedies for a violent cold and for tedious coughs. 'We have the common Maidenhair, which grows on old walls, and which will act as a laxative medicine; whilst idiots are said to have taken it remedially, so as to recover their senses.'

In Basutoland the leaf is smoked for colds in the head and chest.

Catalan: Falsia roja—; *English*: Common Spleenwort, Maidenhair—; *French*: Polytric des officines—; *German*: Rotes Frauenhaar—; *Languedoc*: Fau capillero—; *Spanish*: Politrico—; *Suto*: Lehorometso—; *Tamil*: Mailakkondei—.

2. **Asplenium falcatum** Lam. is found in the western mountains of the Madras Presidency, in Ceylon, the Malay Peninsula, Australia, Polynesia, South Africa and its islands.

In Goa and Malabar the plant is used as an alterative in cases of prolonged malarial fever.

Bombay: Pana—; *Goa*: Pandan—; *Malayalam*: Nelapannamaravara—.

3. **Asplenium ruta-muraria** Linn. is common in Afghanistan and extends through Kashmir and Baltistan as far as West Nepal, with a gap between West Kashmir and Kumaon. It is found in Turkish Armenia, the Ural Mountains to Caucasus, Turkestan, Tibet, and Siberia, Algiers, the Cape Colony, and many parts of the United States. It is distributed all over Europe: from Arctic Regions, to Spain and Portugal, Italy, the Mediterranean Isles, Greece, and Turkey.

This small herb is still used to some extent in England as a deobstruent and expectorant.

'The Wall Rue is likewise good for them that have a cough, or are shortwinded, or be troubled with stitches in the sides. It stayeth the falling or shedding of the hair, and causeth them to grow thick, fair, and well coloured. Also it helpeth ruptures in children.'

The Fern has been called Tentwort, from its use as a specific or sovereign remedy for the cure of rickets, a disease once known as 'the taint'.

In China it is considered a valuable medicine for the treatment of pulmonary diseases.

Catalan: Falsia blanca, Ruda de rata—; *Chinese*: Koui Ts'ang—; *English*: Tentwort, Wall Rue—; *French*: Capillaire blanc, Doradille des murailles, Rue des murailles, Sauve-vie—; *German*: Weinkraeutel—; *Spanish*: Calantrillo blanco mayor—.

4 **Asplenium adiantum-nigrum** Linn. is found in Afghanistan and Kashmir, at 5,000-8,000 ft., extending to Dalhousie and Chamba. It occurs in Java, the Sandwich Islands, Azores, Canaries Madeira, Cape Verde Isles, St. Helena, Mascarene Isles, Cameroon Mounts, Algeria, Abyssinia, Natal, the Cape Colony. It is distributed all over Europe—the United Kingdom, Norway and

Denmark, through Holland, Belgium, Germany, France, Switzerland, Austria, Spain, Portugal, Mediterranean Region, Greece, Turkey— and Northern Asia—Armenia, Syria, Arabia, Persia, and Siberia—.

Yunanists describe the plant as bitter, diuretic, laxative, lessening inflammations and abating hiccough, producing sterility in women. They consider it useful in the treatment of ophthalmia and diseases of the spleen.

A decoction or syrup of the fronds is used as an expectorant, pectoral, and emmenagogue in Europe.

The rhizome is used as an anthelmintic in Basutoland.

Catalan: Falsia negra—; *English*: Black Spleenwort—; *French*: Capillaire noir—; *La Reunion*: Capillaire noir—; *Spanish*: Capilera negra—; *Suto*: Lehorometso—.

5. **Asplenium furcatum** Thunb. is common on the higher western mountains of South India, and in Ceylon at 5,000-7,000 ft. It is found in Tropical America, the Polynesian Islands, Australia, Cape Colony, Abyssinia, the Mascarene Islands, and Canaries.

The rhizome is used as an anthelmintic in Basutoland.

Suto: Lehorometso—.

ATHYRIUM.

This cosmopolitan genus numbers 120 species.

Athyrium filix-foemina Roth. is found all along the Himalayas from Afghanistan to Sikkim, at 6,000-13,000 ft., and extends to Sind and the Bombay Presidency. It occurs throughout Europe from Lapland, Russia and Scandinavia to Spain, Portugal, Italy, Greece, Crete, and the Caucasus. From Japan and Kamschatka it crosses to Sitka and Labrador, Canada, British Columbia, the United States, Cuba, Caraccas and Venezuela. It is also met with in the Azores and Macaronesian Isles, Algeria, the mountains of Cameroon, Abyssinia, and Natal.

The rhizome is supposed to possess vermifugal properties similar to those of the male fern for which it is used as an occasional substitute.

In Germany the rhizomes of this fern are frequently found mixed with those of the true male fern.

Catalan: Falguera femella—; *Spanish*: Helecho hembra—.

BLECHNUM.

This genus consists of about 60 species, mostly tropical American.

Blechnum orientale Linn. is found throughout India, Ceylon and the Malay Peninsula, up to 6,000 ft. in the south, and 4,000 ft on the Himalayas. It extends to the Malay Islands, China, Polynesia, and Australia.

The rhizome is used in China as an anthelmintic.

Cantonese: Kun chung—; *Chinese*: Kuan Chung—; *Malaya*: Koon choong—.

BOTRYCHIUM.

A cosmopolitan genus consisting of about 40 species, three of which are credited with medicinal properties.

I. Stipes erect, smooth, cylindrical, hollow, succulent; vernation plicate or folded straight; the fertile branch clasped by the sterile before unfolding, pinnate or bipinnate ... ... 1. *B. lunaria.*

II. Stipe 1-2 in. long; petiole of the sterile segment 2-8 in. long, the latter 3-6 in. each way; fertile peduncle up to 18 in. long, generally considerably over-topping the sterile segment 2. *B. ternatum.*

III. Stipes 3-8 in. long; sterile portion not prolonged beyond the fertile spike 4-12 in. each way; fertile branch of the rachis springing from the base of the sterile portion or from the middle of it ... ... 3. *B. virginianum.*

1. **Botrychium lunaria** Sw. is profuse in Afghanistan at 9,000-10,000 ft., common from Kashmir to Sikkim ascending up to 13,000 ft., and extending to the Karakoram Range in Western Thibet. It is found in South Australia, Tasmania, New Zealand, Japan, Kamschatka, Greenland, Newfoundland, Canada, British Columbia, New York, Lake Superior, Colorado, Patagonia, Ice-land, Arctic Russia, Livonia, Lithuania and Caucasia, the British Isles, Spain, Italy, Switzerland and the Mediterranean Isles.

The plant is considered a good vulnerary. It is also much used in dysentery.

'This herb is cold and drying, and is available for wounds both outward and inward. The leaves boiled in red wine, and drank, stay immoderate courses and the whites. It stays bleeding, vomiting, and other fluxes. It helps all blows and bruises, and consolidates all fractures and dislocations. It is good for ruptures, and is put into oils and balsams to heal fresh and green wounds.'

English: Moonwort—; *French*: Herbe aux serpents, Petite lunaire—; *German*: Walpurgiskraut—; *Italian*: Vindicta—.

2. **Botrychium ternatum** (Thunb.) Sw. is found in the Himalayas from the Punjab to Nepal and Sikkim, and has been reported from Mount Abu. It extends to Tasmania and New Zealand, Japan, Siberia and Kamschatka, Nootka and Hudson's Bay Territory, Canada, New England, westwards to California, Washington, and southwards to Florida, hence to New Grenada. In Europe it occurs in Lapland, Hungary, and the Pyrenees.

The plant is used as a vulnerary in China; and the root is prescribed in dysentery.

Chinese: Yin Ti Chueh—; *Indo-China*: Am dia guyet—.

3. **Botrychium virginianum** Sw. occurs in the Himalayas from the Punjab to Sikkim. It is found from Norway to Austria, in Ecuador and Brazil, from New Brunswick to Florida, and westwards to Arizona and the Pacific Coast—British Columbia southwards to Oregon.

The fleshy root is used by American Indians in application to cuts and bruises.

Pacific Coast: Grape Fern, Moonwort, Rattlesnake Fern—.

Ceratopteris.

The only species known, **C. thalictroides** Brong., is found throughout India, Ceylon, and the Malay Peninsula up to 3,000 ft. elevation; common in tanks, ditches, and swampy places, or even dry ground during the rains. It extends to the Great Coco Island, the Nicobar Islands, Java, the Philippine Islands, Japan, China, Hongkong, South Florida, Mexico, and the West Indies southwards to Brazil. It also occurs in Arabia Felix and South-Eastern Arabia; and is found in Tropical West Africa, Angola, and Madagascar.

The plant is used in China as a tonic and styptic.

Chinese: Shui Chueh—.

Cheilanthes.

The genus consists of 120 species, inhabiting tropical and temperate regions.

C. hirta Swartz. is used medicinally in South Africa.

Cheilanthes tenuifolia Sw. is common in the Madras Presidency, in the plains and on low hills up 4,000 ft. It occurs in Bengal, in the plains of Assam, Chittagong, Dacca, Chota Nagpore, in Khasia up to 5,000 ft., Sikkim, the Malay Peninsula, and the Malay Islands. It extends to China, Australia, New Zealand, Polynesia, and Uruguay.

The Santals prescribe a preparation from the roots for sickness attributed to witchcraft or the evil eye.

Santali: Dodhari, Nanha—; *Tasmania*: Parsley Fern—.

Cibotium.

This genus contains 10 species, inhabiting tropical America, Polynesia, and Asia.

Cibotium barometz Link. (=*C. glaucum* Bedd.) is found in Mishmi, Assam, and Tavoy, extending to the Malay Islands and South China.

The drug consists of the lower part of the caudex, reddish brown in colour, in longitudinal slices, covered on the outside with golden brown moniliform hairs suggesting the fur of an animal. The hairy rhizome thus sometimes resembles a lamb, hence the specific name *barometz*, Russian for lamb. It is the Agnus Scythicus, or Tartarian or Vegetable Lamb, a drug of great repute alluded to in medical works of the sixteenth and seventeenth centuries.

The root is employed in China as a tonic, and is said to exercise a special action on the genito-urinary organs. It is also given for lumbago.

In Annam the stems are considered tonic and styptic. The rhizome, like other fern roots, is used as a vermifuge.

The golden brown hair from the base of the fronds is used in Malaya, Java and Sumatra for stanching wounds. It has the power of causing rapid coagulation of blood, and, when properly used, of mechanically arresting hemorrhages from capillaries. It has been much used in the physiological laboratories of Europe and America, and was employed in huma. medicine during the Middle Ages.

Annam: Bach chi, Cau quyet, Cau thank, Cau tich, Cay cu lan, Cay cu li, Cay cu lon, Cay ku lien, Cay tien phuong vi, Cuong lu, Cu seoga, Kim Mao, Phu can—; *Cantonese*: Kau Tsek—; *Chinese*: Kou Chi, Toei Choun—; *French*: Agneau de Scythie, Agneau de Tartarie, Chien roux, Poil de singe—, *Malaya*: Kow chiak—.

CYSTOPTERIS.

This genus consists of 5 species distributed through the temperate regions of the world.

Cystopteris fragilis Bern. is found in Afghanistan and on the Himalayas from Kashmir to Sikkim ascending to 10,000-15,000 ft. It extends to Australasia, New Zealand, the Sandwich Isles, the Arctic and temperate regions of North America, California, Mexico, and through the whole length of the Andean Chain. It is found in the West Indies, and everywhere in Europe from Iceland and Novaya Zemlya and Spitzbergen in the Arctic Regions to Spain, Sicily, Cyprus and the Caucasus. It also occurs in the Lebanon, Persia, Kurdistan, Siberia, Manchuria, Kamschatka, Tibet, and North China. It is distributed to Madeira, Fernando Po. Abyssinia, and South Africa.

The Sutos of Basutoland use a decoction of the rhizome as an anthelmintic enema.

Suto: Lehorometso—.

DRYMOGLOSSUM.

This tropical genus numbers about 10 species.

Drymoglossum carnosum Hook. is common in Nepal, Sikkim, and Bhutan at 2,000-5,000 ft. elevation. It extends to China and Japan.

The fronds are pectoral, diuretic, and astringent. They are used in China in urinary calculus and rheumatism.

Chinese: Lo Yen Ts'ao—.

DRYNARIA.

This genus consists of 20 species, inhabitants of the Palaeotropics.

Drynaria quercifolia J. Sm. (=*Polypodium quercifolium* Linn.) is found throughout India, in the plains or very low down on the mountains, on trees or rocks.

The Ayurvedists describe the rhizome as bitter, tonic, astringent to the bowels, and useful in the treatment of typhoid fever.

The plant is commonly used in the treatment of phthisis, hectic fever, dyspepsia, and cough.

Ilocano: Capcapa—; *Malayalam*: Pannakilhannumaravala—; *Marathi*: Ashvakatri, Basingh, Wandurbashing—; *Pampangan*: Gona, Tibatib—; *Sanskrit*: Ashvakatri—; *Tagalog*: Pacpaclauin, Paipaiamo—; *Visayan*: Cabcab, Cabcaban, Cabcabun—.

DRYOPTERIS.

This genus numbers 250 species found distributed throughout the northern temperate regions.

It is probable that all of the species of this genus possess more or less anthelmintic properties. The following are commonly used in China—*D. sophoroides* O. Kuntze—; in Japan—*D. crassirhizoma* Nakai—; in North America—*D. marginalis* (Linn.) Asa Gray, *D. rigida* Underw.—; in South Africa—*D. athamantica* (Ktze) O. Kuntze, *D. inaequalis* O. Kuntze.— But by far the most famous species is *D. filix mas* (Linn.) Schott, one of the oldest drugs known, and one which is still retained by all pharmacopoeias.

Dryopteris filix-mas (Linn.) Schott.—MALE FERN—.

The synonyms for the male fern are extraordinarily numerous, and the following are among the better known or more commonly occurring:—*Aspidium filix-mas* of many authors, *A. wildeanum* Goeppert, *Dryopteris filix-mas* (Linn.) Schott, *Lastrea filix-mas* Presl., *Lophodium filix-mas* Newm., *Nephrodium filix-mas* Rich., *Polypodium filix-mas* Linn., *P. nemorale* Salisb., *Polysti-hum abbreviatum* DC., *P. durum et induratum* Schur., *P. filix-mas* Roth., *Tectarea filix-mas* Cavan. Since the term *Dryopteris* was first used by Amman in 1739, and applied in 1763 by Adamson, as the name of the genus to which the term *Aspidium* was applied in 1800 by Swartz, the use of the generic term *Dryopteris* would seem to be necessitated by the laws of botanic nomenclature.

The Male Fern is of very wide distribution, occurring in America from Greenland, westwards and southwards, along the Rocky Mountains and Andes to Peru. It is found throughout Europe, in North Asia, eastwards to China and Japan. It is also distributed to Abyssinia, the Azores and Macaronesia. In the Indian region it is generally confined to considerable elevations—6,500-11,000 ft.—on the mountains, from the Lowari Pass to Kashmir, Chamba, Kullu, the Simla Region, and Garhwal.

The male fern has been recommended as a local application in eczema and acne. It is, however, as a vermifuge that the drug is better, if not exclusively, known. It is as such that it is mentioned in the works of Dioscorides, Theophrastus, Galen and Pliny, and by some of the earlier modern writers. It is now used in medicine almost solely for the purpose of getting rid of various intestinal parasites, especially the *tapeworm*. As to its value in other forms of helminthiasis there is difference of opinion. Some authors report as many as 75 per cent of cures in cases of *hookworm* infection, but other workers state that the drug has been found absolutely without value in this infection. There has been a great deal of uncertainty as to the efficacy of male fern, especially in tropical climates, and many observers have found it to be not only inefficient and unreliable, but a toxic and a dangerous remedy.

A large amount of chemical work has been done and a number of substances have been isolated from the drug, but the chemical nature of the different constituents is still not clear. This is not surprising as these compounds are unstable bodies and undergo chemical changes in the dry rhizome as well as in the preparations of the oleoresin and the ethereal extract. That is one of the reasons why such widely controversial results have been obtained regarding their physiological activity.

No doubt that the rhizomes of other species of fern are frequently substituted for the 'official', and that in the dried state it is difficult to distinguish them; but which is the official form? A look at herbariums will convince any one that a heterogeneous mass of plants is named *filix-mas*. A talk with the leading pteridologists of the day will further bring home the astounding information that experts are not agreed that even the European forms of *filix-mas* all belong to the same species!

Catalan: Falguera mascle—; *Danish*: Bregne—; *Dutch*: Varen Kruid—; *English*: Male Fern, Shield Fern—; *French*: Fougère mâle—; *German*: Farnkraut, Farnwurzel, Wurmfarn—; *Hungarian*: Erdei pajzsikapapany—; *Italian*: Felce maschio—; *Naples*: Fielici, Filici—; *Norwegian*: Bregne—; *Pacific Coast*: Aspidium, Basket Fern, Bear's Paw, Knotty Brake, Male Fern, Male Shield Fern, Sweet Brake—; *Piedmont*: Fales, Feles, Fleis—; *Portuguese*: Dentebrura, Feto macho—; *Potenza*: Fivece—; *Puglia*: Fidvitti—; *Reggio*: Felsa mas'c—; *Sardinia*: Filighee mas'cia, Filixi maschin—; *Spanish*: Helecho macho—; *Swedish*: Ormbunk, Traejon—; *Treviso*: Felese mas'cio—; *Turkish*: Serhasi müzekker—.

GLEICHENIA.

This genus consists of 25 species inhabiting tropical and subtropical regions.

Gleichenia dichotoma Willd. is found in the mountains of Southern India and Ceylon, up to 6,000 ft., Kumaon, Nepal, Sikkim, and Bhutan, Khasia Hills up to 5,000 ft., Sylhet, Pachmarhi, Tenasserim, South Andamans, the Malay Peninsula, Sumatra, East Timor, Tropical Australia, Japan, America—.

In Annam the rhizome is used as an anthelmintic. In Madagascar the fronds are given as a cure for asthma.

Annam: Hac cot mang, Hac cot phuong, Tieu ly bach—; *Madagascar*: Ampangantsirika—.

HELMINTHOSTACHYS.

The only species, **H. zeylanica** Linn., is found in South India, in the western forests in swampy places up to 3,000 ft. elevation. It also occurs in Ceylon, about Colombo and other parts of the western and southern provinces, in the plains of Bengal to Assam and Cachar. It extends to the Malay Peninsula, the Malay Islands, the Philippine Islands, Tropical Australia, and New Caledonia.

It is regarded in the Moluccas as a mild aperient.

Annam: Quan trong—; *French*: Osmonde de Ceylan—.

HEMIDICTYUM.

This genus consists of 2 species found mostly in temperate regions.

Hemidictyum ceterach Linn. (=*Asplenium ceterach* Linn. =*Ceterach officinarum* Willd.) occurs on the rocks overhanging the

Karriah River in the Kurram Valley, but is rare there. It is very common in Kashmir, Punjab, and Garhwal ascending up to 9,000 ft. It is found in many parts of Germany, in Switzerland, the Tyrol, Hungary, Dalmatia, the Caucasus, Belgium, France, Spain, Italy, Greece. In Britain it is to be seen in all or nearly all the southern, northern, and western counties; 'in Somersetshire and Devonshire it is especially abundant, in Scotland it is much less frequent' chiefly in the west and south-west; it is frequent in Ireland, though local. It extends to Palestine and Persia, the Canaries, Madeira and Cape Verde Islands, Morocco, Algiers, Abyssinia, and the Cape of Good Hope.

'No herbe maie be compared therewith for his singular virtue to help the sicknesse or grief of the spleen', says one of the oldest Herbals. 'It is generally used against infirmities of the spleen, helps the strangury, and wastes the stone in the bladder, and is useful against the jaundice and the hiccough', states Culpeper.

The plant is considered diuretic and astringent. It is still occasionally used in France for diseases of the urinary tract. The rhizome is used medicinally in several parts of Europe for enlargement of the spleen, incontinence of urine, calculus, jaundice, and malaria.

Dioscorides mentions the use of a decoction of the plant in vinegar for enlargement of the spleen, and also the local application of a plaster made of the leaves steeped in wine. Women were not allowed to use it as it was supposed to cause sterility. Pliny ordered 'it should not be given to women, because it bringeth barrenness'.

The rusty-coloured scales under the blades were at one time used as a cure for gonorrhoea.

Catalan: Dauradella, Herba daurada—; *English*: Finger Fern, Miltwaste, Rusty Back, Scaly Fern, Spleenwort, Stone Fern—; *French*: Cétérach officinal, Dauradille, Doradille, Doradille cétérach, Doradille d'Espagne, Herbe dorée, Scolopendre vraie—; *German*: Milzfarn—; *Portuguese*: Douradinha, Escolopendra—; *Spanish*: Capilera dorada, Ceterach, Doradilla—.

Lygodium.

This genus numbers 22 species to be found in tropical and subtropical regions.

Two of the species are credited with medicinal properties:—

Pinnules 8-12 in. long, 6-12 in. broad ... ... 1. *L. flexuosum.*
Pinnules smaller, 4-8 in. long, about as broad ... 2. *L. japonicum.*

1. **Lygodium flexuosum** Sw. (=*L. pinnatifidum* Sw.) is very common in the Dehra Dun, Kumaun, Sajahanpur, Gorakhpur; it is abundant throughout the plain in Bengal up to 5,000 ft., and also on both sides of the Madras Presidency up to about 4,000 ft. It extends to Ceylon, the Malay Peninsula, the Philippine Islands, North Australia, Angola, and the Guinea Coast.

The plant is commonly used as an expectorant.

In Tirhut the fresh root is boiled with mustard oil and used externally in rheumatism, sprains, scabies, ulcers, eczema, and cut wounds. It is particularly useful as a local application to carbuncles.

In Indo-China an infusion of the plant is used in blennorrhagia; it is said to be a good lactagogue.

Indo-China: Bong bong, Duong vong, Thach vi day—; *Malayalam*: Vallipanna—; *Tirhut*: Kalazha—.

2. **Lygodium japonicum** Sw. occurs abundantly in North India from Kashmir to Sikkim and Bhutan from 2,000-7,000 ft. elevation. It is rare in the western mountains of South India. It is distributed to Ceylon, Java, the Philippine Islands, China, Japan, and North Australia.

The plant is used as an expectorant in China. It is also used in haematuria and blennorrhagia.

Chinese: Hai Chin Sha—; *Indo-China*: Hai kim sa—.

Ophioglossum.

This genus consists of 30 species inhabiting tropical and temperate regions.

Ophioglossum vulgatum Linn. occurs in North India from Chamba State to Sikkim ascending up to 9,000 ft. on Mount Hattu and 2,000 ft. below Darjeeling; it is also found in Chota Nagpore on the Parasnath Mount at an altitude of 2,500 ft. It extends to Japan, the Sandwich Islands, Australia, and New Zealand. It is met with in North America, from Quebec and Ontario southwards to Florida and California, Kentucky, Tenessee, Texas and Arizona to Alaska. In Europe it inhabits Lapland, the British Isles and almost all other countries to Caucasia. In Africa it occurs in the Azores, Abyssinia, the Guinea Coast, Angola, St. Helena, Zambesiland, Cape Colony, and the Mascarene Isles.

A preparation from this plant, known as the 'green oil of charity', is in request in England as a vulnerary and remedy for wounds.

The plant is held in France and Spain as a vulnerary of great repute.

The plant yields a mucilaginous and astringent decoction which is used in angina in La Reunion. The fronds are considered tonic and styptic, and used in contusions, wounds, and haemorrhages.

A warm decoction of the rhizome is used by the Sutos as a lotion for boils.

Catalan: Llansa de Cristo, Llengua de serp—; *English*: Adder's Tongue, Christ's Spear—; *French*: Herbe à daucune, Herbe sans couture, Lance de Christ, Langue de serpent, Luciole, Ophioglosse, Ophioglosse commune, Petite serpentaire, Serpentine—; *Hausa*: Mashinzomo—; *La Reunion*: Herbe un coeur, Herbe paille-en-queue, Langue de serpent—; *Spanish*: Lengua de serpiente—; *Suto*: Mmadiyo, Tsebe-ngwe, Tseyananyane—.

Osmunda.

This genus consists of 10 species distributed throughout the temperate and tropical countries of the world.

Fronds 2-4 ft. long, 1 ft. or more broad, bipinnate ... 1. *O. regalis*.
Fronds 1-2 ft. long, 8-12 in. broad, simply pinnate ... 2. *O. claytoniana*.

1. **Osmunda regalis** Linn. is common on the western mountains of South India at the higher elevations; it is found in the

Himalayas from the Chamba State to Sikkim and Bhutan, but has become very rare in the Simla region; it is common, or at least frequent, in the Khasia District up to 4,000-6,000 ft., in the Central Provinces, and in the Bombay Presidency. It extends to South and West China, Hongkong, Japan, Canada and the Saskatchewan to Brazil. In Europe it is distributed over the British Isles, Sweden and Russia to Spain, Italy, Turkey, and to Siberia. It is met with in the Azores, Algeria, Tunis, Abyssinia, Angola, Central Africa, Nyassaland, Zambesiland, Natal, Cape Colony, and the Mascarene Isles.

The plant is a well known tonic and styptic, and is still very much employed all over Europe. It is used for rickets in England and in France.

The white centre of the root, boiled in some kind of liquor, was supposed good for persons wounded, dry-beaten, and bruised, or that have fallen from some high place.

The root or rhizome stamped in water or gin till 'the liquor becometh a stiff mucilage, has cured many deplorable pains of the back, that have confined the distracted sufferers close to bed for several weeks.' This mucilage was to be rubbed over the vertebrae of the back each night and morning for five or six days together. Also for rickets, 'take of the powdered roots with the whitest sugar, and sprinkle some thereof on the child's pap, and on all his liquid foods. It maketh a noble remedy without any other medicine.'

The tender sprigs of the plant at their first coming are 'good to be put into balmes, oyles, and healing plasters'.

This is much more effectual than the other ferns, both for inward and outward uses, says Culpeper. It is accounted singularly good in wounds, bruises, or the like; the decoction to be drunk, or boiled into an ointment of oil, as a balsam or balm, and so it is singularly good against bruises, and bones broken, or out of joint, and gives much ease to the colic and splenetic diseases; as also ruptures and burstings.

In Guinea an extract is prepared and is used externally for rheumatism and internally for intestinal griping.

English: Flowering Fern, Osmond Royal, Osmund-the-Waterman, Royal Fern, Royal Flowering Fern, Water Fern—; *French*: Fougère aquatique, Fougère fleurie, Fougère royale, Osmonde, Osmonde fleurie, Osmonde royale—; *Fulah*: Kolo kouli—; *German*: Koenigsfarn, Traubenfarn—; *Hova*: Ampanga-fenakoho—; *Indo-China*: Vi—; *Languedoc*: Fougeiroux—: *Spanish*: Helecho acuatico, Helecho florido, Helecho real—.

2. **Osmunda claytoniana** Linn. is found growing in the Himalayas from Kashmir to Bhutan at 6,000-10,000 ft. elevation, in the Khasia Mountains up to 4,500-6,000 ft. It also occurs in Canada, Newfoundland, and throughout the United States.

The rhizomes are used as an adulterant for Male Fern in the American market.

PELLAEA.

This genus consists of 40 species found in subtropical regions.

P. calomelanos Link., *P. consobrina* Hook., *P. hastata* (Thunb.) Prantl., *P. involuta* Bkr. are used medicinally in South Africa.

Pellaea calomelanos Link. is found in the Sirmur State, Dehra Dun, Garhwal, Kumaon, ascending up to 5,000 ft. It extends to Angola, Abyssinia, Somaliland, the Ruenzori Mounts, Zambesi-land, Mashonaland, Natal, Cape Colony, and La Reunion.

The Sutos use the rhizome as an anthelmintic, and smoke the leaf for colds in the head and chest.

Suto: Lehorometso, Pata-lewana, Pata-mawa—.

PLEOPELTIS.

This genus numbers 35 species distributed throughout tropical and subtropical regions.

Pleopeltis lanceolata Linn. is found on the Nilgiris and higher mountains on the west side of the Madras Presidency, and extends to Assam and Ceylon. It occurs in Tropical America and the West Indies, South Africa and its islands, St. Helena, and the Sandwich Islands.

In Mexico a tea made from the fronds is taken to cure the itch.

PTERIS.

A cosmopolitan genus numbering 160 species.

P. aquilina Linn. is used medicinally in Europe, China, La Reunion—; *P. multifida* Poir. in China—; *P. leptophylla* Sw., *P. palmata* W., *P. pedata* Sw. in Brazil—; *P. buchanani* Bkr. in Basutoland—.

Pteris aquilina Linn. is very common in the Himalayas, common in the Khasia Hills at 2,000-8,000 ft. It extends to the Deccan, the Madras Presidency, Ceylon, Tenasserim, the Malay Peninsula, and eastwards to Australia and New Zealand. It is universal in Europe, except in the extreme north, and never an Alpine plant; its range in the British Isles is said to agree closely with that of corn cultivation, and in the Scottish Highlands it never rises above 2,000 ft. It is found in the Cameroon Mountains and in Abyssinia. In general it may be said to thrive all round the world, both within the Tropics, and in the North and South Temperate Zones.

The rhizome is reputed astringent and anthelmintic.

A decoction of the rhizomes and fronds has been given in chronic disorders arising from obstructions of the viscera and spleen.

'For thigh aches smoke the legs thoroughly with Fern braken.'

'The roots being bruised and boiled in mead, or honeyed water, and drunk, kills both the broad and long worms in the body, and abates the swelling and hardness of the spleen. The green leaves eaten, purge the belly, and expel choleric and waterish humours that trouble the stomach. They are dangerous for women with child to meddle with by reason they cause abortions. The roots bruised and boiled in oil, or hog's grease, make a very profitable ointment to heal wounds, or pricks gotten in the flesh. The powder of them used in foul ulcers, dries up their malignant moisture, and causes their speedier healing.'

Cantonese: K'uet—; *Catalan*: Falguera femella—; *Chinese*: Chueh—, *Dutch*: Groote varen, Varen—; *English*: Bracken, Brakes, Female Fern—;

French: Filipode, Fougère à l'aigle, Fougère commune, Fougère femelle, Fougère impériale, Fruchière, Ptéride—; *German*: Adlerfarn, Farmkrautweiblein, Fluegelfarn, Jesuschristuswurzel—; *Indo-China*: Guyet—; *Ireland*: Fern of God—; *Languedoc*: Feuvé—; *Malaya*: Keat—; *Malayalam*: Tavi—; *Malta*: Bracken, Eagle Fern, Felce aquilina, Felce capannaja, Felicilla, Felicita—; *New Caledonia*: M'Baoue—; *Portuguese*: Feto—; *Punjab*: Dio, Kakei, Kakhash, Lungar—; *Roumanian*: Navalnic pajuriu, Spinarea lupului—; *Russian*: Paporotnik—; *Saora*: Manmarda—; *Spanish*: Helecho hembra—; *Tamil*: Parnai—.

STENOLOMA.

S. chinensis Bedd. (=*Davallia tenuifolia* Hook.) is found in the western mountains of the Madras Presidency at 3,000-6,000 ft.; in the Himalayas from the Simla region to Bhutan at 1,000-7,000 ft., plentiful eastwards; it is common in the Khasia Hills at 1,000-3,000 ft. Central Provinces, Pachmarhi, Ceylon, the Malay Peninsula, Japan. It extends to Polynesia and the Mascarene Islands, being common everywhere.

It is administered internally for chronic enteritis in Mauritius.

Chinese: Wu Chiu—; *English*: Parsley Fern—; *Mauritius*: Petite fougère—.

THE MEDICINAL AND POISONOUS GRASSES OF INDIA.

BY

J. F. CAIUS, S.J., F.L.S.

The GRAMINEAE are one of the largest families of plants, including about 480 genera and 5,880 species. They are found in all climates and situations, but are most numerous in temperate regions.

The medicinal and poisonous grasses of the world belong to 78 genera:—AGROPYRON (temperate regions of Europe and Asia); AMPHILOPHIS (tropical Asia and Africa); ANATHERUM (Brazil); ANDROPOGON (tropics of both hemispheres; South Europe, North America, temperate Asia); ANTHEPHORA (Africa); ANTHOXANTHUM (temperate regions); ARISTIDA (all warm dry countries); ARTHRAXON (tropics and subtropics of the Old World); ARUNDINELLA (tropical Asia and America); ARUNDO (Mediterranean region; East Indies, Malay Archipelago, Central and South America, South Africa, Madagascar, New Zealand); AVENA (Mediterranean region; temperate region of the eastern hemisphere); BAMBUSA (Eastern Asia, Australia); BECKMANNIA (China); BRACHIARIA (both hemispheres, mostly African); BROMUS (temperate regions of northern hemisphere, South America); CENCHRUS (tropical and subtropical regions); CEPHALOSTACHYUM (North India, Malaya, Madagascar); CHLORIS (all parts of the world in tropical and subtropical regions); CHRYSOPOGON (Old World tropics, West Indies); COIX (East Indies, Ethiopia; warm regions); CROSSOTROPIS (Africa); CYMBOPOGON (Old World tropics); CYNODON (Australia, cosmopolitan); CYNOSURUS (Mediterranean region; Europe, Canaries); DACTYLOCTENIUM (Tropics); DENDROCALAMUS (South-eastern Asia); DESMOSTACHYA (India to Syria and North Africa); DIARRHENA (China); DIGITARIA (all warm countries); ECHINOCHLOA (warm regions of both hemispheres); ELEUSINE (all warm countries); ELIONURUS (tropical and subtropical regions of both hemispheres); ELYMUS (northern and southern temperate regions); ERAGROSTIS (all warm and temperate regions); ERIANTHUS (mainly tropical); FESTUCA (temperate regions); GLYCERIA (North America); GYNERIUM (South Brazil, Argentina); HETEROPOGON (tropical Africa and Asia to Mediterranean); HIFROCHLOA (temperate regions); HORDEUM (temperate regions; Europe, North Africa, Asia, North and South America); HYGRORHIZA (India, Ceylon, Tongking); IMPERATA (tropical or warm regions of both hemispheres); LOLIUM (temperate regions; Europe, Asia, North Africa); LOPHATHERUM (Indo-Malaya, China, Japan); MANISURIS (tropical countries; India); MELICA (subtropical and temperate regions; Europe, Asia, Africa, America); MELINIS (Africa); MISCANTHIDIUM (tropical Africa south of the Equator, South Africa); NASTUS (Madagascar); OLYRA (Africa); ORYZA (tropical and subtropical regions of both hemispheres); ORYZOPSIS (northern hemisphere); OXYTENANTHERA (Africa, India, Malaya); PANICUM (tropical and subtropical regions); PASPALUM (all warm regions); PENNISETUM (most warm dry regions; Africa); PHALARIS (Mediterranean region; South Africa, California to Chile); PHRAGMITES (almost the whole world); PHYLLOSTACHYS (Eastern Asia); POA (temperate regions); POGONARTHRIA (Africa); POLYTOCA (tropical Asia, Australia); SACCHARUM (tropical and subtropical regions of both hemispheres); SASA (China); SECALE (temperate regions); SETARIA (warm and temperate regions); SORGHUM (warm and temperate regions; Africa, India); SPOROBOLUS (tropical and subtropical regions of both hemispheres; America); STENOTAPHRUM (tropics of both worlds); STIPA (tropical and subtropical regions of both hemispheres); THELEPOGON (India to tropical Africa); THEMEDA (tropical and subtropical regions of the world); THYSANOLAENA (Indo-Malayan region); TRITICUM (Mediterranean region; temperate regions); VETIVERIA (tropics of the Old World); ZEA (all warm countries); ZIZANIA (North America, North-Eastern Asia).

The medicinal and poisonous grasses of India belong to 49 genera:—AGROPYRON, AMPHILOPHIS, ANTHOXANTHUM, ARISTIDA, ARTHRAXON, ARUNDINELLA, ARUNDO, AVENA, BAMBUSA, BROMUS, CENCHRUS, CEPHALOSTACHYUM, CHLORIS, CHRYSOPOGON, COIX, CYMBOPOGON, CYNODON, DACTYLOCTENIUM, DENDROCALAMUS, DESMOSTACHYA, ECHINOCHLOA, ELEUSINE, HETEROPOGON, HORDEUM, HYGRORHIZA, IMPERATA, LOLIUM, LOPHATHERUM, MANISURIS, ORYZA, OXYTENANTHERA, PANICUM, PASPALUM, PENNISETUM, PHALARIS, PHRAGMITES, PHYLLOSTACHYS, POLYTOCA, SACCHARUM, SETARIA, SORGHUM, STENOTAPHRUM, STIPA, THELEPOGON, THEMEDA, THYSANOLAENA, TRITICUM, VETIVERIA, ZEA.

I. Spikelets spicate, all unisexual, male spikes in terminal panicles or continuous with the female spikes.
- a. Fruiting spikelets enclosed in a strong nut-like polished bract ... ... COIX.
- b. Fruiting spikelets having all the inner glumes concealed within the greatly enlarged hardened outer POLYTOCA.
- c. Fruiting spikelets densely crowded on a cylindric spongy rachis ... ... ZEA.

II. Spikelets homo- or heterogamous, 1-2-flowered, solitary or 2- rarely 3-nate, on the internodes of an articulate spike or raceme.
- Spikelets 2-nate; lower involucral glume globose ... MANISURIS.

III. Spikelets with a terminal perfect flower, and one or more imperfect male or neuter below it; rachilla not produced beyond the perfect flower.
- a. Panicle contracted or spiciform; spikelets awnless PHALARIS.
- b. Panicle spiciform; glumes 6, III and IV neuter, awned, stamens 2 ... ... ANTHOXANTHUM.

IV. Spikelets panicled; rachilla not produced beyond glume III; floral glumes rigid or hard, awned.
- a. Glume III narrow, awn 3-flowered ... ... ARISTIDA.
- b. Glume III narrow, awn entire, grain terete ... STIPA.

V. Spikelets homogamous, in compound racemes or panicles. Lower involucral glume not sunk in a hollow of the rachis.
- a. Rachis not fragile. Spikelets in a cylindric silvery thyrsus ... ... IMPERATA.
- b. Rachis fragile. Spikelets in a thyrsus of spiciform racemes ... ... SACCHARUM.

VI. Spikelets many, in solitary digitate or fascicled spikes, usually heteromorphous. Lower involucral glume not sunk in a hollow of the rachis. Margins of the lower involucral glume of sessile spikelet not inflexed.

a. Spikes 2-nate or digitate; spikelets 1-flowered ... ARTHRAXON.

b. Spikes digitate; spikelets 2-flowered; glume I tubercled ... ... THELEPOGON.

VII. Spikelets heterogamous, 1-flowered, 2-rarely 3-nate on the whorled articulate branches of simple or compound racemes or panicles.

a. Spikelets in two superposed series; upper series of heterogamous pairs, and a terminal male, lower a whorl of 4 persistent males or neuters ... THEMEDA.

b. Spikelets 2-nate, or the uppermost 3-nate, spicate or panicled.

i. Sessile spikelets of all the pairs alike in sex and form throughout the spikes.

¶ Joints of rachis many, stout or slender, opaque VETIVERIA.

¶¶ Joints of rachis few; spikes articulate on the obliquely truncate bearded tips of the branches of the panicle ... ... CHRYSOPOGON.

¶¶¶ Joints of rachis very few, filiform, opaque; lodicules ciliate; grain dorsally compressed SORGHUM.

¶¶¶¶ Joints of rachis and pedicles of upper spikelets linear compressed with usually a central translucent channel and thickened margins ... AMPHILOPHIS.

ii. Lowest one or more sessile spikelets differing from all those above it in sex or form.

¶ Spikes binate on the peduncle which is enclosed in a spathiform sheath; sessile spikelet dorsally compressed ... ... CYMBOPOGON.

¶¶ Spikes solitary, lower pairs of spikelets very unlike the upper; upper sessile spikelets cylindric ... ... HETEROPOGON.

VIII. Spikelets 2-flowered; upper flower bisexual, lower male or neuter, rarely both fertile.

a. Spikelets dorsally flattened, base not thickened; glumes 3 with very rarely a minute 4th ... PASPALUM.

b. Spikelets panicled or spicate; lower floral glume not beaked, upper floral glume crustaceous ... PANICUM.

c. Spikelets innumerable, very minute, hairy, densely crowded in the capillary branches of a very large panicle ... ... THYSANOLAENA.

d. Spikelets persistent on the pedicels, cr glumes I and II separately deciduous, IV deciduous; usually awned, awn bent ... ... ARUNDINELLA.

e. Glumes awned from the entire acute or acuminate tip or caudate or cuspidate-acuminate. Racemes dense, more or less secund ... ... ECHINOCHLOA.

f. Spikelets articulate on their pedicels, each surrounded by an involucel of bristles ... ... SETARIA.

g. Spikelets persistent on their pedicels, at least the fertile, or deciduous with their pedicels.

i. Spikelets in involucelled deciduous fascicles.

¶ Involucel of bristles ... ... PENNISETUM.

¶¶ Involucel of spines connate at the base ... CENCHRUS.

ii. Spikelets not involucelled, 2-seriate on a flat subarticulate rachis ... ... STENOTAPHRUM.

IX. Spikelets 2- or more-flowered; rachilla produced.
Spikelets 2-6-flowered; flowering glumes awned; awn subterminal or dorsal ... AVENA.

X. Spikelets panicled, 2-many-flowered; glumes very narrow, flowering glumes penicillate.
a. Rachilla elongate, glabrous; flowering glumes silkily hairy ... ... ARUNDO.
b. Rachilla very short; flowering glumes glabrous; callus with long silky hairs ... ... PHRAGMITES.

XI. Spikelets 1-many-flowered, 1-2-seriate on the rachis of a simple spike or on the long slender branches of a simple panicle. Leaves broad, tessellately nerved.
Spikelets 1-flowered, secund and articulate at the base; upper glumes empty, convolute with setiform tips ... ... LOPHATHERUM.

XII. Spikelets usually panicled, 2- or more-flowered; floral glumes 5-many-nerved, upper empty or 0; styles short.
Spikelets panicled, 3-many-flowered; floral glumes mucronate or awned, dorsally rounded, 5-9-nerved, top of ovary lobed villous ... ... BROMUS.

XIII. Inflorescence various; spikelets 2-many-flowered; flowering glumes 1-3-nerved, entire, 3-toothed, 3-lobed, or 3-awned. Floral glumes entire; acute or acuminate ... ... DESMOSTACHYA.

XIV. Spikelets 1- or more-flowered, biseriate and secund on an inarticulate spike or on the spiciform branches of a slender panicle; flowers all or the lower only bisexual.
a. Spikes digitate, 1-flowered, upper imperfect flower absent ... ... CYNODON.
b. Spikes solitary, spicate or racemed; spikelets 2- or more-flowered, upper flower imperfect ... CHLORIS.
c. Spikes digitate or whorled; spikelets 3-6-flowered, densely crowded, awnless ... ... ELEUSINE.
d. Spikes terminating with a sharp point; upper involucral glume and floral glumes rigidly mucronate or shortly awned ... ... DACTYLOCTENIUM.

XV. Spikelets 1-flowered, articulate on their pedicels and deciduous from them; palea 1-3-nerved, stamens 6 or fewer.
a. Spikelets bisexual, awned; glumes 2, narrow, thin HYGRORHIZA.
b. Spikelets bisexual, awned or not; glumes 4; I and II minute or setaceous ... ... ORYZA.

XVI. Spikelets sessile, singly or in clusters; florets 1 or more.
a. Flowering glumes 5-9-nerved, lateral nerves not conniving; short or ending in teeth or awns ... TRITICUM.
b. Flowering glumes 5-7-nerved, lateral nerves conniving or confluent with the single terminal awn AGROPYRON.
c. Spikelets solitary at the nodes of the spike, 3-many-flowered; glumes coriaceous, lower involucral absent, except in the terminal spikelet; floral glumes 5-many-nerved ... ... LOLIUM.
d. Spikelets in groups of 3 at the nodes of a dense spike; floral glumes 5-nerved ... ... HORDEUM.

XVII. Shrubby or arboraceous grasses. Leaves flat jointed on the sheath.
a. Pericarp thin, adnate to the seed.
i. Stamens 3; spikelets 1-4-flowered, spicate; empty glumes 2-3 ... ... PHYLLOSTACHYS.

ii. Stamens 6; filaments free; paleas entire or slightly cleft, all 2-keeled ... .. BAMBUSA.

iii. Stamens 6; filaments connate. Spikelets few-flowered, palea of upper flower absent or glume-like, not keeled ... ... OXYTENANTHERA.

b. Pericarp crustaceous; seed free; spikelets 2-many-flowered; palea 2-keeled, lodicules none; ovary hirsute at the top ... ... DENDROCALAMUS.

c. Fruit oblong, beaked, pericarp thick, separable. Spikelets 1-flowered, bracteate, crowded in globose heads ... ... CEPHALOSTACHYUM

AGROPYRON.

This genus numbers 45 species, inhabiting mainly the temperate region of Europe and Asia.

Agropyron repens Beauv. (=*Triticum repens* Linn.) is found in Kashmir and Western Tibet, ascending up to 14,000 ft. in the Karakoram Range. This grass is a native of Europe, where it is found freely in road-sides, fields and waste places. The commercial supplies come largely from England and Germany. The plant now abounds in meadows and cultivated grounds in the northern United States of America, where it is often very troublesome as a weed. 'This grows in gardens and arable lands, as an infirmitie or plague of the fields, nothing pleasing to husbandmen; for after the field is plowed, they are constrained to gather the roots together with harrowes and rakes, and being so gathered and laid upon heaps, they set them on fire lest they should grow againe.'

'Although that Couch-grasse be an unwelcome guest to fields and gardens, yet his physicke vertues do recompense those hurts. Couch-grasse healeth green wounds.'

The plant has been employed in England from remote times as a vulnerary, and to remove difficulties of urination. 'Being boiled and drunk it openeth obstructions of the liver and gall, and the stoppings of urine, and easeth the griping pains of the belly, and inflammations; wasteth the matter of the stone in the bladder, and the ulcers thereof also. The seed doth more powerfully expel urine and stayeth laxes and vomiting. The distilled water alone, or with a little worm-seed, killeth the worms in children.

'The way of use is to bruise the roots, and having well boiled them in white wine, drink the decoction; it is opening, but not purging, very safe; 'tis a remedy against all diseases coming of stopping, and such are half those that are incident to the body of man; and although a gardener be of another opinion, yet a physician holds an acre of them to be worth five acres of carrots twice told over.'

In modern days an infusion of the root is generally regarded as a soothing diuretic, helpful to the bladder and kidneys. Formerly, this was a popular drink to purify the blood in spring. The decoction may be made from the whole fresh plant, or from the dried root sliced, two to four ounces being put in a quart of water, reduced to a pint by boiling. A wineglassful of this may be given for a dose. It certainly palliates irritation of the urinary passages, and helps to relieve against gravel. A liquid extract is also dispensed by the druggists, of which one or two teaspoonfuls are given in water.

In Europe the decoction is used to a considerable extent as a diluent and slightly nutritious drink. The French specially value it for its stimulating fragrance of vanilla and rose perfumes.

The rhizome is officinal in the pharmacopeias of France, Switzerland and Turkey. It is diuretic, demulcent and aperient. It exerts an influence upon the genito-urinary organs, and it is used internally as a demulcent drink for irritable bladder, and also for cystitis. It is one of the least stimulant of the remedies of its class, and may be employed very freely. It is also recommended in gout and rheumatism.

The rhizome contains grape sugar, fruit sugar, a sugar of undetermined identity, gum and inosite. It is also said to contain glucosides, one of which yields vanillin on hydrolysis, also a carbohydrate resembling inulin which has been named *triticin*.

Brahui: Milinj—; *Catalan*: Agram, Gram—; *Chinese*: Se Mao Ts'ao—; *Dutch*: Kweek—; *English*: Couch Grass, Dog Grass, Quack-grass, Quick Grass, Quilch, Twitch Grass, Witch Grass—; *French*: Agram, Agropyre rampant, Auge, Blé sauvage, Chiendent, Chiendent des boutiques, Chiendent officinal, Chiendent ordinairé, Froment rampant, Herbe à deux bouts, Laitue de chien, Petit chiendent, Sainte neige, Tranuge, Vagon, Wagon—; *German*: Ackergras, Ackermannswurzel, Bagenwurzel, Fadenwurzel, Fegwurzel, Flechtgras, Graswurzel, Haarstrang, Hundsgraswurzel, Hundsruecken, Knotengras, Kriechweizen, Landdreck, Paeden, Peden, Peyer, Poeden, Queck, Rebel, Ribel, Schnur, Schweinegras, Sehnengras, Spulwurz, Tuerkisches Gras, Weisswurz, Wreeten, Wuemgras, Wurmgras, Zweckenwurzel—; *Greek*: Agrostis—; *Italian*: Caprinella, Dente canino, Graminga, Granacina—; *Polish*: Perz—; *Portuguese*: Grama—; *Provence*: Grame—; *Roumanian*: Albeiu, Iarba caineasca, Pir—; *Russian*: Eja, Pirei, Porei—; *South Africa*: Couch Grass, Dog Grass, Triticum—; *Spanish*: Grama, Grama de las boticas—; *Swedish*: Quick hwete—; *Tasmania*: Couch—; *Turkish*: Ayrikotu—.

Amphilophis.

This genus consists of 25 species or more, distributed throughout tropical Africa and Asia, many of them closely allied and difficult to differentiate.

Amphilophis odorata A. Camus (=*Andropogon odoratus* Dna. Lisboa) is found in the Deccan, where it is considered carminative.

Bombay: Ushadhana—.

Anthoxanthum.

The genus includes 4 or 5 species, natives of Europe and temperate Asia; one being found also in North America and Australia.

Anthoxanthum odoratum Linn. is a native of Europe and Central and Northern Asia occasionally found in the hill districts of India, but only as introduced by man or as an escape from cultivation. It is found in North Africa and the Atlantic Islands. Naturalised in North America it occurs as a meadow grass in pastures from Washington to California.

This grass is specially provocative of hay fever and hay asthma with persons liable to suffer from these distressing ailments. Accordingly a medicinal tincture is made from it with spirit of wine, and if some of the same is poured into the open hand-palms for the volatile aroma to be sniffed well into the nose and throat, immediate relief is afforded during an attack. At the same time three or four drops of the tincture should be taken as a dose with water, and repeated at intervals of twenty or thirty minutes, as needed.

The flowers contain *coumarin*, and their volatile pollen impregnates the atmosphere in early summer. But, strange to say, while the plants growing in spring have a delightfully characteristic odour and a pleasant aroma, those flowers which appear late in autumn emit a most nauseating odour.

English: Pig Grass, Sweet-scented Grass, Sweet-scented Spring Grass, Sweet-scented Vernal Grass, Sweet Vernal Grass—; *French*: Flouve odorante—; *German*: Ruchgras—; *Italian*: Paleino odoroso, Paleo—; *Uruguay*: Grama de olor—.

ARISTIDA.

It is assumed that this genus includes about 300 species, natives of warm countries.

Aristida adscenscionis Linn. is found throughout the plains and low hills of India, Burma and Ceylon, ascending to 8,000 ft. in Kashmir; it is distributed over most warm countries.

In Madagascar an ointment consisting of lard and the ashes of the flowers is used topically for itch and ringworm.

Hova: Horombavy, Pepaka—; *Southern Nigeria*: Oka olongo—.

ARTHRAXON.

The genus consists of 20 species distributed throughout the tropics and subtropics of the Old World.

Arthraxon ciliaris Beauv. is found in the hilly districts of India from Kashmir eastwards to Burma, ascending the Himalaya to 5,500 ft., and southwards to the Nilgiri Hills and Ceylon. It extends to southern and central China, and Shantung.

The plant is used medicinally in various parts of China.

Bombay: Turade, Turde—; *Chinese*: Chin Ts'ao—.

ARUNDINELLA.

This genus consists of about 55 species, found mainly in Tropical Asia and America.

Arundinella brasiliensis Raddi occurs throughout the hilly parts of India, from the Himalaya, ascending to 7,000 ft., to Burma and Ceylon. It is distributed to China, Malaya, Australia, South Africa, South Europe, tropical America.

The natives of Basutoland use the plant in compounding many medicines and in making a lotion for washing wounds.

Bombay: Dundi, Kotir—; *Suto*: Mahlakamane—.

Arundo.

This genus includes 12 species, natives of the Mediterranean region, and distributed to the East Indies and the Malay Archipelago, Madagascar, New Zealand, Central and South America.

Arundo donax Linn. is found in the Lower Himalaya from Kashmir to Nepal, ascending to 3,500 ft., from the Punjab to Sylhet, the Naga Hills up to 5,000-8,000 ft., Burma, the Circars, Coorg, the Nilgiri Hills. It extends westwards to Europe, North Africa and North Asia.

The rhizomes are much used in decoction as an emollient and diuretic. They are said to stimulate the menstrual discharge and diminish the secretion of milk.

Catalan: Canya—; *Egypt*: Busfarisk, Bus haggai, Buz haggni, Ghab, Gub-rumy, Gasab—; *English*: Common Reed, Giant Reed, Great Reed, Reed, Spanish Cane—; *French*: Canne de Provence, Donax, Grand roseau—; *Greek*: Kupho Kalamo—; *Indo-China*: Cong, Cuoi, Dich lo chuc—; *Italian*: Canna—; *Maltese*: Kasba—; *Portuguese*: Canna—; *Spanish*: Caña comun—; *Sokoto*: Gabara, Machara, Wutsiyar giwa—; *Yemen*: Kassab—.

Avena.

This genus numbers about 55 species found chiefly in temperate regions.

A. fatua Linn., *A. sativa* Linn., *A. sativa* var. *orientalis* Hook. f. are used medicinally in Europe; *A. fatua* Linn. is also used in China.

1. Ligule short, very obtuse, up to 3 mm. long ... *A. fatua.*
2. Ligule truncate, 1-3 mm. long ... ... *A. sativa.*

1. **Avena fatua** Linn. is found in the Punjab, North-Western Himalaya, and Western Tibet, in cultivated fields, ascending to 9,000 ft.; also in Sikkim Himalaya up to 9,000 ft. It is distributed to temperate Europe, Northern Africa and Northern Asia.

In Europe the seeds are used for their emollient, refrigerant, and diuretic properties.

In the Punjab the seed is believed to produce poisonous and deleterious effects.

Arabic: Bakhragan, Basheruk, Ghallash sheytany, Khafur, Shair-el-ghul, Sabus ssabbaras, Zommeyr—; *Catalan*: Cugula—; *Chinese*: Ch'iao Mai—; *Egypt*: Chafur, Sammeir—; *English*: Drake, Wild Oat—; *French*: Avéneron, Averon, Avoine bouffe, Avron, Coquiolle, Couyonne, Folle avoine, Pied de mouche—; *German*: Windhafer—; *Greek*: Agrio genima—; *Hindi*: Gandal, Ganer, Jei, Kuljud—; *Iraq*: Dosar, Shofan, Sunaislah—; *Italian*: Lippa—; *Jhalawan*: Gandamkao—; *Kurdish*: Gia brishi, Kaltas—; *Languedoc*: Coughioulo—; *Mosul*: Hurtuman—; *Punjab*: Ganerjei, Gozang, Kasamm, Upwa, Yupo—; *Roumanian*: Odos—; *Russian*: Jivoi ovios—; *Spanish*: Avena loca, Cula—; *Turkish*: Chodar—.

2. **Avena sativa** Linn. is cultivated in Northern India, from Bengal to the Indus; and in the Himalaya up to 12,000 ft. It was known to the ancients, and is now cultivated in all civilized countries; but its original locality has not been satisfactorily ascertained. It grows wild in Sicily, and is said to have been seen in the Island of Juan Fernandez, off the Coast of Chili.

'Common Otes put into a linnen bag, with a little bay salt quilted handsomely for the same purpose, and made hot in a frying pan, and applied very hot, easeth the pain in the side called the stitch.' Oats fried with bay salt, and applied to the sides, take away the pains of stitches and wind in the sides of the belly.

'A poultice made of the meal of oats, and some oil of bays added, helps the itch and leprosy; as also the fistulas of the fundament, and dissolves hard imposthumes.' A yeast poultice made by stirring oatmeal into the grounds of strong beer is a capital cleansing and healing application to languid sloughing sores.

The meal of oats boiled with vinegar and applied, takes away freckles and spots in the face, and other parts of the body. 'Ote-meale is good for to make a faire and wel-coloured maid to look like a cake of tallow, especially if she take next her stomacke a good draught of strong vinegre after it.'

As a food oatmeal is apt to undergo some fermentation in the stomach, and to provoke sour eructations. Furthermore, it is somewhat laxative, because containing a certain proportion of bran which mechanically stimulates the intestinal membranes; and this insoluble bran is rather apt to accumulate.

American doctors have taken of late to extol the Oat when made into a strong medicinal tincture with spirit of wine, as a remarkable nervine stimulant and restorative: this being especially valuable in all cases where there is a deficiency of nervous power, for instance, among over-worked lawyers, public speakers, and writers.

The tincture is ordered to be given in a dose from ten to twenty drops, once or twice during the day, in hot water, to act speedily; and a somewhat increased dose in cold water at bed-time so as to produce its beneficial effects more slowly then. It proves an admirable remedy in nervous prostration and exhaustion after all febrile diseases, and in sleeplessness from nervous exhaustion. It seems to exert a very beneficial action upon the heart muscles and on the urinary organs, speedily relieving spasmodic conditions of bladder and ureter.

Catalan: Sibada—; ***Dutch***: Haver—; ***English***: Groats, Haws, Oat, Oats—; ***French***: Avoine, Avoine cultivée, Avoine noire—; ***German***: Haber, Hafer—; ***Greek***: Bromi—; ***Hindustani***: Jai—; ***Hova***: Varintsoavahy—; ***Hungarian***: Zab—; ***Italian***: Avena, Vena—; ***Languedoc***: Arracho, Aveino, Civada, Civado, Sibado—; ***Malta***: Oats, Avena, Hafur—; ***Polish***: Owies—; ***Portuguese***: Avea—; ***Punjabi***: Javi—; ***Roumanian***: Oves—; ***Russian***: Ovios—; ***Spanish***: Avena—.

Avena sativa var. **orientalis** Hook. fil. is characterised by the presence of bristly hairs at the base of glume III. It is found in North-Western India, the Soane Valley, and Murshidabad.

The seeds are used in Spain as an emollient, refrigerant, and diuretic.

BAMBUSA.

This genus consists of about 73 species, occurring in Eastern Asia, Malaya, and one in Australia.

63.

B. arundinacea Retz. is used medicinally in Guiana, *B. barbata* Trin. in Madagascar, *B. vulgaris* Schrad. in Indo-China.

B. blumeana Schult. is used as a poison in Malaya.

I. Arborescent or shrubby. Stems not spinous ... 1. *B. vulgaris*.

II. Arborescent. Stems spinous.

a. Stem-sheaths long blade with felted hairs within. Empty glumes 0-2 ovate lanceolate acute or mucronate, many-nerved ... 2. *B. arundinacea*.

b. Stem-sheaths short. Empty glumes 2-3 ovate acute, 3-7 nerved ... ... 3. *B. blumeana*.

1. **Bambusa vulgaris** Schrad. is cultivated throughout tropical India and in all hot countries.

The roots, the shoots, the bark and the leaves are all used medicinally in Indo-China.

The roots and the shoots are considered emollient, diuretic, diaphoretic; they are used to purify the blood, and they are commonly administered for obstruction, retention of urine, indefinite pain and syphilis. Mixed with equal parts of tobacco and betel leaves, the fresh root is macerated in oil and applied externally to hard tumours and in cirrhoses.

The astringent bark is used in haemorrhage, excessive menstruation, nausea and vomiting.

The leaves are cooling and emollient; a decoction is given in fevers, melancholia, cough and sore throat.

In Cambodia the nodes are considered bechic, astringent, tonic and depurant; they are prescribed in bronchitis, blennorrhagia, leucorrhoea, metrorrhagia and the yaws.

Bengal: Basini bans—; *Bombay*: Kallak, Vansa kalaka—; *Cambodia*: Asey, Russey—; *English*: Yellow and Green Striped Bamboo—; *French*: Bambou des jardins—; *Indo-China*: Long dau chuc, Tam vong, Tre tau, Tre vuon—; *Malay*: Buluh minyak has, Buluh pan—; *Sinhalese*: Una—.

2. **Bambusa arundinacea** Retz. is found throughout the plains and low hills of India, Burma and Ceylon, wild and cultivated. It is cultivated in the lower Himalaya and in the valleys of the Ganges and Indus.

The hollow stems of this bamboo often contain a silicious and crystalline substance known in the bazaars of India as *tabashir*. This is largely used by Hindus and Mahomedans who prize it very highly as a stimulant and aphrodisiac. It is a demulcent, tonic, aphrodisiac and pectoral, which has been found useful in cough, consumption, asthma and fever. In combination with other astringent medicines it is given in chronic dysentery and internal haemorrhages.

The most efficacious application for dislodgement of worms in ulcers is a poultice made by pounding the young shoots of the bamboo. The juice is first poured on the vermin, and the ligneous mass is applied and secured by a bandage.

Yunani practitioners consider the root a good tonic; they apply the ash to ringworm, bleeding gums and pains in the joints. In the Tamil country the root is considered diluent, the bark is used as a cure for eruptions, the leaves are given to stimulate the menstrual flow, and the tabashir is prescribed in paralysis and for flatulence.

A belief in the emmenagogue properties of the leaves is common alike in India and in China. A decoction of the leaf-bud is said to be very effective. The juice of the leaves with aromatics is given in blood vomiting. With black pepper and common salt the leaves are used to check diarrhoea in cattle.

The juice of the flower is dropped in the ear for earache, and in deafness.

A decoction of the joints is supposed to have an action on the uterus, and is used by females after delivery to cause a free flow of the lochial discharge. The same part of the plant pounded with water is applied to inflamed joints.

The Ayurvedists describe the seeds as acrid, sweet, aphrodisiac and fattening. They use them freely in biliousness, urinary discharges. and all forms of poisoning.

Mhaskar and Caius have shown experimentally that the bark, the seeds and the tabashir are equally useless in the antidotal treatment of either snake bite or scorpion sting.

Annam: Tre, Tre pheo—; *Arabic*: Qasab—; *Assam*: Bnah, Kata, Koto—; *Bengal*: Bans, Bansh, Behurbans, Kautabans, Ketua—; *Bombay*: Dougi, Kalak, Mandgay, Padhai—; *Burma*: Kyakatwa—; *Canarese*: Bambu, Bidaru, Bidiru, Bidirumale, Bidru, Bidungalu, Bidungulu, Biduru, Dongi, Elubidiru, Gale, Hebbidiru, Hebbidru, Hennubidiru, Kalale, Karira—; *Cantonese*: T'in Chuk Wan, T'in Chuk Wong—; *Catalan*: Bambu—; *Central Provinces*: Kattang—; *Chinese*: T'ien Chu Huang, T'ien Chu Yuen—; *Chittagong*: Bariala—; *Deccan*: Bambu, Bhans, Chansa—; *English*: Spiny Bamboo, Thorny Bamboo—; *French*: Bambou, Bambou commun, Canne bambou—; *French Guiana*: Bambou—; *Garo*: Whahkanteh—; *German*: Bambus—; *Gond*: Katiwadur—; *Gujerati*: Toncor, Wans—; *Haldwani*: Kanwas—; *Hasada*: Katangarimad—; *Hindi*: Bans, Kantabans, Kattang, Magarbans, Malbans—; *Italian*: Bambu, Canna indiana—; *Java*: Singkara—; *Kadir*: Mula, Mulai—; *Kolami*: Katammadh, Katanga—; *Konkan*: Kalak, Padhai—; *Konkani*: Conoqui, Kanaki, Vellu—; *Korku*: Mad—; *Kumaon*: Kanta bans—; *Lambadi*: Bambu—; *Lao*: Mai pai—; *Madras*: Ponteveduru—; *Magahi*: Wanah—; *Malaya*: Thian chook wong—; *Malayalam*: Illi, Kampu, Kaniyaram, Karmmaram, Mula, Mulmulam, Pattil, Tejanam, Trinadhvajan, Valiyamula, Venu—; *Marathi*: Kallak—; *Mundari*: Katanggamad, Katanggarindd—; *Naguri*: Katangmad—; *Panch Mahals*: Vas—; *Persian*: Nai—; *Portuguese*: Bambu espinhoso, Spodio—; *Provence*: Boulou—; *Punjab*: Magae, Nal—; *Roumanian*: Bambu—; *Russian*: Bambuk—; *Sanskrit*: Bahupallava, Brihattrina, Dhanurdruma, Dhatushya, Dridhagranthi, Dridhakanda, Dridhapatra, Duraruha, Kamatha, Kantaki, Kantalu, Karmmara, Kichaka, Kilati, Kishkuparva, Kugshirandhra, Mahabala, Maskara, Mrityubija, Navagragandha, Phalantaka, Purvayoni, Pushpaghataka, Shataparva, Shatpadalaya, Suparva, Suparvan, Tajana, Tejana, Trinadhvaja, Trinaketu, Trinaketuka, Tvachisara, Tvakasara, Vadaniya, Vansha, Vanya, Venu, Yavaphala—; *Santali*: Mat—; *Sinhalese*: Kattuuna, Una—; *Spanish*: Bambu, Caña de Indias para bastones, Mambú—; *Tagalog*: Canavangtotoo—; *Tamil*: Ambal, Ambu, Aril, Bongu, Iraivarai, Kalai, Kambu kilai, Kisagam, Kulaimungil, Masukkaram, Mirittusam, Mudangal, Mulai, Mullumungil, Mundlaveduru, Mundul, Mungil, Nadimungil, Nedil, Nettil, Palandam, Panai, Pandil, Pasy, Periamungil, Perumungil, Peruvarai, Sabam, Sanagi, Sey, Tandu, Tattai, Tulai, Tumbu, Valai, Vannigaruppam, Varaimungil, Vedir, Velam, Venu, Veral, Vey, Veyal, Vindil—; *Telugu*: Bongu, Bongudu, Bonguveduru, Kichakamu, Maskaramu, Mullaveduru, Mudusuveduru, Pentiveduru, Trinadhvajamu, Veduru—; *Thana*: Kalak, Katestokar, Padai—; *Tongking*: Tre gai—; *Tulu*: Bedru—; *Upper Godavery*: Kanka—; *Urdu*: Bansa—; *Uriya*: Bendobaunso, Kontabanso, Kontabaunso—; *Uruguay*: Bambú, Caña.

3. **Bambusa blumeana** Schult. is indigenous, and also cultivated to some extent, in the open sandy country of the Malay Peninsula, at Pekan in Pahang, at Perlis and Kanga in Kedah. It extends to the Malay Islands.

The very fine hairs found on the sheath leaves of the young sprouts are prominent among Malay poisons. They are used either alone or in combination with other things.

Malay: Buloh duri, Buluh duri—.

Bromus.

This genus numbers about 40 species, found mostly in the temperate regions of the Northern hemisphere and of South America, several in the high mountains of the tropics.

B. catharticus Vahl., a Peruvian plant, is said to be purgative, and *B. purgans* Linn. of Canada is reputed emetic.

Bromus mollis Linn. is found in North-West India Europe, the Mediterranean region as far as Persia, Western Siberia, Japan and the Atlantic Islands. It is naturalised in North and South America.

The grains of this plant bring on giddiness in man and the quadrupeds, and are fatal to poultry.

English: Bull-grass, Goose-corn—; *Italian*: Spigolina—; *Maltese*: Hortan, Hurtan—.

Cenchrus.

This genus consists of about 25 species, found in the warm regions of both hemispheres, and in temperate North America.

Cenchrus catharticus Del. occurs in the Punjab and in the Upper Gangetic Plain. It is found in Perim Island, at the mouth of the Narbada River, and is distributed over Arabia, and tropical Africa.

In La Reunion a decoction of the fruits is taken as a diuretic and pectoral.

Bombay: Baront, Kukar—; *Hausa*: Karangiya—; *La Reunion*: Gratte-cul, Pagode—; *Rajputana*: Bharut—; *Sudan*: Cram-cram, Heskanit—.

Cephalostachyum.

This genus numbers 8 species inhabiting North India, Malaya and Madagascar.

Cephalostachyum capitatum Munro is found in the Sikkim and Bhutan Himalaya at 2,000-8,000 ft., and also in the Kasia, Jaintea and Naga Hills.

The leaves are much used medicinally in Madagascar. The infusion, 60 leaves for a litre of water, is given as tonic, anthelmintic, stomachic and carminative. An ointment consisting of 20 parts of ash and 30 parts of lard or suet is applied for the itch and for rheumatism.

Betsimisaraka: Volozatsy, Volozevona—; *Lepcha*: Pa-yang, Pa-yong—; *Madagascar*: Volo, Volotsangana—.

CHLORIS.

This genus contains 75 species found in the tropical and subtropical regions of both hemispheres.

C. petraea Thunb. yields hydrocyanic acid.

Chloris virgata Sw. is found abundantly in Rajputana, Upper and Lower Gangetic Plains, southwards to Bihar, Central and Southern India, Burma, Western Tibet. It extends westwards to Algeria, and is distributed over China, Mongolia, Tropical and South Africa and America.

A decoction of the plant or of its roots is used by the Xosas of South Africa as an addition to baths for the treatment of colds and rheumatism.

Rajputana: Gharaniagas—; *Nigeria*: Kafar gauraka, Kafar fakar, Sawun gauraka—; *Telugu*: Uppugaddi—; *Transvaal*: Sweet Grass, Zoet Grass—; *Xosa*: uMadolwana—.

CHRYSOPOGON.

This genus consists of about 18 species found in the hot parts of the Old World, a few extending into the temperate zones; one in Florida and Cuba.

Chrysopogon serrulatus Trin. occurs in the Western Himalaya, from Kashmir to Nepal, ascending to 6,000 ft. It is found in the hilly districts from the Punjab to Bihar and Orissa, Burma and southwards to Ceylon. It extends to Indo-China, Afghanistan, Portuguese East Africa, Madagascar, Rhodesia and South Africa.

The plant is said to yield hydrocyanic acid. The seeds are a popular vermifuge in Annam.

Annam: Bong co, Co may, Co mu, Nam hoang neu—; *Indo-China*: Chiendent grenille—; *Jhalawan*: Kaj—; *Kharan*: Kohigayab—; *Kila Saifulla*: Sabba—; *Nushki*: Kuj—; *Pab*: Kaj—; *Quetta*: Kohigayab—.

COIX.

The genus includes 5 or 6 species, natives of the East Indies and Ethiopia, one being broadly dispersed in warm regions.

Coix lachryma-jobi Linn. is found throughout the hotter and damper parts of India and Ceylon, wild and cultivated—not wild in Ceylon. It is distributed over tropical Asia and is cultivated in Africa and America.

Yunani doctors use the seed as a tonic and diuretic. In Western practice the seeds are given as a tincture or decoction in catarrhal affections of the air passages and in inflammation of the urinary passages. In Tongking they are considered a good blood purifier and an excellent diuretic.

The kernels deprived of their shells are used as a food and medicine throughout China, Malaya, Indo-China, the Philippine Islands, and in La Reunion. They make an excellent diet drink for invalids, and have diuretic and cathartic properties. They are also employed for lung and chest complaints.

Among the Santals the root is given in strangury and in the menstrual complaint known as 'silka'.

An infusion of the parched and ground grains is used instead of tea by the Japanese. The Chinese consider the toasted seeds to be tonic and diuretic; they prescribe them in blennorrhagia.

The root is one of the ingredients in Vagbhata's snake remedy Ashtanga-agada; but Mhaskar and Caius have shown experimentally that the drug is not an antidote to snake venom.

Arabic: Damudud—; *Ashanti*: Akrokosaibia—; *Assam*: Koamonee, Sohriu—; *Balaghat*: Gurlu—; *Bengal*: Gurgur, Jilgurgur, Julgurgur, Kunch—; *Bombay*: Gurgur, Kardia, Kassaibija, Kassar, Keruch, Ranjondhla, Ranmaka—; *Bundelkhand*: Ganddula, Garun—; *Burma*: Cheik, Kalithi, Kyeikphun, Kyeit, Sakyeik—; *Cachar*: Jhonki—; *Central Provinces*: Galbi, Ganddula, Kasei—; *Chanda*: Gadi, Galbi, Kasei—; *Chinese*: I I Jen, Kiai Li, T'u I Mi, Y I Mi—; *Cutch*: Dhamra—; *Egypt*: Dimma ayub—; *English*: Christ's Tears, Job's Tears—; *Fiji*: Sila—; *French*: Larme de Job, Larmillé des Indes—; *Gond*: Kasai—; *Gujerati*: Kasai, Ranzondlo—; *Hindi*: Baru, Dabhir, Ganduta, Garahadua, Gargaridhan, Garun, Gulbigadi, Gurlu, Jilgurgur, Julgurgur, Kaiya, Kansh, Kasei, Sanklu, Sankhru, Sankru—; *Igorotte*: Agda—; *Indo-China*: Bo bo hoang, Cuom gao, Neang vong, Y di—; *Jaintia Hills*: Sohriu—; *Japan*: Judsudama, Tomugi—; *Karen*: Be, Bema—; *Khasia Hills*: Sohriu, U sohriew—; *Konkani*: Ranzondlo, Ranzonnalo—; *La Reunion*: Job—; *Lepcha*: Kun-dap, Tung-lin—; *Lushai Hills*: Mim—; *Malay*: Jilai batu, Mulai tikus, Ringuiringui—; *Malaya*: Yee mai, Yee yin—; *Malta*: Dmuh ta Giobb, Hara tac Ciaul, Job's Tears, Lacrima di Giobbe, Zibeg tal curum—; *Manipur*: Changmimkhombi, Mim, Mung—; *Marathi*: Ranjondhala, Ranmakkai—; *Matheran*: Kasai, Ranmaka—; *Mexico*: Acayacoyotl—; *Mount Abu*: Dabhir—; *Mundari*: Bakrihoreng, Horeng, Loeonghoreng—; *Naga Hills*: Kasi, Kesi, Koasangti, Kudhati, Kudhiathia, Sikrakravu, Sotsa—; *New Caledonia*: E'Houa—; *North-West Provinces*: Sankru—; *Philippines*: Alimodias—; *Poona*: Jondhali—; *Porto Rico*: Camandula—; *Portuguese*: Lagrimas de Job—; *Punjab*: Sanklu—; *Rajputana*: Dabhir—; *Ranchi*: Danghir—; *Sabathu Hills*: Sanklu—; *Saharanpur*: Baru—; *Samoa*: Maniusniu, Samasama—; *Sanskrit*: Gaberu, Gavedhu, Gavedhuka, Gavedu, Gojivha, Gundraguttha, Jargadi, Kshudra, Kunta—; *Santali*: Jargadi—; *Sema*: Akiti—; *Seoni*: Galu—; *Sierra Leone*: Boukon, Boukori, Ewaruwura, Kali bugi—; *Sinhalese*: Karibu, Kikirindi, Kikirindimana, Kukirrindi—; *Spanish*: Lagrimas de San Pedro—; *Tagalog*: Tegbe, Tigbi—; *Tamil*: Kattu-kundamani—; *Twi*: Owu-ammang-mankang-m'asaim—; *Uruguay*: Lagrima de Cristo, Lagrima de Job, Lagrima de Maria—; *Visayan*: Adlay—.

CYMBOPOGON.

This genus consists of about 36 species distributed throughout the tropical, and more rarely the subtropical, regions of the Old World.

C. winterianus Jowitt is officinal in Germany.

C. schoenanthus Spreng. is used medicinally in China, Guinea, Madagascar and Guiana; *C. nardus* Linn. in Cambodia, Guinea and Madagascar; *C. citratus* Stapf. in the Gold Coast; *C. excavatus* Stapf., *C. marginatus* Stapf., *C. validus* Stapf. in South Africa.

A. Basal leaf-sheaths ultimately loosened and curled; blades flat; raceme-fascicles compound ... 1. *C. jwarancusa.*

B. Basal leaf-sheaths in dense tufts, tightly clapping, thickened below; blades more or less filiform and flexuous, except when very short; raceme-fascicles more or less simple ... ... 2. *C. schoenanthus.*

C. Sessile spikelets lanceolate or ovate- or obovate-lanceolate; back flat. Lowest pedicel of raceme scarcely stouter than the upper. All the spikelets awnless ... ... 3. *C. nardus.*

D. Sessile spikelets linear to lanceolate-linear, awnless; back distinctly concave in the lower part; panicle usually loose; branches slender, the ultimate branchlets more or less nodding; spathes long and narrow; hairs of joints and pedicels rather spreading ... ... 4. *C. citratus.*

1. **Cymbopogon jwarancusa** Schult. (=*Andropogon jwarancusa* Jones) is found in the outer hill zone of the United Provinces, Kumaon, Garhwal, and westwards as far as Peshawar, Jodhpur and Jaisalmer, Sind, Bihar. It occurs in Western Tibet, ascending to 8,000 ft., and extends westwards to North Africa.

The plant is described by both Ayurveda and Yunani writers.

This grass is used to purify the blood, and in coughs, chronic rheumatism and cholera. It is recommended as a valuable aromatic tonic in dyspepsia, especially that of children. It is also used as a stimulant and sudorific in gout, fever and rheumatism.

The flowers are astringent and are used to stop haemorrhage.

When a Baluchi has fever, a large quantity of this grass is collected and made into a bed. He lies on it and at once begins to sweat.

Arabic: Izkhair—; *Baluchi*: Haveh—; *Bengal*: Gandhavena, Ibharankusha, Karankusa—; *Bombay*: Ibarankusa, Iwarankusa, Izkhir, Kurankusa—; *Brahui*: Have, Haveh—; *Canarese*: Kari lavancha—; *Gujerati*: Jalavalo, Khadajala, Pilovalo—; *Hindi*: Bur, Ghatyari, Ibharankusha, Karankusha, Khavi, Khawi, Khoi, Lamjak, Panni, San, Solara—; *Marathi*: Izkir, Lavaja, Pivalavala—; *North-West Provinces*: Bad, Ganguli, Misiyaban, Piriya—; *Persian*: Gurgiyah—; *Punjab*: Bur, Ghatyari, Ibharankusha, Karankusha, Khavi, Khawi, Khoi, Lamjak, Panni, San, Solara—; *Sanskrit*: Amrinala, Avadahaka, Avadataka, Dirghamula, Isthakapathika, Jalashaya, Jwarankusa, Laghu, Lamajjaka, Laya, Nalada, Sevya, Shighra, Sunala, Sunila—; *Tongking*: Co de—.

2. **Cymbopogon schoenanthus** Spreng. (=*Andropogon schoenanthus* Linn. = *A. laniger* Desf.) is found in the hotter parts of India, wild or cultivated, from the Punjab to Burma and southwards to Travancore and Ceylon. It extends to Afghanistan and through North Africa to Morocco; it also occurs in Eritrea, Socotra and Somaliland.

Ayurvedists describe the grass as acrid, bitter and pungent; useful in fevers, bronchitis, pains, leprosy, heart and throat troubles, and epileptic fits in children.

In Mahomedan medicine it is prescribed as a diuretic, emmenagogue and deobstruent, also as a stimulant to promote perspiration and expectoration.

A decoction of the plant is given as a febrifuge. It is used in China against oppression and spasmodic dyspnoea.

A strong infusion of the grass is commonly used in Egypt as an antispasmodic and diuretic, and is reputed to have valuable results, especially in promoting the passage of urethral stones.

The oil is considered a powerful stimulant when applied externally. It is used as a liniment in chronic rheumatism and neuralgia, and is believed to have the property of curing baldness.

The root and the stem are popular snake and scorpion remedies; but Caius and Mhaskar have demonstrated experimentally that both of them are useless in the antidotal treatment of either snake bite or scorpion sting.

Arabic: Izkhir, Mahareb—; *Bengal*: Agyaghas, Gandhabena, Kamakarpura—; *Bicol*: Baliyoc—; *Bombay*: Rohisha, Rosegavat—; *Cairo*: El-mahareb—; *Canarese*: Vasanchullu—; *Ceylon*: Camel Grass—; *Chinese*: Cha Yang, Che Yang, Mao Hsiang, Yang Mao Ts'ao—; *English*: Camel Grass, Camel Hay, Geranium Grass—; *French*: Barbeau musqué, Chiendent musqué, Foin de chameau, Gramen oriental, Jonc odorant, Pâture de chameau, Schoenanthe, Schoenanthe odorant—; *French Guiana*: Citronnelle—; *Fort Sandeman*: Sargarah—; *Gujerat*: Roshdo, Roshgas, Rusaghas, Rush, Rushgas—; *Hassan*: Ezkryr—; *Hedjaz*: M'haz, Ulfaz udwiyah—; *Hindi*: Bujina, Ghandhabela, Ghatyari, Gundbel, Khavi, Mirchiagand, Musel, Palakhari, Rohisha, Rousaghas, Saundhiya—; *Hova*: Fiahana, Fiahina, Fiehana, Verofehana, Veromanitra—; *Italian*: Fieno di camelo—; *Jhalawan*: Hawai—; *La Reunion*: Citronnelle—; *Las Bela*: Pui—; *Malayalam*: Sirek—; *Marathi*: Rohisha, Rosegavat, Rushgavat—; *Matheran*: Rusha—; *New Caledonia*: N'Dowi—; *North-West Provinces*: Bujina, Palakhari—; *Persia*: Gorgiyah—; *Philippines*: Paja de meca, Raiz de moras, Salaid—; *Punjab*: Ranus, Rauns—; *Sanskrit*: Bhustrina, Bhuti, Bhutika, Devajagdha, Dhupagandhika, Dhyama, Dhyamaka, Katrina, Paura, Putimugdala, Rohisha, Rohishatrina, Saugandhika, Shyamaka, Sugandhatrinashita, Sushitala—; *Shahrig*: Sargarah—; *Siwaliks*: Mirchiagard—; *Spanish*: Esquenanto, Paja de camello, Paja de la Meca—; *Tagalog*: Salai, Salay, Tanglad—; *Tamil*: Kamachipillu, Kavatampillu, Vassinapillu—; *Teheran*: Azkar—; *Telugu*: Kamachikussu—; *Visayan*: Tanglad—.

3. **Cymbopogon nardus** Linn. (=*Andropogon nardus* Linn.) occurs throughout the hotter parts of India, Burma, the Malay Peninsula, and Ceylon, wild or cultivated, common everywhere. It extends to tropical Asia, Africa and Australia.

The infusion of the leaves is used as a stomachic and carminative.

The oil is stimulant, carminative, antispasmodic and sudorific. It is used as rubefacient.

In Cambodia the flowers and the leaves are considered good against cough, and sudorific; the roots are used as a diuretic, sudorific and antiperiodic.

Bengal: Kamakher, Kurankosha—; *Burma*: Singoumia—; *Cambodia*: Sakrey, Slek Rey—; *Canarese*: Gandahanchikhaddi, Kamakshihullu—; *Deccan*: Ganjni—; *English*: Citronella Grass—; *Hindi*: Ganjni, Ganjnikaghas, Iwarankosa, Pustburn—; *Hova*: Verofehana—; *Indo-China*: Citronnelle, Hung nghe huyen, La sa, Satrey—; *Malay*: Sereh wangi—; *Malayalam*: Chorapulla, Kamakshipulla—; *Marathi*: Ganjni, Usadhana—; *Sinhalese*: Maana, Mana, Pengirimana, Watusaewendara—; *Spanish*: Espicanardo espurio—; *Tamil*: Kavattampillu, Kamachipillu, Mandappillu, Sukkunaruppillu, Sunnarippillu—; *Telugu*: Kamakshikasuvu, Kamachikussu, Kamanchigaddi—; *Zulu*: ubungu—.

4. **Cymbopogon citratus** Stapf. (=*Andropogon citratus* D.C.) is only known in the cultivated state. Probably of Indian origin it is now widely distributed over the tropics of both hemispheres.

Ayurvedists describe the grass as pungent, bitter, sharp and hot. They say it is a good laxative, an aphrodisiac and anthelmintic; useful in bronchitis, leprosy and epileptic fits.

An infusion of the leaves is a popular sudorific, stimulant, antiperiodic, very frequently given in simple catarrh. As a fever drink it has great effect in inducing a remission or intermission by bringing on sweat; and it is especially successful with weakly subjects or when the fever is of a typhoid type. It is also used as a medicinal vapour bath in mild cases of fever, or as an inhalation.

The roots and tender leaves are sometimes given with black pepper in cases of disordered menstruation and in the congestive and neuralgic forms of dysmenorrhoea.

In flatulent and spasmodic affections of the bowels, and in gastric irritability, the oil is a remedy of value. In cholera it proves very serviceable, not only by allaying and arresting the vomiting, but by aiding the process of reaction as well. Applied externally, it forms an excellent embrocation in chronic rheumatism, neuralgia, sprains and other painful affections. Externally it is also used for the ringworm.

In the Gold Coast the leaves are often boiled in water like tea, and the liquor is drunk to cure fever. It is sometimes put into hot bath water, and the patient stands in the hot vapour given off. This is also said to be a good cure for fever.

Bengal: Gandhabena—; *Burma*: Sabalin—; *Canarese*: Kavanche hullu, Majjigehullu, Purhalihulla, Vasanehullu—; *Chinese*: Mao Hsiang—; *Colombia*: Limoncillo—; *Deccan*: Hazarmasalah—; *English*: Lemon Grass, Melissa Grass—; *French*: Chiendent citronnelle, Verveine des Indes Orientales—; *Gujerati*: Lilacha, Lilicha—; *Hindi*: Gandhatrina—; *Indo-China*: Huong mao thao, La sa, Sacrey, Say sa—; *Java*: Sireh—; *Malay*: Sereh makan—; *Malayalam*: Shambharapulla, Vasanappulla—; *Marathi*: Hirvacha, Olancha, Olencha—; *Persian*: Chaekashmiri, Hazarmasalah—; *Portuguese*: Herba cheirosa, Capim de Cheiro—; *Sanskrit*: Abichhatraka, Atigandha, Badhira, Badhiradhvanibodhana, Bhustrina, Bhutina, Chhatra, Gochhalaka, Guchhala, Guhyabija, Gundardha, Jambukapriya, Karenduka, Kutimbaka, Malatrinaka, Punsvavighraha, Putigandha, Roshisha, Samalambi, Shringaroha, Sugandha—; *Sinhalese*: Penquin, Saira—; *Spanish*: Grama de limon, Limoncillo—; *Tagalog*: Tanglad—; *Tamil*: Karapurappillu, Vasanappillu—; *Telugu*: Chippagadi, Nimmagaddi, Vasangaddi—; *Venezuela*: Citronera, Limonera, Malojillo—.

CYNODON.

Of the 7 species which constitute this genus, 1 is cosmopolitan, the others Australian and South African.

C. dactylon Pers. is used medicinally in Madagascar, La Reunion and South Africa; *C. hirsutus* Stent. in Basutoland.

C. bradleyi Stent. yields hydrocyanic acid.

Cynodon dactylon Pers. is cosmopolitan and distributed through all warm countries. It is found throughout India, Burma and Ceylon, ascending to 5,000 ft. in the Himalaya.

A decoction of the root is chiefly used as a diuretic in South India and in Chota Nagpore. In Mysore it is employed for secondary syphilis.

The cold infusion of the root often stops bleeding from piles.

The roots crushed and mixed with curds are a Deccan remedy for chronic gleet.

The expressed juice of the plant is an Ayurveda specific for hysteria, epilepsy and insanity.

The expressed juice of the plant, however, is a popular astringent commonly used as an application to fresh cuts and wounds and given internally in cases of chronic diarrhoea and dysentery.

It is also diuretic and used in cases of dropsy. It has moreover been found useful in catarrhal ophthalmia.

In the Konkan the grass is prescribed in compound decoctions with more active drugs for the cure of dysentery, and abnormally profuse menstruation. A white variety, which appears to be only a diseased state of the plant, is used medicinally by the native Hindu practitioners. It is acidulous and is given to check vomiting in bilious complaints.

A preparation of the plant is applied by the Santals in a parasitic disease, which attacks the spaces between the toes.

In Madagascar the whole plant, or the rhizome alone, is applied specially in gout and rheumatic affections.

An infusion of the plant is taken as a tonic in Venezuela.

Europeans in the Transvaal use the plant for heartburn. It is taken bruised and mixed with sodium bicarbonate and other substances. The bruised plant alone is applied as a styptic to wounds.

The Xosas use a decoction as a lotion for sores and swellings.

Mhaskar and Caius have shown that the plant is not an antidote to either snake venom or scorpion venom.

Afrikaans: Batawiese kweek, Fynkweek, Kwagga, Kwaggakweek, Kruisgras, Oostindiese kweek—; *Antsianaka*: Arampandrotra—; *Arabic*: Nagil—; *Baluchi*: Milinu—; *Bengal*: Dub, Dubla, Durba—; *Betsileo*: Kindrese—; *Bombay*: Harala, Haryeli—; *Canarese*: Garikehullu, Karkeri hullu, Kudi garikai—; *Catalan*: Agram, Gram—; *Central Provinces*: Dhupsa, Hariali—; *Chinese*: Ko Oung—; *Cuba*: Grama—; *Cutch*: Chhabar, Chhabbar—; *Egypt*: Nediil, Negil, Nishil, Nisiil—; *English*: Bahama Grass, Bermuda Grass, Couch Grass, Creeping Panic Grass, Devil's Grass, Doab Grass, Dog's Tooth Grass, Doob Grass, Doorwa, Doub Grass, Dun Grass—; *French*: Chiendent pied de poule, Gros chiendent, Herbe des Bermudes, Pied de poule—; *Greek*: Agria, Agriada—; *Guam*: Grama—; *Gujerat*: Dhro, Dhrokad, Gharo, Khabbal—; *Hausa*: Kiri kiri, Tsar-Kiyar zomo, Tsirkiar zomo—; *Hawaii*: Manienie—; *Hindi*: Dhoboghas, Dub, Dubra, Durba, Huriyale, Kalighas, Kabbar, Ramghas—; *Hova*: Fandrotararana—; *Indo-China*: Co chi trang, Co ong, Hanh nghi chi—; *Iraq*: Thaiyil—; *Jhalawan*: Char, Chobi, Godmaz—; *Kolami*: Dhobighas, Dubi—; *Konkani*: Dhurva, Harialy—; *Languedoc*: Limaouça—; *Lao*: Ya Pet—; *Las Bela*: Sabah—; *Loanda*: Capim dos caballos—; *Loralai*: Sabah—; *Madagascar*: Kindresy—; *Malta*: Capriola, Couch Grass, Doub Grass, Gramigna, Nigem—; *Marathi*: Dhurva, Durva, Harialy, Karala—; *Menabe*: Fandrahana—; *Mundari*: Dubila, Dubilatasad, Dubitasad—; *Nairobi*: Uganda Grass—; *Nisirabad*: Kabb—; *North-West Provinces*: Duba, Kalighas, Ramghas—; *Pacific Coast*: Bermuda Grass, Dog's Tooth Grass, Indian Couch Grass, Scotch Grass, Wire-Grass—; *Panjgur*: Shez—; *Punjab*: Daurva, Dun, Dubra, Kabbar, Khabbal, Talla, Tilla—; *Rajputana*: Dob, Nilldub—; *Rarotonga*: Mati—; *Sadani*: Dublaghas—; *Sakalave*: Fandrotsana, Fandrotsarana—; *Sanskrit*: Amari, Amrita, Ananta, Anuvallika, Asitalata, Bahuvirya, Bhargavi, Bhutahantri, Dhurta, Dhurva, Durmara, Gauri, Guna, Harasalika, Harita, Haritali, Jaya, Kachharuha, Mahaushadhi, Mahavari, Mangala, Nanda, Niladurva, Ruha, Sahsravirya, Saumya, Shadvala, Shambhavi, Shanta, Shashpa, Shataparva, Shatagranthi, Shatavalli, Shatmula, Shita, Shitakumbhi, Shitala, Shiva, Shiveshta, Shyama, Tiktaparva, Vamini, Vijaya—; *Santali*: Dhobighas, Dubi—; *Sind*: Chhabar, Chhabbar, Chibbur—; *Sokoto*: Kirikiri, Tsarkryarzomo, Tsirikiarzomo—; *South Africa*: Bahama Grass, Bermuda Grass, Bermuda Quickgrass, Devil's Grass, Doab Grass, Dub Grass, Fine Couchgrass, Florida Grass, Germiston Grass, Scotch Grass—; *Spanish*: Grama común—; *Suto*: Mohlwa, Morara—; *Tagalog*: Cautcauran, Colatay, Malit—; *Tamil*: Arugampillu, Hariali—; *Tasmania*: Indian Couch—; *Telugu*: Ghericha, Gurka hariali—; *Trans Indus*: Burawa—; *Transvaal*: Quagga Quick—; *Upper Godavery*: Haryali—; *Urdu*: Dub—; *Uriya*: Dubbulaghanso—; *Uruguay*: Gramilla, Gramilla colorada, Gramilla del tiempo, Pata de perdiz—; *Virginia*: Wire Grass—; *West Indies*: Bahama Grass—; *Xosa*: uQuaqaqa—; *Zulu*: isiNandi, umFulwane, uNgwengwe—.

DACTYLOCTENIUM.

The genus consists of 5 species inhabiting warm regions, one widely spread throughout the tropics.

Dactyloctenium aegyptium Rich. (=*Eleusine aegyptiaca* Desf.) is spread throughout the low country of India and Burma, to Malacca and Ceylon. It is also known from Egypt, Nubia, Eritrea, Arabia and other countries of Tropical Asia and America. It was introduced into the New World from the warm regions of the Old.

The grains are used medicinally by the Mundas of Chota Nagpore; they are parched in an earthern vessel and consumed in small doses for three to eight days by women who after childbirth suffer from belly-ache.

A decoction of the seeds is renowned in Africa as an alleviator of pains in the region of the kidney, and its herbaceous parts are applied externally for the cure of ulcers.

Annam: Co chi trang—; *Australia*: Egyptian Finger Grass—; *Bombay*: Mhar, Nachani, Nagli, Natchni, Raj—; *Bundelkhand*: Makamakna, Tipakia—; *Central Provinces*: Chikara, Chotamandiya, Mathna, Utesirkum, Utesirla—; *Egypt*: Im-es-salib, Rigl-el-herbayeh—; *English*: Goose Grass—; *Hausa*: Gudegude—; *Hindi*: Makra, Makri—; *Malayalam*: Kavarapullu—; *Mundari*: Bulungcuri, Bulungkode, Bulungruti, Bulungrutui—; *Natal*: Inane—; *Philippines*: Salai maya—; *Punjab*: Bhobra, Chimbari, Chubrei, Karmadhana, Madana, Madhana—; *Rajputana*: Malicha, Maligha, Mansa—; *Sadani*: Nonmarua—; *Santali*: Suntubukrui—; *Sinhalese*: Putatana—; *South Africa*: Coast Grass, Duck Grass—; *Tagalog*: Alam—; *Tamil*: Sodi, Tamida—; *Telugu*: Muttengapillu—; *Transvaal*: Egyptian Kweek-grass, Natal Kweek-grass—; *United States*: Comb Fringe Grass, Crowfoot Grass, Little Crowfoot—; *Uriya*: Kakuriya—; *Zulu*: isiNandi, isInane, uNgwengwe—

DENDROCALAMUS.

This genus numbers 24 species, which inhabit the Indo-Malayan region, the Philippine Islands, China and Africa.

Dendrocalamus strictus Nees is found in the dry hills of Northern India, from Garhwal to Nepal and southwards from the Punjab to Bihar, the Konkan, Central and South India, Burma and Great Cocos Island. It extends to Java

The leaves are boiled in water for convalescents to bathe in.

A decoction of the leaves is given to aid parturition; the juice in two-ounce doses taken frequently is used as an abortifacient.

In Arabia the joints are made into a decoction and used as a medicine to procure abortion.

The minute hairs found on the sheath leaves are used by the Malays as a poison; it is given either alone or in combination with other things.

Baigas: Bhiru—; *Bengal*: Karail—; *Bombay*: Bas, Bassa, Kaban, Udha, Vassa—; *Burma*: Myinwa—; *Canarese*: Bidiru, Gandubediru, Kibbidary, Kiribidiru—; *English*: Male Bamboo, Solid Bamboo—; *Gond*: Halpa, Vadur, Veddar—; *Gujerat*: Nakorvans, Narvans, Vans—; *Hindi*: Bans, Banskaban,

Banskhurd, Kopar, Lakdibans, Narbans—; *Khond*: Maringi—; *Kolami*: Burumadh, Burumat, Mathan, Saring—; *Kumaon*: Bans—; *Kurku*: Indo—; *Lambadi*: Vasi—; *Malay*: Buloh batu, Buloh berang, Buloh tempat—; *Malayalam*: Arinkantam, Cheriyamula, Kalmula, Kalmungil, Karinalimula, Karinkana—; *Marathi*: Bans, Bharilvel, Bhovarlit, Vela—; *Mundari*: Birmad, Burumad, Janggimad—; *Palamow*: Bukhar—; *Pandratola*: Kark—; *Reddi*: Kondaveduru—; *Sanskrit*: Vansha, Venu, Yavaphala—; *Santal*: Burumadh, Burumat—; *Saora*: Kondaveduru—; *Tamil*: Karanai: Kalmungil, Kattumungil, Mungil, Sinnamungil, Sirumungil, Siruvari—; *Telugu*: Chittiveduru, Gattiveduru, Kankaveduru, Penveduru, Potuveduru, Rativeduru, Sadanapaveduru, Sannaveduru, Veduru—; *Thana*: Bundi, Manwel—; *Tulu*: Lavakiri, Panjibedru—; *Uriya*: Saliabans, Saliabhanso, Salimbo, Salimbobaunso, Salimbo bhanso, Sanobaunso—.

DESMOSTACHYA.

The only species of the genus is **D. bipinnata** Stapf. (=*Eragrostis cynosuroides* Beauv.). It is found in the plains of India, from Peshawar and Sind to Burma and southwards, in hot dry places. It extends westwards to Syria, Egypt, Nubia and East Tropical Africa.

The culms are diuretic and stimulant. In the Konkan they are prescribed in compound decoctions with more active drugs for the cure of dysentery, and profuse menstruation.

Afghanistan: Drab, Kuthag—; *Bengal*: Kusha—; *Bolan*: Drab—; *Bombay*: Darbh—; *Brahui*: Drak—; *Bundelkhand*: Dabvi—; *Central Provinces*: Chir, Dabhat, Kusha—; *Egypt*: Ambarfe, Halfa, Hamarte, Gash—; *Gazechah*: Kuthag—; *Gujerati*: Dabha, Darabha—; *Hindi*: Dab, Davoli, Durva—; *Kani*: Kuthag—; *Kila Saifulla*: Sparmaghaz—; *Libya*: Halfa—; *Marathi*: Darbha—; *North-Western Provinces*: Dab, Daboi, Dhab, Kush—; *Punjab*: Dab, Dhab, Dib, Drab, Drabh, Kusa—; *Sanskrit*: Barhi, Darbha, Durbha, Garbha, Hrasva, Kurava, Kusha, Kutha, Kutupa, Pavitra, Suchyagra, Yajnabhushana—; *Shahrig*: Dab—; *Sibi*: Drab—; *Telugu*: Aswalayana, Dabha, Darbha, Durpa, Kusadarbha—; *Turbat*: Ding, Drab—.

ECHINOCHLOA.

This genus consists of 20 to 25 species distributed throughout the warm regions of both hemispheres.

E. colona Link. var. *frumentacea* Blatter and McCann, and *E. crus-galli* P. Beauv. are used medicinally in China.

1. Upper involucral glume and lower floral glume equally acute or cuspidate ... ... *E. colona*.
2. Upper involucral glume and lower floral glume cuspidate or produced into an awn, the latter more than the former. Ligule absent ... *E. crus-galli*.

1. **Echinochloa colona** Link var. **frumentacea** Blatter & McCann (=*Panicum frumentaceum* Roxb.) is cultivated over the greater part of India, up to 6,500 ft on the Himalaya.

The plant is said to be a good remedy useful in biliousness and constipation.

America: Billion Dollar Grass—; *Baluchi*: Samako—; *Bengal*: Samrashama, Sanwa, Saon, Shama, Shamula, Syamadhan—; *Bihar*: Sama, Sanwan, Sawan—; *Bombay*: Bavto, Sama, Samuka, Sanwak, Shama—; *Canarese*: Same, Save—; *Central Provinces*: Sawa, Sema—; *Chinese*: Shan Tzu—; *Deccan*: Kangra, Kathi, Kathli, Sanwa, Saon, Sawa Shama, Shamula—; *English*: Japanese Barnyard Millet—; *Garhwal*: Jhungara—; *Gujerat*: Samo, Samoghas, Janglisamak—; *Hindi*: Samak, Sanwa, Sawa, Sawan, Shama, Shamula—; *Kashmir*: Karin, Soak—; *Khondmals*: Jhari—; *Kumaon*: Jhangora,

10

Koni, Kungni, Mandira—; *Malay*: Padi burong—; *Marathi*: Janglisama, Samul—; *North-Western Provinces*: Jhungara, Saman, Sawan—; *Oudh*: Sama, Samei, Sanwan, Sawan, Sawanbhadeha—; *Persian*: Bajri—; *Punjab*: Chandra, Sama, Samuka, Sanwak, Sawank, Soak—; *Sanskrit*: Avipriya, Rajadhanya, Shyama, Shyamaka, Sukumara, Tribija, Trinabijottama—; *Sind*: Saon, Saron—; *Sinhalese*: Welmarukku—; *Tamil*: Kudraivallipillu, Railpillu—; *Telugu*: Bontachamalu, Bontashama, Bonthshama, Chama, Chamalu, Pala oodalu, Sawa—; *Uriya*: Samu—.

2. **Echinochloa crus-galli** P. Beauv. (=*Panicum crus-galli* Linn. =*P. crus-corvi* Linn.) is common throughout the greater part of India and Malaya; as a weed it is found throughout the warm temperate countries of the northern hemisphere. It is somewhat rare in the tropics of Africa and the New World, and south of the Tropic of Cancer.

The plant is used to check haemorrhage, and is prescribed for diseases of the spleen

Australia: Barnyard, Cockspur Grass—; *Baluchi*: Samo—; *Bengal*: Burashama, Dul—; *Betsileo*: Ahibary, Aibary—; *Bombay*: Borur, Jiria, Pakud, Pakushama, Sawank, Todia, Tor—; *Canarese*: Kadu dabhai hullu—; *Central Provinces*: Baribhodore, Bharta, Datia, Kondabuttamgodi—; *Chinese*: Pai—; *Dutch*: Vingergras—; *Egypt*: Dineyb—; *English*: Panic Grass—; *French*: Crête de coq, Ergot de coq, Millard, Panis des marais, Panis pied de coq Patte de poule, Pied de coq—; *German*: Hahnenfussfennich—; *Greek*: Kechri—; *Gujerat*: Adbausamo—; *Hindi*: Samak, Sanwak—; *Hova*: Farimanga, Tsimparifarifolsy, Tsimparifarifary, Tsimparifarimango, Tsimparifarimena—; *Iraq*: Dahnan, Danan, Dukhain—; *Italian*: Cresta de gallo—; *Malta*: Panickgrass, Cock's shin Grass, Panicastrella, Xrika—; *Marathi*: Sama—; *Mundari*: Camatasad, Iri—; *North-Western Provinces*: Dhand, Jalsawank—; *Punjab*: Barasanwak, Bharti, Dhand, Jarotha—; *Rajputana*: Horma, Sama, Siwan—; *Roumanian*: Bujorul—; *Sadani*: Sauna, Sawa—; *Sakalave*: Karangy—; *Sanskrit*: Jalasamaka—; *Spanish*: Daza, Mijo—; *Tagalog*: Dauadaua—; *Teulgu*: Peddawundu—; *United States*: Barnyard Grass—: *Uruguay*: Pata de gallo—.

ELEUSINE.

This genus consists of 10 species distributed over the warm regions of the eastern hemisphere, one being widely spread throughout the tropics.

E. coracana Gaertn. is used medicinally in South Africa, *E. indica* Gaertn. in Guiana and Cambodia.

1. Spike stout, often incurved, pubescent at base. Seed globose ... ... *E. coracana*
2. Spike slender, nearly glabrous at base. Seed oblong, obtusely trigonous ... ... *E. indica*.

1. **Eleusine coracana** Gaertn. is cultivated throughout India and in many parts of tropical Africa and tropical Arabia.

Ayurveda authors describe the grain as acrid, bitter and sweet and cooling. They consider it tonic, and recommend it in biliousness and blood diseases.

The grain is generally used as an astringent.

In South Africa the Tongas and Shangaans use it along with *Plumbago zeylanica* Linn. as an internal remedy for leprosy.

Abyssinia: Talban—; *Bengal*: Marua, Murha—; *Bombay*: Nagli, Nangli—; *Canarese*: Raggi, Ragi—; *Egypt*: Bishna—; *Eritrea*: Dagussa—; *French*: Coracan, Eleusine coracana—; *Golungo Alto*: Luco, Lucu—; *Gujerati*: Bavtonagli, Navtonagli—; *Hadramaut*: Dokhn—; *Hausa*: Tamba—; *Hindi*: Makra,

Mandua, Marua, Maud, Murua, Rotka—; *Katagum*: Tamba—; *Khasi*: U'raitruh—; *Kolami*: Kode—; *Konkani*: Gonddo, Nachno—; *Malayalam*: Chettipulli, Mutami—; *Marathi*: Nachiri, Nagli—; *Mundari*: Dasaikode, Dumbakode, Kode—; *Natal*: Upoko—; *Nile*: Ooleyzei—; *Northern Nigeria*: Tomba—; *North-West Himalayas*: Koda, Kodon, Kodra, Kutra—; *North-West Provinces*: Makra, Mandua, Marua, Rotka—; *Oudh*: Makra, Mandua, Marua, Rotka—; *Persian*: Mandwah—; *Portuguese*: Nachinim—; *Punjab*: Chalodra, Koda, Kodon, Kodra, Mandal—; *Sanskrit*: Bahupatraka, Bhuchara, Guchha, Kadhina, Kanisha, Lanchhana, Maliyasa, Narttaka, Nrityakunda, Ragi, Raji, Rajika—; *Santali*: Kode—; *Shangaan*: Liphokho—; *Sind*: Nachni, Nangli—; *Sinhalese*: Kurakka—; *Tamil*: Kalvaragu, Kapai, Kayur, Ragi—; *Telugu*: Chodalu, Ponassa, Ragi, Ragulu, Sodi, Tamidelu—; *Unyoro*: Bolu—; *Uraon*: Kodai—; *Uriya*: Mandia—; *Zambesi*: Marumbi, Mitchinin—; *Zanzibar*: Vimbe—; *Zulu*: uPoko—.

2. **Eleusine indica** Gaertn. is found growing throughout the plains of India and Ceylon. It is common in the tropics of the Old World. It is naturalized in the Mediterranean region, and is said to have been introduced in the New World.

The whole plant, but more especially the root, is considered sudorific and febrifuge in Cambodia. It is much used in liver complaints.

In Guiana a decoction of the plant is given to children for convulsions.

Bundelkhand: Gurchawa—; *Burma*: Hsengnomyeet, Singnomyet—; *Cambodia*: Choeung Kras—; *Central Provinces*: Godchabba, Gurragadi, Kakariya, Madanya, Malghi, Mandial—; *Cuba*: Pata de gallina—; *French Guiana*: Pied de poule—; *Golungo Alto*: Pé de galinha—; *Guam*: Umog—; *Gujerat*: Adbaunagli—; *Hausa*: Chiyawartuji, Tuji—; *Hindi*: Malankuri—; *Katagum*: Chiyawar tuji, tugi—; *Kolami*: Bir Kode—; *Kumaon*: Mandavi—; *Lagos*: Ese-kana-kana, Gbegi—; *Malay*: Rumput sambau—; *Marathi*: Ran nachani—; *Mundari*: Kodai, Sukari kodai—; *North-Western Provinces and Oudh*: Gadha, Gadhacharwa, Gathamandwi, Jhingri, Jhinjhor, Lijhar, Makraila—; *Philippines*: Baquisquisan—; *Rajputana*: Mandwa—; *Sesuto*: Moseli—; *Shahrig*: Chhabal—; *Sierra Leone*: Osoekurunyi—; *Sinhalese*: Walkurakkan—; *South Africa*: Crowfoot, Goose Grass, Wild Rapoko Grass—; *Tagalog*: Sabongsabongan, Sambale—; *Tamil*: Thipparagi—; *Telugu*: Karuchodi, Karsodi, Kuror—; *Uganda*: Kasibanti—; *United States*: Goose Grass, Yard Grass—; *Uriya*: Nandia—; *Zulu*: uMunyankomo, umNyankomo, uPoko.

Heteropogon.

This genus consists of about 6 species found in the tropical and subtropical regions of the whole world.

H. contortus Roem. & Schult. is used medicinally in China and South Africa.

Heteropogon contortus Roem. & Schult. (=*Andropogon contortus* Linn.) is found in all warm regions. It is distributed throughout India, Burma and Ceylon to the Straits of Malacca, ascending the Himalaya to 5,000 ft. It is common throughout the whole Tropical Africa, and the adjacent islands, ascending in Eritrea to over 8,800 ft. It extends all over Africa, the Mediterranean region, and the tropical and subtropical regions generally.

The root is a stimulant and diuretic.

The Sutos of South Africa use the plant with *Tribulus terrestris* Linn. for the treatment of rheumatism in the hands

Afrikaans: Steek Grass—; *Bihar*: Chorant—; *Bombay*: Bal, Bale, Daklisuckal, Italisuckal, Musaneh, Pendrisuckali, Suckal, Suckali kussal, Suckeri—; *Canarese*: Kari vunugada hullu, Sunkari hullu—; *Chinese*: Ti Chin—; *English*: Spear Grass, Wild Oats—; *Ga*: Akorsorfeng, Ananugangi—; *Gujerat*: Dhabsuliyum—; *Hausa*: Bunsurundaji—; *Hawaii*: Pili—; *Hindi*: Kher, Saral, Shurighas, Shurval—; *Ho*: Saiyu—; *Kharwar*: Chorant—; *Kohlu*: Barwu—; *Marathi*: Kant egawta—; *Matheran*: Bal, Kusal, Pandrisukal—; *Mundari*: Sauri—; *Ormara*: Abdarka—; *Sambalpur*: Sukla—; *Santali*: Saurighas—; *Sesuto*: Selokana—; *Shahrig*: Barwaz—; *Somaliland*: Aggar—; *Tamil*: Karunsipullu, Panipullu, Usipullu—; *Telugu*: Dubbagasarigaddi, Eddigaddi, Kaserigaddi, Yedda, Yeddi, Yerragoyi—; *Uriya*: Dauria, Sinkola, Sinkolo—; *Zulu*: isiTupe—.

HORDEUM.

This genus numbers about 25 species, natives of temperate regions, mainly found in the northern hemisphere.

Hordeum vulgare Linn. is cultivated chiefly in North India and up to 13,000 ft. in the Himalaya. It is widely cultivated in all temperate regions.

Both Ayurveda and Yunani writers credit barley with anti-bilious, antispasmodic, and antifebrile properties, and recommend its use in the treatment of bronchitis.

Barley is demulcent, and easy of digestion, and is for these reasons used in the dietary of the sick. A powder of the parched grains is much employed in the form of a gruel in cases of painful and atonic dyspepsia.

In Patna the ashes of the leaf are employed in the formation of cooling sherbets. The ashes of the stalks are prescribed for indigestion in the plains of the Punjab.

From the earliest times barley has been employed to prepare drinks for the sick, especially in feverish disorders, and for sore lining membranes of the chest. Barley in the form of the decoction popularly known as *barley water*, affords a mucilaginous drink very valuable in all cases requiring demulcent treatment. Pearl barley is the form usually preferred for the preparation of the decoction, made by pouring four pints of boiling water on two Troy ounces of pearl barley and boiling away to two pints, and straining. It is especially used in infant feeding, as it seems to prevent the formation of large milk curds by its colloidal character.

Pearl barley is the seed deprived of all its investments and afterwards rounded and polished in a mill. It is in small round or oval grains, having the remains of the longitudinal furrow of the seeds, and of pearly whiteness. It is wholly destitute of hordein, and abounds in starch, with some gluten, sugar and gum. This is the proper form of barley for medicinal use.

Honey may be added beneficially to the decoction of barley for bronchial coughs. Barley water with gum arabic, one ounce of the gum dissolved in a pint of the hot decoction, is a very useful drink to soothe irritation of the bladder and of the urinary passages.

Barley meal is formed by grinding the seeds previously deprived of their husk. It may be made into a coarse, heavy, hard bread, which in some countries is much used for food, though apt to purge.

In China barley sprouts are especially prepared for medicinal use. This is done by moistening the grains with water, allowing them to germinate, and then drying them in the sun. The sprouts are rubbed off and the grain ground into flour. The sprouts enter into a number of prescriptions given for infantile complaints.

The germinated barley or malt with the radicle is used in China as peptic, stomachic, lenitive, demulcent and expectorant. Preparations of malt have acquired some reputation of late years in Europe and America, since they are more demulcent and nutritious than those of the unmalted barley.

Barley is officinal in France, Portugal, Spain and the United States of America.

Afghanistan: Jao, Jaoshirin, Jaotush—; *Arabic*: Shaair, Shair—; *Armenian*: Kari—; *Ashkobi*: Arbus—; *Bagwana*: Brehnajau—; *Behar*: Jowakhar—; *Bengal*: Jab, Jao, Jau—; *Bhotia*: Nas—; *Bombay*: Jav, Satu—; *Brazil*: Cevada, Cevada sancta—; *Burma*: Muyau—; *Canarese*: Javegodhi—; *Catalan*: Ordi, Ordi comú—; *Chinese*: Kung Mai, No Mai, Ta Mai—; *Danish*: Byg—; *Deccan*: Satu—; *Dutch*: Gerst—; *Egypt*: Shair—; *English*: Barley—; *Finland*: Ohva—; *French*: Béchette, Blé d'Egypte, Blé de mai, Epeautre, Epente, Epinte, Espigan, Orge, Orge commune, Orge grosse—; *German*: Garsten, Geste—; *Greek*: Krithari, Krithi—; *Gujerati*: Jau, Jav, Ymvah—; *Harboi Hills*: Jau, Sa, Urbusha—; *Hindi*: Jau, Jav, Jawa, Suj—; *Hungarian*: Arpa—; *Indo-China*: Dai mach, Lua mach nha, Ta me, Thoc mach nha—; *Iraq*: Sha'ir—; *Italian*: Farro, Orzo, Scandella, Spelita, Spelta—; *Jhalawan*: Jau, Sa, Urbusha—; *Kila Saifulla*: Jau, Sa, Urbusha—; *Konkani*: Cevad, Jav—; *Kurdish*: Ju—; *Languedoc*: Espeulto, Espigan, Feraje hordi, Ordi, Ordigal—; *Lapland*: Kordne—; *Lassa*: Soah—; *Malaya*: Mai ngai—; *Malta*: Barley, Orzo, Xghei, Xghei tal birra, Xghei tal mazza—; *Marathi*: Cevad, Jav, Java, Satu—; *Nepal*: Tosa—; *North-West Provinces*: Indarjau, Jau, Yurk—; *Persian*: Jao—; *Polish*: Jenczmien—; *Portuguese*: Cevada—; *Punjab*: Buza, Chak, Chang, Chung, Jaon, Jau, Jawa, Nai, Thanzatt—; *Roumanian*: Orz—; *Russian*: Jetschmen—; *Sanskrit*: Akshata, Dhanyaraja, Divya, Hayapriya, Hayeshta, Kanchuki, Medhya, Pavitradhanya, Praveta, Shaktu, Shvetashunga, Sitashuka, Sitrishuka, Tikshnashuka, Turagapriya, Yava, Yavaka—; *Sharig*: Jau, Sa, Urbusha—; *Spanish*: Cebada, Cebada común—; *Swedish*: Biugg—; *Tamil*: Barliyarisi, Barliyarishi—; *Tartary*: Arpah—; *Telugu*: Barlibiyam, Dhanyabhedam, Pachchayava, Yava, Yavaka, Yavala—; *Turkish*: Arpa—; *Urdu*: Jav—; *Yemen*: Schair—.

Hygrorhiza.

H. aristata Nees., the only species of this genus, is found in the upper and lower Gangetic Plains, Assam, Chittagong, Pegu, the Deccan and Ceylon, extending to Tongking. It occurs either floating on the surface of water, or creeping on wet ground.

Ayurveda describes the seeds as sweet and acrid, oleagenous, digestible and cooling. They are astringent to the urinary tract, and useful in biliousness. They cause constipation and flatulence.

Bengal: Uridhan—; *Bombay*: Deobhat, Urodhan—; *Canarese*: Jyarahumedhe—; *Gujerati*: Vanti—; *Hindi*: Janglidal, Tili, Tini—; *Malayalam*: Nirvallipullu—; *Marathi*: Deobhata—; *North-Western Provinces*: Parsal, Passahi, Passai, Passari, Tinni—; *Punjab*: Pastal—; *Sanskrit*: Aranyadhanya, Aranyajali, Munidhanya, Nivara, Prasadhika, Trinadhanya, Trinodbhava, Vanavrihi—; *Sinhalese*: Gojabba—; *Tamil*: Valli pullu—

IMPERATA.

This genus consists of 5 tropical species found in the warm regions of both hemispheres, but chiefly American.

Imperata arundinacea Cyrill. (=*I. cylindrica* Beauv.) is found in the hotter parts of India, from the Punjab southwards and eastwards, to Malacca and Ceylon. It grows in all warm countries and often covers enormous areas. When entering cultivated land it may become a serious pest.

The roots are used as an emollient in Cambodia, mostly in the fumigation of piles.

The roots are much appreciated in China for their restorative, tonic, haemostatic and antifebrile properties.

The Sutos of South Africa use the root in making a medicine for chest colds in children, while the Zulus regard it as a specific for hiccough.

Australia: Blady-grass—; *Baluchi*: Drug—; *Bengal*: Ulu—; *Bhabar*: Shiro—; *Bombay*: Dhub—; *Burma*: Thek-kay nyen—; *Cambodia*: Sbeou—; *Canarese*: Sanna dabbai hullu—; *Cantonese*: Pak mau kan—; *Chinese*: Mao Tsao, Pai Mao Ken—; *Hausa*: Tofa, Toha—; *Hindi*: Dabh, Siru, Ulu—; *Hova*: Tenina, Tenona, Tsevoka—; *Indo-China*: An do bach mao, Co danh, Mao ken, Tranh—; *Iraq*: Halfah, Shismallah—; *Kolami*: Chero—; *Malay*: Alang alang, Lalang—; *Malaya*: Pak mau kan—; *Mundari*: Huringciru—; *North-Western Provinces*: Shiro—; *Punjab*: Siru, Ulu—; *Sakalave*: Manevica—; *Sanskrit*: Balbajamu, Barhissu, Darbha—; *Santali*: Chero—; *Simla*: Dab—; *Sinhalese*: Iluk—; *Southern Nigeria*: Spear Grass—; *Suto*: Mohlabalerumo, Mohlorumo—; *Tamil*: Dharbai pul, Tharpaipillu—; *Telugu*: Barumbiss, Dharba, Modavagaddi—; *Uganda*: Lusanke—; *Upper India*: Bharwai, Sil, Sir, Usirh—; *Zulu*: umThente—.

LOLIUM.

This genus includes 6 species, natives of Europe, temperate Asia, and North Africa; but naturalised in many countries.

Many of the species are reputed to possess a poisonous principle.

Lolium temulentum Linn. is an annual weed of cultivation, also occasionally met with as a wayside weed and in waste ground, common in the plains and hills of the Upper Gangetic Plain, the Punjab, the North-West Provinces ascending to 4,000-6,000 ft. in the Himalaya, and Sind. It extends throughout Europe and Western Asia, has been found in the island of Madura and in North Africa, and occurs as an introduced plant in the United States and Australia.

The seeds or grains were used medicinally by the ancient Greeks and Romans. The old sages, however, supposed it to cause blindness, and with the Romans, *lolo victitare*, to live on Darnel, was a phrase applied to a dim-sighted person. Gerard says, 'The new bread wherein Darnel is eaten hot, causeth drunkenness'. A general trembling, followed by inability to walk, hindered speech, and presently profound sleep with subsequent headache and vomiting, are the symptoms produced by Darnel when taken in a harmful quantity.

'As it is not without some vices', says Culpeper, 'so hath it also many virtues. The meal of Darnel is very good to stay gangrenes and other such like fretting and eating canker and putrid sores; it also cleanseth the skin of all leprosies, morphews, ringworms, and the like, if it be used with salt and raddish roots. And being used with quick brimstone and vinegar, it dissolveth knots and kernels, and breaketh those that are hard to be dissolved, being boiled in wine with pigeon's dung and linseed. A decoction thereof made with water and honey, and the places bathed therewith is profitable for the sciatica. Darnel meal applied in poultice draweth forth splinters and broken bones in the flesh. The red darnell boiled in red wine and taken, stayeth the lax and all other fluxes and women's bloody issues and restraineth urine that passeth away too suddenly.'

Darnel meal is still recommended as a sedative poultice, and is still believed to cure freckles.

Catalan: Juy, Sisanya—; *Devonshire*: Eaver, Iver—; *Dutch*: Bolderik, Dolik, Dravik—; *English*: Bearded Darnel, Cheat, Darnel, Jura, Ray Grass, Rye Grass, Wary, Bragge, Cockle, Dornel, Dragge, Drake, Drank, Dravick, Droke, Drunk, Drunken Plant, Eaver, Ivray, Jum, Lover's Steps, Ray, Riely, Rivery, Sturdy—; *French*: Ivraie, Ivraie annuelle—; *German*: Lolch, Taumelkorn—; *Hindi*: Machni—; *Iraq*: Hanaitah, Ruwaitah, Shailam—; *Italian*: Gioglio, Loglio—; *Maltese*: Sicrana—; *Portuguese*: Joio—; *Roumanian*: Neghina, Selbata, Zizanie—; *Russian*: Plevel, Pshenetz—; *Somersetshire*: Devon Ever—; *Spanish*: Joyo, Zizaña—; *Sussex*: Crop—; *Uruguay*: Joyo, Trigollo, Zizaña—.

Lophatherum.

This genus consists of 2 species inhabiting Indo-Malaya, China and Japan.

Lophatherum gracile Brongn. is found in Tropical Himalaya from Sikkim eastwards, the Khasia and Naga Hills, and Ceylon up to 4,000 ft. It is common along forest paths in the whole Malay Peninsula from Singapore to Kedah, from the plains to 5,000 ft. altitude in the Thaiping Hills and Pahang. It extends to the waste land in China, Japan, Java, Amboyna and New Guinea.

In China the leaves are considered to have antifebrile and diuretic properties.

Annam: Co may, Dam chuc diep—; *Chinese*: Tan Chu—; *Malay*: Rumput jarang, Rumput Kelurat, Rumput ubi bulush—; *Sino-malau*: Tham chook—.

Manisuris.

The one species of this genus, **M. granularis** Sw., is found throughout the hotter parts of India, from the Punjab eastwards to Burma, and southwards to Ceylon. It occurs in most tropical countries.

Anislie says that in Behar the plant is prescribed internally in conjunction with a little sweet oil, in cases of enlarged spleen and liver.

Ajmere: Kangni—; *Berar*: Ratop—; *Chanda*: Agimaligadi—; *Gujerat*: Kasium Kasiunghas—; *Hindi*: Kangni, Trinpali—; *Naguri*: Buruburuludiatasad, Ludiatasad—; *Rajputana*: Dhaturoghas—; *Sanskrit*: Palanggini—; *Sierra Leone*: Ardande, Fesifesi, Guisqui—; *Udaipur*: Dhaturoghas—.

ORYZA.

This genus numbers about 17 tropical species extending to the subtropical regions of both hemispheres.

Oryza sativa Linn. is widely cultivated. It occurs indigenous in the marshes of Rajputana, Sikkim, Bengal, the Khasia Hills, Central India, the Circars and Pegu. It extends to Australia.

Ayurveda describes the rice grain as acrid, sweet, oleagenous, tonic, aphrodisiac, fattening, diuretic and useful in biliousness.

In India rice is used variously in sick diet. It is invariably ordered as the safest and best food in all dysenteric complaints. Boiled rice, when hot, is used as a poultice. Rice-water is recommended as an excellent demulcent, refrigerant drink in febrile and inflammatory diseases, dysuria and other affections requiring this class of remedies. This decoction is also given as an enema in affections of the bowels.

In Cambodia the husk of the grain is considered anti-dysenteric. The roasted grain, mixed with an equal amount of palm-sugar, is prescribed in *Strychnos* poisoning. Boiled in water and then dried in the sun, the grain enters into the composition of remedies for leprous ulcers.

Certain varieties of specially prepared grains are used medicinally in China and Malaya. Malted rice is used as a peptic, carminative and tonic.

In Europe the grain has long been considered to exercise pectoral virtues, and useful for persons troubled with lung disease, and spitting of blood, as in pulmonary consumption. 'The grain stays laxes and fluxes of the stomach and belly, especially if it be parched before it is used, and hot steel quenched in the milk wherein it is boiled, being somewhat drying and binding. The flour of the rice has the same property, and is put into cataplasms to repel humours from flowing to the place, and also to women's breasts to stay inflammations.'

Rice bread and rice cakes, simply made, are very light and easy of digestion. The gluten confers the property of rising, on dough or paste made of rice flour. But as an article of sustenance rice is not well suited for persons of fermentative tendencies during the digestion of their food, because its starch is liable to undergo this chemical change in the stomach.

The grain is officinal in Austria, Belgium, France, Germany, Great Britain, Holland, Italy, Portugal, Spain, Switzerland and Turkey.

Amblaw: Fala—; *Annam*: Lua nep, Lua nui, Lua song lon, Lua te, Lua toc—; *Arabic*: Arruz, Arz—; *Armenian*: Priusch—; *Belgaum*: Bhatta—; *Bengal*: Chal, Chanvol, Dhan, Oridhana—; *Bombay*: Bhatta, Dangar—; *Bouru*: Hala, Halai, Pala—; *Broach*: Dangar—; *Burma*: Chan, Saba, San—; *Cambodia*: Srau damnop, Srau khasai, Srau prapeai vea—; *Canarese*: Akki, Batta—; *Catalan*: Arros—; *Central Provinces*: Deaodhan—; *Ceram*: Fala—; *Chinese*: Ch'en Lien Mi, Hsien, Keng, Tao, Tao Tzeu—; *Cochin China*: Lua—; *Danish*: Riis—; *Deccan*: Chanval—; *Dutch*: Rijst—; *East Indies*: Paddy—; *Egypt*: Arus, Rus, Ruz—; *English*: Rice—; *Ewe*: Molung, Morli, Morlu—; *Fanti*: Omo—; *Fatehpur*: Phasai—; *French*: Riz—; *Ga*: Omong—;

Gambia: Mannow—; *German*: Reiss—; *Greek*: Oryza, Oryzion, Oryzon—; *Guam*: Fae, Fai, Faria—; *Gujerati*: Chokha—; *Hausa*: Shinkafa—; *Hazara*: Shali—; *Hindi*: Chaval, Deodhan, Dhan, Pusui—; *Hungarian*: Riskasa—; *Indo-China*: Khao, Nganh, Nhu—; *Iraq*: Shilib, Ruzz—; *Italian*: Riso—; *Japanese*: Ko, Kome, Motsj—; *Java*: Bai—; *Jhang*: Munji—; *Kashmir*: Dein, Tani—; *Khmer*: Srau, Srou, Srur—; *Kolami*: Baba, Birbaba—; *Konkani*: Bhat—; *Krepi*: Morli—; *Krobo*: Omong—; *Kurdish*: Chaltuk, Pishik—; *Laos*: Khao chao, Khao hai, Khao loi, Khao nieu—; *Madagascar*: Vary—; *Malay*: Pady—; *Malayalam*: Ari, Nellu—; *Marathi*: Bhat, Tandula—; *Matabello*: Feha—; *Mon*: Sro—; *Mount Abu*: Garri, Sal—; *Mundari*: Baba—; *Mysol*: Fas—; *Mysore*: Bhatta, Nellu—; *North-West Provinces*: Chanwal, Dhan, Jarhan, Lehi, Munji, Pusai—; *North Queensland*: Kineyah—; *Oudh*: Dhan, Pashai, Passari, Tinni—; *Palaung*: Rekao, Sakao, Takao—; *Pandran*: Kandahari, Khisumbhuz, Wilaiti—; *Partabgarh*: Sathi—; *Persian*: Biranj—; *Peshawar*: Shol—; *Philippines*: Bolahan, Palai—; *Polish*: Ryz—; *Portuguese*: Arroz—; *Punjab*: Dhan, Munji, Shalian, Tai—; *Rajputana*: Garri—; *Rampur*: Phasai—; *Roumanian*: Orez—; *Russian*: Psheno, Ris, Sarachinskoe psheno—; *Sanskrit*: Arunya, Ashuvrihi, Dhanya, Nivara, Shali, Tandula, Vrihi—; *Santal*: Uri, Urihoro—; *Sema*: Aghi—; *Shahrig*: Shali—; *Sind*: Chanwar, Sari, Sugdasi—; *Sinhalese*: Goyan, Hal, Uruwi—; *Spanish*: Arroz—; *Sunda*: Pare, Pari;— *Swedish*: Ris—; *Tagalog*: Bigas, Binambang, Bolohan, Dumali, Lamuyo, Malagguit, Palay, Quinanda, Tangi—; *Tamil*: Arishi, Arisi, Nellu—; *Tartary*: Dugu—; *Tayabas*: Nilomot—; *Telugu*: Biyam, Dhanyamu, Errajilama, Nevaridhanyamu, Ouri, Urlu, Vadlu, Vudlu—; *Tobu*: Shali—; *Turkish*: Pirinj, Pirins—; *Twi*: Aimong, Mong—; *Uriya*: Balunga, Chaul, Dhan, Dhanno, Rabana—; *Xong*: Ruko—; *Zehri*: Pirkalanari—.

Oxytenanthera.

This genus includes 16 species occurring in India, Burma, Malaya and Africa.

Oxytenanthera sinuata Gamble is found south of the Malay Peninsula in Johor, and Negri Sembilan; but nowhere very common.

The very fine hairs found on the sheath leaves of the young sprouts are prominent among Malay poisons. They are frequently used with pounded glass and set up a train of symptoms like that of a chronic pseudo-dysentery.

In Kelantan bamboo hairs are sometimes mixed with the juice obtained from *Caryota mitis* Lour. and an extract of toad. This toad extract is made from the common brown toad by allowing it to decompose in a bamboo cylinder containing a little water; after an interval of seven days this is smeared or sprinkled over wearing apparel, and the ingredients are said to set up an incurable and painful skin disease like a ringworm in appearance.

The hairs are mixed with the ground bark of *Wickstroemia ridleyi* Gamble and decayed copra, and then thrown into rivers to stupefy fishes.

Malay: Buloh minyak—.

Panicum.

This genus is a heterogeneous assemblage of about 400 species, mostly natives of warm regions, a few being found in temperate climates of the northern hemisphere, one or two being almost cosmopolitan.

Many species are much valued as fodder-grasses but a few are said to be toxic to the animals.

I. Annuals.

- a. Spikelets $\frac{1}{6}$-$\frac{1}{5}$ in. long. Stems tufted, 2-4 ft. long, stout. Leaves 6-12 in. long ... 1. *P. miliaceum.*
- b. Spikelets $\frac{1}{10}$-$\frac{1}{8}$ in. long. Stem 1-3 ft. long, rather slender. Leaves 1-2 ft. long ... 2. *P. miliare.*

II. Perennials.

- a. Spikelets $\frac{1}{10}$ in. long, laxly clustered on the branches; glume IV smooth. Leaves 6 by $\frac{1}{6}$ in. Panicle 6 in. high ... ... 3. *P. antidotale.*
- b. Spikelets $\frac{1}{11}$-$\frac{1}{10}$ in. long, large, scattered; glume IV finely rugulose. Leaves 1-2 ft. by $\frac{1}{2}$-1 in. Panicle 1-2 ft. high ... 4. *P. maximum.*
- c. Spikelets small, very numerous, crowded: glume IV smooth. Leaves 5-12 in. by $\frac{1}{4}$-$\frac{1}{2}$ in. Panicle 7-12 in. long ... ... 5. *P. sarmentosum.*

1. **Panicum miliaceum** Linn. is cultivated or naturalized throughout the hotter parts of India, Africa and other hot countries. It is grown in various parts of India, even up to 10,000 ft. on the Himalaya, but nowhere to any great extent.

Ayurveda writers describe the plant as sweet and acrid, causing biliousness and indigestion.

At Shoran, in Baluchistan, the plant is used as a cure for gonorrhoea.

Arabic: Dokhu, Worga, Worglo—; *Ashkobi*: Azhaum, China—; *Bagwana*: Peonprish—; *Bengal*: China—; *Bihar*: China, Chinh, Chinna—; *Bombay*: Bansi, Chenah, China, Chirwa, Phikar, Rali, Sama, Sawa, Vari, Varikaanu, Varisava, Wadi, Worga—; *Bundelkhand*: Bansi, Phikai, Rali—; *Canarese*: Baragu, Bilibaragu, Karibaragu, Save—; *Chinese*: Chi, Shu—; *Deccan*: Sava, Sawi, Shamakh, Wari—; *Dutch*: Gierst—; *Egypt*: Dokhn—; *English*: Broom Corn Millet, Chena Millet, Common Millet, Indian Millet—; *French*: Mil, Mil en branches, Mil commun, Mil d'Inde, Petit mil, Millet, Millet commun, Millet à grappes, Millet rond, Millet rouge—; *German*: Hirse—; *Gujerati*: Chino, Samli, Vari—; *Hindi*: Chena, China—; *Iraq*: Dukhn—; *Kashmir*: Chinwa—; *Kila Saifulla*: Azhdan—; *Ladak*: Tzede—; *Lepcha*: Mung-cher—; *Marathi*: Barag, Sava, Vari—; *North-Western Provinces*: Chehna, Chinwa Chirwa, Kuri, Sawanchaitwa, Sawanjethwa—; *Persian*: Arzan—; *Portuguese*: Milho miudo—; *Punjab*: Anne, Chena, China, Chini, Salan, Salar—; *Quetta*: Gamh—; *Roumanian*: Malaiu, Meiu—; *Russian*: Proso—; *Sanskrit*: Anu, China, Chini, Rad, Varaka, Vrihibheda—; *Shoran*: Chabor—; *Sind*: Chinu—; *Sinhalese*: Mainairi—; *South Africa*: Indian Buffalo Grass—; *Tamil*: Kadaikanni, Katakanai, Samai, Varagu—; *Telugu*: Barigalu, Varagalu, Warigalu, Worga, Wuragi—; *Tobu*: Azhdun—; *Uriya*: Rala—; *Yemen*: Kossayb, Mileb—.

2. **Panicum miliare** Lam. is cultivated or naturalized throughout India and Ceylon. It is cultivated in the tropics.

According to some Ayurveda practitioners the plant acts as a nervine stimulant and tonic.

Bengal: Gondula—; *Bombay*: Badi, Bagad, Burburi, Kulti, Poi, Warai—; *Canarese*: Bilisamai hullu, Shamai—; *Central Provinces*: Chika—; *English*: Little Millet—; *Hasada*: Bicagurulu—; *Hindi*: Kungu, Kutki, Savan—; *Indo-China*: Co gang, Co ong—; *Khond*: Gundli, Pani kuhuri—; *Malayalam*: Shama—; *Mundari*: Arabende, Saramecadlongurulu—; *Naguri*: Hendegudulu—; *North-Western Provinces*: Kutki, Mighri—; *Punjab*: Kutki—; *Rajputana*: Nillashama—; *Ranchi*: Mota gundli—; *Santali*: Gundli—; *Sinhalese*: Meneri—;

Tamil: Chamai, Perusamai, Samai, Shamai—; *Telugu*: Chamalu, Gangasamalu, Nallachamalu, Nellashama, Nellashamalu, Sani, Savai—; *Uriya*: Luniva—.

3. **Panicum antidotale** Retz. grows abundantly in the Punjab, the Upper Gangetic Plain, the Western Peninsula and Ceylon. It extends to Afghanistan, Africa and Australia.

The smoke of the burning plant is used for fumigating wounds, also as a disinfectant in small-pox.

In Madras it is given for throat affections.

A report from Hissar states that this grass is grazed only when green, as it afterwards becomes bitter or saltish, and poisonous to cattle.

Baluchi: Bershonk, Gomaz, Gumazg—; *Bombay*: Barigagli, Barn, Barwari, Ghamar, Gharam, Girni, Git, Male, Mangrur, Sera, Shamukha—; *Gujerat*: Dun, Dusghas, Dusto—; *Hindi*: Ghemor, Gunara, Jamur—; *Kulanch*: Gomaz—; *Mand*: Dariagi—; *North-Western Provinces*: Gamur, Ghamor—; *Punjab*: Baru, Garm, Ghamrur, Ghamur, Gharam, Ghirri, Girui, Mangrur—; *Pushtu*: Male, Shamukha—; *Rajputana*: Bangagli, Banvari, Gramna—; *Sadani*: Bende—; *Santali*: Layogundli—; *Shahrig*: Gunj—; *Sibi*: Gam—; *Sinhalese*: Krimisastru—; *Tamil*: Nassiampillu, Pinisupillu—.

4. **Panicum maximum** Jacq. is a native of Africa, cultivated in many parts of India. It occurs throughout South Africa, the Mascarene Islands, Madagascar and in Yemen. It has been introduced and is widely spread in America.

This grass is considered one of the best horse-fodders, but it may cause fatal colic if given in too large a quantity or when wet.

It is said to be responsible in South Africa for the production of 'dikoor', a disease which affects young sheep, four to twelve months old, which are running on old lands.

Ala: Ikbo mili—; *Brazil*: Capim de Angola, Capim de Colonia—; *Burmese*: Nau-Ka-thau-hau—; *Canarese*: Gini hullu—; *English*: Guinea Grass—; *Ewe*: Kogbe—; *Fanti*: Nykekyer—; *Ga*: Nto—; *Hindi*: Gini ghaus—; *Orange Free State*: Blousaad—; *Seychelles*: Fataque—; *Sinhalese*: Ratatana—; *Southern Nigeria*: Ikbo mili—; *Tamil*: Giniopillu—; *Transvaal*: Buffelgras—.

5. **Panicum sarmentosum** Roxb. is found in Assam, Sylhet, Cachar, and the Khasia Hills, ascending to 5,000 ft., Chittagong, and Burma. It is very abundant in the Malay Peninsula in thickets and open edges of woods, often forming great masses scrambling over bushes and ascending to tree-tops. It extends to the Malay islands, Tongking, and China.

In Malaya the roots are chewed with betel nuts as an aphrodisiac.

Malay: Rumput jangut ali, Rumput kulubong—.

PASPALUM.

This is a plymorphic genus including over 200 species, scattered through the tropics of both hemispheres, but most abundant in America.

I. Spikelets oblong or lanceolate acute or acuminate.
   Margin of rachis minutely toothed ... 1. *P. sanguinale.*

II. Spikelets orbicular or broadly oblong.
   a. Spikelets ciliate from the margins of the upper glume ... ... 2. *P. conjugatum.*
   b. Spikelets not ciliate ... ... 3. *P. scrobiculatum.*

1. **Paspalum sanguinale** Lam. is found throughout India, in moist and dry situations, ascending the Himalaya to 6,000 feet. It is cultivated in the Khasia Hills. This grass is one of the commonest weeds in all tropical and warm countries, becoming less frequent in Central Europe.

The plant is known to yield hydrocyanic acid.

Berar: Chikhar:—; *Central Provinces*: Korkoljodi—; *Egypt*: 'Iraq-en-Najil, Tayyin—; *Hindi*: Takri, Takria—; *Khasi*: U'rai-shan—; *North-Western Provinces*: Charamara, Kewai—; *Punjab*: Baratakria, Dubra, Farw, Motikhabbal, Takkri—; *Rajputana*: Hen—; *Suto*: Mmoyane—; *Trans-Indus*: Khurash—; *United States*: Brab Grass—.

2. **Paspalum conjugatum** Berg. is common in the hotter parts of America from the Gulf States southwards; also in the Indo-Malayan region, Lower and Upper Guinea, and Polynesia.

In some places of the Gold Coast the grass is boiled and used in the bath to cure faintness.

Ashanti: Nsuorhwea—; *Hawaii*: Hilo Grass—; *Jamaica*: Sour Grass—; *Malay*: Rumput kerbau—; *Malaya*: Buffalo Grass—; *Montserrat*: Crab Grass—; *Sierra Leone*: Yane, Yani—; *Singapore*: Green Grass—; *Twi*: Nsorhwea—; *West Indies*: Sour Grass—.

3. **Paspalum scrobiculatum** Linn. is found wild or cultivated throughout the hotter parts of India, from the Punjab eastwards and southwards to Singapore and Ceylon. It is distributed through the Tropics of the Old World.

Ayurveda writers describe the plant as sweetish and bitter, tonic, and antidotal to poisons, useful in the treatment of ulcers; it causes constipation and flatulence, upsets the physiological balance of the body, and leads to hallucinations and dysuria.

Mohammedan authors consider it styptic, useful in inflammation and in diseases of the liver, liable to cause constipation and temperature.

Though used as food by a very large number of people in India, the grain is decidedly a narcotic poison, and its deleterious effects are well known in the various parts of the country where this crop is grown. The symptoms of poisoning resemble those caused by datura, and are more severe in cattle than in man, due no doubt to the animals eating the grain and husk, and also to the absence of vomiting, an effect which almost always takes place in man.

Sushruta prescribes the plant in combination with other drugs for the treatment of scorpion sting, but Caius and Mhaskar have shown experimentally that this grass is not an antidote to scorpion venom.

Australia: Ditch Millet, Hureek—; *Bengal*: Khodoadhan, Kodoadhan—; *Bihar*: Koda, Kodai—; *Bombay*: Harik, Kodra, Kodri, Kodro, Kodroakora, Pakod, Pakodi—; *Canarese*: Arikel, Haraka hullu, Harik—; *Central Provinces*: Kodie, Kodo—; *Ceylon*: Koda Millet—; *Gujerati*: Kodra, Kodro, Meya—; *Hausa*: Tumbijaki, Tumbin jaki—; *Hindi*: Koda, Kodaka, Kodava, Kodon—; *Kolami*: Garakode—; *Konkani*: Pacodd, Pacoll—; *Kumaon*: Kodo, Kodra, Kodram—; *Marathi*: Harik, Kodra, Kodru—; *Mundari*: Birjane, Loeongjane, Pirijane, Tasadjane—; *North-West Provinces*: Koda, Kodon, Kodram—; *Porebunder*: Kodo—; *Punjab*: Kodon, Kodra—; *Sanskrit*: Koddara, Kodrava, Koradusha, Kordrava, Kuddala, Madanagraka, Uddala, Vanakodrava—; *Santali*: Janhe—; *Sinhalese*: Amu, Karalamu, Walamu—; *Southern Nigeria*:

Ikbonta—; *Tagalog*: Paragis, Sabungsabungan—; *Tamil*: Karuvaragu, Varagu, Varaku—; *Telugu*: Allu, Alu, Arikalu, Arike, Arikelu, Aruga, Kiraruga, Nita arigaddi, Pataarige—; *Urdu*: Kodon—; *Uriya*: Khoddi, Kodus—; *Zulu*: isAmuyisane—.

PENNISETUM.

This genus numbers about 40 species which are found in most warm countries, particularly in dry regions.

Annual	1. *P. spicatum.*
Perennial	2. *P. compressum.*

1. **Pennisetum spicatum** Roem. & Schult. (=*P. typhoideum* Rich.) is cultivated in numerous forms in India, northern and tropical Africa, and southern Europe.

Ayurvedists consider the plant heating, aphrodisiac in women, and tonic; it is useful in diseases of the heart; it improves the appetite for food and the relish of it, but it causes flatulence.

In Guinea the root is reputed toxic to man and animals.

In South Africa the Sutos use the plant as a medicine for cows which have been incompletely delivered at parturition.

Arabic: Dookhr, Duku—; *Ashanti*: Ewio—; *Bihar*: Gahuma, Jondhariya—; *Bombay*: Bajera, Bajra, Bajri—; *Canarese*: Sajje—; *Egypt*: Dok, Hema, Gusab—; *English*: Bulrush Millet, Cumboo Millet, Pearl Millet, Spiked Millet—; *Eritrea*: Bultuc—; *Ewe*: Gbekul, Lu—; *Ga*: Ngma—; *Gambia*: Sannio—; *Hausa*: Damro, Dauro, Gero, Maiwa—; *Hindi*: Bajera, Bajra, Bajri, Kasajonar, Lahra—; *Kolami*: Tuti—; *Krobo*: Ngma—; *Kumaon*: Bajra—; *Malayalam*: Kampam, Mattari—; *Mandingo*: Sannio—; *Matabele*: Um-velli-velli—; *Msutu*: Leeuja—; *North-Western Provinces*: Bajra, Bajra tangunanwa, Bajri, Lahra—; *Nupe*: Gero—; *Punjab*: Bajra, Bajza—; *Rajputana*: Bajra—; *Sanskrit*: Agradhanya, Nali, Nalika, Nilakana, Nilasasya, Sajaka, Varjari, Varjarika—; *Santali*: Lendha—; *Sesuto*: Nyalothie—; *Shahrig*: Bajari—; *Sind*: Bajaro—; *Sudan*: Dukhu—; *Tamil*: Kambu—; *Telugu*: Gantelu, Peddaganti, Sajja, Sajjalu, Sazza—; *Transvaal*: Leeuja—; *Tripoli*: Shessab—; *Uriya*: Bajramula, Gantiva—; *Zambesi*: Mishuera—; *Zanzibar*: Mawali—; *Zulu*: Amabile, Nyaloti, Nyawoti—.

2. **Pennisetum compressum** R. Br. (=*P. japonicum* Trin.) is found in Burma, Taong Dong, the Sham Hills up to 4,000 ft. It extends to China, Japan, Tongking and Australia.

This grass is used as a tonic in China.

Chinese: Lang Wei Ts'ao.

PHALARIS.

This genus consists of 10 species, natives of the Mediterranean region, but widely dispersed as weeds; one species occurs in the boreal region and in South Africa, and another is found from California to Chile.

Phalaris arundinacea Linn. is found in Kashmir by the Wular Lake at an altitude of 5,550 ft. It is fairly common along the pond edges and stream banks, and in wet places in Eastern Washington, but is infrequent in California. It extends to the northern temperate and Arctic regions.

This grass is reported poisonous.

English: Gardeners' Garters—; *Pacific Coast*: Daggers, Reed Canary-grass, Ribbon-grass—.

PHRAGMITES.

This genus includes 2 species, one cosmopolitan, and one in Argentine.

Phragmites maxima Blatter & McCann (=*P. communis* Trin.=*P. roxburghii* Steud.=*P. karka* Trin.) is cosmopolitan.

Ayurvedists describe the plant as sweet and acrid, cooling, aphrodisiac; useful in biliousness, urinary troubles, vaginal and uterine complaints, erysipelas, and diseases of the heart.

The root is regarded as cooling and diuretic by the Chinese. It is considered diuretic and sudorific in Spain.

The plant is said to be rich in sugar. A manna, known as California Manna or Father Picolo's Manna, is mentioned as being deposited on the plants growing in California and collected by the Indians. It is believed that the manna is a saccharine deposit caused by aphides.

Arabic: Bus—; *Bengal*: Nal—; *Bihar*: Narkat—; *Burma*: Kaing—; *Canarese*: Hulugalagu, Hulugilahullu, Hulugilu—; *Catalan*: Canya borda, Canya de escombretas, Canyis, Canyisos—; *Chinese*: Lu—; *Egypt*: Buz haggny, Buz hagney, Ghab, Ghabrihy, Haggn—; *English*: Nodding Reed—; *French*: Cannette, Petit roseau, Roseau à balai, Roseau aquatique, Roseau des marais—; *Garhwal*: Bichhra—; *Gujerati*: Nairi, Nali—; *Hausa*: Machara, Wachiar giwa, Wutsiyar giwa—; *Hindi*: Nainarakula, Nal, Narkul, Nulanara—; *Hova*: Barorata, Katsaoka, Volotara—; *Indo-China*: Kiem, Lo can, Sai, Say—; *Iraq*: Qasab—; *Irish*: Giolcach—; *Italian*: Canela de Ema, Canna palustre, Cannuccia—; *Katagum*: Machara, Wachiar giwa, Wutsiyar giwa—; *Kolami*: Jankai—; *Kumaon*: Karka, Khaila, Khailuwa, Nal—; *Lagos*: Ifu—; *Languedoc*: Rouza, Rouzo—; *Loralai*: Nal, Nar—; *Libya*: Buzzam—; *Malay*: Gudabong—; *Malayalam*: Nadam, Nalam, Nannana, Naval—; *Malta*: Common Reed, Spire Reed, Canna di palude, Canna da spazzole, Kasbiet irrih—; *Marathi*: Deonala, Nala—; *Mexico*: Carrizo—; *Pacific Coast*: Common Reed—; *Punjab*: Bagnarri, Dila, Nai, Nal, Nar, Nara, Naria—; *Pushtu*: Drumbi, Dwarena, Ghwarga—; *Roumanian*: Rogoz—; *Sanskrit*: Dhhamana, Nada, Nala, Potagala, Shunyamadhya—; *Shahrig*: Nal—; *Sokoto*: Gabara—; *Spanish*: Cañeta, Garrizo—; *Tagalog*: Tambo—; *Tamil*: Perunanal—; *Telugu*: Kikkasagaddi, Mettantisa, Nagasvaramu, Peddarellu, Puvvuguttigaddi—; *Turkish*: Gamish—; *Uniyoro*: Mataetae—; *Uriya*: Nolo—; *Zhob*: Nal—.

PHYLLOSTACHYS.

This genus consists of about 20 Eastern Asiatic species.

Phyllostachys bambusoides Sieb. & Zucc. is found in Upper Assam on the Mishmi Hills. It inhabits China and Japan.

The plant is much used medicinally in China, where the root is considered tonic and the sprouts parasiticidal. The silicious concretion or *tabashir*, is a demulcent, tonic, aphrodisiac and pectoral, commonly administered in cough, asthma, consumption and fever.

Chinese: Chu, Chu Huang, Chu Sun, Hsien Jen Chang, Kuei Ch'ih, Suan Sun—; *Indo-China*: Son truc, Truc—.

POLYTOCA.

This genus consists of 8 species spread over tropical Asia and Australia.

Polytoca barbata Stapf. is found in the hot and damp parts of India, from the Punjab to Assam and Munnepore, and southwards to Ceylon, extending to Java.

According to Ayurveda the plant is bitter and sweet, tonic, laxative and aphrodisiac. It is reputed useful in strangury, burning sensations, phthisis, vesical calculi, biliousness, diseases of the blood and tendency to haemorrhage.

Balghat: Kadpi—; *Bengal*: Gurgur, Keshaghansa—; *Canarese*: Kajalu—; *Central Provinces*: Kadpi—; *Chanda*: Kirmagilaramgadi—; *Gond*: Karpia—; *Guejrati*: Kansado—; *Hindi*: Kansa—; *Konkan*: Kasada—; *Marathi*: Kasai, Varival—; *Sanskrit*: Amarapushpaka, Ashvabala, Chamarapusha, Darbhapatraka, Ikshugandha, Ikshura, Ishika, Kanda, Karmamula, Kasekshu, Kasha, Nadeya, Niraja, Potagala, Sharada, Shiri, Sukanda, Vanahasaka—; *Telugu*: Ghellagadi—; *Uriya*: Phultainr—.

SACCHARUM.

This genus includes 8 tropical and subtropical species.

I. Hairs on callus much exceeding the spikelet. Glumes I and II not dorsally villous.
  a. Culms not leafy above, under .7 in. diameter. Leaves under .8 in. width. Glumes I and II ciliate ... ... 1. *S. spontaneum.*
  b. Culms densely leafy above, over 1 in. diameter. Leaves over 1 in. width. Glumes I and II glabrous ... ... 2. *S. officinarum.*

II. Hairs on callus of sessile spikelet shorter or not much longer than the spikelet. Glumes I and II often dorsally villous. Nodes of culm not bearded. Sheaths not hirsute. Glumes I and II dorsally villous.
  a. Foliage not glaucous. Culms densely leafy above. Sessile spikelet shorter than internodes ... ... 3. *S. arundinaceum.*
  b. Foliage glaucous. Culms not leafy above. Sessile spikelet longer than internode of rachis ... ... 4. *S. munja.*

1. **Saccharum spontaneum** Linn. is found throughout the warmer parts of India and Ceylon, ascending to 6,000 ft. in the Himalaya. It is distributed from the south of Europe to the warm regions of the Old World, and extends to East Australia.

Ayurveda practitioners commonly use this grass as a substitute for *Polytoca barbata* Stapf. Sanskrit authors credit the two plants with the same medicinal properties.

Bombay: Bochri, Dharbi, Dhub, Kan—; *Bengal*: Kagara, Kans, Kas, Kash, Kashiya, Khagra, Khansi—; *Burma*: Thekkaygyee, Thetkiakyn—; *Canarese*: Darbhe, Hodakehullu, Mutullahullu—; *Central Provinces*: Kans, Khan, Padar—; *English*: Thatch Grass, Wild Sugar Cane—; *Gujerat*: Kans, Kansado, Kansadoghas—; *Hasada*: Karetasad, Karitasad—; *Hausa*: Abokin, Kyamo, Kyamo kibiya, Kyamro, Kyauro, Kyauro kibiya, Sheme, Sansari—; *Hindi*: Kagara, Kans, Kas, Khansi, Khugura, Kosa, Kus—; *Kolami*: Puyal—; *Kumaon*: Jasha, Jhaush, Kash—; *Lao*: Pong—; *Lepcha*: Brung—; *Malayalam*: Nannana—; *Marathi*: Kagara, Kas—; *Mundari*: Kariba—; *Naguri*: Kasitasad—; *North-Western Provinces*: Kans, Kansa, Kansi—; *Oudh*: Khagar, Rara—; *Punjab*: Kahi, Kanh, Kans, Sarkara—; *Rajputana*: Kans, Kash, Kashi—; *Sadani*: Kasighas—; *Sanskrit*: Ikshugandha, Kasa, Kahsa, Khaggara—; *Sind*: Kahu, Khan, Khau—; *Tamil*: Achabaram, Anjani, Eruvai, Kosangam, Kucham, Kumil, Kurbagam, Nanal, Nanarbul, Nanmugappul, Peykkarumbu,

Sangabidam, Saravanam, Sarupparasi, Sasabaram, Sugattan, Suvedasaram, Tittiru, Tittiruchi, Tuttam, Vedasam—; *Telugu*: Billugaddi, Kakicheraku, Kakiveduru, Koregadi, Rasalamu, Rellugaddi, Vetticharaku—; *Uriya*: Chhataiagaso, Inkoro, Kaso, Khhodi, Pothhorokhhodi—.

2. **Saccharum officinarum** Linn., probably a native of South Asia, is grown everywhere in India.

Ayurveda describes the sugar cane as sweet, oleagenous, diuretic, tonic, cooling and aphrodisiac. It recommends it as a useful remedy in fatigue, thirst, intestinal troubles, anaemia, erysipelas and leprosy. The cane, however, is indigestible, and likely to interfere with the heat factor in the body causing inflammations and ulcers.

Yunani describes the cane as sweet, laxative, diuretic, fattening and aphrodisiac. It considers it useful to purify the blood, good for the lungs, but bad for the liver.

The roots are used as a cooling and diuretic medicine; and the stem is said to be useful as a remedy for cough.

Sugar cane enters into the composition of Cambodian medicines used for the treatment of ulcers of the skin and mucous membranes. A decoction of the stem is given in diarrhoeas of childhood.

The sugar cane is recognised in Portugal, Spain, Switzerland and the United States as the officinal source of sucrose or saccharose.

The Hindus set a great value upon sugar, and in medicine it is considered by them as nutritious, pectoral and anthelmintic.

In Arabian works on Materia Medica, sugar is described as detergent and emollient. Many writers speak of it as attenuant and pectoral. It has also been supposed to have virtues in calculus complaints.

In the Punjab sugar is considered heavy, tonic and aperient, useful in heat delirium and disorders of the bile and wind.

In cases of poisoning by copper, arsenic, or corrosive sublimate, sugar has been successfully employed as an antidote; and white sugar finely pulverised is occasionally sprinkled upon ulcers with unhealthy granulation.

The root, the bark, the stem and the leaf, treacle and sugar, have all been boomed as snake-bite remedies. Mhaskar and Caius have, however, shown experimentally that, whether administered internally or applied externally, every part of the sugar cane is useless in the treatment of snake-bite.

Ada: Afunu—; *Annam*: Mia—; *Arabic*: Kasabishakar, Kasibshakar, Qasabussakar—; *Awina*: Fofongu—; *Behar*: Katari, Ketari, Khusiyar, Ukh, Ukh, Ukhi—; *Bengal*: Ak, Ganna, Ik, Kajuli, Kulluar, Kushiar, Puri, Uk, Ukyo—; *Bombay*: Ganderi, Gol, Serdi, Us—; *Brazil*: Canna, Canna de assucar, Tacomaree, Viba—; *Burma*: Keyan, Kyan—, *Cagayan*: Agbo—; *Cambodia*: Ampeou, Ampou—; *Canarese*: Ikshu, Ikshudanda, Ingolu, Kabbu, Kantara, Kantaraka, Madhura, Marakabbu, Pundra, Rasadali, Rasala, Rastale, Tanigarbu, Trinaraja—; *Catalan*: Cana dolsa, Cana de sucra—; *Ceylon*: Karambu—; *Chinese*: Kan Che, Sha T'ang, Shih Mi—; *Cochin China*: Mia—; *Deccan*: Ganda, Us—; *Dutch*: Suiker riet—; *Easter Island*: To—; *Egypt*: Ghah, Qassab sukkar—; *English*: Sugar Cane—; *Ewe*: Fofongu, Bogleng—; *Fanti*: Ahwerenkakraba—; *Fiji*: Dovu, Vico—; *French*: Canamelle, Cannamelle, Canne de Batavia, Canne de la Chine, Canne d'Haïti, Canne à sucre, Roseau à sucre—; *French Guiana*: Canne à sucre—; *Ga*: Sheng—; *German*: Zuckerrohr—; *Guam*: Tupo, Tupu—; *Gujerati*: Naisakar, Serdi, Sheradi, Sherdi, Uns—; *Gulf States*: Ribbon Cane, Sugar Cane—; *Hausa*: Deke, Karansariki, Rake—; *Hindi*: Ganna, Ikh, Kajuli, Khulua, Kumad, Naishakar, Rikhu, Uk, Ukh—; *Hova*: Fary—; *Iraq*: Qasab sukkari—; *Japanese*: Kansia—; *Java*: Tebu—; *Kano*: Rake—; *Konkani*: Uny, Uss—; *Krepi*: Bogleng, Boglengbiei, Boglengfe, Boglengyibor—; *Korbo*: Ahleu—; *Kumaon*: Rikhu—; *Kurdish*: Gamish-i-shikir—; *La Reunion*: Canne—; *Madagascar*: Fary—; *Malay*: Tebu, Tibu mira, Tubu—; *Malayalam*: Darbheshu, Ikshu, Kantarakam, Karimpu, Madhutrinam, Vellakarimpu—; *Marathi*: Aos, Kabbo, Us, Usa—; *Mundari*: Gurdanda, Gurkatauri, Gurkosear—; *Nepal*: Akali, Chaku, Uk—; *Newar*: Tu—; *New Caledonia*: Ariva, Arolam, Boiepe, Boinlioua, Delenole, Dilou, Dogangueni, Gadenadeboui, Goreate, Jate, Kabopolenouen, Kiaboue, Kinemaite, Kondimoua, Koubala, Maiou, Mebouangue, Mengou, Migao, Moene, Moindiene, Moueouete, Ngala, Niemba, Ouali, Ouane, Oudiepe-ait, Ouen, Ouen ebail, Ouen mangia, Ouen ou poudendate, Paiambou, Paieme, Pidiak, Pobone, Poilote, Schimate, Sthiabanghi, Tangalite, Thsiogan, Tilibi, Tshiambo—; *North-West Provinces*: Ganna, Ikh, Ikhari, Kanthirikhu, Punarikhu, Rikhu, Ukh, Ukhari—; *Parbuttiah*: Ghenra—; *Persian*: Naishakar—; *Philippines*: Tubo—; *Portuguese*: Canna de assucar, Canna doce—; *Punjab*: Canna, Ikh, Kamad, Kamand, Khand, Paunda, Shakarsurkh—; *Roumanian*: Trestie de zahar—; *Russian*: Saharnyi trastnik—; *Sakalave*: Fisika—; *Samoa*: Tolo—; *Sanskrit*: Adhipatra, Asipatra, Bhurirassa, Dirghachhada, Gandidi, Gudada, Gudadaru, Gudakashtha, Gudamula, Gudatrina, Ikshu, Ikshura, Kantara, Kantaraka, Karkotaka, Khadgapatraka, Koshakara, Madhutrina, Madhuyashit, Maharasa, Mrityupushpa, Payodhara, Pundraka, Rasala, Rasalu, Sastra, Sukumasaka, Trinadhiya, Vansha, Vipularasa, Vrishya—; *Santali*: Akh, Ikhshu—; *Sind*: Kamand—; *Sinhalese*: Uk, Ukgas—; *Sokoto*: Arakke—; *Spanish*: Caña de azucar, Caña dulce, Cañamiel—; *Tagalog*: Tubo—; *Tahiti*: To—; *Tamil*: Angarigai, Asibattiragam, Ikku, Kalai, Kannal, Karumbu, Madudirunam, Paruvayoni, Pundaram, Ukkiragandam, Ukkiragandi, Velam, Vengarumbu—; *Telugu*: Arukanupulakranuga, Cheraku, Cherakubhedamu, Cherakubodi, Cherakudubu, Ikshupu, Inju, Kantaramu, Kanupulacheraku, Lavucheraku, Pottikamucheraku, Tellacheraku, Tiyyamranu, Tunta, Vamsukamu—; *Tongking*: Mia, Mia co ke, Mia lau, Mia ly—; *Tulu*: Karumbu—; *Twi*: Ahwereu—; *Urdu*: Cana—; *Uriya*: Aku, Gudodaru, Ikhyu—; *Visayan*: Quilaba—; *Yemen*: Muddardjend—.

3. **Saccharum arundinaceum** Retz. is found throughout the plains and low hills of India, extending into China.

The root is demulcent and diuretic.

Bengal: Teng—; *Burma*: Phoungga—; *Canarese*: Abbe, Baragu, Hodakai hullu, Lekhinihullu, Munja, Munji, Nala, Rellu, Sara—; *Ceylon*: Elephant Grass—; *English*: Devil Sugar Cane, Reedy Sugar Cane, Wild Sugar Cane—; *Malaya*: Elephant Grass, Tebrau—; *Malayalam*: Mekhalapullu, Munja, Sarappullu—; *Punjab*: Sarkanda—; *Rajputana*: Sarpat—; *Sanskrit*: Gundra, Munja, Sara, Tejanaka—; *Sinhalese*: Rambuk—; *Tamil*: Elhudugiranaanal, Munji, Pi karumbu—; *Telugu*: Adavicheruku, Bramhamekhalamu, Gundra, Kondakanamu, Munjagaddi, Mungamu, Nadamu, Polagaddi, Ponika, Ponugu, Saramu—; *Uriya*: Kantosoro, Soro—.

4. **Saccharum munja** Roxb. (=*S. ciliare* Anders.) is abundant over the greater part of North-West India, especially in the Punjab, extending to the Upper Gangetic Plain.

Ayurveda describes the stem as sweet and acrid, cooling, and aphrodisiac; useful in thirst, burning sensations, erysipelas, diseases of the blood, urinary complaints, and eye troubles.

In the Punjab the root is burnt near women after delivery, its smoke being considered beneficial. The root is also used as a fumigant for burns and scalds.

Ajmere: Sara, Sarpat—; *Bengal*: Mucha, Ramshara, Sar, Sara, Sarpata, Shar—; *Bolan River*: Kash—; *Bombay*: Kerpa, Munj, Sar—; *English*: Munj

11

Grass—; *Hasada*: Mail—; *Hindi*: Munj, Munja, Ramsar, Sara, Sarkanda, Sarkara, Sarpat, Sarpatta—; *Marathi*: Mole—; *Mundari*: Hatumail—; *Naguri*: Mahil—; *North-West Provinces*: Ikar, Patawar, Sarhar, Sarkanda—; *Oudh*: Palwa—; *Pishin*: Surghashae—; *Punjab*: Kanda, Kharkana, Sarjbar, Sarkara—; *Quetta*: Surghashae—; *Sanskrit*: Bahupraja, Bana, Bhadramunja, Brahmanya, Chakshuveshtana, Darbhavhaya, Dridhatrina, Durmula, Ikshukanda, Maunji, Munja, Munjanaka, Munjata, Ranjana, Shakrabhanga, Shara, Shiri, Sthuladarbha, Sumekhala, Tejana, Tejanavhaya, Trinakhya, Vaniraka—; *Santali*: Sar—; *Telugu*: Gundra, Ponika—; *Tharu*: Kandikhar, Karai—; *Trans-Indus*: Darga, Karre—; *Turbat*: Dil—.

SETARIA.

This genus numbers about 100 species, found in tropical and warm temperate regions.

S. italica Beauv. and *S. viridis* Beauv. are used medicinally in China; *S. plicata* T. Cooke in La Reunion; *S. sulcata* Raddi in Zululand.

I. Leaves more or less plicate. Perennial. Culm reaching 5 ft. ... ... 1. *S. plicata.*

II. Leaves flat, not plicate.
- a. Culm 2-5 ft. high ... ... 2. *S. italica.*
- b. Stem 1-2 ft. high ... ... 3. *S. viridis.*

1. **Setaria plicata** T. Cooke is found throughout the moister hilly parts of India, from Kumaon in the Himalaya eastwards, ascending to 5,000 ft. in Sikkim, the Shan Hills in Burma, the Nilgiris up to 6,000 ft., the Malay Peninsula, and Ceylon. It extends to the Malay Islands and China.

The plant is used in La Reunion as an emollient diuretic.

La Reunion: Trainasse—; *Malay*: Akar pimpan—.

2. **Setaria italica** Beauv. is cultivated throughout India, and up to 6,000 ft. in the Himalaya. It is found in most warm, temperate and tropical countries.

According to Ayurvedic texts the plant is sweet and acrid, fattening, aphrodisiac, and sedative to the gravid uterus. It is useful in dispelling burning sensations, and in healing fractures. It produces flatulence.

It is a popular domestic remedy for alleviating rheumatic pains and the pains of parturition.

The grain is said to act as a diuretic and astringent, and to be of use externally in rheumatism.

When taken as the sole food Indian Millet is sometimes apt to produce diarrhoea. It has long been regarded as unsatisfactory for horses unless fed sparingly.

Andamans: Tanahal—; *Arabic*: Dukhn—; *Bengal*: Bertia, Chena, Kakni, Kakun, Kangni, Kangu, Kauni, Kirakang, Kirang, Kora, Rala, Rawla, Tangan—; *Bombay*: Kakun, Kang, Kangni, Korakang, Korakangu, Vavani—; *Bonai*: Kango—; *Bundelkhand*: Kakun—; *Burma*: Puki, Pyounglaykouk, Zami—; *Canarese*: Karibiragu, Naoni, Navanai, Navani, Vavani—; *Catalan*: Cúa de guilla, Cúa de guinea, Panis, Panissa—; *Cebu*: Mijo—; *Central Provinces*: Kungni, Rala—; *Chinese*: Liang, Shu, Su—; *Cochin China*: Cay khe—; *Deccan*: Bertia, Chena, Kakni, Kakun, Kangni, Kangu, Kauni, Kirakang, Kiranj, Kora, Rala, Rawla, Tangan—; *English*: Boer Manna, Foxtail Millet, Hungarian Grass, Italian Millet, Millet—; *French*: Mil à épi, Mil d'Italie, Millet en épi, Millet des oiseaux, Panis d'Italie, Panouil, Panouque Penille—; *Gujerati*: Kang, Karang—; *Hindi*: Bertia, Chena, Kakni, Kakun,

Kalakangni, Kanghuni, Kangni, Kangu, Kauni, Kirakang, Kiranj, Koni, Kora, Rala, Rawla, Tangan—; *Ilocano*: Bicacao, Bucacao—; *Iraq*: Dukhn—; *Italian*: Panico—; *Kashmir*: Pingi, Shali—; *Khasi*: U'rai-shoh—; *Kolami*: Haraba, Marang Kukru—; *Khondmals*: Arga—; *Konkani*: Kangu—; *Kumaon*: China, Gandra, Kangni, Koni, Mandira, Mandua, Murhoa, Shungura—; *Lynngam*: Jrai—; *Malay*: Iskoi—; *Malayalam*: Navana, Tauna, Tena—; *Malta*: Italian Millet, Panico—; *Marathi*: Chenna, Kang, Kangu, Rala, Rale—; *Mundari*: Irba—; *North-Western Provinces*: Kangni, Tangun—; *Pampangan*: Borona—; *Persian*: Arzun, Gal—; *Punjab*: Chanwalkangni, Chiurr, Gal, Husketkangni, Kangni, Khauni, Kher, Kusht, Shak, Shali—; *Pushtu*: Gal—; *Sanskrit*: Chinaka, Kangu, Kanguni, Kangunika, Pitatandula, Priyangu—; *Santali*: Erba—; *Sema*: Asüh—; *Sind*: Kirang—; *Sinhalese*: Tanahal—; *Spanish*: Mijo menor, Panizo—; *Tagalog*: Daua, Dava—; *Tamil*: Tenai, Tennai—; *Telugu*: Kora, Koralu—; *Uriya*: Kangu, Tangun—; *Visayan*: Daoa, Daua, Dava—.

3. **Setaria viridis** Beauv. is found in temperate Himalaya and Western Tibet, ascending to 11,000 ft.; but it is in no way common anywhere. It is also rare on the plains of India. It has been reported from the Nilgiri Hills. It is distributed throughout the temperate and subtropical regions of the Old World.

The plant, crushed and mixed with water, is used in China as an external application for bruises.

Chinese: Kou Wei Ts'ao—.

SORGHUM.

This genus consists of about 35 species spontaneous in the tropical and subtropical regions of both hemispheres, very few extending into the temperate zone. One group of forms is widely cultivated in the tropics, particularly in Africa and India.

A form of *S. guineense* Stapf. is grown for medicine in Upper Guinea; its grain is said to give colic to animals.

Perennial ... ... 1. *S. halepense.*
Annual ... ... 2. *S. vulgare.*

1. **Sorghum halepense** Pers. (=*Andropogon halepensis* Brot.) is common all over India, Burma and Ceylon in cultivated and uncultivated ground. It is distributed over most warm countries.

The seeds are demulcent and diuretic.

The grass is a good fodder both for grazing and for hay, but it is held to have injurious effects if eaten when too young or when the plants are stunted by drought. It has been reported from the Punjab that cattle after eating it had developed fatal head affections; and reports from various parts of Iraq indicate that the plant may cause the death of sheep or goats. This property is also well known in Australia, the United States of America and many other countries.

Banda: Bajra, Bara, Barru—; *Bengal*: Kalamucha—; *Berar*: Kartal—; *Bhabar*: Buru, Rikhon—; *Bombay*: Bara, Barru, Batal, Dacle, Kartal—; *California*: Arabian Millet, Evergreen, Means Grass, Millet—; *Canarese*: Kadukambu-hullu—; *Ceylon*: Johnson-weed, Sorghum-weed—; *Chanda*: Gallajari, Gudi, Paddajalla—; *Egypt*: Gerrau, Hashish-el-faras—; *English*: Aleppo Grass, Evergreen Millet—; *French*: Herbe de Para—; *Greek*: Kalamagra—; *Hindi*: Baru—; *Iraq*: Halaiyan, Sifrand—; *Kashmir*: Braham—; *Kohlu*: Baran—; *Kumaon*: Bikhonda—; *Malta*: Aleppo Millet-grass, Cannarecchia, Dente cavallino—; *Merwara*: Bowari—; *Punjab*: Baru, Barwa, Braham—; *Pushtu*:

Barua—; *Sakalave*: Fembamboloky—; *Shahrig*: Baran—; *South Africa*: Cuba Grass, Johnson Grass—; *Tamil*: Kadu cholam—; *Telugu*: Gaddijanu—; *United States*: Cuba Grass, False Guinea Grass, Johnson's Grass, Mean's Grass, Mears Grass—; *Uruguay*: Gramillon, Pato ruso, Sorgo de Alepo—.

2. **Sorghum vulgare** Pers. (=*Andropogon Sorghum* Brot.) is widely cultivated in India. It is cultivated in numberless forms in the tropical and subtropical regions, particularly in the Old World, and in the warmer parts of the temperate zones of both hemispheres.

According to Ayurveda writers the grain is cooling, and aphrodisiac, improving both appetite and relish for food. It is useful in the treatment of general disorders, diseases of the blood, piles, ulcers, tumours. It is highly indigestible, and causes constipation.

The seeds are diuretic and demulcent.

American Negroes like the decoction of the seeds as a remedy for urinary, bladder and kidney, complaints.

The adventitious shoots growing out of derelict stumps are fatal to cows and goats. In Guranjwala, Gujerat, and Shahpur districts, it is said to be poisonous until the rains are over, when the cattle eat it with impunity.

Afghanistan: Jaor, Jaoriturkimani, Jawars, Jowar, Kiosagi—; *Arabic*: Dakkn, Dhura, Dhurat, Jawars, Taam, Zura—; *Ashanti*: Atokoor—; *Babian Shahrig*: Jowari, Targhar—; *Baghwana*: Dhutar, Turi—; *Bengal*: Jowar, Juar, Kasajonar, Kurbi—; *Betsileo*: Variampemby, Varifemba—; *Bhabar*: Junali—; *Bombay*: Burru, Jaundri, Joar, Jondla, Jowari, Kangra—; *Burma*: Pyoung—; *Canarese*: Jolah, Kenjol, Nirgol, Shalu, Yengara—; *Catalan*: Melca, Menca, Mill africá—; *Central Provinces*: Jowar, Phag, Thuthera—; *Ceylon*: Cholum, Durra, Great Millet, Guinea Corn—; *Chinese*: Kao Liang, Shu Shu—; *Deccan*: Jondla, Jowari—; *Egypt*: Dokhn, Dura, Dura-belledi, Durra, Kaydi—; *English*: Broom-corn, Guinea Corn, Great Millet, Indian Millet—; *Ewe*: Fo—; *French*: Balai, Balai de jonc, Balai d'eau, Blé de Guinée, Houque à balais, Maïs de Guinée, Mil d'Italie, Mil d'Inde, Millet d'Afrique, Millet à balais, Grand Millet noir, Millet de Turquie, Gros mil, Mil d'Ethiopie, Sorgho, Sorgho d'Afrique, Sorgho à balai, Sorgho commun—; *Ga*: Akoko—; *Gambia*: Bassiqui, Bassiwulima, Kous, Manio—; *German*: Moorhirse, Sorghum—; *Gold Coast*: Guinea Corn—; *Gujerati*: Jowar, Juar, Sundia—; *Hindi*: Janera, Joar, Jondla, Jowari, Juar, Jundri, Juvarijondhla, Kanggni—; *Hova*: Ampemby—; *Iraq*: Dhurah, Idhrah baidha—; *Khasi*: U Krai—; *Kolami*: Gangai—; *Konkani*: Juar—; *Krepi*: Fo—; *Krobo*: Koko—; *Kumaon*: Jowar, Junali—; *Kurdish*: Zurat spi—; *Languedoc*: Millangue, Millanque—; *Lao*: Ya pong—; *Las Bela*: Jowari—; *Lepcha*: Ra-Kong, Tsung-Kong—; *Makran*: Mohammadisa, Sohro—; *Malayalam*: Chavela—; *Mal Paharia*: Sissua—; *Malta*: Dariseed, Durrah, Karabocc, Melica, Saggiwa—; *Marathi*: Jondhala, Juari, Kadval, Shalu—; *Mundari*: Gangae, Ganggai—; *North-West Provinces*: Bajrahopanwa, Chotijuar, Juar, Junri—; *Oudh*: Bajrahopanwa, Chotijuar, Juar, Junri—; *Portuguese*: Milho—; *Punjab*: Bajrahupanwa, Chari, Chotijuar, Chotijunri, Joar, Junri, Ka—; *Russian*: Sorgo—; *Sakalave*: Bakaka, Moramа—; *Sanskrit*: Dirghamala, Dirghashara, Ikshupatraka, Kshetrekshu, Raktakhumah, Shikhari, Vrittatandula, Yavanala—; *Santali*: Juar—; *Sarakhala*: Tarighara—; *Sema*: Atsünákhi—; *Sinhalese*: Karaliringu—; *South Africa*: Kaffir Corn, Kafir Corn—; *Spanish*: Alcandia, Saina—; *Tamil*: Cholum—; *Telugu*: Bondajanu, Janu, Jonna, Jonnalu, Kondajanu, Tellajanu—; *Turkish*: Baiyadh dari—; *Twi*: Atoko, Awi, Kokorte—; *Uruguay*: Maiz de Guinea, Sorgo—; *Visayan*: Batad—; *Zulu*: Amabele, Imfe—.

STENOTAPHRUM.

This genus consists of 7 species, found mainly on the shores of the tropical and subtropical seas, four being confined to Madagascar and the neighbouring islands.

Stenotaphrum glabrum Trin. occurs in the plains throughout India. It is found in Lower and Upper Guinea, also in South Africa from Cape Town to Natal, on the American shores of the Atlantic from South Carolina to the La Plata and in the Pacific from Southern Mexico to Australia. It has been introduced in Southern France and Italy.

The rhizome is diuretic, and the decoction is used in Brazil as a diuretic and sudorific.

America: St. Augustine Grass—; *Australia*: Buffalo Grass—; *Bahia*: Grama da Praya—; *Bermuda*: Crab Grass—; *Florida*: Mission Grass—; *Jamaica*: Pimento Grass—; *La Reunion*: Chiendent de boeuf—; *Mauritius*: Herbe bourrique—; *Natal*: Tweek Grass—; *St. Helena*: English Wire Grass, Mat Grass—; *Transvaal*: Tweek Grass—.

STIPA.

This is a genus of over 100 species, spread over the temperate and tropical regions of both hemispheres.

S. inebrians Hance, *S. sibirica* Lam., and *S. viridula* Trin. are actively poisonous; *S. vaseyi* Scribner is a very active narcotic.

I. Leaves smooth; ligule oblong. Callus short, conical. Anther-tips bearded ... ... 1. *S. sibirica.*

II. Leaves scaberulous: ligule lanceolate. Callus scabrid, top bearded. Anther-tips naked ... 2. *S. capillata.*

1. **Stipa sibirica** Lam. is found in the Western temperate Himalaya, from Kashmir to Kunawar up to 8,000-9,000 ft. altitude, and to the Black Mountains. It extends to Afghanistan, Siberia and Korea.

This grass is actively poisonous and is reported to have caused death in horses and other domestic animals.

2. **Stipa capillata** Trin. is found in Western Himalaya, ascending up to 11,000 ft. in Kashmir. It extends westwards to Spain, Dahuria, China.

This grass frequently kills sheep, not, however, by a direct poisonous action, but by its glumes working through the skin into the vital organs.

THELEPOGON.

The one representative of this genus, **T. elegans** Roth., occurs in Central India, the Konkan and Malabar, extending to tropical Africa.

In Northern Nigeria it is fed to horses as a tonic. It is described as very bitter.

Hausa: Dataniya, Dwatana—; *Katagum*: Dandata—.

THEMEDA.

This genus numbers about 15-16 species, all natives of the tropical and subtropical regions of the Old World, mostly Indo-Malayan.

Themeda triandra Forsk. is found in the hotter drier parts of India, from Upper Bengal to Travancore, ascending the Himalaya to 3,000-4,000 ft. It also occurs in Burma and Ceylon, ascending to 4,000 ft. It is one of the commonest grasses of the drier regions of Africa, often forming extensive associations. It extends to Australia and all the warm regions of the Old World.

This plant is said to be used medicinally in China. It has been found to yield hydrocyanic acid.

Canarese: Bettanchi hullu, Gondamanchi hullu, Thodda anji hullu—; *Indo-China*: Quan—; *Natal*: Insinda—; *South Africa*: Rooi Grass—; *Tamil*: Erigai thattu pullu—.

THYSANOLAENA.

T. procera Mez. (=*T. agrostis* Nees=*T. acarifera* Nees and Arnot.), the only species of this genus, is found in subtropical Himalaya, from Kumaon eastwards and the Khasia Hills at altitudes of 4,000-5,000 ft., in Behar on the Parasnath, in the Deccan, Burma, Nicobar Islands, and the Malay Peninsula, where it abounds in dry banks in the hills at 1,000 ft. altitude and upwards from Penang to Malacca. It extends eastwards to New Guinea.

A decoction of the root is used in Chota Nagpur as a mouth-wash during fever.

Birja: Jopono—; *Gond*: Phulbari—; *Hasada*: Durhitasad, Duritasad—; *Indo-China*: Dong trung ha thao, Dot, Le, Ong anh, Xay—; *Khond*: Saderi, Saperi—; *Kolami*: Dodri, Gara-jonor—; *Malay*: Buluh tubaro—; *Mundari*: Garajapono, Garajono—; *Naguri*: Durhitasad—; *Santali*: Bushnia, Karsar—; *Saora*: Kondachipuru—; *Uriya*: Phulosoro—.

TRITICUM.

This genus consists of 20 species, natives of the Mediterranean region and of Western Asia.

T. aestivum Linn. is used medicinally in Europe, and three of its races—*T. amyleum* Ser., *T. durum* Desf., *T. spelta* Linn.—are used as tonics in Brazil.

Triticum aestivum Linn. (=*T. sativum* Lam.) is cultivated in many parts of Northern India and the Deccan Peninsula, especially in the North-West, and up to 13,000 ft. in the Himalaya and Tibet.

Ayurveda writers describe wheat as cooling, oleagenous, tonic, aphrodisiac, laxative, and fattening, improving appetite and relish for food. They held it as a useful medicine in general disorders of health; but they admitted that it was hard to digest.

'The fine floure', says Gerard, 'mixed with the yolke of an egge, honey and a little saffron, doth draw and heale byles and such like sores, in children and in old people, very well and quickly. . . . Slices of fine white bread laid to infuse or steepe in Rose water, and so applied unto sore eyes which have many hot humours falling into them, doe easily defend the humour, and cease the paine. The oyle of wheat pressed forth between two plates of hot iron, healeth the chaps and chinks of the hands, feet and fundament, which come of cold, making smooth the hands, face or any other part of the body.'

And Culpeper: 'The oil pressed from wheat by means of plates of iron or copper, heals tetters and ringworms, if used warm. The green corns chewed, and applied to the bites of mad dogs, heals them; wheat-bread poultices made with red wine, and applied to hot, inflamed, or blood-shot eyes, helps them. Hot bread poultices applied three days together, heals Kernels in the throat. Wheat flour mixed with juice of henbane stays the flux of humours to the joints, if laid thereto; or mixed with the yolk of an egg, honey or turpentine, draws, cleanses, and heals boils, plague sores, or foul ulcers. It is more useful for food than medicine; though a poultice made of it with milk, easeth pains, ripens tumours and imposthumations; and a piece of toasted bread dipped in wine, and applied to the stomach, is good to stay vomiting.

Wheaten flour, bran, starch, and bread-crumb are still very much in demand as popular medicines in Europe and in all places inhabited by European settlers or their descendants.

The flour, which is officinal in France, is esteemed as an external application in erysipelas, burns, scalds, and various itching and burning eruptions. A mixture of flour and water is used as an antidote in cases of poisoning by salts of mercury, copper, zinc, silver and tin, and by iodine.

Bread is formed by making flour into a paste with water, with the addition of yeast, setting it aside to ferment, and then exposing it to the heat of an oven. A poultice is often made by boiling bread with milk; its effects are similar to those of a flaxseed poultice although it does not hold the heat so well. Bread-crumb is useful for giving bulk to pills, and for making poultices.

Bran is laxative, and may be used with advantage to prevent costiveness. Bran bread made from the unsifted flour, is an excellent laxative article of diet in some dyspeptic cases, and, owing to its freedom from starch, in diabetes. Bran poultices are useful for many purposes. Bran is sometimes used in the form of a decoction or infusion, as an emollient bath, and also internally as a demulcent.

Starch prepared from wheaten flour is officinal in all pharmacopoeias. It is employed for sprinkling over inflamed surfaces, to absorb acid secretions, and prevent excoriation. Mucilage of starch is valued in pharmacy for many purposes, and medicinally as a demulcent. In surgery it is occasionally employed for stiffening bandages.

Gluten is a term popularly known in connection with the names of 'gluten flour' and 'gluten bread', used and recommended especially in connection with the dietary of diabetics when starchy foods are to be administered in small amounts or altogether prohibited. These preparations are legion and in many instances are sold under claims which are as false and misleading as they are cruel in the hope of which they hold out to the sufferer from the disease.

In China the grains are roasted and are considered useful in colliquative sweating, especially in tuberculosis in women.

Afghanistan: Ganam, Gandam—; *Afrikaans*: Koring—; *Arabic*: Burr, Hintah—; *Ashkobi*: Spinghanam, Wasaghanam—; *Bagwana*: Dayak, Kandahari, Kub. Shorawaki, Shuthardandan, Sundia, Trimali—; *Bengal*: Gam, Giun, Gom—; *Bombay*: Gahu, Ghawn, Ghawutghum, Gium, Gohum, Kapale, Marghum—; *Brazil*: Trigo. Trigo candeal, Trigo tremez—; *Burma*: Giyonsaba Gyungsaba—; *Canarese*: Godhi—; *Catalan*: Blat menut, Xeixa, Xexa—

Central Provinces: Ghubot, Seonikar—; *Chinese*: Ch'ao, Cheng Ping, Chiang, Hsiao Mai—; *Deccan*: Gohun—; *Dutch*: Tarwe—; *English*: Wheat—; *French*: Blé, Blé cultivé, Blé marcel, Blé marcet, Blé de mars, Blé d'été, Blé trémois, Froment, Froment cultivé—; *German*: Weizen—; *Greek*: Sitari—; *Gujerati*: Gawn, Ghavum, Govum—; *Hadramaut*: Burr—; *Harboi Hills*: Ghanam, Kholam, Shirukh—; *Hausa*: Alkama—; *Hindi*: Gehub, Gehun, Giun, Kunak—; *Hova*: Lafarina, Varimbazaha—; *Hungarian*: Buza—; *Italian*: Frumento, Grano—; *Jhalawan*: Geroli, Sundia—; *Kharan*: Pashmik—; *Khuzdar*: Geroli—; *Kila Saifulla*: Ghanam, Sarghanam, Spinghanam—; *Kirta*: Barkhani, Sungsillah—; *Konkani*: Gahum—; *Languedoc*: Blad, Bladet, Blat, Bled—; *Malayalam*: Gendum, Kotanpam—; *Malta*: Spring Wheat, Grano d'estate, Marzuolo, Civitella, Tomnija—; *Marathi*: Gahum, Gahung—; *Mastung*: Ghanam, Kholam—; *Michi*: Shruk, Tokar, Tomar, Tro—; *Nigeria*: Common Wheat, Egyptian Wheat, Miracle Wheat, Mummy Wheat, Wheat—; *North-Western Provinces*: Gehun—; *Oudh*: Gehun—; *Pab*: Khisankah—; *Persian*: Gandum—; *Portuguese*: Trigo—; *Punjab*: Do, Dro, Gehun, Kanak, Nis, Rozatt, To, Zud—; *Quetta*: Pashmik—; *Roumanian*: Griu—; *Russian*: Pshenitza—; *Sadani*: Gehom, Gohom, Gom—; *Sanskrit*: Arupa, Bahudugdha, Godhuma, Kshiri, Mlenchhabhojana, Nistusha, Rasala, Saman, Sumana, Yava, Yavana—; *Saruna*: Khalam—; *Shahrig*: Boja, Gandun, Ghanam, Kholam—; *Sinhalese*: Tiringu—; *Sohrab*: Pesur, Puzho, Shuthardandan—; *Spanish*: Trigo candeal, Trigo chamorro, Trigo común, Trigo jejar, Trigo marzal—; *Tamil*: Godumai, Godumbaiyarisi—; *Telugu*: Godumulu—; *Wad*: Sundia—; *Zulu*: uKolo, uKolweni—.

a. **Triticum durum** Desf.—*Portuguese*: Trigo durasio—.

b. **Triticum spelta** Linn.—; *Arabic*: Hunta, Qamh, Qamh m'ghejir, Qamh ne ejghe, Qamh stejri—; *Catalan*: Escanya major, Espelta—; *Egypt*: Hunta Qamh—; *French*: Ampeutre, Engrain, Froment rouge, Grande épeautre, Ingrain—; *Spanish*: Escanda, Espelta—.

c. **Triticum amyleum** Ser.—*Brazil*: Trigo branco—.

Vetiveria.

This genus consists of 7 species, distributed throughout the tropics of the Old World; one has been introduced into the New World.

Vetiveria zizanoides Stapf. (=*Andropogon muricatus* Retz =*A. squarrosus* Cooke) is found practically over the whole of India, and eastwards to Burma. It is very common in every part of the Coromandel Coast, and in Bengal and Burma, where it meets with a low, moist, rich soil, especially on the banks of the water-courses. It covers large tracts of waste land in Cuttack. It inhabits the plains of the Punjab and North-West Provinces, and ascends into Kumaon, 1,000-2,000 ft. in altitude. It is cultivated in Rajputana and Chota Nagpur. It occurs occasionally in Malay gardens, and extends to the West Indies, Brazil, and Lower Guinea. So far it has been found in the wild state only in tropical Asia.

Ayurvedists describe the root as bitter, cooling, stomachic, astringent and an antidote to poisons. They recommend its use in burning sensations, bilious fevers, sweats, foul breath, thirst, strangury, ulcers, and diseases of the blood.

Yunanists hold the root as cooling to the brain, bitter and soporific. They recommend its use in spermatorrhoea, headache, and diseases of the blood.

An infusion of the root is given as a febrifuge, and a powder in bilious complaints. It is regarded as stimulant, sudorific, stomachic, and refrigerant.

The essence is used as a tonic.

A paste of the pulverised roots in water is also used as a cooling external application in fevers.

In Guinea the infusion of the roots is used as a tonic and an emmenagogue. In China it is chiefly used to remove dandruff.

The root and stem are Ayurveda cures for snake-bite and scorpion-sting, but Mhaskar and Caius have shown experimentally that neither is an antidote to either snake or scorpion venom.

Arabic: Izkhir, Usir—; *Bengal*: Bena, Khaskhas, Khor, Kuskus—; *Bombay*: Khasakhasa, Vala—; *Burma*: Miyamoe—; *Canarese*: Dhappa sajjai hullu, Kadu karai, Lavancha—; *Chinese*: Ko Cha—; *Cutch*: Vala—; *Deccan*: Khaskhas—; *English*: Cuscus, Khuskhus, Koosa—; *French*: Chiendent des Indes, Vétiver—; *Gujerati*: Valo—; *Hindi*: Bala, Balah, Bena, Ganrar, Kaskas, Khas, Onei, Panni, Usir—; *Indo-China*: Chuoc mat, Huong bai—; *Kolami*: Sirom—; *La Reunion*: Vétiver—; *Malay*: Akar wangi—; *Malayalam*: Ramachchamver, Vettiver—; *Marathi*: Vala—; *Mundari*: Birnijono, Sirum, Sirumjono—; *Oudh*: Tin—; *Persian*: Bikhiwala, Khas—; *Philippines*: Moro, Raiz de Moro—; *Punjab*: Panni—; *Sadani*: Birni—; *Sanskrit*: Abhaya, Amrinala, Avadaha, Dahaharana, Gandhadhya, Haripriya, Indragupta, Ishtakapatha, Jalamoda, Jalashaya, Jalavasa, Rambhu, Katayana, Laghubhaya, Lamajjaka, Nalada, Ranapriya, Samagandhika, Sevya, Shishira, Shitamulaka, Sugandhimula, Ushira, Vira, Virabhadra, Virana, Viratarn, Vitanamulaka—; *Santali*: Sirom—; *Sinhalese*: Saivandera, Savandramul—; *Tamil*: Ilamichamver, Vakil, Vettiver, Vilhalver, Viranam, Virkel, Viyal—; *Telugu*: Avurugaddiveru, Kassavu, Kuruveru, Lamajjakamuveru, Vettiveru, Vidavalliveru—; *Tharu*: Khatra—; *Urdu*: Khas—; *Uriya*: Bena—.

ZEA.

The only species of this genus is of American origin, in cultivation in all warm countries of the world.

Zea mays Linn. is widely cultivated in India.

According to Hindu writers the grain is nutritive, demulcent, emollient, diluent, and absorbent; but, it is liable to produce flatulence.

It is considered by Mahomedan writers to be resolvent, astringent, and very nourishing; they consider it to be a suitable diet in consumption and a relaxed condition of the bowels. It is analgesic, and a decoction is recommended as a hip-bath for piles.

In the Konkan an alkaline solution is prepared from the burnt cobs and is given in lithiasis. In the Central Provinces the centre of the cob, deprived of the seeds and reduced to an ash by burning, is given in combination with common salt in bronchial catarrh and whooping cough.

In Europe and in the United States of America the grain is much used as a valuable article of diet for invalids and children. Corn bread contains more nourishment than wheaten bread, and is a better diet for persons suffering from diseases of the liver, or of the kidneys.

In Greece the silky stigmata are used in decoction in diseases of the bladder, a custom which also obtains in the Philippine Islands, where the plant is considered diuretic. The styles and stigmas have now been generally recognised as diuretic, anodyne, and demulcent. Under the popular name of 'corn silk', a liquid extract is sold as a remedy in irritable conditions of the bladder with turbid and irritating urine; it has a marked diuretic action,

Zea, or the stigmas and styles, is employed in acute and chronic cystitis, and in the bladder irritation of uric acid and phosphatic gravel. It has also been employed in gonorrhoea, and has been affirmed to be a useful diuretic and even cardiac stimulant in the dropsy of heart disease.

The meal is made into emollient poultices. It is also made into mush by boiling, but its exclusive use gives rise to a skin disease.

In Cambodia the grains are prescribed in angina, and the stigmas in paludism.

Animals are sometimes poisoned from eating the green plant. Infants fed on Indian corn flour grow up rickety; it contains only about eighteen grains of proteic substance to the pound. The male flowers are said to contain hydrocyanic acid. The pollen is a possible cause of hay fever.

Accra: Abblay—; *Afghanistan*: Jaoari, Jaori, Jaorikhurdani—; *Afrikaans*: Mielie—; *Annam*: Bap ngo, Lua ngo—; *Arabic*: Durahkisan, Durahshani, Hintaherunu, Khalavan, Khandaruz, Zurratulmakkah—; *Ashanti*: Aburow—; *Awuna*: Akple—; *Bengal*: Bhutta, Janar, Kokka—; *Bombay*: Barjoar, Buta, Macka, Makai—; *Brazil*: Zaburro—; *Burma*: Pyaungbu—; *Cambodia*: Paut, Put—; *Canarese*: Goinjol, Makkai jola, Mekkejola, Musukujola—; *Catalan*: Blat de moro—; *Ceylon*: Cholum—; *Chinese*: Pang Tzeu, Yu Shu Shu—; *Chota Nagpur*: Jonar—; *Cochin China*: Bap ngo, Lua ngo—; *Dardanelles*: Kalamasitaro—; *Deccan*: Makkajari, Makkajowari—; *Dutch*: Mais—; *Egypt*: Durra—; *English*: Indian Corn, Maize—; *Ewe*: Akple, Blikple—; *Fanti*: Abru, Aburow—; *Fiji*: Sila ni papalagi—; *French*: Blé de Barbarie, Blé d'Espagne, Blé de Guinée, Blé d'Inde, Blé d'Italie, Blé de Rome, Blétture, Blé de Turquie, Froment des Indes, Graine de Turquie, Maïs, Maïz, Mil d'Espagne, Gros millet des Indes, Troquet, Truquiet, Turquie—; *Ga*: Able—; *Garhwal*: Junala, Mungari—; *German*: Tuerkische Korn, Tuerkisher Weizen—; *Gold Coast*: Agbahdor—; *Guam*: Maeis—; *Gujerati*: Makkari—; *Hasada*: Jonra—; *Hausa*: Masara—; *Hindi*: Barajuar, Bhotta, Bhuta, Bhutta, Jawdra, Junri, Kukri, Makai, Makka—; *Hova*: Kasobotso, Katsamanga, Katzaha, Tsako, Tsakotsako—; *Indo-China*: Bao, Bap, Cennaron, Khao phot, Mali, Ngo, Ngoc thuc thu, Put, Sali—; *Iraq*: Indhrah safra, Idhrat-ash-sham—; *Italian*: Grano siciliano, Grano turco, Melicatto Meliga—; *Kashgar*: Conac—; *Khasi*: U riew hadem—; *Kila Saifulla*: Makai, Maki—; *Konkani*: Maeo, Zonallo, Zondllo—; *Krepi*: Adakple, Blikple, Kple, Kpledzi, Watsikple—; *Krobo*: Blaifo—; *Kumaon*: Bhutta, Junala, Mukni—; *Kurdish*: Gumnah shami, Zurat zard—; *Languedoc*: Artho, Avari, Avati, Blamari, Blarama, Garouilhé, Garouillet, Millaral, Millargo, Millargou—; *Laos*: Khao phot, Khot—; *Lepcha*: Kun-tsong—; *Lynngam*: Soh rikhaw—; *Madagascar*: Katsabazaha, Katsakandevolahy, Sako—; *Malayalam*: Cholam, Makka cholam—; *Malta*: Frumentone, Granoturco, Indian Corn, Kamh irrum, Maize—; *Marathi*: Bonda, Maka—; *Menomini*: Apesapimin—; *Moldavia*: Popusoiu—; *Mundari*: Gorajonra, Gurulujonra, Jondra, Jonra, Loeongionra—; *Naguri*: Jondra—; *North-West Provinces*: Barajuar, Bhutta, Junri, Maka, Makai, Makka—; *Persian*: Bajri, Gaudumemakkah, Koshahemakki—; *Philippines*: Borona, Maiz—; *Portuguese*: Milho, Milho grosso—; *Punjab*: Barajuar, Chhale, Juar, Kukri, Kuthi, Mak, Makkei, Makki—; *Rajputana*: Mukkha—; *Roumanian*: Porumb—; *Russian*: Kukuruva, Mais—; *Sanskrit*: Kandaja, Mahakaya, Makaya, Samputantastha, Shikhalu, Yavanala—; *Santal*: Jondra—; *Sarakhala*: Makkai, Maki—; *Sema*: Kolakiti—; *Shahriq*: Badaghariuari—; *Sierra Leone*: Amarsk Kabe, Nvin—; *Sind*: Barajuar, Makkai—; *Sinhalese*: Bada iringu, Tringu—; *South Africa*: Mealie Corn, Mealies—; *Spanish*: Maiz, Trigo de las Indias, Trigo de Turquia, Zara del Perú—; *Suto*: Poone—; *Tamil*: Makkasholam, Turaka cholam—; *Telugu*: Makkanzonnalu, Mokkajanna, Mokka jonnalu—; *Tobo*: Makai—; *Tongking*: Bap ngo, Lua ngo—; *Transylvania*: Cucuruz—; *Turkish*. Misir, Sari dari—; *Twi*: Aburow—; *Urdu*: Makai—; *Uriya*: Buta, Maka, Mokha—; *Yemen*: Rumi—; *Zomba*: Chimanga—.

99.

THE MEDICINAL AND POISONOUS ORCHIDS OF INDIA

BY

J. F. CAIUS, S.J., F.L.S.

The ORCHIDACEAE include about 610 genera with over 8,000 species, mostly natives of tropical and subtropical countries, less frequent in temperate countries, a few being found in cold mountainous regions.

The medicinal and poisonous Orchids of the world belong to 38 distinct genera:—ACAMPE (Indo-Malaya, Africa); ACERAS (Mediterranean); ANACAMPTIS (Europe, North Africa); ANGRAECUM (La Reunion, Mauritius); ANSELLIA (Tropical Africa); APLOSTELLIS (Old World tropics and subtropics); BLETIA (Tropical America); CORALLORHIZA (North America); CREMASTRA (Japan); CYMBIDIUM (Tropical Asia, Africa, Australia); CYPRIPEDIUM (Northern cold and temperate regions, Mexico); CYRTOPODIUM (America); DENDROBIUM (Tropical Asia to Japan, Australia, Polynesia); DESMOTRICHUM (Indo-China, Malay Islands); EPIDENDRUM (Tropical America); EULOPHIA (Africa, Indo-Malaya); GALEANDRA (Tropical America); GASTRODIA (Indo-Malaya, Japan, China, Australia); GOODYERA (Europe, Asia, Mascarene Islands, North America); GOVENIA (Tropical America); GYMNADENIA (Europe, Asia); HABENARIA (whole world); HETAERIA (Africa, Indo-Malaya, Polynesia); LIMODORUM (Mediterranean); LISSOCHILUS (Africa); LUISIA (Indo-Malaya, Polynesia); MICROSTYLIS (cosmopolitan; chiefly tropical); MONADENIUM (Africa); NEOTTIA (Temperate Europe, Temperate Asia); OBERONIA (Old World Tropics, Japan); OPHRYS (Europe, Mediterranean); ORCHIS (Mediterranean, Temperate Europe, Asia, North America, North Africa); PLATANTHERA (North America, Tropical Asia, Africa); RHYNCHOSTYLIS (Indo-Malaya); SACCOLABIUM (Africa, Indo-Malaya, China); VANDA (Asia, Australasia); VANILLA (All tropical countries); ZEUXINE (Tropical Asia, Africa).

The medicinal and poisonous Orchids of India belong to 15 distinct genera:—ACAMPE, CYMBIDIUM, DENDROBIUM, DESMOTRICHUM, EULOPHIA, HABENARIA, HETAERIA, LUISIA, OBERONIA, ORCHIS, RHYNCHOSTYLIS, SACCOLABIUM, VANDA, VANILLA, ZEUXINE.

A. Anther 1, opercular. Pollinia waxy, 1-4 in each cell:
- 1. Column footless. Lip entire or lobed, with a nectar depression at base ... ... OBERONIA.
- 2. Column with a long foot. Lip adnate to the produced foot of the column, contracted at the base or clawed:
  - a. Flowers fascicled ... ... DESMOTRICHUM.
  - b. Flowers solitary or in fascicles or racemes ... DENDROBIUM.

B. Anther 1, posticous. Pollinia waxy, usually 2 or 4:
- 1. Scape usually leafless. Petals like the dorsal sepal; lip gibbous or saccate, rarely spurred ... EULOPHIA.
- 2. Sepals and petals subequal, spreading; lobes of the lip embracing the unwinged column. Leafy stems short, pseudobulbous ... ... CYMBIDIUM.

10

3. Sepals and petals fleshy, widely spreading from a narrow base. Flowers large, in simple racemes. VANDA.
4. Sepals and petals subequal. Columns short stout, no foot. Flowers small pink or white crowded in dense racemes showy; lip entire ... ... RHYNCHOSTYLIS.
5. Sepals and petals widely spreading. Column without appendages. Flowers small, in simple or branched racemes; lip 3-lobed with callus in the spur mouth ... ... SACCOLABIUM.
6. Sepals and petals thick, concave. Flowers small, crowded in a short rigid, simple or branched peduncle; lip obscurely 3-lobed, no callus ... ACAMPE.
7. Sepals subequal or the depressed dorsal smaller. Petals narrower often longer, spreading. Column much shorter than the lip, truncate. Flowers few in short lateral spikes. Lip sessile on column base, divided into basal and apical parts ... LUISIA.

C Stem not bulbous. Anther 1, posticous, opercular or erect and persistent. Pollen granular, powdery or in small masses:

1. Sepals and petals subequal, spreading. Column long, anther broad ... ... VANILLA.
2. Sepals free, dorsal with the petals cohering in a hood; lip with dilated end. Column very short with lamellar or conic processes in front ... ZEUXINE.
3. Upper sepal usually adnate to petals pubescent; lip without a dilated end. Column rather large with long rostellar arms ... ... HETAERIA.

D. Anther 1, posticous, erect, inclined or reflexed. Pollinia 1 rarely 2 in each cell, granular, produced into short caudicles. Terrestrial herbs:

1. Lip spurred. Glands of the pollinia both in one pouch ORCHIS.
2. Lip spurred, sepals equalling or exceeding the petals. Flowers spicate or racemose; rostellum not elongate ... ... HABENARIA.

It is chiefly as sources of SALEP that orchids have become therapeutically famous, and it is as sources of salep that some species are still retained in the pharmacopoeias of Austria, Belgium, Germany, Holland, Japan, Norway, Portugal, Russia and Switzerland.

Salep is the name given to the tubers or tuberous roots of various orchids, collected in the flowering season, immersed in boiling water for a short time, and dried. At present the salep of European commerce is prepared chiefly in the Levant, but to some extent in Germany and other parts of Europe. The tubers are collected from wild plants growing in Asia Minor, South-west Asia, France and Germany. The salep of the Indian bazars is produced in the hills of Afghanistan, Baluchistan, Persia and Bokhara; a part of it is obtained from the Nilgiri Hills and also from Ceylon.

From ancient times salep has been considered to possess great invigorating virtues, and has hence been extensively prescribed both in Europe and the east for recruiting the exhausted vitality of aged and enervated persons. An infusion or decoction was taken generally in England as a beverage before the introduction of tea and coffee. Sassafras chips were sometimes added for giving the drink a flavour.

Though its medieval and oriental reputation as an aphrodisiac has been irretrievably lost in the West, salep is still considered highly nutritive and a most useful article of diet for those who suffer from chronic diarrhoea. All scientific opinion, however, concurs in the belief that salep is devoid of medicinal virtue and that it is highly doubtful if its nourishing properties are as great as they are supposed to be.

In the east, nevertheless, faith in the efficacy of the drug is still unshaken and the dry tuber has an immense reputation as an aphrodisiac, a nervine restorative, and a fattener. It is much used by *vaids* and *hakims* in conjunction with other nervine tonics. It is also considered a very nourishing article of diet, and is given mixed with milk and flavoured with spices and sugar. In fact salep is so very highly esteemed and is in such great demand that 'imitation salep' is largely prepared for the Indian trade. Wealthy Orientals are known to have paid a handsome price for pounded potatoes and gum. 'Royal Salep', said to be much used as a food in Afghanistan, has been identified as the product of a liliaceous plant, *Allium macleanii* Baker.

The Order is remarkable for the absence of poisonous properties; only in so far, however, as trifles may be neglected, for three species are considered toxic. The natives of Portuguese East Africa say that a liquid extract of *Monadenium lugardae* N. E. Br. is poisonous and produces vomiting; the leaves and stem of the Malayan *Vanilla griffithii* Reich. f. contain an irritating latex; reports of violent dermatitis in workmen handling vanilla beans, *Vanilla planifolia* Andrews, are not uncommon.

It is no less interesting to note that four orchids are used in the treatment of various forms of insanity. Natives in the Mpika district of Northern Rhodesia administer an infusion of the leaves and stems of *Ansellia humilis* Bull. as a remedy for madness; American Indians give a decoction of the Swamp Orchid, *Simodorum giganteum* Thunb., to combat manias; on the Malabar Coast the powdered flowers of *Vanda spathulata* Spreng. are said to abate frenzy; and in France the fruit of *Vanilla planifolia* Andrews is prescribed in order to overcome melancholia.

ACAMPE.

This genus contains 12 species, mostly Indo-Malayan and Chinese, a few African.

Acampe wightiana Lindl. (=*Saccolabium praemorsum* Hook. f. =*S. Wightianum* Hook. f.) is found throughout Bengal, the Western Peninsula, and Ceylon.

The plant is a bitter tonic. It is used in rheumatism.

Canarese: Marabale—; *Malayalam*: Taliyamaravala—; *Marathi*: Kanbher—.

CYMBIDIUM.

The 30 species which constitute this genus inhabit Africa, Asia, Australia, extending as far as Japan.

Cymbidium aloifolium Swartz occurs in Nepal, the Terai, tropical Himalaya, eastwards to Sikkim, Western and Southern India, Ceylon.

The plant is said to be emetic and purgative.

English: Boat Orchid—.

Dendrobium.

Epiphytic herbs belonging to about 750 species and distributed throughout Tropical Asia, Japan, Australia and Polynesia.

D. ceraia Lindl., *D. moniliforme* Swartz, *D. nobile* Lindl. are used medicinally in China; *D. reptans* Franch. and Sav. is similarly used in Japan.

Two Indian species are credited with therapeutic properties:

1. Stems tufted, 12-18 in. long, usually slender. Flowers in lateral and terminal racemes 3-6 in. long *D. ovatum.*
2. Pseudobulbs tufted, cylindric or dilate upwards, 2 in. long. Flowers 1 or 2 from terminal tuft of bracts ... ... *D. pumilum.*

1. **Dendrobium ovatum** (Willd.) Kranzl. (=*D. chlorops* Lindl.) is found on the Western Ghats and the Western Coast of the Madras Presidency.

The plant is an emollient. The juice of the fresh plant is given internally for all kinds of stomach ache; it excites the bile and acts as a laxative to the intestines.

Malabar: Maravar—; *Marathi*: Nagli—.

2. **Dendrobium pumilum** Roxb. is very common on orchard trees in the Malay Peninsula whence it extends to Burma and Borneo.

The Malays use it as a drug in dropsy.

Malay: Sakat kalumbai—.

Desmotrichum.

The genus consists of about 30 Indo-Malayan species.

Desmotrichum fimbriatum Bl. (=*Dendrobium macraei* Lindl.) inhabits the Western Ghats of the Bombay and Madras Presidencies, Ceylon, Sikkim, the Khasia Hills, Burma, the Malay Peninsula, whence it spreads to Siam, Borneo and the Philippine Islands.

Sanskrit writers describe the plant as cold, mucilaginous, light, strengthening, and as a remedy for disorders of the bile, blood and phlegm. The whole plant is used in decoction along with other drugs supposed to have similar properties. The fruit is said to be an aphrodisiac.

The plant is prescribed by Sushruta, in combination with other drugs, for the treatment of snake-bite and scorpion-sting; but Mhaskar and Caius have proved experimentally that it is not an antidote to either snake-venom or scorpion-venom.

The plant is occasionally used alone as a stimulant and tonic. As a tonic it is given in debility due to seminal losses.

Bengal: Jibai, Jibanti—; *Gujerati*: Jivanti, Radarudi, Wajhanti—; *Hindi*: Jiban, Joivanti, Sag—; *Marathi*: Jivanti—; *Sanskrit*: Bhadra, Jiva, Jivabhadra, Jivada, Jivani, Jivaniya, Jivanti, Jivapatri, Jivapushpi, Jivavardhini, Jivavrisha, Jivdatri, Jivya, Kanjika, Kshudrajiva, Madhushvasa, Madhusrava, Mangalya, Mrigaratika, Payaswini, Pranada, Putrabhadra, Raktangi, Shakashreshtha, Shashashimbika, Shringati, Srava, Sukhankari, Supringala, Yashaskari, Yashasya—.

EULOPHIA.

This genus numbers 200 species distributed through the warm countries of the Old World; they are especially numerous in Africa and Indo-Malaya.

E. arenaria Bohn., *E. flaccida* Schltr., *E. hians* Spreng., *E. robusta* Rolfe are used medicinally in South Africa.

Among Eulophias which furnish salep to the Indian bazaars may be mentioned *E. campestris* Wall., *E. herbacea* Lindl., *E. nuda* Lindl. and *E. pratensis* Lindl.

A. Column produced into a foot. Flowers large; lateral sepals inserted on the spur of the lip ... ... *E. nuda.*

B. Column not produced into a foot:

1. Scape 7-12 in. long, sheathed at intervals by loose membranous bracts. Sepals linear-lanceolate. Petals narrower than the sepals, oblanceolate ... *E. campestris.*
2. Scape 1-3 ft. long with appressed acute sheaths. Sepals elliptic-oblong. Petals shorter and broader than the dorsal sepal, elliptic-oblong, obtuse. Lip broader than long ... ... *E. pratensis.*
3. Scape stout, 2-3 ft. high with large acuminate sheaths. Sepals linear-lanceolate. Petals oblong-elliptic, obtuse or subacute. Lip longer than broad ... ... *E. herbacea.*

1. **Eulophia campestris** Wall. is found in the sub-Himalayan tracts of Rohilkhand and North Oudh, Nepal, Sikkim, Chittagong, Bengal, Upper Burma, Baluchistan and Afghanistan.

Ayurvedists consider the tuber tonic, stomachic, aphrodisiac and alterative; they commonly give it to stimulate appetite and to purify the blood in heart troubles.

Yunani doctors credit the tuber with tonic, aphrodisiac and astringent properties; they consider it useful in stomatitis, purulent cough and paralytic conditions.

The plant furnishes a salep which is much prized by Indians as a tonic and aphrodisiac.

Arabic: Khusyu-uth-thalab—; *Bengal*: Salibmisri, Sungmisrie—; *Gujerati*: Salum—; *Hindi*: Salibmisri—; *Marathi*: Salamishri—; *Nepal*: Hattipaila—; *Persian*: Sungmisri—; *Punjab*: Salibmisri—; *Sanskrit*: Amrita, Amritodbhava, Jiva, Jivani, Pranabhrita, Pranada, Sudhamuli, Virakanda—; *Santal*: Bongataini—; *Urdu*: Salabmisri—.

2. **Eulophia herbacea** Lindl. is found in the Western Himalaya and Western Peninsula; it is met with in Kanara, Konkan and Gujerat.

The tuber furnishes an esteemed kind of salep.

3. **Eulophia nuda** Lindl. occurs in Tropical Himalaya from Nepal eastwards to Sikkim, in Chota Nagpur, Assam, the Khasia Hills, Manipur, Burma, the Western Peninsula and Ceylon.

Hindu practitioners use the tuber in tuberculous glands of the neck, tumours, bronchitis and diseases of the blood; they say it gives appetite and prescribe it accordingly.

The pounded fresh plant is applied to boils and abscesses to promote maturation and suppuration.

The powdered tuber is said to be an antidote for poisoning; it is administered internally to those suffering from intestinal worms.

The plant is a source of salep.

Bengal: Budbar—; *Hindi*: Ambarkand, Gourma—; *Marathi*: Ambarakand, Bhuikakali, Manakanda—; *Sanskrit*: Balakanda, Granthidala, Kandalata, Malakanda, Panktikanda, Trishikhadala—.

4. **Eulophia pratensis** Lindl. is apparently endemic in the Western Peninsula; it is found in the pasture-lands of the Deccan, in the Sholapur District and in Belgaum.

The tubers are applied externally and given internally to remove scrofulous glands in the neck. They are also administered internally as a worm remedy. They make up a large portion of the salep of Western India.

Marathi: Satavari—.

HABENARIA.

The genus contains about 500 species, widely diffused through tropical and subtropical regions.

An infusion of the root of *H. foliosa* Reichb. fil. is used as an emetic by the Zulus.

Habenaria commelinifolia Wall. occurs in the outer ranges of Western Himalaya from the Punjab to Kumaon up to 5,000 ft., extending eastwards to Parasnath, Chota Nagpur, Bihar and Upper Burma, Central India, Konkan, Western Ghats, Deccan, North Kanara and Bababudan Hills.

The plant is a source of salep.

Sadani: Jadu, Jaitjadu—.

HETAERIA.

The 40 species which make up this genus inhabit Africa, Indo-Malaya and Polynesia.

Hetaeria obliqua Bl. is common in the woods of the Malay Peninsula, extending thence to Sumatra and Borneo.

The plant is used by the Malays for poulticing sores.

Malay: Poko tumbak hutan—.

LUISIA.

The genus numbers about 15 species, mostly tropical Asiatic, a few Australian.

Luisia tenuifolia Bl. inhabits the Western Peninsula and Ceylon.

The pounded plant is used as an emollient. It is applied as a poultice to boils, abscesses and tumours.

OBERONIA.

The genus consists of about 60 species, tropical Asiatic, Australian and Pacific.

Oberonia anceps Lindl. is common throughout the Malay Peninsula; it extends to Tenasserim and Java.

The Malays use the leaves for poulticing.

Malay: Sakat lidah buaya—.

ORCHIS.

The 70 species of this genus are found in Europe, temperate Asia, North Africa and America.

The following are used in Europe for the preparation of salep:— *O. bifolia* Linn., *O. fusca* Linn., *O. hircina* Linn., *O. latifolia* Linn., *O. laxiflora* Lam., *O. maculata* Linn., *O. mascula* Linn., *O. militaris* Linn., *O. morio* Linn., *O. pyramidalis* Linn., *O. sambucina* Linn., *O. simia* Lam.

Orchis latifolia Linn. occurs in the Western Temperate Himalaya from Nepal to Kashmir, and Western Tibet at altitudes between 8,000 and 12,000 ft. It extends to Northern Asia and through Afghanistan to North Africa and the Atlantic. It usually grows in damp meadows and marshes.

The tubers are expectorant and astringent. They yield salep.

English: Marsh Orchid—; *Spanish*: Palma Christi—.

RHYNCHOSTYLIS.

The genus consists of 2 Indo-Malayan species.

Rhynchostylis retusa Bl. is distributed throughout India, Burma, the Malay Peninsula, Siam, Java, the Philippine Islands and Ceylon.

The fresh plant is used as an emollient.

SACCOLABIUM.

This genus comprises about 200 species, inhabitants of Africa, Indo-Malaya and China.

Saccolabium papillosum Lindl. is found in Bengal and the lower Himalaya Mountains from Sikkim eastwards, Assam, the Gangetic Delta and Tenasserim.

The root is considered a specific for rheumatism. It is commonly used as a substitute for sarsaparilla and for *Vanda tessellata*.

In the Konkan the roots are used as a cooling remedy.

Marathi: Nakuli, Rasna—; *Sanskrit*: Gandhata, Nakuli, Rasna—.

VANDA.

The genus consists of about 30 species found in Tropical Asia and Australia.

Two Indian species are credited with therapeutic properties

1. Flowers 1.2 in. diam., golden yellow ... ... *V. spathulata.*
2. Flowers 1.5-2 in. diam., tessellated with brown ... *V. tessellata.*

1. **Vanda spathulata** Spreng. occurs in the Western Peninsula from Malabar to Travancore and Ceylon.

On the Malabar Coast the juice of the plant is supposed to temper the bile and abate frenzy. The golden yellow flowers, reduced to powder, are given in consumption, asthma and mania.

Malayalam: Ponnamponmaraiva—.

2. **Vanda tessellata** Hook. (=*V. Roxburghii* R. Br.) is found in Bengal, Chota Nagpur, Bihar, the Central Provinces, the Western Peninsula, Travancore and Ceylon.

Ayurvedists consider the root antipyretic and an antidote for poisoning; they use it in dyspepsia, bronchitis, inflammations, rheumatic pains, hiccough, tremors and diseases of the abdomen.

Yunani practitioners hold it to be laxative, and tonic to the liver and the brain; they prescribe it for bronchitis, piles, lumbago, toothache and boils on the scalp; they further say that it lessens inflammation and heals fractures.

Rasna root is commonly prescribed in a variety of forms for rheumatism and allied disorders. It enters into the composition of several medicated oils for external application in rheumatic pains and diseases of the nervous system.

In Chota Nagpur, the leaves pounded and made into a paste are applied to the body during fever, and the juice is introduced into the aural opening for inflammation of the middle ear.

The plant is occasionally given as a cure for snake-bite. Combined with other drugs it forms some of Charaka's and Sushruta's snake and scorpion remedies. Caius and Mhaskar have, however, shown experimentally that whether single or compounded it is not an antidote to either snake- or scorpion-venom.

Arabic: Kharkittan—; *Bengal*: Nai, Rasna—; *Bombay*: Rasna—; *Canarese*: Bandanike—; *Gujerati*: Rasno—; *Hindi*: Banda, Nai, Perasara, Persara, Rasna, Vanda—; *Marathi*: Rasna—; *Sanskrit*: Atirasa, Bhujangakshi, Chhatráki, Dronagandhika, Elaparni, Gandhanakuli, Muktarasa, Nakuleshta, Nakuli, Palankapa, Rasadhya, Rasana, Rasna, Rasya, Sarpagandha, Shreyasi, Sugandha, Sugandhimula, Surasa, Suvaha, Vandaka, Vriksharuha, Yuktarasa—; *Santal*: Darebanki—; *Telugu*: Chittiveduri, Kanapabadanika, Mardaru, Vadanika—; *Urdu*: Banda—.

VANILLA.

The genus consists of about 20 species; to be found in all tropical countries.

Vanilla planifolia Andrews and other species furnish the well-known vanilla pods of commerce, used as a flavouring ingredient and sometimes as a tonic,, carminative and aphrodisiac. The pod of *V. planifolia* is official in the pharmacopoeias of Austria, France, Japan, Portugal, Switzerland and Turkey.

Vanilla griffithii Reichb. fil. is common in the forests of the Malây Peninsula and the Karimun isles.

The leaves and stem, which contain an irritating latex, are mashed and applied to the hair for strengthening it.

The flowers pounded with water are rubbed on the body for fever.

Malay: Akar punubal, Telinah kerbau bukit—.

Zeuxine.

This genus numbers about 20 species, tropical Asiatic and African.

Zeuxine strateumatica Schlechter (=*Z. sulcata* Lindl.) is found throughout the greater part of India up to 5,000 ft. on the outer Himalayan ranges; it also occurs in the Malay Peninsula and Ceylon, and is distributed to Afghanistan. China, Japan, Java and the Philippine Islands.

The tubers are used as salep.

Bengal: Shwethuli—.

THE MEDICINAL AND POISONOUS AROIDS OF INDIA.

BY

J. F. CAIUS, S.J., F.L.S.

The ARACEAE include about 107 genera with over 1,000 species, most of which are natives of tropical regions and the rest of temperate regions, in both hemispheres. Most of them contain an acrid juice, which is sometimes highly irritating and poisonous.

The medicinal and poisonous Aroids of the world belong to 42 genera:—ACORUS (northern temperate regions, South-Eastern Asia); AGLAONEMA (Tropical Asia and Africa); ALOCASIA (Tropical Asia); AMORPHOPHALLUS (Tropical Asia and Africa); ANADENDRUM (Malaya); ANCHOMANES (Tropical Africa); ANTHURIUM (West Indies, Central and Southern America); ARISAEMA (Asia, Abyssinia, America); ARISARUM (Mediterranean region); ARUM (Europe, Northern Africa, Western Asia); BIARUM (Asia Minor, Syria, Northern Africa); CALADIUM (South America, Indo-Malaya); CALLA (Europe, Northern Asia and America, Philippine Islands); COLOCASIA (Tropical Asia); CRYPTOCORYNE (Indo-Malaya); CULCASIA (Tropical Africa); DIEFFENBACHIA (Central and South America, Japan); DRACONTIUM (South America, Java); DRACUNCULUS (South Europe, Canary Islands); EPIPREMNUM (Malaya, Polynesia); HOMALOMENA (Asia, South America); LAGENANDRA (India, Ceylon); LASIA (Indo-Malaya); MONSTERA (Tropical America); MONTRICHARDIA (Tropical America); ORONTIUM (Northern America); PHILODENDRON (Tropical America); PINELLIA (China, Japan); PISTIA (Tropics and subtropics except Polynesia and Macronesia); PLESMONIUM (India); POTHOS (Indo-Malaya, Madagascar); REMUSATIA (Tropical Africa, Indo-Malaya); RHAPHIDOPHORA (Indo-Malaya); RICHARDIA (Southern Africa); SAUROMATUM (palaeotropics); SCINDAPSUS (Indo-Malaya); STYLOCHITON (Tropical and Southern Africa); SYMPLOCARPUS (Northern America); SYNANTHERIAS (Southern India, Ceylon); TYPHONIUM (Indo-Malaya); TYPHONODORUM (Madagascar); XANTHOSOMA (West Indies, Tropical America).

The medicinal and poisonous Aroids of India belong to 21 genera:—ACORUS, AGLAONEMA, ALOCASIA, AMORPHOPHALLUS, ANADENDRUM, ARISAEMA, COLOCASIA, CRYPTOCORYNE, EPIPREMNUM, HOMALOMENA, LAGENANDRA, LASIA, PISTIA, PLESMONIUM, POTHOS, REMUSATIA, RHAPHIDOPHORA, SAUROMATUM, SCINDAPSUS, SYNANTHERIAS, TYPHONIUM.

I. Spadix with or without an appendage.
  A. Leaves entire or lobed. Stamens free; ovary 1-celled; ovules orthotropous.
    1. Spadix appendaged.
      a. Marsh herbs. Ovaries spirally disposed ... LAGENANDRA.
      b. Terrestrial tuberous herbs.
        i. Leaves compressed, lobed ... ... ARISAEMA.
        ii. Leaves pedatipartite, appearing after the flowering ... ... SAUROMATUM.
        iii. Leaves ovate or lobed, coetaneous with the flowers ... ... TYPHONIUM.

2. Spadix without an appendage.
   a. Floating stemless herb; leaves obovate ... PISTIA.
   b. Submerged or aquatic herbs; leaves ovate or lanceolate ... ... CRYPTOCORYNE.

B. Tuberous herbs large. Leaves appearing without flowers, solitary, very large, 3-lobed or pinnatisect. Ovules anatropous.
1. Spadix without an appendage; neuters very large PLESMONIUM.
2. Spadix appendaged.
   a. Appendage large, naked; neuters absent ... AMORPHOPHALLUS.
   b. Appendage slender, naked; neuters below the male ... ... SYNANTHERIAS.

C. Herbs not tuberous; stem creeping, erect or subterranean. Leaves undivided, cordate or peltate.
1. Limb of spathe refracted, open; ovules parietal. REMUSATIA.
2. Limb of spathe erect.
   a. Ovules many, parietal; fruit subcapsular ... COLOCASIA.
   b. Ovules few, basal; fruit drupaceous, red ... ALOCASIA.

II. Spadix without an appendage.

A. Flowers monoecious. Perennial herbs, flowering and leafing at the same time. Spadix shorter than the spathe. Fruit enclosed in the spathe tube.
1. Spathe eventually deciduous; fruit a red drupe. AGLAONEMA.
2. Spathe persistent; fruits very small, not drupaceous or red ... ... HOMALOMENA.

B. Flowers bisexual.
1. Climbers often woody.
   a. Perianth none, or a thin semicircular limb.
      i. Drupes distinct, red ... ... ANADENDRUM.
      ii. Fruits crowded, not drupaceous or red.
         ¶ Ovary 1-celled, 1-seeded. Leaves ovate to oblong-lanceolate entire ... ... SCINDAPSUS.
         ¶¶ Ovary 1-2-celled, many-ovuled. Berries confluent. Seeds oblong ... ... RHAPHIDOPHORA.
         ¶¶¶ Ovary 1-celled, 1-seeded. Leaves entire or lobed. Seeds kidney-shaped ... ... EPIPREMNUM.
   b. Perianth lobes 6. Spadix flowering upwards. Leaves distichous ... ... POTHOS.
2. Large marsh plant with stout stem, prickly erect. Spadix, much shorter than the long spathe, flowering downwards, sessile. Leaves lobed. LASIA.
3. Stemless rhizomatous plant; rhizome aromatic. Leaves ensiform. Spathe none ... ... ACORUS.

ACORUS.

The genus consists of 2 species, inhabiting the northern temperate regions and South-Eastern Asia.

A. calamus Linn. is used medicinally in India, Malaya, Indo-China, China, Europe and South Africa; *A. gramineus* Soland in Malaya, Indo-China and China.

Midrib of leaves stout ... ... 1. *A. calamus*.
Midrib of leaves absent ... ... 2. *A. gramineus*.

1. **Acorus calamus** Linn. is found throughout India and Ceylon, in marshes, wild or cultivated, ascending the Himalaya up to 6,000 ft. in Sikkim. It occurs in all European countries except Spain, and becomes more abundant eastwards and in Southern Russia, Northern Asia Minor and Southern Siberia, China and Japan, Indo-China and Malaya. It is also met with in the

Northern United States of America, where it appears to be indigenous.

The rhizome is described in the Nighantas as pungent, bitter, heating; emetic, laxative, diuretic, carminative, anthelmintic; improves appetite, voice, throat; good for diseases of the mouth; useful in abdominal pains, inflammations, fevers, epilepsy, bronchitis, delirium, hysteria, dysentery, tumours, thirst, loss of memory, rat-bite, worms in the ear.

According to Mahommedan writers the rhizome has a very bitter sharp taste; it is laxative, expectorant, carminative, alexiteric, tonic to the brain, emmenagogue; useful in general weakness, stomatitis, toothache, inflammations, pains in the liver and the chest, kidney troubles, leucoderma.

The aromatic rhizome or rootstock is considered emetic in large doses, and stomachic and carminative in smaller doses. It is a simple useful remedy for flatulence, colic or dyspepsia, and a pleasant adjunct to tonic or purgative medicines. It is also used in remittent fevers and ague by the native doctors, and is held in high esteem as an insectifuge, especially for fleas.

The root has been employed in medicine since the time of Hippocrates. By the moderns it is successfully used in intermittent fevers, even after Cinchona bark has failed, and it is certainly a very useful addition to Cinchona. It is also a useful adjunct to bitter and stomachic infusions.

It is commonly used to allay distressing cough. A small piece of the dried rootstock kept in the mouth acts better than many cough lozenges. It produces a warm sensation in the mouth and a beneficial flow of saliva. The candied root is sold as a favourite medicine in every Indian bazaar. The Indians of the Hudson's Bay Territory also use the rootstock as a cough medicine.

The Mundas of Chota Nagpur apply the ground rhizomes to the cranial suture of feverish babies.

In Ceylon the rootstock is used in bowel complaints. It has been found extremely useful in the dysentery of children, and also in bronchitic affections. The Europeans of South Africa use it as a carminative and as a diarrhoea remedy.

The Meskwaki Indians of Wisconsin use it as a physic, and in combination with other medicines as a cure for burns, while the Menomini give it for cramps in the stomach.

The root is supposed by the Chinese to affect the heart and lungs and to be beneficial for cancer. In general it is taken as a restorative for the body and spirits.

Culpeper says: 'The spicy bitterness of the root of this plant bespeaks it as a strengthener of the stomach and head, and therefore may fitly be put into any composition of that intention. The root preserved may with good success be used by itself, and both the Germans and Turks are very fond of it, and reckon it a great preservative against infectious air, which makes them commonly eat a piece of the preserved root fasting.'

Even nowadays the candied rhizome is widely employed in Europe, and the Turks use it as a preservative against contagion.

Moodeen Sheriff experimented with the drug and wrote: 'The

rhizome is emetic, nauseant, antispasmodic, carminative, stimulant and insecticide. As an emetic it is more nauseant and depressant than Ipecacuanha, and it is therefore useful in most of the diseases in which the latter is indicated, including dysentery. It is one of the two vegetable drugs in this country which act efficiently as emetics in so small a dose as 30 grains. It should not be used in more than 35 grains, and in 40 grains its action is very violent and obstinate. It is a good remedy in asthma, to relieve which, it should be first used in pretty large or nauseant doses (15 to 20 grains) and then repeated every 2 or 3 hours in smaller or expectorant doses (10 grains) till relieved. Among other diseases which are most benefited by this drug are bronchial catarrh, hysteria, neuralgia and some forms of dyspepsia. The rhizome can also be used in the form of a tincture or an infusion.'

Later Koman gave a decoction to several patients suffering from indigestion and found it to be efficacious.

The rhizome is officinal in Austria, Germany, Holland, Hungary, Italy, Norway, Portugal, Russia, Sweden and Switzerland.

Caius and Mhaskar have found the rhizome useless for the antidotal treatment of snake bite and scorpion sting.

Afrikaans: Kalmoes—; *Annam*: Thach xuong bo, Xuong bo—; *Arabic*: Vaj, Vash—; *Assam*: Bach—; *Bengal*: Bach, Shwetbach—; *Burma*: Linhe—; *Canarese*: Baje—; *Chinese*: Che Ts'ang Pou, Choui Ts'ang Pou, Pai Ch'ang, Ts'ang P'ou—; *Cutch*: Vekhanda—; *Deccan*: Gandkilakri, Vach—; *Dutch*: Kalmus, Zwanenbrood—; *English*: Bastard Water Flag, Calamus, Cinnamon Sedge, Myrtle Grass, Myrtle Sedge, Sweet Cane, Sweet Flag, Sweet Myrtle, Sweet Root, Sweet Rush—; *French*: Acore, Acore aromatique, Acore odorant, Acoro vrai, Acori, Acrois, Galanga des marais, Roseau aromatique, Roseau odorant—; *German*: Ackermagen, Deutscher Ingber, Gewuerzkalmus, Kalmus, Karmes, Karmsen, Kaumeles, Kolmas, Magenwurz, Nagenwurz, Schiemen, Schienenzurz, Wechel, Zehrwurz—; *Gujerati*: Gandhilovaj, Godavaj, Vekhand—; *Hindi*: Bach, Ghorbach, Gorbach, Igir, Kussebbewa—; *Italian*: Acoro, Acoro aromatico, Acoro odoroso, Calamo, Calamo aromatico, Canna odorifera, Erba cannella, Erba di Venere, Erba venerea—; *Jhalawan*: Kull—; *Kashmir*: Vahi—; *Malaya*: Cheong fu, Deringu, Jeringu—; *Malayalam*: Vashampa—; *Marathi*: Vekhand—; *Menomini*: Apaxkiu utcipa, Enausapokotcikun, Weke, Wiabiskitcipa kakasapotcikun—; *Meskwaki*: Mishinepisuni—; *Mundari*: Otehajam—; *Norfolk*: Gladdon—; *Norwegian*: Kalmus—; *Pacific Coast*: Beewort, Calamus-root, Flag-root, Myrtle Flag, Sweet Cane, Sweet Cinnamon, Sweet Flag, Sweet Myrtle—; *Pampangan*: Bueng—; *Persian*: Agar, Agreturki—; *Portuguese*: Calamo aromatico, Canna cheirosa—; *Prairie Potawatomi*: Sinipisiwun—; *Punjab*: Bariboj, Wach—; *Roumanian*: Spervian—; *Sanskrit*: Bhadra, Bhutanashini, Bodhaniya, Galani, Golomi, Ikshuparni, Jalaja, Jatila, Kanga, Kshudraptri, Lomasha, Mangalya, Rakshoghni, Shadagrantha, Shataparvika, Schleshmaghni, Smarani, Tikshna, Tikshnapatra, Ugra, Ugragandha, Vacha, Vijaya—; *Sinhalese*: Wadakaha, Wadakahavasambu—; *South Africa*: Myrtle Flag, Sweet Flag, Sweet Sedge—; *Spanish*: Acoro, Acoro verdadero—; *Swedish*: Kalmus—; *Tagalog*: 'Lubigan—; *Tamil*: Vashambu—; *Telugu*: Vadaja, Vasa, Wasa—; *Urdu*: Bacha—.

2. **Acorus gramineus** Soland occurs in the Sikkim Himalaya up to 6,000 ft., and in the Khasia Hills. It extends to China and Japan.

The root has stimulant, tonic, antispasmodic properties, and is used in China as a diaphoretic, an insectifuge and an insecticide. In Cochin-China it is considered a good substitute for Sweet Flag.

Annam: Thach truong bo—; *Chinese*: Ch'ang P'u, Shih Ch'ang P'u, Shui Ch'ang P'u, Wai Ch'ang P'u—; *Malaya*: Soi cheong phoo—.

AGLAONEMA.

The genus includes about 20 species, Tropical Asiatic and African.

Aglaonema angustifolium N. E. Br. var. **undulatum** Ridl. is found in the Malay Peninsula, where the natives use a decoction of the roots for fever and dropsy.

Malay: Penggehe, Sagut, Sumpuh bulau, Sumpuh kring—.

ALOCASIA.

The genus numbers 45 species, found in Tropical Asia.

A.	Leaves broadly ovate-cordate, repand, nerves 5-6 pairs; petiole short stout	4. *A. montana.*
B.	Leaves large ovate, deeply sagittately cordate, repand; petiole stout, transversely clouded ...	2. *A. indica.*
C.	Leaves triangular-sagittate, shortly acuminate, about $\frac{1}{3}$ as broad as long; petiole long slender ...	1. *A. denudata.*
D.	Leaves broadly ovate-sagittate, repand; basal lobes rounded, connate for $\frac{1}{10}$th of their length; petiole 2-4 ft., base sometimes as thick as the arm ...	3. *A. macrorrhiza.*

1. **Alocasia denudata** Engl. (=*A. singaporensis* Linden) is common in hedges and woods all over the Malay Peninsula, extending to Lingga and Borneo.

The acrid juice is used as a poison in Kelantan.

Malay: Berar Kitan, Keladi chandek, Keladi rimau, Keladi ular.

2. **Alocasia indica** Schott is cultivated in India and in most parts of the tropics.

Sanskrit writers describe the rootstock as pungent, fragrant, cooling; useful in inflammations, leprosy, anasarca, diseases of the abdomen and spleen.

Medicinally it is said to be useful in anasarca. The flour obtained by pounding the dried stems boiled with rice flour until all the water has evaporated, is given to the patient, and no other food allowed except milk. The ash, mixed with honey, is used in cases of aphthae.

As a food taken frequently, it seems to act as a mild laxative and diuretic. In piles and habitual constipation it is useful.

The tubers chopped fine, tied in a cloth and heated, are used as a fomentation in rheumatism.

The acrid juice of the petioles is much used as a common domestic remedy on account of its styptic and astringent properties. The petiole is slightly roasted and the juice expressed. Purulent discharge from the ears in children is often stopped by a single application.

The juice of the fresh petiole has no action in the symptomatic treatment of scorpion sting (Caius and Mhaskar).

Assam: Mankachu—; *Bengali*: Manguri, Mankachu—; *Betsimisaraka*: Saombia—; *Canarese*: Manaka—; *English*: Acrid Taro, Giant Taro—; *Fiji*: Dranu, Via—; *Goa*: Curcas—; *Guam*: Piga—; *Hindi*: Mankanda, Mankandu, Mankashu, Manna—; *Malay*: Birah, Keladi sebaring—; *Marathi*: Alu—; *Mundari*: Mansaru—; *Philippines*: Biga, Elephant's Ears—; *Samoa*: Ape,

Ta'amu—; *Sanskrit*: Brihachhada, Chhatrapatra, Mahapatra, Mana, Manaka, Sthalapadma, Vistirnaparna—; *Sinhalese*: Rata-ala—; *Tagalog*: Badiang, Gabingonac—; *Visayan*: Badiang—.

3. **Alocasia macrorrhiza** Schott is found wild and cultivated throughout tropical and subtropical India. It is cultivated in the tropics generally.

The root is a mild laxative and diuretic.

It is numbered among the recorded fish-poison plants in Australia. The milky juice of the lower part of the stem is used in Queensland as an antidote to the stings of *Laportea gigas* Wedd., the Giant Nettle.

Chinese: Hai Yu—; *English*: Acrid Taro, Giant Taro—; *Easter Island*: Kape—; *Guam*: Piga—; *Hawaii*: Ape, Apü—; *New Caledonia*: Alendiete, Baouen, Diamote, Ouagan, Pera, Taro—; *Philippines*: Biga—; *Queensland*: Banganga, Bargadga, Cunjevoi, Hakkin, Nargan, Pitchu—; *Rarotonga*: Kape—; *Samoa*: Ape, Ta'amu—; *Sinhalese*: Habarala—; *Tahiti*: Ape—.

4. **Alocasia montana** Schott occurs in the Northern Circars, and has been found in Java.

The root is said to be used to poison tigers (Roxburgh).

AMORPHOPHALLUS.

This genus includes about 90 species, spread over tropical Asia and Africa.

A. rivieri Durieu is used medicinally in China; *A. dracontioides* N. E. Br., *A. flavovirens* N. E. Br., and *A. johnsonii* N. E. Br. are used in the Gold Coast

Style many times longer than the ovary ... ... 1. *A. campanulatus*.
Style absent ... ... 2. *A. prainii*.

1. **Amorphophallus campanulatus** (Roxb.) Bl. is largely cultivated throughout the plains of India and Ceylon. It occurs in open pastures, in grassy fields, and on the borders of woods all over the Malay Peninsula, whence it extends to Siam, Sumatra, Java and the Andamans.

Sanskrit writers describe the tuber as dry, acrid, pungent; increasing both appetite and taste; stomachic, constipating; useful in piles, enlargement of the spleen, tumours, asthma, bronchitis, vomiting, abdominal pain, blood diseases, elephantiasis; causes itching sensation; harmful in leprosy, leucoderma. The tuber is covered with a layer of earth, roasted in hot ashes, and administered with the addition of oil and salt. Several confections are also used. The dried tubers, peeled and cut into segments, are sold in the shops.

The corm and the seeds are used as irritants and relieve the pain of rheumatic swellings when applied externally. The corm is considered a hot carminative in the form of a pickle. The tuber contains a large quantity of farinaceous matter, mixed with acrid poisonous juice, which may be extracted by washing or heat. When fresh, it acts as an acrid stimulant and expectorant, and is used in acute rheumatism. It is supposed to have restorative powers and is in much request. It is considered serviceable in haemorrhoids.

The root is used in ophthalmia and applied to boils. It is also used as an emmenagogue.

The raw tuber, well ground, is rubbed on swellings of the extremities by the Mundas of Chota Nagpur. If eaten, it produces unbearable itching in the mouth and throat, and it may prove fatal.

Bengal: Ol—; *Bombay*: Janglisuran—; *Burma*: Wa—; *Cagayan*: Bagang—; *Cutch*: Janglisuran—; *Deccan*: Kanda—; *Hasada*: Hada, Hatuhada—; *Hindi*: Kanda, Ol, Zaminkand—; *Ilocano*: Carot, Corot—; *Jolo*: Bagong—; *Kolami*: Ol—; *Konkani*: Suma, Surna—; *Malayalam*: Karunakarang—; *Marathi*: Suran—; *Matheran*: Suran—; *Naguri*: Haluhada, Ol—: *Persian*: Zaminkand—; *Punjab*: Nagphannigand, Zamin gand—; *Sanskrit*: Arsaghna, Arshoghna, Bahukanda, Durnamari, Kanda, Kandala, Kandarha, Kandashurana, Kandi, Kandula, Kandvardhana, Kanthalla, Olla, Rutchyakanda, Sthulakandaka, Sukandi, Suvitra, Tivrakantha, Vatari—; *Tagalog*: Apon, Pungapung, Tocodlangit—; *Tamil*: Karunaikkalang, Karunaikkilhangu—; *Telugu*: Daradakandagadda, Ghemikanda, Kanda, Kandagodda, Manchikanda, Polikanda—; *Uriya*: Olna—; *Visayan*: Anto, Oroy, Pamangguilon, Pungapung, Tocodlangit—.

2. **Amorphophallus prainii** Hook. fil. is common in the forests of the Malay Peninsula, especially by the forest paths. It extends to Sumatra.

The acrid juice is used as a poison by the Malays. Mixed with *Antiaris* latex it is one of the Sakai dart poisons.

Malay: Sikir, Loki, Lokie—; *Sakai*: Begung—.

ANADENDRUM.

The genus consists of 4 or 5 Malayan species.

Anadendrum montanum Schott var. **longirostris** Hook. fil. is common in the Malay Peninsula, where it is found in the low country growing low down on tree trunks in thick forest. It extends to Tenasserim, Java and Borneo.

The roots are used medicinally by the Malays.

Malay: Akar asam tebing darat, Akar merayan, Akar sugunja, Akar cumbung—.

ARISAEMA.

The genus includes 105 species, inhabiting Asia, Abyssinia and America.

A. japonicum Schott, *A. ringens* Schott, *A. thunbergii* Blume are used medicinally in China; *A. thunbergii* Blume is used in Indo-China; *A. triphyllum* Schott in North America.

A. Leaves trifoliolate. Limb of spathe ovate-lanceolate, incurved caudate-acuminate ... ... 2. *A. speciosum*.
B. Leaves pedatisect. Limb of spathe incurved, broadly cymbiform, acuminate ... ... 3. *A. tortuosum*.
C. Leaflets whorled. Limb of spathe suberect, cymbiform, long caudate-acuminate ... ... 1. *A. leschenaultii*.

1. **Arisaema leschenaultii** Blume is found in the Nilgiris, Travancore and Ceylon.

The roots are used medicinally by the Sinhalese.

Sinhalese: Walkidaran—.

2. **Arisaema speciosum** Mart. occurs in the temperate Himalaya from Kumaon to Sikkim and Bhutan.

In Hazara, the root is stated to be poisonous; in Chumba, it is applied pounded to snake-bites. In Kulu, where the tuber is given to sheep for colic, the fruit is said to have deleterious effects on the mouth when eaten by children.

The root is not an antidote to snake venom (Mhaskar and Caius).

Punjab: Kiralu, Kirikikukri, Sampkikhumb—; *Sinhalese*: Walkidaran—.

3. **Arisaema tortuosum** Schott is found in the Sikkim Himalaya, Manipur, Bengal and the Western Peninsula.

It is stated to have poisonous qualities. In Kulu, the seeds are said to be given with salt for colic in sheep. The roots are used to kill the worms which infest cattle in the rains.

Kharwar: Dain—; *Nepal*: Birbanka—; *Punjab*: Don, Gurin, Jangosh, Kirakal, Kirkichalu—; *Santali*: Tuyajondra—.

COLOCASIA.

The genus numbers 7 species, natives of tropical Asia, one of which is cultivated or naturalised in many subtropical countries.

Colocasia esculenta (Linn.) Schott (=*C. antiquorum* Schott) is found wild and cultivated throughout the hotter parts of India and Ceylon. It is cultivated in all hot countries.

The pressed juice of the petioles is styptic, and may be used to arrest arterial haemorrhage. It is sometimes used in earache and otorrhoea, and also as an external stimulant and rubefacient. The juice expressed from the leaf-stalks is used with salt as an absorbent in cases of inflamed glands and buboes. The juice of the corm is used in cases of alopaecia. Internally, it acts as a laxative, and is used in cases of piles and congestion of the portal system, also as an antidote to the stings of wasps and other insects. The tubers chopped fine, tied in a cloth and heated, are used as a fomentation in rheumatism.

The corm is used by the Mundas as a remedy for body-ache.

Caius and Mhaskar have demonstrated that the juice of the corm is without action in the symptomatic treatment of scorpion sting.

Accra: Mankani—; *Angami Naga*: Dzu, Kirth—; *Annam*: Khoai nuoc—; *Arabic*: Kalkas, Kur, Gulgas—; *Ashanti*: Kooko—; *Bengal*: Ashukuchu, Bunkuchu, Charkuchu, Guri, Kachhu, Kachu, Kalokuchu, Kuchu—; *Bombay*: Kachualu, Terem—; *Brazil*: Tayoba de São Thome—; *Burma*: Mahuyapein—; *Cagayan*: Gabi—; *Canarese*: Kesavedantu, Keshavanagadde, Shamegadde—; *Carolines*: Chaua—; *Chinese*: Yu, Yu T'eou—; *Cuba*: Malanga—; *Deccan*: Arvi, Chamkurekagaddah—; *Egypt*: Culcas, Kolkus, Golkas—; *English*: China Potato, Coco, Coco Yam, Eddoes, Egyptian Arum, Kopeh, Scratch Coco, Taro—; *Fanti*: Kooko, Mankani—; *Fiji*: Taro—; *French*: Chou caraïbe, Colocase, Colocasie, Gingembre d'Egypte—; *Fufulde*: Tandauje—; *Gold Coast*: Coco Yam—; *Guam*: Sune, Suni—; *Hausa*: Gwaza, Kamu, Koko yam—; *Hindi*: Arvi, Arwi, Ashukachu, Auri, Avois, Gagli, Ghoya, Ghuian, Ghuiya, Ghuya, Ghwiya, Gorikachu, Kachalu, Kachalu arvi, Kachu—; *Japanese*: Akaimo, Imo, Midsu-imo, To-no-imo—; *Kangra*: Arbi, Gandiali, Kachalu—; *Khasi*; Ka shiriw—; *Kolami*; Pichgisaru—; *Konkani*; Allum—; *Languedoc*;

Farrao—; *Le Reunion*: Songe, Songe du pays, Sonze de Chine, Sonze de Maurice, Sonze noire, Sonze du pays—; *Lynngam*: Chew—; *Madagascar*: Songe—; *Malay*: Keladi china, Keladi hudang, Keladi telor—; *Malayalam*: Chempakizhanna, Kaladi—; *Malta*: Coco, Cocoaroot, Colocasia, Aro di Egitto, Ghorghas—; *Marathi*: Alu—; *Marquesas*: Tao—; *Mexico*: Quequexquic, Quequeste—; *Mundari*: Birsaru, Kucusaru—; *New Caledonia*: Barenik, Coboné, Dadi, Diali, Diamboilaté, Doboua, Jabouak, Jalapé, Kandié, Kandiéren, Kavé, Kiamoan, Néré, Ouagapé, Ouaoua, Oumon, Ounégaté, Paricraouté, Pobo, Tanmaouté, Taro, Tianaboué, Tiaouné, Tiréné—; *Pampangan*: Gabi, Gandus, Gavay, Lagway—; *Panama*: Oto—; *Philippines*: Badiang, Dagmay—; *Polynesia*: Kalo, Talo, Taro—; *Porto-Rico*: Yautia malanga—; *Punjab*: Alu, Gagli, Ghuyan, Givian, Kachalu, Kachalu arbi, Kasauri, Rab—; *Sadani*: Bonsaru—; *Sanskrit*: Kachchi, Kachu, Kachwi—; *Sema Naga*: A'i—; *Singapore*: Gahala, Tadala—; *Sinhalese*: Gahala, Habarala, Kandala, Tadala—; *South Carolina*: Carib Cabbage—; *Spanish*: Aro de Egipto, Name de Canarias, Name de Egipto, Yame de Canarias, Yame de Egipto—; *Sunda*: Talas, Taloes—; *Tagalog*: Gabi, Gabingpola, Gabingsilangan, Gabynasiboyas, Gabynasiniboyas, Gandus, Gavay, Lagvay—; *Tamil*: Shamakkilangu, Shemakkalenga—; *Telugu*: Chamadumpa, Chamagadda, Chamakura, Chamakuru, Chema, Shamathumpa—; *Twi*: Kooko—; *Uriya*: Saru—; *Visayan*: Abalong, Apipi, Biga, Dagmayngaapipi, Dagmayngabolilao, Dafimayngainitlog, Dagmayngaquinson, Dagmayngatapol, Gabi, Gabingmorada, Ganyngaguinatos, Galiang, Gandus, Gavay, Guinatos, Lagway, Quimpoy—; *West Africa*: Coco Yam—; *West Indies*: Coco, Eddo, Eddoes, Dasheen, Tania, Tanier—; *Yoruba*: Koko—.

CRYPTOCORYNE.

The genus consists of 40 Indo-Malayan species.

Cryptocoryne spiralis Fisch occurs from Khandesh to North Kanara, Calicut, Coromandel, Ceylon and Bengal.

It is a well-known drug in Ceylon, where it is employed by the native doctors in decoctions, in combination with other drugs as a remedy for infantile vomiting and cough, and in the case of adults for abdominal complaints and fever.

The plant is considered a substitute for Ipecacuanha. It contains neither emetine nor cephaeline (Chopra).

Tamil: Nattativadayam—; *Telugu*: Nattativasa—.

EPIPREMNUM.

The genus consists of 14 Malayan and Polynesian species.

Leaves very large entire ... ... 1. *E. giganteum*.
Leaves deeply cut into acuminate lobes ... ... 2. *E. mirabile*.

1. **Epipremnum giganteum** Schott is common on rocks and trees in the Malay Peninsula, whence it spreads over to Siam and Cochin-China.

The powdered spadices are used to mix with the dart poison by some of the Sakai tribes.

Malay: Rengut, Ringhut, Ringut—.

2. **Epipremnum mirabile** Schott is found in Martaban, Tenasserim, China, the whole of Malaya, Australia and Polynesia.

The plant is said to have anthelmintic properties,

HOMALOMENA.

The genus includes 30 Indo-Malayan and tropical American species.

A. Spathes red ... ... 3. *H. rubescens.*
B. Spathes green.
 a. Spathe 3 in. oblong, acute at base and top. Basal lobes of leaves divaricate ... ... 1. *H. aromatica.*
 b. Spathe 1-1¼ in. linear-oblong. Base of leaves rounded ... ... 2. *H. ovata.*

1. **Homalomena aromatica** Schott is found in Assam and Chittagong.

The large rhizome, which is invested with the old withered leaf-scales, bears numerous white long rootlets issuing from its surface, and is said to be held in high estimation by the natives as an aromatic stimulant.

Bengal: Kuchugundubi—.

2. **Homalomena ovata** Hook. fil. is found in Bukit, Timah and Singapore, whence it extends to Borneo.

The plant is used medicinally by the Malays.

Malay: Asam tikus, Kelamoyiang, Kumayan, Kumayan Jantan—.

3. **Homalomena rubescens** Kunth. is found in the Sikkim Himalaya, the Khasia Hills and Chittagong, extending to Java.

The plant is used as a poison by the Malays. It enters into the composition of 'ipoh' and is thrown into rivers to poison the water.

LAGENANDRA.

The genus consists of 5 species, spread over India and Ceylon.

Lagenandra ovata (Linn.) Thwaites (=*L. toxicaria* Dalz.) is a marsh plant occurring from the Konkan to North Kanara, Mysore, Coorg, Cochin, Travancore, and Ceylon where it is common in the Central Provinces.

The plant is said to have insecticidal properties. It is considered very poisonous in Southern India.

Southern Konkan: Vatsanabh—.

LASIA.

The genus consists of 3 Indo-Malayan species.

Lasia aculeata Lour. (=*L. heterophylla* Schott) is a stout spinous marsh plant occurring from tropical Sikkim Himalaya, Assam, Bengal and Burma, southwards to Ceylon and the Malay Peninsula, and extending to the Malay Islands and China.

The root is highly esteemed by the Santals as a remedy for affections of the throat (Campbell).

Among the Mundas the petioles, ground and mixed with water, are given to drink to cattle affected with throat disease.

The leaves and roots are a common remedy for piles in Ceylon.

Bengal: Kantakachu—; *Burma*: Zayap—; *Malay*: Bekil, Gli-gli—; *Mundari*: Janumsaru—; *Santal*: Rantasaru—; *Sinhalese*: Kohilla, Mahakshilla—; *Telugu*: Kantakachoramu, Mulasari—.

PISTIA.

The only species of this genus, **Pistia stratiotes** Linn., is a stemless stoloniferous floating herb found in still sweet water throughout India and Ceylon. It occurs in the tropics and subtropics, with the exception of Polynesia and Macronesia. In the Malay Peninsula it is commonly cultivated by the Chinese to feed pigs on.

The Ayurvedists consider the plant to be cooling and laxative; useful in diseases of the blood and tuberculous glands.

The Yunanists describe the root as bitter, diuretic, good for wounds, inflammations and burns.

The plant is cooling and demulcent, and is given in dysuria. It is said to destroy bugs most effectively.

The root is laxative and emollient.

The leaves are made into poultices and applied to haemorrhoids. Mixed with rice and cocoanut milk they are given in dysentery, and with rose-water and sugar in cough and asthma.

The ashes are applied to ringworm of the scalp.

The Mundas use the juice of the plant medicinally in ear complaints.

The whole plant has long been employed in ancient Chinese prescriptions. It is applied to boils, syphilitic eruptions and in many skin complaints.

A decoction of the leaves is used in La Reunion as a diuretic and prescribed in diseases of the urinary tract. Made into pills the leaves are used in syphilis.

The natives of the Gambia use the plant as an anodyne eye-wash: the leaves are soaked in cold water. If the plant comes in contact with the body it produces a sharp stinging sensation, like a nettle sting, and the pain continues for some minutes.

An oil was prepared by boiling the juice of the leaves in cocoanut oil and used externally in chronic skin diseases. Relief was obtained in a number of cases (Koman).

Arabic: Satar-al-tayutas—; *Ashanti*: Ntanoa, Todia—; *Awuna*: Aflo—; *Bengal*: Pana, Takapana—; *Bombay*: Gondala, Prishni—; *Cagayan*: Aluluan—; *Canarese*: Antaragange—; *Chinese*: Fou Ping—; *Deccan*: Anterghunga—; *Egypt*: Hay-adem-el-ma—; *English*: Water Lettuce, Water Soldier—; *Ewe*: Aflo—; *Fanti*: Ntangtangaba—; *Ga*: Taitraimantai—; *Gambia*: Eyepity—; *Gujerati*: Jalakumbhi—; *Hausa*: Kainuwa—; *Hindi*: Jalkhumbi, Takapana, Unterungha—; *Hova*: Tsinkafonkafona—; *Ilocano*: Loloan—; *La Reunion*: Pensée d'eau, Pourpier de Madagascar—; *Madagascar*: Azafo, Hazafo ; *Malaya*: Fow phing, Kambiang, Kiambang—; *Malayalam*: Koddapail—; *Marathi*: Gondala, Jalamandri, Prasni—; *Mundari*: Jhalkumbhi—; *Sadani*: Jhalkumbhi—; *Sanskrit*: Akashamuli, Ashakumbhi, Daladhaka, Jalavalkala, Khali, Khamulika, Kumbhika, Kumuda, Kutrina, Paniyaprishthaja, Parni, Prashni, Shvetaparna, Untareitamara, Varimuli, Variparni—; *Sinhalese*: Atiudayan, Dayaparandella—; *Tagalog*: Quiapo—; *Tamil*: Agasatamarai—; *Telugu*: Akasatamara, Antaratamara, Koddapail, Nirubudiki—; *Tongking*: Beo cai—; *Twi*: Ntaya—; *Urdu*: Jalakumbhi—; *Uriya*: Baujhanjhe—; *Uruguay*: Lechuguita de agua, Ropollito de agua—; *Visayan*: Cayapo, Louanlouan,

PLESMONIUM.

The genus is represented by **P. margaritiferum** Schott, a tuberous herb flowering before leafing and found in Bengal.

The country-people in Goa use the crushed seed to cure toothache; a small quantity is placed in the hollow tooth and covered with cotton; it rapidly benumbs the nerve; they also use it as an external application to bruises on account of its benumbing effect.

Among the Mundas the raw tuber, well ground, is rubbed on swellings of the extremities. The tubers are eaten. If they are not boiled in several successive kettlefuls of water, or together with tamarind leaves or tamarind pulp, they keep a noxious acidity which affects the throat and mouth with unbearable itching. It is said that the itching has sometimes proved fatal. Acid drinks are a counterpoison. The tamarind added ought, therefore, not to be spared.

Biru: Bonggajorena—; *Goa*: Azomut, Uzomut—; *Hasada*: Birhada, Hada—; *Naguri*: Hada, Tonanghada—; *Oudh*: Kharhar—.

POTHOS.

The genus includes 60 species, inhabitants of tropical Asia, Australia, Polynesia and Madagascar.

Bracts large concave thickly coriaceous ... ... 1. *P. cathcartii.*
Bracts very small ... ... 2. *P. scandens.*

1. **Pothos cathcartii** Schott occurs in the tropical Himalaya from Kumaon to Bhutan, Assam, the Khasia Hills, Manipur and Burma.

In Lakhimpur the leaves, fried in ghee, are eaten to cure various pains.

Lakhimpur: Hathidenkiya—.

2. **Pothos scandens** Linn. is found throughout India, on walls and tree trunks, from Bengal eastwards to Burma, and southwards to Singapore, and from the Konkan to Ceylon, the Andamans and the Nicobar Islands, extending to the Malay Islands and China.

The bruised stem and leaves are mixed with ox-urine and applied to the wounds in snake-bite; and an aqueous extract of the fresh stem and leaves is given internally (Roberts).

The stem and leaves have no effect in the treatment of snake-bite, whether taken internally or applied externally (Mhaskar and Caius).

Badaga: Arkaburu—; *Canarese*: Adikabiluballi—; *Malayalam*: Anapparuva—; *Siamese*: T'kep—; *Sinhalese*: Potawel—.

var. **cognatus** Engler, known in Malaya as 'Juloh-juloh', is used medicinally by the Malays. The powdered leaves are applied to the body to cure small-pox; the stem, cut up with camphor, is smoked like tobacco for asthma.

REMUSATIA.

The genus consists of 2 species, inhabiting tropical Asia and Africa.

Remusatia vivipara Schott is found in the subtropical Himalaya from Kumaon to Sikkim, the Khasia Hills and Burma, Chota Nagpur, the Western Ghats, Mysore and Ceylon. It extends to Cochin-China, Java and tropical Africa.

The root is made into an ointment with turmeric and used as a remedy for itch; and the juice with cow's urine is considered to be alexipharmic (Rheede).

Marathi: Rukhalu—; *Matheran*: Rokhalu—.

RHAPHIDOPHORA.

The genus includes 60 Indo-Malayan species.

Rhaphidophora pertusa Schott (=*Scindapsus pertusus* Schott) is a tall climber found in the Deccan Peninsula, the Coromandel Coast, Malabar, Ceylon and the Malay Islands.

The juice of the plant with black pepper is given to people who are bitten by the Russell's Viper. The juice, with that of the roots of *Croton oblongifolium*, and of the fruit of *Momordica charantia* is also applied to the bitten part.

This treatment is ineffective (Mhaskar and Caius).

Jolo: Mamao—; *Malayalam*: Anachukiri, Ilattimaravala—; *Marathi*: Ganeshkanda—; *Tagalog*: Pacpaclauin—; *Tamil*: Ilattimaravalai—; *Telugu*: Enuganalleru—; *Visayan*: Dayia—.

SAUROMATUM.

The genus consists of 4 palaeotropical species.

Sauromatum guttatum Schott occurs in the Punjab, the Gangetic Plain, the Western Himalaya, Chota Nagpur, the Bombay Presidency and Burma; extending to Sumatra.

The tubers are used as a stimulating poultice.

Bombay: Loth—; *Central Provinces*: Bhasamkand—.

SCINDAPSUS.

The genus consists of 20 Indo-Malayan species.

Spathe green without yellow within, 4-6 in. oblong subcylindrical beaked. Spadix greenish-yellow ... 2. *S. officinalis*.
Spathe white or pale yellow, 2-2½ in. boat-shaped acuminate narrow long-beaked. Spadix white ... 1. *S. hederacea*.

1. **Scindapsus hederacea** Schott is found throughout the Malay Peninsula, climbing on rocks and trees in the low country.

A decoction of the stem is used by the Malays for rheumatism.

Malay: Akar lubang alah, Akar ular—.

2. **Scindapsus officinalis** Schott occurs in the tropical Himalaya from Sikkim eastwards, Bengal, Chittagong, Burma and the Andamans.

The ripe fruit is described in the Nighantas as pungent, sharp, heating, anthelmintic, aphrodisiac, galactagogue; sharpening hearing and appetite; regulating the bowels; useful in dysentery, asthma and troubles of the throat.

Mahommedan writers speak of the fruit as aphrodisiac, cardiotonic; useful in ozoena and bronchitis.

The dried fruit is commonly used as a stimulant, diaphoretic, and anthelmintic.

Among the Santals the fruit is applied externally for rheumatism.

A decoction of the sliced fruit was tried in cases of asthma and found to act as an expectorant; it did not diminish the severity of the fit (Koman).

Bengal: Gajapipal, Gajapipul—; *Canarese*: Doddahippali, Gajahippali—; *Deccan*: Hattipipli—; *Dehra Dun*: Poriabel—; *Gujerati*: Mottopiper—; *Hindi*: Braipipli, Gaja pipal, Gajapipli, Maidah, Pippaljhhanca—; *Malayalam*: Anattippali, Attittippali—; *Marathi*: Thorapimpli—; *North-West Provinces*: Gajpippali, Hath, Ungliya—; *Sanskrit*: Chavyaphala, Chhidravaidehi, Dirghagranthi, Gajakrishna, Gajapippali, Gajavha, Ibhakana, Iboshana, Kapivalli, Karipippali, Kolavalli, Kunjarapippali, Shreyasi, Tejasi, Vartuli, Vashira—; *Santali*: Darejhapak, Dharejhapak—; *Tamil*: Anaittippili—; *Telugu*: Enugatippali, Gajapippali—; *Urdu*: Gajapippali—; *Uriya*: Girudhuni, Gojopippoli, Odisimo—.

SYNANTHERIAS.

The genus is limited to 1 species, **S. sylvatica** Schott, found in the Deccan Peninsula, from the Northern Circars to the Konkan, and Ceylon.

The plant is pungent, anthelmintic, heating; improves taste; useful in tumours, pains, piles; causes biliousness.

The country people use the crushed seed to cure toothache. A small quantity is placed in the hollow tooth and covered with cotton; it rapidly benumbs the nerve; they also use it as an external application to bruises on account of its benumbing effect. In the Konkan, the seeds rubbed into a paste with water are applied repeatedly to remove glandular enlargements. The taste of the fruit is intensely acrid; after a few seconds it causes a most painful burning of the tongue and lips, which lasts, for a long time, causing much salivation and subsequent numbness.

Goa: Uzomut—; *Marathi*: Wajramuta—; *Sanskrit*: Aranyasurana, Chitrakandaka, Shvetasurana, Sitasurana, Surendra, Vanja, Vanakanda, Vanaolla, Vanya—.

TYPHONIUM.

The genus includes 25 Indo-Malayan species.

Typhonium trilobatum (Linn.) Schott is found in the Eastern and Western Peninsula and Ceylon, in Bengal, Assam, Chittagong, Burma, the Malay Peninsula. It extends to Siam, Cambodia, Tongking, Borneo and Java.

The roots are exceedingly acrid, and used in poultices; and also applied externally to the bite of venomous snakes, at the same time given inwardly about the size of a field bean. It is certainly a most powerful stimulant

The acrid principle is very volatile; and by the application of heat, or by simple drying, the root becomes innocuous or even a wholesome article of diet As an article of food, it relaxes the bowels and thereby relieves haemorrhoids,

The tubers eaten with bananas cure stomach complaints.

Mhaskar and Caius have shown experimentally that the root is not an antidote to snake venom.

Bengal: Ghetkochu—; *Hasada*: **Cakad**, Nirbisi, Niribisi—; *Malayalam*: Chena—; *Naguri*: Najompicki—; *Sinhalese*: Panuala—; *Tamil*: Karkarunaik-kilhangu, Karunaikkilhanga—; *Telugu*: Duradakandagadda, Kandagadde—.

THE MEDICINAL SPIDERWORTS OF INDIA.

BY

J. F. CAIUS, S.J., F.L.S.

The COMMELINACEAE are perennial herbs generally distributed in the Tropics and Subtropics, a few in China and Japan, the Southern United States and in Australia. The family includes 30 genera with about 400 species.

Many species possess an abundant mucilage, which is alimentary after being cooked; and the tuberous rhizomes of some contain, besides mucilage, starch, which adds to their nutritive qualities. The species credited with medicinal properties belong to 8 genera:—ANEILEMA (Tropics); COMMELINA (Tropics and Subtropics); CYANOTIS (Tropical Asia and Africa); FLOSCOPA (Tropics); PALISOTA (Tropical Africa); POLLIA (Old World Tropics and Subtropics); TINANTIA (Mexico to Brazil); TRADESCANTIA (Tropical and North America).

The medicinal Spiderworts of India belong to 4 genera:—ANEILEMA, COMMELINA, CYANOTIS, FLOSCOPA.

I. Capsule loculicidal. Stamens 3, perfect; with 1-3 staminodes.
   1. Cymes solitary, included in a spathe. Ovary 3-celled; cells, 1-2 ovuled ... ... COMMELINA.
   2. Cymes naked, panicled, rarely in a spathe ... ANEILEMA.

II. Capsule loculicidal. Stamens 6, perfect.
   1. Cymes from imbricating bracts, scorpioid, or 1-2-few-flowered. Corolla tubular below ... ... CYANOTIS.
   2. Flowers panicled. Stem erect ... ... FLOSCOPA.

ANEILEMA.

This genus numbers 85 species, tropical and subtropical, chiefly Asiatic.

In the Gold Coast Colony the leaves of *A. beninense* Kunth. and *A. ovato oblongum* P. Beauv. are used in enemas to cure constipation.

Four Indian species are used medicinally either in India or in Malaya.

I. Cells of ovary 3-8-ovuled; cells of capsule 1-3- or more-celled.
   1. Stout plant with large leaves; seeds 3 or 4 in a cell, smooth ... ... 2. *A. lineolatum*.
   2. Tufted herb; seeds 5 or 6 in a cell, superposed, sharply 3-gonous ... ... 4. *A. scapiflorum*.

II. Cells of ovary 2-ovuled; cells of capsule 2-seeded. Small herb, decumbent ... ... 3. *A. nudiflorum*.

III. Cells of ovary 1-ovuled; cells of capsule 1-seeded. Stout herb ... ... 1. *A. conspicuum*.

1. **Aneilema conspicuum** Kunth. is found in the Patkai Hills of Assam, whence it extends to Burma and Siam, Penang. Malacca, Sumatra and Java.

The plant is used medicinally by the Malays, and is said to be a powerful emmenagogue.

Malay: Poko tampoh kalin—.

2. **Aneilema lineolatum** Kunth. occurs throughout Tropical India; in the Western Ghats from the Konkan to Travancore; and from Sikkim, Bhotan and the Khasia Hills up to 2-4,000 ft., to Burma, Malaya and the Malay Islands.

The plant is used by the Malays as an abortifacient.

Malay: Poko mati hidup—.

3. **Aneilema nudiflorum** Br. is to be found throughout India; from the North-West Himalaya ascending to 6,000 ft. eastwards and southwards to the Khasia Hills, Burma, the Deccan, Travancore and Ceylon. It is a common weed in waste ground and dry sandy places throughout the Malay Peninsula. It is also met with in China.

The herbage cooked in oil is employed in the treatment of leprosy and ringworm.

Malay: Rumput kurunit, Rumput lipah lumbu, Rumput sarang tupai, Rumput tapah burong—.

4. **Aneilema scapiflorum** Wight is a native of the Himalaya. It is spread from the Upper Gangetic plain to Bhotan, Travancore, Tenasserim, Ceylon and the Shan Hills.

The root is said to have astringent and tonic properties. It is considered useful in headache, giddiness, fever, jaundice and deafness. It is also regarded as an antidote to poisons, and as a cure for snake-bite.

The root-bark dried in the shade is said to have been employed with benefit in asthma. It is commonly used in colic, piles and infantile convulsions. It is a remedy of great repute for impotence and spermatorrhoea and also for incontinence of urine.

The dried powder is much used in combination. Mixed with sugar it is employed as an aphrodisiac. With the juice of the *tulsi* leaves it is given for pains in the kidneys, and is one of the chief remedies used by the Hakims in spermatorrhoea. With long pepper it is prescribed in bites of mad dogs, both internally and externally.

Mhaskar and Caius have shown experimentally that the drug is not an antidote to snake venom.

Bengal: Kureh—; *Gujerati*: Sismulia—; *Hindi*: Siyahmusli—; *Persian*: Musli-e-riyah—.

COMMELINA.

This genus includes 115 tropical and subtropical species.

Many species are used medicinally in various parts of the world:—*C. communis* Linn. in China; *C. pallida* Willd. and

C. tuberosa Linn. in Mexico; *C. deficiens* Herbert in Brazil; *C. barbata* Lam. and *C. benghalensis* Linn. in La Reunion; *C. madagascarica* C. B. Clarke in Madagascar; *C. africana* Linn. and *C. benghalensis* Linn. in South Africa; *C. nudiflora* in the Gold Coast.

The species used therapeutically in India number five.

I. Capsule 3-celled; two cells with 2, and one with 1 seed.
  1. Leaves elliptic to ovate-lanceolate, 2 to 3 in. long, glabrous; seeds cylindric reticulate, black ... 2. *C. nudiflora.*
  2. Leaves linear-lanceolate, 3 to 6 in. long, nearly glabrous; seeds globose, smooth, truncate, black powdered with white ... ... 4. *C. salicifolia.*
  3. Leaves elliptic or ovate, blunt, 1 to 3 in. long, pubescent; seeds closely pitted ... ... 1. *C. benghalensis.*

II. Capsule 3-celled, subequally 3-valved.
  Leaves lanceolate oblique acuminate, many-nerved, 6 in. long; seeds ellipsoid compressed, lead-coloured, margins often marbled ... ... 3. *C. obliqua.*

III. Capsule 2-celled.
  Leaves ovate-lanceolate acuminate, 3 in. long, scabrid pubescent; seeds ellipsoid rugose, straw-coloured or brown ... ... 5. *C. suffruticosa.*

1. **Commelina benghalensis** Linn. occurs in the peninsula of India generally, and is fairly common everywhere. It is also found in Ceylon, Burma, the Malay Peninsula, China and Tropical Africa.

The plant is described by Ayurveda as bitter, useful in leprosy, and a good regulator of the nervous system.

In La Reunion the plant is reputed emollient and mucilaginous. It is much used in the form of a decoction as a drink, a lotion or a bath.

The Sutos prepare from it a medicine for treating barrenness in women.

Bengal: Kachradam, Kanchara, Kanchura, Kanshira, Kanuraka—; *Canarese*: Hittagani—; *English*: Day Flower, Dew Flower—; *Gujerati*: Mhotunshishmuliyun—; *Hindi*: Kanchara, Kanuraka—; *La Reunion*: Grosse herbe d'eau, Grosse trainasse—; *Malay*: Rumput mayiam—; *Marathi*: Kena—; *Pampangan*: Biasbias—; *Philippines*: Alibangon, Anagalide azul, Uligbonggon—; *Punjab*: Chura, Kanna, Khanna—; *Sanskrit*: Kanchata: Marishajalaja, Paniya, Tanduliya—; *Santal*: Kana arak—; *Sind*: Chura, Kanna, Khanna—; *Sinhalese*: Deyamainaireya, Diyameneriya—; *Spanish*: Anagalide azul—; *Suto*: Khotswana—; *Tagalog*: Alicbanggon—; *Tamil*: Kanangakarai—; *Telugu*: Nirukassuvu, Vennadeviruka—; *Twi*: Onyame bewu na mahu—; *Visayan*: Cabilao, Sabilao, Sabilaonggalabaan—.

2. **Commelina nudiflora** Linn. is very common in waste land everywhere. It is frequent in Bengal and extends to Burma, the Malay Peninsula and Ceylon. It is distributed to most of the tropical and subtropical countries: Africa, Madagascar, the Seychelles, Mauritius, the Sandwich Islands, Australia, etc.

The leaves are used by the Malays for poulticing sores.

The natives of the Gold Coast Colony use the plant to cure a disease called 'okwaha'—a swelling in the groin—producing very itchy spots and sores and swellings. The leaves are pounded, mixed with the seeds of *Leea guineensis* G. Don. and those of long pepper. The mixture is then put in a plantain leaf, which has first been warmed to prevent it from splitting too easily, applied to the affected parts, bound, and kept there for three days. The application first affords relief; and when it is removed on the fourth day it is found that the swelling bursts.

Ashanti: Onyame bewu na mawu—; *Awuna*: Agbenokui nokui, Agbormaku maku—; *English*: Day Flower, Dew Flower—; *Ewe*: Agbenkui nokui, Agbormaku maku—; *Fanti*: Nyame bewu ansang na mewu—; *Ga*: To lilairy—; *Gujerati*: Shishmuli, Shishmuliyun—; *Hausa*: Balasa, Balasaya, Kunungura—; *Hindi*: Kanshura—; *Malay*: Pulau aur, Rumput kukupu, Tapak eti—; *Marathi*: Kina, Velichibhaji—; *Nzima*: Nyamele wua ngwosu—; *Philippines*: Aligbangon—; *Sanskrit*: Katsapriya, Koshapushpi—; *Spanish*: Anagalide azul—; *Twi*: Onyame bewu na mawu—.

3. **Commelina obliqua** Ham. is common over the low moist parts of India, flowering during the rainy season chiefly. It also occurs on the lower Himalaya, ascending up to 7,000 ft. in altitude. It is distributed to Ceylon and Burma, and hence to the Malay Peninsula where it is mostly found in the shady spots of the forests.

The root is a refrigerant and laxative, useful in strangury and costiveness. It is used in vertigo, fevers and bilious affections, and as an antidote to snake-bites.

Mhaskar and Caius have demonstrated that the root is not an antidote to snake venom.

Bengal: Jatakanchura, Jatakanshira—; *Bijnor*: Kana, Korna—; *Hindi*: Kana, Kanjuna—; *Kumaon*: Kanjura—.

4. **Commelina salicifolia** Roxb. is common in wet places in the peninsula of India, especially in Bengal, Coromandel and Bombay. It is distributed to Burma and to the Malay Peninsula where it is met with in abandoned rice-fields. It also occurs in Java.

The plant is used in dysentery and insanity.

Bengal: Languli, Panikanchira—; *Hasada*: Madtonggaara, Madtonggata-sad—; *Hindi*: Jalpipari—; *Sanskrit*: Jalapippali, Languli, Langulu—; *Santali*: Bir kana arak'—.

5. **Commelina suffruticosa** Bl. is a native of Bengal. It is found in Tropical India from Nepal, Sikkim and Bengal to Central India and the Malay Peninsula. It is distributed to the Malay Islands.

The Santals apply the root to sores.

Santali: Dare orsa—.

CYANOTIS.

This genus consists of 35 palaeotropical species, two of which are used medicinally in India.

1. Cymes enclosed in biseriate falcate imbricating bracteoles ... ... 2. *C. tuberosa.*
2. Cymes from the leaf sheaths axillary ... ... 1. *C. axillaris.*

1. **Cyanotis axillaris** Schult. f. is a herbaceous annual, met with in many parts of India, in the plains from the Upper Gangetic valley to Assam and southwards to Ceylon. It may be found in rice-fields in some parts of Malaya. It is distributed to Siam, China and Australia.

The Indians occasionally drink a decoction of the plant for dropsy.

On the Malabar Coast it is considered as a useful remedy in tympanitis.

In Behar it is used as an external application in cases of ascites, especially when mixed with a little oil.

Bombay: Itsaka—; *Hindi*: Baghanulla, Soltraj—; *Jolo*: Hauli—; *Tamil*: Nirpulli—; *Telugu*: Golagandi—.

2. **Cyanotis tuberosa** Schult. f. is found in Chota Nagpore, the Deccan Peninsula—on the west side from the Konkan to Travancore—and in Ceylon.

The root is given in long-continued fevers, and also for worms in cattle.

Santali: Hodojerengarak, Meromchunchi—.

Floscopa.

The genus consists of 15 tropical and subtropical species.

Floscopa scandens Lour. occurs throughout tropical India, in swamps from Eastern Nepal, Sikkim and the Khasia Hills to Travancore, Ceylon and Malacca, where it is common in muddy ditches. It is distributed to Eastern Asia and tropical Australia.

In Lakhimpur (Assam) the juice of the stem is put into sore eyes.

Lakhimpur: Karahimlu—; *Malay*: Rumput kumpai tikus, Rumput tapak eti—.

THE MEDICINAL AND POISONOUS CAMPIONS OF INDIA.

BY

J. F. CAIUS, S.J., F.L.S.

The CARYOPHYLLACEAE are herbs, rarely undershrubs. They mostly inhabit the extra-tropical regions of the northern hemisphere, extending to the Arctic regions and to the tops of the highest Alps. They are rarer in the southern hemisphere, and still more so in the tropics, where they are almost confined to the mountains. They form 80 genera with about 1,300 cosmopolitan species.

The medicinal and poisonous campions and pinks of the world belong to 23 genera:—AGROSTEMMA (Mediterranean region); ALSINE (northern hemisphere); ARENARIA (northern temperate regions); BUFONIA (Mediterranean region); CERASTIUM (northern temperate regions); CUCUBALUS (northern temperate regions); DIANTHUS (Europe, Asia, Africa; mostly Mediterranean); DRYMARIA (tropical and southern temperate regions); GYPSOPHILA (Europe, Asia; especially eastern Mediterranean); HERNIARIA (Mediterranean; Europe, South Africa); HOLOSTEUM (northern temperate regions); LYCHNIS (northern temperate regions); MELANDRIUM (northern hemisphere; South Africa, South America); POLLICHIA (tropical and South Africa); POLYCARPAEA (cosmopolitan); POLYCARPON (cosmopolitan); SAGINA (northern temperate regions); SAPONARIA (northern temperate regions, chiefly Mediterranean); SILENE (northern temperate regions, chiefly Mediterranean); SPERGULA (temperate regions); SPERGULARIA (cosmopolitan); STELLARIA (cosmopolitan); TISSA (cosmopolitan).

The medicinal and poisonous campions of India belong to 13 genera:—ARENARIA, CERASTIUM, CUCUBALUS, DIANTHUS, DRYMARIA, HOLOSTEUM, LYCHNIS, POLYCARPAEA, POLYCARPON, SAPONARIA, SILENE, SPERGULA, STELLARIA.

A. Calyx gamosepalous, 4-5 lobed. Petals clawed and stamens inserted on an elongate gynophore (rarely sessible). Styles free. Stipules absent.

I. Capsule dehiscent; styles 2. Seeds pelate; hilum facial; embryo straight.
Calyx bracteate, striate ... ... DIANTHUS.

II. Capsule bursting by short or long valves. Hilum lateral; embryo annular.

1. Calyx veined. Capsule shortly 4-valved; styles 2 ... ... SAPONARIA.
2. Calyx 10-∞-nerved. Capsule partially celled, shortly 3-6-valved; styles 3 ... SILENE.
3. Calyx 10-nerved. Fruit fleshy; styles 3 ... CUCUBALUS.
4. Calyx 10-nerved. Capsule shortly 4-5- or 8-10-valved; styles 5. ... ... LYCHNIS.

B Sepals free, or connate at the base only. Petals subsessile and stamens inserted on an annular disk, rarely perigynous. Styles free.

I. Stipules absent. Sepals free. Stamens hypogynous.

1. Petals jagged. Capsule cylindric or conic, 6-valved. Styles 3. Seeds many, compressed ... ... HOLOSTEUM.
2. Petals notched or entire. Capsule cylindric or conic, 8-10-valved. Styles 3-5, opposite the sepals. Seeds many ... ... CERASTIUM.
3. Petals bifid. Capsule globose ovoid or oblong. Styles 3-5. Seeds few or many. STELLARIA.
4. Petals entire or lacerate. Capsule globose ovoid or oblong. Styles usually 2-3. Seeds many ... ... ARENARIA.

II. Stipules scarious. Sepals free. Petals entire. Styles 3 or 5. Capsules 3- or 5-valved ... ... SPERGULA.

C. Sepals free. Petals subsessile and stamens inserted on an annular disk. Styles 2-3, combined. Stipules scarious, rarely obsolete.

I. Sepals not keeled. Petals 2-6-fid. ... ... DRYMARIA.

II. Sepals keeled. Petals entire. Style 3-fid ... ... POLYCARPON.

III. Sepals scarious, not keeled. Stigma 3-toothed ... POLYCARPAEA.

ARENARIA.

The genus consists of about 100 species, all natives of cold and temperate regions.

Arenaria serpyllifolia Linn. is found in the subtropical and temperate Himalaya from Nepal to Kashmir, in Rohilkhand and the Punjab at 10,000-11,000 ft., in Western Tibet at 11,000-13,000 ft. It extends to Afghanistan, temperate Europe and Asia.

The herb has long been used in China for bladder diseases. It is considered very valuable for calculus troubles and acute and chronic cystitis.

Chinese: Li Ke Ts'ao—*English*: Lloyd's Sandwort, Sandwort, Thyme-leaved Sandwort—.

CERASTIUM.

The genus numbers about 100 species, all natives of temperate and cold regions.

C. arvense Linn. is used medicinally in Europe.

Cerastium vulgatum Linn. occurs throughout the temperate and subalpine regions of India and Ceylon, ascending to 15,000 ft. in Sikkim and Western Tibet. It extends to temperate Northern Europe and Asia.

The plant is used in Spain as a mild refrigerant.

English: Chickenweed, Mouse-ear, Mouse-ear Chickweed, Murion—.

Cucubalus.

The one species of this genus, **C. bacciferus** Linn., occurs in the temperate Himalaya from Kashmir at 5,000-8,000 ft. to Sikkim at 8,000-12,000 ft.; in the Khasia Hills, at the Boya Pani at 5,000 ft. It is distributed westwards to South Europe, to Central Asia and to West Siberia.

A decoction of the herb is used in Spain to check haemorrhage.

English: Berry-bearing Campion, Bladder Campion—; *French*: Behen baccifère, Coulichon—; *Spanish*: Oreja de vecino, Pamphlina de bayas, Purga cabras—.

Dianthus.

The genus includes about 250 species, natives of the northern temperate zone exclusive of Western North America; one of Siberia extending into Arctic America.

D. armeria Linn., *D. barbatus* Linn., *D. carthusianorum* Linn., *D. caryophyllus* Linn., *D. plumarius, D. sylvestris* Wulf. are used medicinally in Europe; *D. superbus* Linn. in China; *D. crenatus* Thunb., *D. scaber* Thunb. in South Africa.

1.	Leaves very narrow, rigid, slender, mid-rib very thick, margin much thickened, finely toothed ...	*D. anatolicus.*
2.	Leaves channelled above, bent-back, with smooth margins, linear	*D. caryophyllus.*
3.	Leaves broad and nearly flat or slightly trough-shaped, 3-5-nerved	*D. chinensis.*

1. **Dianthus anatolicus** Boiss. is found in the Western Himalaya, Western Tibet, Kashmir, and extends to Armenia.

It is used as an antiperiodic in intermittent fevers.

Persian: Kantariyun—.

2. **Dianthus caryophyllus** Linn. is generally supposed to be a native of the Mediterranean. It occurs in Baltistan, the Punjab and Kashmir at 7,000-8,000 ft., extending westwards to Europe.

The flowers are used medicinally in Spain and North America; they are considered cardiotonic, diaphoretic and alexiteric. They are nervine and antispasmodic.

The whole plant is used as a vermifuge in China.

Annam: Cu mach—; *Catalan*: Clavellina—; *English*; Carnation Pink, Clove Pink, Wild Carnation—; *French*: Œillet commun, Œillet des fleuristes, Œilllet giroflée Œillet des jardins, Œillet à ratafia—; *Roumanian*: Garoafa—; *Russian*: Gvasdika—; *Spanish*: Clavelina comun—.

3. **Dianthus chinensis** Linn. is a native of China and Japan cultivated in Indian gardens.

A broth of the fresh new leaves is given in infantile diarrhoea. The old dried leaves are pounded and used in diseases of the eye to clear the eyesight.

The dried flowering plants are used in China as a diuretic, anthelmintic, and abortifacient.

In Indo-China the dried flowers and buds are given as a diuretic for indigestion and in difficult labour; they are also prescribed in kidney troubles.

In Malaya the herb is used as a remedy for gonorrhoea.

English: China Pink—; *French*: Œillet de Chine, Œillet de la régence, Œillet rouge—; *Indo-China*: Thach truc, Truc tu hoa—; *Portuguese*: Cravina—.

DRYMARIA.

The genus includes about 40 species; one in tropical Asia and Africa, one in Australia, the rest in America.

D. cordata Willd. and *D. glandulosa* Bartl. are used as cooling medicines in French Guiana.

Drymaria cordata Willd. is found in tropical and sub-tropical India and Ceylon, extending westwards to the Punjab and ascending the Himalaya to 7,000 ft. in Sikkim. It occurs in the Malay Peninsula, Java, Sumatra, Papua, tropical Africa and America.

In French Guiana it is eaten as a salad for its cooling properties.

French Guiana: Mignonnette—.

HOLOSTEUM.

The genus consists of 6 species inhabiting the northern temperate regions.

Holosteum umbellatum Linn. is found in Kashmir, and in cultivated fields in the Quetta valley; it extends to western Asia, North Africa and Europe.

It is reputed to have refreshing and slightly demulcent properties.

English: Umbellate Chickweed—.

LYCHNIS.

The genus is made up of 10 species, natives of Arctic and Temperate northern regions, and of the Andes of South America.

The following are used medicinally in Europe—*L. chalcedonica* Linn., *L. coronaria* Desr., *L. dioica* Linn., L. *flos-cuculi* Linn., *L. githago* Lam.—; in China—*L. grandiflora* Jacq.—; in Indo-China—*L. fulgens* Fisch., *L. Senno* Sieb. and Zucc.—; in South Africa—*L. githago* Lam.—; in North America—*L. alba* Mill.—.

Lychnis coronaria Desr. is abundant in Kashmir: Gadsar, roadsides and dry places; near Shirazia Bagh, on top of hill in rocky and grassy soil; Dachigan Rukh; below Gulmarg, wooded hill-side at 8,000 ft. It extends to West Asia and South Europe.

An extract or a decoction of the root is used in Spain for infraction of the lymph glands of the mesentery, and for diseases of the lungs and the liver.

Catalan: Pessiganassos—; *French*: Coquelourde—; *Spanish*: Candelaria, Claveles lanudos, Coronaria purpurea, Guantes de reina, Rosa de Grecia—.

POLYCARPAEA.

This genus consists of 30 cosmopolitan species.

P. arenaria Gagnep. is used medicinally in Indo-China, *P. corymbosa* Lam. in Malaya and Transvaal, *P. linearifolia* DC. in Nigeria.

Polycarpaea corymbosa Lam. is a cosmopolitan herb. It is found in the Western Peninsula and Ceylon, Central and North-Western India, Sind, ascending the Western Himalaya to 7,000 ft.; it also occurs in Burma and the Malay Peninsula.

The herb is useful in strangury, urinary calculi, boils, inflammatory swellings, and ulcers. Its ash mixed with pepper is applied externally to boils and ulcers (Ayurveda).

The pounded leaves are used as poultice, warm or cold, over boils and inflammatory swellings.

In Pudukotah the herb is administered both externally and internally as a remedy for the bites of venomous reptiles. In Porebunder the pounded leaves are used for bites from animals, and given with molasses in the form of a pill in jaundice.

In Malaya the drug is found in the shops as flowering heads, arising from silvery cymes, with portions of the stem and tomentose leaves, and is used as a demulcent and astringent.

Whether administered internally or applied externally the herb is useless in the treatment of snake-bite (Mhaskar and Caius).

Chinese: Pai T'ou Weng—; *Hausa*: Bakin suda—; *Malaya*: Pak thow yoong—; *Porebunder*: Okharad—; *Rajputana*: Zutaniakhand—; *Sanskrit*: Bhisatta, Okharadi, Tadagamritikodbhava—; *Santal*: Janhenanjur—; *Tamil*: Nilaisedachi—; *Telugu*: Bommasari, Rajuma—.

POLYCARPON.

A small genus of 7 species, distributed throughout all the warm and temperate climates of the world.

Polycarpon indicum Merrill is found in the hotter parts of India in fields and waste places; in Indo-China and tropical Africa.

In Indo-China an infusion of the roasted leaves is given for cough following upon a fever, more particularly measles.

Indo-China: La coc man—.

SAPONARIA.

The genus includes 20 species, chiefly Mediterranean and Western Asiatic.

The following are used medicinally in Europe—*S. ocymoides* Linn., *S. officinalis* Linn., *S. Vaccaria* Linn.—; in China and Indo-China—*S. Vaccaria* Linn.—; on the Pacific Coast: *S. officinalis* Linn.

The root of *S. officinalis* Linn. is official in France. A toxic glucoside, saporubin, has been extracted from it.

Saponaria Vaccaria Linn. is a cosmopolitan weed of cultivation found throughout India.

The Yunani practitioners use the plant for enlargement of the spleen, dysmenorrhoea, and ulcers; the leaves are commonly given for scabies and itch.

The mucilaginous sap of the plant is considered to have febrifuge and tonic properties in long continued fevers of a low type.

In Indo-China the plant is used as a mild depurative. In China it is used in furunculosis and scabies.

The leaves and roots contain from 8 to 15 per cent of saponin. The seeds contain a sapotoxin, and are somewhat toxic to stock.

Arabic: Gafis, Gleylé—; *Bengal*: Sabuni—; *Brahui*: Karari—; *Chinese*: Wang Pu Liu Hsing—; *Egypt*: Ful-el-'arab—; *English*: Cockle, Cow Basil—; *French*: Blé de vache, Copatte, Saponaire à vaches, Saponaire des vaches—; *German*: Wilder Weid, Kuhkraut—; *Hindi*: Musna—; *Hova*: Savonimbary—; *Indo-China*: Vuong bat luu hanh—; *Iraq*: Kharaz bint-al-fallah, Ziwan—; *Italian*: Cetino, Mezzettino—; *Malta*: Cow Basil, Cow-herb, Cetino, Mazzettino—; *Mosul*: Ziwan—; *Persian*: Gafis, Guligafis—; *Santali*: Musna—; *Sind*: Musna—; *South Africa*: China Cockle, Cockle, Cow basil, Cow foot, Glond, Soapwort, Spring cockle—; *Spanish*: Teta de vaca—; *Uriya*: Tilothi—.

SILENE

A genus of 320 species, natives of temperate Europe and Asia chiefly Mediterranean.

The following are used medicinally in Europe—*S. apetala* Willd., *S. cerastoides* Linn., *S. conoidea* Linn., *S. Cucubalus* Wibel, *S. gallica* Linn., *S. hirsuta* Lag., *S. inaperta* Linn., *S. italica* Pers., *S. nicaeensis* All., *S. nocturna* Linn., *S. nutans* Linn., *S. otites* Sm., *S. rupestris* Linn., *S. saxifraga* Linn., *S. viscosa* Pers.—; in Indo-China—*S. macrosolen* Steud.—; in South Africa—*S. Burchellii* Otth.; *S. capensis* Otth.—; in North America—*S. laciniata* Cav, *S. noctiflora* Linn., *S. stellata* Ait., *S. virginica* Linn.

A. Perennial.
  Petals imbricate in bud. Calyx inflated, 20-nerved, reticulately veined ... ... *S. Cucubalus.*

B. Annuals.
  Petals contorted in aestivation. Calyx not recticulately veined.
  I. Calyx 20-, 30-, or 60-nerved, in fruit conical from a broad base. Petals pink ... ... *S. conoidea.*
  II. Calyx 10-nerved. Flowers in unilateral racemes.
    1. Face of the seeds smooth, rather concave, back deeply grooved with two wavy wings. Petals absent or minute, bright red ... ... *S. apetala.*
    2. Face of the seeds flat with an ear-shaped depression, back not grooved. Petals red or white ... ... *S. gallica.*

1. **Silene Cucubalus** Wibel is found in the temperate Himalaya at 5,000-11,500 ft., from Nepal to the Indus. It is distributed over temperate Asia, Europe and North Africa.

In Spain the plant is considered emollient, and it is used in baths or as a fumigant; the juice is prescribed in ophthalmia.

Baleares: Trons—; *Catalan*: Botets, Colitxos, Culissos, Cuniells, Curibells, Esclafidors—; *English*: Bladder Campion, Inflated Catchfly, White Ben, White

Bottle—; *French*: Behen blanc, Behen commun—; *German*: Klatschnelke, Leimkraut, Schaummohn—; *Italian*: Behen bianco, Bubbolini, Crepa terra—; *Maltese*: Kaskeiza—; *Spanish*: Alcanducea, Cascabelillo de Canarias, Collejas, Conejera, Polemonio pratense—.

2. **Silene conoidea** Linn. occurs in Kashmir and Baluchistan; from Kumaon to the Indus and from Oudh to the Punjab; in West Tibet at Ladakh. It extends westwards to the Atlantic Ocean.

The plant is used therapeutically in Spain as a substitute for *S. Cucubalus.*

Harboi Hills: Dardiri—; *Kalat*: Gardi—; *Kharan*: Guto—; *Spanish*: Claveletas de barranco—.

3. **Silene apetala** Willd. is found in Peshawar; it extends through Afghanistan and Persia to South Europe and North Africa.

In Spain the juice is used in eye troubles; the plant is used in baths or as a fomentation for its emollient properties.

4. **Silena gallica** Linn. is found in Garhwal and in the Carnatic; it is common everywhere in the Mediterranean region.

The plant is used in Spain as a substitute for *S. Cucubalus.*

Spanish: Carmentilla—.

Spergula.

A very small genus of 3 European and Asiatic species.

Spergula arvensis Linn. is found in cultivated fields in various cool parts of India and throughout the northern hemisphere.

It is used as a diuretic in Colombia.

Bogotá: Abrojito—; *Dutch*: Spurrie—; *English*: Beggar-weed, Bottle Brush, Cowquake, Dodder, Dother, Farmer's Ruin, Franke, Granyagh, Lousy Grass, Make-beggar, Mountain Flax, Pick Pocket, Pick Purse, Poverty Weed, Sandweed, Spurrey, Spurry, Toad Flax, Yarr, Yarrel, Yawr, Yur—; *French*: Espargoule, Espargoute, Espargoutte, Spargoule, Spargoute, Spergoute, Spergule, Sperjule—; *German*: Spergel—; *Roumanian*: Granat—; *Russian*: Taritsa—; *Spanish*: Espergula—.

Stellaria.

A genus of about 100 species, dispersed over the whole world, but most abundant in cold and temperate regions.

A. Annual.
- Sepals free to the base. Petals shorter than the calyx or absent. Stamens 3-10, hypogynous. Styles 3. Capsule with 4-6 entire valves ... ... *S. media.*

B Perennials.
- I. Sepals free to the base. Petals with diverging lobes. Stamens hypogynous. Styles 5. Capsule with 5 bifid valves ... ... *S. aquatica.*
- II. Sepals more or less connate at the base into an obconic tube. Petals minute or absent. Stamens 10, perigynous. Styles 3. Capsule with 6 entire valves ... ... *S. uliginosa.*

1. **Stellaria media** Linn. is found in cultivated and waste places, roadsides and edges of streams throughout the Punjab and temperate regions of India, ascending in the Himalya to 12,000 ft.

and in Western Tibet to 14,500 ft.; in the Mediterranean basin, throughout Europe and northern Asia, and carried out as a weed to all the temperate and colder regions of the globe.

The plant is employed in plaster for broken bones and swellings, as it is supposed to be binding and cooling.

The herb is used in Spain as a vulnerary, astringent and resolvent. On the Pacific Coast it is reputed refrigerant, demulcent and alterative.

Catalan: Burrissol, Murrons, Murrons blanchs, Murrons de canari, Picagallina, Picapoll—; *Dutch*: Muur—; *Egypt*: Gezäze—; *English*: Chickweed, Common chickweed—; *French*: Alsiné, Alsiné des oiseaux, Mouron blanc, Mouron des oiseaux, Mouron d'hiver, Morsgeline, Morsgeline—; *German*: Augentrostgras—; *Italian*: Budellina, Centonchio, Centoncolo, Morgellina, Paperina—; *Malta*: Common Chickweed, Budellina, Centocchio, Paperina, Harira bajda—; *Pacific Coast*: Alsine, Chickenweed, Common Chickweed, Satin-flower, Tonguegrass, White-bird, Winter-weed—; *Roumanian*: Limba vrabiei—; *Russian*: Makritsa—; *Spanish*: Pamplina, Pamplina de canarios, Pajarera, Yerba gallina, Yerba pajarera—.

2. **Stellaria aquatica** Scop. occurs in the temperate Himalaya from Nepal to Marri, at 4,000-8,000 ft. It extends to Northern and Western Asia, Europe, and North Africa.

In Indo-China a decoction of the leaves is given as a galactagogue to nursing mothers suffering from want of milk.

In China the plant is used for the treatment of fistulae.

Chinese: Fan Lu—; *English*: Chickweed, Great Chickweed, Water Chickweed—; *French*: Stellaire aquatique—; *Indo-China*: Phon lau—; *Malta*; Water Starwort, Budellina d' acqua—.

3. **Stellaria uliginosa** Murr. is found in wet places of the temperate and alpine regions at 4,000-10,000 ft.; in the temperate and alpine Himalaya from Kashmir to Sikkim, here up to 16,000 ft.; in the Khasia Hills at 5,000-6,000 ft. It is generally distributed over the temperate regions of the northern hemisphere.

A decoction of the leaves is used as a galactagogue in Indo-China.

English: Bog Stitchwort—; *Indo-China*: Tuoc thiet thao—.

THE MEDICINAL AND POISONOUS CROWFOOTS OF INDIA.

By

J. F. Caius, s.j., f.l.s.

The Ranunculaceæ are annual or perennial herbs, rarely shrubs or climbers. The family includes 40 genera with about 700 species, distributed all over the world, abundant in temperate and cold regions, rare in the tropics and usually at high altitudes.

This is a very important family from our point of view, as owing to the fact that a poisonous, acrid, and narcotic principle prevails in varying degree throughout the Order, there is scarcely one plant which can be regarded as harmless, while some are deadly. However, most crowfoots lose their acrid qualities on drying or cooking. Their roots, when perennial, contain, besides the acrid, a bitter extractive principle, present in various proportions, with a volatile oil, which render them drastic and emetic. Their seeds are acrid; some contain both a fixed and a volatile oil, and are aromatic.

The various members which have been chemically examined have been found to contain (1) *alkaloids*—aconitine, ajacine, ajaconine, atisine, benzaconine, berberine, bikhaconitine, canadine, clematine, coptine, cynoctodine, damascenine, delcosine, delphinine, delphinoidine, delphisine, deltaline, hydrastine, indaconitine, isopyrine, japaconitine, jervine, jesaconitine, lappaconitine, lycaconitine, myoctonine, neopelline. palmatisine, pseudaconitine, septentrionaline, staphisagroine—; (2) *volatile acrid and bitter principles*—anemonin, cimicifugin, nigellin—; (3) *acids*—aconitic, anemonic, isoferulic, valerianic—; (4) *crystalline or amorphous glucosides*—adonidin, delphinin, helleborein, helleborin, paeonin—; (5) *saponins, tannins, resins, oleoresins, sugars* and *fats*.

The better known medicinal and poisonous crowfoots of the world belong to the following 24 genera:—Aconitum (Northern temperate regions); Actaea (Northern temperate regions); Adonis (Northern palaeotemperate regions); Anemone (cosmopolitan); Aquilegia (Northern temperate regions); Caltha (temperate regions); Cimicifuga (Northern temperate regions); Clematis (cosmopolitan); Clematopsis (Nigeria, Tanganyika, Congo, Angola); Coptis (Northern temperate and arctic regions); Delphinium (Northern temperate regions); Eranthis (Northern palaeotemperate regions); Helleborus (Europe; Mediterranean region); Hydrastis (Japan; North America); Isopyrum (Europe, Asia, North America); Knowltonia (South Africa); Myosurus (temperate regions); Nigella (Europe; Mediterranean region); Paeonia (Europe; Asia; western North America); Pulsatilla (Central Europe; Siberia; East Indies); Ranunculus (cosmopolitan); Thalictrum (Northern temperate regions); Trollius (Northern temperate and arctic regions); Zanthorhiza (Atlantic North America).

The medicinal and poisonous crowfoots of India belong to the following 14 genera:—ACONITUM, ACTAEA, ADONIS, ANEMONE, AQUILEGIA, CALTHA, CIMICIFUGA, CLEMATIS, COPTIS, DELPHINIUM, NIGELLA, PAEONIA, RANUNCULUS, THALICTRUM.

A. Erect or climbing shrubs. Leaves opposite. Sepals petaloid, valvate. Carpels 1-ovuled; ovule pendulous. Fruit of many achenes
   Petals o ... ... CLEMATIS.

B. Herbs. Leaves radical or alternate. Sepals petaloid, imbricate. Carpels 1-ovuled; ovule pendulous. Fruit of many achenes
   I. Petals o, flowers involucrate ... ... ANEMONE.
   II. Petals o, flowers not involucrate ... ... THALICTRUM.
   III. Petals 5-16, with no nectariferous pit ... ... ADONIS.

C. Herbs. Leaves radical or alternate. Sepals herbaceous, imbricate. Carpels 1-ovuled; ovules erect. Fruit of many achenes
   Sepals 3-5, deciduous, petals usually 5 ... ... RANUNCULUS.

D. Herbs or undershrubs. Leaves alternate. Sepals petaloid or herbaceous, imbricate. Carpels several, many-ovuled. Fruit of 1 or more follicles (berried in ACTAEA)
   I. Flowers regular solitary or panicled
      a. Petals o. Leaves undivided ... ... CALTHA.
      b. Petals small, carpels long-stalked ... ... COPTIS.
      c. Petals small, unguiculate. Carpels sessile. Flowers pale blue ... ... NIGELLA.
      d. Petals as large as the sepals, all spurred ... AQUILEGIA.
   II. Flowers regular, racemed
      a. Carpels solitary, berried when ripe ... ACTAEA.
      b. Carpels 4-8, follicular when ripe ... CIMICIFUGA.
   III. Flowers irregular
      a. Posterior sepal spurred ... ... DELPHINIUM.
      b. Posterior sepal vaulted ... ... ACONITUM.

E. Herbs or undershrubs. Sepals herbaceous imbricate. Carpels several, several-ovuled, girt by a fleshy disk. Fruit of coriaceous few-seeded follicles ... ... PAEONIA.

ACONITUM.

The genus numbers 110 species distributed over the northern temperate regions of the globe.

Aconites are at once highly poisonous and most valuable medicinal plants. The following species are used medicinally in Europe—*A. Anthora* Linn., *A. Cammarum* Linn., *A. Lycoctonum* Linn., *A. Napellus* Linn., *A. variegatum* Linn.—; in Japan and China—*A. Fischeri* Reichb., *A. Lycoctonum* Linn., *A. Napellus* Linn., *A. uncinatum* Linn.—; in North America—*A. Fischeri* Reichb., *A. Napellus* Linn., *A uncinatum* Linn.—; in Indo-China—*A. Fischeri* Reichb., *A. Kusnezoffii* Reichb., *A. Lycoctonum* Linn., *A. Napellus* Linn.

Those aconite alkaloids which have been fully examined belong to two well-defined groups:—(1) aconitines, which are highly poisonous—aconitine and bikhaconitine from *A. Napellus*, indaconitine from *A. chasmanthum*, japaconitine from Japanese hondo aconite roots, pseudaconitine from *A. deinorrhizum* and *A. Balfourii*, jesaconitine from Japanese bushi aconite roots, lappaconitine and lycaconitine from *A. Lycoctonum*—; (2) atisines, which are not toxic —atisine from *A. heterophyllum*, palmatisine from *A. palmatum.*

Nothing is known of the properties and uses of 11 species recognized as Indian. The other 13 Indian Aconites are arranged according to the structure of their roots, on the authority of Stapf:—

A. Root perennial, long, fusiform usually breaking up at length into cord-like anastomosing or free strands; old plants often with several stems from the collar ... 1. *A. luridum.*

B. Roots biennial, paired, tuberous; each tuber producing normally one simple or rarely branched stem

- I. Stem erect, rarely ascending, never twining
  - 1. Seed-angles winged, faces smooth or almost so (not transversely lamellate)
    - a. All the leaves distinctly to very long-petioled, cordate-orbicular or reniform in outline, deeply dissected
      - i. Tubers 2-3.5 cm. long, fracture in the dry state horny or cartilaginous, brown (at least the part outside the cambium); taste slightly bitter, followed by a tingling sensation; carpels 5, glabrous or nearly so, never tomentose ... ... 2. *A. chasmanthum*
      - ii. Tubers 0.5-2.5 cm. long, fracture in the dry state almost farinaceous, pure white; taste indifferent or slightly sweetish, not followed by any tingling sensation; carpels densely tomentose; honey-gland subterminal ... ... 3. *A. violaceum.*
    - b. All the leaves with the exception of the lowest shortly petioled to sub-sessile, cordate-ovate, coarsely crenate ... ... 4. *A. heterophyllum*
  - 2. Seeds with hyaline wavy transverse lamellae
    - a. Cambium of tubers discontinuous, forming in cross-section circular or tangentially flattened to horseshoe-shaped strands
      - i. Nectary-hood very slender with a minute lip; carpels quite glabrous; taste of tubers intensely and pure bitter, not followed by any tingling sensation ... 5. *A. palmatum.*
      - ii. Nectary-hood wide with a rather large lip; carpels hairy; taste of tubers rather indifferent, followed by a strong and persistent tingling sensation
        - † Inflorescence and carpels greyish pubescent; leaves divided almost to the base, ultimate divisions narrow; carpels usually 3 ... ... 6. *A. deinorrhizum.*
        - †† Inflorescence and carpels spreadingly yellowish tomentose; leaves rather less divided than in the preceding species, ultimate divisions broad; carpels 5 ... ... 7. *A. Balfourii.*
    - b. Cambium of tubers continuous, forming in cross-section a more or less sinuous ring
      - i. Secondary sieve-strands of mother-tubers not encased in sclerenchymatic sheaths; lip of nectary widened from the base upwards; leaf-segments not or slightly divaricate

† Leaves divided to in the inner, to $\frac{3}{4}$ or less in the outer incisions, ultimate divisions rather broad; carpels 5; follicles 10-18 mm. long

¶ Intermediate leaf-division rhomboid-cuneate, sparingly and coarsely inciso-crenate; nectary-hood much leaning forward, slightly widened at the top, scarcely gibbous; carpels glabrous or nearly so; follicles 14-18 mm. long ... ... 8. *A. Falconeri.*

¶¶ Intermediate leaf-division mostly ovate in outline, copiously inciso-crenate or dentate; nectary-hood slightly leaning forward, distinctly gibbous at the top; carpels villous; follicles about 10 mm. long ... ... 9. *A. spicatum.*

†† Leaves divided almost to the very base, ultimate divisions narrow; carpels mostly 3; follicles 18-25mm. long ... ... 10. *A. laciniatum.*

ii. Secondary sieve-strands of mother-tubers encased in sclerenchymatic sheaths, forming persistent fibres; lip of nectary narrow; leaf-segments conspicuously divaricate ... ... 11. *A. ferox.*

II. Stem very slender, twining

1. Flowers on recurved pedicels (up to 5 cm. long), nodding; helmet conic-ovate in profile ... 12. *A. Elwesii.*
2. Flowers on straight or almost straight pedicels (up to 7.5 cm. long); helmet depressed, semi-orbicular in profile ... ... 13. *A. lethale.*

1. **Aconitum luridum** Hook. fil. & Th. is found in the Himalaya from Eastern Nepal to Chumbi, between 12,000 and 14,000 ft. The root is reputed poisonous.

Bengal: Bish, Butsnabbish—; *Bhoti*: Tchendook—; *Bombay*: Butchnab—; *Darjeeling*: Soorja banshi—; *Hindi*: Mahoor—; *Nepal*: Atisingeeabish, Bikh, Bish, Bishnak—; *Sikkim*: Suya-banshi—; *Telugu*: Ativassa—.

2. **Aconitum chasmantum** Stapf inhabits the subalpine and Alpine zone of the Western Himalaya, from Chitral and Hazara to Kashmir, between 7,000 and 12,000 ft.

The root contains, according to Dunstan, very small proportions of aconitine (1897). Later the roots were examined by Dunstan and Andrews, and found to contain the crystalline alkaloid indaconitine (1905). But an alkaloid, obtained by Bauer and Radjhan, did not appear identical with indaconitine (1931).

Hazara: Mohri—; *Jhelum*: Mohri, Piun—; *Kashmir*: Ban-bal-nag—.

3. **Aconitum violaceum** Jacquem. occurs in the Alpine zone of the Himalaya from Gilgit to Kumaon, between 10,000 and 15,000 ft.

The roots are eaten by the hillmen of Kanawar as a pleasant tonic

Kachung: Tilia kachang—; *Kashmir*: Jokser, Jokser me-hin-pat-val—; *Sutlej*: Tilia kachang—

4. **Aconitum heterophyllum** Wall. is common in the subalpine and alpine zone of the Himalaya, from the Indus to Kumaon, from 6,000 to 15,000 ft.

Ayurvedists describe the root as a bitter tonic, hot, stomachic, and digestive. They recommend it as a remedy in fevers, diarrhoea, dyspepsia and cough, and also as an aphrodisiac and an astringent. They consider it useful in the treatment of bites and stings from poisonous animals.

Yunanists use the drug much in he same way as the Ayurvedists. Thus the root is a common Indian medicine employed as a mild and bitter tonic, aphrodisiac, antiperiodic and tonic.

The Central Indigenous Drugs Committee (1901) while admitting that the drug is a good bitter and tonic, declared it to be worthless as an antiperiodic. Koman (1919) reported having cured a case of acute dysentery and one of chronic enteritis. Caius and Mhaskar proved experimentally that it is not an antidote to snake venom (1931) and that it is useless in the treatment of scorpion sting (1932).

The alkaloid *atisine* has been obtained from the root.

Bashahr: Patis, Patris—; *Bhoti*: Ais—; *Bombay*: Atis, Atvika—; *Canarese*: Atibaje—; *Cutch*: Ativista—; *Gujerat*: Atavakhanikali, Atavishnikali, Ativakh, Ativish—; *Hindi*: Atis, Atvika—; *Jhelum*: Chitijari, Sukhi-hari—; *Kashmir*: Hongisafed, Mohandigujsafed, Mohundguji, Narmada, Nerr-mada—; *Lahul*: Bonga—; *Leh*: Ais—; *Marathi*: Atavish—; *Persian*: Batis, Vajjeturki—; *Punjab*: Bonga, Chitijari, Patis, Patris, Sukhihari—; *Sanskrit*: Amrita, Aruna, Ataicha, Atisaraghni, Ativisha, Bhangura, Bhringi, Ghunavallabha, Kashmira, Madri, Mahaushadham, Mahoshadha, Mridvi, Prativisha, Pravisha, Shishubhaishyajja, Shokapaha, Shringi, Shringika, Shuklanda, Shveta, Shvetakanda, Shvetavacha, Shyamkanda, Upvisha, Vira, Virupa, Visha, Visharupa, Vishva—; *Tamil*: Atividyam—; *Telugu*: Ativasa—.

5. **Aconitum palmatum** D. Don is found in the alpine Himalaya of Nepal, Sikkim, and the adjoining part of South Tibet, from 10,000 to 16,000 ft.

The non-poisonous root is tonic and antiperiodic; it is given for rheumatism and diarrhoea. Mixed with pepper it is used internally as a remedy for pains in the abdomen, diarrhoea and vomiting, and intestinal worms; externally it is used as an application for rheumatism.

The root contains the alkaloid *palmatisine*.

Bombay: Bikhma, Vakhma, Wakhma—; *Gujerati*: Vakhamo—; *Hindi*: Bikhma, Bishawa, Bishma, Bisma—; *Sikkim*: Gniong-rip-gmok, Setobikhoma—.

6. **Aconitum deinorrhizum** Stapf occurs in the alpine Himalaya of Bashahar.

The root contains the poisonous alkaloid *pseudaconitine*.

Bashahar: Doodhiya moura, Maura, Mohra—; *Gujerat*: Dudhio vachhanag—; *Hindi*: Mahoor, Mahore, Maur, Maura bikh, Mora, Moura bikh, Muhura—; *Kashmir*: Dudhia bish—; *Punjab*: Dudhia bish—; *Ravi*: Dudhia moura—; *Sutlej*: Kala mohra, Mahoor, Mahore, Maur, Maura bikh, Mohra, Mora, Moura bikh, Muhura—.

7. **Aconitum Balfourii** Stapf occurs in the subalpine and alpine Himalaya from British Garhwal to Nepal.

The tubers contain the poisonous alkaloid *pseudaconitine*.

Darma: Gobriya—; *Garhwal*: Banwa—; *Hindi*: Teliya, Teliya bachnag, Teliya bis—; *Lahore*: Bachnab—; *Nepal*: Gobari—; *Ralam*: Bhanva, Dhaulia, Dhumuriya, Fatkia, Gobaria, Gobriya, Jhirina, Kawriya, Phatkia, Phutkia, Ratwa, Tiliya—.

8. **Aconitum Falconeri** Stapf inhabits the alpine and subalpine zone of the Himalaya of Garhwal.

The root is evidently poisonous.

Hindi: Bikh, Bis, Mahoor, Mahore, Maur, Maura bikh, Meethatellia, Mora, Moura bikh, Muhura—; *Tihri Garhwal*: Mohra—.

9. **Aconitum spicatum** Stapf is found in the alpine zone of the Himalaya of Sikkim and Chumbi, between 10,000 and 12,000 ft.

This species is the principal source of the 'Bikh' or 'Bish' of the Calcutta market. The toxic principle is the alkaloid *bikhaconitine*.

Bengal: Dagra, Dakra, Katbish, Kathbish—; *Bombay*: Bachnab—; *Canarese*: Vasanabhi—; *Cutch*: Bachnaga—; *Deccani*: Lal bachnag—; *Gujerati*: Shingadiovachnag, Vachnag, Vachlanag—; *Hindi*: Bachhnag, Bachnag, Bachnak, Bechnak, Bikh, Bish, Bis, Kalabachhnag, Singya—; *Lepcha*: Ning—; *Malayalam*: Valsanabhi—; *Sanskrit*: Sringibish, Vatsanabha—; *Sikkim*: Bikh, Gniong, Gniongmot, Gniongning, Kalabikoma, Shoddukmot—; *Sinhalese*: Vachanabhi—; *Tamil*: Navi, Sennabi, Vachanavi—; *Telugu*: Nabhi, Vasanabhi—.

10. **Aconitum laciniatum** Stapf occurs in the subalpine and alpine Himalaya of Sikkim and adjoining Tibet, between 10,000 and 14,000 ft.

It forms, together with *A. spicatum*, the article known as 'Nepal Aconite'.

Sikkim: Kalobikhmo—.

11. **Aconitum ferox** Wall. is found in the alpine Himalaya of Nepal.

The drug is chiefly employed in India in the treatment of leprosy, fever, cholera, and rheumatism.

It is a very effective medicine in various diseases, acting as a narcotic sedative, regarded as heating and stimulant, useful in fever, cephalalgia, affections of the throat, dyspepsia, and rheumatism. It is much used as an external application, the root being formed into a paste and spread upon the skin in neuralgia, boils, etc. Internally it is chiefly used in the treatment of chronic intermittent fevers.

Ayurveda prescribes that the root be specially prepared before use. To this end the tough horny material is soaked in either urine or milk from the cow and exposed to the sun until it becomes so soft as to be easily pierced by a pin. The process requires three or more days and the urine or milk has to be renewed every day. Bose (K. C.) has shown that this treatment is attended with very important modifications of the physiological properties :—tincture of Aconite at first slows the heart-rate, lowers the blood pressure, and increases the peripheral circulation, later the heart-rate is accelerated and the blood pressure raised; after treatment of the root with cow's urine the tincture increases the rate and systole of the heart, the blood pressure, and the peripheral circulation, and the effects persist for a very long time. Mhaskar and Caius have confirmed those findings and demonstrated, moreover, that if the root is treated with cow's milk, instead of urine, the above changes are much more pronounced.

The root, either alone or in combination with other drugs, has been much recommended for the treatment of snake-bite and scorpion-sting; but Caius and Mhaskar have proved experimentally that the drug is not an antidote to snake venom and is useless in the treatment of scorpion-sting.

A preparation of the root is much used in all the hilly districts in India to poison arrows. The toxic principle is the alkaloid *pseudaconitine*.

Arabic: Bish—; *Assam*: Bish—; *Bengal*: Bish, Butsnabbish, Katbish—; *Bombay*: Butchnab—; *Canarese*: Vasanabhi, Vatsanabhi—; *Cutch*: Buchnaga—; *Gujerat*: Shingadiovachnag, Vachhanag, Vachnag—; *Hindi*: Bachhnag, Bachnak, Bis, Bish, Mahoor, Mithazahar, Singya, Singyabis, Teliyabis—; *Lepcha*: Lung-ji nying, Nyine, Nying—; *Malayalam*: Vatsanabhi—; *Marathi*: Bachnag, Vachnag—; *Nepal*: Atisingeeabish, Bikh, Bish, Bishnak—; *Persian*: Bishnag, Zher—; *Sanskrit*: Aheya, Amrita, Bhugara, Brahmaputra, Darada, Gara, Garada, Garala, Ghora, Halahala, Haridra, Jangala, Jangula, Jivanaghata, Kakola, Kalakuta, Kasakula, Kishala, Kshveda, Nila, Pradipana, Pranahara, Raktashringika, Rasa, Rasayana, Shanklakeya, Shringi, Sowrashtrika, Tikshna, Vatsanabha, Visha—; *Sinhalese*: Vachanabhi—; *Tamil*: Vashanavi—; *Telugu*: Ativasa, Naghi, Vasanabhi—.

12. **Aconitum Elwesii** Stapf is found in the alpine Himalaya of north-eastern Sikkim.

The root is reputed poisonous.

Lepcha: Tukschak gniong—.

13. **Acconitum lethale** Griff. occurs in the higher parts of the Mishmi mountains.

The root contains the non-poisonous alkaloid *palmatisine*. It is, however, believed to be the source of the celebrated 'Bhi' or 'Bis' poison of the Mishmis.

Mishmi: Bee, Bih, Bhi, Bis—.

ACTAEA.

This genus consists of 10 species distributed over the northern temperate regions.

A. alba Mill., *A. eburnea* Rydb., and *A. spicata* Linn. are used medicinally in North America by the Meskwaki, Pillager Ojibwe, and Forest Potawatomi Indians of the State of Wisconsin.

Actaea spicata Linn. occurs in the temperate Himalaya from Hazara to Bhutan. It is distributed to Northern Asia, Europe, and North America.

The root is an active emeto-purgative, acrid, parasiticide: it is held in repute as an expectorant and antispasmodic. It must be administered with caution.

Canadian doctors administer the root in snake-bite; and it is said to be attended with much success in the treatment of nervous diseases, rheumatic fever, chorea and lumbago. The berries were formerly used internally for asthma and scrofula, and externally for skin complaints.

The Pillager Ojibwe and the Forest Potawatomi Indians of Wisconsin make a tea from the root, to be drunk by women after childbirth; it is meant to clear up the system. White men use it for stomach troubles, and as a substitute for *Cimicifuga racemosa* Nuttall. It has been used in cases of ovarian neuralgia, uterine tenderness, sub-involution, and as a substitute for digitalis in the treatment of fatty or irritable heart. It affords some relief for a headache which is due to eye strain.

In some parts of Europe the powdered leaves, stems, and flowers are used as an insecticide. The toxic principle is referred to as 'oil of baneberry'.

Mhaskar and Caius have shown that the root is useless as an antidote to snake-bite.

Catalan: Herba de San Cristófol—; *English*: Baneberry, Bugbane, Grapewort, Herb Christopher, Toadroot—; *French*: Actée, Actée en épi, Christophoriane, Faux hellébore noir, Herbe aux poux, Herbe de saint Christophe—; *German*: Christophskraut, Christophswurz—; *Italian*: Barba di capra—; *North America*: Coral-and-pearl, European Baneberry, Grapewort, Herb Christopher, Poison-berry, Red Baneberry, Red-berry Snakeroot, Red Cohosh, Snake-berry, Toad-root—; *Ojibwe*: Wîckobidjibîk—; *Potawatomi*: Wasgobîdjibikeok—; *Spanish*: Cristobalina, Yerba de San Cristobal—.

ADONIS.

The genus numbers 10 species inhabiting the northern palaeotemperate regions, chiefly southern Europe and western Asia.

A. aestivalis Linn., *A. autumnalis* Linn., *A. microcarpa* DC., *A. vernalis* Linn. are used medicinally in Europe; *A. vernalis* Linn. is similarly used in North America.

The herb of *A. vernalis* Linn. is official in Austria, Holland, Italy, Russia, Spain, and Switzerland.

Adonis aestivalis Linn. extends from temperate Europe to Asia; it is found in the Western Himalaya from Peshawar to Hazara and Kumaon.

The whole plant is used as a cardiotonic and diuretic in Euope. The flowers are considered laxative, diuretic, and lithothriptic.

Baluchi: Buchki—; *Catalan*: Ull de perdiu—; *French*: Adonide d'été, Rougeotte—; *German*: Feuerroeschen—; *Languedoc*: Rubissa—; *Reggio*: Pe d'esen—; *Romagna*: Chiga sanev, Erba corta, Fior d'Adone, Gattapozzia—; *Russian*: Goritzvyet Krasnee—; *Sardinia*: Ogu de boi—; *Spanish*: Ojo de perdiz—; *Treviso*: Gioze de sanguse, Stiantamalanni—; *Waziri*: Chadangul, Lavang—.

ANEMONE.

This cosmopolitan genus numbers 100 species.

The following are used medicinally in Europe—*A. alpina* Linn., *A. coronaria* Linn., *A. hepatica* Linn., *A. narcissiflora* Linn., *A. nemorosa* Linn., *A. pratensis* Linn., *A. pulsatilla* Linn., *A. ranunculoides* Linn., *A. stellata* Lam., *A. sylvestris* Linn., *A. trifolia* Linn., *A. vernalis* Linn.—; in North America—*A. acutiloba* Laws.,

A. cylindrica A. Gray, *A. dichotoma* Linn., *A. parviflora* Michoe., *A. patens* Linn., *A. pratensis* Linn., *A. thalictroides* Linn., *A. virginiana* Linn.—; in Japan and China—*A. cernua* Thunb.—; in Indo-China—*A. cernua* Thunb., *A. dichotoma* Linn., *A. hepatica* Linn., *A. japonica* Sieb. and Zucc., *A. nikoensis* Max.—; in South Africa—*A. caffra* Harv., *A. transvaalensis* Prantl.—.

The leaves and flowers of *A.. nemorosa* Linn. are official in Portugal, those of *A. pulsatilla* Linn. in France and Portugal.

A. multifida Poir. and *A. nemorosa* Linn. are reputed poisonous in America.

In Siberia the juice of *A. altaica* Fisch., *A. narcissiflora* Linn., *A. parviflora* Mich., *A. ranunculoides* Linn., *A. reflexa* Steph. is used to poison arrows.

Anemonin, or anemone camphor, has been obtained from *A. nemorosa, A. pratensis, A. pulsatilla,* and *A. transvaalensis.*

1. Achenes strigose, rarely glabrous. Radical leaves 3-partite ... ... *A. obtusiloba.*
2. Achenes nearly orbicular with an incurved beak. Leaves palmately 5-partite ... ... *A. narcissiflora.*

1. **Anemone obtusiloba** D. Don occurs in temperate and alpine Himalaya from Kashmir to Sikkim, between 8,000 and 15,000 ft.

In Hazara the pounded root, which is acrid, is mixed with milk and given internally for contusions. In Bessahir it is said to be used as a blister, but to be apt to produce sores and scars.

The seeds, if given internally, produce vomiting and purging. The oil extracted from them is used in rheumatism.

Jaunsar: Ageli—; *Kumaon*: Kakriya, Ratanjota, Rattanjog—; *Punjab*: Padar, Ratanjota, Rattanjog—.

2. **Anemone narcissiflora** Linn. is found in Kashmir. It is distributed to the Alps of Southern and Central Europe, Western Asia, Siberia, and North-Western America.

The plant is toxic and is used as a poison in most places where it is found growing.

AQUILEGIA.

The genus consists of 75 species occurring in the northern temperate regions.

A. alpina Linn., *A. Kitaibelii* Schott, *A. parviflora* Ledeb., *A. pyrenaica* DC., *A. vulgaris* Linn. are used medicinally in Europe; *A. canadensis* Linn., *A. formosa* Fisch., *A. vulgaris* Linn. are similarly used in North America.

Aquilegia vulgaris Linn. is found in the Western temperate and subalpine Himalaya; it is distributed over temperate Europe and Asia.

In Europe the plant and the seeds are credited with aperient, resolvent, diuretic, and diaphoretic properties. The leaves are generally used in lotions with good success for sore mouths and throats. A tincture is employed to strengthen the gums.

'Tragus saith that a dram of the seed taken in wine with a little saffron openeth obstructions of the liver and is good for the yellow jaundice, if the person after taking thereof be laid to sweat well in bed. The seed also taken in wine causeth a speedy delivery of women in childbirth; if one draught suffice not let her drink a second, and it will be found effectual. The Spaniards used to eat a piece of the root thereof in a morning fasting, many days together, to help them when troubled with stone in the veins or kidneys.'

The plant contains a cyanogenetic glucoside.

Catalan: Ausellets, Corns, Perfet amor—; *Dutch*: Akelei—; *English*: Blue Starry, Boots-and-shoes, Capon's-feather, Capon's-tail, Cock's-foot, Colombine, Colourbine, Columbine, Cullavine, Culverkeys, Culverwort, Curranbine, Dove's-foot, Granny's Night-cap, Hawk's-feet, Hen and Chickens, Lady's shoes, Lady's slippers, Snapdragon, Sowdwort, Two faces under a hat—; *French*: Ancholie, Ancholie commune, Ancholie des jardins, Ancholie sauvage, Ancholie vulgaire, Ancolie, Bonne-femme, Clochette, Colombine, Cornette, Eglantine, Fleur-églantine, Galantine, Gants de Notre-Dame, Gélantine, Manteau royal—; *German*: Acklei, Adlerblue, Adlersblume, Aglei, Akelei, Falsche Glockenblume, Glockenblume, Harlekinsblume, Jowisblume, Jupiterblume, Unsererliebenfrauen-handschuh—; *Italian*: Amor mascosto, Aquilegia, Aquileia, Aquilina—; *Roumanian*: Caidarusa, Cinci coade, Clopotei cornuti—; *Russian*: Vadasbor—; *Spanish*: Aquileña, Aquileña, Manto real, Pajarilla, Pajaritas—; *United States*: Boots-and-shoes, Cock's-foot, Culverwort, Dove's-foot, European Columbine, Garden Columbine, Granny's-nightcap, Hawk-feet, Lady's-shoes—.

CALTHA.

The genus numbers 20 species inhabiting the temperate regions of the world.

Caltha palustris Linn. is found in western temperate Himalaya from Kashmir to Nepal, between 8,000 and 10,000 ft. It occurs in temperate Europe and North America.

Every part of the plant is strongly irritant, and cases are on record of serious effects produced by rashly experimenting with it. The juice is efficacious in curing warts.

Internally, the boiled young sprouts were used by the Iroquois in rheumatism and neuritis.

In the Pyrenees and in Germany the young buds are pickled like capers.

A tincture made from the whole plant when in flower may be given in cases of anaemia, in small, well-diluted doses.

Catalan: Flor del mal d'ulls—; *English*: Bassinet, Big Watercup, Blob, Boots, Bublican, Bull-flower, Bulls' Eyes, Butter-blob, Carlock-cups, Chirms, Christ's Eyes, Claut, Colt's Foot, Crow Cranes, Crazy, Dandelion, Drunkard, Fire o'Gold, Golden Cup, Goldins, Golds, Gollin, Great Butter-flower, Ground Ivy, Halcups, Horse-blobs, John Georges, Johnny Cranes, Kingcob, King Cup, Leopard's Foot, Mare-blebs, Mare-blob, Marsh Horsegowl, Marsh Mallow, Marsh Marigold, May Blobs, Mayflower, Meadow Bouts, Meadow Bright, Meadow Gowan, Moll-blob, Open Gowan, Palsywort, Publicans and Sinners, Soldiers' Buttons, Water-blobs, Water-bouts, Water Buttercup, Water Caltrops, Water Dragon, Water Goggles, Water Gowan, Wild Bachelor's Buttons, Yellow Gowan—; *French*: Populage, Souci d'eau, Souci des marais—; *German*: Butterblume, Dotterblume, Kuhblume, Sumpfdotterblume—; *Italian*: Farferugine—; *Menomini*: Wäsasipu 'Kwa—; *North America*: American Cowslip, Cowslip, Cow's Lip, Marsh Marigold—; *Ojibwe*: O'îgîte'bûg—; *Punjab*: Baringu, Mamiri—; *Spanish*: Calta, Centella palustre, Yerba centella—.

CIMICIFUGA.

This genus consists of 12 species, which inhabit northern temperate regions.

C. foetida Linn. is used medicinally in Europe, China, and Indo-China; *C. racemosa* (L.) Nuttall in North America, where the dried rhizome and roots are official.

Cimicifuga foetida Linn. inhabits temperate Himalaya from Kashmir to Bhutan, between 7,000 and 12,000 ft. It is found in Siberia and Eastern Europe.

The plant is a nerve depressant.

In Europe the root is considered a mild emeto-purgative. In China and Indo-China it is used as an antiperiodic and sudorific, and prescribed in rheumatic affections, dropsy, the early stage of phthisis, and chronic bronchial diseases.

In Siberia the root is used to drive away bugs and fleas.

Annam: Thang ma—; *Chinese*: Sheng Ma—; *English*: Bugbane—; *French*: Actée fétide, Chasse-punaise, Cimicaire, Herbe aux punaises—; *German*: Wanzenkraut—; *Punjab*: Jiunti—; *Spanish*: Yerba de las chinches—.

CLEMATIS.

A cosmopolitan genus numbering 220 species, rare in the tropics.

This genus is acrid and poisonous; the leaves and fresh stems, if bruised and applied to the skin, cause vesication; in some species the seeds have purging properties.

The following species are used medicinally in Europe—*C. alpina* Mill., *C. flammula* Linn., *C. integrifolia* Linn., *C. recta* Linn., *C. vitalba* Linn., *C. viticella* Linn.—; in Indo-China—*C. apiifolia* DC., *C. chinensis* Retz., *C. dioica* Linn., *C. florida* Thunb., *C. paniculata* Thunb., *C. patens* Morr. and Dec., *C. smilacifolia* Wall., *C. virginiana* Linn., *C. vitalba* Linn.—; in China—*C. apiifolia* DC., *C. chinensis* Retz., *C. paniculata* Thunb., *C. recta* Linn.—; in North America—*C. crispa* Linn., *C. ligusticifolia* Nutt., *C. viorna* Linn., *C. virginiana* Linn.—; in Jamaica—*C. dioica* Linn.—; in Guinea and Nigeria—*C. grandiflora* DC., *C. Thunbergii* Steud.—; in the islands of Mauritius and La Reunion—*C. mauritiana* Lam.—; in Madagascar—*C. Bojeri* Hook., *C. mauritiana* Lam., *C. saxicola* Bojer and Hills.—; in Southern Africa—*C. brachiata* Thunb., *C. stanleyi* Hook., *C. stewartiae* Burtt-Davy—.

A. Achenes with long feathery styles. Flowers in axillary fascicles; pedicels 1-flowered ... ... 1. *C. napaulensis.*

B. Achenes with long feathery tails. Flowers in axillary panicles

- I. Filaments glabrous, connective of anthers not produced
  - a. Glabrous ... ... 3. *C. smilacifolia.*
  - b. Silky ... ... 2. *C. triloba.*
  - c. Young parts pubescent ... ... 4. *C. gouriana.*
- II. Filaments hairy, connective of another not produced ... ... 5. *C. graveolens.*

1. **Clematis napaulensis** DC is found in teemperate Himalaya from Garhwal to Bhutan. In Kumaon throughout, between 4,500 and 7,000 ft., in shady moist ravines.

In Kanawar the leaves are said to act deleteriously on the skin.

Kumaon: Ghantiali, Jai, Kanguli—; *Punjab*: Birri, Pawanne, Wandak—.

2. **Clematis triloba** Heyne inhabits the Bombay Presidency: Konkan and Deccan, Western Ghats.

It is applied externally to boils and itch, and to parasitic worms.

It is prescribed for the treatment of snake-bite; but Mhaskar and Caius have found every part of the plant equally useless in the antidotal treatment of snake-bite.

Bombay: Moravela, Moriel, Morvel, Ranjae, Ranjai—; *Canarese*: Morhari—; *Gujerat*: Morvel—; *Hindi*: Churahar, Dhantiali, Murhari—; *Marathi*: Moravela, Morbel, Ranjai—; *Matheran*: Morvel, Ranjat—; *Porebunder*: Trekhdovelo—; *Sanskrit*: Murva, Devashreni, Devi, Dhanurmala, Dhanurguna, Dridhasutrika, Gokarni, Laghuparnika, Madhulika, Madhurasa, Madhushreni, Madhusrava, Morata, Piluparni, Prithakaparni, Snigdhaparni, Snuva, Tejani—; *Sind*: Maruva—.

3. **Clematis smilacifolia** Wall. occurs in Kumaon, the Himalaya from Sikkim eastwards up to 5,000 ft., Assam, Ganjam, Konkan, Kanara and Deccan of the Bombay Presidency. It extends to Indo-China, Java, Borneo, and the Philippine Islands.

In Indo-China a decoction of the roots is used for aching of the muscles.

Canarese: Hottuhambu—; *Indo-China*: Day ong lao, Khoua kau san, Linh tien—; *Lepcha*: Krang rik—; *Malayalam*: Vatiyampu—; *Sinhalese*: Narawel—; *Telugu*: Gurraputige—.

4. **Clematis gouriana** Roxb. is found in the Punjab hills, the Western Himalaya up to 5,000 ft., the hilly districts throughout India between 1,000 and 3,000 ft., and Ceylon. It is distributed to Java and the Philippine Islands.

The leaves of the fresh stems, if bruised and applied to the skin, cause vesication.

Bombay: Moriel, Morvel, Ranjai—; *Canarese*: Telejadari—; *Dehra Dun*: Belkangu, Belkum, Belkun—; *English*: Indian Traveller's Joy—; *North-Western Provinces*: Belkangau, Belkun—; *Saora*: Pannedang, Pedutivva—; *Uran*: Golarang—; *Uriya*: Boromojhanti, Idiya—; *Visayan*: Calupad, Tolangsangbanug—.

5. **Clematis graveolens** Lindl. inhabits the Western temperate Himalaya to Kumaon, between 6,000 and 11,000 ft., and extends to Baluchistan, where it is considered poisonous.

Baluchi: Hushokawal—.

COPTIS.

This genus consists of 10 species which are found in the northern temperate and arctic regions of the world.

C. anemonaefolia Sieb. and Zucc. is used medicinally in Malaya, Indo-China, Japan, and North America; *C teeta* Wall. in India, Malaya and China; *C. trifolia* Salisb. in North America.

Berberine has been isolated from *C. teeta* and *C. trifolia*.

The dried rhizome of the species growing in Japan is official in the Japanese pharmacopoeia.

Coptis teeta Wall. grows in the temperate parts of the Mishmi mountains, east of Assam.

The root is an all-round tonic and stomachic medicine. It has been found to produce excellent effects in debility, convalescence after fevers and other debilitating diseases, atonic dyspepsia, and in mild forms of intermittent fevers.

It is used in Sind, Assam, and other parts of India for restoring the appetite, being given in infusion; also as an appliation on sores and for inflammation of the eyes.

It contains berberine and coptine.

Assam: Mishmeeteeta, Tita—; *Bombay*: Mamiran, Mishmitita—; *Cantonese*: Choon lin—; *Chinese*: Huang Lien—; *English*: Coptis, Gold Thread—; *French*: Coptide— *Hindi*: Mamira, Mamiran, Mishmitita—; *Malaya*: Choon lin—; *Sind*: Mahmira—; *Sinhalese*: Pitakarosana—.

DELPHINIUM.

The genus numbers 150 species, natives of temperate regions in the northern hemisphere.

The flowers are acrid, bitter, astringent; the seeds are emetic, cathartic, anthelmintic, and insecticidal.

The following species are used medicinally in Europe—*D. ajacis* Linn., *D. consolida* Linn., *D. elatum* Linn., *D. grandiflorum* Linn., *D. hybridum* Willd., *D. laxiflorum* DC., *D. montanum* DC., *D. orientale* Gay., *D. palmatifidum* DC., *D. peregrinum* Linn.. *D. pubescens* DC., *D. staphisagria* Linn., *D. triste* Fisch.—; in Indo-China—*D. ajacis* Linn.—; in North America—*D. ajacis* Linn., *D. consolida* Linn., *D. staphisagria* Linn.—.

Deltaline has been isolated from the whole plant of *D. occidentale* S. Wats., and delcosine from *D. consolida* Linn.

Ajacine, and ajaconine have been isolated from the seeds of *D. ajacis;* delphinine, delphisine, delphinoidine, and staphisagroïne have been obtained from the seeds of *D. staphisagria.*

D. consolida Linn. in flower is official in Portugal; the seeds of *D. staphisagria* Linn. are official in France, Holland, and Portugal.

I. Spur subulate
   1. Radical leaves 5-15 cm. diam., orbicular, divided to the base or nearly so, 5-9-parted. Follicles 3. ... 1. *D. denudatum.*
   2. Radical leaves 3-4 cm. diam., suborbicular, 5-7-lobed. Follicles 5 ... ... 2. *D. caeruleum.*
   3. Radical leaves 10-15 cm. diam., orbicular or reniform, 5-7-lobed or -parted. Follicles 3 ... 3. *D. elatum.*

II. Spur inflated, conical ... ... 4. *D. Brunonianum.*

1. **Delphinium denudatum** Wall. occurs in the Western temperate Himalaya from Kashmir to Kumaon, between 8,000 and 12,000 ft.,

Ayurveda doctors consider the bitter root as cooling, vulnerary, and alexiteric. They recommend its use for diseases of the blood, and for the treatment of snake-bite and scorpion-sting.

Yunanists describe the root as a stimulant in conditions of debility, a brain tonic useful in insanity, an analgesic good in tooth ache and in painful piles, an antidote to snake and scorpion venoms. The drug promotes appetite, but it is strongly binding and a purgative should be administered with it.

In Bashahr the root is used for tooth ache, and also as an adulterant for aconite.

Mhaskar and Caius have shown that the root is not an antidote to either snake or scorpion venom.

Arabic: Judwar, Mahferfin—; *Bhutia*: Nirbisi—; *Canarese*: Nirvishi—; *Gujarat*: Nirvishi—; *Hindi*: Judwar, Nirbishi, Nirvisi—; *Jaunsar*: Main—; *Marathi*: Nirvishi—; *Nepal*: Nilobikh—; *Persian*: Zadwar—; *Punjab*: Judwar—; *Sanskrit*: Apavisha, Avisha, Nirvisha, Vishabhava, Vishaha, Vishahantri, Vishavairini, Vivisha—; *Simla*: Manila—; *Sind*: Judwar—; *Urdu*: Jadwar—.

2. **Delphinium caeruleum** Jacq. is found in the alpine Himalaya from Kumaon to Sikkim, up to 17,000 ft. in the Eastern Himalaya.

The root is applied to kill the maggots in the wounds of goats.

Punjab: Dakhangu, Dhakangu—.

3. **Delphinium elatum** Linn. spreads over the western temperate Himalaya from Kashmir to Kumaon, between 10,000 and 12,000 ft.

The whole plant, but more especially the seed, is emetic, aperient, diuretic, and anthelmintic.

This plant is found growing in Europe where the seeds are used as insecticides, and much employed in the treatment of itch and other skin eruptions. The flowers are astringent and used in troubles of the eye.

The toxic principles are delphinine, delphisine, and delphinoidine.

4. **Delphinium Brunonianum** Royle grows in the alpine region of Western Tibet at an altitude of 14,000 ft.

The juice of the leaves of this plant is used in Kurram to destroy ticks in animals, chiefly when they affect sheep. In Leh it is considered so poisonous that the dew from the leaves falling on grass is said to poison cattle and horses.

Garhwal: Kasturi—; *Kumaon*: Nepari—; *Ladakh*: Ladara—; *Pangi*: Mundwal—; *Ravi*: Sapfalu—; *Sutlej*: Laskar, Liokpa, Panni, Ruskar, Spet, Supalu—.

NIGELLA.

The genus includes 16 species, natives of the Mediterranean region and Central Europe, one of them cultivated and run wild in India.

The following are used medicinally in Europe—*N. arvensis* Linn., *N. damascena* Linn., *N. hispanica* Linn., *N. sativa* Linn.—; in Syria—*N. orientalis* Linn.—; in Algeria and the Hadramaut—*N. arvensis* Linn.—; in Malaya and North America—*N. sativa* Linn.—.

Nigella sativa Linn. is found in the Punjab and Bihar, sometimes cultivated and an occasional weed of cultivation. It extends to the Mediterranean region.

The seeds when bruised are strongly scented, and they are used in Europe as emmenagogues, diuretics, and galactagogues. They are carminative and are an excellent adjunct to purgative draughts. Put among woollens, furs, etc., they prevent the attack of insects.

In the Presidencies of Bengal and Bombay the seeds are given to nursing women to promote the secretion of milk. In the Punjab they are reckoned stimulant, and are administered by hakeems in milk for rheumatism and cough. It is said they were at one time used in place of pepper.

The oil expressed from the seeds is employed for the same purposes as the seed.

In eruptions of the skin the seeds reduced to powder and mixed with sesamum oil are much used in India as an external application.

Koman found the seeds useful in mild cases of puerperal fever.

The seeds have been recommended for the treatment of snake-bite and scorpion-sting; but Mhaskar and Caius have demonstrated that they are not an antidote to snake venom, and that they are useless in the treatment of scorpion sting.

Afghanistan: Shewadaru, Siyahdaru—; *Arabic*: Habbat-us-suda, Hab-es-souda, Kamun-e-avad, Shoonez, Shuniz—; *Bengal*: Kalajeera, Kalijira, Kalzira, Mugrela—; *Bombay*: Kalenjira, Kalonji—; *Burma*: Samonne—; *Canarese*: Karejirage, Karijirigi, Karimsiragam—; *Deccan*: Kulanjan—; *Egypt*: Habb-suda, Hubsindee, Kamnun iswid, Niyeh—; *English*: Black Cummin, Fennel Flower, Nutmeg Flower, Roman Coriander, Small Fennel—; *French*: Cumin noir, Faux cumin, Nielle de Crète, Nielle romaine, Nigelle cultivée, Nigelle romaine, Quatre épices, Toute épice—; *German*: Schwarzkuemmel, Gemeiner Schwarzkuemmel, Roemischer Schwarzkuemmel—; *Gujerati*: Kalonjijirum—; *Hebrew*: Qesah—; *Hindi*: Kalajira, Kalonji, Mugrela—; *Iraq*: Habbat-soda—; *Italian*: Cinnamonea, Cuminella, Erba spezie—; *Kashmir*: Tukmigandna—; *Malaya*: Ku sheng—; *Malayalam*: Karunchirakam, Karunshiragam—; *Mauritius*: Cumin noir, Nigelle—; *North America*: Black Caraway, Black Cumin, Fitch, Roman Coriander—; *Persian*: Shuniz, Siahudanah, Siyahbiranj—; *Punjabi*: Kalvanji—; *Russian*: Tchernushka—; *Sanskrit*: Bashpika, Kalajaji, Karavà, Karavi, Krishnajiraka, Kunchi, Kunchika, Kunjika, Musavi, Prathvikà, Prithivi, Frithu, Prithuka, Prithivi, Sthuljiraka, Sushavi, Upakunchiraka—; *Sinhalese*: Kaluduru—; *Tamil*: Karunjiragam—; *Telugu*: Nellajeelakara, Nullajilakara—.

PAEONIA.

This genus numbers 15 species distributed over Europe, Asia, and the western part of North Africa.

The following are used medicinally in Europe—*P. corallina* Retz., *P. Moutan* Sims, *P. officinalis* Linn., *P. peregrina* Mill., *P. tenuifolia* Linn.—; in Japan, China, and Indo-China—*P. albiflora* Pall., *P. Moutan* Sims—; in California—*P. Brownii* Dougl., *P. officinalis* Linn.—.

Paeonia emodi Wall. is found in the Western Himalaya from Kashmir to Kumaon.

The tubers of this plant are highly esteemed by hakeems as a medicine for uterine diseases, colic, bilious obstructions, dropsy, epilepsy, convulsions and hysteria. *Ud-salap* is generally given to children as a blood-purifier.

In Booner, the root in combination with other drugs is a favourite remedy for bruises, sprains, etc.

An infusion of the dried flowers is highly valued as a remedy for diarrhoea.

The seeds are emetic and cathartic.

Bhutia: Bhumamadiya, Yetghas—; *English*: Himalayan Paeony, Paeony Rose—; *Hindi*; Udsalap—; *Kashmir*: Mid—; *North-Western Provinces*: Chandra—: *Punjab*: Mamekh—; *Urdu*: Udasaliba—.

RANUNCULUS.

The genus includes 300 species distributed all over the world, but most numerous in temperate or cool regions.

The following species are used medicinally in Europe— *R. acris* Linn., *R. arvensis* Linn., *R. asiaticus* Linn., *R. bulbosus* Linn., *R. falcatus* Linn., *R. Ficaria* Linn., *R. Flammula* Linn., *R. glacialis* Linn., *R. gramineus* Linn., *R. hederaceus* Linn., *R. Lingua* Linn., *R. monspeliacus* Linn., *R. muricatus* Linn., *R. pedatus* W. & K., *R. parviflorus* Linn., *R. repens* Linn., *R. sceleratus* Linn., *R. Thora* Linn., *R. trichophyllus* Chaix, *R. trilobus* Desf.—; in China—*R. acris* Linn., *R. sceleratus* Linn.—; in Indo-China—*R. Langsdorfii* DC., *R. pensylvanicus* Linn. f., *R. sceleratus* Linn., *R. ternatus* Thunb.—; in North America—*R. abortivus* Linn., *R. bulbosus* Linn., *R. Flammula* Linn., *R. multifidus* Pursh., *R. sceleratus* Linn.—; in Madagascar—*R. pinnatus* Poir.—; in South Africa—*R. capensis* Thunb., *R. pinnatus* Poir., *R. pubescens* Thunb.—.

A. Water plant. Flowers white ... ... 1. *R. trichophyllus.*
B. Land plants. Flowers yellow
  I. Achenes smooth or granular, not muricate or tubercled, shortly beaked
    1. All leaves undivided, lanceolate, entire or remotely toothed ... ... 2. *R. Lingua.*
    2. Leaves 3-partite. Achenes many in an oblong head, not margined ... ... 3. *R. sceleratus.*
    3. Leaves ternatisect. Achenes smooth, glabrous, margined .. .. 4. *R. pensylvanicus.*
  II. Achenes tubercled or muricate
    1. Sepals appressed to and half the length of the petals ... ... 5. *R. arvensis.*
    2. Sepals reflexed, rather shorter than the petals ... 6. *R. muricatus.*
  III. Achenes gibbous at the side with a long terminal beak. 7. *R. falcatus.*

1. **Ranunculus trichophyllus** Chaix is found in the plains of the Punjab, in the Western Himalaya from the Indus to Kumaon ascending to considerable elevations, in Waziristan, Baluchistan and Afghanistan whence it extends to North Africa and Europe and spreads over the temperate parts of both hemispheres.

The amount of toxic principle is so small as to be negligible and, in Europe, the plant is given in intermittent fevers, rheumatism, and asthma.

English: Water Crowfoot, Water Fennel—.

2. **Ranunculus Lingua** Linn. occurs in Kashmir; it is found throughout the temperate zone of the northern hemisphere.

In Europe the leaves are applied as a blister to the wrists in rheumatism.

English: Great Spearwort, Sparrow Weed, Sparrow Crowfoot—; *French*: Grande douve, Renoncule lancéolée—; *German*: Grosser Hahnenfuss—.

3. **Ranunculus sceleratus** Linn. is found in Mount Abu, Sind, Waziristan, Northern India, the warm valleys of Himalaya, and Bengal; it extends to Siam and the temperate zone of the northern hemisphere.

The plant is an emmenagogue and a galactagogue; it is much used in the treatment of skin diseases.

The fresh plant is poisonous, and produces violent effect if taken internally. The bruised leaves form an application to raise blisters, and may also be used to keep open sores caused by vesication, or by other means.

The leaves are still used occasionally in Europe as a vesicant. When made into a tincture with spirit of wine, and given in small diluted doses, the plant proves curative of stitch in the side, and of neuralgic pains between the ribs, likewise of pleurisy without feverishness.

In Indo-China the seeds are given as a stomachic, to destroy foul breath, to reduce abscesses, and in kidney troubles.

The plant contains an acrid volatile oil, an inert resin, and a narcotic principle called *anemonin.*

Arabic: Kafessaba—; *Catalan*: Gata rabiosa—; *Chinese*: Shih Lung Juei—; *Danish*: Puggepebe—; *Dutch*: Water haanewoet—; *Egypt*: Zaghalanta, Zaghlul—; *English*: Celery-leaved Buttercup, Celery-leaved Crowfoot, Marsh Crowfoot, Water Celery—; *French*: Grenouillette aquatique, Grenouillette d' ean, Herbe sardonique, Mort aux vaches, Renoncule des marais, Renoncule scélérate—; *German*: Blasenziehender Hahnenfuss, Boese Ranunkel, Boesewicht, Froscheppich, Giftranunkel—; *Indo-China*: Thach long noi—; *Italian*: Appio riso, Erba sardonica, Pie corvino, Ranocchiella aquatica, Ranunculo di palude—; *Kumaon*: Shim—; *North America*: Biting Crowfoot, Blisterwort, Cursed Crowfoot, Marsh Crowfoot, Water Celery—; *Persian*: Kabikaj—; *Portuguese*: Ranunculo mataboi—; *Spanish*: Ranunculo malvado—; *Tirhut*: Polica—.

4. **Ranunculus pensylvanicus** Lin. fil. is found in the Upper Gangetic Plain, the Nepal Terai, the Khasia hills up to 6,000 ft.; it extends to North China, Amur-land, and North America.

The plant is used to raise blisters.

Indo-China: Tieu hoi hoi toan—.

5. **Ranunculus arvensis** Linn. occurs in the Western Himalaya from Kashmir to Kumaon, in Mount Abu, in Afghanistan, extending to temperate Asia, Siberia, Europe, and North Africa.

In Europe the plant is used in intermittent fevers, gout, and asthma.

As a fodder it frequently produces symptoms of irritant poisoning.

The leaves are said to contain hydrocyanic acid.

Baluchi: Zardphul—; *Boya*: Kanjarvaliai—; *Egypt*: Zaghalantah—; *English*: Corn Buttercup, Corn Crowfoot, Crow's-claws, Devil-on-both-sides, Devil's-claws, Devil's Coach-wheel, Devil's Currycomb, Dill-cup, English Stavesacre, Field Crowfoot, Goldweed, Gye, Hard-iron, Hedge-hog, Hellweed, Horse Gold, Hungerweed, Jack-o'-both-sides, Joy, Peagle, Prickle-backs, Scratchbur, Starveacre, Yellowcup, Urchin Crowfoot, Yellow Crees—; *French*: Bassinet des champs, Picot, Piguereaux—; *Malta*: Devil's Claws, Hedgehog—; *Miram Shah*: Zirgulac—; *Punjab*: Chambul—; *Spanish*: Coronilla de la Virgen, Gata rabiosa, Yerba del amor—; *Waziri*: Peri makhlak, Sarsundan—.

6. **Ranunculus muricatus** Linn. occurs in the Punjab and Punjab Himalaya, Kashmir, Western Asia, Europe, and North America.

In Europe the plant is used in intermittent fevers, gout, and asthma.

Egypt : Zaghlil—; *Spanish* : Abrojos a cuatro—.

7. **Ranunculus falcatus** Linn. occurs in Kashmir, Punjab, Baluchistan, Western Asia, and South-eastern Europe.

The plant when pounded and applied to the skin produces blisters.

Baluchistan : Wahwashu—; *Pushtu* : Karamghundai—.

THALICTRUM.

The genus consists of 80 species which inhabit the northern hemisphere.

The following are used medicinally in Europe—*T. angustifolium* Linn., *T. aquilegifolium* Linn., *T. flavum* Linn.—; in Indo-China—*T. aquilegifolium* Linn., *T. flavum* Linn., *T. foliolosum* DC.—; in North America—*T. polycarpum* S. Wats.—; in South Africa—*T. minus* Linn.—.

Achenes short, oblong, ribbed, subsessile. Leaves ternately or pinnately decompound.

1. Stem 4-8 ft. glabrous. Achenes usually 2-5, small, oblong ... ... *T. foliolosum.*
2. Stem 2-4 ft. erect. Achenes 5-8, oblong ... *T. minus.*

1. **Thalictrum foliolosum** DC. is found throughout the Himalaya between 5,000 and 8,000 ft., the Khasia hills between 4,000 and 6,000 ft., Burma, and Siam.

The root combines tonic and aperient properties, and has been found useful in convalescence after acute diseases, in mild forms of intermittent fevers, and in atonic dyspepsia.

The root is largely used as an *anjan,* or application for ophthalmia in Afghanistan and throughout India.

In the Punjab, the root is used as a purgative and diuretic.

Arabic : Mamiranchini—; *Bengal* : Gurbiani—; *Bombay* : Mamiran—; *Hindi* : Mamira, Pilijari, Pinjari, Shuprak—; *Kashmir* : Chaitra—; *Kumaon* : Barmat, Penglajari, Pilajari—; *Persian* : Mamiranchini—; *Punjab* : Chireta, Chitramul, Gurbiani, Keraita, Mamira, Pashmaran, Phalijari—.

2. **Thalictrum minus** Linn. is found in the inner valleys of the Temperate Himalaya, and in West Tibet between 9 000 and 12,000 ft.; it is distributed to Northern Asia, Europe, Abyssinia and Southern Africa.

An infusion of the leaf or a decoction of the root is used by the Sutos for fever.

Sosuto : Lefokotsane, Tloro-ya-ngwale—.

MEDICINAL AND POISONOUS PLANTS OF INDIA: *MAGNOLIADS, DILLENIADS, ANONADS, MENISPERMADS, BERBERIDS.*

BY

J. F. CAIUS, S.J., F.L.S.

I

The MAGNOLIACEAÆ are trees or shrubs, sometimes climbing, often aromatic. They are chiefly natives of the tropical and temperate Asiatic mountains and United States, a few are Australian.

There are about 70 species belonging to 10 genera, 9 of which are used medicinally in various parts of the world:—DRIMYS (South America; New Zealand to Borneo); EUPTELEA (Japan to Bengal); ILLICIUM (Atlantic North America; Asia); KADSURA (tropical Asia, China, Japan); LIRIODENDRON (North America); MAGNOLIA (Asia, North America); MICHELIA (tropical Asia, China); SCHIZANDRA (tropical and warm temperate Asia, North America); TALAUMA (eastern Asia, South America, West Indies).

Many members yield essential oils.

Among the products isolated the following may be mentioned:—(1) *cyanogenetic glucosides* and *saponins*; (2) *crystalline compounds of alkaloidal* (?) *nature*—tulipiferine—; (3) *bitter principles*—liriodendrin, magnolin, shikimin—; (4) *toxic principles*—sikimitoxin—: (5) *tannins* and *resins*—; (6) *mucilage*.

The medicinal Magnoliads of India belong to the three genera ILLICIUM, KADSURA, and MICHELIA.

A. Erect trees or shrubs.
  Stipules absent. Carpels in one whorl .. ... ILLICIUM.
  Stipules conspicuous. Gynophore stalked ... ... MICHELIA.

B. Climbing shrubs.
  Carpels of fruit capitate ... ... KADSURA.

ILLICIUM.

The genus consists of 20 species distributed over India, China, Japan, and Atlantic North America.

The fruits are aromatic, stimulant, and carminative.

The following species are used medicinally in China—*I. anisatum* Linn.—; in Indo-China—*I. anisatum* Linn., *I. Griffithii* Hook.f. & Th., *I. verum* Hook.f.—; in Malaya—*I. verum* Hook. f.—; in the Philippine Islands—*I. anisatum* Linn.—; in North America—*I. floridanum* Ell., *I. parviflorum* Michx.—

Sikimitoxin, a highly toxic substance, has been isolated from *I. anisatum*.

The fruit of *I. verum* Hook. f. or the essential oil from it is official in Austria, Brazil, France, Germany, Great Britain, Hungary, Italy, Portugal, Russia, Spain, Sweden, Switzerland, and the United States.

The essential oil from *I. anisatum* Linn. is official in Mexico.

Illicium Griffithii Hook. fil. & Th. occurs in Bhutan and the Khasia Hills between 4,000 and 5,000 feet.

The fruit is aromatic, stimulant, and carminative.

French: Faux badianier—; *Indo-China*: Dai hoi nui—.

Kadsura.

This genus includes 8 species, which inhabit tropical Asia, China, and Japan.

K. japonica Juss. is used medicinally in Japan.

Kadsura scandens Bl. is common in the forests of the Malay Peninsula and the Malay Islands at low elevations.

A decoction of the root is used by the Malays for rheumatism.

Malay: Akar dama daura, Akar kapala patong—.

Michelia.

The genus consists of 15 species inhabiting tropical Asia, and China.

The barks are bitter and aromatic, stimulant and antiperiodic.

The following are used medicinally in Indo-China—*M. Champaca* Linn.; in Malaya and Java—*M. Champaca* Linn., *M. montana* Blum.

1. Leaves 15-25 cm. long. Flowers yellow. Sepals and petals 15 or more ... ... *M. Champaca.*
2. Leaves 5-10 cm. long. Flowers white or with a tinge of yellow. Sepals and petals 9-12 ... *M. nilagirica.*
3. Leaves up to 19 cm. long. Flowers white. Sepals and petals 8 ... ... *M. montana.*

1. **Michelia Champaca** Linn. occurs wild in the Eastern sub-Himalayan tract and lower hills up to 3,000 feet, Assam, Burma, Western Ghats and South India; it is much cultivated in various parts of India and Burma. It is distributed to Yunnan, Indo-China, Siam, and Malaya—not in the Malay Peninsula where, however, it is common in gardens.

The root is bitter and demulcent; it is also considered purgative.

The dried root and root-bark, mixed with curdled milk is useful as an application to abscesses, clearing away or maturing the inflammation. In the form of an infusion it is a valuable emmenagogue.

The bark has febrifugal properties; it is a stimulant, expectorant, and astringent. It is described in Ayurveda as bitter with a sharp acrid taste, causing warmth in the abdomen, destroying poisons, removing worms, facilitating micturition and sweating, useful in bile and blood affections.

A decoction of the bark was tried in sixteen mild cases of chronic gastritis. After using it for several days the patients obtained considerable relief from the pain and discomfort in the epigastrium with which they were troubled, but none of the cases was completely cured (Koman).

Ayurveda practitioners use the leaves in combination with other drugs to remove the foetid odour of vaginal discharges. In Dacca the juice of the leaves is given with honey in cases of colic.

According to Ayurveda the flowers are bitter, stomachic, diuretic; they remove bilious conditions; they are good in leprosy, skin diseases, and ulcers. Yunani doctors consider their smell a good stimulant; they use the flower as an expectorant and look upon it as useful in cough and rheumatism.

The flowers and fruits are considered bitter and cool remedies, and are used in dyspepsia, nausea, and fever. The flowers mixed with sesamum oil form an external application, which is often prescribed in vertigo; they are also applied to foetid discharges from the nostrils. They are useful as a diuretic in renal diseases and in gonorrhoea. They are considered stimulant, antispasmodic, tonic, stomachic, and carminative.

The perfumed oil prepared from the flowers is a useful application in cephalalgia, ophthalmia, and gout.

The seeds and fruit are said to be useful for healing cracks in the feet.

The flowers and the fruit in combination with other drugs are recommended as an antidote to snake and scorpion venoms, but Caius and Mhaskar have shown experimentally that both the flower and the fruit are useless in the antidotal and symptomatic treatment of snake bite and scorpion sting.

Annam: Dam bac, Su Nam—; *Assam*: Phulchopa, Phulsopa, Titasoppa—; *Bengal*: Champa, Champaka, Chumpa, Kotu—; *Bombay*: Champa, Chapha—; *Burma*: Changal, Paranyam, Saga—; *Canarese*: Champaka, Kendasampige, Kolasampige, Sampage, Sampaghy, Sampige—; *Deccan*: Champa—; *English*: Golden Champa, Yellow Champa—; *French*: Champac—; *Gujerati*: Champo, Pitochampo, Raechampo, Sonchampa—; *Hindi*: Champ, Champa, Champac, Champaca—; *Indo-China*: Kim cuong moc—; *Kathiawar*: Pilochampo, Rayachampo, Sachochampo—; *Konkani*: Champa, Champo, Pudchampo—; *Lao*: Chum pi—; *La Reunion*: Champac—; *Malay*: Bongasjampacca, Mangliet—; *Malayalam*: Champakam— *Marathi*: Kudchampa, Pivalachampa, Sonachampa—; *Matheran*: Champaka, Pilachampa, Souchampa—; *Mundari*: Champabadaru—; *Nepal*: Aulechamp, Ouliachamp—; *North Western Provinces*: Champa—; *Philippines*: Champaca, Champaga—; *Portuguese*: Champo—; *Punjab*: Chamba, Chamoti, Champa, Chamuti—; *Sadani*: Campa—; *Sanskrit*: Anjana, Atigandhaka, Bramaratithi, Bhringmohi, Chambunala, Champaka, Champeya, Deepapushpa, Gandhaphali, Hemanga, Hemapushpa, Hemapushpaka, Hemavha, Kamabana, Kanchana, Katu, Kumara, Kusuma, Kusumadhipa, Kusumadhirata, Nagapshpa, Patichampaka, Peetapushpa, Punyagandha, Rajachampaka, Shatapadatithi, Shitala, Shitalachchada, Sthiragandha, Sthirpushpa, Subhaga, Sukumara, Surabhi, Svarnachampakapaka, Svarnapushpa, Ugragandha, Vanadapika, Vanadeepa, Vanamalika, Varalbdha—; *Sinhalese*: Champak, Hapu, Sapu—; *Tagalog*: Champaga, Sampac, Sampaka, Tsampaka—; *Tamil*: Amariyam, Sambagam, Sembagam, Shampangi, Vandumarmalar—; *Telugu*: Champakmu, Champeyamu, Gandhaphali, Gangaravi, Hemangamu, Hemapushpamu, Kanjanamu, Sampangi, Sampega—; *Tulu*: Champaka, Sampay—; *Uriya*: Chompa, Chompoko—.

2. **Michelia nilagirica** Zenk. is found in the shola forests of the Nilgiris, Anamalais and Pulneys above 5,000 feet.

The bark is used as a febrifuge.

Canarese: Bilisampage, Doddasampage, Sampana, Sapage—; *Hindi*: Pilachampa—; *Marathi*: Pilachampa—; *Sinhalese*: Walsapu—; *Tamil*: Kattuchambagam, Nilagirishambagam, Shambagam—.

3. **Michelia montana** Bl. is found in the Himalaya, the mountains of Perak and Redah, and in Java.

The bark is a bitter tonic useful in fevers.

II

The DILLENIACEÆ are trees or shrubs, chiefly natives of the southern hemisphere. Tropical America and Asia possess about an equal number of species; they are rare in Africa, and hitherto none have been found in South Africa or temperate South America.

4 of the 16 genera are reputed medicinal :— CUROTELLA (tropical America, West Indies); DAVILLA (tropical America, West Indies); DILLENIA (Indo-Malaya); TETRACERA (tropics; chiefly America). Two of these—DILLENIA and TETRACERA—are represented in India.

I. Woody climbers; filaments dilated upwards. Carpels 2 to 5 ... ... TETRACERA.
II. Trees; filaments not dilated.
Follicles not expanding nor dehiscing enclosed in the thickened pulpy sepals; seeds not arillate ... DILLENIA.

DILLENIA.

This genus includes 20 Indo-Malayan species.

In Indo-China the calyces of *D. Baillonii* Pierre, *D. indica* Linn., and *D. ovata* Wall. are made into jellies and cooling drinks; the bark of *D. ovata* is used medicinally in Cambodia.

1. Flowering with the leaves; flowers large, 6 to 9 in. across, white; fruit very large ... ... *D. indica.*
2. Flowering before leaves; flowers 2 to 5 in. wide, yellow; leaves 4 in. long, oblong, blunt, velvety beneath ... ... *D. ovata.*

1. **Dillenia indica** Linn. occurs from Nepal to Assam, extending to the Malay Peninsula and Indo-China; southwards it spreads over to Ceylon.

A syrup of the juice of the unripe fruit allays coughs, assists expectoration, and cures angina and aphthae.

The juice of the fruit, mixed with sugar and water, is used as a cooling beverage in fevers, and as a cough mixture. The fruit is slightly laxative, and is apt to induce diarrhoea if too freely indulged in.

The bark and the leaves are astringent. The bruised bark is applied as a cataplasm in arthritis.

The Malays use the pulp of the fruit as a hair-wash.

Assam: Chalita, Otengah—; *Bengal*: Chalita, Chalta, Hargesa, Ruvya—; *Bombay*: Karambel, Mothakarmel, Mothekaramala—; *Burma*: Thabyu, Thibuta, Zinbrun, Zinpyunngan—; *Canarese*: Bettadakanagal, Bettakanigala, Ganagalu, Kadkanagula, Kanagala, Kanigala, Neyitaku—; *Deccan*: Mutakurmul—; *Garo*: Panpui—; *Gujerati*: Karmbal, Otaphal—; *Hindi*: Chalta, Chaltr, Girnar—; *Indo-China*: Dok shan, So ba—; *Kachin*: Masang—; *Kolami*: Korkotta—; *Konkani*: Corombol—; *Lepcha*: Kyangmozhu, Phamsikol, Phan-se kung—; *Magahi*: Chauralesi, Thapru—; *Malay*: Chimpuh, Simpoh—; *Malayalam*: Chalita, Punna, Syalita, Valapunna—; *Marathi*: Karmbel, Motakarmal, Motakarmbal—; *Monghyr*: Chilta—; *Mundari*: Korkotadaru, Kurkutadaru—; *Nepal*: Panchkule, Panchphal, Ramphal—; *Sanskrit*: Bhavya, Ruvya—; *Santal*: Korkot, Korkotta—; *Sinhalese*: Hondapara, Wampara—; *Taleing*: Carllow—; *Tamil*: Akku, Ugakkay, Uva, Uvav, Uvatteku—; *Telugu*: Kalinga, Peddakalinga, Uvva—: *Uriya*: Chalota, Oao, Ou, Rai, Uau—.

2. **Dillenia ovata** Wall. is found in the Eastern Peninsula, Malaya, Cambodia, Indo-China.

In Cambodia the astringent bark is given for diarrhoea in the form of a strong infusion.

Cambodia: Knang krepeu, Plou—; *Cochinchina*: Thu dau mot—.

TETRACERA.

The genus consists of 25 tropical species.

The following are used medicinally in the Philippine Islands—*T. indica* Merrill, *T. macrophylla* Wall.—; in Nigeria and Guinea—*T. alnifolia* Willd., *T. potatoria* Azfl.—; in Guiana—*T. alnifolia* Willd., *T. aspera* Willd., *T. ovalifolia* DC.—; in Brazil—*T. oblongata* D.C.—.

1. Follicles 3-5-seeded.
   Leaves 2-5 in. long up to 2 in. wide, toothed, glabrous except the nerves beneath ... *T. indica*.
2. Follicles 1-2-seeded.
   Leaves 3-5 in. long, oblong or lanceolate. Sepals silky inside, glabrous outside, broadly oval ... *T. laevis*.
3. Follicles 1-seeded.
   Leaves 5-7 in. long, 3-6 in. wide. Sepals silky inside, puberulous outside, obovate obtuse, edges ciliate ... ... *T. macrophylla*.

1. **Tetracera indica** Merrill (=*T. assa* DC.) is found in Eastern Bengal and the Eastern Peninsula from Chittagong to Singapore. It extends to Siam, Cochin-China, Java, and the Philippine Islands.

In the Philippines an infusion of the plant is used internally for pulmonary haemorrhage, and externally as a gargle for the treatment of aphthae.

Malay: Ampelas, Mempelas minyak, P'las payah—; *Tagalog*: Malacatmon—.

2. **Tetracera laevis** Vahl is found in the Western Peninsula, the forests of Malabar, Ceylon, Java and Borneo.

In Malabar a decoction of the leaves is mixed with rice gruel and given for the treatment of aphthae.

Malayalam: Piripu—; *Tamil*: Anaittichal—.

3. **Tetracera macrophylla** Wall. occurs in the Eastern Peninsula from Penang to Singapore, extending to Sumatra.

An infusion of the plant is used in the Philippine Islands as a drink for haemoptysis, and as a gargle for aphthae.

Malay: Ampelas gajah, Ampelas lidah kuching, Ampelas rimau, Ampelas rimbah, Mempelas gajah, Mempelas lidah kuching, Mempelas rimau, Mempelas rimbah—; *Tagalog*: Malacatmon—.

III

The ANONACEÆ are trees or shrubs, often climbing and frequently aromatic. There are 80 genera comprising 820 species, nearly all tropical, especially of the Old World.

The medicinal and poisonous Anonads of the world belong to 22 genera: ALPHONSEA (tropical Asia): ANAXAGOREA (tropical

Asia, America); ANONA (Tropics; chiefly America); ARTABOTRYS (palaeotropics); ASIMINA (Cuba, Mexico, Eastern United States); BOCAGEA (tropical America and Asia); CANANGIUM (tropical eastern Asia to Australia); DESMOS (tropical Asia); ENANTIA (West Africa); GONIOTHALAMUS (tropical Asia); HABZELIA (tropical Asia); HEXALOBUS (tropical Africa, Madagascar); MELODORUM (Tropics); MONODORA (tropical Africa, Madagascar); OROPHEA (Indo-Malaya); OXYMITRA (palaeotropics); POLYALTHIA (palaeotropics); ROLLINIA (tropical America); SAGERAEA (Indo-Malaya); UNONA (tropical Asia, Africa); UVARIA (warm regions); XYLOPIA (Tropics).

The medicinal and poisonous Anonads of India belong to 12 genera:—ALPHONSEA, ANAXAGOREA, ANONA, ARTABOTRYS, CANANGIUM, DESMOS, GONIOTHALAMUS, MELODORUM, OROPHEA, POLYALTHIA, SAGERAEA, UVARIA.

A. Petals 2 in 2 rows, one or both rows imbricate in bud. Stamens many. Anther-cells concealed by a flat appendage
- 1. Sepals imbricate. Trees or shrubs. Ovules 6-8. Torus flat ... ... SAGERAEA.
- 2. Sepals valvate. Climbers. Ovules many ... UVARIA.

B. Petals valvate in bud, flat, or base only concave. Inner petals similar to outer. Stamens many, with overlapping appendages. Ovaries indefinite, rarely few
- 1. Petals connivent at concave base, covering stamens and ovaries. Climbers with hooked peduncles. Flowers rather large ... ... ARTABOTRYS.
- 2. Petals flat. Carpels indehiscent
  - a. Trees. Ovules many in 2 rows. Stamen-appendages acute ... ... CANANGIUM.
  - b. Trees. Ovules 1-2 ... ... POLYALTHIA.
  - c. Shrubs, half climbing. Ovules 6 in 1 row. Petals sometimes connate and reduced to 2 ... DESMOS.
  - d. Shrubs. Fruit follicular dehiscent ... ... ANAXAGOREA.

C. Petals valvate, outer spreading; inner concave connivent different, arching over stamens. Stamens and pistils numerous and anther appendages overlapping except *Orophea*. Trees. Inner petals clawed.
- 1. Inner petals connivent, not vaulted. Flowers rather large ... ... GONIOTHALAMUS.
- 2. Inner petals vaulted, longer than outer. Stamens 6. OROPHEA.

D. Petals valvate, thick, connivent; outer long narrow or broad, inner small but similar.
- 1. Trees or shrubs. Ovules solitary. Fruit fleshy of many connate carpels ... ... ANONA.
- 2. Climbers. Ovules 2 or more. Outer petals broad. Torus convex ... ... MELODORUM.

E. Petals valvate, outer often very small. Stamens definite loosely imbricate. Ovules 1 or many. Flowers usually solitary. Trees or shrubs.
Ovaries indefinite. Petals all larger than sepals saccate at base ... ... ALPHONSEA.

ALPHONSEA.

The genus consists of 9 to 10 Indo-Malayan species.

Alphonsea ventricosa Hook. fil. & Th. occurs in Assam and Chittagong and extends to the Andamans and Penang.

The plant contains a poisonous alkaloid.

ANAXAGOREA.

The genus numbers about 8 species, natives of tropical Asia and America.

Anaxagorea Scortechinii King is found in the woods of the Malay Peninsula.

The Malays put the seeds among clothes.

Malay : Bunga lerak, Pali monyet—.

ANONA.

The genus includes 70 tropical species, especially American.

The following species are used medicinally in Senegal—*A. senegalensis* Pers.—; in Nigeria—*A. cherimolia* Mill., *A. senegalensis* Pers.—; in Guinea, Oubanghi-Chari and the Gold Coast—*A. muricata* Linn., *A. senegalensis* Pers.—; in Indo-China—*A. muricata* Linn., *A. squamosa* Linn.—; in the Philippine Islands—*A. muricata* Linn., *A. reticulata* Linn., *A. squamosa* Linn.—; in North America—*A. muricata* Linn., *A. palustris* Linn., *A. spinescens* Mart., *A. squamosa* Linn.—; in California—*A. cherimolia* Mill., *A. muricata* Linn.—; in Porto Rico and Jamaica—*A. muricata* Linn.—; in Guiana—*A. Ambotay* Aubl.—; in Brazil—*A. cherimolia* Mill., *A. Marcgravii* Mart., *A. muricata* Linn., *A. palustris* Linn., *A. Pisonis* Mart., *A. reticulata* Linn., *A. spinescens* Mart., *A. squamosa* Linn.—.

A. muricata Linn., *A. reticulata* Linn., and *A. squamosa* Linn. are used medicinally in India

Fruit tubercled ... ... 1. *A. squamosa.*
Fruit smooth, slightly areolate ... ... 2. *A. reticulata.*
Fruit bearing numerous fleshy spines ... ... 3. *A. muricata*

1. **Anona squamosa** Linn. is a native of the West Indies, now cultivated throughout India.

The root is considered a drastic purgative, and is administered in acute dysentery. It is also employed internally in depression of spirits and in spinal diseases.

The astringent bark is used as an antidiarrhoeal cure in Cambodia.

An infusion of the leaves is considered efficacious in prolapsus ani of children; and the bruised leaves with salt make a cataplasm to induce suppuration.

In Brazil they are applied as a poultice over boils and ulcers. In Gambia, the West Indies, Central and South America they are used to kill lice and to prevent bed bugs, etc.

The ripe fruit is medicinally considered a maturant; and when bruised and mixed with salt, is applied to malignant tumours to hasten suppuration. The seeds contain an acrid principle fatal to insects, and the dried unripe fruit, powdered and mixed with gram flour, is used to destroy vermin.

In Chota Nagpur the seeds are crushed and used for destroying worms in the wounds of cattle.

The seeds are a powerful irritant of the conjunctiva.

Annam: Mak khieb, Mang cau ta, Qua na, Tiep—; *Antilles*: Cachimant, Hattier, Pomme-canelle—; *Arabic*: Ambeberesch, Beresch, Chermisch, Sariphа—; *Assam*: Ata, Katal—; *Bengal*: Ata, Luna, Meba, Sitaphal—; *Brazil*: Ata, Pinha, Frutta de Condessa—; *Burma*: Auza—; *Cambodia*: Tiep srok—; *Canarese*: Amritaphala, Duranji, Sitaphala—; *Ceylon*: Anoda—; *Cuba*: Anon—; *Deccan*: At, Sitaphul—; *English*: Custard Apple, Sugar apple, Sweet Sop—; *Ewe*: Eevunyikleng—; *Fanti*: Apre—; *French*: Anone écailleuse, Ate, Attier, Cachiman, Guanabane, Hattier, Marie baise, Pomme canelle—; *Ga*: Ngawyei, Nangwi, Ngaasie—; *Guam*: Atis—; *Gujerati*: Anan, Anuram, Seetaphul, Sitaphal—; *Hasada*: Sarupa—; *Hindi*: At, Ata, Atasitaphal, Sariphа, Shariphal, Sitaphal—; *Ibo*: Mbubo-ago—; *Indo-China*: Phan le chi—; *Kolami*: Mandal, Nenwa—; *Konkani*: At, Ath—; *Krobo*: Hangbue—; *Lambadi*: Sitaphal—; *La Reunion*: Attier—; *Malay*: Mannapapuwa, Srikaya—; *Malayalam*: Antacheecha, Attachchakka, Sirpa, Sitapalam, Sutakanni—; *Marathi*: At, Sitaphal—; *Mexico*: Anona, Anona blanca—; *Mundari*: Borordaru, Neoa, Sampa, Sarupa—; *Nagpuri*: Neoa—; *Nepal*: Sharifal—; *North-Western Provinces*: Behli, Sharifa—; *Persian*: Kaj, Sharifah—; *Portuguese*: Ateira, Fructo de Conde—; *Puerto Rico*: Anon, Anona con escamas, Atis—; *Punjabi*: Sharifa—; *Sanskrit*: Agrimakhya, Atripya, Bahubijaka, Gandagatra, Krishnabija, Sitaphala, Subha, Suda, Vaidehivallabha—; *Sakalave*: Konkony—; *Santal*: Mandargom—; *Sind*: Sharifa—; *Sinhalese*: Anoda, Atta—; *Sumatra*: Sirikayu—; *Tagalog*: Ates—; *Tamil*: Atta, Sitapalam—; *Telugu*: Gandagatramu, Sitapandu, Sitaphalamu—; *Timne*: Momina—; *Tulu*: Amritakay—; *Twi*: Apre, Bororfo nyankonga—; *Uriya*: Ato, Sitapholo—.

2. **Anona reticulata** Linn. is a native of the West Indies, now cultivated in India.

The bark is a powerful astringent, and is used as a tonic by the Malays. In China it is used in the treatment of inflammation of the eye.

In Brazil the leaves are used as a maturant.

In the West Indies and in Central and South America the green fruit is much used as an antidysenteric and an anthelmintic.

A. de Santos has isolated from the bark of the trunk an alkaloid which he has named *anonaine*.

Antilles: Cachimen, Coeur de boeuf, Corossol réticulé, Mamilier—; *Bengal*: Luvuni, Nona—; *Bombay*: Ramphal—; *Burma*: Awza—; *Canarese*: Ramaphala, Ramphal—; *Ceylon*: Anona—; *Chinese*: Meng Pa—; *Cuba*: Anona, Mamon—; *Deccan*: Ramphal—; *English*: Bullock's Heart, Custard Apple, Netted Custard Apple—; *French*: Anone en réseau, Cherimolier, Mamilier—; *Goa*: Anona—; *Guam*: Anonas—; *Gujerati*: Ramphal—; *Haiti*: Guanabano—; *Hindi*: Anta, Luvuni, Nagnewa, Nona, Ramphal—; *Hova*: Voankobohobo—; *Konkani*: Anon—; *La Reunion*: Anone, Coeur de boeuf—; *Malay*: Manua, Nona, Nona Kapri—; *Malayalam*: Manilanilam, Parankichchakka, Ramachchita—; *Marathi*: Ramphal—; *Mexico*: Anona, Ilama—; *Mundari*: Nagneva—; *Philippines*: Anonas—; *Portuguese*: Anona—; *Puerto Rico*: Anona de redecilla, Corazon—; *Sakalave*: Hobohobo—; *Sanskrit*: Krishnabija, Lavali, Lavani, Mriduphala, Raktatvatch, Ramawhaya, Ramphala, Vasanta—; *Santal*: Gom—; *Seychelles*: Coeur de boeuf—; *Sinhalese*: Anoda—; *Tamil*: Aninuna, Ramachita, Manilayatta—; *Telugu*: Ramaphalamu, Ramasitaphalamu—; *Uriya*: Barhial, Neua, Ramopholo, Ramositapholo—; *Venezuela*: Rinon—.

3. **Anona muricata** Linn. is a native of tropical America, now common in the tropics of the Old World.

In La Reunion the root is considered antispasmodic and parasiticidal; the leaves are given in fevers, and also used in the form of a poultice to produce suppuration; the flower buds and the flowers are considered an excellent remedy for cough; the unripe fruits when dried and powdered are given in chronic dysentery, and they are used for aphthae in the form of a decoction; the seeds are valued for their astringent and emetic properties.

In Jamaica a decoction of the root is said to be an antidote against fish-poison.

In Southern California and Tropical America the bark is used as a drastic purgative.

In Brazil the leaves are steeped in hot water or ground with oil, and used as a maturant. An infusion of the leaves is used as a remedy for dysentery in Porto Rico. In general the leaves are regarded as a useful remedy for fever and dysentery. In Guinea they are pounded and applied fresh to cicatrice wounds. In Gambia they are used to get rid of bed bugs.

The seeds are emetic and astringent. In Southern California and Tropical America they are said to be poisonous and they are used as a parasiticide and an insecticide; they are used to poison fish.

Akim: Abrorfontunkum—; *Antillas*: Anona de puntitas—; *Antilles*: Anone en bouclier, Cachiman épineuse, Corossol montagne, Grand corossol, Sapadille—; *California*: Guanabana, Sour-sop—; *Canarese*: Mulluramphala—; *Cartagena*: Anona de broquel, Catuche, Guanabana—; *English*: Sour sop—; *Ewe*: Vo, Voti, Votsi—; *Fanti*: Apre—; *French*: Corrossolier—; *Fula*: Dukumeporto—; *Ga*: Aluguntungung, Nkrangmrobe—; *Guam*: Laguaná—; *Indo-China*: Mak khieb thet, Mang cau xiem, Tiep parang—; *Krepi*: Yevunyakle—; *Krobo*: Alukutum—; *La Reunion*: Corossol—; *Malayalam*: Mullanjakka, Vilattinuna—; *Mexico*: Anona, Anona amarilla, Cabeza de negro, Catuche, Guanabano—; *Peru*: Guanabano—; *Philippines*: Goyabrano, Guanabano—; *Sakalave*: Kaoraosaly—; *Sinhalese*: Katuanoda—; *Tamil*: Mulluchitta, Pulippala—; *Tschaudjo*: Alola—; *Twi*: Aduatungnkungm, Apre, Deboo, Nkrangmrobe—.

ARTABOTRYS.

This genus consists of 30 species distributed over the tropics of the Old World.

A. odoratissimus R. Br. is used medicinally in Indo-China, the Philippine Islands, the Malay Archipelago; *A. suaveolens* Bl. in Java.

1. Limb of petals broad, lanceolate or elliptic-oblong ... 1. *A. odoratissimus*.
2. Limb of outer or all the petals linear, narrow-oblong, or subclavate ... ... 2. *A. suaveolens*.

1. **Artabotrys odoratissimus** R. Br. is largely cultivated in India, Ceylon, Java and China.

A decoction of the leaves is given for cholera in some of the islands of the Malay Archipelago.

Bombay: Vilayatichampa—; *Canarese*: Kandalisampage, Manoranjanballi—; *Deccan*: Madanmast, Madmanti—; *Hindi*: Champa—; *Indo-China*: Day cong chua—; *Malayalam*: Madanakameswari, Manoranjitam—; *Philippines*: Ilang-ilang de China—; *Sanskrit*: Harachampaka, Nllachampaka, Phalasampenga—; *Southern California*: Climbing Vine, Climbing Ylang-ylang—; *Tagalog*: Alagalag sonson, Alangilang sonson—; *Tamil*: Manoranjidam—; *Telugu*: Manoranjidamu, Muddasampenga, Phalasampanga, Sakalaphalasampanga—; *Uriya*: Kalomuro, Monasocompa—.

2. **Artabotrys suaveolens** Bl. is found in Sylhet, Chittagong, Mergui, the Malay Peninsula and the Malay Islands.

The Malays use the leaves to prepare an aromatic infusion whose good effects have been extolled in the treatment of cholera.

Malay: Akar chenana, Durie carban—.

CANANGIUM.

This genus consists of 3 species inhabiting tropical Eastern Asia and extending through Burma to New Guinea and the Philippine Islands.

1. Leaves glabrous. Flowers over 2 in. long ... *C. odoratum*.
2. Leaves ovate to orbicular, woolly beneath, deciduous. Flowers large ... ... *C. latifolium*.

1. **Canangium odoratum** Baill. is cultivated throughout India. From Ava and Tenasserim it extends to Java.

The flowers yield the 'ilang-ilang' of perfumes. 'Cananga Oil' consists of the early portions of the distillate.

The oil is used as an application in cephalalgia, ophthalmia, and gout.

Burma: Kadapnyan, Kadatnyan—; *Canarese*: Apurvachampaka—; *English*: Ylang-ylang—; *French*: Bois de Bananen, Bois de lance bâtard, Caneng aromatique—; *Guam*: Alangilang—; *Ilocano*: Alangilang, Ilangilang—; *Jolo*: Angilang—; *Malay*: Cananga, Ilangilang, Kenanga—; *Rarotonga*: Moto-oi—; *Samoa*: Moso'oi—; *Sinhalese*: Wanasapu—; *Tagalog*: Alangilang, Ilangilang—; *Tamil*: Karumugai, Maladi, Maramanoranjidam, Sadi, Sirusambagam—; *Telugu*: Apurvachampakamu—; *Visayan*: Alangilang, Ilangilang—.

2. **Canangium latifolium** Pierre is distributed over Burma, the Malay Peninsula, and Cambodia.

In Indo-China the wood is considered a febrifuge.

Indo-China: Su tay, Tai nghe, Tom xui—.

DESMOS.

This genus numbers 20 species inhabiting tropical Asia.

Flower opposite leaf. Leaves glabrous.
1. Pedicel 4 to 6 in. long; leaves hardly glaucous beneath ... ... *D. cochinchinensis*.
2. Pedicel 1 to 2 in. long; leaves quite glaucous beneath *D. chinensis*.

1. **Desmos cochinchinensis** Lour. is found in Assam, the Malay Peninsula, and Cochin-China.

The Malays use a decoction of the roots for fever.

Malay: Akar sugi-sugi, Larak salai—.

2. **Desmos chinensis** Lour. occurs in East Himalaya, Burma, the Malay Peninsula and Islands, and Cochin-China.

The Malays use a decoction of the plant for dysentery.

Malay: Akar darah, Akar mariam, Akar singa, Kenanga hutan, Poko sikenchong—,

GONIOTHALAMUS.

The genus includes about 50 species distributed over tropical Asia.

Goniothalamus macrophyllus Hook. fil. is common in woods all over the Malay Peninsula whence it extends to Sumatra and Java.

The wood is aromatic. The plant is used medicinally by the Malays.

Malay : Bongsoi, Penawar hitam—.

MELODORUM.

The genus consists of about 40 species found in tropical Asia, Africa, and Australia.

Melodorum Kingii Boerl. is found in the forests of Malacca and Perak.

A decoction of the flowers is used by the Malays for dyspepsia.

Malay : Poko achar—.

OROPHEA.

This genus consists of 30 Indo-Malayan species.

Inner petals vaulted.
Stamens 12 ; stem hairy ; carpels globose ... ... 1. *O. setosa.*
Stamens 6 ; stem hairy ; carpels globose ... ... 2. *O. hirsuta.*

1. **Orophea setosa** King is found in the woods of Perak and Negri Sembilan from 500 or 1,200 ft. altitude.

A decoction of the roots is used by the Malays in malaria.

Malay : Pialu—.

2. **Orophea hirsuta** King inhabits Perak, where it is however rare.

The Malays chew the plant with betel-nut in cases of cough.

Malay : Supucha pelandok—.

POLYALTHIA.

The genus includes 70 palaeotropical species ; mostly in tropical Asia, a few in Africa and Australasia.

P. Oliveri Engl. is used medicinally in Sierra Leone, Libéria, and parts of Cameroon.

1. Leaves narrow, lanceolate, undulate ... ... 1. *P. longifolia.*
2. Leaves oblong ... ... 2. *P. simiarum.*

1. **Polyalthia longifolia** Benth. and Hook. fil. is cultivated throughout the hotter parts of India.

The bark is used as a febrifuge in the Balasore District of Orissa.

Bengal : Debdaru, Devadar, Devadaru— ; *Bombay* : Asok, Asoka, Asopalav, Asupal, Asupala, Devadaru— ; *Canarese* : Putrajivi— ; *Ceylon* : Maraillupai— ; *English* : Indian Fir, Mast Tree— ; *Gujerati* : Ashopalo— ; *Hindi* : Asok, Debdari, Deodar, Devadar, Devadaru, Devidari— ; *Konkani* : Assok, Devandaru— ; *Madras* : Nettilingam— ; *Malayalam* : Aranei, Aruna, Ashokam, Ashvattam, Chorani, Hemapushpam, Vanjolam— ; *Sanskrit* : Devadaru, Putrajiva— ; *Tamil* : Asogam, Asuvattai, Kalgoli, Kasubam, Kolikkudi, Nettilingu, Pundi, Ravadam, Saribam, Selai, Sendu, Tevadaram, Vansulam— ; *Telugu* : Asokamu, Asvattamu, Devadaru— ; *Tulu* : Asoka— ; *Uriya* : Debodaru, Asoka, Oswottho, Putikashto—.

2. **Polyalthia simiarum** Benth. and Hook. fil. is found in Orissa, Pegu Yoma, Martaban and Tenasserim.

The bark is used in Orissa as a cure for scorpion sting.

Bhuia: Champa—; *Modesia*: Khari—; *Nepal*: Khutti, Labshi—; *Orissa*: Mongai, Ojarh, Mojarh—; *Santali*: Dighibentia—.

SAGERAEA.

The genus consists of 6 Indo-Malayan species.

Sageraea laurifolia Blatter occurs in South Konkan.

The leaves have a pungent, astringent, and bitter taste. In the Konkan they are used for fomentation.

Bombay: Andi—; *Canarese*: Sagare—; *Konkani*: Sageree—; *Marathi*: Harkinjal, Sageri, Undie—.

UVARIA.

The genus numbers 100 species distributed over the Tropics of the Old World.

V. Chamae P. Beauv. is used medicinally in Gold Coast, Sierra Leone, Togo, Lagos, and Senegal; *V. Afzelii* Scott Elliot in Liberia; *V. calamistrata* Hance in Indo-China.

1. Stamens all cuneate ... ... 1. *U. Narum.*
2. Outer stamens flat ... ... 2. *U. dulcis.*

1. **Uvaria Narum** Bl. occurs in the Bombay Presidency—Konkan and North Kanara—, and the Madras Presidency—forests of the Western Ghats from South Kanara to Travancore, and hills of Salem up to 4,000 feet.

The oil obtained from the roots by distillation, as well as the root, are used medicinally in various diseases. The root is fragrant and aromatic, and the bruised leaves smell like cinnamon.

Canarese: Kariballi, Unamini—; *Malayalam*: Narampanal—; *Matheran*: Narampanal—; *Tamil*: Pulichan—; *Tulu*: Pandel—.

2. **Uvaria dulcis** Dunal occurs in Burma, the Malay Peninsula, Java and the Philippines.

The bark of the root has astringent, stimulant, and alterative properties.

Malay: Pisang-pisang hitam—; *Visayan*: Dalaganum, Dalagao—.

IV

The MENISPERMACEAE are scandent or twining shrubs, found chiefly in the tropics of the Old and New World. They number 65 genera and 360 species. Many of them figure among the poisonous plants used in making arrow and dart poisons.

The medicinal and poisonous Menispermads of the world belong to 21 genera:—ABUTA (tropical South America); ANAMIRTA (Indo-Malaya); ANOMOSPERMUM (Brazil, Guiana); BURASAIA (Madagascar); CISSAMPELOS (tropical and subtropical regions); COCCULUS (tropical and subtropical regions); COSCINIUM (Indo-Malaya); CYCLEA tropical Asia); FIBRAUREA (tropical and subtropical Asia); JATRORRHIZA (tropical Africa); MENISPERMUM (eastern Asia, Atlantic North America); PACHYGONE (eastern Indo-Malaya); PENIANTHUS (western tropical Africa); PERICAMPYLUS (eastern Indo-Malaya); SPHENOCENTRUM (western tropical Africa); SPIROSPERMUM (Madagascar); STEPHANIA (palaeotropics); STRYCHNOPSIS (Madagascar); TILIACORA (Indo-Malaya); TINOMISCIUM (tropical Asia); TINOSPORA (Palaeotropics).

The following are among the products isolated:—(1) *alkaloids*—bebeerine, isobebeerine, beta-bebeerine, berberine, buxine, chondrodine, curarine, menispermine, para-menispermine, pelosine, sangoline, tiliacorine—; (2) *quaternary bases* (none of which has been obtained in a free state)—columbamin, jatrorrhizin, palmatin—; (3) *crystalline bitter principles*—anamyrtin, columbin, picrotoxin, picrotoxinin, trilobin—; (4) *acids*—columbic, picrotoxic—.

The medicinal and poisonous Menispermads of India belong to 12 genera:—ANAMIRTA, CISSAMPELOS, COCCULUS, COSCINIUM, FIBRAUREA, JATRORRHIZA, PACHYGONE, PERICAMPYLUS, STEPHANIA, TILIACORA, TINOMISCIUM, TINOSPORA.

A. Flowers trimerous. Ovaries usually 3. Drupes with a subterminal, rarely ventral or basal style-scar. Seeds oblong or subglobose
1. Sepals 6. Stamens 6 or 3, often connate ... ... JATRORRHIZA.
2. Sepals 6. Petals 6. Filaments free ... ... TINOSPORA.
3. Sepals 9. Petals 6. Filaments free ... ... TINOMISCIUM.
4. Sepals 6. Petals 0. Filaments free ... ... FIBRAUREA.
5. Sepals 6. Filaments all connate ... ... ANAMIRTA.
6. Sepals 9. Outer filaments free ... ... COSCINIUM.

B. Flowers trimerous. Ovaries usually 3. Drupe with a subbasal, rarely subterminal style-scar. Seed horseshoe-shaped. Albumen copious. Embryo slender, cotyledons linear or slightly dilated
1. Petals 6, minute. Ovaries 3-12. Style subulate ... TILIACORA.
2. Petals 6. Ovaries 3-6. Styles subulate ... ... COCCULUS.
3. Petals 6. Ovaries 3. Styles forked ... ... PERICAMPYLUS.

C. Flowers 3-5-merous. Ovaries usually solitary. Drupe with a subbasal style-scar; endocarp dorsally muricate or echinate. Seeds horseshoe-shaped. Embryo linear; cotyledons appressed
1. Sepals 6-10, free. Petals of male flower and female 3-5, free ... ... STEPHANIA.
2. Sepals 4, free. Petals of male 4, connate, of female 1 CISSAMPELOS.

D. Flowers usually trimerous. Ovaries usually 3. Drupes with a subbasal or ventral style-scar. Seed curved, hooked or inflexed.
Sepals, petals and stamens 6 each ... ... PACHYGONE

ANAMIRTA.

The only Indian species of this genus, **A. paniculata Colebr.** is found in the Khasia Hills, Assam, Eastern Bengal, from Orissa and the Bombay Konkan to Ceylon; it extends through the Malay Archipelago to New Guinea.

According to Yunani authors the berry is slightly bitter, expectorant, demulcent, and carminative; useful in the treatment of rheumatism; poisonous to fish.

The bitter berries are sometimes used in the form of an ointment. This ointment is employed as an insecticide to destroy pediculi, and in some obstinate forms of chronic skin diseases. Koman prepared an ointment containing 80 grains of finely powdered seeds to one ounce of vaseline, and used it as an external application in ringworm; cases of one to two months' duration were cured by a few days application, but chronic cases did not derive benefit.

The fresh leaves are used in Bengal as a snuff in the treatment of quotidian ague.

In Ceylon the bruised fresh bark is applied to the bitten part in cases of snake-bite. But Mhaskar and Caius have found that the bark, leaves, and berries are equally useless as an external application in the treatment of snake-bite.

The seeds are used in the nightsweats of phthisis.

The jungle tribes of the Malay Peninsula use the plant to poison their arrows and their kriss.

Both its poisonous properties and any therapeutic virtues the plant may contain depend upon picrotoxin of which the fruit is the source. The plant also contains two minor alkaloids of unknown constitution, menispermine and paramenispermine, both physiologically inactive.

Arabic: Mahijehreh—; *Bengal*: Kakamari—; *Bombay*: Kakaphala, Kakphal, Vatoli—; *Burma*: Hong—; *Calcutta*: Bacaenkaphal—; *Canarese*: Kakamari, Kakkisoppugade—; *Deccan*: Kakmari—; *Dutch*: Indiaansche rezies, Koklus—; *English*: Crow Killer, Fish Berry, Fish Killer, Indian Berry, Indian Cockles, Louse Berry, Oriental Berries, Poison Berry—; *French*: Arbre à enivrer, Bois à enivrer, Bois enivrant, Bois ivrant, Coque du Levant, Coque Levant, Herbe à tous maux, Pareire à feuilles rondes—; *German*: Fischkoerner baum, Kokkelskoernerbaum, Tollkoernerbaum—; *Gujerati*: Jermae, Kakaphula, Kakmari—; *Hindi*: Jermae, Kakmari—; *Indo-China*: Ben nau, Day tao, Seg Dom—; *Italian*: Coccole d'India—; *Konkan*: Garudphul, Kadul—; *Malay*: Dawonnboelann, Tubabidji—; *Malayalam*: Anakrytu, Garalaphala, Garaphala, Kantakakonnuveli, Kantakakunavam, Karantakam, Miunannu, Machattinkaya, Nanninkuru, Pola, Pullukunavam—; *Oceania*: Lacton, Lagton, Libtang, Soma, Suma, Tubaflava—; *Pampangan*: Balasin, Bayati, Lactang, Lanta, Lictang, Lingtangbaguin, Suma, Tuba—; *Persian*: Mahijehreh—; *Philippines*: Abutra, Pangmavan—; *Portuguese*: Coca de Levante, Fruta matapeixe—; *Sanskrit*: Garalaphala, Kakahva, Kakamari, Kakanashika—; *Sinhalese*: Pangan, Tittawel—; *Spanish*: Coca de Levante, Coca levantina—; *Tagalog*: Balasin, Bayati, Lactang, Lanta, Lictang, Lingtangbaguin, Suma, Tuba—; *Telugu*: Kakamari, Kakichempoo, Koditige, Tippatige—; *Tulu*: Chipulu—; *Urdu*: Mahijehreh—; *Uriya*: Kalabiti Nai, Kaumari—; *Visayan*: Balasin, Bayati, Lactang, Lagtal, Lanta, Lictang, Lintangbaguin, Suma, Tuba—.

CISSAMPELOS.

This genus includes 21 species inhabiting the warm regions of the world.

The following are used medicinally in the Philippine Islands and Indo-China—*C. Pareira* Linn.—; in the Antilles and Guiana—*C. Pareira* Linn.—; in South America—*C. caapeba* Linn., *C. Pareira* Linn.—; in Brazil—*C. glaberrima* St. Hil., *C. ovalifolia* DC.—; in the Gold Coast and in Nigeria—*C. mucronata* A. Rich., *C. owariensis* P. Beauv.—; in Southern Africa—*C. angustifolia* Bucch., *C. capensis* Thunb., *C. mucronata* A. Rich., *C. Pareira* Linn., *C. torulosa* E. Mey.—; in Mauritius—*C. mauritiana* Pet. Th.—.

Cissampelos Pareira Linn. occurs throughout tropical and sub-tropical India; and throughout the warm parts of Asia, East Africa, and America.

The root is the part most esteemed; it has an agreeable, bitterish taste, and is considered a valuable stomachic. It is frequently prescribed in the later stages of bowel complaints, in conjunction with aromatics. It is given for pains in the stomach and for dyspepsia, diarrhoea, dropsy and cough; also for prolapsus uteri. It is applied externally in snake-bite and scorpion-sting.

Among the Mundas of Chota Nagpur the root, ground and mixed with water, is used against stomach-ache and diarrhoea, especially against infantile diorrhoea.

In Hausaland the bitter root is sold as a medicine for many purposes.

The Chuanas, Subias, and Kobas of South Africa drink an infusion of the powdered root for stomach-ache. The Filabusi natives and the Xosas drink a similar preparation for snake-bite. The Pedis use a decoction of the root as a wash for children who have pimples over their body.

In Madagascar the root is considered diuretic, emmenagogue, and antipyretic; it is given for urinary gravel.

In La Reunion the bitter root is considered tonic, stomachic, and diuretic; and it is used in the treatment of bladder troubles.

In French Guiana the roots are used as a diuretic in cases of dysuria and calcular nephritis. An infusion is expectorant and is considered a potent alexiteric.

The root acts as an antiseptic of the bladder and is used in chronic inflammation of the urinary passages. In Yoruba it is used as an anti-emmenagogue.

The leaves are said to have a peculiarly cooling quality, and they are used locally in cases of unhealthy sores and sinuses. In the Gold Coast they are applied to abscesses. The Xosas apply a paste of the leaf to wounds.

The root in combination with other drugs is an Ayurveda cure for snake-bite and scorpion-sting. Mhaskar and Caius have, however, demonstrated that the root is not an antidote to snake venom and that it is useless as an external application in the treatment of scorpion-sting.

The root contains the alkaloid bebeerine. It is official in Portugal.

Annam: Moi tron, Tiet re—; *Antsianaka*: Voaravinaviavy—; *Ashanti*: Akuraso—; *Batangas*: Calancalamayan—; *Bengal*: Akanadi, Nemuka, Nimuka, Tejomalla—; *Betsileo*: Vahemboatavo—; *Betsimisaraka*: Vahifotsy—; *Bombay*: Pahadmul, Pahadvel, Venivel—; *Canarese*: Padavali—; *Cebu*: Batangbatang—; *Chuana*: Mokaekae—; *Dehra Dun*: Parhe, Purhe—; *East Africa*: Kinukadjio —; *English*: False Pareira brava, Ice Vine, Velvet-leaf—; *Filabusi*: M'cessie —; *French*: Faux pareira brava, Liane à glacer l'eau, Liane à serpents—; *French Guiana*: Pareira brava, Paria brava—; *Garhwal*: Pahari, Parhe—; *Goa*: Parayel—; *Gujerati*: Venivel—; *Hausa*: Fiyaka, Jibdakassa, Judarkas, Piyaka—; *Hindi*: Akanadi, Dakhnirbissi, Harjeuri, Harjori, Pari—; *Hova*: Vahivory, Voriravina—; *Ilocano*: Cuscusipa—; *Java*: Areujtjantjuan—; *Koba*: Mokaekae—; *Kolami*: Pitu-singh, Ranu-sed—; *La Reunion*: Liane blanche, Pareira brava—; *Lepcha*: Tamshaprip—; *Madagascar*: Ravinbury, Vahenosy—; *Malaya*: Akar mumpanang, Gasinggasing, Gegasing—; *Malayalam*: Kattuvalli, Patuvalli—; *Marathi*: Padavali, Padavel, Pahadvel, Paharmul, Paharvel—;

Mexico: Oreja de raton—; *Mundari*: Cutulutur, Dirikakaru, Haruarajaite, Huringpitusing—; *Nepal*: Batulpoti—; *North Western Provinces*: Dakhnirbisi, Duknirbissi—; *Oudh*: Harjewri—; *Pedi*: Lepeta—; *Philippines*: Calaad, Chanchae, Sansao—; *Porebunder*: Kardhiyunbang—; *Porto Rico*: Bejuca de mono—; *Portuguese*: Abutua, Pareira brava—; *Punjab*: Bat, Batindupath, Bel, Katori, Parbik, Pataki, Tikri, Zakhmihaiyat, Zucumyeat—; *Sakalave*: Hamafana—; *Sanskrit*: Ambashtha, Ambashthika, Avidhakarni, Brihattikta, Chchinnaveshika, Devi, Ekashthila, Kuchela, Kucheli, Laghupatha, Mahanjasi, Malati, Malavi, Papacheli, Papehelika, Patha, Patika, Prachina, Prachinambastika, Pratanini, Rasa, Ruchishya, Shishira, Shreyasi, Sthapini, Sushhira, Tiktapushpa, Trishira, Trivrita, Uthika, Vallika, Vara, Varatikta, Vatsadini, Venivalli, Vidhakarni, Vridhakarnika, Vriki, Vrittaparni—; *Santal*: Tejomalla—; *Saora*: Paterutivu—; *Sind*: Belpath, Katori—; *Sinhalese*: Deyamitta, Diyamitta, Weniwela—; *Spanish*: Butua, Pareira brava—; *Subia*: Mokaekae—; *Tagalog*: Chinchaochinchauan, Gulangulamanan, Sansaosansansaosan, Yemoumohan—; *Tamil*: Appatta, Puimushti, Punaittitta, Punmushti, Puttutiruppi, Sina, Titta, Tuvan, Tuvigaba, Vattattiruppi—; *Telugu*: Adivibankatige, Pata, Visaboddi—; *Tharu*: Batulpati—; *Twi*: Aportororkungma—; *Uriya*: Okanobindhi—; *Uruguay*: Butua, Pareira brava—; *Visayan*: Hampapari, Himpapara, Pari, Sampapari—; *Yoruba*: Je-in-joko, Jo-ko-je—.

COCCULUS.

This genus consists of 11 species found in all warm countries.

C. laurifolius DC. and *C. trilobus* DC. are used medicinally in Indo-China and Malaya, *C. Thunbergii* DC. in Japan, and *C. pendulus* Diels in the Sudan and in Senegal.

I. Scandent shrubs.
  1. Leaves suborbicular, glabrous; panicles large ... 4. *C. macrocarpus*.
  2. Leaves deltoid or hastate, villous; panicles short ... 1. *C. hirsutus*.
  3. Leaves oblong or trapezoidal, at length glabrous. Male flowers fascicled, female solitary ... 2. *C. pendulus*.

II. A shrub or small tree ... ... 3. *C. laurifolius*.

1. **Cocculus hirsutus** Diels occurs in tropical and subtropical India from the foot of the Himalayas to South India and Ceylon and Pegu; it extends to Southern China, Arabia, and tropical Africa.

The root is generally used as a refrigerant, and also as a gentle laxative. It has been extensively used as an alterative in chronic rheumatic and venereal diseases.

A decoction of the fresh roots, with a few heads of pepper, in goat's milk, is administered for rheumatic and old venereal pains; half a pint every morning is the dose. It is reckoned heating, laxative, and sudorific.

In the Konkan, the roots, rubbed with Bonduc nuts, are administered as a cure for belly-ache in children. In bilious dyspepsia, they are given in 6 *massa* doses, with ginger and sugar.

In Sind, the root and leaves are used in headache and neuralgic pains.

The juice of the leaves, mixed with water, has the property of coagulating into a green jelly-like substance, which is taken internally, sweetened with sugar, as cure for gonorrhoea.

In Baluchistan the mucilage is used to cure spermatorrhoea, taken in milk; it is used for coughs and to put on to sore eyelids and to soften breasts

Arabic: Haddal, Herrije, Luah, Schirwai—; *Baluchistan*: Afaband, Zamur—; *Bengal*: Huyer—; *Bombay*: Parvel, Vasanvel—; *Canarese*: Dagadiballi, Dusariballi, Sugadiballi, Yadaniballi—; *English*: Broom Creeper, Ink Berry—;

Gujerati: Vevati, Vevdi—; *Hansot*: Vachhana—; *Hindi*: Chireta, Dier, Faridburti, Hier, Jalayamani, Jamtikibel—; *Jaisalmer*: Bajarbel—; *Konkan*: Vanatiktika—; *Marathi*: Hunder, Parvel, Tana, Vasanavela, Vasanel, Vasanvel —; *Matheran*: Tan, Vasanyel—; *Nimar*: Bochan—; *North-Western Provinces*: Patha—; *Oudh*: Karsane—; *Persian*: Faridbutti—; *Porebunder*: Vadhinovelo—; *Sanskrit*: Patalagarudi, Chchilihinda, Dirghakanda, Dirghavalli, Dridhakanda, Dridhalata, Garudi, Mahahala, Mahamula, Mochakabhida, Sauparni, Somavalli, Tiktanga, Vanatiktika, Vasandi, Vasantitikta, Vatsadani—; *Sind*: Kursan, Zamir—; *Tamil*: Kattukodi—; *Telugu*: Chipurutige, Dusaritige, Katlatige—; *Urdu*: Faridbutti—; *Uriya*: Musakani—.

2. **Cocculus pendulus** Diels occurs in Sind, Baluchistan, Waziristan, the Punjab Plains to the Northern Circars, Kathiawar, Deccan, Carnatic to Tinnevelly. It extends through Afghanistan and Arabia to tropical and subtropical Africa.

The root is used in Sind and Afghanistan in the treatment of intermittent fevers and as a substitute for *Anamirta paniculata*. It is used by the natives in Senegal and in the French Sudan for the cure of periodic fevers.

The flowers are added to food, and an infusion of the plant is used in Senegal to assist in removing thorns from the feet.

The Arabs make an intoxicating drink from the fruits.

Heckel and Schlagdenhauffen have found in the root about 2 per cent. of pelosine, and about 3 per cent of a new crystalline alkaloid 'sangoline'. The root also contains columbin.

Arabic: Ssag-el-ghorab, Turrach—; *Baluchi*: Zamor—; *Brahui*: Zamor—; *Egypt*: Lebakh-el-gebel—; *Falor*: Tiati—; *Gujerat*: Parwatti—; *Jaisalmer*: Pilwan—; *Jodhpur*: Pilwan—; *Punjab*: Illarbillar, Parwatti, Vallur, Vehri—; *Sind*: Ullarbillar—; *Senegal*: Sangol—; *Songhai*: Liligui—; *Telugu*: Dusaratige—; *Tukulor*: Safatou—; *Waziri*: Jadhai, Parwatiae—; *Wolog*: Mboum sehet, Mboum tiéré—.

3. **Cocculus laurifolius** DC. occurs in subtropical Himalaya from Nepal to Jammu up to 5,000 feet, in the Western Ghats of the Madras Presidency, Eastern Bengal, and Burma. It is distributed to South Cochin-China, South China, Java, Formosa, and Japan.

The jungle tribes of the Malay Peninsula use the plant to poison their arrows and darts.

China: Wu Yåo—; *Dehra Dun*: Tildhara, Tilphara—; *Garhwal*: Tildhara—; *Japan*: Kansirowujak, Vjaku, Wujak—; *North-Western Provinces*: Tilphara—.

4. **Cocculus macrocarpus** W. and A. is found in the Konkan, North Kanara, the South Mahratta Country, and China.

The powdered leaves are taken in milk as a cure for biliousness, gonorrhoea, and syphilis.

Matheran: Vatoli, Wat-yel, Watan-yel—; *Tamil*: Kottaiyachachi—.

COSCINIUM.

The genus consists of 6 Indo-Malayan species.

Coscinium fenestratum Colebr. occurs in South India and Ceylon, extending to Sumatra.

The root is extensively used in Ceylon as an efficient bitter tonic, and is viewed as a very good substitute for Columba. It has also antiseptic properties to a great extent, and can be used for dressing wounds and ulcers.

The wood is valued as a bitter tonic by the Sinhalese. It is also much used as a cure for tetanus.

A decoction of the stem is given internally in cases of bites from monkeys, snakes, brahmin-lizards and geckos. Mhaskar and Caius have found experimentally that the stem is not an antidote to snake venom, whether colubrine or viperine.

The decoction of the bark was administered in doses of one to two ounces three times a day to several patients suffering from malarial fever, but no beneficial effects were noticed (Koman).

The plant is used by the Sakais in the preparation of their dart poisons.

Tummin Katti and Shintre (1930) have analysed the alcoholic extract of the stems and they have reported *inter alia* a considerable amount of a mixture of alkaloids.

Bengal: Haldigach—; *Canarese*: Doddamaradarasina, Maradarashina, Maramanjali—; *Deccan*: Jhadihaladi, Jharkihaldi—; *English*: Calumba Wood, Ceylon Calumba Root, Columbo Wood, False Columbo, Tree Turmeric—; *Malay*: Akar kunyit, Kunyit babi—; *Malayalam*: Haridram, Maramannal—; *Marathi*: Venivel—; *Sakai*: Tol—; *Sanskrit*: Daruharidra, Darvi, Pitadru—; *Sinhalese*: Bangwellgetta, Venivel, Woniwol—; *Tamil*: Atturam, Imalam, Kadari, Manjalkodi, Maramanjal, Pasamantram, Sanniyam, Seyebasam, Tiyaram, Udaravi, Udubadi—; *Telugu*: Manupasupu—; *Tulu*: Maramanjali—.

FIBRAUREA.

The genus consists of 4 species found in tropical and subtropical Asia.

Fibraurea chloroleuca Miers is very common in hedges and woods from Singapore to Penang. It extends to Tavoy and to the Malay Islands and Celebes.

The plant is used medicinally in Cochin-China and in the Malay Peninsula.

The root is used as a diuretic. The bitter stem is considered a valuable tonic, and is given in intermittent fevers and hepatic troubles.

Indo-China: Day vang giang—; *Malay*: Akar kinching kerbau, Akar kuning—.

JATRORRHIZA.

The genus consists of 2 species, natives of tropical Africa.

J. strigosa Miers is used medicinally by the Bakwiri and others in Cameroon.

Jatrorrhiza palmata Miers is cultivated in some parts of India.

The root is much used in Zambezi, Mozambique, Madagascar, and Indo-China as a bitter tonic and stomachic. By the natives of Mozambique it is used in dysentery and various other diseases.

It is a mild bitter free from astringency. It is useful in functional atonic conditions of the digestive organs, especially with other tonics, aromatics, or cathartics.

The root contains three physiologically active bases: columbamin, jateorhizin, and palmatin.

J. palmata Miers. is official in all pharmacopoeias. The difference in botanical origin is merely one of synonymy; and there is considerable confusion in the spelling of both the generic and specific names.

Austria: Kalumba—; *Belgium*: Colombo—; *Bombay*: Colombo—; *Brazil*: Calumba—; *Denmark*: Kolumbo—; *England*: Calumba, Colombo—; *Finland*: Kolumbo—; *France*: Colombo—; *Germany*: Kolombo—; *Holland*: Calumba—; *Hungary*: Calumba—; *Indo-China*: Phong Ky—; *Italy*: Colombo—; *Japan*: Calumba—; *Mexico*: Columbo—; *Norway*: Kolumbo—; *Portugal*: Calumba—; *Russia*: Kolombo—; *Spain*: Colombo—; *Sweden*: Kalumba—; *Switzerland*: Colombo—; *Tamil*: Kolumbu—; *Telugu*: Kalamba—; *Turkey*: Guvercin, Kalumba—; *United States*: Calumb, Calumba, Columbo—; *Uriya*: Kolombo—.

PACHYGONE.

This genus consists of 11 Indo-Malayan species.

Pachygone ovata Miers inhabits South India and Ceylon.

The dried fruit is used to destroy vermin and to stupefy fish.

Burma: Ngupyu—; *Ceylon*: Kadukkodi—; *Tamil*: Kadukkodi—.

PERICAMPYLUS.

This genus consists of 6 Indo-Malayan species.

Pericampylus glaucus Blatter occurs in Sikkim, Khasia, Assam, Pegu, Martaban, Tenasserim, Mergui, Chittagong, the Nicobars, and the Malay Peninsula. It extends to the Malay Archipelago, eastwards to the Moluccas, Cochin-China, Laos, Tongking, South China and Formosa.

The roots have long been held in great repute among snake-charmers in India as an antidote to the bites of poisonous snakes; but Mhaskar and Caius have shown that they are not an antidote to snake venom.

Bengal: Barakkanta—; *Cochin-China*: Day loi tien—; *Java*: Aroygeureung —; *Nepal*: Lahara, Pipalpati—; *Sumatra*: Currung—.

STEPHANIA.

The genus numbers 32 species, found in the Old World, chiefly in tropical Africa, China, and the Malayan Islands.

S. abyssinica Walp. and *S. Dinklagei* Diels are used medicinally in Yorula, Liberia and Sierra Leone.

1. Leaves ovate or subdeltoid, peltate ... ... *S. hernandifolia.*
2. Leaves broadly ovate or suborbicular, at the base rotund ... ... *S. glabra.*

1. **Stephania hernandifolia** Walp. is found on the western and eastern coasts of the Indian Peninsula, in Cachar, Sikkim, Eastern Bengal, Assam, and Penang; it is distributed over Siam and the Malay Archipelago to Australia.

The bitter root enters into the composition of a good many Ayurveda preparations as a substitute for that of *Cissampelos Pareira* Linn.

It is regarded as bitter, astringent, easily digestible and useful in fever, diarrhoea, urinary diseases, dyspepsia, etc.

The extract acts as a strong poison on frogs (Bancroft).

Bengal: Agnadnemuka, Akanadi—; *Burma*: Sha ma say nway—; *China*: Chien Chin T'eng—; *Hindi*: Akanadi—; *Java*: Areujgeureung, Ojotminjak—; *Malayalam*: Patakilannu, Patavalli—; *Mundari*: Marangpitusing—; *Nepal*: Tambarki—; *Sadani*: Gaipari—; *Sanskrit*: Ambashtha, Patha, Vanatiktika—; *Saora*: Duvgyatige—; *Sinhalese*: Lunuketigawel, Lupuketigawel—; *Uriya*: Musakani, Nimukha, Okanobhindi, Sondhimali—; *Zulu*: umTambane—.

2. **Stephania glabra** Roxb. occurs in the Himalaya from Simla to Sikkim, in the Khasia Hills, Assam, and Tenasserim.

Roxburgh states that the acrid root is used medicinally in Sylhet.

In Cochin-China it is used in pulmonary tuberculosis, asthma, dysentery, and fever.

Annam: Cu binh voi, Cu mot, Tu nhien—; *Dehra Dun*: Parah, Purha—; *Garhwal*: Gindaru—; *Nepal*: Brakulilahara, Nimilahara, Tambarki—; *Tongking*: Day moi tron—.

TILIACORA.

The genus includes 15 species, which inhabit India and tropical Africa.

Tiliacora acuminata Miers occurs from Bengal to Orissa and Konkan, Ceylon, Singapore, Java, and Cochin-China.

The root rubbed between stones and mixed with water, is given as a drink for the cure of venomous snake-bites; but, according to Mhaskar and Caius, the root is not an antidote to snake venom.

Bengal: Tiliakora, Tiliakoru—; *Canarese*: Kuri—; *Hindi*: Bagamushada, Karraath, Karwanth, Rangoe—; *Malayalam*: Vallikkanniram—; *Oudh*: Karwanth, Rangoe—; *Telugu*: Kappatige, Nagamushini, Nallatige, Pataveru, Tigemushidi, Vettichitramulamu—; *Uriya*: Ralajati noi, Kolichiti—.

TINOMISCIUM.

This genus consists of 8 Malayan species.

Tinomiscium petiolare Miers is very common in forests and secondary jungle, along roadsides, etc., from Singapore to Kedah. It is distributed to Tongking, Sumatra, and Borneo.

The Malays use the plant as a drug in rheumatism.

Malay: Akar langkap, Akar lempoyang, Akar lumpang, Akar mumbulu Akar nasi—.

TINOSPORA.

The genus numbers 40 species distributed over most of the palaeotropical regions, but chiefly in Indo-Malaya.

The following species are used medicinally in Senegal—*T. bakis* Miers.—; in Indo-China—*T. cordifolia* Miers., *T. crispa* Miers., *T. malabarica* Miers.—; in the Philippine Islands—*T. crispa* Miers.—; in Java—*T. Rumphii* Boerl.—.

I. Endocarp distinctly tuberculate
  1. Leaves pilose on both sides or at least beneath ... 1. *T. malabarica*.
  2. Leaves glabrous ... ... 2. *T. crispa*.

II. Endocarp rotund-ovoid, ribbed on the back, otherwise almost smooth ... ... 3. *T. cordifolia*.

1. **Tinospora malabarica** Miers is found in Bengal, Khasia, Assam, Orissa, Konkan, Kanara, nearly all the districts of the Madras Presidency, and Ceylon.

The plant has tonic properties.

In China and Tongking the fresh leaves and the stem are used in the treatment of chronic rheumatism.

The whole plant is used medicinally in Cambodia. Fumigations are recommended for piles and ulcerated wounds. Medicated baths are prepared in cases of liver complaint.

Almora: Gureh—; *Annam*: Khoan can dang—; *Bengal*: Bara padmagulancha, Padmagaluncha—; *Cambodia*: Kambaur—; *Ceylon*: Bukinda, Walkinda—; *Garhwal*: Gileh—; *Hindi*: Giloe, Gulancha, Gureh—; *Malaya*: Bara sarasati lat—; *Marathi*: Gulvel—; *Ramnagar*: Gurja—; *Sinhalese*: Bukinda—; *Tamil*: Potchindil—; *Tongking*: Day dan xuong—; *Uriya*: Guduchi, Gulochi, Podmogulochi—.

2. **Tinospora crispa** Miers occurs in Sylhet, Assam, and Burma. It is cultivated in the Malay Peninsula and Ceylon. It is found throughout the Malay Archipelago.

The whole plant is exceedingly bitter, and is employed by the Malays in the cure of intermittent fevers.

In Indo-China the stem is considered a febrifuge of great importance. As a tonic it is said to give as good results as quinine. There is a widespread opinion that it is an excellent tonic during convalescence from exhausting diseases.

In the Philippine Islands it is considered to be a panacea to be applied to all bodily afflictions. It is given in general debility, in chronic rheumatism, in malarial fevers.

The jungle tribes of the Malay Peninsula use the plant in the preparation of their arrow and dart poisons.

Annam: Day Ki nin, Day than thong—; *Cambodia*: Bandaul pech, Bora phet—; *French*: Liane-quinine—; *Malay*: Toba, Tuba, Tubabidyi—; *Sinhalese*: Tithakinda, Tittakinda—; *Sunda Islands*: Andawali—; *Tagalog*: Macabuhai—; *Tongking*: Thuoc sot ret—; *Visayan*: Paliavan, Panavan, Pangiavan—. The Indian names are the same as for *T. cordifolia*.

3. **Tinospora cordifolia** Miers is found throughout tropical India, Burma, the Andamans, and Ceylon.

The stem is one of the most popular drugs in India, being held in high repute by Ayurveda and Yunani doctors alike.

The stem is a bitter stomachic; stimulates bile secretion; causes constipation; tonic; allays thirst, fever, burning sensation, vomiting; diuretic; enriches the blood; cures jaundice; useful in skin diseases; the juice is useful in diabetes, vaginal and urethral discharges, low fevers, and enlarged spleen (Ayurveda).

Stem bitter; appetiser, stomachic, tonic, antipyretic, expectorant; good in cough, jaundice, giddiness, vomiting, piles, anaemia, chronic fever; renews the blood; mixed with sesame oil it is useful for massaging the body (Yunani).

An infusion of the powdered stem is used as an alterative and tonic, and has enjoyed the reputation among ancient Hindu writers of being an aphrodisiac.

In Ceylon the stem is used in fevers, skin diseases, jaundice, and syphilis. A decoction of the stem is given internally, and the bruised fresh leaves are applied externally for the bites of flies and the sting of bees.

The starch obtained from the roots and stems of the plant is similar to Arrow-root in appearance and effect. It answers not only as a remedial medicinal agent in chronic diarrhoea and some forms of obstinate chronic dysentery, but it is also a valuable nutrient, when there is intestinal irritability and inability to digest any kind of food. I have myself had experience of the usefulness of this starch. I think that this drug is useful where there is an acid diarrhoea, due to an acidity of the intestinal canal or acid dyspepsia. It is useful in relieving the symptoms of rheumatism. There is another preparation of this plant, the *succus* (juice), freshly prepared from the fresh plant. It acts as a powerful diuretic. It is prescribed by ancient Hindu physicians in gonorrhoea with advantage (Kirtikar).

The Mundas of Chota Nagpur apply the ground plant on fractures.

In certain parts of India the plant is regarded as a specific for the bites of poisonous insects and venomous snakes. The juice and decoction of the root are applied to the part bitten, poured frequently into the eyes and administered internally by mouth at intervals of half an hour.

The majority of Sanskrit authors prescribe the root and stem in combination with other drugs as an antidote to snake-bite and scorpion-sting. Mhaskar and Caius have, however, shown experimentally that the root and stem are useless in the antidotal treatment of snake-bite; the root is also useless as a collyrium or as an external application to the bitten part; the root is equally useless in the treatment of scorpion-sting.

A tincture was tried in mild cases of malaria and was found to be useless in such cases. The aqueous extract was tried in the low chronic fever of kala-azar and also in diabetes, but it was not found to possess the virtues attributed to it. . . . The drug was given, another fair trial in the form of a liquid extract in several cases of malarial fever, both in children and adults. Its action was found to be very slow. The drug had to be administered for several days before even mild cases were cured. In chronic cases it did not do any good, although it is a reputed medicine in the Hindu Pharmacopoeia for all fevers (Koman).

Arabic: Gilo—; *Bengal*: Gadancha, Giloe, Gulancha, Guluncha, Gunchi, Nimgilo—; *Bombay*: Ambarvel, Gharol, Giroli, Guloe, Gulwel—; *Burma*: Singomone, Sinzamanne—; *Canarese*: Amrytaballi, Madhuparne, Uganiballi—; *Cantonese*: Fun khu hang—; *Central Provinces*: Gulwel—; *Ceylon*: Chintil—; *Chinese*: K'uan Chu Hsing—; *Deccan*: Gulbel, Gulo, Gulvel—; *French*: Culancha—; *Goa*: Amritvel, Amrityel—; *Gujerati*, Gado, Galo, Gulo, Gulwel—; *Hansot*: Galavel—; *Hindi*: Giloe, Gulancha, Gulbel, Gulel, Guloh, Gurach, Gurcha—; *Indo-China*: Day than thong—; *Kathiawar*: Galo, Galonowelo—; *Khond*: Gursilai—; *Konkani*: Amontevel, Ekoiss givanem—; *Kumaon*: Gulancha, Gurcha—; *Lepcha*: Kantherrik—; *Malaya*: Foon kan thang, Sarasati lat—; *Malayalam*: Amrytu, Peyamrytam, Sittamrytu—; *Marathi*: Ambarvel, Gharol, Giroli, Gulavela, Gulaveli, Guloe, Gulvel—; *Mundari*: Harajora, Harajuri, Harjora—; *Nepal*: Garjo—; *Persian*: Gulbel—; *Punjab*: Batindu, Garham, Garum, Gilo, Gilogularich, Zakhmihaiyat—; *Reddi*: Korapattatige—; *Sadani*: Harjora—; *Sanskrit*: Amrita, Amritalata, Amritavallari, Amritavalli, Bhishakapriya, Chakralakshana, Chakrangi, Chandrahasa, Chandrapasa, Chchinna,

Chchinnaruha, Chchinnodbava, Chchinnodhana, Dhira, Goraksha, Guduchi, Guluchi, Jivanthika, Jivantiha, Jwaranashini, Jwarari, Kundalini, Kundalli, Madhuparni, Madhuparnika, Nagakumarika, Nirjara, Pamrodhara, Pittaghni, Rasayani, Shyama, Somalatika, Somavalli, Surakrita, Tantri, Tantrika, Uddhara, Vara, Vataraktari, Vatsadani, Vayastha, Vishalya—; *Sikkim*: Gurjo—; *Sinhalese*: Galuchi, Rasakinda, Rassakinda—; *Tamil*: Amridavalli, Amudam, Asasi, Kaippuchindil, Kunali, Narsindil, Niraidarudian, Paganrai, Padalamulam, Parivai, Pattigai, Sadi, Sagadundam, Sagamuli, Silam, Sindil, Sivandi, Sivedai, Ubavam, Vallikkandam, Vayamadu—; *Telugu*: Duyutige, Guduchi, Iruluchi, Jivantika, Madhuka, Manapala, Somida, Tellatippatige, Tippatige—; *Tharu*: Gulach—; *Tulu*: Amrytaburu—; *Uriya*: Gulancha, Gulochi—.

V.

The BERBERIDACEÆ are glabrous herbs or shrubs, sometimes climbing. They number 200 species included in 12 genera. They grow in the temperate regions of the northern hemispheres, in tropical mountains, in the Andes and extra-tropical South America. They are absent from tropical and South Africa, Australia and New Zealand.

The medicinal and poisonous Berberids of the world belong to 10 genera:—ACHLYS (Japan; Pacific North America); BERBERIS (Northern hemisphere; South America); BONGARDIA (Eastern Mediterranean); CAULOPHYLLUM (North-eastern Asia; North-America); DIPHYLLEIA (Atlantic North America; Japan); EPIMEDIUM (northern temperate regions); LEONTICE (northern temperate regions); MAHONIA (southern hemisphere; cultivated); NANDINA (China, Japan); PODOPHYLLUM (northern temperate regions).

In general the root and bark are purgative; the bitter bark is tonic and antiperiodic or depurative and cooling; the sour fruit is astringent and antiscorbutic.

The following are among the products isolated:— (1) *alkaloids*—berbamine, berberine, oxyacanthine—; (2) *resins*—podophylloresin, podophyllotoxin—; (3) *colouring matter*—podophylloquercetin—; (4) *acids*—malic, citric—.

The medicinal and poisonous Berberids of India belong to 4 genera—BERBERIS, BONGARDIA, MAHONIA, PODOPHYLLUM.

A. Stem o or erect. Flowers hermaphrodite. Carpel 1. Seeds usually small
  I. Leaves unequally pinnate. Leaflets opposite ... MAHONIA.
  II. Leaves simple, fascicled in the axils of 3-5-partite, rarely simple spines ... ... BERBERIS.
  III. Leaves simple, palmate. Ovules many ... ... PODOPHYLLUM.
B. Glabrous herb with tuberous root. All the leaves radical ... ... BONGARDIA.

BERBERIS.

This genus includes 190 species inhabiting the northern hemisphere, and South America.

About forty species of *Berberis* are used medicinally, and they all seem to have similar therapeutical properties. Their stems and barks are bitter tonics and mild laxatives.

There are three well-defined medicinal groups:—(1) The Rocky Mountain group, including *B. aquifolium* Pursh.; (2) the Asiatic group, which includes *B. aristata* DC.; (3) the European group, which includes *B. vulgaris* Linn.

The following deserve special mention in Europe—*B. aetnensis* Presl., *B. vulgaris* Linn.—; in China—*B. Sieboldii* Miq., *B. vulgaris* Linn.—; in Indo-China—*B. asiatica* Roxb., *B. Lycium* Royle, *B. japonica* R. Br., *B. vulgaris* Linn.—; in North America—*B. aquifolium* Pursh., *B. nervosa* Pursh., *B. pinnata* Lag., *B. valgaris* Linn.—; in Central America—*B. trifoliatus* Moric.—; in Colombia—*B. glauca* DC.—.

Berberine has been isolated from *B. aetnensis, B. aquifolium, B. buxifolia, B. glauca, B. nervosa, B. vulgaris*. Berbamine and oxyacanthine are contained in the root barks of *B. aquifolium* and *B. vulgaris*.

A. Ripe fruit red. Inflorescence a simple raceme. Leaves usually toothed, lower pedicels 8-25 mm. long ... 1. *B. vulgaris*.
B. Ripe fruit blue black. Inflorescence a simple raceme
 1. Leaves glossy green not glaucous beneath ... 2. *B. aristata*.
 2. Leaves pale glaucous beneath
  a. Leaves 1.3-3.8 cm. broad, secondary nerves prominent above ... ... 4. *B. asiatica*.
  b. Leaves 7.5-18 mm. broad, secondary nerves not prominent above ... ... 3. *B. lycium*.

1. **Berberis vulgaris** Linn. is found in the Western Himalaya from Kashmir to Nepal up to 12,000 feet. It also occurs in Western Tibet, and is generally distributed over temperate Asia and the greater part of Europe and Northern Africa.

The root-bark and the stem-bark have the same composition and possess the same medicinal properties.

Barberry bark is tonic, purgative, and antiseptic. As a bitter stomachic tonic, it proves an excellent remedy for dyspepsia and functional derangement of the liver, regulating the digestive powers, and if given in larger doses, acting as a mild purgative and removing constipation.

The bark is used for all cases of jaundice, general debility and biliousness, and for diarrhoea. It possesses febrifuge powers and is used as a remedy for intermittent fevers. It also forms an excellent gargle for a sore mouth. A good lotion for application to cutaneous eruptions has also been made from it.

In the Punjab, the root bark is used as diuretic, and for the relief of heat, thirst and nausea. It is considered astringent, refrigerant and antibilious. In small doses it is tonic, in larger cathartic. In the form of decoction, it is useful in scarlet fever and brain affections.

In Baluchistan the roots are boiled in water and the decoction given both to men and cattle for internal injuries. It is said to be good for fever, especially high fever.

On the Pacific Coast of North America the root-bark is considered tonic and aperient; the leaves are used as an antiscorbutic.

The berries are official in France. They contain citric and malic acids, and possess astringent and antiscorbutic properties. They are useful in inflammatory fevers, especially typhus, also in bilious disorders and scurvy; and in the form of a jelly they are very refreshing in irritable sore throat, for which also a Syrup of Barberries made with water, proves an excellent astringent gargle.

In England the fruits are macerated in about twelve times their quantity of water, to which a little fennel seed has been added, and the liquid is used as a cooling draught in cases of fever.

The Egyptians are said still to employ a diluted juice of the berries in pestilential fevers. In Ceylon the ripe fruits are used in the preparation of cooling, astringent, and antiscorbutic drinks which are given in febrile diseases and diarrhoea.

In Persia the consumption of the berries is said to remove itch and other skin complaints.

Arabic: Ambarbaris, Ambar-us-shahab—; *Baluchi*: Karoskae, Korae, Zrolg—; *Bombay*: Zarishk—; *Brahui*: Zarch—; *Catalan*: Coralets—; *Dutch*: Barbarine, Berberis, Kweekdorn, Zuurdoorn—; *English*: Barbaryn, Barberry, Barboranne, Berber, Berberry, Guild, Jaundice Berry, Maiden Barberry, Pepperridge, Piperidge, Piperidge Bush, Piperidge Rilts, Pipperidge-bush, Piprage, Woodsore, Woodsour, Woodsower Tree—; *French*: Berbéris, Chivafou, Epine-aigrette, Epine-vinette, Verre-vinette, Vinette, Vinettier—; *German* Basselbeere Berberitze, Berberstrauch, Bramelbeere, Bromelbeere, Dreidorn, Erbselbeere, Erbsele, Essigdorn, Farsbeere, Gallhageldorn, Gelbbeere, Passelbeere, Peiselbeere, Reifbeere, Reisbeere, Reissbeere, Reisselbeere, Rhabarberbeere, Sauch, Sauerdorn, Sauerrach, Saurach, Spitzbeere, Versuchsbeere, Weinaengleinstrauch, Weinling, Weinnaegelein, Weinshade, Weinschaerl, Weinzaepfel, Wuetscherling, Zwackholzbaum—; *Greek*: Axokanda—; *Hamadan*: Zirishk—; *Hattu*: Chamchur, Chochar—; *Hindi*: Zarishk—; *Italian*: Berberi, Berbero, Crespigno, Crespino, Spina santa, Trespino—; *Jaunsar*: Chatrod—; *Pacific Coast*: Barberry, Common Barberry, European Barberry, Garden Barberry, Guild-tree, Jaundice-berry, Sowberry, Wood-sour—; *Persian*: Bedana, Cutch, Karoskai, Zarishk—; *Portuguese*: Berberis—; *Punjab*: Chachar, Chochar, Kashmal, Zirishk—; *Pushtu*: Karoskai—; *Roumanian*: Dracila, Lemn galben, Macris de riuri—; *Russian*: Barbaris—; *Sibi*: Karoskae, Zrolg—; *Spanish*: Acetin, Agracejo, Agracillo, Agraz, Agrecillo, Alarguez, Alguese, Azlacristo, Berberos—; *Toba Hills*: Karoskae, Zrolg—; *Urdu*: Ambar—; *Zarghun*: Karoskae, Zrolg—; *Zhob*: Korae, Zrolg—.

2. **Berberis aristata** DC. occurs in the Himalaya from Chota Banghal to Nepal, between 6,000 and 10,500 feet.

The wood, root-bark and extract of Indian Barberry have been used in Hindoo Medicine from a very remote period. Its properties are said to be analogous to those of turmeric. Indian Barberry and its extract, *rasot,* are regarded as alterative and deobstruent, and are used in skin diseases, menorrhagia, diarrhoea, jaundice, and above all in affections of the eyes. Sarangdhara recommends a simple decoction of Indian barberry to be given, with the addition of honey in jaundice. In painful micturition from bilious or acrid urine, a decoction of Indian barberry and emblic myrobalan is given with honey. A decoction of the root-bark is used as a wash for unhealthy ulcers, and is said to improve their appearance and promote cicatrization. *Rasot,* mixed with honey, is said to be a useful application to aphthous sores.

A decoction of the root bark in doses of one to two ounces was given to several patients for malarial fever and was found to be beneficial, the effect being very slow (Koman).

Arabic: Aargis, Ambarbaris, Hhodhub, Itrarahh, Messuk—; *Bagi*: Chochar, Kambra, Tharmala—; *Bhutia*: Tsema—; *Canarese*: Bagisutra—; *English*: Indian Barberry, Tree Turmeric—; *Garhwal*: Kingora—; *Greek*: Lykion indikon—; *Hindi*: Chitra, Chotra, Darhald, Hooshish, Kashmal, Kashmar, Rasvat, Rusot—; *Jaunsar*: Kashmoi—; *Malayalam*: Maradarisina, Maramanjal—; *Narkand*: Chochar, Kambra, Tharmala—; *Nepal*: Chitra, Chutro—; *North-Western Provinces*: Chitra—; *Persian*: Chitra, Zirishk—; *Punjab*: Chitra,

Kasmal, Kemloo, Simlu, Sumlu, Tutrum—; *Sanskrit*: Daruharidra, Darvi, Kata, Katankati, Kateri, Pitadaru, Suvarnavarna—; *Simla*: Kammul, Kashmal, Kaumul—; *Tamil*: Mullukala, Usikkala—.

3. **Berberis lycium** Royle is found in the Punjab Himalaya between 3,000 and 9,000 feet, and in Kumaon at 2,500-8,500 feet.

The root is highly esteemed as a febrifuge and as a local application in eye diseases.

An extract prepared by digesting in water sliced pieces of the root, stem, and branches, is called *rusot*, and is used advantageously in cases of ophthalmia.

The leaves are administered in Baluchistan as a cure for jaundice.

In Indo-China the fruit is given as a tonic in troubles of the kidney.

The tincture was found to be useless as a remedial agent in fevers of malarial origin (Central Indigenous Drugs Committee).

Arabic: Ambarbarin, Hooziz, Hoozizindi—; *Baluchistan*: Koroski, Zarch—; *Cutch*: Kasmal—; *Hindi*: Chitra, Kushmul—; *Indo-China*: Cau tu, Ky tu—; *Jaunsar*: Chahoi, Daruhaldli, Kashmal—; *North-Western Provinces*: Kushmul—; *Persian*: Unjebar—; *Simla*: Chochar, Chotra, Kashmal, Kasmal—; *Sind*: Kushmul—; *Waziri*: Khadavanai. Sakazie, Sarghazie—.

4. **Berberis asiatica** Roxb. is found in the dry valleys of the Himalaya, from Garhwal to Bhutan, between 2,000 and 8,000 feet. It also occurs in Bihar, Mount Abu, and Afghanistan.

The plant is used for the same purpose and in the same ways as *B. aristata*. It is reputed useful in the treatment of snake-bite and scorpion-sting; but Mhaskar and Caius have shown the root, stem, and gum are all equally useless for the purpose.

Almora: Kilmoru—; *Bengal*: Daruharitra—; *Canarese*: Maradarisina—; *Dehra Dun*: Kingora—; *Garhwal*: Kingora—; *Gujerat*: Daruhaldar—; *Hindi*: Daruhaldi, Sumlu—; *Jaunsar*: Kishornoi—; *Kumaon*: Kilmora, Kilmoru—; *Marathi*: Daruhaldi—; *Nepal*: Chitra, Kissie, Matekissie—; *Persian*: Darhuld, Daruhuld—; *Ranikhet*: Kilmoru—; *Sanskrit*: Daruharidra, Darupita, Darunisha, Darvi, Dvitiyabha, Haimavati, Haridra, Hemakanti, Hemkranta, Kaliyaka, Kamini, Kapitaka, Karkatini, Karnavati, Kashtha, Kashtharajani, Katankateri, Marnmari, Nirdishta, Pachampacha, Parjani, Parjaniya, Pita, Pitachandâna, Pitadaru, Pitadru, Pitatvaka, Pitika, Sthirraga—; *Simla*: Kammula, Kashmala, Kaumula—.

BONGARDIA.

B. Rauwolfii C. A. Mey. The only species of this genus occurs in Baluchistan, whence it spreads to Afghanistan, Persia, Transcaucasus, Syria, Palestine, Bithynia, and the Greek Islands.

In Baluchistan the leaves are used as a cure for sore eyes in horses.

Ghandoba: Puccatutuka—; *Kotra*: Shrin—.

MAHONIA.

This genus includes 50 species, which inhabit the southern hemisphere. Many are cultivated as ornamental shrubs.

In North America the berries of *M. fascicularis* DC. are considered a cooling laxative medicine.

Mahonia napaulensis DC. is found in the temperate Himalaya at 4,000-8,000 feet, from Garhwal to Bhutan; in the Khasia Hills between 4,000 and 5,000 feet; in Mergui; and in the Nilgiri mountains between 5,000 and 8,000 feet.

The berries are considered diuretic, and demulcent in dysentery.

Chamba: Kandlu, Sharor—; *Garhwal*: Gurm, Haldia—; *Jaunsar*: Khoru —; *Nepal*: Chatri, Jamnemunola, Milkissie—; *Punjab*: Amudanda, Chiror—; *Ramnagar*: Bankhilmana—; *Travancore*: Maranthu—.

PODOPHYLLUM.

The genus consists of 5 species distributed over the northern temperate regions.

The rhizome of and resin from *P. emodi* Wall. are official in Great Britain; the dried rhizome and roots of *P. peltatum* Linn. are officially recognized in Belgium, Brazil, Denmark, France, Germany, Great Britain, Holland, Italy, Japan, Norway, Portugal, Russia, Spain, Sweden, Switzerland, Turkey, United States; the resin from *P. peltatum* Linn. in Austria, Finland, Hungary and Yugoslavia.

Podophyllum emodi Wall. is found in the interior ranges of the Himalaya from Hazara and Kashmir to Sikkim.

The constituents of the Indian podophyllum and of the American podophyllum (*P. peltatum* Linn.) are identical. The chief constituent is podophyllotoxin. An uncrystallizable resin, podophylloresin, has also been isolated.

The drug was administered in the form of a tincture to six cases and was found to possess all the properties that the podophyllin of the British Pharmacopoeia possessed. It acts as a hepatic stimulant and cholagogue purgative (Koman).

Gujerat: Venivel—; *Hindi*: Bakrachimyaka, Bhavanbakra, Papra, Papri, Nibishi, Pilijati—; *Kashmir*: Banwangan—; *Marathi*: Podwel, Patvel—; *Punjab*: Banbakri, Bankakra, Bankakri, Chimyaka, Chyakri, Gulkakri, Gulkakru, Kakra, Papri, Wanwangan—.

THE MEDICINAL AND POISONOUS SPURGES OF INDIA.

BY

J. F. Caius, s.j., f.l.s.

The Euphorbiaceæ represent almost all habits of growth, and exhibit a high degree of adaptability to varying environments. This large family includes 220 genera, and about 4,000 species distributed all over the world, except in arctic and antarctic regions. About half the species inhabit equatorial America; they are much rarer in temperate America. In the Old World they are more abundant in the Mediterranean region and in temperate Asia than between the tropics.

The properties of the spurgeworts are as distinctive as their botanical characters, and the ancients had so accurately recognized this, that all plants with a 3-coccus fruit were regarded by them as hurtful and suspicious. The members of the family secrete a very acrid milky juice, varying in strength with the species, the organ of the plant, and the climate. In some this juice is one of the most deleterious of poisons, in others its acridity is so far neutralized by mucilage and resins as to reduce it to a simple purgative and diuretic. Some species are slightly narcotic-acrid, others aromatic. The albumen usually contains a fixed bland oil, without the acridity which is found in the embryo and the integuments of the seed. It is to a liquid resin and a volatile principle that all the properties of the spurgeworts are due; they are thus strongest in alcoholic tinctures, but are dissipated or weakened by the application of heat. The root of the Manihot offers a remarkable example: there is scarcely a more poisonous juice than this, yet the action of fire converts the plant into very wholesome food.

Enormous though it be, the research work purporting to deal with the juice has not yet been crowned with success, and the literature regarding the toxic principles isolated from spurges is but one great mass of confusion.

The medicinal and poisonous spurges of the world belong to 76 genera:—Acalypha (tropics and subtropics of both hemispheres); Alchornea (all tropics); Aleurites (Asia and Pacific); Andrachne (tropics and subtropics); Anthostema (Madagascar; tropical Africa); Antidesma (palaeotropics); Aporosa (tropical Asia); Baliospermum (Indo-Malaya); Bischofia (Indo-Malaya, Pacific Islands); Breynia (Africa, Asia, Polynesia); Bridelia (Africa, Asia, Australia); Buxus (temperate regions of the Old World); Chrozophora (Mediterranean region, Asia, Africa); Cicca (Asia); Cleistanthus (Africa, Indo-Malaya); Cnesmone (India, Malay Peninsula, Java, Sumatra); Croton (tropics and subtropics); Dalechampia (tropics); Daphniphyllum (tropical Asia); Dichostemma (tropical Africa); Drypetes (Africa, Indo-Malaya, tropical America); Elaeophorbia (tropical

Africa); EREMOCARPUS (California); ERYTHROCOCCA (tropical Africa); EUPHORBIA (subtropical and warm temperate regions); EXCOECARIA (Old World tropics); FLUEGGEA (Old World tropics); GELONIUM (Asia, Africa); GLOCHIDION (tropical); HEVEA (Brazil, Guiana); HIPPOMANE (warm America, West Indies); HOMONOIA Indo-Malaya); HURA (tropical America); HYMENOCARDIA (Indo-Malaya, Africa); ICHTHYOTHERE (Brazil); JATROPHA (America, Malaya); JOHANNESIA (Brazil); JULOCROTON (Central and South America); MABEA (Central America, Brazil, Guiana); MACARANGA (tropics of the Old World); MAESOBOTRYA (tropical Africa); MALLOTUS (Asia, Africa); MANIHOT (South America to Mexico); MANNIOPHYTON (tropical Africa); MAPROUNEA (Brazil, Guiana, West Africa); MAREYA (tropical Africa); MERCURIALIS (Europe, Western Asia, Japan); MICRODESMIS (Asia, Africa); NECEPSIA (tropical Africa); OLDFIELDIA (tropical Africa); OMPHALEA (tropical Africa and America); OPHTHALMOBLAPTON (Brazil); PARIS (Europe, northern Asia); PEDILANTHUS (Central and South America, West Indies); PETALOSTIGMA (Australia); PHYLLANTHUS (tropics); PSEUDOLACHNOSTYLIS (tropical Africa); PUTRANJIVA (India); PYCNOCOMA (tropical Africa, Comoro Islands, Madagascar); RICHERIA (tropical America); RICINODENDRON (tropical Africa); RICINUS (tropical Africa); SAPIUM (all tropics); SAUROPUS (Indo-Malaya); SEBASTIANIA (America, Africa); SPONDIANTHUS (tropical Africa); STILLINGA (America, Africa, Mauritius, Madagascar, Fiji); SYNADENIUM (tropical and South Africa, La Reunion); TETRACARPIDIUM (tropical Africa); TETRORCHIDIUM (tropical America and Africa); TOXICODENDRUM (South Africa); TRAGIA (Asia, Africa); TREWIA (India); TRIGONOSTEMON (Indo-Malaya); TRITAXIS (Indo-Malaya); UAPACA (tropical Africa, Madagascar).

The medicinal and poisonous spurges of India belong to 36 genera :—ACALYPHA, ALCHORNEA, ALEURITES, ANDRACHNE, ANTIDESMA, APOROSA, BALIOSPERMUM, BISCHOFIA, BREYNIA, BRIDELIA, BUXUS, CHROZOPHORA, CICCA, CLEISTANTHUS, CROTON, EUPHORBIA, EXCOECARIA, FLUEGGEA, GELONIUM, GLOCHIDION, HIPPOMANE, HOMONOIA, HURA, JATROPHA, MACARANGA, MALLOTUS, MANIHOT, MICRODESMIS, PHYLLANTHUS, PUTRANJIVA, RICINUS, SAPIUM, SAUROPUS, SEBASTIANIA, TRAGIA, TREWIA.

A. Cells of ovary 2-ovuled.

I. Male flowers of one pedicelled stamen numerous surrounding a single-pedicelled female, all enclosed in a cup-like involucre ... ... EUPHORBIA.

II. Flowers apetalous; sepals 4 in male, imbricate; sepals 4 or 6 in female.

a. Leaves alternate ... ... HURA.

b. Leaves opposite ... ... BUXUS.

III. Sepals 1-2-seriate. Petals minute or absent. Stamens 1-2-seriate. Ovary 2-many-celled; cells 2-ovuled.

a. Leaves alternate, undivided, quite entire. Flowers petaliferous. Stamens 3-6.

1. Calyx valvate, stamens in a column.

i. Fruit a drupe; ovary 2-celled ... BRIDELIA.

ii. Fruit capsular; ovary 3-celled ... CLEISTANTHUS.

2. Calyx imbricate, stamens free. Flowers small. Disc of 10 glands. Herb ... ... ANDRACHNE.

9

b. Leaves alternate, rarely subopposite, undivided, quite entire. Sepals imbricate. Petals absent. Stamens 3 or more. Flowers solitary or in axillary fascicles.
   1. Fruit dry capsular of 3 bivalved cocci.
      i. Herbs, shrubs or trees; styles free or united below; disc present, of scales or a ring ... PHYLLANTHUS
      ii. Shrubs or trees; styles long united in a column, rarely free; disc absent ... GLOCHIDION.
   2. Fruit of 6 cocci; epicarp fleshy white ... FLUEGGEA.
   3. Fruit baccate of 3 to 6 hard cocci in a fleshy epicarp.
      i. Sepals 4; stamens 4; fruit large; tree ... CICCA.
      ii. Male flowers turbinate or hemispheric, lobes minute; fruit small; shrubs or treelets ... BREYNIA.
      iii. Male flowers rotate or discoid, lobes minute; fruit medium; shrublets ... ... SAUROPUS.

c. Leaves alternate, undivided, entire or serrate. Sepals imbricate. Petals absent. Stamens few or many. Styles or stigmas 2-3, dilated. Fruit a drupe.
   Stamens 2-3. Ovary 2-3-celled. Drupe 1-celled PUTRANJIVA.

d. Leaves alternate, trifoliolate. Sepals imbricate. Petals absent. Fruit a berry.
   Sepals 5. Stamens 5 ... ... BISCHOFIA.

e. Leaves alternate quite entire. Styles or stigmas usually very minute. Fruit various.
   1. Fruit drupaceous; disc circular pulvinate or conic; seeds laterally flattened reticulate; filaments long ... ... ANTIDESMA.
   2. Fruit capsular dehiscent; flowers in catkins; filaments long ... ... APOROSA.

B. Cells of ovary 1-ovuled.

I. Perianth double, of calyx and corolla. Stamens 4-10; filaments free. Ovary 1-3-celled. Fruit a small drupe.
   Flowers in axillary panicles. Petals imbricate. Stamens 5-10 ... ... MICRODESMIS.

II. Perianth single, or of the male, or of both sexes, double. Stamens 1-2-seriate. Ovary 2-3-celled. Inflorescence terminal or axillary.

a. Flowers in terminal 2-3-chotomous cymes; cymes unisexual, or with the central flower female, petaliferous.
   1. Leaves digitately nerved or lobed. Stamens many. Fruit capsular ... ... JATROPHA.
   2. Leaves palmilobed. Juice usually milky. Stamens 10 in two whorls ... ... MANIHOT.
   3. Leaves digitately nerved. Stamens 8-20. Fruit a drupe ... ... ALEURITES.

b. Flowers in terminal androgynous spikes or racemes. Petals usually villous. Stamens strongly incurved in bud alternate with the sepals. Capsule of 3 cocci. CROTON.

c. Flowers in axillary racemes. Stamens erect in bud. Petals present. Sepals valvate. Anther-cells connate. Styles bifid. Fruit capsular. Stellately tomentose herbs or shrubs ... ... CHROZOPHORA.

d. Flowers in axillary rarely terminal spikes racemes or panicles, apetalous. Calyx valvate in bud.
   1. Filaments free; anthers erect, 2-celled, cells united by their base only. Styles very long fimbriate or lacerate. Female flowers in large bracts ... ... ACALYPHA.
   2. Filaments free; anthers 2-celled or 4-locellate; cells oblong or globose, laterally attached by a narrow or broad connective.
      i. Stamens 4 to 10. Anthers 2-celled. Filaments slender, anther-cells oblong parallel. ALCHORNEA.

ii. Stamens very numerous. Anthers 2-celled.
† Leaves opposite. Anther-cells parallel. Fruit fleshy or capsular ... ... TREWIA.
†† Leaves opposite or alternate. Stamens very many, central in the flower; anther-cells usually globose, adnate to the often broad connective ... ... MALLOTUS.
iii. Stamens one or more central; anthers 3-4-locellate; anther-cells subglobose. Styles entire. Fruit capsular ... ... MACARANGA.
3. Filaments variously connate in bundles.
i. Male and female flowers in axillary spikes. Staminal bundles indefinite. Capsule unarmed ... ... HOMONOIA.
ii. Flowers in terminal panicles. Leaves palmate, serrate. Staminal bundles indefinite ... RICINUS.

e. Flowers in axillary or leaf-opposed clusters, rarely panicled or racemed, apetalous. Sepals of male imbricate or calyx shortly toothed. Stamens numerous, central in the flower.
1. Male-flowers fascicled. Sepals 5. Fruit smooth, indehiscent ... ... GELONIUM.
2. Male-flowers in panicled cymes. Sepals 4-5. Capsule of three 2-valved cocci ... BALIOSPERMUM.

f. Flowers in axillary spikes or racemes, apetalous. Male calyx valvate. Styles connate. Fruit capsular.
Racemes androgynous. Male calyx 3-5-partite; female calyx of 5 sepals. Stamens 1-3. Styles spreading above ... ... TRAGIA.

III. Perianth single. Calyx of male minute and open in bud, or obsolete.
a. Herb with very short leaf opposed bisexual spikes. Annual ... ... SEBASTIANIA.
b. Trees or shrubs.
1. Racemes terminal. Male calyx terete, 2-3-lobed. SAPIUM.
2. Racemes lateral or terminal. Calyx terete, 3-partite. Stamens 3 ... ... EXCOECARIA.
3. A tree. Flowers monoecious. Male calyx 2-3-lobed ... ... HIPPOMANE.

ACALYPHA.

The genus includes 320 tropical and subtropical species.

The following species are used medicinally in the Philippine Islands—*A. indica* Linn.—; in Madagascar—*A. spiciflora* Burm. f.—; in La Reunion—*A. indica* Linn., *A. integrifolia* Willd.—; in Zululand—*A. peduncularis* Meissn.—; in Southern Rhodesia—*A. petiolaris* Krauss—; in Gold Coast and Nigeria—*A. ciliata* Forsk.—; in Oubanghi-Chari—*A. ornata* Hochst.—; in Indo-China —*A. Evrardii* Gagnep., *A. indica* Linn.—; in the Malay Peninsula —*A. siamensis* Gage—; in India—*A. fruticosa* Forsk., *A. hispida* Burm., *A. indica* Linn., *A paniculata* Miq. .

A. Herb. Leaves long-petioled, ovate or rhombic-ovate, crenate-serrate ... ... 3. *A. indica.*

B. Shrub.
a. Leaves ovate-acute or subacute, crenate, glabrous and glandular beneath ... ... 1. *A. fruticosa.*
b. Leaves broadly rhombic-ovate, at the base shortly cuneate, at the apex cuspidate-acuminate, subacutely and coarsely serrate ... ... 2. *A. hispida.*

c. Leaves long-petioled, ovate-acuminate ... 4. *A. paniculata.*
d. Leaves sub-coriaceous rhomboid narrowed to blunt base, bluntly serrate, tip blunt ... 5. *A. siamensis.*

1. **Acalypha fruticosa** Forsk. occurs in the Deccan Peninsula, Ceylon, Pegu, Martaban and Ava. It extends to Arabia and tropical Africa.

The leaves are attenuant and alterative, and an agreeable stomachic in dyspepsia and other ailments. The dose of the infusion of the leaves as ordered by the Vaidyas in Southern India is half a teacupful twice in the day.

The leaves are a Chinese remedy for colic and diarrhœa.

In Arabia the leaves are macerated in water, and children with pustules washed in the liquid.

Arabic: Bortam—; *Canarese*: Chinni—; *Chinese*: Tch'a Iang—; *Deccan*: Chinni—; *French*: Ricinelle—; *Hadie*: Defran—; *Surdud*: Shohat—; *Tamil*: Sinni—; *Telugu*: Chinni—; *Uahfat*: Ansthat—; *Yemen*: Bortom, Defran—.

2. **Acalypha hispida** Burm. is cultivated in Indian gardens.

The leaves are beaten up with green tobacco leaf and an infusion of rice, and applied to inveterate ulcers.

The flowers, boiled in water or administered in the form of a conserve, are said to be specific in diarrhœa and similar disorders.

Malayalam: Vattattali—.

3. **Acalypha indica** Linn. occurs throughout the hotter parts of India and Ceylon. It extends to the Philippine Islands and tropical Africa.

The plant is used as expectorant as a substitute for senega. It has also a diuretic action. It is a useful remedy for bronchitis, asthma and pneumonia; also for rheumatism. It was formerly employed as a purgative and anthelmintic.

The root, bruised in hot water, is employed as a cathartic, and the leaves as a laxative in decoction mixed with common salt. The leaves are used in scabies and, mixed with lime, in other cutaneous diseases.

The powder of the dry leaves is given to children in worm cases, also a decoction prepared from the leaves with the addition of a little garlic. The juice of the same part of the plant, together with that of the tender shoots, is occasionally mixed with a small portion of margosa oil, and rubbed on the tongues of infants for the purpose of sickening them and clearing their stomachs of viscid phlegm.

The expressed juice of the leaves is in great repute, wherever the plant grows, as an emetic for children, and is safe, certain and speedy in its action. Like Ipecacuanha, it seems to have little tendency to act on the bowels or to depress the vital powers, and it decidedly increases the secretion of the pulmonary organs.

A decoction of the leaves is given in ear-ache; a cataplasm of the leaves is applied as a local application to syphilitic ulcers, and as a means of relieving the pain of snake-bite and the irritation caused by the bite of the centipede.

Fresh leaves ground into a paste, made into a ball to the size of a large marble, and introduced into the rectum, are very useful in relieving obstinate constipation of children.

The plant is used in congestive headaches: a piece of cotton is saturated with the expressed juice and inserted into each nostril, relieving head symptoms by causing hæmorrhage from the nose. The powder of the dry leaves is used in bed sores and wounds attacked by worms.

The leaf of this plant is said to be a parasiticide and is applied externally, ground with common salt or quicklime or lime juice. A paste of the leaves with lime juice was prescribed for cases of ringworm (different varieties of cutaneous tinea). In chronic cases it had no effect, but in recent cases it did some good (Koman).

Caius and Mhaskar have had nothing but disappointment with the plant. They found it quite ineffective as an anthelmintic; and the leaves proved useless as an external application in the treatment of snake-bite.

Arabic: Harram-ed-dhibbel—; *Bengal*: Kanda-amadu, Muktajari, Murkanta, Shwetbusunta—; *Bombay*: Khokali, Khokli, Kuppi—; *Canarese*: Chalmari, Kuppi—; *Chinese*: T'ie Han Ts'ai—; *English*: Indian Mercury—; *French*: Bois queue de rat—; *Godavari*: Murkandachettu, Murukonda—; *Gujerati*: Vanchhikanto—; *Hindi*: Khokali, Khokhla, Khokli, Kuppi—; *Indo-China*: Tai tuong xanh—; *Malay*: Rumput lislis—; *Malayalam*: Kuppamani—; *Marathi*: Khajoti, Khokli, Kupi—; *Mauritius*: Herbe chatte, Ortie de l'Inde—; *Philippines*: Bugos—; *Porebunder*: Dadro—; *Sanskrit*: Arittamunjariye—; *Sinhalese*: Kupamenya, Kuppamaniya, Waelkupamenya—; *Tamil*: Kuppaimeni, Kuppamani—; *Telugu*: Kuppaichettu, Kuppintaku, Murkandachettu—; *Uriya*: Indramaris—.

4. **Acalypha paniculata** Miq. is found in the Deccan Peninsula, from Malabar southwards, and in Ceylon; it ascends to 3,000 feet. It is distributed over to Java and tropical Africa.

It is often used as a substitute for *A. indica*.

Aguku: Odukwe—; *Ibo*: Odukwe—; *Yoruba*: Ilewu—.

5. **Acalypha siamensis** Gage is found in sandy open country on the east coast and in the north of the Malay Peninsula, whence it spreads over Siam.

The Malays use the leaves for making tea.

Malay: Tumput—.

ALCHORNEA.

The genus numbers about 50 tropical species, rarest in Asia.

A. cordifolia Muell.-Arg. and *A. hirtella* Benth. are used medicinally in tropical West Africa.

The root of *A. floribunda* Muell.-Arg has properties similar to those of Indian hemp and is much used in the Belgian Congo as a stimulant and narcotic.

Alchornea rugosa Muell.-Arg. is found at Amherst in Burma, in the Andaman Islands and Malacca, extending eastwards to China, the Malay Archipelago and Bouton Straits,

The seeds are used as a purgative in Indo-China.

Indo-China: Bo met, Bo net, Chua mot, Dat mot, Muoi, Soi giai—; *Lao*: Khang poi—.

ALEURITES.

The genus consists of 6 species, inhabiting warm Asia.

A cordata Steud. is used medicinally in China; *A. moluccana* Willd. in India, the Philippine Islands, Madagascar, Brazil and Guiana.

Aleurites moluccana Willd. is typically a seashore plant. It is found wild and planted in the Malay Peninsula, on the sea-coasts and on limestone rocks at 1,000 to 1,500 feet altitude. It is distributed over the Malay and Polynesian islands. The tree is cultivated in India, and very much so in South India.

The oil obtained from the kernels by expression, has been found, in doses from one to two ounces, to act as a mild and sure purgative, producing in from three to six hours after ingestion free bilious evacuations. It was found to approach nearly to castor oil in the mildness and certainty of its operation, but superior to it as having neither taste nor smell, and as producing its cathartic action without any nausea.

In Madagascar and La Reunion the leaves, heated with a flat iron, are applied several times in succession to articulations in cases of acute rheumatism.

Arabic: Jouzebarri, Khasifehindi—; *Bengal*: Akola, Akrot, Jangliakrot—; *Bombay*: Akrod, Jaiphul, Jangliakrot, Janglieranda, Japhal, Jelapa—; *Brazil*: Noz de Bancoul, Noz da India—; *Burma*: Tosikyasi—; *Canarese*: Akroda, Natakrodu—; *Caroline Islands*: Raguar—; *Chinese*: Shih Leih—; *Cutch*: Akrota—; *English*: Belgaum Walnut, Bengal Walnut, Camiri Nut, Candleberry Tree, Candle Nut, Indian Walnut, Singapore Nut—; *French*: Aleurit des Molluques, Bancoulier—; *Guam*: Lumbang, Raguar—; *Gujerati*: Akhoda, Jangli-akhroda—; *Hawaii*: Eboc, Kukui—; *Hindi*: Akola, Akrot, Jangliakrot—; *Hova*: Bakoly, Savoga, Savokara—; *Ilocano*: Balucanag—; *Jamaica*: Country Walnut—; *Kemirih-Malay*: Buaktas—; *La Reunion*: Bancoul, Bancoulier—; *Madagascar*: Mofotra—; *Malay*: Buahkeras, Kanieri—; *Malayalam*: Akrottu, Akshotam, Karankolam—; *Marathi*: Akhod, Japhala, Ranakhrot—; *Mauritius*: Bancoul—; *Persian*: Chaharmaghz-e-hindi, Girdagan-e-hindi—; *Samoa*: Lama —; *Sanskrit*: Akharota, Akhota, Akshota, Asphotaka, Gudashaya, Kandarala, Karparala, Kaureshta, Madanabhaphala, Parvatiya, Phalasneha, Pritakchhada, Bekhaphala, Svadumajja, Vrittaphala—; *Shan*: Kanyin, Mak man yau, Mak yau lik—; *Sinhalese*: Kakkuna, Ratakekuna, Tel-kekuna—; *Tagalog*: Baguilumban, Balocanad, Balucanad, Calumban, Capili, Lumban, Lumbang—; *Tamil*: Nattakkar-ottu—; *Telugu*: Natakrotu, Uduga—; *Urdu*: Akharut—; *Uriya*: Akuroto—; *Visayan*: Rumbang—.

ANDRACHNE.

This genus consists of 15 species, broadly distributed over the tropical and temperate regions of both hemispheres.

A. ovalis Müll.-Arg. is used in South Africa as an anthelmintic and insecticide.

Andrachne cordifolia Müll.-Arg. is found at altitudes of 5,000-8,000 feet in the Central and Western temperate Himalaya, from Nepal westwards to Murree and Afghanistan.

The plant is poisonous, and the twigs and leaves are said to kill cattle when browsed in the early morning on an empty stomach.

Beas: Chirmutti, Mutkar, Pin—; *Chenab*: Barsu—; *Garhwal*: Bhatula—: *Jaunsar*: Bharti, Bhartoi—; *Jhelum*: Gurguli, Kurkni—; *Punjab*: Gurguli, Kurkni, Kurkuli—; *Ravi*: Barotri, Madare—; *Sutlej*: Tsatin—.

ANTIDESMA.

This genus includes 90 palaeotropical species.

A. madagascariense Linn. is used medicinally in La Reunion; *A. membranaceum* Muell.-Arg. in Liberia; *A. venosum* E. Mey. in South and tropical West Africa; *A. zeylanicum* Lam. in Ceylon; *A. ghaesembilla* Gaertn. in Cambodia; *A. bunius* Spreng, in India.

A. Ovary tomentose or pubescent ... ... 2. *A. ghaesembilla.*

B. Ovary glabrous.
 a. Leaves 3-6 in. Spikes stout often 4-5 in. ... 1. *A. bunius.*
 b. Leaves 1-3 in. Spikes slender 1-1½ in. ... ... 3. *A. zeylanicum.*

1. **Antidesma bunius** Spreng. is a small evergreen tree found throughout the hotter parts of India, from the Nepal and Sikkim Terai and Assam, southwards to Singapore and Penang; and from Parusnath in Behar to Ceylon. It extends to the Malay Islands and China.

The acid leaves are used in snake-bites; and, when young, they are boiled and used in syphilitic cachexia.

Mhaskar and Caius have demonstrated that the leaves are not an antidote to snake venom.

Burma: Kywepyisin—; *Canarese*: Nayikute—; *Lepcha*: Kantjer—; *Malayalam*: Cherutali, Nulittali—; *Marathi*: Amati—; *Nepal*: Himalcheri—; *Pampangan*: Bignay—; *Philippines*: Bubbugnay—; *Sinhalese*: Karawellakebella—; *Tagalog*: Bignai—; *Tamil*: Nolaidali—; *Telugu*: Anepu, Janupolari—; *Visayan*: Bugnai—.

2. **Antidesma ghaesembilla** Gaertn. is a small tree found in the tropical Himalaya, from Simla to Bhutan and southwards to Ceylon, Singapore, Perak and Penang. It extends to Siam and China, the Malay Islands, Australia and Africa.

The bark, the leaves, and the wood are used medicinally in Cambodia. The bark is considered astringent and tonic. The leaves are finely crushed in a mortar and applied to the fontanel of newly-born babes, and to the head of children suffering from dengue or from a cold with cephalalgia. The wood, or better the young stems with their bark, are used as an emmenagogue.

Cambodia: Dangkiep kdam—; *Canarese*: Pulimpurase, Pullampurasigida—; *Central Provinces*: Jhondri—; *Malay*: Balng ayam, Gunchiak—; *Malayalam*: Cheriyannatam—; *Mundari*: Koetang, Koetangnuetasura, Matasuradaru—; *Telugu*: Janupulisaru, Kondapolari, Lona, Nallaballi, Polari—; *Uriya*: Nihanahari—.

3. **Antidesma zeylanicum** Lam. occurs in the southern Deccan Peninsula, and is common in Ceylon.

A decoction of the leaves is used as a snake remedy; but Mhaskar and Caius have shown experimentally that it is useless as an antidotal as well as a symptomatic treatment.

Sinhalese: Hinembilla—; *Tamil*: Nolaidali—,

APOROSA.

The genus consists of 35 Indo-Malayan species.

Aporosa lindleyana Baill., a much-branched evergreen tree, occurs in the Deccan Peninsula from the Konkan southwards, and is abundant in Ceylon.

A decoction of the root is given in jaundice, fever, headache, insanity, and seminal loss.

Canarese: Sarali—; *Malayalam*: Kotili, Vetti, Vittil—; *Sinhalese*: Baranaembilla, Kebella—; *Tamil*: Kotili, Vittil—; *Tulu*: Sarali—.

BALIOSPERMUM.

This genus consists of 10 Indo-Malayan species.

Baliospermum montanum Müll.-Arg. (=*B. axillare* Bl.) is a stout subherbaceous leafy shrub. It is found in the tropical and subtropical Himalaya, from Kashmir to Bhutan; in Assam and the Khasia Hills down to Chittagong, Pegu, Tenasserim, Burma, and Penang; in the Deccan Peninsula, from Berar and the Konkan to Travancore. It also occurs in Siam and Java.

This plant is much used in Hindu medicine, where purgation is indicated, the root being generally prescribed. This is described as 'pungent, heating; purgative, anthelmintic, diuretic, alexiteric; useful in pains, diseases of the skin and of the abdomen, piles, wounds, enlarged spleen, itching, inflammations, anæmia, leucoderma, jaundice'.

The seeds are used as a drastic purgative, but in overdoses are an acro-narcotic poison; they are sometimes used as a substitute for *Croton Tiglium*. They are also used externally as a stimulant and rubefacient. The oil is a powerful hydragogue cathartic, and is useful for external application in rheumatism. Madden states that to the east of the Sutlej the leaves are in high repute for wounds, and the sap is believed to corrode iron. The root is considered cathartic, and is used in dropsy, anasarca, and jaundice.

A decoction of the leaves is said to be useful in asthma.

Caius and Mhaskar have found the root equally useless in the treatment of snake-bite and of scorpion-sting.

Arabic: Habbussalatinebarri, Habbussalatinesahrai—; *Bengal*: Danti, Hakum, Hakun—; *Berar*: Boa—; *Bombay*: Dantimul, Jamalgota—; *Burma*: Natcho—; *Canarese*: Danti, Kaduharalu—; *Cutch*: Dantimul, Jamalgota—; *Gujerati*: Dantimul, Jamalgota—; *Hindi*: Dante, Danti, Hakum, Hakun—; *Lepcha*: Poguntig—; *Malayalam*: Dantika, Katalavanakku, Nakadanti, Neervalam, Niratimuttu—; *Marathi*: Danti—; *Melghat*: Danti, Jangli jamalgota—; *North-Western Provinces*: Janglijamalgota—; *Persian*: Bedanjirekhatai—; *Sanskrit*: Amukheti, Anukula, Bhadra, Danti, Dantika, Erandapatri, Erandaphala, Gunapriya, Jayapala, Kumbhi, Madhupushpa, Makunaka, Nagadanti, Nagasphota, Nepala, Nikumbhi, Nishalya, Nishkumbha, Pratyakparni, Raktadanti, Rechani, Ruksha, Shighra, Shwetaghanta, Shyenaghanta, Taruni, Udumbaraparni, Upachitra, Varahangi, Vishalya, Vishodhini—; *Tamil*: Kattamanakku, Niradimuttu, Nirettimuttu, Peyamanakku—; *Telugu*: Ettadundiga, Kanakapata, Kondamudamu, Nelajidi—; *Uriya*: Donti—.

BISCHOFIA.

The only species of this genus, **B. javanica** Bl., is a large evergreen tree occurring in the sub-Himalayan forests and outer hills, spreading over Chota Nagpur, Assam, Burma, Chittagong, and the Western Peninsula. It is found in the Malay and the Pacific Islands.

In Assam the juice of the leaves is considered a cure for sores.

Assam: Uriam, Uriana—; *Bombay*: Bok—; *Burma*: Aukhyu, Pogaungza, Yepadauk—; *Cagayan*: Toogan—; *Canarese*: Gobranerul, Goparimgurulle, Govarnellu, Nigurulle, Nile, Nilimara—; *Dehra Dun*: Paniala—; *English*: Java Cedar, Vinegar Wood—; *Formosa*: Ka-tang—; *Garhwal*: Kaen, Kanji, Kot semla—; *Haldwani*: Pankoen—; *Hindi*: Bhillar, Irum, Kain, Kein, Kotsemba, Paniala, Pankain—; *Igorrote*: Tuel—; *Ilocano*: Quitaquita—; *Kadir*: Naunal, Nira, Tirippu—; *Kumaon*: Koen—; *Lepcha*: Sa-nong kung—; *Malayalam*: Nira, Tirippu—; *Malkot*: Billar—; *Mundari*: Daruhajam, Hajam—; *Nilgiris*: Red Cedar—; *Oudh*: Irum—; *Philippines*: Canarem, Dueg, Tua—; *Santali*: Pader—; *Saora*: Panta—; *Tagalog*: Dampol, Toa, Toob, Tooc, Toog—; *Tamil*: Madagirvembu, Milachadayan, Milachittyan, Tondi—; *Telugu*: Nalupumushti—; *Tharu*: Areng—; *Uriya*: Dingiri—; *Visayan*: Tooc—.

BREYNIA.

This genus consists of 12 species, found in Africa, Asia and Polynesia.

Calyx greatly enlarged in fruit	1. *B. patens.*
Calyx not or scarcely enlarged in fruit	2. *B. rhamnoides.*

1. **Breynia patens** Rolfe is found in the tropical Himalaya from Nepal to Mishmi, in Assam, Chittagong, Burma, the Western Peninsula, and Ceylon.

The plant is astringent to the bowels. The juice of the stem is given in conjunctivitis.

Gujerati: Kamboi, Kedakamboi, Khedakamboi—; *Hindi*: Kalamahomad, Kambhi—; *Malayalam*: Peruniruri—; *Marathi*: Kalichikali—; *Porebunder*: Kali-kamboi, Kamboi—; *Sanskrit*: Bahupraja, Bahupushpa, Devadaru, Kamboji—; *Saora*: Kintaipude—; *Telugu*: Devadari—; *Uriya*: Deulopohora, Jajan, Medhokotahotoru—.

2. **Breynia rhamnoides** Müll.-Arg. is found throughout tropical India, Ceylon, and the Malay Peninsula; extending to the Malay and the Philippine Islands, and China.

The bark is astringent. In Behar the dried leaves are smoked like tobacco, in cases in which uvula and tonsils are swelled.

Canarese: Bamari, Gandupachcheri, Huli, Kadunugge, Karisuli, Kempehuri, Kempuhili, Suli—; *Ceylon*: Manippulanti—; *Hindi*: Surasaruni—; *Malay*: Suruyian—; *Malayalam*: Chuvannaniruri, Kattuniruri, Pavilappula—; *Oudh*: Tikkar—; *Sanskrit*: Aruin—; *Santali*: Kadrupala, Karki—; *Sinhalese*: Gaskayila—; *Tamil*: Kattuniruri, Manippulanji, Pavalappul, Pavalappula, Pamari, Sigappuppula—; *Telugu*: Ettaballi, Ettapurugudu, Nagalipachchari, Pagadapu, Pagadurugudu, Vellari, Yellari—; *Uriya*: Jajan—.

BRIDELIA.

This genus includes 60 species, spread over the tropics and subtropics of the Old World.

B. cambodiana Gagnep. is used medicinally in Cambodia, *B. schlechteri* Hutch. in South Africa; *B. atroviridis* Muell.-Arg., *B. ferruginea* Benth. and *B. micrantha* Baill. are used in Nigeria and in Gold Coast; *B. montana* Willd. and *B. retusa* Spreng, in India.

Leaves membranous	1. *B. montana.*
Leaves rigidly coriaceous	2. *B. retusa,*

1. **Bridelia montana** Willd. is found in the sub-Himalayan tract from the Jhelum eastwards; it also occurs in the Khasia Hills, the Central Provinces, Bihar, Orissa, and Upper Burma.

The root and the bark are much used as an astringent medicine in Bombay and Goa.

The plant has been credited with anthelmintic properties.

Assam: Kaisho—; *Bombay*: Asano—; *Canarese*: Kittoe, Sidigulige—; *Cutch*: Asana—; *Garhwal*: Gaya—; *Goa*: Faturfoda—; *Gujerati*: Asano—; *Hindi*: Geia, Gondni, Kargnalia, Khaja, Kusi—; *Khond*: Balli—; *Kumaon*: Kurgnulia—; *Marathi*: Asana—; *Mundari*: Karakadaru—; *Nepal*: Geio—; *Saharanpur*: Gondni—; *Tamil*: Vengaimaram—; *Telugu*: Gundubigalu, Panchavoni, Pantangi, Pantegi, Pantiga, Patunga, Sannakodari, Tellavegisa—; *Uriya*: Marda, Nuanali—.

2. **Bridelia retusa** Spreng. is found throughout India, and Ceylon, Burma and Malacca.

The root and the bark are valuable astringents.

Hindu practitioners in Western India use the bark for the removal of urinary concretions.

The bark is used as a liniment with gingelly oil in rheumatism.

Ajmere: Lamkana—; *Assam*: Kohir—; *Banswara*: Angnera—; *Bengal*: Kasai, Kosai—; *Betul*: Kasai, Sankana—; *Bhil*: Asana, Gunjan, Katiain—; *Bhumij*: Karika—; *Bombay*: Asauna, Assana, Phatarphod—; *Burma*: Seikche, Seikchi, Seikgyi, Tseichyi—; *Canarese*: Asana, Garige, Goje, Gojji, Guju, Gurige, Gworgie, Havugandha, Kogyamunji, Maguva, Muljane, Mullahonne, Mullubenga, Mulluhonne, Nasinagandha, Nasinage, Phatterphodi, Siruhonne—; *Central Provinces*: Karka, Kassei, Kassi—; *Chittagong*: Kamkui—; *Deccan*: Sun—; *Dehra Dun*: Gaya, Khaja—; *Garhwal*: Gauli, Gaya, Khaja—; *Garo*: Kashi—; *Gond*: Kassei—; *Gujerati*: Asana, Ekalakanto, Monj—; *Hindi*: Gauli, Kaj, Kajja, Kassi, Khaja—; *Ho*: Kaka—; *Kalagarh*: Goli—; *Kharwar*: Kaj, Kaji, Kanj—; *Khond*: Rugendi—; *Kolami*: Kaka, Kharaka—; *Konkani*: Kongy—; *Kurku*: Karka, Karkha—; *Lambadi*: Makhiro—; *Lepcha*: Pengji—; *Malayalam*: Chalaru, Komanji, Mullukayani, Mulluvenya, Mulluvenna—; *Mandevi*: Haklo—; *Marathi*: Asana, Asanakutgi, Gunjan, Kantakauchi, Katiain, Katiyen, Kutki, Palapasana, Patharaphoda, Phatharaphoda—; *Matheran*: Asana, Hasana, Palehasan, Phatarphod—; *Mechi*: Katakuchi—; *Monghyr*: Kuj—; *Mundari*: Karaka—; *Nepal*: Geio—; *Nimar*: Kasai, Sankana—; *Portuguese India*: Bentka—; *Punjab*: Mark, Pathor—; *Rajbanshi*: Nanda—; *Rajputana*: Angnera, Lamkana—; *Ramnagar*: Goli—; *Reddi*: Annemu—; *Saharanpur*: Ekdania, Gondni—; *Sanskrit*: Asana, Ekadivī, Ekavira, Mahavira, Sakridvira, Suvarika—; *Santali*: Kadrupala—; *Saora*: Annemu—; *Sinhalese*: Kattakaala, Ketakala, Ketakele—; *Tamil*: Adamarudu, Asuvai, Kadugai, Komanji, Malaivengai, Mullumarudu, Mulluvengai, Mulvengai, Sem, Singattan, Siruvengai—; *Telugu*: Annemu, Bontavegita, Duriyamaddi, Ettapattaka, Kodaratti, Koramaddi, Koramanu, Kosengi, Peddayanemu, Poramanu, Selaka—; *Tharu*: Khooj—; *Tinnevelly*: Adamaradu—; *Tulu*: Mullugunje—; *Uriya*: Kasai, Kasi, Kosai, Kosi, Panikasi—.

BUXUS.

The genus consists of 25 species, inhabiting the temperate regions of the Old World.

B. balearica Lam. is used medicinally in Majorca, *B. sempervirens* Linn. in Europe and North America.

Buxus sempervirens Linn. occurs in the temperate Himalaya, from Kumaon to Simla and Bhutan at 5,000-9,000 feet, and on the Salt Range in the Punjab. Westwards it extends to North Africa, and to western and southern Europe, where it abounds in limestone districts. Northwards it extends to western Siberia, Turkestan, China and Japan.

Yunani practitioners use both the leaves and seeds:—'The leaves are good for headache, pain, prolapsus ani. The seeds are bitter, astringent, tonic to the heart and brain; used in stomatitis, to dry the bad humours of the liver'.

The wood in its native countries is considered diaphoretic, being given in decoction as an alterative for rheumatism and secondary syphilis. It is used as a substitute for guaiacum in the treatment of venereal disease when sudorifics are considered to be the correct specifics.

Boxwood has been found narcotic and sedative in full doses; emetico-cathartic and convulsant in overdose. The tincture was formerly used as a bitter tonic and antiperiodic, and had the reputation of curing leprosy.

A volatile oil distilled from the wood has been prescribed in cases of epilepsy. The oil has been employed for piles and also for toothache.

The leaves, which have a nauseous bitter taste, have sudorific, alterative, and cathartic properties; they are given in powder, in which form they are also an excellent vermifuge.

Various extracts and perfumes were formerly made from the leaves and bark. A decoction was recommended as an application to promote the growth of the hair.

Dried and powdered, the leaves are still given to horses for the purpose of improving their coats. The powder is regarded by carters as highly poisonous, to be given with great care. In Devonshire, farriers still employ the old remedy of powdered Box leaves for bot-worms in horses.

In some parts of Europe Box is one of the active ingredients in a cure for the bite of a mad dog.

Camels readily eat the leaves; but they are poisoned by them.

The bark of the root is officinal in Portugal.

Afghanistan: Shandalaghune—; *Arabic*: Bakas—; *Catalan*: Boix—; *Dutch*: Boxboom, Palm, Palmboom—; *English*: Box Tree, Dudgeon, Evergreen Box—; *French*: Bois bénit, Buis, Buis commun, Ozanne—; *German*: Buchsbaum—; *Greek*: Pyxos—; *Italian*: Bosso, Bossolo, Busso, Buxo, Mortella—; *Jaunsar*: Papri, Sansadu—; *Kashmir*: Chikri—; *Kumaon*: Papri—; *Mashudi*: Shamshath—; *North America*: Box-tree, Bush-tree, Common Box, Dudgepn—; *North-Western Himalayas*: Chikri, Papri, Poppar, Sansadu, Shamshad—; *Persian*: Shamshad—; *Portuguese*: Buxo—; *Punjab*: Chikri, Papar, Paprang, Papri, Papur, Shamshad, Shumaj—; *Roumanian*: Cimisu, Cimsu, Merisor tureesc—; *Russian*: Buksus, Samshit—; *Spanish*: Boj—; *Swedish*: Buxbom—; *Urdu*: Shamshad—.

CHROZOPHORA.

This genus consists of 11 species, natives of the Mediterranean region, tropical Asia and tropical Africa.

C. plicata A. Juss. is a poison with acrid properties. *C. senegalensis* A. Juss. is used medicinally in Northern Nigeria.

A prostrate herb. Stamens 15 in two whorls. Ovary and capsule with stellate tomentum and silvery scales. 2. *C. rottleri.*

A procumbent stellately woolly herb. Ovary and capsule with stellate tomentum but without silvery scales ... 1. *C. prostrata.*

1. **Chrozophora prostrata** Dalz. is found throughout India: the Punjab, the Upper Gangetic Plain, Bihar, Gujerat, Konkan, Deccan, North Kanara, Sind, Central and South India. It extends westwards to North Africa and Spain.

The ashes of the root are given to children for cough. The leaves are considered depurative. The seeds are used as purgative.

The Santals mix the root with that of *Carissa Carandas* for blistering purposes.

Bengal: Khudiokra—; *Chinese*: Pa Teou Nan—; *French*: Bancoulier—; *Gujerati*: Betookharada, Okharada—; *Hindi*: Shadevi, Sonballi, Subali—; *Punjab*: Nilakrai, Nilkhanti, Putkhanda—; *Sanskrit*: Suryavarta—; *Santal*: Pangonari—; *Sind*: Shadevi, Sonballi, Subali—; *Tank*: Nealboti—; *Telugu*: Guruguchettu, Lingamiriyam—.

2. **Chrozophora rottleri** A. Juss. occurs through the whole of India. It extends to Afghanistan and thence to the Mediterranean region.

The plant has emetic, drastic, and corrosive properties. The seeds are used as a cathartic

Afghanistan: Kapochist—; *English*: Turnsole—; *French*: Morelle à indigo—; *Gujerat*: Kalookharad—; *Hindi*: Shadevi, Sonballi, Subali—; *Marathi*: Suravarta—; *Punjab*: Kukronda, Nilan, Tappalbuti—; *Sanskrit*: Suryavarta—; *Sind*: Shadevi, Sonballi, Subali—; *Spanish*: Tornasol—.

CICCA.

C. disticha Linn. (=*Phyllanthus distichus* Müll.-Arg.) is cultivated in India and the Malay Peninsula. It occurs in the Malay Islands and in Madagascar.

The root and the seed are cathartic. The fruit is acid and astringent.

The juice of the root-bark is used as a poison, often for criminal purposes. It produces headache, sleepiness, and death accompanied by severe abdominal pains.

Bengal: Hariphal, Hariphul, Loda, Noari—; *Bombay*: Harparawri, Raiavala—; *Burma*: Thimbawzibyu, Thinbawniyu, Thinbozihpyu—; *Canarese*: Aranelli, Karinelli, Kirinelli, Kirunelli, Rayaranelli—; *English*: The Country Gooseberry, Star Gooseberry—; *French*: Brignolier—; *Goa*: Cherambola—; *Hindi*: Chatmeri, Harfarauri, Nuree. Nurphal—; *Konkani*: Rojanvalli, Rosanvalli—; *La Reunion*: Cherimbelier—; *Malay*: Chermela—; *Malayalam*: Arinelli, Nellipuli—; *Mauritius*: Chérimbolier—; *Pampangan*: Iba—; *Sanskrit*: Ghana, Komalavalkala, Lavali, Lavalippala, Lavani, Pandu, Skandhaphala, Srigdha, Sugandhamula—; *Sinhalese*: Ratanelli—; *Tagalog*: Bangquiling, Banquilin, Iba, Yba—; *Tamil*: Arunelli, Sadadaram, Sadagam—; *Telugu*: Rachayusirika, Ratsavusiriki—; *Tulu*: Karinelli—; *Urdu*: Harpharuri—; *Uriya*: Aurakuli, Bungarada, Narokoli—; *Visayan*: Bangquiling, Layohan, Poras—.

CLEISTANTHUS.

The genus includes 110 African and Indo-Malayan species.

Cleistanthus collinus Benth. occurs in Bihar, Chota Nagpur the Satpura Range, the Chanda District and the Western Peninsula. It is rare in Ceylon.

The plant is poisonous. An extract of the leaves and fruit acts as a violent gastro-intestinal irritant,

In Chota Nagpur the fruit and bark are employed to poison fish; the latter is also considered a useful application in cutaneous diseases. For severe headache, the head and upper part of the body are bathed in water in which the leaves have been steeped. The Mundas throw the leaves, dried and pounded, into pools and ponds to kill fish.

Berar: Ghara—; *Canarese*: Badedarige, Bodauaraga, Kadagargari, Kodasige—; *Central Provinces*: Ganari, Garar, Garari—; *Gond*: Garari—; *Hasada*: Parasudaru—; *Hindi*: Garari, Garrar, Karla—; *Ho*: Pasu—; *Kharwar*: Kargali, Kergali—; *Khond*: Kadise—; *Kolami*: Larshuter, Parasu, Pas, Pasu—; *Malayalam*: Nilappala—; *Marathi*: Garari—; *Mundari*: Kargelung, Parasu—; *Naguri*: Kargelodaru, Kargelungdaru—; *Sanskrit*: Indrayava, Kaudigam, Kutaja, Nandi—; *Santali*: Kargali, Kargalli—; *Tamil*: Nilaippalai, Odaichi, Odan, Odishi, Odu, Odugu, Oduppai, Oduvan—; *Telugu*: Kadise, Korshe, Korsi, Vadise—; *Uriya*: Karada, Koroda, Korora—.

CROTON.

This genus includes 600 species, inhabiting all tropical and subtropical countries.

The 'Crotons' of gardens do not belong to this genus, but to the genus *Codiaeum* of Papua and Polynesia.

The following species are used medicinally in China, Malaya, the Philippine Islands—*C. Tiglium* Linn.—; in Indo-China— *C. argyratus* Bl., *C. Cumingii* Müll.-Arg., *C. maieuticus* Gagnep., *C. oblongifolius* Roxb., *C. poilanei* Gagnep., *C. Tiglium* Linn., *C. tonkinensis* Gagnep.—; in North America— *C. californicus* Müll.-Arg.—; in Mexico — *C. dioicus* Cav., *C. draco* Cham. and Schlecht., *C. niveus* Jacq., *C. vulpinus* Sesse and Moq.—; in Venezuela and Colombia— *C. leptostachyus* H. B. and K., *C. malasubo* Karst.—; in Peru— *C. suberosus* H. B. and K.—; in Brazil— *C. antisiphiliticus* Mart., *C. campestris* St. Hil., *C. fulvus* Mart.—; in Guiana— *C. origanifolius* Lam.—; in the West Indies— *C. cascarilla* Benn., *C. eluteria* Benn., *C. flavens* Linn., *C. glabellus* Linn., *C. lucidus* Linn.—; in Gambia and Nigeria— *C. amabilis* Muell.-Arg., *C. lobatus* Linn.—; in Gold Coast— *C. lobatus* Linn., *C. penduliflorus* Hutch—; in East Africa— *C. elliotianus* Baill., *C. gubouga* S. Moore; in Southern Africa— *C. gratissimus* Burch., *C. gubouga* S. Moore, *C. sylvaticus* Hochst.—; in Madagascar— *C. macrobuxus* Baill.—.

A. Inflorescence lepidote.
- a. Leaves densely lepidote beneath
  - 1. A tree ... ... 1. *C. argyratus.*
  - 2. A shrub.
    - i. Petiole ½-1½ in. Stamens 15-18. Capsule ½ in. long, broadly oblong ... ... 5. *C. reticulatus.*
    - ii. Petiole ¾ in. Stamens 12-15. Capsule ¼ in. long, scaly ... ... 3. *C. Cumingii*
- b. Leaves glabrous when old ... ... 4. *C. oblongifolius.*

B. Inflorescence stellately tomentose or glabrous.
- a. Leaves 3-5-plinerved at the base, scabrid or softly pubescent with stellate hairs beneath ... 2. *C. caudatus*
- b. Leaves strongly triple-nerved ... ... 6. *C. Tiglium.*

1. **Croton argyratus** Bl. is a small evergreen tree, common in the forests and woods of Burma, Malaya, Sumatra, Java, and Borneo.

In Indo-China the leaves are infused for a digestive tea.

Indo-China: Bac la, Day mang—; *Malay*: Cherit badak, Mungke, Sim-mungke—.

2. **Croton caudatus** Geisel is found in the Eastern Himalaya, in Assam and Bengal down to the Deccan, Malacca, and Ceylon; extending to Java and the Philippine Islands.

The leaves are applied as a poultice to sprains.

In Lakhimpur the young leaf buds are powdered with the leaves of *Caesalpinia sappan* and used for liver diseases.

Bengal: Nanbhantur—; *Lakhimpur*: Latamahundi—; *Lepcha*: Tackcha-brik—; *Malay*: Akar guroh periat, Akar tuku takal—; *Uriya*: Pholancokriti, Sanaushnota, Wusta—.

3. **Croton Cumingii** Müll.-Arg. is a straggling shrub found on rocks, usually limestone, throughout the Malay Peninsula, Siam, Indo-China, and the Philippine Islands.

The plant is used medicinally in Annam.

Indo-China: Co chu lon—.

4. **Croton oblongifolius** Roxb. is found in Sylhet, Bengal, Chota Nagpur, the Central Provinces, Burma, the western Peninsula, and Ceylon.

The seeds and fruits are purgative.

The Goanese and the inhabitants of Southern Konkan administer the bark in chronic enlargements of the liver and in remittent fever. In the former disease, it is both taken internally and applied locally. As an application to sprains, bruises, rheumatic swellings, etc., it is in great request.

In the Southern Konkan, it has a reputation as a remedy in snake-bites.

The Santals use the bark and root as a purgative and as an alterative in dysentery.

The roots enter into the composition of pills used by the Mundas of Chota Nagpur for chronic rheumatism. The root bark is given in small doses as a purge; a larger quantity is said to be poisonous.

The leaves are used externally in Cambodia for liver complaints and scabies. The leaves are often boiled in water, and the liquid applied hot to the itch.

The root, bark, and seeds are equally useless in the antidotal and symptomatic treatment of snake-bite; the root is useless as an external application (Mhaskar and Caius).

Assam: Burmaparokupi—; *Bengal*: Baragach, Putri—; *Bombay*: Ganasuar, Gunsur—; *Burma*: Theyin, Thityin—; *Cambodia*: Trapung—; *Goa*: Gonsur—; *Hindi*: Buragach—; *Indo-China*: Cui den, Tra poung—; *Kharwar*: Bhain swan—; *Koderma*: Maisonda—; *Kolami*: Konya, Kuli, Kurti, Kutikonyer, Poter—; *Lohardugga*: Putri—; *Malay*: Kote, Putol—; *Mal Paharia*: Putol—; *Marathi*: Ganasura, Gunsur—; *Mundari*: Kuti, Kutidaru—; *Nepal*: Ach—; *Oudh*: Arjunna—; *Patna*: Chucka—; *Sanskrit*: Bhutankusa—; *Santali*: Gote—; *Tamil*: Milgunari—; *Telugu*: Bhutalabhairi, Bhutankusamu—; *Tharu*: Mahson—; *Uran*: Poter—; *Uriya*: Masundi—.

5. **Croton reticulatus** Heyne is found in the Deccan Peninsula, from the Konkan southwards. It also occurs in the Central Province of Ceylon.

The bark is used as a bitter stomachic.

Bombay: Panduray—; *Marathi*: Pandhari, Pandharisala—; *Matheran*: Pandhri, Pandurai—.

6. **Croton Tiglium** Linn. is found naturalized and cultivated in Bengal, Assam, and southwards to Burma, Malacca, and Ceylon. It extends to the Malay Islands and Ceylon.

According to Ayurveda the fruit and the seed are purgative; and the oil from the seed is purgative, carminative, useful in diseases of the abdomen, mental troubles, convulsions, fever, insanity, inflammations, bronchitis.

According to Yunani practitioners the seeds have a bitter bad taste, causing a burning sensation; cathartic, expectorant, emetic; good in sore eyes, excessive phlegm, leucoderma.—The oil is cathartic, tonic; removes pus and bad matter from the body.

The seeds and the oil are officinal and their properties are well known: irritant, rubefacient, cathartic. The seeds are officinal in Finland and Portugal; the oil is officinal in Austria, Belgium, Beazil, Denmark, France, Germany, Great Britain, Japan, Mexico, Norway, Russia, Spain, Switzerland, the United States, and Yugoslavia.

In Lakhimpur the seeds are ground in water, and the infusion is used to kill insect pests.

The fruits are employed by Dayaks in Borneo to poison fish. The root is used in Kelantan as an abortifacient.

On account of their drastic purgative properties the seeds and oil are regarded by the Chinese as entirely poisonous. The bark is used as a tonic in Annam.

Caius and Mhaskar gave two adult males one minim of croton oil in one dose. The patients had several motions, but passed no worms though they were ultimately found to have been harbouring 356 hookworms. These same workers found that the seeds are not an antidote to snake venom, and that they are useless as a collyrium in the treatment of snake-bite or as an external application to the sting of the scorpion; they also demonstrated that the leaves have no action when applied externally to the site of the bite from a poisonous snake.

Annam: Ba dau—; *Arabic*: Batu, Dand, Datun, Habbussalatin—; *Bengal*: Jamalgota, Jaypal, Jhayal, Rechuk—; *Bombay*: Jaipa, Jaipal, Jamalgota—; *Burma*: Kanako—; *Canarese*: Danti, Japala, Jayapala, Nepala—; *Chinese*: Pa Teou, Pa Teou Seou, Pa Tou—; *Deccan*: Jamalgota—; *English*: Purging Croton—; *French*: Bois des Molluques, Bois de Pavana, Bois de Pavanna, Bois purgatif, Graine de Tilly, Graine de Moluques, Petit pignon d' Inde—; *Gujerati*: Nepal—; *Hamadan*: Habb-el-salatin—; *Hindi*: Jamalgot, Jamalgota— *Indo-China*: Ba dau, Ba tau, Bat khlot, Cong khoi—; *Italian*: Grana tiglio—; *Japan*: Hazu—; *Java*: Cheraken—; *Lakhimpur*: Kanibhi—; *Malay*: Chemkian, Chengkian—; *Malaya*: Bori, Pai Tow—; *Malayalam*: Dantibijam, Katalavanakku, Nirvalam—; *Marathi*: Jamalagota, Jepal, Jeyapal—; *Mauritius*: Croton tilly, Tilly—; *Persian*: Badanjirekhatai, Bidendjirehkhatai, Dund, Habbdilmaluk, Habbelkhatai—; *Portuguese*: Croton tiglio—; *Sanskrit*: Danti, Jayapala, Nepala—; *Sinhalese*: Jayapala, Nepalam—; *Tagalog*: Camaisa, Tuba, Tubacamaisa—; *Tamil*: Kattukkattai, Naganam, Nagandi, Nervalam, Nigumbam, Nirvalam, Sambari, Sayabalam, Sevalangottai, Siduram, Sittudu, Tendi—; *Telugu*: Nepalamu, Nepalavemu—; *Tulu*: Berada, Byaribittu—; *Uriya*: Jaipalo Joyopalo, Konika—; *Visayam*: Camarandag, Macasla, Tubasabuquid—.

EUPHORBIA.

The genus comprises 750 species, inhabiting chiefly subtropical and warm temperate countries. It is very various in habit. Some species have a fleshy angular stem, furnished with spines, resembling that of a Cactus; these species yield by incision a resinous juice which is strongly drastic, and used externally as a vesicant. The other Euphorbias, which have a normal stem and leaves, have a purgative milky juice; some are considered efficacious in syphilitic cachexia; others are emetic, or slightly astringent and aromatic, or astringent and slightly narcotic.

The following species are used medicinally in Europe— *E. amygdaloides* Linn., *E. Chamaesyce* Linn., *E. Characias* Linn., *E. Cyparissias* Linn., *E. Esula* Linn., *E. Gerardiana* Jacq., *E. helioscopia* Linn., *E. hyberna* Linn., *E. Lathyris* Linn., *E. Myrsinites* Linn., *E. palustris* Linn., *E. Peplis* Linn., *E. Peplus* Linn., *E. resinifera* Berg., *E. serrata* Linn.—; in the Mediterranean region— *E. aleppica* Linn., *E. Apios* Linn., *E. dendroides* Linn., *E. Peplis* Linn., *E. spinosa* Linn.—; in Arabia— *E. antiquorum* Linn., *E. officinarum* Linn.—; in China— *E. adenochlora* Morr. and Dcne., *E. helioscopia* Linn., *E. humifusa* Willd., *E. Lathyris* Linn., *E. pekinensis* Rupr., *E. Sieboldiana* Morr. and Dcne.—; in Indo-China: *E. antiquorum* Linn., *E. Atoto* Forst., *E. Chamaesya* Linn., *E. edulis* Lour., *E. hirta* Linn., *E. hypericifolia* Linn., *E. neriifolia* Linn., *E. parviflora* Linn., *E. rosea* Retz, *E. thymifolia* Linn., *E. Tirucalli* Linn.—; in Malaya— *E. Lathyris* Linn., *E. pekinensis* Rupr., *E. Sieboldiana* Morr. and Dcne.—; in the Philippine Islands— *E. hirta* Linn., *E. neriifolia* Linn., *E. Tirucalli* Linn.—; in Australia— *E. Drummondii* Boiss.—; in Fiji— *E. hirta* Linn.—; in North America— *E. albomarginata* Torr. and Gray, *E. coralloides* Linn., *E. corollata* Linn., *E. hirta* Linn., *E. humistrata* Engl., *E. Ipecacuanha* Linn., *E. Lathyris* Linn., *E. maculata* Linn., *E. marginata* Pursh., *E. ocellata* Durand, *E. Preslii* Guss., *E. serpyllifolia* Pers.—; in the West Indies— *E. buxifolia* Lam., *E. linearis* Retz, *E. pulcherrima* Willd.—; in Mexico— *E. glyptosperma* Linn., *E. prostrata* Ait., *E. pulcherrima* Willd.—; in Peru— *E. laurifolia* Lam., *E. pulcherrima* Willd.—; in Chile— *E. portulacoides* Linn.—; in Brazil— *E. brasiliensis* Lam., *E. caecorum* Mart., *E. cotinifolia* Linn., *E. heterodoxa* Muell.-Arg., *E. hypericifolia* Linn., *E. papillosa* St. Hil., *E. phosphorea* Mart.—; in Guiana— *E. cotinoides* Miq., *E. hirta* Linn., *E. hypericifolia* Linn., *E. punicea* Sw.—; in the West Indies— *E. hirta* Linn., *E. pulcherrima* Willd.—; in the Canary Islands and in Guinea— *E. balsamifera* Ait., *E. canariensis* Linn.—; in West Africa— *E. aegyptiaca* Boiss., *E. balsamifera* Ait., *E. convolvuloides* Hochst., *E. hirta* Linn., *E. kamerunica* Pax, *E. lateriflora* Schum. and Thonn., *E. leonensis* N. E. Br., *E. Poissonii* Pax, *E. polycnemoides* Hochst.—; in Abyssinia— *E. cerebrina* Hochst.—; in Ethiopia— *E. officinarum* Linn.—; in Madagascar— *E. anagalloides* Bak., *E. Bojeri* Hook., *E. emirnensis* Bak., *E. erythroxyloides* Bak., *E. hirta* Linn., *E. splendens* Boj., *E. trichophylla* Bak. —; in La Reunion— *E. dissimilis* Cordem., *E. goliana* Lam., *E. heyneana* Boiss., *E. hirta* Linn.,

E. hypericifolia Linn., *E. Peplus* Linn., *E. prostrata* Ait., *E. thymifolia* Burm., *E. Tirucalli* Linn.—; in Southern Africa— *E. abyssinica* J. F. Gmel., *E. basutica* Marl., *E. bupleurifolia* Jacq., *E. clavarioides* Boiss., *E. helioscopia* Linn., *E. ingens* E. Mey., *E. mauritanica* Linn., *E. procumbens* Mill., *E. restituta* N. E. Br., *E. sanguinea* Hochst. and Steud., *E. striata* Thunb., *E. truncata* N. E. Br., *E. virosa* Willd.—.

E. Candelabrum Tremaut, *E. mauritanica* Linn., *E. striata* Thunb., and *E. virosa* Willd. are used as arrow poisons by the Bushmen of Southern Africa; *E. sapini* de Wild. is similarly used by the tribes of Oubanghi-Chari.

The latex of *E. canariensis* Linn. is officinal in Portugal and Spain; that of *E. resinifera* Berg. in Austria, Belgium, Brazil, Denmark, Finland, France, Germany, Hungary, Italy, Norway, Portugal, Russia, Spain, Sweden, Switzerland, Turkey, and Yugoslavia.

A. Herbs, rarely shrubby below; prostrate or ascending. Leaves all opposite. Glands 4-5, usually furnished with a membranous petaloid limb.

- I. Erect or decumbent branching herbs or shrubs, with leaves $\frac{1}{3}$-$\frac{2}{3}$ in. long or thereabouts. Involucres in axillary and subterminal cymes or crowded in the uppermost leaves. Limb of gland conspicuous though small.
  - a. Glabrous. Stem stout swollen at the nodes ... 2. *E. Atoto.*
  - b. Glabrous or sparsely pubescent
    - 1. Annual. Seeds smooth or with broad shallow transverse pits ... ... 7. *E. hypericifolia.*
    - 2. Perennial. Seed rugose ... ... 12. *E. rosea.*
  - c. Hispid with copious crisped hairs ... ... 6. *E. hirta*
- II. Prostrate, rarely erect plants, with leaves $\frac{1}{6}$-$\frac{1}{3}$ in. long, rarely more. Involucres solitary, or in very short subsessile cymes, usually crowded along one side of the stem, or of short lateral branches. Limb of gland obscure or absent.
  - a. Stems hispidly hairy. Capsules pubescent or hirsute.
    - 1. Leaves crenulate. Capsules pubescent. Seeds furrowed ... ... 16. *E. thymifolia.*
    - 2. Leaves entire. Capsules hirsute. Seeds pitted ... ... 4. *E. granulata.*
  - b. Stems glabrous. Capsules glabrous.
    - 1. Leaves nearly as broad as long. Seeds smooth ... ... 9. *E. microphylla.*
    - 2. Leaves from a very unequal base, elliptic-rhomboid, very obtuse ... ... 14. *E. sanguinea.*
  - c. Stems glabrous or short-hairy ... ... 19. *E. turcomanica.*

B. Shrubs or trees with fleshy terete ribbed angled or flattened stems and branches. Leaves absent or alternate or the upper opposite

- I. An erect unarmed shrub or tree. Branches spreading, cylindric, scattered, whorled or clustered. Leaves absent or few small linear-oblong ... ... 17. *E. Tirucalli.*
- II. Erect shrubs or trees, with stout angled ribbed or winged branches that are crenate or nodose at the angles or wings, and there provided with 2 or 3 stipular prickles.
  - a. Pairs of stipular spines on tubercles or swellings of the branchlets ... ... 10. *E. neriifolia.*

10

b. Pairs of stipular spines on flat brown or black corky areas, not on swellings of the branchlets. 11. *E. nivulia.*
c. Pairs of stipular spines in the sinuses of the thick sinuate wings .... 1. *E. antiquorum.*
d. Pairs of stipular spines on the margins of the thick obtuse undulate wings ... ... 13. *E. Royleana.*
e. Pairs of stipular spines on the lobes of the lobulate wings ... ... 18. *E. trigona.*

C. Herbs, rarely shrubby below, usually erect. Leaves alternate, upper opposite.
I. Involucral glands transversely oblong with rounded margins.
a. Perennial herbs.
1. Capsule smooth. Plant quite glabrous ... 15. *E. Thomsoniana.*
2. Capsule warted. Plant glabrous or sparsely hairy ... ... 8. *E. longifolia.*
b. Annuals. Capsule smooth. Seeds pitted ... 5. *E. helioscopia.*
II. Involucral glands truncate retuse or 2-cornate. Annuals. Seeds oblong leprous ... ... 3. *E. dracunculoides.*

1. **Euphorbia antiquorum** Linn. occurs throughout the hotter parts of India and Ceylon, in dry places, ascending to 2,000 feet. It is common all over Burma, being often cultivated for hedges. It is found on the limestone rocks of the Malay Peninsula, where the plant is also cultivated occasionally.

In the Nighantas the plant is described as purgative, pungent, digestive, bitter and heavy, and is said to be useful in constipation, flatulent distention, tumours, swellings, abdominal enlargements, rheumatism, spleen, leprosy, mania and jaundice.

A plaster, prepared from the roots and mixed with asafœtida, is applied externally to the stomachs of children suffering from worms. The bark of the root is purgative, and the stem is given in decoction in gout.

The juice, which flows from the branches, is used as a purgative to relieve pain in the loins. It is an acrid irritant in rheumatism and tooth-ache. When taken internally, it acts as a drastic purgative. It is also employed in nervine diseases, dropsy, palsy, deafness and amaurosis. It is a popular application to warts and other cutaneous affections.

The juice is prescribed as a purge and deobstruent in those visceral obstructions and dropsical affections which are consequent on long-continued intermittent fever. Externally, mixed with margosa oil, it is applied to limbs which have become contracted from rheumatism.

In Bombay the root is mixed with country liquor to make it more intoxicating, and the juice is used to kill maggots in wounds, and is dropped into the ear to cure earache, a practice common to many parts of India. In the Konkan the stem is roasted in ashes, and the expressed juice, with honey and borax, given in small doses to promote the expectoration of phlegm.

The Mundas of Chota Nagpur use the milky juice to poison and catch fish. The juice is also used as a purge: mixed in a raw egg it gives as many stools as there are drops of the juice in the mixture.

A preparation from this plant is given as a cure for cough by the Santals.

The hard woody part of the stem is used medicinally by the Malays.

In Cambodia the latex is used as an emetic; the core of the plant is administered in dysentery and in febrile conditions. Throughout Indo-China the milky juice is reputed emetic, purgative, and hydragogue.

The Hindus employ the root mixed with black pepper as a medicine for the cure of snake-bites internally and externally. Mhaskar and Caius have experimented with both the root and the milky juice, and they have found that, whether given internally or applied externally, those drugs are useless in the treatment of snake-bite.

Annam : Xuong rong— ; *Arabic* : Zaqqum-e-hindi— ; *Belgaum* : Tindarisend— ; *Bengal* : Bajvaran, Lariyadaona, Nagri, Narasij, Tekatasij— ; *Burma* : Pyathal, Shasoung-pyathal, Shazanoji, Tazaung— ; *Cambodia* : Chanlat day— ; *Bombay* : Bontakalli, Bontegalli, Bontekalli, Chaturagalli, Jadekalli, Mundugalli— ; *Ceylon* : Chaturakkalli— ; *Deccan* : Tidharisend, Tindharisend— ; *English* : Spurge Cactus— ; *Gujerati* : Tandharisend— ; *Hasada* : Siddaru— ; *Hindi* : Narashij, Sayord, Tidharasend— ; *Indo-China* : Chan leat day, Thanh long, Xuong rong— ; *Malayalam* : Chaturakalli, Kalli, Ratakkali— ; *Marathi* : Naraseja, Narsej— ; *Michi* : Shidu— ; *Mundari* : Eteke, Etete, Siddaru, Marangeteke— ; *Persian* : Zaqqum-e-hindi, Zaqunniya-e-hindi— ; *Sadani* : Sid— ; *Sanskrit* : Sihunda, Simhunda, Snuhi, Tridharaka, Vajra, Vajrakantaka, Vajri— ; *Santal* : Etke, Etkec'— ; *Sinhalese* : Daluk— ; *Tamil* : Amudangam, Kalli, Kandiravam, Kaniravam, Kodiravam, Mavirukkam, Murittargalli, Sadurakkalli, Saduchi, Sunakkudam, Tiruvargalli, Vachiram, Vachirangam, Valangalli— ; *Telugu* : Bommajemudu, Bontajemudu, Bontakalli, Peddajemudu, Simhundumu— ; *Tulu* : Darekhalli— ; *Urdu* : Zakum— ; *Uriya* : Dokahanasiju, Dokanasij, Lohasiju— ; *Yemen* : Gholak, Kelah—.

2. **Euphorbia Atoto** Forst. is found on the Malabar Coast, from Canara southwards. It is common on the sandy seashores of the Malay Peninsula, the Andaman Islands, and Ceylon. It is distributed to the Malay Archipelago, China, Australia, and the Pacific Islands.

In Indo-China women use the milky juice as an emmenagogue, and also as an abortifacient. According to Sallet, when prepared in a certain way and cooked with honey over the water bath the latex is a good remedy for tuberculosis.

In New Caledonia sea-water in which the plant has been soaked and malaxated is very much used as a purgative.

Indo-China : Thuoc doi—

3. **Euphorbia dracunculoides** Lam. is met with throughout India in the plains and low hills. It extends westwards to Arabia and tropical Africa.

The fruit is used to remove warts.

Bengal : Chhagulpuputi, Jychee— ; *Hindi* : Chagulputputi— ; *Punjab* : Kangi, Richni, Sudab— ; *Rajputana* : Bamburi— ; *Santal* : Parwa— ; *Telugu* : Tillakada—.

4. **Euphorbia granulata** Forsk. inhabits the Western Peninsula, Chota Nagpur, Malwa, the Rajputana Desert, the Punjab. Sind, Afghanistan, Arabia, Egypt, and the Canary Islands.

In Las Bela the plant is said to purify the blood.

Arabic: Lebbaejde, Lebbejn, Libbein, Melachene—; *Jhalawan*: Gwandar—; *Kharan*: Shirgonah—; *Las Bela*: Khirwal—; *Rajputana*: Dudeli—; *Santali*: Kantha arak—.

5. **Euphorbia helioscopia** Linn. is a common field-weed in spring throughout the Punjab plains and the Siwalik tract, ascending to 7,000 feet in the outer Himalaya. It has been introduced into the Nilghiris. It spreads over Afghanistan, and westwards to the Atlantic. It also occurs in Japan.

The plant is used as a hydragogue cathartic, and the juice is applied to remove warts. The juice is sometimes, though improperly, used to cure sore eyelids, causing in many instances intolerable pain and inflammation.

In the Punjab the milky juice is applied to eruptions, and the seeds are given with roasted pepper in cholera. In Waziristan the juice is used in herpes.

In Sind the juice is used in the form of a liniment in neuralgia and rheumatism, and the root is employed as an anthelmintic.

The latex is successfully used for the removal of warts in South Africa.

Afrikaans: Melkbos, Melkgras, Wolfsemelk—; *Arabic*: Rumude, Ssaade—; *Baluchistan*: Gurbagund, Sohrbulok, Zahrihk—; *Catalan*: Lletera, Lleterola, Mal d'ulls—; *Chinese*: Tse Ch'i—; *English*: Cat's milk, Churn-staff, Sun Spurge, Wartwort—; *French*: Lait de couleuvre, Omblette, Réveille matin, Tithymale—; *Hindi*: Hirruseeth, Mahabi—; *Iraq*: Khannaq-ad-dijaj, Um-al-halib—; *Italian*: Esula, Titimaglio, Titimalo, Tutumaglio—; *Kharan*: Gurbagund—; *Malta*: Cat's Milk, Spurge, Sun Spurge, Erba calenzuola, Tenghoud—; *Punjab*: Chatriwal, Dudal, Gandabute, Kulfadodakak—; *South Africa*: Milkweed, Spurge—; *Spanish*: Lechetrezna, Mirasol—; *Waziri*: Sagergotiae—.

6. **Euphorbia hirta** Linn. (=*E. pilulifera* Linn.) is a common weed in waste ground, found throughout the hotter parts of India from the Punjab eastwards, and southwards to Ceylon and Singapore. It inhabits most of the tropical and subtropical countries.

The dried herb is stimulant, expectorant, and diuretic. It has been frequently reported useful for bronchitis and asthma affections.

The juice of the plant is given in dysentery and colic, and the milk applied to destroy warts. A decoction is used in asthma and chronic bronchial affections.

The plant is chiefly used in the affections of childhood, in worms, bowel complaints, and cough. It is sometimes prescribed in gonorrhœa.

The root is given by the Santals to allay vomiting, and the plant to nursing mothers when the supply of milk is deficient or fails.

The plant is widely used in West Africa as a medicine. In Gold Coast it is ground and mixed with water for use as an enema for constipation. The white juice is used by women to increase the flow of milk. The leaves are used in curing sores; their juice is sometimes squeezed into the eyes to cure eye trouble.

Similarly in Liberia the plant is given along with lime juice as a laxative, and the latex is applied for conjunctivitis; a thorn or other foreign body is extracted with the help of the latex, which is also applied locally as an antidote to arrow poison or to assist extraction of the head.

The herb is very much used in La Reunion as an astringent in chronic diarrhœas and dysenteries. It is also applied topically to ulcers, oedemas, and phlegmons. The juice is given for aphthae. It is considered tonic, narcotic, and antiasthmatic.

A popular astringent and haemostatic in the Philippine Islands. It is used in the treatment of dysenteries in Annam.

The plant is much used in Guiana as a febrifuge. As a poultice it is applied to abscesses and inflamed glands.

The plant is known in Australia as 'Queensland asthma herb'. It is used there as a remedy for coughs, bronchial and pulmonary disorders, but more especially in paroxysmal asthma.

There is no doubt that an extract of this plant has a sedative effect on the mucous membrane of the respiratory and genito-urinary tract. It has been found by me very beneficial in cases of asthma. I have been using a tincture of it in my private practice in diseases of the genito-urinary tract and in chronic bronchitis and asthma. The result has been very satisfactory (Koman).

According to Dikshit and Kameshwar Rao experiments conducted on cats, dogs, and rabbits, with and without anaesthesia, show that the drug has got a fairly marked action on the respiratory system, depressing the respiration and producing a well marked dilatation of the bronchioles. The drug produces a local irritation of the stomach when given by mouth in larger doses and produces nausea and vomiting. If injected intravenously in animals, however, it has got a depressant action on the movements of the intestines. There is an immediate cessation of the automatic movements of the intestines and a relaxation of the tone of the muscle. The cardiovascular system is depressed, and an intravenous injection leads to a fall of blood pressure chiefly due to the depression of the heart. Perfusion experiments show a depression of the heart. Other systems of the body like the genito-urinary system are not markedly affected (18th Ind. Sc. Congress; Nagpur, 1931).

Australia : Queensland Asthma Herb—; *Awuna* : Ahinkodze, Notsigbeng—; *Bengal* : Barakerui, Burakeru, Burokeruee—; *Betsileo* : Aidinono—; *Bombay* : Nayeti—; *Canarese* : Achchegida—; *Central Provinces* : Gordon—; *Ceylon* : Madduppachcharichippillu, Palavi—; *English* : Asthma Herb, Asthma-weed, Australian Asthma Herb, Euphorbia, Pill-bearing Spurge, Snake Weed—; *Ewe* : Notsigbee—; *Fiji* : Do ni osi—; *French* : Euphorbe à globules, Euphorbe à pilules, Herbe à Jean Renaud, Herbe Jean Robert, Herbe à serpents, Madelone plat, Mal famée, Malnommée, Pilulier, Poil de chat—; *French Guiana* : Grande malnommée, Petite malnommée rouge, Malnommée—; *Guam* : Golondrina—; *Gujerati* : Dudeli, Dudh, Nagladudheli, Ratidudheli—; *Hausa* : Nonankurchiya—; *Hindi* : Bahidudhi, Dudhi, Laldudhi—; *Indo-China* : Co sua lon la, Ka ta, Phi duong thao, Vu sua—; *Kolami* : Pusitoa—; *Konkani* : Dudurly—; *Kwang Tung* : Yu Tchou Ts'ao—; *La Reunion* : Jean Robert—; *Lepcha* : Sung-gryong muk—; *Malay* : Ambin jantan, Gelang susu, Kurumak susu—; *Malayalam* : Nelapalai—; *Mano* : To a gbondo—; *Marathi* : Dudhi, Dudnali, Govardhan, Mothidudhi, Nayati—; *Mauritius* : Jean Robert—; *Mundari* : Kodasing—; *Nzima* : Akubaa—; *Oloke Meji* : Emi-ile, Emile—; *Pampangan* : Bolobotones, Magatas, Malismhalis,

Sisiohan, Sisiwhan—; *Philippines*: Batabotonis, Botobotones, Golandrina—; *Porebunder*: Mothidudheli, Motodakardumaro, Ratodakardumaro—; *Sanskrit*: Nagarjuni, Pusitoa—; *Sinhalese*: Budadakiriya, Dadakiriya, Kiritala—; *Susu*: Boxeforotai, Nyalefoxe, Nyayafoke—; *Tagalog*: Batobatonis, Gatasgatas, Sayican—; *Tamil*: Amumpatchaiyarissi, Patchaiyarissi—; *Telugu*: Bidarie, Nanabala, Nanabiyan, Reddinanabrolu—; *Timne*: Ebit, Yankara-ebit—; *Twi*: Ahinkodge—; *Visayan*: Bovi, Buyayava, Gatasgatas, Tavava—; *Wolof*: Homguelem—; *Yoruba*: Egele, Ege-ile, Emi-ile—.

7. **Euphorbia hypericifolia** Linn. is common throughout the hotter parts of India and Ceylon, and occurs up to 4,000 feet on the Himalaya. It is found in the tropics generally, with the exception of Australia and Pacific Islands.

An infusion of the dried leaves is a remedy in dysentery, diarrhœa, menorrhagia, and leucorrhoea, and it affects the system as an astringent and feeble narcotic.

It is given with milk to children in colic.

In Guiana cataplasms of the plant are applied in cephalalgia. In La Reunion the plant is considered drastic, but rarely used as such; it is better known as an astringent and styptic.

In Annam a decoction of the plant is used in measles and in gravel.

Arabic: Melachene—; *Bombay*: Nayeti—; *Brazil*: Erva de Santa Luzia—; *French Guiana*: Marie claire—; *Gujerati*: Dudheli—; *Hindi*: Dudhi, Dudhikalava, Hakshardana—; *Indo-China*: Chanleat, Co sua, Vu sua—; *La Reunion*: Herbe Saint-Jean, Jean Belan—; *Marathi*: Dhaktidudhi, Dudhi, Dudhmogra—; *Mauritius*: Herbe colique, Herbe mal lévé—; *Porebunder*: Dhakradunmaro, Dudheli—; *Punjab*: Hazardana—; *Rajputana*: Dudeli—; *Sanskrit*: Dugdhika—; *Sinhalese*: Eladadakiriya—.

8. **Euphorbia longifolia** D. Don has been reported from Thoucote in Nepal. It is also reported that the juice is applied to fistulous sores in Kashmir and the Punjab.

9. **Euphorbia microphylla** Heyne is found in Bengal, Bundelkhand, Southern India, Burma and Java.

In Chota Nagpur, a preparation of this plant, along with that of *Cryptolepis Buchanani* is given to nursing mothers when the supply of milk fails or is deficient.

Bengal: Chotokeruee—; *Gujerati*: Dudhi—; *Marathi*: Dudhi—; *Porebunder*: Ekphulidudheli—; *Santali*: Dudhiaphul—.

10. **Euphorbia neriifolia** Linn. is common in rocky places throughout the Western Peninsula. It is cultivated elsewhere in India, as also in Ceylon, Burma, Baluchistan and the Malay Islands.

The lant is a reputed Ayurveda medicine. The Nighantas describe it as 'bitter, pungent; laxative, carminative, alexipharmic; improves the appetite; useful in abdominal troubles, bronchitis, tumours, loss of consciousness, delirium, leucoderma, piles, inflammations, enlargement of the spleen, anaemia, ulcers, fevers.—The milk is pungent, laxative; good for abdominal troubles, tumours, leucoderma.—The leaves are heating, carminative; improve the appetite; good for tumours, pains, inflammations, abdominal swellings.

The Yunanists consider it a good substitute for *E. Tirucalli.*

The milky juice is used as a purgative and a rubefacient. It is prescribed as a cathartic and deobstruent in visceral obstructions and dropsical affections consequent on long continued fever. It enters into the composition of most of the drastic purgatives.

The juice is empolyed in earache; mixed with shoot it is applied to the eye in ophthalmia; mixed with margosa oil it is used as an application in rheumatic affections.

On the Western coast the bark of the root, boiled in rice water with the addition of arrack, is considered useful in dropsy.

In the Rajputana Desert the milky juice is used as a cure for coughs, and is applied to the skin as a blister.

The leaves are used as a diuretic in Indo-China.

The juice of the leaves is a popular cure for earache in the Philippine Islands.

A succus consisting of equal parts of the juice of this plant and simple syrup was prepared and administered in doses of 10 to 20 minims three times a day in cases of asthma, and was found to give relief to the fits of that disease (Koman).

The milk and the root enjoy an almost universal repute as remedies in snake-bite. Mhaskar and Caius have, however, found the two drugs useless in the antidotal and symptomatic treatment of snake-bite and scorpion-sting, and equally useless when applied topically to the bite or the sting.

Annam: Xuong rong rao—; *Arabic*: Wurzizuker—; *Bengal*: Hijdaona, Mansasij, Patashij, Shij—; *Berar*: Thuar, Thuhar—; *Bombay*: Minguta, Newarang, Thor—; *Burma*: Shasaung, Shasoung, Shazawnminna, Zizaung—; *Canarese*: Elegalli, Yalekalli—; *Deccan*: Kuttekijibhkapatta, Kuttekijibhkasend, Newrung, Putteunkesaynd, Thoor—; *Goa*: Nevulkanta—; *Gujerati*: Thor, Tuaria—; *Hindi*: Munsasej, Pattonkisend, Putteun, Sehund, Senhur, Sij, Thohar, Thor—; *Ilocano*: Carambuaya—; *Indo-China*: Giang lam, Xuong rong, Xuong rong rao, Xuong rong ta—; *Kolami*: Etke—; *Konkani*: Nivalkantem, Nivalkanti—; *Malay*: Sesudu—; *Malayalam*: Ilakalli, Kalli—; *Marathi*: Mingut, Nevagunda, Newrang, Niwarung—; *Nimar*: Thuar, Thuhar—; *Pampangan*: Bait, Sorogsorog, Sorosoro—; *Philippines*: Lengua de perro—; *Portuguese*: Cardeira—; *Punjab*: Gangichu—; *Rajputana*: Thor—; *Sanskrit*: Patrasnuhi, Pattakari, Snuhi, Svarasana—; *Santali*: Etke—; *Sind*: Minaguta, Nivadunga, Thohur, Thoor—; *Sinhalese*: Paluk, Patuk—; *Tagalog*: Bait, Sorogsorog, Sorosoro—; *Tamil*: Ilaikkalli, Kalli, Manjevi, Nadangi, Naynakki—; *Telugu*: Akujemudu—; *Tulu*: Irekalli—; *Urdu*: Zakum—; *Uriya*: Siju—.

11. **Euphorbia nivulia** Ham. is found in dry rocky places throughout Bengal, South India, and Bengal. It is often planted for hedges.

The Yunanists use the plant as a substitute for *E. Tirucalli.*

The juice of the leaves is used internally as a purgative and a diuretic; mixed with nim oil it is applied externally in rheumatism. The warm juice is a good cure for earache, and it is occasionally rubbed over the eyes to remove dimness of sight.

On the Western Coast the root bark is boiled in rice water and arrack, and given in dropsy.

The pulp of the stem, mixed with green ginger, is given to persons **bitten by mad dogs.**

Bengal: Sij—; *Berar*: Thuar, Thuhar—; *Burma*: Shasoung—; *Canarese*: Dubbakalli, Dundukalli, Gutagalli, Yellakalli—; *Deccan*: Patteoon—; *Dehra Dun*: Thor, Thuor—; *Gujerati*: Thorkantalo—; *Hasada*: Eteke—; *Hindi*:

Katathohar, Senhur, Sij, Thor, Thuhar, Thura—; *Jaunsar*: Suru—; *Kolami*: Etke, Etki—; *Lambadi*: Motataria—; *Malayalam*: Ilakalli—; *Marathi*: Kaudenivdung, Newrang, Nivdung, Niwarung—; *Mundari*: Burueteke—; *Nimar*: Thuar, Thuhar—; *Porebunder*: Kautalio, Thorkantalo—; *Rajputana*: Raj, Tor—; *Sanskrit*: Patrasnuhi, Sehuda, Svarasana, Vajradruksha, Vajri—; *Santali*: Etke—; *Tamil*: Ilakalli, Manjevi, Nanangalli—; *Telugu*: Akujemudu, Akukalli, Bonthajamudu—; *Urdu*: Zakum—; *Uriya*: Kolosiju, Siju—.

12. **Euphorbia rosea** Retz is found in the Deccan Peninsula, from the Carnatic to Tranquebar, on the coast. It is common near the sea in Ceylon and Indo-China. It also occurs in Afghanistan.

The plant is used as a drastic purgative in Indo-China. The leaves and seeds are considered to be a good worm remedy.

Indo-China: Co sua, Nohn la—.

13. **Euphorbia Royleana** Boiss. is common on dry rocky hillsides of the outer Himalaya from the Indus to Kumaon, ascending to 6,000 feet. It occurs also on the Salt Range.

The acrid milky juice possesses cathartic and anthelmintic properties.

Beas: Chu, Chunga, Surs—; *Chenab*: Chula—; *Dehra Dun*: Thor, Thuor—; *Garhwal*: Surai—; *Hindi*: Senhur, Shakarpitan, Thar, Thor—; *Jaunsar*: Suru—; *Jhelum*: Suli—; *Kumaon*: Sihund, Sohund—; *Punjab*: Shakarpitan, Thar, Thor—; *Rajputana*: Thor—; *Ravi*: Chun—; *Salt Range*: Tordanda—; *Sutlej*: Suro, Tsui—.

14. **Euphorbia sanguinea** Hochst. occurs in Baluchistan, spreading over to Afghanistan, tropical Arabia and Africa.

When used for medicinal purposes it is prepared like 'bhang' (Hotson).

The Sutos use it as an application to sore nipples in suckling mothers.

Brahui: Meshir—; *Pretoria*: Spurge—; *Suto*: Kxamamaswana, Selwe, Tatampoi-e-nyenyane—.

15. **Euphorbia Thomsoniana** Boiss. is found in Kashmir between 10,000 and 12,000 feet.

The crushed root-stocks are employed by the natives of Kuram as detergents for washing the hair, and, when boiled, are given as purgatives.

Kashmir: Hirtiz—.

16. **Euphorbia thymifolia** Linn. is common in paths and gravelly spots in the plains and lower hills of India, Ceylon, and the Malay Peninsula; ascending up 5,500 feet in Kashmir. It is found in all hot countries except Australia.

The dried leaves and seeds are slightly aromatic and astringent. They are given in worm cases and in certain bowel affections of children in the Tamil country. In Northern India they are considered stimulant and laxative. They are commonly used as a purgative in Sind.

The Santals use the root as a remedy for amenorrhœa.

In the Konkan the juice is used to cure ringworm. Among the Mundas of Chota Nagpur it is employed to stop diarrhœa; half a handful of the whole plant being pounded and drunk in a mixture with water and sugar-candy.

In Indo-China the plant is used as a cathartic; the seeds and leaves are considered anthelmintic.

In Arabia the juice of the plant is made into a paste with wheat flour, and administered as a purgative in the form of pills. The bruised plant is applied to wounds.

In La Rèunion the herb is used as an astringent in diarrhœa and dysentery.

The expressed juice or the powdered plant is administered internally with wine as a remedy for snake-bite, and it is applied externally to the part bitten; but whether given internally or applied externally the plant is equally useless in the treatment of snake-bite (Mhaskar and Caius).

Arabic: Rummid—; *Bengal*: Dudiya, Snweetkeruee, Swetkerua, Swetkirui—; *Bombay*: Nayata, Nayeti—; *Ceylon*: Ammanpachcharichippillu, Chittirapalavi—; *Gujerati*: Nahanidudheli—; *Hindi*: Chhotidudhi, Chothaduhi—; *Indo-China*: Co sua, Nhon la—; *La Reunion*: Rougette—; *Malay*: Segan padang—; *Marathi*: Ghakdidudhi, Lahandudhi, Mothidudhi—; *Mauritius*: Petite rougette—; *Mundari*: Kdoasingjaite—; *Nasirabad*: Khiri—; *Persian*: Gazardanah—; *Porebunder*: Jhinkidudheli, Jhinkodhakardhumaro—; *Punjab*: Baradodak, Dhoduk, Hazardana—; *Sanskrit*: Laghudugdhika, Raktavindachada—; *Santali*: Nanhapusitoa, Nanhapusituar—; *Sind*: Dhoduk, Hazardhana—; *Sinhalese*: Bindadakuriya—; *Tamil*: Chinamampatchaiyarisi, Sittrapaiadi—; *Telugu*: Bidurunadabiyam, Reddivarimanubala—; *Yemen*: Rummid—.

17. **Euphorbia Tirucalli** Linn., a native of Africa, has become naturalised in the drier parts of Bengal, the Deccan, South India, and Ceylon; elsewhere it is largely cultivated for hedges, and in Berar is much grown to shelter young mango plants from direct sunlight.

According to Yunani practitioners the juice of this plant is purgative, carminative; useful in gonorrhœa, whooping cough, asthma, dropsy, leprosy, enlargement of the spleen, dyspepsia, jaundice, colic, tumours, stone in the bladder.

The fresh milky juice is applied to warts and used as a rubefacient embrocation in rheumatism. It is good alterative in syphilis, and a good application in neuralgia. In the Konkan it is given as a purge; and the charcoal, which is very light, is used in making pastilles. In Goa it is used to poison fish. The milky juice is a dangerous drug to handle, causing severe and tenacious conjunctivitis.

In Cambodia the plant is used to stupefy fish.

The root is used in La Reunion as a vesicant, more rarely as an emeto-cathartic.

The bark is used as a vesicant in Java, and is also applied to fractures.

In some parts of India a decoction of the root is given in certain cases of colic, and the milky juice mixed with melted butter is prescribed as a purge. The juice is also applied topically in neuralgia.

In Malaya the latex is inserted into cuts in bullocks to form the elevated recognition marks.

Annam: Cay san ho, San ho—; *Arabic*: Azfur-zukkum, Zaqqum-e-hindi—; *Baluchi*: Dedar—; *Bengal*: Gul, Lankasij, Latadaona—; *Bombay*: Niwal, Sehund, Seyr, Shera, Tej, Thor, Thora—; *Burma*: Shashoungleknyo, Shasoungleknyo, Tazaunglekhnyo—; *Canarese*: Bontakalli, Bottugalli, Kadunevali, Kalli, Kodukalli, Kolugalli, Kolukalli, Nandugalli, Pachanegalli, Yellegulla—; *Cochin-China*: Cay san ho xanh—; *Deccan*: Barkisend, Karikisend, Send—; *English*: Milk Bush, Milk Hedge, Indian Tree Spurge—; *Goa*: Nival—; *Gujerati*: Dandalithora, Pardeshithora, Thordandalio—; *Hindi*: Konpahlsehnd, Sair, Sehud, Sehund, Sendh, Shirthohar, Sij—; *Indo-China*: Ramok, San ho xanh, Thanh san ho, Thap nhi, Xuong kho—; *Java*: Kayoo-oorb—; *Lambadi*: Taria—; *La Reunion*: Laine sans feuilles, Tirucalli—; *Malayalam*: Guda, Katerumakkali, Kolkalli, Tirukalli—; *Marathi*: Nevli, Nirval, Nivla, Niwarung, Seyr, Seyrteg, Shar, Shera, Tej, Thuvar, Vajraduhu—; *Mauritius*: Calli—; *Melghat*: Thuar, Thuhar—; *Nimar*: Thuar, Thuhar—; *Persian*: Shirtothar, Zaquniya-e-hindi—; *Porebunder*: Kharsanithora, Thora—; *Sanskrit*: Bahukshira, Dandasruha, Ganderi, Snuka, Trikuntaka, Vajradruma—; *Santali*: Siju—; *Sind*: Sair, Sij, Thora—; *Sinhalese*: Navahandi, Thovar—; *Tagalog*: Catuit—; *Tamil*: Kalli, Kiri, Kombukkalli, Pachankalli, Parchanu, Tirukkalli, Tiruvatti—; *Telugu*: Chemudu, Jemudu, Kadujemudu, Kalli, Kanjiemudu, Sannajemudu, Tirukalli—; *Tigrinia*: Qoutscheh—; *Tulu*: Kodukalli—; *Urdu*: Zakum—; *Uriya*: Lonkasiju—; *Yemen*: Dahan, Rummid—; *Zulu*: uMunde, umIululu—.

18. **Euphorbia trigona** Haworth inhabits dry rocky hills in the Deccan, and probably other parts of India. It is found in the Andaman Islands and the Moluccas.

The fresh milky juice is a direct irritant both when taken internally and applied externally. Taken in very small quantities, it is a drastic purgative.

Telugu: Kattimandu—.

19. **Euphorbia turcomanica** Boiss. is found in Baluchistan, whence it extends to Persia and Mesopotamia.

In Baluchistan the plant is boiled and used as a medicine for gripe.

Brahui: Meshir—

EXCOECARIA.

The genus consists of 30 species, inhabiting the tropics of the Old World.

A deciduous shrub ... ... 1. *E. acerifolia.*
A small evergreen tree ... ... 2. *E. agallocha.*

1. **Excoecaria acerifolia** Fr. Didrichs. is found at 5,000-6,000 feet in the western and central Himalayas from Nepal to Kumaon. It also occurs in the Khasia Hills.

The Bhutias of East Kumaon use the leaves as a remedy for rheumatism.

Garhwal: Dudhila—; *Kumaon*: Basingh—.

2. **Excoecaria agallocha** Linn. is a common small evergreen tree of the Coast and tidal forests of India, Burma, the Malay Peninsula, the Andaman Islands, the Malay Archipelago, North Australia, New Caledonia to Tonga.

The milky juice, which exudes from the bark of this tree when green and fresh, is very acrid and injurious to the eyes, hence it is called 'the blinding tree of India'.

A decoction of the leaves is occasionally given by Hindu doctors in epilepsy, in the quantity of a quarter of a tea-cupful twice daily. This decoction is also used as an application to ulcers.

From the lower part of the trunk and roots, a soft, light, reddish suber is obtained, which is sold by the itinerant medicine men of Western India, under the name of *Tejbul,* as an aphrodisiacal tonic.

The natives of Eastern Australia, as well as those of New Guinea, use the milky juice to cure certain ulcerous chronic diseases, e.g., leprosy.

In Fiji, it is employed for the cure of leprosy, its mode of application being very singular. The body of the patient is first rubbed with green leaves; he is then placed in a small room and bound hand and foot, when a small fire is made of pieces of the wood of this tree from which rises a thick smoke; the patient is suspended over this fire, and remains for some hours in the midst of the poisonous smoke and under the most agonizing torture, often fainting. When thoroughly smoked, he is removed, and the slime is scraped from his body; he is then scarified and left to await the result. In some cases he is cured, but frequently the patient dies under the ordeal.

The milky latex obtained from the bark when it is green is used by Malays as an irritant poison; it is so acrid when fresh that it blisters the skin. In Kelantan it is given internally as a poison.

Andamans: Yekin—; *Australia*: Blind-your-eyes, River Poisonous Tree—; *Bengal*: Gangwa, Gengwa, Geogheria, Geor, Geria, Goria, Uguru—; *Bombay*: Geva—; *Burma*: Kayau, Kayaw, Tayau, Tayaw—; *Canarese*: Hara, Haro—; *English*: Blinding Tree, Milky Mangrove—; *Fiji*: Sinu-gaga—; *Formosa*: Hsiang shu—; *Hindi*: Gewa—; *Igorrote*: Ali—; *Indo-China*: Gia, Ngai, Soen gia, Tia mu, Va, Ya—; *Malay*: Aquila, Baboeta, Babooter, Bebuta, Bubuta, Butabuta, Garamataboeta, Noto—; *Malayalam*: Katappala, Karmmatti, Komatti—; *Marathi*: Geva, Phungali, Suran, Surind, Surund—; *Marinduque*: Diladila—; *Pampangan*: Butabuta, Himbabao—; *Queensland*: Balavola karping—; *Samoa*: Toto—; *Sanskrit*: Agaru, Uguru—; *Sinhalese*: Talakiriya, Telakiriya, Tellakwiya—; *Spanish*: Arbol del ciego, Buta de Filipinas—; *Tagalog*: Botabota, Buta, Butabuta—; *Tamil*: Agadil, Agi, Ambalatti, Ambalavirukkam, Tillai, Perundillai—; *Telugu*: Chilla, Tella, Tilla—; *Uriya*: Ghona, Gowan—; *Visayan*: Alipata, Himbabao, Lipata, Siac—.

FLUEGGEA.

The genus consists of 6 species, found in the tropics of the Old World.

Branchlets stout ending in spines ... ... 1. *F. leucopyrus.*
Branchlets slender unarmed ... ... 2. *F. virosa.*

1. **Flueggea leucopyrus** Willd. occurs in Sind, the Punjab Plain, Burma, the Western Peninsula from Canara southwards. It is abundant in Ceylon.

The juice of the leaves, or the leaves made into a paste with tobacco, are used to destroy worms in sores.

The plant is said to be a fish poison.

Afghanistan: Perapastawane—; *Canarese*: Bilchuli, Gudahale, Huli—; *Central Provinces*: Challamanta, Salemanta—; *Goa*: Parpo—; *Gujerati*: Shinavi—; *Hindi*: Suffaidmahommad—; *Malayalam*: Perinklavu—; *Marathi*: Kandekuvana, Pandharphali—; *Matheran*: Pandharphali—; *Merwara*: Halepan, Salipan—; *Nepal*: Achal—; *North-Western Provinces*: Aintha, Hartho—; *Porebunder*: Tumari—; *Punjab*: Bata, Bhathi, Gargas, Girk, Girthan, Kakun, Vanuthi—; *Rajputana*: Halepan, Salepan—; *Sanskrit*: Apiyadruma, Bhuriphali, Panduphali, Svetakamboja—; *Sind*: Kiran—; *Sinhalese*: Hinkatupila—; *Tamil*: Irubulai, Mudbulanji, Mulluppulatti, Pula, Pulanji, Varadbul, Varadbula, Vedbula, Vellaippula, Vellaippulanji—; *Telugu*: Challamunta, Mekarayi, Puli, Tellapulisaru, Tellapuliyudu, Tellapurugudu—.

2. **Flueggea virosa** Baill. (=*F. microcarpa* Bl.) occurs throughout India, Ceylon, and the Malay Peninsula. It extends to the Malay Islands, China, Australia, and tropical Africa.

The juice of the leaves, or the leaves made into a paste with tobacco, are used to destroy worms in sores.

The West Ashantis use the roots to cure gonorrhœa. They boil the whole plant in water and wash themselves in the water to 'become strong'.

In Rhodesia the root is used as an aphrodisiac.

The Ewe people of Togoland use the leaves to cure constipation. For this purpose they are boiled and the water is drunk.

In Indo-China the astringent bark is used to stupefy fish.

Arabic: Dabalab, Hamrur, Hemmele, Hennet-el-bagar, Kartjikartji—; *Ashanti*: Nkangnaa—; *Bedde*: Zandanu—; *Bombay*: Kantepuwan, Pandharphali—; *Buldana*: Pittor—; *Burma*: Konchiny a—; *Canarese*: Belahuli, Bilisuli, Gudahale—; *Chindao*: Umsosoti—; *Dagomba*: Susuwuluga—; *Dehra Dun*: Rithoul—; *Ewe*: Esre, Hesre, Hlese—; *Fulani*: Chamal, Chambe, Chame—; *Ga*: Gbekebii able-tcho—; *Goa*: Papo—; *Gujerati*: Shaunavi—; *Hausa*: Gussu, Gwiwar kare, Tsa, Tsugawun kare, Tswa—; *Hindi*: Dalme, Patala—; *Indo-China*: Bong ne, Bong no, Cang pa, Mai ten, Nhat giep thu, Xeur nam—; *Kumaon*: Athaina—; *Kanuri*: Shimshim—; *Lambadi*: Jarigaro—; *Lepcha*: Hik-ti bi—; *Malayalam*: Perinklavu—; *Marathi*: Kodarai—; *Melghat*: Pithondi—; *Merwara*: Halepan, Salipan—; *Nimar*: Jhonri—; *Sanskrit*: Apiyadruma, Bhuriphala, Dhusara, Nilishila, Panduphali, Patali, Vrittabijaka—; *Saora*: Jungjang—; *Sierra Leone*: Tigwi—; *Sokoto*: Tsa, Tswa—; *Tamil*: Irubulai, Pula, Varadbula, Vedbula, Vellaippula—; *Telugu*: Mekarayi, Sulamunta, Tellapuli, Tellapurugudu—; *Uriya*: Janjingi, Kanilehya—; *Yoruba*: Iranje—.

GELONIUM.

The genus includes 20 Asiatic and African species.

Gelonium multiflorum A. Juss. occurs in Bengal and the Circars, reaching northwards to the foot of the Sikkim Himalaya. It extends from Chittagong to Tenasserim and Malacca, and is found in the woods of Upper and Lower Burma, and the Malay Peninsula. It is distributed over the Malay Islands, Siam, and China.

The bark is considered a good tonic for the gums, and is prescribed for gingivitis in Cambodia; it is also used as a purgative in hepatic troubles.

Burma: Setahanbaya—; *Cambodia*: Tromung sek—; *Hindi*: Ban naringa—; *Malay*: Buah punei, Lampun hitam, Mingantok, Ruas-ruas—; *Telugu*: Sarugata—.

GLOCHIDION.

The genus includes 160 species spread over tropical Asia and Polynesia.

G. rubrum Bl. is used medicinally in Indo-China.

Leaves 2-4 in. thin. Anthers 3 ... ... 1. *G. hohenackeri.*
Leaves 4-8 in. coriaceous. Anthers 5-6 ... ... 2. *G. zeylanicum.*

1. **Glochidion hohenackeri** Bedd. is found in Chota Nagpur and Orissa, and in the Western Peninsula; it is very common in the Konkan.

The bark is given medicinally when the stomach revolts against food.

Canarese: Banavara, Nirchalli, Nirchelli, Nirjani, Sullai—; *Malayalam*: Kuluchan—; *Marathi*: Bhoma—; *Uriya*: Baniakandhum, Chikni, Kalchia—

2. **Glochidion zeylanicum** A. Juss. occurs in the Western Peninsula.

The bark is used as a stomachic.

Canarese: Banda, Savregidda—; *Malayalam*: Nirvetti—; *Tamil*: Kumbalam—; *Telugu*: Itepulla—.

HIPPOMANE.

H. mancinella Linn. is a native of tropical America, introduced nto Indian gardens.

The bark and the berries yield a highly poisonous milky acrid juice. It is a violent irritant and powerful cathartic, diuretic, and vesicant. The least drop applied to the eye will cause blindness for some days; the smoke from the wood when burnt will also seriously affect the eyes.

The juice is much used in Cuba for tetanus and in Guiana for worms in children. Indians use it to poison their arrows.

Cayenne: Figuier—; *Cuba*: Manzanillo—; *English*: Manchineal Tree—; *French*: Arbre de mort, Arbre poison, Bois de lait, Figuier vénéneux, Mancenillier, Noyer vénéneux—; *German*: Manzenillerbaum—.

HOMONOIA.

The genus consists of 4 Indo-Malayan species.

Homonoia riparia Lour. is a small rigid evergreen shrub, found on the rocky and stony river-beds of the Sikkim Himalaya at an altitude of 1,000 to 2,000 feet; also in Assam, the Khasia Hills, and southwards to Burma, Tenasserim, the Malay Peninsula, the Andaman Islands, and Bundelkhund; and in the Deccan Peninsula from the Konkan southwards. It is also common in Ceylon up to an elevation of 2,000 feet. It extends to Siam, Cochinchina, Sumatra, Java, and the Philippine Islands.

The root is given for ulcers. It is laxative and diuretic, and a decoction is given in piles, stone in the bladder, gonorrhœa and syphilis

In Cambodia an infusion of the wood is given in paludism. The whole plant is considered depurative.

Burma: Momaca, Yactagyiben, Yactakyee—; *Cambodia*: Rey tuk—; *Gond*: Sundeh—; *Indo-China*: Bo song, Rey tuk, Ri ri—; *Konkan*: Sherana—;

Kumaon: Kandayar—; *Kurku*: Jeljambu—; *Lao*: Kai—; *Lepcha*: Mongthel, Mung-thel kung—; *Malayalam*: Katalluri—; *Mundari*: Garahuri, Garahuru—; *Nepal*: Kholaruis—; *Philippines*: Lumanaja, Mimbre—; *Sanskrit*: Kshudrapashanabheda, Pashanabedaka—; *Santali*: Gurjor, Sunukui—; *Tagalog*: Agoyoy, Agucuc, Balanti, Dumanai, Lumanai—; *Telugu*: Cheppunjerinjal, Taniki—; *Uriya*: Jamla—; *Visayan*: Mayagos, Mayoyos, Miayos.

HURA.

The genus consists of 2 tropical American species.

Hura crepitans Linn., indigenous in tropical America, has been introduced into India from Jamaica.

The juice of the bark is a Brazilian remedy for leprosy. It is considered caustic and irritant in Guiana and Mexico, and is said to cause blindness. At Guadas the latex is used as a fish poison.

In Mexico and Guiana the leaves are much used for chronic pains; the fresh seeds are emetic and purgative and in great demand; the oil from the seeds is said to be a useful though very drastic purgative.

The bark and the latex are officinal in Portugal.

Canarese: Retidani—; *English*: West India Sandbox Tree—; *French*: Arbre du diable, Sablier, Sablier élastique—; *French Guiana*: Maman cacao, Sablier, Sablier élastique—; *Indo-China*: Ba dau tay, Diep tay, Ngo dong—; *Mauritius*: Sablier—; *Portuguese*: Assacu, Oassacu—; *Surinam*: Postentree—; *Tamil*: Mullarasomam—; *Telugu*: Simaburuga—; *West Indies*: Sandbox Tree—; *Youri-Taboca*: Oassacu—.

JATROPHA.

This genus includes 160 tropical and subtropical species, chiefly American.

The following species are used medicinally in Arabia— *J. glandulifera* Roxb.—; in Cambodia, Indo-China, the Philippine Islands and Guiana— *J. Curcas* Linn., *J. multifida* Linn.—; in Mexico— *J. macrorhiza* Benth.—; in South America and the West Indies— *J. Curcas* Linn., *J. gossypifolia* Linn., *J. multifida* Linn.—; in Gold Coast— *J. Curcas* Linn., *J. gossypifolia* Linn.—; in West Africa, Madagascar, La Reunion— *J. Curcas* Linn.—; in South Africa— *J. capensis* Sond., *J. hirsuta* Hoch., *J. Zeyheri* Sond.—.

A. Evergreen tree or large shrub
   1. Leaves simple and ovate or 3-5-lobed ... ... 2. *J. glandulifera.*
   2. Leaves long-petioled, orbicular-cordate or 3-5-lobed or angled ... ... 1. *J. Curcas.*

B. Shrubs
   1. Leaves entire or 3-lobed, base cuneate, lobes entire. 5. *J. nana.*
   2. Leaves 5-lobed or -partite, lobes glandular-serrulate ... 3. *J. gossypifolia.*
   3. Leaves long-petioled, orbicular, palmately cut into many narrow entire or lobulate caudate acuminate segments ... ... 4. *J. multifida.*

1. **Jatropha Curcas** Linn. is an evergreen shrub, indigenous to America, but cultivated in most parts of India, especially on the Coromandel Coast and in Travancore. It is a common hedge-plant in the Konkan, and in the Malay Peninsula.

According to Ayurvedic practitioners 'the fruit and the seed are anthelmintic; useful in chronic dysentery, thirst, urinary discharges, abdominal complaints, biliousness, anæmia, fistula, and diseases of the heart'. The oil is reckoned a valuable external application to itch, herpes, chronic rheumatism, and sores or wounds.

The leaves are applied as a rubefacient and discutient, and a decoction of them is said to excite the secretion of milk in women. The viscid juice which flows from the stem upon incision is painted over cuts and wounds to check bleeding and promote healing.

In Goa the root-bark is applied externally in rheumatism. In the Konkan it is rubbed with a little asafoetida and given with buttermilk in dyspepsia and diarrhoea; the fresh stems are used as a tooth brush to stop bleeding from the gums.

The roasted nuts are used as a purgative by the Mundas, one nut producing three or four stools.

In Cambodia the leaves are considered insecticidal, the latex styptic, the oils from the seeds abortive.

The leaves are extensively used in the Cape Verde Islands, in the form of decoction and cataplasm to the mammae, as a lactagogue. The Bakwiri of Cameroons add a decoction of the young leaves to beer as a diuretic for rheumatism.

In Nigeria a decoction of the leaves with native natron is used by women as a wash for a month before childbirth. In Gambia the leaves are used to make a mouth-wash. In Southern Nigeria they are a remedy for jaundice, applied by rectal injection.

In West Africa the sap of the plant is used to cure toothache. It is also used as a styptic, for stopping bleeding. It is said to blind the eyes. The bark of the roots is ground and used as a dressing for sores.

In Western Ashanti the leaves are burned and the ashes applied to guinea worms, which are said to come out quickly in consequence.

The seeds are used in native Hausa medicine; the oil extracted from them is used as an application for sores in domestic stock. In Gabon the seeds are ground and mixed with palm oil to kill rats.

The oil from the seeds is applied topically in Guinea in rheumatism, herpes, and pruritus. The juice of the plant and the pounded leaves are applied to wounds and refractory ulcers.

In Gold Coast the leaves are ground and used with pounded palm nuts to make a mixture which is used as an enema for weak babies to strengthen them. They are also used for smoking bedbugs out of a house. They are often crushed in water and used as an enema. The juice of the leaves is squeezed into water, lemon or lime added, and the liquid used as a bath for curing fever. When the young leaves are boiled, the liquor is drunk to cure fever also. The seeds are commonly used as a very effective purgative. Sometimes they are used in curing eye trouble.

Indians in Natal use the seeds as a purgative.

The wood is used in Madagascar and Guiana as an anti-diarrhœic; the latex is applied to decayed teeth and to wounds, and is used as a styptic; the roots are given as emetic and purgative.

The seed is a Brazilian anthelmintic; the principle is contained in the embryo.

The blacks of Rio Nunez saponify the oil with the ashes of the Papaw, and use the preparation to heal the wounds caused by circumcision.

A toxalbumin, curcin, is present in the seeds.

Ada: Kwadidicho—; *Agolo*: Ubolu, Ubwolu—; *Agona*: Kaneadua—; *Akim*: Nkrangye dua—; *Akwapim*: Aborotortor—; *Anang*: Mbubok—; *Antilles*: Fève d'enfer, Herbe du Bon Dieu, Herbe du diable Mancenillier bénit, Médicinier bénit, Médicinier des Barbades, Noix américaine, Pignon de Barbarie, Pignon d'Inde—; *Antsianaka*: Valavenola, Voanongo—; *Arabic*: Dandebarri, Dandenahri, Habb el meluk—; *Ashanti*: Akanedua—; *Awka*: Elu—; *Awuna*: Krorti—; *Bambara*: Baga-ni, Iridingue—; *Baoule*: Propo—; *Bengal*: Baghbarinda, Bagbherenda, Bonbheranda, Erandagachh, Paharierand, Safedind—; *Benin*: Aru-ebo—; *Bete*: Badaguigui—; *Betsimisaraka*: Kizika—; *Bhil*: Ratan—; *Bobo-Dioula*: Manan-naga—; *Bobo-Fing*: Daya-naga—; *Bombay*: Irundi, Jaiphal, Jepal, Kurikarlu, Maraharalu, Mogalieranda, Yerand—; *Brazil*: Figo do inferno, Mandubiguasu, Munduyguasu, Pinhao do Paraguay, Pinhao de purga, Pinhao paraguay, Pinheiro do inferno, Pinheiro de purga—; *Burma*: Kesugi. Thinbankyekku, Thinbaukyeksu, Thinbawkyetsu—; *Cambodia*: Lohong khvangsa—; *Canarese*: Adaluharalu, Bettadaharalu, Bettaharalu, Doddaharalu, Dundiga Kadandla, Kaduharalu, Karnochchi, Maraharalu, Maraharlu, Maraharuli, Marayavudula, Parangiharalu, Turukaharalu—; *Cape de Verde*: Pulguiera—; *Ceylon*: Kaddamanakku—; *Chinese*: Ma Fong Chou—; *Cuba*: Piñon botija—; *Deccan*: Ehanduejot, Erundi, Jangli-yarandi—; *Efik*: Eto-mkpa—; *Egypt*: Habb-el-meluk—; *English*: Barbados Nut, Physic Nut, Poison Nut, Purging Nut—; *Ewe*: Babatsi, Gbomagboti, Kpoti, Krorti—; *Fanti*: Aburokyiriaba, Adadze—; *Formosa*: T'ung yu-shu—; *French*: Grand médicinier, Grand pignon d'Inde, Gros ricin, Haricot du Pérou, Manioc bâtard, Médicinier, Oignon d'Inde, Pignon des Barbades, Pignon d'Inde, Pourghère, Pulguère, Purghère, Ricin d'Amérique—; *French Guiana*: Médicinier—; *French Guinea*: Barané—; *Fulani*: Kidi, Koladebbe, Kolakolaje, Kwolkwolaje—; *Ga*: Kplukacho—; *Gaboon*: Ogombo—; *Gbari*: Kwotewi—; *German*: Purgirnuss, Schwarzelrechnuss—; *Golunga Alto*: Mupuluca—; *Guam*: Tubatuba—; *Gujerati*: Jamalgota, Jangliarandi, Ratanjota—; *Hasada*: Kulaiaradaru—; *Hausa*: Binidazugu, Chinidazugu—; *Hindi*: Bagberenda, Bagbherenda, Baghrandi, Bhernda, Jangliarandi, Pharierand, Safedarand, Safedind—; *Hova*: Tanantanambazaha, Tanantanamposy, Tanantanankisoa—; *Ibibio*: Etomkpo—; *Ibo*: Buluolu, Okweni, Olulu-idu, Owulu-idu—; *Ilocano*: Tavatova, Tawatawa—; *Indo-China*: Ba dau me, Ba dau nam, Coc daû, Dau me, Dong thu, Kuang, Lohong, Vao, Vong dau ngo—; *Java*: Dijarak—; *Kolani*: Kulajara, Kulejera, Totkabendi—; *Konkani*: Erond, Erondd, Jemphal—; *Kontagora*: Binidazugu—; *Krepi*: Abrortorto—; *Krobo*: Kitigblaicho, Kutugblaicho—; *Lambadi*: Ranniarendero—; *La Reunion*: Pignon d'Inde—; *Madagascar*: Kinampotsy, Kinanafotsy, Kinapotsy, Kinopotsy—; *Malay*: Jarak b'landa—; *Malayalam*: Kattavanakku—; *Malinke*: Bagha, Baha—; *Mandingo*: Baga, Bagauro, Bagha—; *Marathi*: Mogalieranda, Ranayerandi—; *Mauritius*: Pignon d'Inde—; *Melghat*: Mogali yerand—; *Mende*: Katawi, Kata-wului—; *Mexico*: Avellanes purgantes, Sangregaod—; *Misahohe*: Wabati—; *Mozambique*: Sassi—; *Mundari*: Jadabindi, Totkabindi—; *Naguri*: Kulabindidaru, Totkabindidaru—; *Nepal*: Kadam—; *Nzima*: Taprika, Tapuleka—; *Ibu*: Okwata, Ubwolu—; *Orissa*: Baigab—; *Pahouin*: Ofossn tang—; *Panama*: Arbol santo, Coquillo, Jaquillo, Piñon de purga, Tapate—; *Persian*: Dandebarri, Dandenahri—; *Philippines*: Bolongcauit, Casta, Cator, Kator, Tavatava, Tuba—; *Porto Rico*: Tartago—; *Portuguese*: Grão malucco, Grão muluco, Pinhao de purga, Pulza, Purgheira, Purgueira, Ricino maior—; *Quittah*: Abrortorto—; *Sakalave*: Savoa, Valavelo, Voanongo—; *Samoa*: Puavai—; *Sanskrit*: Akhuparnika, Chitra, Dravanti, Kananeranda, Mushikaparni, Nyagrodhi, Parvateranda, Pratyakshreni, Randa, Shanbari, Sutashreni, Vrisha—; *Santali*: Bhernda—; *Saora*: Peddanepalemu—; *Serere*: Tabanani—; *Sierra Leone*: Bagauro—; *Sinhalese*: Endaru, Erandu, Rata-endaru, Velendaru—; *Soussou*: Baranè, Barhane, Baxane—; *Spanish*: Arbol de los Piñones de Indias, Piñoncillo—; *Tagalog*: Tatataba—; *Tamil*: Adalai, Kadalamanakku, Kaitta, Kattamanakku, Kattukkottai, Kuribaravuni, Naligadi, Nikkurottam, Tiravadi, Vellaiyamanakku—; *Telugu*: Adaviyamudamu, Katiyamudamu, Kondamudamu, Nepalemu,

Peddanepalemu, Pepalemu—; *Timne*: Esigororopetr—; *Tivi*: Igadam—; *Tulu*: Kadalambudu—; *Twi*: Aborotortor, Akanedua, Nkrangyedua—; *Uriya*: Baigoba, Norokokalo—; *Visayan*: Casla—; *Wolof*: Tuba—; *Yoruba*: Botije, Botuje, Botuje-ubo, Lobotuje, Olobontuje, Shenrijun, Ubo—.

2. **Jatropha glandulifera** Roxb. is common near villages in Bengal, Burma, the Northern Circars, and the Deccan, rare in Oudh and the Punjab. One of the most abundant of hedge plants in the lower provinces, but also prevalent on village waste lands.

The juice of the plant is used in various parts of the country as an escharotic to remove films from the eyes.

The root brayed with water is given to children suffering from abdominal enlargements. It purges, and is said to reduce glandular swellings.

The fixed oil from the seeds has purgative properties. It is applied to sinuses, ulcers, foul wounds, ringworm, and also in rheumatism and paralysis.

Arabic: Abab, Kharub—; *Bengal*: Lalbherenda—; *Bombay*: Janglierandi, Undarbibi—; *Canarese*: Kariturukahalugida, Totlagida—; *Hindi*: Janglierandi, Undarbibi—; *Kolami*: Verendi—; *Konkani*: Ranerandi—; *Malayalam*: Atala, Nakadanti—; *Marathi*: Jangliarandi—; *Persian*: Baide anjira—; *Sanskrit*: Nikumbha—; *Tamil*: Adalai, Eliyamenakku, Erikkaraikattamanaku, Kattamanaku, Puliyamanakku—; *Telugu*: Dundigamu, Dundigapu, Nelaamida, Nelanepalemu, Nepalemu, Nelayamudamu, Vettiyamudamu—; *Urdu*: Jangli eranda—; *Uriya*: Simanorokokalo—.

3. **Jatropha gossypifolia** Linn. is a native of Brazil naturalized in many parts of India. It is now established in Burma and Singapore.

A decoction of the bark is used as an emmenagogue.

The leaves are applied to boils and carbuncles, eczema and itches.

The seeds act as an emetic; but they are said to cause insanity.

In the pith of the old thick stems a yellowish-brown substance is found which is sold in Gold Coast medicine markets. It is put into a clean cloth and squeezed into the nostrils, causing the patient to sneeze and effectively curing headache.

In Gold Coast the leaves and the seeds are used as purgatives. The leaves are boiled and used in the bath to cure fever. Their juice is used to cure sores on the tongues of babies.

Antioquia: Tuá-tuá—; *Ashanti*: Kaagya—; *Bambara*: Baga, Bagha, Baha, Santanan—; *Bengal*: Lalbherenda—; *Benin*: Oru-ebo—; *Canarese*: Kariturukaharalu—; *Ewe*: Babatsi, Gbomagboti, Nakrakpoti—; *Fanti*: Aburokyiraba, Akandedua—; *Ga*: Engmelbii, Kpitikpitsho—; *Hindi*: Bhernda, Verenda—; *Ibo*: Akimbogho—; *Kolami*: Bhernda, Verenda—; *Lagos*: Lobotuje pupa—; *Malayalam*: Simayavanakku—; *Malinke*: Baga, Bagha, Baha—; *Mandingo*: Baga, Bagha, Baha—; *Mende*: Katawi, Kata-wului—; *Oloke-meji*: Olobontije-pipa—; *Panama*: Frailecillo, Frailejon, Purga de fraile, Purga de Huane—; *Santali*: Bhernda, Verenda—; *Tamil*: Adalai, Kattamanakku, Simaiyamanakku—; *Telugu*: Nepalemu, Simanepalemu—; *Uriya*: Rangakalo—; *Venezuela*: Tuá-tuá—; *Yoruba*: Botuje pua, Lapalapa pupa, Lobotuje, Olobontuje—.

4. **Jatropha multifida** Linn. is a native of South America, cultivated and naturalized in various parts of India.

The seeds are regarded as a powerful purgative. In French Guiana they are used fresh as a purgative and emetic.

In Cambodia the leaves, the latex, and the oil from the seeds are used medicinally. The leaves are used in scabies; the latex is applied to wounds and ulcers; the oil is used both internally and externally as an abortifacient.

In Indo-China the dried root is given as a decoction for indigestion and colic; it is also prescribed as a tonic, in orchitis and in oedema.

Benin: Iboisa—; *Brazil*; Flor de coral, Pinhoen—; *Cambodia*: Lohong khvang kraham—; *Canarese*: Sime auvdala, Vilayatiharalu—; *English*: Coral Plant, French Physic Nut, Purging nut, Small Physic Nut—; *French*: Arbre corail, Médicinier bâtard, Médicinier d'Espagne, Noisetier purgatif, Petit médicinier—; *French Guiana*: Bouquet corail—; *Ilocano*: Mana—; *Indo-China*: Bach phu tu—; *Lagos*: Ege—; *Mauritius*: Arbre corail—; *Sanskrit*: Bhadradanti, Brihaddanti, Dugdhagarbha, Guchhphala, Jayavaha, Jyotishka, Virechani, Vishabhadra—; *Spanish*: Avellana purgante de Santo Domingo, Piñol de Cumaña, Purgante de España—; *Tagalog*: Mana, Tuba—; *Tamil*: Kattunervalam, Malaiyamanakku—.

5. **Jatropha nana** Dalz. is found in the Deccan, in stony places near Poona, Bombay, and throughout the Konkan.

The juice is employed as a counter-irritant in ophthalmia.

Marathi: Kirkundi—.

MACARANGA.

The genus includes 180 species, occurring in the tropics of the Old World.

M. huraefolia Beille is used medicinally in tropical West Africa.

Leaves deltoid-ovate, peltate, entire. Evergreen tree, 50-60 ft. ... ... 1. *M. indica*.
Leaves peltate, entire. Small resinous tree ... 2. *M. peltata*.

1. **Macaranga indica** Wight is found in Sikkim up to 3,000 feet, the Mishmi and Khasia Hills, the Deccan Peninsula, the Andaman Islands, and Ceylon.

The gum is applied to sores.

Kumaon: Ramalo—; *Malayalam*: Puthatamara—; *Malkot*: Papri—; *Sinhalese*: Kenda—; *Tamil*: Vuttuttamara—.

2. **Macaranga peltata** Muell.-Arg. is found on the hills of Orissa and the Circars, and on the Ghats from the Konkan to Travancore. It is abundant in Ceylon, up to 3,000 feet.

The gum, powdered and made into a paste, is reckoned a good external application for venereal sores.

Bombay: Chanda—; *Canarese*: Chandakanne, Chandkal, Chenthakanni, Chinthakanni, Jatiyuppalige, Kanchupranti, Kanjupparanti, Upaligi, Uppalige, Upparanti—; *Ceylon*: Vaddakkanni—; *Kadir*: Pimugam—; *Kanara*: Chandkal—; *Konkani*: Chandori—; *Malayalam*: Pattavanni, Uppila, Vatta—; *Marahi*: Chanda, Chandoda, Chandora, Chandwar—; *Matheran*: Chandara—; *Mayurbhanj*: Tabhari—; *Mysore*: Chenthakanni—; *Sinhalese*: Bukenda, Kenda, Patkenda—; *Tamil*: Vatta, Vattikanni, Vattittutti—; *Telugu*: Boddi, Kondajaphara, Kondatamara, Pulichinjalamu—; *Uriya*: Gondaguria, Piania—.

MALLOTUS.

The genus includes 120 species, found in the tropics of the Old World.

M. apelta Müll.-Arg., *M. cochinchinensis* Lour., *M. furetianus* Müll.-Arg., *M. philippinensis* Müll.-Arg. are used medicinally in Indo-China; *M. japonicus* Müll.-Arg. is used in Japan; *M. philippinensis* Müll.-Arg. in the Philippine Islands; *M. oppositifolius* Müll.-Arg. in Nigeria and Gold Coast.

Capsule echinate or woolly; stamens 45 to 100 ... 1. *M. cochinchinensis.*
Capsule not spiny; stamens 40 to 50 ... ... 2. *M. philippinensis.*

1. **Mallotus cochinchinensis** Lour. is very common in the secondary jungle and woods of Burma, the Malay Peninsula, the Malay Islands, Cambodia, Indo-China, and China.

The root and the fruit are applied topically to bruises in Indo-China.

Indo-China: Bai bai, Bet, Bo, Bong bet, Giay, Plan—; *Malay*: Balik angin—.

2. **Mallotus philippinensis** Müll.-Arg. is found throughout tropical India; along the foot of the Himalaya from Kashmir eastwards, ascending to 5,000 feet; all over Bengal and Burma, Singapore, and the Andaman Islands; and from Sind southwards to Ceylon. It is distributed to China, the Malay Islands, New Guinea, and Australia.

The leaves are bitter, cooling; give appetite; cause flatulence and constipation. The glands on the fruit are pungent, heating; purgative, anthelmintic, vulnerary, detergent, maturant, carminative, alexiteric; heal ulcers and wounds, tumours, stone in the bladder; useful in bronchitis, diseases of the abdomen, enlargement of the spleen (Ayurveda).

The glands and hairs on the fruit are bitter; anthelmintic, styptic; lessen intestinal pain; useful in scabies, ring-worm, and other skin diseases (Yunani).

Among the Mundas of Chota Nagpur the root, well ground, is rubbed on the painful parts in articular rheumatism. The root is used internally by the Arabs for leprosy, and in solution to remove freckles and pustules.

The leaves, fruit, and root with honey are applied to poisoned bites, bruises, ring-worm, pimples and freckles.

In Katha, Burma, the seeds are ground to a paste, and applied to wounds and dah cuts.

The powder prepared from the fruit is used as an anthelmintic, vermifuge and purgative medicine. It is also said to possess cathartic properties. This powder, which is known as 'Kamala' and consists of the small glands and hairs on the ripe fruits, is officinal in Austria, Germany, Hungary, Italy, Japan, Portugal, and Switzerland.

In most European and American works it is stated that Kamala has long been used in India in the treatment of tapeworm; but neither literature nor tradition bear this out. It would appear from results obtained recently under controlled conditions that the value of the drug as an anthelmintic has been greatly overrated. As for its antidotal properties they have been experimentally disproved by Mhaskar and Caius in the case of cobra and daboia venoms.

Akola: Kunkumphal—; *Almora*: Roli—; *Arabic*: Kampileh, Kinbil—; *Assam*: Gangai, Jaggaru, Puddum—; *Banda*: Kamela, Reoni, Roli—; *Bengal*: Kamalagundi, Kamila, Tung—; *Berar*: Kuku, Sendri—; *Betul*: Kunkuma, Rauni, Rori—; *Bhil*: Shendrya—; *Bijnor*: Kamela, Ruinia—; *Birbhum*: Dholasinua—; *Bombay*: Kamala, Kamela, Kapela, Kapila, Ruhin, Shendri—; *Bundelkhand*: Rori—; *Burma*: Tanthieden, Tawtheeteng, Tawthidin—; *Canarese*: Chandrahittu, Ettunalige, Honne, Hullichellu, Kapila, Kesari, Kesarimavu, Kunkuma, Kunkumada, Kunkume, Munnaga, Punnaga, Purushatunga, Surahonne, Suraparni, Suvarnakesari, Vasare—; *Central Provinces*: Chamargular, Ningur, Rauni, Rori—; *Ceylon*: Kamalama, Kapila—; *Darjeeling*: Sinduri—; *Deccan*: Kambhal, Kamila, Ruin, Ruli, Wussunthagunda—; *Dehra Dun*: Raini—; *English*: Monkey Face Tree, Spoonwood—; *Garhwal*: Rohni, Roini, Ruina—; *Garo*: Chinderpang, Machugan—; *Gond*: Koku—; *Gujerati*: Kapilo—; *Haldwani*: Roli—; *Hindi*: Kamala, Kambhal, Kambila, Kamela, Kanula, Kamud, Raini, Rauni, Rohni, Roini, Roli, Rora, Ruin, Ruli, Rulu, Wassanthaganda—; *Ilocano*: Buas, Vuas—; *Indo-China*: Aloang lay sa lang, Ba bet, Bae thau, Buom, Canh kiền, Du ma thich, Ema da, Giay rung, Skuoi—; *Jaunsar*: Kambel—; *Kadir*: Manjanai, Maunana, Ponni—; *Kashmir*: Kaimbil—; *Kharwar*: Rori—; *Kolami*: Garasinduri—; *Kumaon*: Rauni, Reru, Riuna, Roli, Ruen—; *Lambadi*: Dholo—; *Lepcha*: Numboonghor, Puroa, Purva, Tukla—; *Lohardugga*: Rori—; *Malayalam*: Chenkolli, Kapila, Kuramatukka, Manjana, Maunana, Piponnakam, Ponnakam, Ponni, Poonagam, Punna, Tavitu—; *Mal Paharia*: Daosindra—; *Marathi*: Shendri, Shindur—; *Matheran*: Asli, Rohen—; *Melghat*: Kuku—; *Merwara*: Senduria—; *Michi*: Baraiburi, Sindurpong—; *Mundari*: Garisinduri—; *Nepal*: Safedmallata, Sinduria—; *Nimar*: Kunkuma, Rauni, Rori—; *North-Western Provinces*: Purvahung, Sinduria—; *Oudh*: Rohni—; *Persian*: Kampileh, Kanbela—; *Peshawar*: Kambalia—; *Punjab*: Kamal, Kambal, Kamela, Kumila, Reini, Reun, Rulya—; *Queensland*: Poodgee-poodgera—; *Ramnagar*: Roli—; *Sanskrit*: Bahupushpa, Chandra, Kampilla, Kampillaka, Kapila, Karkatha, Kesara, Laghupatraka, Lohitanga, Madhuka, Nadivasa, Pikaksha, Punnaga, Punnagakesara, Punnama, Raktachurnaka, Raktanga, Raktaphala, Ranjaka, Rechanaka, Rechani, Rechi, Rochana—; *Santali*: Rora—; *Shan*: Tawthadin—; *Sinhalese*: Hamparandella, Hamparila, Kamala—; *Tagalog*: Banato, Sala—; *Tamil*: Avam, Kabilam, Kabilappodi, Kamala, Kambosam, Kapila, Kopilapodi, Kungumam, Kurangumanjanatti, Manjanai, Suvanagesari, Tavattai, Tiruchalai—; *Telugu*: Adavigubbatuda, Benduruppu, Chendiramu, Kunkuma, Nagarakta, Sinduragunda, Sinduri, Tunga, Vasantagandhamu, Veligarapu—; *Tulu*: Ponne—; *Urdu*: Kalileh—; *Uriya*: Bosontogundi, Kunkumo, Kopilogundi, Punango, Sinduri, Sindurogundi—; *Visayan*: Sala—; *Zambales*: Apuyot—.

MANIHOT.

The genus consists of 150 American species, mostly Brazilian and Mexican.

M. loureirii Pohl, a native of Cochin-China, is used medicinally in China; *M. ultilissima* is used in Brazil, Guiana, and Cambodia.

Manihot ultilissima Pohl is cultivated in various parts of India.

In Cambodia the pounded tuber is applied to ulcerated wounds. It is considered antiseptic in Brazil, and is used to preserve meat; as an ointment it is reputed useful in ulcers of the cornea.

The juice is very poisonous. In Guiana it is boiled down to a syrup and given as an aperient. The fresh rhizome is made into poultice and applied to ulcers.

The fruit is used as a fish poison in California and in Brazil.

The starch is officinal in Holland

Ashanti: Bankye—; *Bajo*: Makwamba, Ngabo, Ngawa, Sanagal—; *Bahamas*: Bay Rush, Bitter Cassava, Cassava, Manioc, Tapioca—; *Bakundu*:

Kasara—; ***Bakwiri***: Mionde—; ***Balondo***: Ewa—; ***Balong***: Sanaga!—; ***Bambara***: Bananiku—; ***Banda***: Ngali—; ***Basari***: Banking, Kambohando—; ***Batanga***: Ewa—; ***Bemoba***: Jangbwanguyuya—; ***Benin***: Bobozi, Ebobozi, Igari, Igari-makole, Ikeripia—; *Betsileo*: Balahazo, Mangahazo—; *Betsimisaraka*: M' Bazaha—; ***Bota***: Makwamba—; ***Brazil***: Mandioca, Maniba, Manioc, Manira—; ***Cambodia***: Kdouch—; ***Canarese***: Kadugenasu, Maragenasu—; ***Central America***: Yuca—; ***Dagomba***: Bankye, Kambonjule—; ***East Africa***: Mhogo—; ***Ebrie***: Bede—; ***Efik***: Ibo iwa, Iwa unenge, Okpo ofop iwa—; ***English***: Bitter Cassava, Brazilian Arrowroot, Cassava, Tapioca Plant—; ***Ewe***: Agbeli—; ***Fanti***: Bankye—; ***French***: Mandioc, Manioc, Manioc doux, Manioque, Pain des negres—; ***French Guiana***: Manioc, Manioc amer, Manioc petit Louis—; ***Fufulde***: Mbai—: ***Fulani***: Bantara, Nyambe—; ***Ga***: Duade—; ***Gold Coast***: Bantshi, Duaday—; ***Guam***: Mandioka, Mandiuka, Mendioka, Mendiuka, Yuka—; ***Hausa***: Baushin kurege, Doyar kundu, Kandirin beguwa, Karaji, Karaza, Rogo—; ***Hindi***: Sakarkanda—; *Hova*: Vomangahazo—; ***Ibo***: Abacha, Abaja, Akbo, Akbonkono, Akbunkolo, Akpu, Akpu-ji, Ji-akpu, Ji-gbo, Oi-akpu—; ***Jamaica***: Cassada—; ***Java***: Hoci dangdua—; *Kabure*: Mbaumhae—; ***Kano***: Rogo—; ***Kanuri***: Garisa—; ***Kissi***: Yambale, Yambale fulalei—; ***Konkomba***: Akombanul—; ***Konno***: Tanga—; *Koranko*: Banake, Tankaso—; ***Krepi***: Gbedi, Gbeli—; ***Krobo***: Agbeli—; *Limba*: Kopintan—; ***Loanda***: Quisaca—; ***Lokko***: Bororanga—; *Losso*: Bondoronde—; *Madagascar*: Kajaha, Kazaha, Matreoka, Ropotra, Tsiveritelo, Vihazo—; *Malinke*: Bananku, Bananku gue, Bara bananku—; *Mamprussi*: Kukulajo—; *Mandingo*: Mandioca, Nyambo—; ***Mano***: Bai, Bei—; ***Mauritius***: Manioc—; *Mbonge*: Kasara—; ***Mende***: Tangei, Rande, Nguwo—; ***Mexico***: Huacamotl—; *Mundari*: Edelsangga—; ***Nigeria***: Bara banankou—; ***North America***: Brazilian Arrowroot, Cassava, Manioc, Tapioca Plant—; ***Nupe***: Rogo, Rogo dzurugi—; *Nzima*: Bairdair—; ***Portuguese***: Mandioca, Pau de farinha—; ***Roumanian***: Manioc—; ***Russian***: Manioc—; ***Sakalave***: Ambazaha, Balafanapaka, Mohogo—; *Sherbro*: Yeke—; ***Shuwa Arabic***: Baghut, Nagasha—; *Sinhalese*: Manyokka—; *Sobo*: Madaka, Medaka—; ***Spanish***: Mandiiba, Maniba del Brasil, Manihot amarga, Manioc amarga, Mondioca, Yuca, Yuca amarga, Yuca brava, Yuca casave de las Antillas, Yuca casavi de las Antillas, Yuca mortifera—; ***Susa***: Yoka, Yoka-fore—; ***Tagalog***: Caomtingcahoy—; ***Timne***: Ayoka Batsanka, Ekanda, Erogbo, Nafange, Yoka-ibi—; ***Tivi***: Rougou, Vambeyon—; *Tschandjo*: Kungola—; ***Twi***: Bankye—; ***Vai***: Dumbai—; ***Visayan***: Camotingcahoy—; *Uruguay*: Mandioca—; ***West Africa***: Mandioca—; *West Indies*: Yuca—; *Wolof*: Gniambi, Nyambe—; ***Yalunka***: Kunde, Kunde-ghrabi-na—; *Yoruba*: Ege—.

MICRODESMIS.

The genus consists of 2 species, one Asiatic, and one African. *M. caseariaefolia* Planch. is used medicinally in Indo-China; *M. puberula* Hook. and *M. Zenkeri* Pax. in tropical West Africa.

Microdesmis caseariaefolia Planch. ocurs in Tenasserim, and is found in woods and forests all over the Malay Peninsula. It is distributed to Borneo and South China.

The fresh sap is used in Indo-China for caries of the teeth.

Indo-China: A luan te he, Chanh oc, Dock ko bang—; *Malay*: Chateng, Chereh rambeh, Kenidei badak, Sigoniah—.

PHYLLANTHUS.

The genus includes 500 species, temperate and tropical, absent from Europe and northern Asia.

The following species are used medicinally in Indo-China— *P. elegans* Wall., *P. Emblica* Linn., *P. Niruri* Linn., *P. reticulatus* Poir., *P urinaria* Linn.—; in China— *P. Emblica* Linn.—; in the Philippine Islands— *P. Niruri* Linn., *P. reticulatus* Poir., *P. urinaria* Linn.;— in Colombia— *P. ichthyomethius* Rusby, *P. salviaefolius* H.B.K.—; in Brazil— *P. conami* Sw., *P. lathyroides* H.B.K., *P*

Niruri Linn.—; in Guiana— *P. conami* Sw., *P. epiphyllanthus* Linn., *P. guyanensis* Klotsch, *P. Niruri* Linn., *P. urinaria* Linn.—; in the West Indies— *P. Niruri* Linn.—; in Lagos and Gold Coast— *P. amarus* Schum. and Thonn., *P. discoideus* Muell.-Arg., *P. floribundus* Müll.-Arg., *P. Niruri* Linn., *P. pentandrus* Schum. and Thonn., *P. reticulatus* Poir., *P. Wildemannii* Beille—; in Madagascar— *P. casticum* Soy.-Will., *P. madagascariensis* Muell.-Arg.—; in La Reunion— *P. casticum* Soy.-Will., *P. Niruri* Linn., *P. phillyreaefolius* Poir., *P. urinaria* Linn.—.

A. A large deciduous tree; bark flaking conchoidally; branchlets slender, pubescent. Leaves equal and symmetrically set ... ... 2. *P. Emblica.*

B. Shrubs or small trees; leaves 1 in. or more long.
  I. Capsule small .1 in. through or less, fleshy. Pedicels slender rather long ... ... 5. *P. reticulatus.*
  II. Capsule long-pedicelled, dehiscent, .1 to .5 in. long. Leaves ovate-lanceolate glaucous beneath, 4 in. long. 1. *P. elegans.*

C. Herbs, occasionally wiry at base; leaves distichous small.
  I. Quite herbaceous; leaves oblong or elliptic obovate; capsule smooth ... ... 4. *P. Niruri.*
  II. Slightly woody at base; leaves linear-oblong; capsule echinate ... ... 7. *P. urinaria.*
  III. Nine inches to 2 ft. tall with long branches, leaves close set very narrow.
    a. Leaves linear or oblong-linear; flowers pedicellate. 6. *P. simplex.*
    b. Leaves cuneate-obovate; flowers sub-sessile ... 3. *P. maderaspatensis.*

1. **Phyllanthus elegans** Wall. is found in Tenasserim, at Mulmein, Mergui, and Amherst; in the Malay Peninsula, on stream banks in Pulau Adang islands and Rawei.

In Indo-China the roots are considered antiperiodic. The juice of the pounded leaves is given to children with aphthae.

In Tongking the leaves are made into a salad and given for fatigue.

Indo-China: Bo ngot, Bu ngot, Hac rien than—.

2. **Phyllanthus Emblica** Linn. is found, wild or planted, throughout tropical and subtropical India; from the base of the Himalaya, from Jummu eastwards, and southwards to Ceylon and Malacca; whence to the Malay Islands and China.

This plant occupies an important place in Hindu Materia Medica, and the medicinal properties of its various parts are described at length in the Nighantas: 'The fruit is acrid, sour, bitter, sweetish; cooling, alexiteric, carminative, alterative, laxative, tonic, antipyretic; useful in burning sensations, vomiting, biliousness, urinary discharges, thirst, leprosy, constipation, inflammations, erysipelas, piles, anaemia, strangury; used in biliousness, thirst, sweats, anura, poisoning, ophthalmia, incipient blindness.—The leaf is used in opthalmia and incipient blindness.—The seed is acrid, sweet; aphrodisiac antipyretic; useful in biliousness, asthma, bronchitis, leucorrhœa, vomiting.'

Mohammedan physicians esteem the plant equally with the Hindus. According to Arabic and Persian authors; 'The flowers are cooling and aperient.—The fruit is acrid, sour, cooling; astringent, tonic, expectorant, vulnerary, laxative, improves the appetite; useful in diseases of the heart, liver complaint, cold in the nose, thirst, piles, biliousness, eye troubles; stops nasal hæmorrhage; purifies the humours of the body.'

The root, the bark, and the ripe fruit are astringent. The flowers are refrigerant and aperient.

The unripe fruit is cooling, diuretic, and laxative. The exudation from the incisions on the fruit is used as an external application in inflammation of the eye.

In the Konkan the juice of the fresh bark with honey and turmeric is given in gonorrhœa.

The leaves are, in Baroda, used as an infusion with fenugreek seeds in cases of chronic dysentery, and are also considered a bitter tonic. In the same locality the milky juice is considered a good application to offensive sores.

The leaves enter into the composition of a Mundari expectorant.

In the Tamil country the calcined fruit is powdered with an equal weight of flower tops of *Mollugo nudicautis* Lam., made into a paste with milk, and given for uterine hæmorrhage during pregnancy.

In Cambodia the leaves are used in the preparation of antithermic lotions and baths, and the decoction of the roots is given in myalgia following upon some febrile condition.

Arabic: Ambliy, Amlaj—; *Assam*: Amlaki, Amluki, Sohmyrlain—; *Bengal*: Ambolati, Amla, Amlaki, Amlati, Amlika, Amulati, Aonla, Aunlah, Yeonlah—; *Betul*: Aonla—; *Bombay*: Amla, Avala, Avalkati, Avla—; *Burma*: Hziphyu, Shabju, Siphiyusi, Tasha, Zibyu, Ziphiyusi—; *Cambodia*: Ngop—; *Canarese*: Amalaka, Chattu, Dadi, Dhanya, Dhatri, Nelli, Sudhe—; *Central Provinces*: Amla, Anla—; *Canton*: Yeou Kan Tse—; *Chinese*: An Mo Le—; *Cuttack*: Alathanda—; *Deccan*: Amla, Owla, Ownla—; *English*: Emblic Myrobalan Tree—; *Garo*: Ambari—; *Gond*: Aunri, Lalla, Milli, Nalli, Nilli, Usir—; *Gujerati*: Amali, Ambala, Ambri, Amla, Bhoza, Bhoza amali—; *Hindi*: Amalachi, Amla, Amlaki, Amlika, Anola, Anuli, Anvula, Anvurah, Anwerd, Aola, Aonla, Aora, Aungra, Aunra, Aunwered, Daula—; *Indo-China*: Bong ngot, Kam lam, Kamlan ko, Yem ma lac—; *Khond*: Durga—; *Kol*: Miral—; *Kolami*: Aura, Miral—; *Konkani*: Anvallo, Dogranvalli, Dogranvallo—; *Kumaon*: Aonla—; *Kurku*: Aunre—; *Lambadi*: Ambla—; *Lao*: Ma kam paum—; *Lepcha*: Pa-om kung, Suom—; *Malayalam*: Amalakam, Nelli—; *Marathi*: Anvala, Aola, Aonli, Avala, Arola, Awla, Bhuiawalai—; *Mauritius*: Embélic—; *Mundari*: Meraldaru—; *Nepal*: Amla—; *Nimar*: Aonla—; *North-Western Provinces*: Amla, Aoula—; *Persian*: Amelah, Amuleh—; *Portuguese*: Mirabolano emblico—; *Punjab*: Ambal, Ambli, Ambul, Amla, Aonla—; *Sanskrit*: Adiphala, Akara, Amalaki, Amamalakam, Amlika, Amraphala, Amrita, Amritaphala, Bahuphali, Dhatri, Dhatrika, Dhatriphala, Jatiphala, Karshaphala, Kayastha, Pancharasa, Rochani, Shanta, Shiva, Shriphala, Shriphali, Tishya, Tishyaphala, Triphala, Vayastha, Vrishya, Vrittaphala—; *Santal*: Meral—; *Savara*: Peddavusirika—; *Sinhalese*: Awusadanelli, Nelli, Nellika—; *Tamil*: Amalagam, Andakoram, Indul, Kattunelli, Nelli, Perunelli, Sirottam, Tattiri, Toppunelli—; *Telugu*: Amalakamu, Amalaki, Nelli, Pullayusirika, Triphalamu, Usirika, Usirikaya, Usiriki—; *Tulu*: Nelli—; *Urdu*: Anwala—; *Uriya*: Aura, Khondona, Onola—.

3. **Phyllanthus maderaspatensis** Linn. is met with throughout the drier parts of India, from Banda through the Deccan Peninsula

to Ceylon. It occurs also in tropical Africa, Arabia, Java, and Australia.

In Southern India an infusion of the leaves is given for headache.

The seeds are used medicinally on account of the mucilage which they afford when soaked in water.

Arabic: Cholf, Marur—; *Gujerati*: Kanochha—; *Hindi*: Hazarmani, Kanocha, Kanochha, Niruri, Sadamanni—; *Matheran*: Kanocha—; *Persian*: Marurshatu—; *Porebunder*: Bakrado—; *Sind*: Niruri, Sadamanni—; *Telugu*: Nalausereki—; *Urdu*: Kanochha—.

4. **Phyllanthus Niruri** Linn. is found throughout the hotter parts of India, from the Punjab eastwards to Assam, and southwards to Travancore, Ceylon, and the Malay Peninsula, ascending the hills to 3,000 feet. It inhabits the tropics generally, except Australia.

Hindu physicians credit the plant with the same therapeutical properties as *P. urinaria*; they, moreover, use the leaves as a diuretic and in menorrhagia.

According to Mahommedan physicians the plant is stomachic, good for sores and in chronic dysentery; the fruit is bitter, useful for tubercular ulcers, wounds, sores, bruises, scabies, and ringworm.

The plant is much used as a diuretic in dropsical affections, gonorrhœa, and other troubles of the genito-urinary tract. It is a South Indian diuretic and antidysenteric. In Indo-China it is reputed to be diuretic, depurant and antisyphilitic. In La Reunion it is very much used in blennorrhagia, dropsy, and diarrhœa.

The fresh root is said to be an excellent remedy for jaundice. In the Konkan it is rubbed down with rice water and given as a remedy for menorrhagia. Bruised with a little water and administered internally with milk, it is a popular Tamil galactagogue. In the Rajputana Desert the root, pounded and mixed with *Commiphora mukul* Engl., is given to camels suffering from indigestion.

The leaves are stomachic. A poultice of the leaves with salt cures scabby affections, and without salt may be applied to bruises and wounds; made with rice water the poultice lessens œdematous swellings and ulcers.

In Gold Coast the leaves are pounded and used to cure gonorrhœa. Parts of the plant are used to cure constipation. The leaves are boiled and the liquor drunk to stop acute pains in the stomach. The chief use of the plant is to allay griping in cases of dysentery.

The milky juice is a good application to offensive sores.

The bark is used as a purgative and in the treatment of various diseases and yields a bitter principle 'phyllanthin', which is a fish poison.

An infusion of the young shoots is given in dysentery.

The infusion of the root and leaves is a good tonic and diuretic, when taken cold in repeated doses.

The decoction of the root and leaves is very bitter and is a favourite remedy among the natives of Santo Domingo and Porto Rico, for the cure of intermittent fevers. I have myself many times proved its efficacy in preventing the expected paroxysm. I was accustomed to employ a tincture made by myself with the whole plant, the dose being two drachms in the morning. Sometimes I repeated the dose, which acted upon the bowels as a slight purgative and this is very useful in inveterate intermittents with infracts of the spleen and liver. The infusion of the root and leaves is a good tonic, and a diuretic when taken cold in repeated doses (A. J. Amadeo; *Pharmaceut. Journ.*, April 28, 1888).

A decoction of the bitter root and leaves is given in Haiti for stomach-ache.

An aqueous extract or a decoction of the fresh roots stem, and leaves is given internally in snake-bite (Roberts); but the root, stem, and leaf, are all useless as an antidotal treatment (Mhaskar and Caius).

Arabic: Mekatkata, Meneckete—; *Ashanti*: Bommaguwakyi—; *Bengal*: Bhujamla, Kiraneli, Sadahazurmani—; *Bombay*: Bhuiavala—; *Brazil*: Erva pombinha—; *Burma*: Miziphiyu—; *Canarese*: Kiranelligida, Kirinelligide—; *Ceylon*: Vellaikkilkaynelli—; *Chinese*: So Cheu—; *Deccan*: Bhuinavalah—; *Efik*: Oyomo-ke-iso-amon-ke-edem—; *French*: Herbe du chagrin—; *Ga*: Ombatoatshi—; *Guam*: Maigo-lalo, Maigu-lalo—; *Gujerati*: Bhonya anmali—; *Hindi*: Bhonya abali, Bhuinanvalah, Jangli amli, Jaramla, Niruri, Sadahazurmani—; *Ibibio*: Oyomo-ke-iso-amon-ke-edem—; *Indo-China*: Cho de—; *Java*: Aijlaunmahaij—; *Krobo*: Ofobiokpab—; *La Reunion*: Petit tamarin blanc—; *Malayalam*: Kirganelli, Kizhanelli, Kizhkkayinelli—; *Marathi*: Bhui avali—; *Mauritius*: Curanellie blanche, Herbe au chagrin—; *Mundari*: Muikoa kantara, Pirikantara—; *Nupe*: Ebogi, Ebo zunmaggi—; *Philippines*: Hierba de San Paulo, Hierba de San Pedro—; *Porebunder*: Bhonya anmari—; *Portuguese*: Erva poubinha—; *Rajputana*: Gugerati bawal—; *Sanskrit*: Adhyanda, Ajata, Ajuta, Amala, Amlika, Amrita, Aphala, Bahupatra, Bahupatri, Bahuphala, Bahupushpi, Bhudatra, Bhudhatri, Bhumyamalaki, Bhuparva, Charati, Chorata, Dalaparshini, Dridhapadi, Hilolika, Jada, Jharika, Jhatamala, Kshetramali, Mala, Nilolika, Putrashronika, Shiva, Sukshmadala, Sukshmaphala, Uchchata, Vishaghni, Vishvaparni, Vitunnaka, Vituntika, Vrishya—; *Seychelles*: Herbe au chagrin—; *Sind*: Niruri—; *Sinhalese*: Pittawaka—; *Spanish*: Yerba de quinina, Yerba de quinono—; *Tamil*: Kilanelli, Kilkkaynelli—; *Telugu*: Nelausirika, Nelavusari—; *Twi*: Bommaguwakyi—; *Urdu*: Bhui amla—; *Uriya*: Bhui aola—; *West Indies*: Petit tamarin blanc, Quinine créole, Yerba de quinino—; *Yemen*: Mekatkata, Meneckete—; *Yoruba*: Yoloba—.

5. **Phyllanthus reticulatus** Poir. is common throughout tropical India, Burma, Malaya, and Ceylon. It extends to the Malay Islands, China, and tropical Africa.

The fruit is astringent to the bowels; useful in inflammations, and diseases of the blood (Ayurveda).

The bark is considered alterative and attenuant, and is prescribed in decoction in the quantity of four ounces or more twice daily.

The leaves are employed as a diuretic and cooling medicine in Sind.

The juice of the leaves is used medicinally in the Konkan. It is made into a pill with camphor and cubebs, which is allowed to dissolve in the mouth as a remedy for bleeding from the gums. It is also reduced to a thin extract along with the juice of other alterative plants and made into a pill with aromatics; this pill is given twice a day, rubbed down in milk, as an alterative in heat of the blood.

In Lakhimpur the juice of the leaves is used for diarrhœa in infants.

In Ashanti the leaves are mixed with palm nuts (*Elæis guineensis*) and made into a broth, which is given to newly delivered women to relieve them.

In Gold Coast the stems are used to cure sore eyes, the juice from them being blown into the eyes.

The Rongas and Europeans of East Africa use the powdered leaf as a local application to sores, burns, and suppurations, and chafing of the skin. The Rongas also apply it to venereal sores.

In Indo-China the plant is used in the treatment of smallpox and syphilis.

Akola: Pithor, Pittur—; *Assam*: Amluki—; *Bengal*: Panjuli—; *Bombay*: Pavana—; *Canarese*: Anamsula, Chippulinellu, Huli, Karesuli, Karihuli, Sannahagesoppu—; *Deccan*: Buinowla, Kalemadhkajhar—; *Fanti*: Nkokobro—; *Gorakhpur*: Sikat—; *Gujerati*: Datwan—; *Hausa*: Alambu, Alambu na tudu, Bak' in alamba, K'alambu—; *Hindi*: Buinowla, Kalemadhkaper, Makhi, Panjoli, Panjuli—; *Indo-China*: No kang pa, No phen den—; *Katagum*: Tsah—; *Konkani*: Kaili—; *Lambadi*: Kombhoiro—; *Malayalam*: Kattuniruri, Kilanelli, Nirnelli, Niruri—; *Marathi*: Pavan—; *Melghat*: Pithor, Pittur—; *Merwara*: Kabonan—; *Nimar*: Kamoi—; *Punjab*: Panjuli—; *Rajputana*: Kabonan—; *Ronga*: Tetenya—; *Sanskrit*: Bahupraja, Bahupushpa, Kamboji, Kambojini, Krishnakambhoji—; *Sind*: Kamohi, Kamohijopun, Kamu, Pikapiru—; *Sinhalese*: Welkyla—; *Tagalog*: Malaiba, Malatinta, Tinatinam, Tintatintahan—; *Tamil*: Abirangi, Karuppuppilanji, Karunelli, Kattukkilanelli, Melanelli, Nirppul, Nirppula, Nirppulanji, Pul, Pula, Pulanji—; *Telugu*: Nallapuli, Nallapurugudu, Nelapurugudu, Pandibarangi, Pulisar, Purugud—; *Timne*: Egbeli—; *Tulu*: Kakesoppu—; *Twi*: Awobe—; *Uriya*: Bonotihudi, Jojangi, Phajoli—; *Visayan*: Matangolang, Sungotolang—; *Yoruba*: Iranje—; *Zamboanga*: Nipin—.

6. **Phyllanthus simplex** Retz is found in the plains and low hills throughout India, from Kumaon to Assam and southwards to Travancore, Ceylon and Malacca, ascending to 6,000 feet. It is distributed to China, the Malay and the Pacific Islands.

The natives use the fresh leaves, flowers and fruit, with cumin seeds and sugar, of each equal parts made into an electuary, for the cure of gonorrhœa, a teaspoonful is given twice a day. The fresh leaves, bruised and mixed with buttermilk, make a wash to cure the itch in children.

The root is used in Chota Nagpur as an external application for mammary abscess.

Gujerati: Motibhonya anmali—; *Indo-China*: Vay oc—; *Marathi*: Bhuiavali, Motibhuiavali—; *Mundari*: Otemeral, Pirimeral—; *Porebunder*: Motibhonya anmari—; *Santali*: Tandimeral—; *Telugu*: Uchchiyusirika.

7. **Phyllanthus urinaria** Linn. is met with throughout India, from the Punjab to Assam, Burma, Malacca, Penang, and Ceylon; it is common in Singapore and Selangor. It inhabits the tropics generally.

The fruit and plant are acrid, sour, cooling, bitter, sweetish; alexipharmic; useful in thirst, bronchitis, leprosy, anæmia, urinary discharges, anuria, biliousness, asthma, hiccough (Ayurveda).

The plant is much used as a diuretic in dropsical affections, also in gonorrhœa and other genito-urinary troubles.

In Chota Nagpur the root is given to sleepless children.

In La Reunion the plant is considered diuretic, sudorific, depurative, and emmenagogue. It is given as a drink in dysentery and cystitis.

In Cambodia the plant is used as a bitter, tonic, astringent, and febrifuge. An infusion is given for liver complaint, diarrhœa and paludism.

Bengal: Hazarmani—; *Burma*: Miziphiyuani—; *Cambodia*: Prak phle—; *Canarese*: Kempukiranelli—; *Ceylon*: Chivappukkilkaynelli—; *Chinese*: Ko Cheu—; *Gujerati*: Kharsadabhonya anmali, Kharsadabhonya ansari—; *Hindi*: Hazarmani, Lalbhuinanvalah—; *Indo-China*: Cho de, Giap ha chau, Kho ham, Prak phle—; *La Reunion*: Petit tamarin rouge—; *Malay*: Ambin buah—; *Malayalam*: Chirukizhukanelli, Chukannakizhanelli—; *Marathi*: Lalmundajanvali—; *Matheran*: Lalbhiuawali—; *Mauritius*: Curanellie rouge, Curanellie urinaire—; *Sanskrit*: Adhyanda, Ajata, Ajuta, Amala, Aphala, Arudha, Bahupatra, Bahuphala, Bahupushpa, Charati, Chotata, Dalasparshini, Drirdhapadi, Hilkolika, Jada, Jharika, Bhatamala, Mala, Nilolika, Putrashronika, Shiva, Sukshmadala, Sukshmaphala Tali, Tamalika, Tamalini, Tamravalli, Uchchata, Vishaghni, Vishvaparni, Vitunnaka, Vituntika, Vrishya—; *Santali*: Badarzhapi—; *Sinhalese*: Binnelli, Ratpittaawka—; *Tagalag*: Ibaibaan—; *Tamil*: Shivappunelli—; *Telugu*: Ettausirika, Yerrausereki—.

PUTRANJIVA.

The genus consists of 4 Indo-Malayan species.

Putranjiva Roxburghii Wall. is found, wild and cultivated, throughout tropical India, from the lower Himalaya in Kumaon eastwards and southwards to Pegu and Ceylon.

The leaves, the fruit, and the stones of the fruit are given in decoction in colds and fevers.

In Siam a medicine for rheumatism is made out of the leaves and fruit.

Bengal: Jiaputa, Putranjiva—; *Bombay*: Jewanputr, Jivputrak, Putajan, Putrajiva—; *Burma*: Badibyu, Egayit, Taukyat, Toukyap, Toukyat—; *Canarese*: Amani, Menasinakale, Putrajiva, Putremjiva—; *Chindwin*: Badibyu—; *Haldwani*: Juti—; *Hazaribagh*: Piten—; *Hindi*: Jiaputa, Jivputrak, Joti, Juti, Patigia, Patji, Putajan, Putijia, Putrajiva, Putranjiva—; *Kumaon*: Juti—; *Lambadi*: Kalo—; *Malayalam*: Pongalam, Ponkolam—; *Marathi*: Jewanputr, Jivputrak, Jiwanputr, Putajan, Putrajwa—; *North-Western Provinces*: Jiaputa, Joti, Juti, Putrajiva—; *Oudh*: Patji—; *Pegu*: Daukyat—; *Punjab*: Jiyaputra, Patajan—; *Ramnagar*: Juti—; *Reddi*: Veku—; *Sanskrit*: Ardhasadhaka, Apatyajiva, Garbhada, Garbhakara, Jivanaputra, Kumarajiva, Mavu, Pavitra, Putrajiva, Putranjiva, Shlipadapaha, Sidhida, Sutajivaka, Sutrajiva, Yashtipushpa—; *Santali*: Pitoj, Pitonj—; *Tamil*: Irukolli, Karupali, Karupilai—; *Telugu*: Kadrojuvi, Kudrajinik, Kudurujivi, Kudurujivir, Mahaputrajivi, Putrajivika—; *Uriya*: Bholokoli, Pohunjona, Poichandia, Poitundia—.

RICINUS.

Ricinus communis Linn. is indigenous to tropical Africa, and is found, wild or cultivated, throughout the Tropics. By cultivation it has been distributed through not only all tropical and subtropical regions, but also in many of the temperate countries of the globe. In the south of England the plant ripens its seeds in favourable situations, and it has been known to come to maturity as far north as Oslo in Norway.

The root is sweetish, heating; carminative; useful in inflammations, pains, ascites, fever, glands, asthma, eructations, bronchitis, leprosy, diseases of the rectum, and the head.—The leaves are useful in intestinal worms, strangury, night blindness, earache; increase biliousness.—The flowers are useful in glandular tumours, anal troubles, vaginal pain.—The fruit is heating and an appetiser; useful in tumours, pains, piles, diseases of the liver and spleen.—The seed is cathartic and aphrodisiac.—The oil is sweetish; cathartic, aphrodisiac, anthelmintic, alterative, useful in tumours, diseases of the heart, slow fevers, ascites, inflammations, typhoid, pain in the back, lumbago, leprosy, elephantiasis, convulsions; causes biliousness (Ayurveda).

The root bark is purgative, alterative; good in skin diseases.—The leaves are galactagogue; good for burns.—The seeds and the oil from them have a bad taste; purgative; useful in liver troubles, pains in the body, lumbago, boils, piles, ringworm, paralysis, inflammations, ascites, asthma, rheumatism, dropsy, amenorrhœa (Yunani).

In Chota Nagpore the bark is ground and applied to burns by the Mundas. A fibre of the bark is tied round the neck to stop vomiting.

The leaf is applied to the head to relieve headache, and is commonly used as a poultice for boils. It is said to possess considerable power as a lactagogue, being applied warm to the breasts. The fresh leaves are used by nursing mothers in the Canary Islands as an external application, to increase the flow of milk. On the other hand it is generally admitted in Indo-China that the leaves act as an antilactagogue, whether applied topically on the breasts or taken internally.

In general the plant is credited with a number of peculiar properties, and is said to be useful in difficult parturition, spasmodic twitchings of the face, facial malformations, and cancer of the stomach.

The seeds and the oil from the seeds are used as a purgative wherever the plant is found growing. The Chinese use the crushed seeds more frequently than they do the oil, the paste is applied to relieve scrofulous sores. In some parts of West Africa a decoction of the leaves is taken to purge.

In Las Bela the oil is expressed and used medicinally; and a fomentation is made with the leaves to cure wounds. At Turbat in Makran it is used as an ointment for sores. At Kotra in Kachhi the leaves are used for fomentations; in Kharan they are bound over boils, and are a good cure.

In French Guinea the leaves are boiled and used as a febrifuge. In native medicine in West Africa the leaves are boiled to form a lotion for fevers, and pounded for application as a poultice to swellings. The leaf is sometimes tied on the forehead for headache. In Togo the leaves of the larger-leaved variety are crushed with cold water to form a useful eye lotion.

An infusion of the leaf is a Zulu remedy for stomach-ache. It is administered orally or as an enema. The Zulus also apply a paste of the root in toothache. In South Africa the root is the basis of several native remedies for toothache.

In Southern Rhodesia the bark is used by natives for stitching up wounds, and as a dressing for wounds and sores. The Chewas boil the root in water which has been previously boiled with the ashes of maize stalks. The sediment from the second boiling is smeared on the teeth and gums to relieve toothache, the material not being swallowed.

The Transvaal Sutos apply the powdered roasted seeds to sores, boils, etc., in children. The foliage is considered emmenagogue, the root-bark purgative, and the leaf useful as a local application in rheumatism. The local application of the leaf to the mammae is said to produce a powerful galactagogic action. In West Africa a decoction is taken internally as a lactagogue and emmenagogue.

The bruised leaves are used for caries of the teeth and given with water for colic in Madagascar.

In La Reunion the leaves are considered lactagogue and are given in infusion or applied to the breasts.

In Guiana the leaves are applied to the breasts to help the secretion of milk. Soaked in vinegar they are applied to the forehad in cases of sunstroke. They act as a powerful sudorific.

A treatment for chancre in Northern Nigeria consists in fumigation by burning stems of *Ricinus* along with those of *Calotropis procera*.

The oil, combined with citron ointment, is used as a topical application in common leprosy.

In West Africa, castor oil is not used as a medicine internally, but externally as an unguent alone or mixed with other substances, such as groundnut or other oil. It is used as an application for parasitic skin-diseases, craw-craw, etc., and as a dressing for sores and parasitic conditions in domestic stock. In South Africa a mixture of castor oil and kerosene is known as a cheap and efficient culicide in the fight against malaria. Castor oil has been found to be almost the only agent which prevents gad-flies from attacking camels for any length of time.

Rendered soft and pulpy by heat, the leaves are applied to the Guinea-worm sore to facilitate its extraction.

Ancient Sanskrit writers prescribed the seed in the treatment of snake-bite and scorpion-sting; but Caius and Mhaskar have shown experimentally that it is not an antidote to either snake or scorpion venom.

The leaves and the seeds are efficinal in Portugal; the seeds and the oil in France and Italy; the oil in Austria, Belgium, Brazil, Denmark, Finland, Great Britain, Holland, Hungary, Japan, Mexico, Norway, Russia, Spain, Sweden, Switzerland, Turkey, the United States, and Yugoslavia.

Afghanistan: Bazanjir, Buzanjir—; *Afrikaans*: Kasterolieboom—; *Almora*: In—; *Amboyana*: Camiri—; *Aowin*: Ateende—; *Arabic*: Charua, Djar, Hurua, Khirurwi, Khirwa, Tebscha, zaejt—; *Assam*: Eri—; *Awuna*: Dzonggdaleng—; *Bafo*: Bokuri—; *Balong*: Bokuri—; *Bambara*; Subarabana—; *Behar*: Anda, Amar, Lenr, Renr—; *Bengal*: Bheranda, Bherenda—; *Benin*: Era ogi—; *Berar*: Yerandi—; *Betsileo*: Kinamena—; *Bombay*: Erendi—; *Brazil*: Carapateiro, Figueira d'inferno, Mamona, Nhambuguasu—; *British Central Africa*: Mbaliki—; *Burma*: Kesu, Kyeksu—; *Cambodia*: Lohong preng—; *Canarese*: Avudala, Avudalu, Chittuharalu, Eranda, Haralu, Manda, Vardhamana—; *Catalan*: Rici, Figuera infernal, Rissino—; *Central Provinces*: Grundi—; *Chinese*: Pi Ma—; *Coimbatore*: Kotteimuttu—; *Danish*: Undertroee—; *Deccan*: Erund, Ind, Rund, Yarand—; *Dutch*: Wonderboom—;

Efik: Etighi-unenge, Eto-aran-ukebe—; *Egypt*: Kharwa, Kiki—; *English*: Castor-oil Plant, Palma Christi—; *Ewe*: Dzegbele, Dzongbati, Kasuwelti, Kasuworliti, Longo—; *Fanti*: Abronkruma, Adedenkuruma, Esusonkuruma, Sunsumnkuruma—; *French*: Avanacoë, Avanacu, Bois de carapat, Carapat, Grand ricin, Palma Christi, Paume Dieu, Pignon de Barbarie, Ricin, Ricin commun, Ricin vulgaire—; *French Guiana*: Palma Christi—; *French Guinea*: Diakoula—; *Fulah*: Diakoula—; *Fulani*: Derre, Diakoula, Kolakolaje gor'de, Zurmaje—; *Ga*: Abromkruma, Adedengkuruma—; *Genoa*: Ricin, Ricinu—; *German*: Wunderbaum—; *Gond*: Nerinda—; *Greek*: Kiki—; *Guam*: Agaliya—; *Gujerati*: Diveli, Diveligo, Tirki—; *Hamadan*: Kercheng—; *Hasada*: Ranggajara, Risajara—; *Hausa*: Chika gida, 'Dan kwasare, Kulakula, Zurma—; *Hindi*: Arand, Arandi, Arend, Arind, Erand, Erandi, Erend, Ind, Rand—; *Hova*: Tanantanankisoa—; *Hungarian*: Ricinus—; *Ibo*: Obwa, Ogba, Ogili-obwa, Ogili-ogba, Ogilisi, Ogiri-aro, Osisi-ogili—; *Ilocano*: Tangantangan, Tavatavangsina, Tawatawasinga—; *Indo-China*: Dau dau, Dù du, Huong thet, Khnhe, Lohong, Miet ma, Thu du tia—; *Iraq*: Khirwa—; *Italian*: Caffee da olio, Catapuzia maggiore, Erba venaria, Fagiolo d' India, Fagiolo romano, Fava d' India, Fico d' inferno, Girasole, Mirasole, Palma Christi, Palmaperta, Ricino—; *Kabure*: Assimballo—; *Kachhi*: Harnoli—; *Kalabari*: Alamba-ngbole—; *Kharan*: Murghpad—; *Kikuyu*: Mubariki—; *Kolami*: Jarabindi—; *Konkani*: Erendi—; *Koranko*: Maraban keyo barambe—; *Kordofan*: Hurna—; *Krepi*: Atornggor, Kasuwelti, Kasuwoliti, Yevu-tongo—; *Krobo*: Kumenglo—; *Kumaon*: Andi, Arandi, Indrendi—; *La Reunion*: Ricin tantan—; *Las Bela*: Hiranr—; *Lepcha*: Hilk bu ruk-lop, Raklop—; *Loanda*: Bafureira—; *Lokko*: Tawabawa—; *Makaran*: Murpad—; *Malabar*: Tonda—; *Malaya*: Jarak, Pee mah—; *Malayalam*: Anandam, Avanaka, Chittavanaku, Erandam, Gandharvahastakam, Kotta, Panjangulam, Pantiyavanakku, Varddhamanam—; *Malta*: Castor-oil Plant, Palma Christi, Ricino, Rionu—; *Marathi*: Erandi, Yarandicha—; *Masai*: Oldule—; *Mauritius*: Palma Christi—; *Mbonge*: Boku-balondu—; *Mende*: Bonde, Ngele-bonei—; *Mentone*: Ricinu—; *Mexican*: Tlaplatl—; *Mozambique*: Ambona—; *Mundari*: Bindidaru, Ranggabindi, Ranggafara—; *Naguri*: Rangga bindidaru, Risabindi, Sutamranggabindi—; *Nandi*: Imanyet—; *Nasirabad*: Harnauli—; *Nepal*: Alha, Areta, Orer—; *Nimar*: Arand—; *North America*: Castor Bean, Castor Plant, Palma Christi—; *North-Western Provinces*: Arand, Bhatreri, Rendi, Reri—; *Nupe*: Kpamfini gulu—; *Pahouin*: N'zoum—; *Pedi*: Mokhura—; *Persian*: Bedanjir, Bedanjira, Garchak, Karchak, Tochme kerchek—; *Portuguese*: Carrapateiro, Mammona, Palma Christi, Ricino—; *Portuguese Africa*: Bafureira—; *Potenza*: Ricin—; *Punjab*: Aneru, Arand, Arind, Bedanjir, Harnauli—; *Pushtu*: Arhand—; *Rajputana*: Arend, Edia—; *Reggio*: Rizoin—; *Roumanian*: Ritina—; *Russian*: Kleshtshexika—; *Sakalave*: Kinana, Tanantanamanga—; *Samoa*: Lama-papalangi—; *San Remo*: Ricin—; *Sanskrit*: Amanda, Amangala, Bhanda, Chankuka, Chitrabija, Chitraka, Dirghadantaka, Eranda, Gandharvahasta, Gandharvahastaka, Ishta, Kanta, Panchangula, Panjangula, Ruvuka, Shukla, Shulashatru, Svehaprada, Taruna, Triputi, Triputiphala, Tuchhadru, Vardhamana, Vatari, Vranaha, Vuka, Vyadatvaka, Vyaghradala, Vyaghrapuchha—; *Santali*: Eradom—; *Sarawan*: Bedanjir—; *Sardinia*: Cacamengiu—; *Sibi*: Harnauli—; *Sicily*: Rigginu—; *Sind*: Ayrunkukri, Heran—; *Sinhalese*: Eudaru, Telendaru—; *South Africa*: Castor Bean, Castoroil Plant, Palma Christi—; *Spanish*: Higuera del diablo, Higuera infernal, Palma Christi, Ricino—; *Sumatra*: Jarak—; *Susu*: Limbiaxule, Mbiakule—; *Suto*: Mokhura—; *Swahili*: Mbariki—; *Swedish*: Undertrard—; *Tagalog*: Lansina, Lingasina, Tangantangan—; *Tamil*: Aimugi, Amanakku, Andagam, Asaram, Attagam, Attamanam, Attugam, Erandam, Kottai, Kottaimuttu, Muttukkottai, Peramanakku, Sanju, Sasambari, Sigandi, Sittaman, Sittamanakku, Sittiram, Tabinjam, Urppulam, Vattamam—; *Telugu*: Amudamu, Chittamudamu, Erandamu, Peramudamu—; *Tigre*: Kella—; *Tigrinia*: Vulleh—; *Timne*: An-fentoel, Lamirenda—; *Tivi*: Harev, Ihurua dzengo, Ijija, Jija, Masev, Sherai jongo, Showara jongo—; *Treviso*: Rizin, Rizino—; *Tschaudjo*: Dendelle, Sau—; *Tuareg*: Fueni—; *Turkish*: Hint—; *Tuscany*: Erba de latte, Erba lattaria, Girasole maggiore, Girasole piccolo, Manoaperta, Meo, Scatapuzia, Zecca, Zeccha—; *Twi*: Abronkruma, Adadenkuruma—; *Urdu*: Eranda—; *Uriya*: Bheronta, Chitroko, Erondo, Gobo, Joda, Kalo, Monto—; *Tindoko*—; *Uruguay*: Ricino, Tartago—; *Verona*: Rizino—; *Waziri*: Randanh—; *Xosa*: umHlavuthwa—; *Yemen*: Jar, Tebsha, Zeit—; *Yoruba*: Ilara, Ilarun, Laa, Lapalapa adengteng, Lara—; *Zulu*: umHlakuva—.

SAPIUM

The genus includes 100 species, inhabiting warm countries.

S. sebiferum Roxb. is used medicinally in China and Indo-China, *S. Grahamii* Prain in Gold Coast. The latex of *S. ellipticum* Pax is in Kenya an ingredient in arrow-poison.

Leaves shortly petioled, elliptic-lanceolate, obtusely acuminate, serrate or subserrate	1. *S. indicum.*
Leaves elliptic or oblong-lanceolate, acuminate, crenate-serrate	2. *S. insigne.*
Leaves long-petioled, ovate, orbicular-ovate or sub-rhombic, quite entire, finely acuminate	3. *S. sebiferum.*

1. **Sapium indicum** Willd. is found in the Sunderbund and in the tidal forests of Tenasserim and Ceylon. It occurs on the sea coasts of the Malay Peninsula, and is distributed to the Malay Islands and New Guinea.

The juice of this tree is reckoned of a very poisonous nature. The taste of the fruit is nauseous beyond description. The seeds are used for intoxicating fish.

Bengal: Batan, Batul, Huru, Hurua—; *Bombay*: Hurna—; *Borneo*: Booroo—; *Hindi*: Hurua—; *Malay*: Gurah, Guring—; *Malayalam*: Karm-matti, Venkshiri—; *Sinhalese*: Kirimakulu, Kirrimakalu—.

2. **Sapium insigne** Trim. is found in the sub-Himalayan tract, from the Beas eastwards to Assam, Chittagong, Burma, and Ceylon, ascending to 4,300 feet.

The whole tree is full of an acrid milk which, when applied to the skin, produces vesication.

Almora: Khin—; *Anamalais*: Garpathola—; *Bombay*: Dudla—; *Canarese*: Kannupade, Kurda, Nanaka—; *Dehra Dun*: Khindra, Khinna, Khinni, Khirni—; *Garhwal*: Khindra, Khinna, Khinni, Khirni, Kinna—; *Hindi*: Khina, Khindra, Khinna, Khiria, Khirni, Khirun, Lendwa, Lienda—; *Kadir*: Karup-puchchulai—; *Marathi*: Dudla, Hure, Ura, Ure—; *North-Western Himalaya*: Boddar, Khinna—; *Punjab*: Bilodar, Biloja, Dudla, Karalla, Ledra—; *Telugu*: Garbhasula—; *Tulu*: Kannupade—.

3. **Sapium sebiferum** Roxb., indigenous to China and Japan, is found as a cultivated plant in various parts of India and elsewhere in warm countries.

The acrid juice is a powerful vesicant. The oil is reputed similar to castor oil.

In China the bark is used as a tonic, and the resin as a purgative.

In Indo-China the root bark in decoction is given in dyspepsia. The fat from the seeds is used in the treatment of skin diseases.

Bengal: Momchina—; *Chinese*: Pa Teou Seou, Wu Chiu—; *Dehra Dun*: Tarcharbi—; *English*: Chinese Tallow Tree—; *Hindi*: Bilaitisissu, Vilayatishis-ham—; *Indo-China*: O cuu moc, Sach, Soi, Soi tia u kiu, Soi trang—; *Mauritius*: Arbre à suif—; *North America*: Chinese Tallow Tree, Tankawang Fat Tree, Vegetable Tallow Tree—; *Saharanpur*: Pahari shisham—; *Sanskrit*: Agaru, Toyapippali—; *Uriya*: Ronojita—.

SAUROPUS.

The genus consists of 20 Indo-Malayan species.

Sauropus quadrangularis Müll.-Arg. is found in Bihar, Chota Nagpur, and the Western Peninsula; it is common in Burma.

The dried leaves are smoked in tonsilitis.

Matheran : Chickli—; *Telugu* : Tellavusirika—.

SEBASTIANIA.

The genus consists of 80 tropical species, mostly American.

Sebastiania chamaelea Müll.-Arg. occurs in Bihar, the Deccan Peninsula, Ceylon, Burma, and the Malay Peninsula. It is distributed to China, the Malay Islands, tropical Australia, and Africa.

The juice of the plant in wine is used as an astringent. A decoction of the plant in clarified butter is considered to be tonic, and is applied to the head in vertigo.

In Indo-China the plant is used in the treatment of syphilis and diarrhœa.

Konkan : Bhui crendi—; *Sinhalese* : Ratpitawakka—.

TRAGIA.

The genus includes 55 tropical and subtropical species.

T. volubilis Linn. is used medicinally in South America and in the West Indies.

T. *angustifolia* Benth. and *T. cordifolia* Benth. are considered toxic in Madagascar.

Tragia involucrata Linn., a common stinging weed, is found in dry places throughout India, from the Punjab and the lower Himalaya of Kumaon, eastwards to Assam and Burma, southwards to Travancore and Ceylon. It extends to China.

The root is considered diaphoretic and alterative; an infusion is given in ardent fever and in itching of the skin.

In the Konkan a paste of the root is used to aid the extraction of guinea-worm; mixed with the juice of *Ocimum sanctum* it is employed as a cure for itchy skin eruptions.

In Chota Nagpur the root is given when the extremities are cold during fever; also for pains in the legs and arms.

The root also forms the basis of an external application in leprosy.

The leaves enter into the preparation of an errhine which is prescribed in cases of headache.

The fruit rubbed over the head with a little water is useful in cases of baldness.

The root of this plant has diaphoretic property and is therefore given in fevers to cause perspiration. A decoction of this root 1 in 10 was found to be useful in relieving bronchitis and the attendant fever (Koman).

The fruit is not an antidote to scorpion venom (Caius and Mhaskar).

Bengal : Bichati, Bichiti—; *Bombay* : Kanchkuri, Khajkhotti—; *Canarese* : Dulagondi, Haligilu, Kiriberalu, Kiruberalu, Kiriturache, Sannaturachi, Turachi—; *Deccan* : Kanchkure—; *Hindi* : Baraanta—; *Ho* : Jipenda—; *Kolami* : Sengelsing—; *Malayalam* : Cherukodithuva, Chorinnanam, Choriyanam, Kotittuva—; *Marathi* : Kanchkuri, Khajkolti—; *Matheran* : Katavi, Kulti—; *Sanskrit* : Duralabha Dusparcha, Dustaparisha, Grahini, Kachchura, Kasaghni,

Samudranta, Virupa, Vrishchikali, Vrishchikapatri—; *Santal*: Sengelsing—; *Tamil*: Ambu, Cherukanjuru, Erumaikkanjori, Kanjori, Kandudi, Kannichi, Karuppukkanjori, Kunasagam, Kurundotti, Punaikkanjori, Samuttirandam, Sendotti, Siruganjori, Tanavaiyadam, Turalobam, Turpparigam, Vellaikkanjori—; *Telugu*: Chinnadulagondi, Dulagondi, Dulagundi, Peddadulagondi, Regadadula, Revatidulagondi, Telumani, Tigedulagondi—; *Tulu*: Pachcherengi, Turuse—; *Uriya*: Bichati, Bichchuati—.

TREWIA.

The genus consists of 2 Indo-Malayan species.

Trewia nudiflora Linn. is common in the hotter parts of India from Kumaon southwards and eastwards to Assam, Malacca, and Ceylon. It extends to Sumatra and Java.

The plant is cooling, tonic, alexiteric; improves taste; removes biliousness (Ayurveda).

The plant is used for the removal of swelling, bile, and phlegm. The root in decoction is given to relieve flatulence and is applied locally in gouty or rheumatic affections.

Bengal: Pitali, Pitoli—; *Bombay*: Bhillauri, Petari, Tumri—; *Burma*: Setkadon, Thitmyoka, Yehhmyot, Yemyot—; *Canarese*: Kadukanji, Kadukere, Kadukumbala, Kanji, Katkumbala, Katkumbla, Padye, Shillauri, Tumri—; *Dehra Dun*: Gamhar, Tumri—; *Haldwani*: Gutel—; *Hindi*: Bhillaura, Gambhar, Gamhar, Khamara, Pindara, Pitoli, Tumri—; *Kolami*: Garalohadaru—; *Kumaon*: Khamara, Tumri—; *Lepcha*: Tungplam—; *Magahi*: Hruprukban—; *Malayalam*: Kanji, Kattukumil, Malankumil, Pamarakkumil, Pambarakmubil—; *Marathi*: Petari, Pitari—; *Monghyr*: Gamhar—; *Mundari*: Garalupung—; *Nepal*: Gamari, Garum, Kurong—; *Oudh*: Bhillaur, Bhillaura—; *Saharanpur*: Dhaulpedda—; *Sanskrit*: Karahata, Karangaha, Pindara—; *Santali*: Gada lopong—; *Tamil*: Attarasu, Attupuvarasu, Kanji, Naykkumil, Sannattuvarai—; *Telugu*: Eruponaku—; *Tharu*: Bilur—; *Uriya*: Monda, Panigambhar, Pitalu, Pithaliya—.

MEDICINAL AND POISONOUS PLANTS OF INDIA: *WATERLILIES, POPPYWORTS, FUMITORIES.*

BY

J. F. CAIUS, S.J., F.L.S.

I.

The NYMPHAEACEÆ are aquatic perennial herbs, with a submerged rhizome. Several members of this family were venerated by the ancients, not only for the magnificence of their flowers and leaves, carpeting the surface of the tranquil waters, but also on account of their utility. Their young rootstocks contain an abundance of starchy, mucilaginous, and sugary matters, which render them nutritive. The flowers, which have a peculiar scent like brandy, are narcotic. The seeds, filled with a floury albumen, are edible. The family consists of 8 genera and 50 species inhabiting the temperate and tropical regions of the globe.

The medicinal waterlilies of the world belong to 5 genera:—BRASENIA (cosmopolitan, except in Europe); EURYALE (south-eastern Asia); NELUMBO (Pennsylvania to Colombia; Asia and Australia); NUPHAR (northern temperate and cold regions); NYMPHAEA (tropical and temperate regions). The genus NUPHAR is not represented in India.

A. Sepals and petals 3 each. Carpels free. Ovules few ... BRASENIA.

B. Sepals 4-6; petals and stamens indefinite. Carpels confluent with one another or with the disk into one ovary; ovules many.

  1. Sepals, petals and stamens half-superior, inserted on the disk which is confluent with the carpels ... NYMPHAEA.

  2. Sepals, petals and stamens superior. Carpels sunk in the torus ... ... EURYALE.

C. Sepals 4-5; petals and stamens indefinite. Carpels irregularly scattered, sunk in pits of the turbinate disk ... NELUMBO.

Nupharine, an amorphous physiologically inactive alkaloid of unknown constitution, has been isolated from the fresh rhizome of *Nuphar luteum* Sibth. & Sm. The seeds of this same plant contain nuphar-tannic acid. The amino-acid, l-asparagin, has been obtained from *Nelumbo nucifera* Gaertn.

BRASENIA.

The only species, **B. Schreberi** J. F. Gmel., is cosmopolitan, except in Europe. In India it is to be found in Bhutan at an altitude of 6,000 feet, and in the Khasia Hills at 4,500 feet.

The leaves are astringent, and have been used in phthisis and dysentery in North America.

In Indo-China the stems and leaves are ground and applied to boils and abscesses. The seeds, in decoction form, allay thirst, neutralise the effects of certain poisons, strengthen the gums; they are considered cooling, digestive, and anti-dysenteric.

China: Ch'un, Shui K'uei, Shun—; *Indo-China*: Chuyen—; *Malaya*: Sooi kwai—; *North America*: Frog-leaf, Water-leaf, Water-lily, Water-shield, Water-target—.

EURYALE.

The only species, **E. ferox** Salisb., extends from Kashmir to Oudh and Eastern Bengal, and thence to China.

The seeds are much used for their tonic, astringent, and de-obstruent properties.

In China an infusion of the seeds is used for ear-ache, and an infusion of the leaves is given after a painful delivery.

Bengal: Makhana—; *China*: Ch'ien Shih—; *English*: Foxnut—; *Hindi*: Machana, Makhana—; *Indo-China*: Ke dau, Khiem—; *Malaya*: Siew sat—; *Persian*: Mukareh, Mukhareh—; *Punjab*: Jewar—; *Sanskrit*: Makhanna, Padma—; *Telugu*: Mellunipadmanu—; *Urdu*: Makhana—; *Uriya*: Kunta-padma—.

NELUMBO.

The only Asiatic species, **N. nucifera** Gaertn. (=*Nelumbium speciosum* Willd.), extends from Persia eastwards to Australia, and is to be found throughout the warmer parts of India.

The plant is used medicinally all over the East. Nearly every part has a distinct name and economic use, and supplies one or more drugs.

The powdered *root* is prescribed for piles as a demulcent; also for dysentery and dyspepsia. It is used as a paste in ringworm and other cutaneous affections.

In Ceylon an aqueous extract of the fresh rootstock of the white-flowered variety is given internally in doses of 1 to 4 ounces for the bites of snakes and tarantulas. It is said to be especially useful in the bites of the cobra.

The flour of the root is used as an arrowroot by the Chinese. It is given in diarrhœa and dysentery; and it is the base of a diet for infants.

The long white rhizomes resemble a string of sausages, each about a foot long and separated from the others by a constricting fibre. These constricting fibres are used medicinally in China and are credited with the property of restoring to health persons suffering from nervous exhaustion. Their special action is regarded as hæmostatic.

The large *leaves* are used as cool bedsheets in high fever with heat and burning of the skin. The milky viscid juice of the leaf and flower stalks is given in diarrhœa.

The *flowers* are used as an astringent in diarrhœa, also cholera, in fever and diseases of the liver; and are also recommended as a cardiac tonic. Koman found that a syrup of the dried flowers in half ounce doses was efficacious in curing mild cases of dysentery.

In the Tamil country the flowers are considered diuretic and cooling. Boiled in milk with an equal amount of flowers of *Eugenia Jambolana* to a pasty consistence, and made into boluses about the size of an areca nut, they are administered to women for false pains in the early months of pregnancy.

The dried red petals are used by the Chinese as a cosmetic application to the face to improve the complexion.

The *stamens* are considered astringent and cooling, useful in burning sensation of the body, bleeding piles and menorrhagia. In bleeding piles they are administered with honey and fresh butter, or with sugar.

The carefully dried and yellow fragrant stamens are an astringent and diuretic remedy in China, where they are also used as a cosmetic. In Tongking they are employed for flavouring and improving the appearance of tea.

The *seeds* are used to check vomiting, and are given to children as a diuretic and refrigerant. They form a cooling medicine for cutaneous diseases and leprosy, and are considered an antidote for poisons.

In China the seeds, roasted and divested of their shells are a favourite article for dessert. Boiled and ground into flour the kernels form a valuable food and medicine, and are considered a general tonic recommended in diarrhœa, spermatorrhœa and erotic dreams.

The *plumule,* caulicle, or germinating embryo of the ripe seed is given in China to reduce high fever, and is said to be useful in the treatment of cholera, hæmoptysis and spermatorrhœa.

A sherbet of the plant is used as a refrigerant in small-pox, and is said to stop eruption; it is also used in all eruptive fevers.

Caius and Mhaskar have reported that no part of the plant is an antidote to either snake or scorpion venom.

Annam : Lien hoa, Sen—; *Arabic* : Karambe-ul-ma, Nilufer, Uss-ul-nilufer—; *Australia* : Aquaie, Pink Water-lily, Sacred Lotus—; *Bengal* : Kombol, Komol, Padama, Padma, Pudmapodu—; *Bijnor* : Besenda—; *Bombay* : Kamala, Kankadi—; *Burma* : Padung ma—; *Cambodia* : Chhuk—; *Canarese* : Kamala, Tavaribija, Tavarigadde—; *China* : Hoa Lien, Lien, Lien Jou, Lien Ou—; *Deccan* : Kungvel—; *English* : Chinese Water-lily, Egyptian Bean, Indian Lotus, Pythagorean Bean, Sacred Lotus—; *French* : Fève d' Egypte, Lis rose du Nil, Lotus sacré, Nénuphar de la Chine—; *Gujerati* : Suriyakamal—; *Hasada* : Kombolba, Upalba—; *Hindi* : Ambuj, Bhe, Kamal, Kanval, Kanwal Kakri, Kaul, Lalkamal, Padam, Padma—; *Ilocano* : Sucao—; *Khandesh* : Dudhamalidakand—; *Konkani* : Salloc—; *Malay* : Seroja, Teratai—; *Malayalam* : Bemtamara, Tamar—; *Marathi* : Kamal—; *Muzaffarnagar* : Pabbin—; *Persian* : Baklakoobtee, Beykhne-el-ufer, Nelofir, Nilufer, Nilufu—; *Poona* : Pandkanda—; *Pondicherry* : Nénuphar—; *Porebunder* : Motunkamal—; *Portuguese* : Flor de loto, Nenuphar—; *Punjab* : Bhen, Kanwalkakri, Pamposh, Pamposh kammal—; *Sanskrit* : Abja, Ambhoja, Ambhoruha, Ambuja, Ambujanma, Ambupadma, Amburoha, Amlana, Arvinda, Asyapatra, Bisakusuma, Bisaprasuna, Drishopadma, Harivetra, Indiralaya, Jalajanma, Kamala, Kawar, Kunja, Kusheshaya, Kutapa, Mahapadma, Mahotpala, Nala, Nalika, Nalina, Padma, Pankaja, Pankeruha, Pathoja, Pundarika, Pushkara, Putaka, Rajiva, Sahsrapatra, Sarojanma, Saroruha, Sarsija, Sarsiruha, Sharada, Sharapadma, Shatapatra, Shri, Shriparna, Shrivasa, Shuklapadma, Sitambuja, Sujala, Tumarasa, Vanashobhana, Varisoha—; *Sind* : Pabban—; *Sinhalese* : Nelumbo, Nelum—; *Tagalog* : Baino, Bayno—; *Tamil* : Ambal, Sivapputamarai, Tamarai—; *Telugu* : Kalung, Erratamara, Tamara—; *Urdu* : Nilufer—; *Uriya* : Padam—.

NYMPHAEA.

The genus consists of 40 tropical and temperate species.

Acrid, astringent, demulcent and diaphoretic properties are attributed to this genus.

The following species are used medicinally in Europe—*N. alba* Linn.—; in Egypt—*N. caerulea* Sav., *N. Lotus* Linn.—; in Nigeria, Sierra Leone, Senegal—*N. Lotus* Linn.—; in Guinea—*N. caerulea* Sav., *N. stellata* Willd.—; in Madagascar—*N. stellata* Willd.—; in China—*N. alba* Linn.—; in Indo-China—*N. alba* Linn., *N. stellata* Willd., *N. tetragona* Giorgi—; in North America—*N. alba* Linn., *N. polysepala* (Engelm.) Greene, *N. odorata* Ait.—.

Four species are used medicinally in India: *alba, pubescens, rubra, stellata.*

A. Flowers white, about 5-11.2 cm. diam. ... ... 1. *N. alba.*
B. Flowers red, pale rose or white, 7.5-20 cm. across ... 2. *N. rubra.*
C. Like *rubra* but leaves more densely pubescent beneath. Flowers smaller, 7.5-10 cm. across ... . 3. *N. pubescens.*
D. Flowers usually pale violet, less commonly blue or purple, 7.5-15 cm. diam. ... ... 4. *N. stellata.*

1. **Nymphaea alba** Linn. occurs in Kashmir, Siberia and Europe.

The mucilaginous and somewhat acrid root and stock are administered in some countries for dysentery. It is an astringent and slightly narcotic medicine, which is reputed to have aphrodisiac properties. In some parts of England the juice is made into an ointment to stimulate the scalp so as to prevent falling out of the hair.

The flowers are reputed to be anti-aphrodisiac, and to have cooling and astringent properties. They are given in Persia for fevers and chest troubles.

An infusion of the flower and fruit is administered in diarrhœa, and as a diaphoretic.

Akwapim: Ntanowa—; *Annam*: Kien thiet—; *Arabic*: Nilufar—; *Bombay*: Pandharenkamal—; *Catalan*: Cadells, Escudet, Escut de riu, Ninifa, Platis-giradors—; *China*: Hung Pai, Lien Hua—; *Danish*: Soeblomster—; *Dutch*: Witte Zeebloem—; *English*: Flatter-dock, Lotus-lily, Nenuphar, Water-rose, White Water-lily—; *French*: Baratte, Blanc d'eau, Burette, Crugeon, Herbe aux plateaux, Lis d'eau, Lis d'étang, Lis des étangs, Lune d'eau, Lunette, d'eau, Nénuphar blanc, Grand nénuphar, Nymphe, Plateau à fleurs blanches, Pyrote, Volant d'eau, Volet, Volet blanc—; *Ga*: Tertermantrer—; *German*: Wasser Lilie, Wasserrose, Weisse Seeblume, Weisse Seerose—; *Greek*: Nymphaia, Sidi—; *Hausa*: Bado—; *Hindi*: Nilofar—; *Indo-China*: Bach lien tu, Hat sen trang—; *Irish*: Duilleoghuidhebhaitighe—; *Italian*: Carfano, Nenufaro bianco, Ninfea, Papeo—; *Kashmir*: Brimposh, Kamud, Nilofar—; *North America*: White Pond Lily, White Water Lily—; *Persian*: Nilufar—; *Polish*: Grzybienie, Wodna lilia—; *Portuguese*: Niniphea branca—; *Punjab*: Kammian—; *Roumanian*: Plumiera alba, Plumuna alba, Pluta, Nufar alb, Titva de apa—; *Russian*: Kubishka, Wodanoi lelei—; *Spanish*: Escudete de rio, Higos de rio, Nenufar blanco, Ninfea blanca—; *Swedish*: Sjoebad—; *Teheran*: Nilafar—.

2. **Nymphaea rubra** Roxb. is common throughout India in the warmer parts.

The powdered rootstock is given in dyspepsia, diarrhœa and piles.

A decoction of the flowers is prescribed in palpitation of the heart.

The flower is much used by Ayurveda practitioners. They prescribe it to allay thirst, cough, and vomting. They use it as a cooling and antibilious medicine; as a blood purifier, a febrifuge, and an aphrodisiac.

Arabic: Nulufar—; *Bengal*: Buroruktokumbal, Chotasundi, Nal, Raktakambal, Saluk, Shaluk—; *Burma*: Kiyani, Kyahphyu—; *Canarese*: Bilitavarai, Nyadalehuvu—; *Deccan*: Alliphul—; *Gujerat*: Kanval, Nilophal—; *Hasada*: Kombolba, Upalba—; *Hindi*: Chotakanval, Kanval, Raktachandana—; *Malayalam*: Ampala—; *Marathi*: Lalakamal, Raktakamal—; *Orissa*: Dhabalakain, Rangkain—; *Persian*: Nilufar—; *Sanskrit*: Alagandha, Alipriya, Alohita, Alpapatra, Aravinda, Arunakamala, Bhadra, Charunalaka, Hallaka, Indivara, Kalharamu, Kamala, Kokanada, Krishnakanda, Kumuda, Kuvalaya, Nilotpala, Raktakairava, Raktakalhara, Raktakamala, Raktakumuda, Raktambhoja, Raktasandhyaka, Raktasaroruha, Raktavarija, Raktavarna, Raktotpala, Ravipriya, Rochana, Sandhyaka, Shonapadma, Somakhya—; *Sind*: Kuni, Puni—; *Sinhalese*: Olu, Otu—; *Tamil*: Allitamarai, Ambal—; *Telugu*: Allitamara, Errakaluva, Kalharamu, Koteka, Tellakalava—.

3. **Nymphaea pubescens** Willd. is found all over India in the warmer parts.

The powdered root is prescribed for piles as a demulcent; also for dysentery and dyspepsia.

The flowers are astringent and cardiotonic.

Bengal: Burro, Shaluk—; *Bombay*: Kummul—; *Hindi*: Bombhur, Kawal zutto—; *Madras*: Alli—; *Persian*: Nelofir—; *Sind*: Koonee—.

4. **Nymphaea stellata** Willd. inhabits the warmer parts of India, and extends to tropical Asia and Africa.

In India the plant is used in the same ways and for the same purposes as *N. rubra*. It is moreover much employed to stimulate and promote the growth of the hair.

In Indo-China the seeds are considered a powerful stomachic and restorative.

In Cambodia the maceration of the leaves is used as a lotion in eruptive fevers. In Madagascar the leaves are applied topically in erysipelas.

In Guinea the infusion of the roots and the stems is considered emollient and diuretic; it is taken against blennorrhagia and diseases of the urinary tract. A decoction of the flowers is considered narcotic and antiaphrodisiac.

Bengal: Chotashaluk, Nilpodma, Nilsaphala, Nilsapla, Sundhi—; *Bijnor*: Bambher—; *Bombay*: Upliakamal—; *Burma*: Kyanyu—; *Cambodia*: Pralit—; *Ceyon*: Niltelolu, Katutelolu, Sudatelolu—; *English*: Blue Water-lily—; *Gujerati*: Nilkamal—; *Hindi*: Bambher, Bhengt, Lilophal, Nilkamal, Nilophal, Nilpadma—; *Hova*: Betsimihilana, Tantama, Tsiazonafo, Voahirana—; *Imerina*: Voalefoka—; *Indo-China*: Cu sung nho, Khiem tuc—; *Jodhpur*: Be—; *Madras*: Karunaydel, Nilotpalam—; *Malay*: Ati-ati payah, Kelipoh, Teratai Kechil—; *Malayalam*: Sitambel—; *Malinke*: Kuludion, Kulukulu—; *Marathi*: Krishnakamal, Poyani—; *Porebunder*: Kalakamal, Kamal, Kumtan—; *Sakalave*: Tatamo—; *Sanskrit*: Asitotpala, Indiwar, Kandota, Kandotha, Kuvalaya, Mridutpala, Nilakamala, Nilaliya, Nilambujanma, Nilapadma, Nilapatra, Nilotpala, Nilpankaja, Nilpatraka, Saugandhika, Utpala, Utpalaka—; *Sinhalese*: Manel, Monch, Nilupul—; *South Africa*: Water-lily—; *Tagalog*: Lauas—; *Telugu*: Nitikulava—; *Uriya*: Subaikain—; *Zulu*: Izibo—.

II.

The PAPAVERACEÆ inhabit the temperate and subtropical regions of the northern hemisphere; but few are met with in the tropics or southern hemisphere. Some species are now dispersed over cultivated ground throughout the world. The family consists of 28 genera and 600 species—annual or perennial herbs, rarely shrubs or small trees.

The members of this Order have sedative, narcotic, stupefying, at times caustic and rubefacient, properties. The seeds are oleaginous, some emetic and cathartic.

The following substances are among the products isolated:—(1) *alkaloids*—α and β—allocryptopines, aporeidine, aporeine, berberine, chelerythrine, chelidonine, homochelidonine, codamine, codeine, cryptopine, dicentrine, glaucidine, glaucine, gnoscopine, lanthopine, laudanidine, laudanine, laudanosine, meconidine, morphine, pseudomorphine, narceine, narcotine, neopine, oxynarcotine, papaveramine, papaverine, protopapaverine, pseudopapaverine, protopine, rheadine, sanguinarine, thebaine, isothebaine, tritopine—: (2) *bitter substances*—chelidoxanthin, glaucopicrin, papaverosin—; (3) *acids*—erratic, fumaric, meconic, rhoeadinic, sulphuric thebolactic—; (4) *fixed oils*; (5) *dyes*; (6) *gums, resins, starch, mucilage*; (7) *neutral principles*—meconin, meconoisin, opionin—; (8) *glucose sugar*; (9) *cyanogenetic glucosides.*

The medicinal and poisonous poppyworts of the world belong to 10 genera:—ARGEMONE (tropical America); BOCCONIA (warm America; West Indies); CHELIDONIUM (Europe; Asia Minor); ESCHSCHOLZIA (Pacific North America); GLAUCIUM (Europe; Asia); HYPECOUM (Mediterranean; Central Asia); MECONOPSIS (northern temperate regions); PAPAVER (Europe, Asia, America, South Africa, Australia); ROMNEYA (California); SANGUINARIA (Atlantic North America).

The medicinal and poisonous poppies of India belong to four genera: ARGEMONE, HYPECOUM, MECONOPSIS, PAPAVER.

A. Capsule usually short, opening by short valves or pores
  1. Stigmas 4 or more, radiating on a sessile disk ... PAPAVER.
  2. Stigmas 4-6, radiating from the top of a depressed style ... ARGEMONE.
  3. Stigmas 4-6, decurrent on the top of the style ... MECONOPSIS.

B. Capsule slender, transversely septate within, breaking up into joints or longitudinally 2-valved ... ... HYPECOUM.

ARGEMONE.

The genus includes 12 species, natives of Tropical America.

A. alba Lestib. is used medicinally in North America, and *A. mexicana* Linn. is so used in every country where it is found growing.

Argemone mexicana Linn., indigenous in tropical America, was introduced in India where it has become naturalized, and where it may be found throughout up to an altitude of 5,000 feet.

Far from being treated as a foreigner the plant occupies a very honourable place in the Materia Medica of both Hindus and Mahommedans. Hakims and vaids are at one in their encomium of this valuable medicinal herb.

The root is an alterative. The Hausas in Nigeria use an infusion of it as a wash for drummers who have to sustain their vigour through an all-night play. Among the Yorubas the root is chiefly used as a children's remedy, as a wash, or worn as a charm, or by fumigation in holding the head over a steaming pot containing this along with other medicines; also as an infusion taken internally by a child who is fretful or frightened at night. Women too are given an infusion to drink at the commencement of the pains of parturition.

The use of the root is attended with benefit in some chronic cases of skin diseases. In La Reunion it is given as a decoction in blennorrhagia; and in French Guiana this same decoction is prescribed for vesicular calculus. In French Guinea the decoction is used as an eye-wash and lotion for inflammatory swellings; also as a mouth-wash for tooth-ache, and taken internally for gleet.

In Gold Coast the pulverised root mixed with onions and applied to the guinea-worm sore is said to be effective in extracting the worm.

In La Reunion and in French Guinea the stem is considered diuretic; a decoction of it is prescribed for vesicular calculus in French Guiana.

The Mandingos of the Gambia use an infusion of the leaves for coughs.

In French Guiana the flowers are considered narcotic. The honey made by bees which visit this flower is said by the Hausas to make men mad.

The latex has slight corrosive properties, and is applied to warts, chancres, etc.

The seeds are laxative, emetic, nauseant, expectorant and demulcent. They are useful in cough and catarrhal affections of the throat and pulmonary mucous membrane, and in pertussis and asthma. Though they do not appear to possess any anti-spasmodic property, they have a distinct control over asthma, apparently from their combined actions of nauseant, emetic, expectorant and demulcent. As their use is often accompanied by more or less vomiting and nausea, they are as a laxative medicine more suited to some pulmonary affections than to other diseases.

In Mexico the seeds are considered an antidote to snake venom. In French Guiana they are used as a cathartic and an emetic.

The smoke from the burning seeds is used in Delhi to relieve tooth-ache; it is also said to be useful in caries of the teeth.

The seeds yield on expression a fixed oil, which has long been in use amongst West India and West Africa practitioners as an aperient. Applied to herpetic and other forms of skin disease, it is reported to exercise a well-marked soothing influence. It is serviceable in some cases in which jalap, rhubarb and castor-oil are indicated, and also in some bronchial and catarrhal affections. In Delhi and in Sind the oil is applied to indolent ulcers and eruptions, also as an external application for headaches.

The yellow juice of this plant is used as a medicine for dropsy, jaundice, and cutaneous affections. It is also diuretic, relieves blisters, and heals excoriations and indolent ulcers. It is held in much esteem as a local application to indolent and ill-conditioned ulcers. It is also used as an external application to the eyelid in conjunctivitis. In the Konkan, the juice with milk is given in leprosy.

In Jodhpur the yellow juice is used in eye affections and rubbed on the body to relieve rheumatic pain. In the West Indies it is reported to be used as a substitute for ipecacuanha.

The seeds are not an antidote to snake venom (Mhaskar and Caius).

The alkaloids of *A. mexicana* have been correctly identified as berberine and protopine (A. C. Santos and P. Adkilen; *Journ. Am. Chem. Soc.*, July 1932).

Annam : Ca gai—; *Arabic* : Bagel—; *Ashanti* : Akusiribie, Kokosekiyatuduru —; *Bengal* : Baroshialkanta, Karikanta, Shialakontha, Shialkanta, Siakanta, Siyalkanta—; *Brazil* : Cardo santo—; *Burma* : Khyaa—; *Canarese* : Balurakkisa, Datturi, Datturi-gidda, Mulludaturi, Parangidatturi—; *Cape Peninsula* : Bathurst Burweed, Steekbossie—; *Chinese* : Lao Chou Li—; *Colombia* : Cardosanto—; *Deccan* : Bharamdandi, Daruri, Farangidhatura, Pila, Piladhatura—; ***English*** : Gamboge Thistle, Jamaica Yellow Thistle, Mexican Poppy, Prickly Poppy, Yellow Mexican Poppy—; *Falor* : Bidjor—; *French* : Chardon bénit des Antilles, Figuier infernal, Pavot épineux, Pavot massue, Pavot du Mexique, Tache de l'oeil—; *French Guiana* : Argémone—; *Fulani* : Nyie gilangeru—; ***German*** : Mexicanische Argemone, Stachelmohn—; *Gujerat* : Darudi—; *Hasada* : Rangainijanum—; *Hausa* : Hakorin kada, Ka ki ruwan Allah, Kamukamu, Kankama, Kankamarka ta bi ka, Kaya geiva, Kurar fatake, Ya ki ruwan Allah—; *Hindi* : Bharbhand, Biladhutura, Brahmadundi, Brahmi, Farangidhutura, Kutaila, Kutila, Satiyanashi, Shialkanta, Suchianas, Ujarkanta—; *Ilocano* : Cachumba, Chicalote, Casubangaso, Casubhangaso—; *Indo-China* : Lao thu lac—; ***Jodhpur*** : Sattiyanasi—; *Kano* : Kwarko, Kwarkoro, Wagiri—; *Kathiawar* : Darudi—; ***Katsina*** : Karanko, Kwarakko, Kwaranko—; *Kolami* : Karikanta, Siyalkanta—; ***Konkani*** : Firingidutro, Redonoxi—; *La Reunion* : Chardon—; *Malayalam* : Brahmadanti—; *Malinke* : Ba, Guame-moni—; *Mandingo* : Tombong suma—; ***Marathi*** : Daruri, Firangidhotra, Kantedhotra, Kontedhotra, Pinvaladhotra—; ***Mexico*** : Cardo santo, Chicalotl, Figo del inferno—; *Mundari* : Bakulajanum—; ***North America*** : Jamaica Yellow Thistle, Mexican Poppy, Mexican Prickly Poppy, Thorn Poppy, Yellow Poppy-thistle—; *North-West Africa* : Devil's Fig, Golden Thistle of Peru, Mexican Poppy, Prickly Poppy—; *North-Western Provinces* : Bharbhurwa, Kantela, Karwah—; *Oudh* : Karwah—; *Portuguese* : Figo do inferno— *Punjab* : Bhatkateya, Bhatmil, Bherband, Kandiari, Katsi, Satyanasa, Sialkanta—; *Sanskrit* : Brahmadandi, Bramhi, Hemadugdha, Hemashikha, Hemavati, Hemavha, Kanchanakchiri, Kanchani, Karhini, Katuparni, Kshirini, Patuparni, Pitapushpa, Rukmini, Srigalakanta, Suvarna, Svarnadagdha, Svarnakshiri, Svarnavha, Tiktadugdha, Yavuchincha—; ***Santali*** : Gokulapanum—; *Sinaloa* : Cardo, Chicalote—; *Sokoto* : Boginahi—; ***Spanish*** : Adormidera espinosa, Cardo santo, Fico del inferno, Figo del inferno—; ***Suto*** : Nthswantsane, Sehlohlo-se-seholo—; *Tagalog* : Diluario, Duluario—; *Tamil* : Bramadandu, Bramarayakash, Kurukkum—; *Telugu* : Brahmadandi— *Twi* : Anansemmenyan—; *Urdu* : Baramdandi—; *Uriya* : Dengbejari, Kantakusham, Sarpuni—; *Wolof* : Homhom bu djiguen, Hompay bu djiguen—; *Yoruba* : Akawn-ekkun, Egunarigbo, Ekan-chun, Kada ka taba yaro—.

HYPECOUM.

The genus includes 18 species of the Mediterranean region, distributed to Western and Central Asia.

1. Exterior petals elliptico-rhomboid, entire. Lobes of leaves linear-awl-shaped ... ... *H. pendulum.*
2. Exterior petals trilobed. Lobes of leaves linear-lanceolate or lanceolate, acute, mostly entire ... ... *H. procumbens.*

1. **Hypecoum pendulum** Linn. is found in Waziristan and Baluchistan, and extends to the Mediterranean region and the south of Europe.

An extract is used as a cooling drink by the Waziris.

Waziri : Margaypal, Zirgulaki—.

2. **Hypecoum procumbens** Linn. occurs in Peshawar, Multan, the Salt Range, and Baluchistan. It spreads over Western Asia as far as the Mediterranean region.

The juice has the same effect as opium. The leaves act as diaphoretic.

In Bavaria the plant is used as a tonic, aperient, and suborific.

English : Horned Cumin—; *Italian* : Cornacchina de grani, Cornacchina verdemare—; *Malta* : Cornacchina, Karn il Moghza—.

MECONOPSIS.

The genus includes 30 species of the northern temperate region.

1. Radical leaves linear-oblong or lanceolate, remotely and irregularly pinnatifid-lobed ... ... 1. *M. aculeata.*
2. Radical leaves pinnatipartite ... ... 2. *M. napaulensis.*

I. **Meconopsis aculeata** Royle occurs in Kashmir, Garhwal and Kumaon ascending up to 11,000-15,000 feet.

The plant, especially the root, is considered narcotic and poisonous.

Jhelum : Guddikum—; *Kumaon* : Kanda—; *Ravi* : Gudi—; *Simla* : Kanta—; *Sutlej* : Kanada—.

2. **Meconopsis napaulensis** DC. is found at 9,000-10,000 feet in Nepal, Sikkim and Bhutan, whence it extends to Sze-chuan.

The root is used as a narcotic.

PAPAVER.

The genus numbers 110 species scattered over Europe, Asia, Australia, America, and South Africa.

The flowers and flower heads are sedative; the milky juice is narcotic, sometimes caustic. Every species of poppy yields opium to a greater or less extent.

The following are used medicinally:—in Europe—*P. alpinum* Linn., *P. Argemone* Linn., *P. dubium* Linn., *P. hybridum* Linn., *P. somniferum* Linn.—; in Indo China—*P. somniferum* Linn.

A. Leaves 5-10 cm., all radical, obovate or oblong, pinnatifid. Flowers orange-yellow. Capsule obovoid, hispid. 7. *P. nudicaule.*

B. Leaves 2-3-pinnatifid. Flowers 2.5-5 cm. diam., scarlet with a black disk. Capsule 13 mm. long, subglobose, hispid. Stigma convex, rays 4-8 ... ... 4. *P. hybridum.*

C. Leaves 1-2-pinnatifid. Flowers 7.5-10 cm. diam., scarlet; pairs of petals unequal. Capsule stalked. Stigma convex, rays 8-12 ... ... 1. *P. Rhoeas.*

D. Leaves 1-2-pinnatifid. Petals scarlet in unequal pairs. Capsule obovoid, glabrous, sessile. Stigma 6-12-rayed ... 2. *P. dubium*

E. Leaves oblong, amplexicaul, lobed, 2-serrate. Flower large, white, purple or scarlet. Capsule 2.5 cm. diam., stalked, globose, glabrous. Stigmatic rays 5-12 ... 5. *P. somniferum.*

F. Leaves bipinnatisect with acute lobes. Petals pale scarlet. Capsule oblong-elliptical, sparingly beset with weak straight bristles ... ... 3. *P. Argemone.*

G. Leaves subcoriaceous, hispid on both sides, pinnatipartite. Flowers large, more than 10 cm. across. Petals 4-6, all scarlet or at the base with a black violaceous spot. Capsule glabrous, glaucous, subglobose, about 2-3 cm. diam. Stigmatic rays 13-15 ... ... 6. *P. orientale.*

1. **Papaver Rhoeas** Linn. is grown in gardens in Gujerat and Northern India. It is common in Kashmir, Tibet, Western Asia, Europe, and North Africa.

There is little to be found in Indian works about this poppy. Ayurvedists credit it with the same properties as *P. somniferum.* Yunanists use the leaves and seeds as tonics in low fevers, and recommend the cooked roots for rheumatism.

Quoth Culpeper of English fame: 'A syrup is made of the seed and flowers, which is useful to give sleep and rest to invalids, and to stay catarrhs and deflusions of rheums from the head into the stomach and lungs, which causes a continual cough, the forerunner of consumption; it helps hoarseness of the throat, and loss of voice, which the oil of the seed does likewise. The black seed boiled in wine, and drunk, stays the flux of the belly, and women's courses. The poppy-heads are usually boiled in water, and given to procure rest and sleep; so do the leaves in the same manner; if the head and temples be bathed with the warm decoction, or with the oil, the green leaves or heads bruised, and applied with a little vinegar, or made into a poultice with barley-water, or hog's-grease, cools and tempers all inflammations, and St. Anthony's fire. It is generally used in treacle and mithridate, and in all other medicines that are made to procure sleep and rest, and to ease pains in the head as well as in other parts. It cools inflammations, and agues. It is put in hollow teeth, to ease the pain; it is also good for the gout'.

Robert Turner says 'how excellent is that flower in diseases of the pleurisic with similar surfeits hath been sufficiently experienced'. But Gerard writes: 'most men being led rather by false experiments than reason, commend the flower against the Pleurisic, giving to drink as soon as the pain comes, either the distilled water, or syrrup made by often infusing the leaves. And yet many times it happens, that the paine ceaseth by that meanes, though hardly sometimes'.

The petals are, no doubt, sudorific; and they have mild sedative properties.

The milk from the capsules has, in a weak form, the same properties as opium; it is used in France as a sedative.

In England the bruised leaves, if applied to a part which has been stung by a bee or a wasp, are said to give prompt relief.

A non-toxic alkaloid, rhoeadine, occurs in all parts of the plant.

Anglo-Saxon: Chebole, Chesebolle, Chybolle—; *Arabic*: Khashkhashulsuda, Nabatulkhashkhash—; *Bengal*: Lalposht—, *Bombay*: Janglimudrika—; *Burma*:

Bhainbinami, Bhinbinami—; *Canarese*: Kempugasgase, Kempukhasakhasi—; *Catalan*: Caputxins, Gallarets, Paparotes, Parama, Pipiripip, Quicaraguichs, Quiquiriquic, Roellas, Rosella, Ruella—; *Chinese*: Li Ch'un Ts'ao—; *Deccan*: Lalkhaskhas—; *Dutch*: Heul Klaproos, Kankerbloem, Kolbloem, Roode Koorenbloem, Wilde Heul Klaproos—; *Egypt*: Qarun, Zaghlil—; *English*: Bledewort, Blind Eyes, Blindy Buff, Blindy-buffs, Canker, Canker Rose, Cheesebowl, Cockrose, Cock's-comb, Collin-hood, Copper-rose, Cop Rose, Corn Flower, Corn Poppy, Corn Rose, Cuprose, Cusk Darnel, Ear-aches, Fireflout, Head Ache, Head Foppy, Head Wark, Joan Silverpin, Lightnings, Maws, Poison Poppy, Pope, Red Weed, Ridweed, Scarlet Poppy, Soldiers, Thunder Bolts, Yedwark—; *French*: Babiot, Babiou, Babot, Babou, Cornrose, Confanon, Coprose, Coq, Coq ponceau, Coquelicot, Coquelicot-pavot, Gravelotte, Gueringuingo, Mahon, Moine, Onder, Pavot coq, Pavot-coquelicot, Pavot des champs, Pavot rouge des champs, Pavot rouge sauvage, Ponceau, Rose de loup—; *German*: Ackerschnalle, Blatsblume, Blutmohn, Boschblume, Fallblume, Feldmangenblume, Feldmohn, Feldrose, Feuerblume, Feuermohn, Flattermohn, Flitschrose, Gaublume, Glatschen, Glatzenblume, Glitsche, Grimmagblume, Haferkrautblaum, Halsrose, Hirnschalblume, Hirnschnalz, Jammerblume, Juffer, Klapperrose, Klatschmohn, Klatschrose, Kohlrose, Kokeschblume, Koklico, Koleblume, Kornmohn, Rote Kornrose, Maagsamen, Makufken, Wilder Mohn, Paterblume, Pflappenrose, Pugerlitze, Schnallen, Schnellerblume, Windmohn—; *Greek*: Roias—; *Gujerat*: Lala, Lalkhaskhas—; *Hindi*: Lal, Lala, Lalpost, Post, Postekebija—; *Hungarian*: Mak—; *Italian*: Bambagella, Bubboline, Citole, Papavero selvatico, Papavero serchione, Rosolaccio, Scitole—; *Languedoc*: Graousella, Graouselle, Roalo, Rouzello—; *Malayalam*: Chovannakashakasha—; *Malta*: Corn Poppy, Poppy, Papavero, Rosolaccio, Pepprin—; *Marathi*: Tambadakhasakhasa—; *Persian*: Gulelalakebija, Khashkhashsiyah, Koknaresurkh—; *Polish*: Maczek—; *Portuguese*: Papoila, Papoila vermelha, Papoula—; *Roumanian*: Mac iepuresc, Mac rosiu, Paparoane, Paparuna—; *Russian*: Mak samosyika—; *Sanskrit*: Raktapostavrikshaha—; *Spanish*: Ababol, Adormidera silvestre, Amapola, Arabol—; *Tamil*: Siguppuppostaka, Sivappugashagasha—; *Telugu*: Erragassagassala, Errapostakaya—; *Turkish*: Gelincik—; *Urdu*: Gulelalakenija, Khashkhashsiyah—; *Waziri*: Kolejhdarai, Zirdgul—.

2. **Papaver dubium** Linn. is found in the Western Himalaya from Hazara to Kashmir and Garhwal. It extends to Afghanistan, Western Asia, Europe and North Africa.

The petals are sudorific.

Two alkaloids, aporeine and aporeidine, have been isolated from this plant. The former is described as a tetanising poison similar to thebaine.

English: Pale-red Poppy, Smooth long-headed Poppy—; *Malta*: Long-headed Poppy—; *Portuguese*: Dormideira, Papoula—.

3. **Papaver Argemone** Linn. is indigenous in Europe in the Mediterranean region. It is cultivated in Indian gardens.

In Spain an infusion or a syrup of the petals is much esteemed as a sudorific.

English: Rough-headed Poppy, Rough long-headed Poppy—.

4. **Papaver hybridum** Linn. occurs in the Punjab in the Jhelum Valley. It is distributed over Central and Western Asia, Europe, and North Africa.

The petals are diaphoretic.

The species contains the non-toxic alkaloid rhoeadine and another base.

Egypt: Abunnom, Aul el num—; *English*: Rough-headed Poppy, Rough round-headed Poppy—; *Malta*: Rough Poppy, Pepprin—.

5. **Papaver somniferum** Linn. is cultivated in the warm and temperate regions of Asia, Europe and North Africa. It is doubtful whether the plant has ever been found wild. In India it is grown here and there throughout the country. In France a form is specially grown on account of its oil-yielding seed.

Opium, the inspissated juice obtained by incision from the immature capsule, is well known for its therapeutic and poisonous properties, and is official in all Pharmacopoeias.

The seeds are used medicinally in Jodhpur, and the capsules are soaked in water which, after being strained, is taken as an intoxicant.

In Indo-China the roasted seeds are considered tonic.

Poppy heads are used in China for diarrhoea, dysentery, and all kinds of fluxes. The seeds ground in cold water are given in diarrhoea and dysentery.

A weak decoction of the dried capsule is good for those complaints which require sedatives, and native practitioners in India administer it as a restrainer in bowel troubles. In Europe, however, the decoction is not given internally, but is employed as an external application to allay pain and soothe.

In Europe syrup of poppy is still often employed to allay cough and likewise as an opiate for children. This syrup is prepared from the capsules. Another syrup, made of the leaves and flowers, is also used in pleurisy and erisypelas.

The green leaves, the fresh flowers, the crushed capsules are all made into various ointments, or used as such for poulticing.

Opium, in combination with other drugs, is recommended for the treatment of snake bite and scropion sting. Caius and Mhaskar have proved experimentally that the drug is not an antidote to snake venom, and is useless in the treatment of scorpion sting.

Arabic: Abunom, Afiun, Bizrulkhashkhash, Khashkhashulbaiza, Khish khash, Qishrulkhashkhash—; *Bengal*: Pasto, Post—; *Bombay*: Aphim, Appo, Khaskhas, Post—; *Burma*: Bhain, Bhainzi—; *Canarese*: Afim, Biligasgase, Gasagase, Khasakhasi—; *Catalan*: Cascall Pintacogues—; *Chinese*: Ying Tzu Su—; *Danish*: Valmue—; *Deccan*: Afim, Khashkhash—; *Dutch*: Heul, Slaapruid—; *Egypt*: Abun-nom, Shikh-shakh—; *English*: Bale-wort, Carnation Poppy, Joan Silverpin, Opium Poppy, Peony Poppy, White Garden-poppy, White Poppy—; *French*: Pavot, Pavot blanc, Pavot des jardins, Pavot à opium, Pavot somnifère—; *German*: Mohn, Saatmohn, Schlafmohn—; *Greek*: Agria, Milkon hymeros—; *Gujerat*: Aphina, Khuskhus, Posta—; *Hamadan*: Khash-khash—; *Hindi*: Afin, Afyun, Kashkash, Pest, Post, Potekebija—; *Indo-China*: A phien, A phu dung, Co tu tuc—; *Italian*: Papavero, Papavero domestico—; *Kachhi*: Doda, Post—; *Kumaon*: Posht—; *Malaya*: Bungapion, Yin soo hock—; *Malayalam*: Afiun, Kashakasha—; *Malta*: Opium Poppy, Poppy, Papavero, Pianta da oppio, Papavru, Xahxieh—; *Marathi*: Aphu, Khuskhus, Posta—; *Nepal*: Aphim ; *Oudh*: Posta—; *Persian*: Afiun, Khashkhash, Khashkhash-sufaid, Koknar, Postekoknar, Tukhmekoknar—; *Portuguese*: Dormideira, Papoula branca—; *Punjab*: Afim, Doda, Khashkhash, Khishkhash, Post—; *Roumanian*: Mac, Mac somnisor, Somnisor—; *Russia*: Mak snotvornyi—; *Sanskrit*: Ahifen, Chosa, Khasa, Khakasa, Ullasata—; *Sinhalese*: Abin, Albinatta—; *Spanish*: Adormidera, Dormidera—; *Swedish*: Vallmo—; *Tamil*: Abini, Gashagasha, Kasakasa, Postaka—; *Telugu*: Abhini, Gasagasala, Gasalu, Kasakasa—; *Turkish*: Hashash—; *Urdu*: Khashkhashsufaid—; *Uruguay*: Adormidera, Amapola—; *Waziri*: Afin, Kolijhdarai—,

6. **Papaver orientale** Linn. is cultivated in Indian gardens. It is common in North Persia, Armenia, and the Mediterranean region.

The petals are sudorific.

A series of alkaloids have been isolated from the plant: thebaine, isothebaine, glaucidine, two phenolic bases, and one non-phenolic base.

7. **Papaver nudicaule** Linn. is found in the Northern Himalaya, up to an altitude of 17,000 feet. It extends through Afghanistan to Northern Asia and to Central and Northern Europe.

The flowers and capsules are mildly diaphoretic.

Waziri: Kakavai—.

III.

The FUMARIACEÆ are often considered as a sub-family or a tribe of the Papaveraceae. They are annual or perennial herbs broadly distributed in temperate and warm regions, mainly in the temperate zone of the northern hemisphere. The family includes 7 genera and about 260 species.

Most fumitories contain in their herbaceous parts mucilage, saline substances, and a peculiar acid or acrid juice, so combined that they are classed among tonic and alterative medicines.

The following have been isolated from members of the Order:—(1) *alkaloids*—bulbocapnine, corybulbine, isocorybulbine, corycavamine, corycavidine, corycavine, corydaline, dehydrocorydaline, corydine, corypalmine, corytuberine, dicentrine, fumarine, protopine, sanguinarine, d-tetrahydropalmatine—; (2) *acids*—fumaric.

The medicinal fumitories belong to three genera: CORYDALIS (Mediterranean region, Europe, Asia); DICENTRA (Asia, North America); FUMARIA (Europe, Asia, Africa; chiefly Mediterranean).

No Indian DICENTRA is credited with therapeutic properties.

1. One outer petal spurred; capsule many-seeded ... CORYDALIS.
2. One outer petal spurred; fruit indehiscent, nut 1-seeded. FUMARIA.

CORYDALIS.

The genus numbers about 140 species, natives of the Mediterranean region, inhabiting Europe and Asia.

The genus is tonic, diuretic, and alterative. The roots are used as emmenagogues and vermifuges.

The following species are employed medicinally in Europe—*C. bulbosa* Pers., *C. claviculata* Pers., *C. fabacea* Pers., *C. lutea* DC., *C. ochroleuca* Koch.; in China—*C. ambigua* Cham. and Sch., *C. incisa* Pers.—; in Indo-China—*C. bulbosa* Pers.—.

Some twenty alkaloids have been isolated from various Corydalis roots.

1. Stems erect. Flowers 2.5 cm. long ... ... 1. *C. Govaniana*.
2. Stems procumbent. Flowers 12 mm. long ... ... 2. *C. ramosa*.

1. **Corydalis Govaniana** Wall. is found in Western Himalaya from Kashmir to Kumaon at 8,000-12,000 feet.

The root is considered tonic, diuretic, alterative, and antiperiodic. It is prescribed in syphilitic, scrofulous, and cutaneous affections.

Bengal: Bhutkesi, Bhutkis—; *Hindi*: Bhutkesi, Bhutkis—; *Punjab*: Bhutkes, Bhutkis—; *Sanskrit*: Bhutakesi—.

2. **Corydalis ramosa** Wall. inhabits the temperate portion of the Himalaya from Kashmir to Sikkim, at 12,000-15,000 feet.

In the Kurram Valley the plant is, like all other plants with yellow sap, employed in the treatment of eye diseases.

Kurram Valley: Mamiran—.

FUMARIA.

This genus includes 46 species, natives of the Mediterranean region, and distributed over Central Europe, Asia, Africa; one being almost cosmopolitan.

The genus is tonic, aperient, and resolvent.

The following are used medicinally in Europe: *F. capreolata* Linn., *F. densiflora* D.C., *F. media* Lois., *F. officinalis* Linn., *F. parviflora* Lam., *F. Schleicheri* Soyer-Willem., *F. spicata* Linn., *F. Vaillantii* Lois.—; in Indo-China—*F. officinalis* Linn., *F. parviflora* Lam.—.

1. Flowers pink with wings of upper petal reflexed upwards, bracts about as long as the pedicel or longer. Flowers purplish pink with wings of upper petal at most obscurely purple. Fruit of moderate size, subtruncate, rarely apiculate ... ... 1. *F. indica.*
2. Flowers white, sometimes flushed with pink, with erect spreading, patent or reflexed wings to the broad upper petal; sepals usually present; outer petals not emarginate; the wings of the upper usually with a purple blotch. Fruit often subacute, rugose ... ... 2. *F. parviflora.*

1. **Fumaria indica** Pugsley is distributed over the greater part of India up to 8,000 feet on the Himalaya. It extends through Baluchistan and Afghanistan to Persia, Turkestan, Songaria and Mongolia.

The plant is diuretic, diaphoretic, and aperient.

An extract is drunk by the Waziris for the purpose of cooling the body, and it is also applied externally for the same purpose.

Waziristan: Lavang, Levanai, Margyapal, Parparie, Sevae—.

2. **Fumaria parviflora** Lam. is found in Baluchistan, the Taurus, and Persia.

The herb is highly esteemed by the Mahommedans in India; it is said to purify the blood and act as a laxative and diuretic. In Persia it is often mixed with chicory for medicinal purposes.

The dried plant is regarded as efficacious in low fever, and is also used as an anthelmintic, diuretic, diaphoretic and aperient, and to purify the blood in skin diseases.

Along with black pepper, it is used in the treatment of ague.

It is used in Europe as an alterative, aperient, and antifebrile drug. It is given in Spain in visceral obstructions, in scorbutic affections, and in various eruptive diseases.

Decoctions of the entire plant were given to cases of mild pyrexia and the result was unsatisfactory (Koman).

The plant contains fumaric acid and the alkaloid fumarine.

Arabic : Baglat-ul-mulk, Bukslat-ul-mulik, Rizud-dejaj, Shahatraja, Shatraraj— ; *Baluchi* : Kahurkah, Kankechk— ; *Bengal* : Bansulpha, Shulpha— ; *Bombay* : Pitpatra— ; *Deccan* : Pitpapara, Shatra— ; *Egypt* : Shahtarag, Shatreyg— ; *English* : Fine-leaved Fumitory, Small-flowered Fumitory— ; *Gujerat* : Khasudlio, Pitpapda, Pittapapado— ; *Hamadan* : Shahtara— ; *Harboi Hills* : Kafutkawa— ; *Hindi* : Banshulpha, Pitpapada, Pitpapara, Pitpapra, Shahatra— ; *Indo-China* : Tug sha tu chinh— ; *Kumaon* : Khairuwa— ; *Marathi* : Pitpapda— ; *Persian* : Shahatra, Shah-tarra, Shatra— ; *Punjabi* : Pitpapra— ; *Pushtu* : Papra, Pitpapra, Shahtara— ; *Quetta-Pishin* : Shatara— ; *Sanskrit* : Araka, Charaka, Kalapanga, Katupatra, Kavachanamaka, Krishnashakha, Ksheparpata, Nakra, Panshu, Panshuparyaya, Parpata, Parpataka, Pragandha, Pittari, Renu, Shita, Shitavallabha, Sutikta, Tikta, Trishnari, Triyashti, Varatikta, Varmakantaka— ; *Sind* : Shahatra, Shatra— ; *Tamil* : Tusa— ; *Teheran* : Shahtara— ; *Telugu* : Chatarasi— ; *Urdu* : Shahatra— ; *Waziri* : Lavang—.

THE MEDICINAL AND POISONOUS CRUCIFERS OF INDIA.

BY

J. F. CAIUS, S.J., F.L.S.

The CRUCIFERÆ are a very important Natural Order, and one which includes many esculents and few poisonous or medicinal plants. They are annual, biennial, or perennial herbs without milky juice. They are dispersed all over the world; reaching, in the polar regions and on the highest mountains, the limits of flowering-plant vegetation. The family consists of 220 genera and 1,900 species; most of these inhabit the South of Europe and Asia Minor; they are rarer in the tropics, in extra-tropical and temperate South America. Several species are cultivated in numberless varieties and forms as vegetables.

The members of the Order are acrid, stimulant, and anti-scorbutic; seeds oleaginous, frequently pungent.

The following have been isolated from different parts of the plants:—(1) *alkaloids*—cheirinine, sinapine—; (2) *aromatic alcohols*—inositol—; (3) *enzymes*—myrosin—; (4) *glucosides*—cheirolin, glucocheirolin, glucotropoeolin, sinalbin, sinigrin—; (5) *acids*—behenic, erucic, fumaric, myronic, oleic, stearic—; (6) *pigments*—quercetin—; (7) *mustard oils*—allyl, benzyl, phenyl ethyl.

The medicinal and poisonous crucifers of the world belong to 45 genera:—ALLIARIA (temperate Europe; Asia); ALYSSUM (Mediterranean; Europe); ARABIDOPSIS (Europe; Western Asia); ARABIS (northern temperate regions; south America); BARBAREA (northern temperate regions); BERTEROA (northern temperate regions of the Old World); BRASSICA (Mediterranean; Europe, Asia); BUNIAS (Mediterranean; Asia); CAKILE (northern temperate regions); CAMELINA (Mediterranean; Europe); CAPSELLA (temperate and sub-tropical regions); CARDAMINE (cosmopolitan; chiefly temperate regions); CHEIRANTHUS (Mediterranean; northern temperate regions); CLYPEOLA (Mediterranean); COCHLEARIA (Europe, Asia Minor); CRAMBE (Mediterranean; Europe, Asia, Polynesia, Patagonia); DENTARIA (northern temperate regions); DESCURAINIA (northern temperate regions; South America); DIPLOTAXIS (Mediterranean; Europe); DRABA (northern temperate and arctic regions; south-western North America); EROPHILA (Mediterranean; Europe); ERUCA (Mediterranean); ERYSIMUM (Mediterranean; Europe; Asia); FARSETIA (Mediterranean); HELIOPHILA (South Africa); HESPERIS (Mediterranean; Europe); IBERIS (Europe, Asia); IODANTHUS (Atlantic North America); ISATIS (Mediterranean; Europe; Asia); LEPIDIUM (cosmopolitan); LUNARIA (Europe); MATTHIOLA (Mediterranean; Europe, South Africa); MORICANDIA (Mediterranean); NASTURTIUM (cosmopolitan); RADICULA (North America); RAPHANUS (Mediterranean; Europe; Java); RAPISTRUM (Mediterranean; central Europe); RORIPA (northern hemisphere); SENEBIERA (subtropical regions; Europe); SESELI (Europe, Africa, Asia, Australia); SISYMBRIUM (northern temperate regions); THLASPI (northern temperate regions); THYSANOCARPUS (California); VESICARIA (Europe); WASABIA (Japan).

The medicinal and poisonous crucifers of India belong to 24 genera :—ALLIARIA, ALYSSUM, ARABIDOPSIS, BARBAREA, BRASSICA, CAPSELLA, CARDAMINE, CHEIRANTHUS, COCHLEARIA, CRAMBE, DESCURAINIA, DRABA, EROPHILA, ERUCA, ERYSIMUM, FARSETIA, ISATIS, LEPIDIUM, MATTHIOLA, MORICANDIA, NASTURTIUM, RAPHANUS, SISYMBRIUM, THLASPI.

A. Pods narrow, long. Seeds usually uniseriate (pods sometimes short in *Nasturtium*)
- I. Sepals narrow, erect; valves without horns or appendages; stigmas erect, connate or decurrent on the style ... ... MATTHIOLA.
- II. Sepals broad or narrow; stigma undivided or shortly bilobed
  - a. Sepals erect, lateral saccate.—Hoary, leafy erect herbs ... ... CHEIRANTHUS.
  - b. Sepals spreading, not saccate; pods tumid; seeds minute, biseriate. Flowers usually yellow ... NASTURTIUM.
  - c. Sepals not saccate; pods 4-angled; seeds uniseriate. Flowers yellow ... ... BARBAREA.
  - d. Sepals not saccate; pods hardly nerved, usually acute; stamens simple. Flowers white or purple ... ... CARDAMINE.

B. Pods short, broad. Seeds usually biseriate (pods sometimes long and seeds uniseriate in *Farsetia*)
- I. Pods 2-celled; sessile, many-seeded. Seeds compressed, often winged, uniseriate ... ... FARSETIA.
- II. Pods 1-2-celled, 2-many-seeded, valves often tumid. Seeds biseriate (sepals never saccate). Flowers white or yellow ......
  - a. Sepals, short, spreading. Glabrous herbs ... COCHLEARIA.
  - b. Stamens often appendaged; pods usually orbicular and 4-seeded.—Hoary herbs ... ... ALYSSUM.
  - c. Stamens not appendaged; pods longer than broad, many-seeded.—Hoary herbs
    - † Petals entire ... ... DRABA.
    - †† Petals bifid ... ... EROPHILA.

C. Pods usually sessile, long, narrow. Seeds usually 1-seriate. Stigma capitate, emarginate or shortly bilobed
- I. Flowers yellow, pedicelled, ebracteate. Pods terete, valves 3-nerved ... ... SISYMBRIUM.
- II. Flowers white. Pods terete, valves 1-3-nerved. Seeds smooth or striate ... ... ALLIARIA.
- III. Flowers yellow, pedicelled, ebracteate. Pods cylindric or flat, valves 1-nerved ... ... DESCURAINIA.
- IV. Flowers white, purple or rose-coloured. Valves of the terete or compressed pod 1-nerved or nerveless ... ARABIDOPSIS.
- V. Flowers yellow, rarely purple, ebracteate. Pods elongate, narrow, compressed tetragonal or terete; valves linear, 1-nerved, frequently keeled ... ERYSIMUM.

D. Pods short or long, cotyledons longitudinally folded or deeply grooved
- I. Stigma capitate, truncate or 2-lobed
  - a. Pods long. Seeds uniseriate ... ... BRASSICA.
  - b. Pods short, turgid, beaked. Seeds biseriate ... ERUCA.
- II. Stigmas connate, forming an erect cone. Pods long. A glaucous large herb or undershrub ... ... MORICANDIA.

E. Pods short, dehiscing throughout their length, compressed laterally at right angles to the septum
   I. Cotyledons incumbent, straight, curved or longitudinally folded
      a. Pods many-seeded, valves not winged ... CAPSELLA.
      b. Pods few-seeded, valves winged or not ... LEPIDIUM.
   II. Cotyledons accumbent, straight. Pods compressed notched; valves winged or keeled ... ... THLASPI.

F. Pods short, indehiscent, not jointed, usually hard
   Pods 1-celled, oblong or linear, wing thick.—Glaucous herbs ... ... ISATIS.

G. Pods long or short, transversely jointed; joints indehiscent or the lower 2-valved or reduced to a pedicel for the upper
   Lower joint of pods a seedless pedicel, upper globose 1-seeded ... ... CRAMBE.

H. Pods long, not jointed, indehiscent, either 1-celled and many-seeded or breaking up into many 1-celled, 1-seeded indehiscent fragments
   Pods long, terete, hollow or septate; seeds globose 2-3-celled ... ... RAPHANUS.

ALLIARIA.

The two species of this genus inhabit Western Asia, Europe, and North Africa.

Alliaria officinalis Andrzej. is found in the Western Himalaya from Kumaon to Kashmir, at 6,000-10,000 feet. From Afghanistan it extends westwards to the Mediterranean, Central and subarctic Europe.

The herb and seeds are esteemed diuretic, diaphoretic, and expectorant, and are used as external application in gangrenous affections, and to promote suppuration, healing of cuts, bruises and ulcers.

In England the leaves used to be taken internally as a sudorific and deobstruent, and externally were applied antiseptically in gangrenes and ulcers. The juice of the leaves taken alone or boiled into a syrup with honey is found serviceable in dropsy.

The plant is to be used fresh, for it loses its properties on drying.

Dutch: Lookkruid, Look zonder look—; *English*: Beggarman's Oatmeal, Cardiacke, Caspere, Eileber, English Treacle, Garlick-roast, Hedge-garlick, Jack-by-the-hedge, Leek-cress, Garlick Mustard, Penny Hedge, Poor Man's Garlic, Poor Man's Mustard, Poor Man's Treacle, Sauce-alone, Swarms—; *French*: Alliaire, Herbe aux ailles, Herbe aux aillets, Herbe aux aulx, Julienne alliaire—; *German*: Knoblauchhederich, Knoblauchkraut, Knoblauchsrauke—; *Greek*: Skordion—; *Italian*: Erba alliaria—; *Pacific Coast*: English Treacle, Garlic Mustard, Garlicwort, Hedge Garlic, Hedge Mustard, Leek Cress, Poor Man's Mustard—; *Portuguese*: Alliaria—; *North America*: Hedge Garlic—; *Russian*: Chessnovitsa—; *Spanish*: Aliaria—.

ALYSSUM.

The genus includes 120 species, natives of Europe, the Mediterranean region as far as Central Asia, the Canary Islands, Madeira; naturalised elsewhere.

Alyssum maritimum Lam. is cultivated in gardens in North-western India.

The plant is commonly used in Spain as an antiscorbutic and diuretic. In Barcelona it is highly esteemed as an astringent for gonorrhoea.

Catalan: Caps blancs, Herba blanca, Sempre en flor—; *English*: Sweet Madwort—; *French*: Alysse, Alysse maritime, Alysse odorant, Alysson, Corbeille d'argent, Corbeille d'or—; *German*: Steinkraut—; *Malta*: Allison, Sweet Alysson, Buttuniera—; *Roumanian*: Galbenus, Lubit—; *Russian*: Taritsa—; *Spanish*: Aliso de mar, Bosqueletes anchos, Bosqueletes de dama, Bosqueletes de plata, Mastuerzo maritimo—.

ARABIDOPSIS.

The genus consists of 11 species distributed over Europe, Asia, and Africa.

Arabidopsis thaliana (Linn.) Heynh. is found in Kashmir at 5,000-10,000 feet, and in Kumaon. It extends to Afghanistan, Western Asia, Central Europe, and the Mediterranean region; it has become naturalized in North and South America, South Africa and Australia.

It is used in Spain and other parts of Europe to cure sores in the mouth.

English: Thale Cress—.

BARBAREA.

The genus consists of 15 species of the northern temperate regions, chiefly European.

Barbarea vulgaris Br. is found in the temperate and subalpine Himalaya and Western Tibet at 6,000-10,000 feet. It is distributed to Western Asia, Europe, North and South Africa, and Australia.

In Spain the leaves are a reputed vulnerary.

In La Reunion the plant is used as a stimulant and antiscorbutic.

Catalan: Herba barbara, Herba barbera, Herba de Santa Barbara—; *English*: Winter Cress, Yellow Rocket—; *French*: Barbarée, Barbarée vulgaire, Cresson de terre, Cresson vivace, Girarde jaune, Herbe aux charpentiers, Herbe de sainte Barbe, Herbe de saint Julien, Herbe de sainte Marguerite, Julienne jaune, Roquette—; *German*: Barbarakraut, Erdkresse—; *La Reunion*: Cresson de terre—; *Pacific Coast*: Bitter Cree, French Cress, Hedge Mustard, Land Cress, Normany Cress, Rocket Cree, St. Barbara's Cress, Winter Cress, Winter Rocket, Yellow Rocket—; *Spanish*: Yerba de los carpinteros, Yerba de Santa Barbara—.

BRASSICA.

The genus numbers 33 species, often polymorphic, chiefly Mediterranean.

The seeds of most of them are stimulant and laxative.

The following species are used medicinally in Europe—*B. campestris* Linn., *B. Napus* Linn., *B. nigra* Koch., *B. oleracea* Linn., *B. Rapa* Linn.—; in Japan—*B. campestris* Linn.—; in China—*B. campestris* Linn., *B. cernua* Thunb., *B. Napus* Linn. var. *chinensis* (Linn.) O. E. Schulz, *B. oleracea* Linn., *B. Rapa* Linn.—; in Indo-China—*B. campestris* Linn., *B. integrifolia* (West.) O. E. Schulz, *B. Napus* Linn. var. *chinensis* (Linn.) O. E. Schulz, *B. oleracea*

Linn.—; in the Philippine Islands—*B. integrifolia* (West.) O. E. Schulz.—.

A. Ovary multi-ovulate (9-45). Pods 1.5-10 cm. long. Beak distinctly conical, sometimes as thick as the pod, often 1- or 2-seeded
  I. Stem-leaves amplexicaul
    1. Flowers not surpassing the buds at flowering
      a. Calyx closed. Petals pale yellow, 1.8-2 cm. long. Pods with undulate margins. Leaves more or less fleshy ... ... 1. *B. oleracea.*
      b. Calyx half-opened. Petals intensely yellow, 0.9-1.8 cm. long. Pods with straight margins. Leaves membranous ... ... 2. *B. Napus.*
    2. Flowers surpassing the buds at flowering ... 3. *B. campestris.*
  II. Stem-leaves sessile or petiolate
    1. Leaves with a plane margin ... ... 4. *B. integrifolia.*
    2. Leaves with a crisp margin ... ... 5. *B. cernua.*

B. Ovary few-ovulate (5-11). Pods 0.8-3 cm. long. Beak very thin, always seedless ... ... 6. *B. nigra.*

1. **Brassica oleracea** Linn. is cultivated in all its numerous varieties all over India. The typical plant is indigenous in Central Europe.

The Cabbage has entered into the Ayurvedic Materia Medica, and is very often prescribed by vaids for a variety of ailments.

The cabbage and its numerous associates are well-known household remedies.

The seeds are diuretic, laxative, stomachic, and anthelmintic. The leaves form a good application in gout and rheumatism. An Irish cure for a sore throat is to tie cabbage leaves round it; and the same remedy is applied in England with hot cabbage leaves for a swollen face.

The red cabbage is most emollient and pectoral. The juice made into syrup, without any condiments, is useful in chronic coughs, and in bronchial asthma. It is considered in France to be highly antiscorbutic.

The leaves of the common white cabbage, when gently bruised and applied to a blistered surface, will promote a free discharge, as also when laid next the skin in dropsy of the ankles. Its juice will cure warts.

Bengal: Kopi—; *Catalan*: Col—; *Chinese*: Kan Lan, Pe Ts'ai, Pao Pao Ts'ai—; *Deccan*: Karam—; *Dutch*: Kool—; *Egypt*: Krumb—; *English*: Cabbage, Colewort, Collet, Sea Cabbage, Wild Cabbage—; *French*: Chou, Navette d'hiver, Rabette—; *German*: Gartenkohl, Kohl—; *Greek*: Krambi—; *Gujerat*: Kobia—; *Hindi*: Band gobi, Karamkalla, Kobi, Kopi—; *Hova*: Laisoa, Laisoamasombika—; *Indo-China*: Cam lam—; *Iraq*: Lahanoh—; *Italian*: Cavolo—; *Kurdish*: Kalam—; *Marathi*: Kholkhol—; *Moldavian*: Curechiu—; *Mundari*: Arakubi, Kubiara, Potomkubi—; *Polish*: Iarmuz, Kapusta—; *Portuguese*: Cauve, Repolho—; *Punjabi*: Band gobi—; *Roumanian*: Varza—; *Russian*: Kapusta—; *Sanskrit*: Dalamalini, Dalasarini, Kebuka, Kechuka, Keluta, Kembu, Kembuka, Pechuka, Pichuni, Polini, Supatra, Swadukanda, Swalpavitapa—; *Spanish*: Berza, Col—; *Swedish*: Kahl—; *Sze-Chwan*: Lien Hoa Pe—; *Tamil*: Gos, Kos—; *Telugu*: Gos.

2. **Brassica Napus** Linn., indigenous in the South Mediterranean region, is cultivated in India.

The root is emollient and diuretic. Its juice is useful in chronic coughs and bronchial catarrh.

The Chinese use the seeds to cure coughs complicated with dyspnoea, and to subdue expectoration due to cold.

Afrikaans: Raap—; *Catalan*: Nap—; *Chinese*: La Pa Tzeu, Lo Pou Tse—; *Danish*: Stekroe—; *Dutch*: Steekraap—; *English*: Coleseed, Navew, Rape—; *French*: Arbi, Chou navet, Navau, Navet—; *German*: Raps, Reps, Ruebsen, Stechruebe—; *Greek*: Bunias—; *Hindi*: Kali sarson, Tori—; *Iraq*: Saljam—; *Italian*: Napo, Navone—; *Malta*: Navone, Neveu—; *Mundari*: Huringmani, Mani, Turimani—; *Persian*: Saljam—; *Polish*: Rzepnica—; *Portuguese*: Nabo—; *Roumanian*: Broajba—; *Russian*: Briukoa, Dikayia Repa, Nyemetskayia Repa—; *Sadani*: Lutumani, Turimani—; *Spanish*: Nabo, Nabo común, Nabo largo—; *Swedish*: Stickrave—; *Uruguay*: Colza, Nabo—.

var. **chinensis** (Linn.) O. E. Schulz is cultivated in India, China, Korea, Japan and Java.

The entire plant is antiscorbutic, arthritic, and resolvent.

The seeds are considered stimulant, stomachic, and laxative.

In Indo-China the leaves and stems are considered to be anti-dysenteric. The seeds are used in gleet and in difficult parturition.

In Chinese drug shops the plant is represented as dry, round, light-coloured cakes of vegetable matter made from the broken leaves. The cakes have been found to contain a large amount of various vitamins (Hsien Wu).

Annam: Cai—; *China*: Sung—; *English*: Chinese Cabbage—; *Indo-China*: an dai—; *Japan*: Onuma—; *Malaya*: Pak chai piang—; *Mundari*: Kubi, Risakubi—.

3. **Bassica campestris** Linn. is naturalized in India. In all probability it is indigenous in the mountainous region of the Mediterranean.

The tuberous roots and the seeds are said to be antiscorbutic.

The seeds yield Oil of Colza, which is officinal in Sweden as *Oleum Rapae.*

Arabic: Chardal—; *Betsileo*: Anambe—; *Brahui*: Mashna muli—; *Catalan*: Colinap—; *Danish*: Raps—; *Dehra Dun*: Banga sarson, Sarson—; *Dutch*: Koolzaad—; *English*: Coleseed, Colza, Field Cabbage, Navette, Swedish Turnip, Wild Navew—; *French*: Chou des champs, Chou colza, Navette—; *German*: Rape—; *Hova*: Ananomby, Anantsonga, Nave—; *Italian*: Colza—; *Malta*: Summer Rape, Wild Cabbage, Rapa selvatica, Caulicelli, Liftia—; *Meerut*: Banga sarson, Sarson—; *Mundari*: Piramani—; *Oudh*: Pila sarson—; *Pacific Coast*: Bargeman's Cabbage, Common Navew, Nape, Rutabaga, Summer Rape, Turnip, Wild Navette, Wild Navew—; *Rohilkhand*: Pila sarson—; *Roumanian*: Rapita—; *Russian*: Colza, Surepitsa—; *Sadani*: Sarso—; *Spanish*: Colinabo, Colza, Nabicol, Nabina—; *Uruguay*: Colza, Nabiza, Nabo—.

var. **rapa** (Linn.) Hartm. is cultivated throughout India. The Ayurvedists distinguish two varieties, the red and the white, and consider the white the better of the two.

The seeds mixed with hot water form an efficient counter-irritant poultice. The oil, combined with camphor, forms an efficacious embrocation in muscular rheumatism, stiff neck, etc.; it is used in dengue fever with benefit, and it is rubbed on the chest in bronchitis.

In Indo-China the roots and the leaves are considered stomachic; the ground fresh leaves are applied to abscesses; the seeds are given in colic.

Afrikaans: Raap—; *Annam*: Lu bu, Tron cu—; *Arabic*: Khardaleasvad—; *Bengal*: Kalisarson, Sadarai, Sanshi, Shalgram, Shurshi, Schwetrai, Sursha,

Sursi—; *Burma*: Amenmiyenzi—; *Canarese*: Kappusasoe, Sarsive—; *Catalan*: Nap de bou, Nap rodo—; *China*: Yun T'ai—; *Cutch*: Surah—; *Deccan*: Kalerayan, Pilerayan—; *English*: Rape, Wild Turnip—; *German*: Reps, Ruebe, Ruebsen, Rurnips—; *Gujerat*: Kalarai, Raira, Sarashire, Sarsawa—; *Hasara*: Dakubi—; *Hindi*: Bangasarson, Baralai, Dain, Dainlai, Jadiya, Jariya, Kalerai, Kalisarson, Khetiya, Lahi, Lahota, Lai, Laita, Pilasarson, Pilirai, Rararada, Rarasarson, Sarson, Sarsonzard, Shalgham, Shetashirsha, Sursi, Teri, Toriya—; *Indo-China*: Man thanh, Vu than—; *Iraq*: Shalghan—; *Kurdish*: Shailam—; *Lepcha*: Kanyang—; *Malayalam*: Karupakatuka—; *Malta*: Swede—; *Marathi*: Kalamohare, Sherasa—; *Mundari*: Dakobi, Piramani—; *Nagpuri*: Dahakobi, Dahakubi—; *Persian*: Sipandanesiyah—; *Punjab*: Gonglu—; *Quetta-Pishin*: Jamboi—; *Rajputana*: Sarson, Sarsu—; *Sanskrit*: Bhutaghna, Bimbata, Duradharsha, Grahagna, Kadamba, Kadambada, Kadambaka, Kalasarshapa, Katukasveha, Katusveha, Rajakshavaka, Rajika, Rakshitaphala, Sarishapa, Sarshapa, Sidhaprayojana, Sidhartha, Tantubha, Tantuka, Tuverika, Ugragandha—; *Sinhalese*: Kaluabbe, Wal-aba—; *Spanish*: Nabo gallego, Turnep—; *Tamil*: Karuppukkadugu—; *Telugu*: Nallaavalu—; *Uthal*: Sarih—.

4. **Brassica integrifolia** (West) O. E. Schulz is much cultivated in India. It is distributed over the Indo-Malayan region, temperate Eastern Asia, Africa, Madagascar, the West Indies, and South America.

The seeds are warming, sudorific, and a well-known aid to digestion. They are much used in internal congestions, in spasmodic, neuralgic, and rheumatic affections, and in morbid states of the cerebro-spinal system.

In medicine the oil is used as an embrocation and is applied to the skin in eruptions and ulcers.

Bengal: Raisarasha—; *Bombay*: Rai, Rajika, Sarson—; *Canarese*: Sasive—; *China*: Pai Chieh—; *Cutch*: Rai—; *English*: Indian Mustard, Rai—; *French*: Moutarde de l'Inde, Moutarde rouge—; *Gujerati*: Rai—; *Hindi*: Badshahirai, Barirai, Barlai, Gohnasarson, Khasrai, Rai, Sarson, Sarsonlahi, Shahzadarai—; *Indo-China*: Ben, Rau cai, Rau mo, Spey top, Thuy thai—; *Kashmir*: Asur—; *Konkani*: Samsvel—; *Malaya*: Pak kai—; *Marathi*: Mohari, Rayan—; *Mundari*: Rai, Raimani—; *Pacific Coast*: Chinese Mustard, Cosson, Indian Mustard, Russian Mustard—; *Philippines*: Mostaza—; *Portuguese*: Mostarda de India, Mostarda vermelha—; *Sadani*: Rai—; *Sanskrit*: Rajika—; *Sinhalese*: Abba—; *Tamil*: Kadugu—.

5. **Brassica cernua** (Thunb.) Forbes and Hemsley is cultivated in India. It is found in China, Japan, and the West Indies.

The leaves are used in Indo-China as an antidysenteric. They are powerfully diaphoretic.

A decoction of the seeds is given in lumbago, cough, and indigestion.

China: Chieh—; *English*: Black Mustard—; *Indo-China*: Gioi—.

6. **Brassica nigra** (Linn.) Koch is cultivated in India and many other countries.

The leaves and seeds are much used by Hindu and Mahommedan practitioners, who credit them with valuable therapeutical properties.

The dried ripe seed is rubefacient, vesicant, emetic, stimulant, tonic and revulsive.

The seeds act as a digestive condiment, if taken moderately. If swallowed whole they operate as a laxative, and for this purpose are sometimes prescribed in dyspepsia and other complaints attended with torpid bowels.

The flour of the seeds is used in the form of a poultice being a useful and simple rubefacient and vesicant. Mustard poultices prove highly serviceable in cases of febrile and inflammatory diseases, internal congestions, spasmodic, neuralgic, and rheumatic affections. Mustard flour in water is highly recommended as a speedy and safe emetic. It is reckoned as a medicine of great value by the Javanese and the Chinese.

The Meskwaki Indians of Wisconsin grind the seed and use the powder as a snuff to cure cold in the head.

The pure fresh oil is a stimulant and mild counter-irritant when applied externally. As such, it is very useful in mild attacks of sore-throat, internal congestion, and chronic muscular rheumatism.

Mustard seed has been recommended by Sanskrit authors as a cure for snake bite and scorpion sting; but Caius and Mhaskar have shown that it has no antidotal value against snake or scorpion venom.

An alkaloid, sinapine, has been isolated from the seeds; also myrosin, sinigrin, inosite, albumins, gums, and colouring matters.

Arabic: Khardal, Khirdal, Khurdal—; *Bengal*: Baisarisha—; *Bogotá*: Rábano, Rábano de canarios—; *Bombay*: Rai, Sarsan—; *Canarese*: Bilesasive, Karisasive, Sasive—; *Danish*: Senep—; *Deccan*: Raiee, Suroo, Sursoo—; *Dutch*: Mosterd, Zwart mosterd—; *Egypt*: Kabar, Khardal, Khodar, Lifsan—; *English*: Black Mustard, Mustard, True Mustard—; *French*: Chou noir, Moutarde. Moutarde grise, Moutarde noire, Navuce rouge, Sénevé, Sénevé noir—; *German*: Gruener Senf, Hollaendischer Senf, Mustert, Schwarzer Senf, Senf—; *Greek*: Napi, Napia, Sinapi, Sinipi—; *Gujerat*: Kalirai, Rai—; *Hindi*: Aslra, Banarasirai, Ghorrai, Jagrai, Kalirai, Lahi, Makrarai, Rai, Sursu, Suru, Taramira, Tira—; *Iraq*: Khardal—; *Italian*: Senapa, Senapa nera, Senapa vera—; *Konkani*: Sansonv—; *Lepcha*: Kun-rang nok—; *Malta*: Black Mustard, Senapa, Senapa nera, Sinapa, Mustarda seuna—; *Meskwaki*: Fitciko, makani—; *Pacific Coast*: Black Mustard, Brown Mustard, Kerlock, Red Mustard—; *Persian*: Sarshaf—; *Polish*: Gorczyka—; *Portuguese*: Mostarda, Mostardeira—; *Potenza*: Lassand—; *Prairie Potawatomi*: Tcotca'mosikûn—; *Reggio*: Senva—; *Roumanian*: Mustar—; *Russian*: Gortchiza—; *Sanskrit*: Asuri, Atitikshva, Jwalanti, Jwalatprabha, Katuasuri, Krimika, Krishnasarshapa, Krishnika, Kshava, Kshavaka, Kshudhabhijanana, Kshujjanika, Kshutabhijanaka, Kshutaka, Madhurika, Raji, Rajika, Raktasarshapa, Raktika, Sarshapa, Tikshnagandha—; *Sind*: Ahor, Kali sursoon, Suraj—; *Sinhalese*: Ganaba, Kalu-aba—; *Spanish*: Mostaza—; *Swedish*: Senap—; *Tamil*: Chirukadugu, Kadugu—; *Telugu*: Avalu—; *Treviso*: Colzat—; *Urdu*: Rai—; *Uruguay*: Mostacilla negra, Mostaza negra—; *Verona*: Ravenele—.

Cardamine.

The genus includes 100 cosmopolitan species, natives of temperate and cold regions, especially of Europe and Western and Central Asia.

The plant is bitter and antiscorbutic. The flowers are diuretic, useful in chorea and asthma.

The following are used medicinally in Europe—*C. alpina* Willd., *C. amara* Linn., *C. asarifolia* Linn., *C. gelida* Schott., *C. hirsuta* Linn., *C. impatiens* Linn., *C. parviflora* Linn., *C. pratensis* Linn., *C. resedifolia* Linn., *C. sylvatica* Linn., *C. trifolia* Linn.—; in Indo-China—; *C. hirsuta* Linn.—; in Northern America—*C. pratensis* Linn.—.

A. Petals linear, erect, very minute, white, often absent ... 2. *C. impatiens*.
B. Petals spreading, 3 times as long as the sepals ... 1. *C. pratensis*.

1. **Cardamine pratensis** Linn. is found in Kashmir. It is distributed to Northern and Western Asia, Europe, Abyssinia, and Noth America.

The plant is considered to be stimulant, diaphoretic, and diuretic. In some parts of Europe the plant is used in the treatment of nervous affections. In Cornwall the flowering tops have been employed for the cure of epilepsy throughout several generations with singular success; though the use of the leaves only for this purpose has caused disappointment.

The seeds contain myronic acid.

Catalan : Creixans de prat, Creixens de prat— ; *Dutch* : Gemeen schuimblad, Koekocksbloem, Pinksterbloem, Weidekers— ; *English* : Apple-pie, Bird's-eye, Bogspinks, Bonny-Bird-Een, Bread-and-Milk, Canterbury Bells, Cuckoo's Bread, Cuckoo-flower, Cuckoo-pint, Cuckoo's Shoes and Stockings, Gilliflower, Gookoo-buttons, Headache, Lady Flock, Lady's Glove, Lady's Smock, Lamb Lakins, Lucy Locket, May Blob, May Flower, Milkgirl, Milk Maids, Paigle, Pigeon's Eye, Pink, Shoes and Stockings, Smell Smocks, Spink, Whitsuntide— ; *French* : Bec à l'oiseau, Bouquet au coup, Cardamine, Cardamine des prés, Cresson amer, Cresson élégant, Cressonnette, Cresson des prés, Cresson sauvage, Passerage sauvage, Petit cresson aquatique— ; *German* : Fleischblume, Gauchblume, Kuckucksschaumkraut, Schaumkraut, Schluesseblume, Wiesenkresse, Wiesenschaumkraut— ; *Greek* : Kardamini, Sisymbrion eteron— ; *Italian* : Cardamantina, Cardamina, Crescione pratense, Nasturzio salvaggio— ; *Languedoc* : Creissouneto, Creissoun de prat— ; *Moldavian* : Stupitul cuerlui— ; *Polish* : Aengkrasse— ; *Roumanian* : Pascele calului, Scuipatul cucului, Spumeala— ; *Russian* : Kress, Serdechnik— ; *Spanish* : Berros de prado, Mastuerzo de prado, Mastuerzo pratense—.

2. **Cardamine impatiens** Linn. occurs in the temperate Himalaya from Kashmir to Sikkim, at 5,000-12,000 feet. It is distributed to Afghanistan, temperate Asia and Europe.

The plant is stimulant and diuretic.

English : Mountain Bitter Cress, Narrow-leaved Bitter Cress.

CAPSELLA.

The genus consists of 4 species, natives of temperate and subtropical regions.

Capsella Bursa-pastoris Medik. is a weed of cultivation; it is found in temperate India in cultivated places.

Plant astringent; seeds stimulant. The herb and its seeds were employed in former times to promote the regular monthly flow in women.

Apparently the uses of the plant are unknown in India.

It has been used in English domestic practice from early times as an astringent in diarrhoea; it was much used in decoction with milk to check active purgings in calves.

A well-known popular remedy in Europe and North-America. It is of prompt use to arrest bleedings and floodings, when given in the form of a fluid extract; and it may be considered a most reliable medicine for staying fluxes of blood. The fluid extract of the herb is also given for dropsy as a diuretic.

The Menomini of Wisconsin use the plant as a cure for poison ivy; the plant is steeped, and the water when tepid is used as a wash.

In Bogota the herb, and more particularly the inflorescence, is considered a cure for blennorrhagia.

In Tongking the seeds are given in troubles of the chest; the whole herb when fresh is used as an hemostatic.

Extracts of Shepherd's Purse have been used as substitutes for ergot. They have been asserted to contain acetylcholine, choline, and possibly tyramine, as well as fumaric acid and inositol; but, according to Wasicky (*Ber. deutsch. Pharm. Ges.* 1922, 32, 142), plants free from fungoid infection are devoid of active constituents.

Baluchistan : Halaku— ; *Catalan* : Bossas de pastor, Bosses de pastor, Sarronets de pastor— ; *Chinese* : Ch'i— ; *Colombia* : Calzoncitos, Pan y quesito— ; ***Dutch*** : Herderstaschje— ; *English* : Bad Man's Oatmeal, Blind-weed, Case-weed, Clapper Pouch, Cocowort, Fat Hen, Lady's Purse, Mother's Heart, Naughty Man's Plaything, Pepper-and-Salt, Pepper-and-Shot, Pick Pocket, Pick Purse, Pick-your-mother's-heart-out, Poor Man's Permacety, Poor Man's Pharmacetty, Purse, Rattle Pouch, Sanguinary, St. James's Wort, Shepherd's Bag, Shepherd's Pouch, Shepherd's Purse, Shepherd's Scrip, Shepherd's Sprout, Tooth-wort, Toywort, Ward-seed, Witches' Pouchers— ; *French* : Bourse à berger, Bourse à curé, Bourse de capucin, Bourse à Judas, Bourse à pasteur, Bourserette, Boursette, Capselle, Fleur de Saint Jacques, Houlette, Malette, Malette à berger, Millefleur, Molette, Molette de berger, Molette des pasteurs, Moufette, Moutarde de Mithridate, Moutarde sauvage, Tabouret, Thlaspi— ; *German* : Bauensent, Beutelschneiderkraut, Blutkraut, Brillenkraut, Daxenkraut, Gaensekresse, Geldbeutel, Grischel, Groeschelkraut, Herzelkraut, Hirtentaeschel, Hirtentaeschlein, Hirtentasche, Klaffer, Klapperkraut, Kleep, Kuespelkraut, Lapatekrokt, Muenselkraut, Nadeldieb, Saeckelkraut, Schaeferkraut, Schedelkraut, Schinken, Taeschelkraut, Taschenblume, Taschendieb, Taschenkraut, Voegelikraut— ; ***Harboi Hills*** : Chambraka— ; ***Indo-China*** : Te, Te thai— ; *Irish* : Clappedepouch— ; ***Italian*** : Borsacchina, Borsa di pastore, Borsapastore, Erba raperina, Millefiori— ; ***Malta*** : Shepherd's Purse, Borsacchina, Borsa di pastore, Gargir il gemel— ; ***Menomini*** : Wisakipûko'sa— ; ***Roumanian*** : Buruiana de figuri, Buruiana viermelui, Iraba rosie, Pungulita, Tascuta ciobanului, Tascutita, Traista ciobanului— ; ***Russian*** : Pastushiya sumka— ; ***Spanish*** : Bolsa de pastor, Botella, Hierba del cangrejo, Hierba del carbonero, Mastuerzo macho, Paniquesillo, Pan y quesillo, Zurrón de pastor— ; *Tongking* : Dinh lich, Dinh luc—.

CHEIRANTHUS.

The genus includes 20 species, natives of the Mediterranean region, Central Europe, the Himalayan region, and North America.

C. Cheiri Linn. is the only therapeutically active species. Its flowers aré officinal in Portugal.

Cheirianthus Cheiri Linn. is cultivated in Indian gardens.

The seeds are tonic, diuretic, expectorant, stomachic, aphrodisiac; good in dry bronchitis, fevers, and injuries to the eye (Yunani).

The flowers, said to be cardiac and emmenagogue, are used in paralysis and impotence. They are boiled in olive oil; and this prepared oil is much used for enemata.

The dried petals are much used in Upper India as an aromatic stimulant.

The *seeds* are also used as an aphrodisiac.

An alkaloid, cheirinine, has been isolated from the leaves and seeds. The seeds contain cheiroline, which appears to exist in the plant as a glucoside. Quercetin has been isolated from the colouring matters of the flowers.

Bengal : Khueri— ; ***Catalan*** : Violer groch— ; ***Dutch*** : Muurbloem— ; ***Egypt*** : Manthur— ; ***English*** : Banwort, Bee Flower, Bleeding Heart, Blood Wall,

Bloody Warrior, Cherisaunce, Chevisaunce, Churl, Geraflour, Gilly-flower, Handflower, Heartsease, Jacks, Jerofileris, Jilliver, July Flower, Keyry, Sweet William, Wallflower, Wallstock-gillofer—; ***French*** : Bâton d'or, Carafée, Casse col, Coquardeau, Giroflée, Giroflée jaune, Giroflée des murailles, Giroflée des murs, Giroflée suissard, Murayer, Murer, Rameau d'or, Ravenelle jaune, Violette giroflée, Violier jaune—; ***German*** : Gelbveiglein, Goldlack, Handblume, Lack, Lackviole, Gelbe Levkoje, Gelbe Viole—; ***Hindi*** : Lahurishubu, Todrisurkh—; ***Iraq*** : Shabbo asfar—; ***Italian*** : Violaciocca gialla, Viola gialle—; ***Languedoc*** : Girouflado fero, Granié—; ***Malta*** : Wallflower, Violaciocca gialla, Gizi safra—; ***Portuguese*** : Goivo amarelho, Violeta amárelha—; ***Punjab*** : Todrisrafarmani, Todrisiyah, Todrisurukh—; ***Roumanian*** : Micsunea—; ***Russian*** : Jeltofiol, Levkoi—; ***Sind*** : Todrisiyah, Todrisurkh—; ***Spanish*** : Alhelí amarillo—; ***Urdu*** : Todrisurkha—.

COCHLEARIA.

The genus consists of 20 species, natives of Northern temperate and Asiatic regions.

Acrid, bitter, antiscorbutic herbs.

The following species are used medicinally in Europe—*C. armoracia* Linn., *C. officinalis* Linn.—; in Brazil—*C. armoracia* Linn.—; North America—*C. officinalis* Linn.

Cochlearia flava Ham. is found in the Gangetic valleys from the Punjab to Bengal.

The plant is used as a febrifuge in Bihar.

CRAMBE.

The genus includes 20 European and temperate Asiatic species; some of them distributed to Polynesia and Patagonia.

In Europe *C. hispanica* Linn. and *C. maritima* Linn. are considered specifics against scrofula.

Crambe cordifolia Stev. is found in Kashmir and Baluchistan, whence it spreads to Afghanistan, Persia, and the Caucasus.

At Hindubagh, in Baluchistan, the plant is used as a cure for itch.

Baluchi : Pilgosch—; ***Harnai*** : Nawarra—; ***Hindubagh*** : Pashai, Skharyae, Skhrai—; ***Khwas*** : Naghour—; ***Kila Saifulla*** : Pursail—; ***Nushki*** : Pilgosch—; ***Waziri*** : Navgra—.

DESCURAINIA.

The genus numbers 43 species; most of them inhabit the cold and temperate regions of the whole of America; some are found in Asia, Europe and Macronesia.

D. sophia (Linn.) Webb. is used medicinally in Europe and in the United States of America.

Descurainia sophia (Linn.) Webb. (=*Sisymbrium Sophia* Linn.) occurs in India from Kashmir to Kumaon up to 14,000 feet, in the Eastern Himalaya, the Salt Range, Peshawar and Baluchistan. It is distributed to Afghanistan, Persia, Mesopotamia, the Mediterranean region, Central Europe, arctic and subarctic Europe, China, Japan, Central Asia. The plant has run wild in South Africa, North and South America, and New Zealand.

The flowers and leaves are astringent and antiscorbutic. An infusion of them is a popular remedy in Spain.

In China and Malaya the drug is regarded as demulcent. laxative, and febrifuge.

In Baluchistan the seeds mixed in syrup are swallowed as a cure for fever. On the Pab Hills in Jhalawan the plant is used as a febrifuge.

In the United States of America the herb is used externally in indolent ulcers, and the seeds are given internally in worms, calculus complaints, etc.

The juice, mixed with an equal quantity of honey or vinegar, has been recommended for chronic coughs and hoarseness, and ulcerated sore-throats. A strong infusion of the herb has proved excellent in asthma, and the seeds formed a special remedy for sciatica.

Arabic: Khubah, Robesetti—; *Catalan*: Herba de la sabiduria, Herba de Santa Sofia—; *Chinese*: T'ing Li—; *English*: Flixweed, Fluxweed—; *French*: Argentine rouge, Herbe des chirurgiens, Herbe de la sagesse, Herbe de sainte Sophie, Mort aux vers, Moutarde de chien, Rhubarbe des paysans, Sagesse des chirurgiens, Sisymbre sophie, Sophie des chirurgiens, Thalictre des marchands, Thalictron, Thalictron des boutiques—; *German*: Besenkraut, Feinblaetrige, Rauke, Schuttrauke, Sophienkraut, Sophienrauke, Wurmkraut—; *Harboi Hills*: Rush—; *Hindi*: Khubkallana—; *Indo-China*: Bo nuong cao—; *Jhalawan*: Rush—; *Kalat*: Rush—; *Kirani*: Khashir—; *Malaya*: Tien theng yan, T'ing li yen—; *Mastung*: Rush—; *Pab Hills*: Rush—; *Persian*: Khakshi—; *Pishin*: Khakshir—; *Quetta*: Khakshir, Khashir—; *Spanish*: Ajenjo de Serifo, Hierba de los cirujanos, Hierba de la sabiduria, Serifio absintio, Sofia de los cirujanos—; *Teheran*: Towdri—; *United States*: Flax weed—; *Urdu*: Khakshir—.

DRABA.

This genus includes 150 species; natives of the northern temperate, arctic, and alpine regions, and of the Andes of South America.

Draba muralis Linn. is found in Kashmir up to 6,000 feet. It extends to Asia Minor, Europe, and North Africa.

The plant is used in Spain as an antiscorbutic.

English: Small Whitlow Grass—.

EROPHILA.

The genus includes 4 species, natives of the Mediterranean region and spread over Europe.

Erophila vulgaris DC. is found in Kashmir at 5,000-6,000 feet. It is distributed to Afghanistan, Western Asia, and Europe.

The plant is considered in Spain as an astringent and vulnerary; it is a popular remedy for the whitlow.

English: Spring Whitlow-grass—; *French*: Drave—.

ERUCA.

The genus consists of 5 species, natives of the Mediterranean region.

E. sativa Gars. is used medicinally in Europe.

Eruca sativa Gars. is indigenous in the Mediterranean; but is cultivated in many parts of India.

The seeds are vesicant, and the whole plant is considered aphrodisiac.

In Europe the young leaves are used as stimulant, stomachic, diuretic, and antiscorbutic. A strong dose will cause vomiting, and may be taken in the place of ipecacuanha. Powdered, the effect is less strong than that of mustard.

Afghanistan: Mandao—; *Arabic*: Buckl—; *Bengal*: Shwetsursha, Suffedshershi, Swetsarish—; *Catalan*: Ruca—; *Egypt*: 'Afin, Gergir, Gery, Shiltam—; *English*: Dame's Rocket, Dame's Violet, Jamba, Purple Rocket, Rocket, Vesper-Flower, White Rocket—; *French*: Roquette, Roquette cultivée, Roquette des jardins—; *German*: Senfkohl—; *Hadramaut*: Buckl—; *Hindi*: Duan, Taramira, Taranuri, Tira—; *Italian*: Ruchetta, Rucola—; *Kumaon*: Chara, Dua—: *Loralai*: Mulai—; *Malta*: Rocket, Ruchetta, Rucola, Aruca, Eruca—; *North-Western Provinces*: Duan, Lalu, Sahwan, Tara, Taramira, Tira—; *Persian*: Jambeh—; *Punjab*: Assu, Jamnia, Tara, Taramira, Usan—; *Sanskrit*: Bhutaghna, Bimbata, Daradharsha, Grahagna, Kadamba, Kadambada, Kadambaka, Rajakshawaka, Rajika, Rakshitaphala, Sarishapa, Sarshapa, Sidhaprayojana, Sidharta, Tantubha, Tantuka, Tuverika, Ugragandha—; *Sind*: Jambeho, Kalasurson—; *Spanish*: Oruga, Raqueta, Roqueta—.

ERYSIMUM.

The genus number 80 species, natives of the Mediterranean region, and inhabiting Europe and Asia.

Erysimum repandum Linn. is found in Kashmir at 5,000-7,000 feet; it is distributed to Persia, and westwards to Eastern Europe and North Africa.

The plant is used in Spain as an antiscorbutic.

The seeds, when placed in water, become coated with transparent mucilage. In Persia they are given in fever; and externally, in the form of a poultice, to relieve pain in the stomach.

Arabic: Khubah—; *Baluchistan*: Kashir—; *Hamadan*: Tukhm-i-khakhshir—; *Persian*: Khakechi—.

FARSETIA.

The genus includes some 20 Mediterranean species, spread over North Africa, Western Asia, and Northern India.

F. ramosissima Hochst. var. *macrocarpa* Schweinf. is used medicinally by the Moors.

1. An erect rigid shrub. Flowers large. Seeds biseriate ... 1. *F. Jacquemontii*
2. Twiggy, covered with denser and finer, closely appressed pubescence. Flowers small. Seeds uniseriate ... 2. *F. Hamiltonii.*
3. A rigid hoary undershrub. Flowers large. Seeds usually biseriate ... ... 3. *F. aegyptiaca.*

1. **Farsetia Jacquemontii** Hook. fil. and Th. is found in Northern India, West Rajputana and Sind, extending to Afghanistan.

The plant has a pleasant pungent taste and is taken as a cooling medicine after pounding. In the Punjab it is considered specific for rheumatism.

Punjab: Faridbuti, Faridmuli, Lathia, Mulei—; *Rajputana*: Kagpilang—.

2. **Farsetia Hamiltonii** Royle occurs in the Punjab, Western Rajputana, Sind and Baluchistan; it extends to Afghanistan, Persia, Arabia and North Africa.

The plant is pounded and taken as a cooling medicine. It is considered specific for rheumatism in the Punjab.

Punjab : Faridbuti—.

3 **Farsetia aegyptiaca** Turra is found in the Punjab. It is distributed to Afghanistan, Arabia, Syria, and North Africa.

In the Punjab the plant is considered specific for rheumatism. It is taken as a cooling medicine after pounding.

Arabic : Djarba, Garba, Garbun, Gjerba, Haifal, Turra—; *Punjab* : Faridburi—.

Isatis.

This genus includes about 50 species, natives of the Mediterranean region, and Central and Western Asia.

Isatis tinctoria Linn. is found in Western Tibet, and is cultivated in many parts of India. It is distributed westwards to Europe and the Canary Islands.

The leaves have been used medicinally in the treatment of jaundice, scurvy, and other ailments; but they are of no value.

California : Ash-of-Jerusalem, Dyer's Weed, Dyer's Woad—; *Catalan* : Herba de Sant Felip, Pastell—; *English* : Dyers' Weed, Dyers' Woad, Woad—; *French* : Gouède, Guède, Guesde, Pastel, Pastel des teinturiers, Vouède—: *German* : Faerber, Waid—; *Italian* : Guado, Isatide, Pastello—; *Roumanian* : Dobrisor, Dobrusor—; *Russian* : Sinilnik, Vaida—; *Spanish* : Añil de Aragón, Hierba pastel, Hierba de San Felipe—.

Lepidium.

The genus includes 100 cosmopolitan species.

The genus is acrid, antiscorbutic, and sternutatory.

The following are used medicinally in Europe—*L. campestre* R. Br., *L. crassifolium* W. K., *L. Draba* Linn., *L. graminifolium* Linn., *L. Iberis* Linn., *L. latifolium* Linn., *L. perfoliatum* Linn., *L. ruderale* Linn., *L. sativum* Linn., *L. virginicum* Linn.—; in West Africa—*L. sativum* Linn.—; in Southern Africa—*L. capensis* Thunb., *L. schinzii* Thel.—; in Brazil—*L. ruderale* Linn.—; in Guiana—*L. virginicum* Linn.—; in North America: *L. medium* Greene.

I. Pods orbicular-ovate, notched at tip, valves winged ... 1. *L. sativum.*
II. Pods transversely oblong, tip entire, valves wingless ... 2. *L. Draba.*
III. Pods elliptic or ovoid, tip entire, valves wingless ...
  1. Leaves fleshy ... ... 3. *L. crassifolium.*
  2. Leaves not fleshy ... ... 4. *L. latifolium.*
IV. Pods elliptic or oblong, tips retuse or notched, valves keeled, wingless or winged at tip only
  1. Valves winged at tip ... ... 5. *L. ruderale.*
  2. Valves almost wingless ... ... 6. *L. perfoliatum.*

1. **Lepidium sativum** Linn. is cultivated throughout India.

The seeds are used by both vaids and hakims on account of their mucilaginous properties. They are prescribed as a laxative, and applied to hurts or sprains as a poultice,

According to Honigberger, the plant in the Punjab was administered in cases of asthma, cough with expectoration, and bleeding piles. The root is used in secondary syphilis and tenesmus.

According to Bellew, the seeds are also considered to be galactagogue in the Punjab, and are administered after being boiled with milk, to cause abortion. O'Shaughnessy found the drug to answer as a gentle and warm aperient.

The leaves are mildly stimulant and diuretic, serviceable in scorbutic diseases.

The seeds are a popular Hausa medicine; preparations are made both for internal and external use. The chief medicinal use is for rheumatic pains and swellings, the seeds crushed and mixed with water as a draught or made into liniments; also a dressing for sores in camels and horses. They are also given for diarrhoea and dysentery.

Afghanistan: Tara tezak—; *Arabic*: Habb-er-reshad, Habrasjat, Half, Harfulabaz, Hurf, Jarjir, Reschad—; *Bengal*: Alevarie, Alleevree, Halim—; *Bombay*: Ahaliva, Asalia—; *Bornu*: Karabou—; *Burma*: Mongnyin, Samungni—; *Canarese*: Allibija, Kurutige—; *Catalan*: Enciamet de caputxi, Herba de la fluxió, Morritort, Murissa—; *Deccan*: Halim—; *Egypt*: Habb-er-reshad, Reshad—; *English*: Common Cress, Garden Cress, Passerage, Pepperwort, Town Cress—; *French*: Anitor, Cresson alénois, Cresson des jardins, Cresson de terre, Nasitar, Nasitor, Nasitort, Passerage cultivée—; *Fulani*: Takere fondu—; *German*: Gartenkresse, Kresse, Pfefferkraut—; *Gujerati*: Asalio, Halim—; *Hausa*: Labsur, Lausur—; *Hindi*: Chansur, Chaunsar, Halim, Halon, Hurf, Malavan, Tara, Tezac—; *Iraq*: Rashad—; *Italian*: Lepidio, Nasturzio—; *Kano*: Lausuru—; *Kanuri*: Lapsur—; *Kaisina*: Lafsu—; *Languedoc*: Cercoucello—; *Malta*: Garden cress, Agretto, Crescione inglese, Habberxa—; *Marathi*: Ahliva—; *Mundari*: Camcur, Cancur, Ilicamcur—; *North-Western Provinces*: Halang, Halim—; *Persian*: Rukhame-ispanda, Tereh tizec—; *Portuguese*: Mastruco—; *Punjab*: Halim, Halon, Shargundei, Tezak—; *Sanskrit*: Ashalika, Bhadra, Chandrashura, Chandrika, Charmahantri, Darakrishna, Drighabija, Kalamesha, Karavi, Nandini, Pashumehankarika, Raktabija, Raktaraji, Sidhaprayojana, Suvasura, Vasupreshpa—; *Sind*: Ahree, Ahreo, Assaria—; *Sokoto*: Lafsur—; *Spanish*: Berro de jardin, Malpica, Mastuerzo, Mastuerzo hortense—; *Tamil*: Aliverai—; *Teheran*: Tukhm-shahi—; *Telugu*: Adalavitulu, Adeli, Adityalu, Adiyalu—; *Urdu*: Halim—; *Waziristan*: Halayo—; *Yemen*: Half—; *Zaria*: Lafsur—.

2. **Lepidium Draba** Linn. is found as a weed of cultivation in the Punjab. It occurs in Persia, Mesopotamia, Caucasus, Europe, and the Mediterranean region.

The plant is said to possess antiscorbutic properties; and to control bleeding, if used raw.

In Loralai the seeds are used as a cure for flatulency, seven or eight being taken at a time.

In Waziristan it is used as a stomachic and tonic. In Europe as an antiscorbutic.

Afghanistan: Bijindak—; *Arabic*: Neffélé—; *Baluchi*: Garbust—; *Brahui*: Garbust—; *Catalan*: Capellans, Palitra—; *English*: Hairy Cress, Hoary Cress—; *French*: Drave—; *Hindi*: Bajindak—; *Iraq*: Jinaibruh, Qunaibrah—; *Ispahan*: Montchih—; *Jhalawan*: Garbast—; *Kalat*: Garbust—; *Kila Saifulla*: Buska—; *Kohlu*: Halia—; *Loralai*: Buski—; *Malta*: Hoary Cress, Cocola, Lattona—; *Mushki*: Garbast—; *Pishin*: Bushka—; *Quetta*: Bushka, Busk—; *Spanish*: Lepidio drava—; *Tabriz*: Khili-wili—; *Tasmania*: Hoary Cress, White Weed—; *Waziri*: Bashka—; *Zhob*: Bushka, Buska—.

3. **Lepidium crassifolium** W. K. is found in Baluchistan and the Oriental region, spreading over to Europe.

The plant is employed as a rubefacient in rheumatism. The seeds are prescribed internally in rheumatism and dropsy.

4. **Lepidium latifolium** Linn. is found in Kashmir. It is distributed to Northern and Western Asia, Europe, and North Africa.

The plant is depurative and antiscorbutic. It is used as a resolvent in skin diseases.

Catalan: Herba de la fluxió, Morritort bort, Morritort salvatge, Piperotis—; *English*: Cress, Dittander—; *French*: Passerage à larges feuilles, Grande passerage—; *German*: Breitblaeterige Kresse, Pfefferkraut—; *Italian*: Mostardina, Peperella—; *Ladak*: Gonguch—; *Portuguese*: Herva pimenteira, Herva serra, Lepidio—; *Spanish*: Lepidio, Lepidio de Plinio, Mastuerzo mayor, Piperitas, Piperitis—.

5. **Lepidium ruderale** Linn. is found in Kashmir at 7,000-13,000 feet. It extends through the Oriental region to Europe. It also occurs in Australia.

The plant is used in impetigo.

Brazil: Mastruco, Menstruco—; *English*: Narrow-leaved Pepperwort—; *Tasmania*: Narrow-leaved Cress—; *Uruguay*: Yuyo de resfrio.

6. **Lepidium perfoliatum** Linn. is found in Baluchistan and Afghanistan, whence through the Oriental region it extends to Southern Europe.

In Europe this plant is popularly believed to be a useful antiscorbutic.

MATTHIOLA.

This genus includes 50 species of the Mediterranean region, inhabiting Europe, South Africa, and the Western Himalaya.

The leaves of the following species were formerly used as emmenagogues in Europe—*M. incana* R. Br., *M. tristis* R. Br., *M. varia* DC.—.

Matthiola incana R. Br. is indigenous in the Mediterranean region and in Western Europe; it is cultivated in Indian gardens.

The Mahommedan authors distinguish between three kinds of seeds; yellow, red, and white; but in practice no distinction seems to be made between them. They are all considered bitter, tonic, diuretic, expectorant, stomachic, aphrodisiac; good in dry bronchitis, fevers, and injuries to the eye.

The seeds are used in infusion in cancer; mixed with wine they are given as an antidote to poisonous bites. In the Punjab and in Sind they are reckoned aphrodisiac.

Mhaskar and Caius have shown experimentally that the seeds are not an antidote to snake venom.

Arabic: Bajar-ul-kumam, Manthur—; *Catalan*: Violé comú—; *Egypt*: Kheyly, Manthur, Manthur madbag—; *English*: Gilliflower, Hoary Stock, Queen Stock, Stock—; *French*: Mathiola—; *German*: Winter Levcofe—; *Indo-China*: Tu la lan hoa—; *Iraq*: Shabbo—; *Malta*: Queen's Stock, Stock, Violaciocca, Gizi—; *Persian*: Todri—; *Punjab*: Todri liba, Todri safed—; *Sind*: Todri safed—; *Spanish*: Aleli encanecido—; *Urdu*: Todri safed—; *Uruguay*: Aleli—.

MORICANDIA.

The genus consists of 10 Mediterranean species.

Moricandia tortuosa Hook. fil. and Th. occurs in the salt plains of the Punjab, west of the Indus.

A decoction of the flowers is rubbed on for eczema in Waziristan.

NASTURTIUM.

The genus consists of 50 cosmopolitan species.

The plant is diuretic, stimulant, and antiscorbutic. The seeds are mildly laxative.

The following are used medicinally in Europe—*N. fontanum* Aschers., *N. palustre* DC., *N. sylvestre* R. Br.—; in La Reunion —*N. fontanum* Aschers.—; in Madagascar—*N. barbareaefolium* Baker, *N. fontanum* Aschers.—; in China—*N. montanum* Wall.—; in Indo-China—*N. fontanum* Aschers., *N. indicum* DC.—; in Brazil —*N. fontanum* Aschers.—; in North America—*N. curvisiliqua* Nutt., *N. fontanum* Aschers., *N. palustre* DC.—.

1. Aquatic, stem creeping and floating ... ... 1. *N. fontanum.*
2. Stems weak, suberect, smooth, slightly hairy ... 2. *N. palustre.*
3. Erect, rough-hairy or sometimes glabrous annual ... 3. *N. indicum.*
4. An annual much resembling *N. indicum* but less robust with larger, smoother, less divided leaves ... 4. *N. montanum.*

1. **Nasturtium fontanum** Aschers. is found in many hill stations, in the Punjab, Waziristan, Baluchistan; and thence to Afghanistan, temperate Asia and Europe.

The fresh plant is widely known for its antiscorbutic and stimulant properties. It is used in Brazil in troubles of the chest.

Amarah: Kobani—; *Brazil*: Agriao—; *Catalan*: Creixans, Creixens, Gredas, Morritort d'aigua—; *Deccan*: Lutputiah—; *Dutch*: Water-kers—; *Egypt*: Gorrat-el-aïn—; *English*: Shamrock, Stertion, Water-cress—; *French*: Cailli, Cresson, Cresson aquatique, Cresson d'eau, Cresson de fontaine, Cresson officinal, Cresson de ruisseau, Nasitord, Santé du corps—; *German*: Braunkersch, Brunnenkresse, Kersche, Quellranke, Wasserraute, Wassersenf, Wiesenkresse—; *Greek*: Sisymbrion—; *Hindi*: Deskandai—; *Hova*: Anandrano—; *Irish*: Biolar—; *Italian*: Agretto—; Crescione, Nasturzio, Sisimbrio, Sisimbro—; *Malta*: Water-cress, Crescione d' acqua, Cresciuni, Sija—; *New-Zealand*: Kowhiti-whiti—; *North-Western Himalayas*: Piriyahalim—; *Portuguese*: Agriao—; *Provençal*: Creiso—; *Roumanian*: Brancuta, Cardama de izvor, Hrenita, Nasturel—; *Russian*: Kress—; *Sinhalese*: Kakutupala—; *Spanish*: Berro, Berro común, Berro de fucute, Mastuerzo acuatico—; *Swedish*: Kiaellkrasse—; *Waziri*: Dalamira, Thermera—.

2. **Nasturtium palustre** DC. occurs in North-Western India, the temperate Himalaya up to 10,000 feet, Assam, and Bengal; it is spread over the temperate regions of both hemispheres.

The fresh plant possesses the antiscorbutic properties of the genus.

China: T'ing Li—; *Egypt*: Manthur, Mufrad—; *New South Wales*: Native Cabbage—; *Pacifiic Coast*: Marsh Cress, Marsh Yellow Cress—.

3. **Nasturtium indicum** DC. is found throughout India in wet places from Ceylon to Kashmir and Mishmi, Bengal, Assam, Chittagong, Tenasserim, Malay Peninsula. It extends to Malaya and China.

In Indo-China the plant is considered diuretic, stimulant, and antiscorbutic. The seeds, which are said to be laxative, are also used in the treatment of asthma.

Indo-China: Cai cot xoi, Dirh lich, Lai hoang, Thuy gioi thai—.

4. **Nasturtium montanum** Wall. occurs in North-Western India up to 7,000 feet, in Sikkim, the Khasia Hills, and Burma. It extends to Java, China, and Japan.

The plant has antiscorbutic properties.

China: Han Ts'ai—.

RAPHANUS.

The genus consists of 8 species, mostly Mediterranean.

R. sativus Linn. is used medicinally in Europe, China, Indo-China, Malaya, and the Malay Archipelago.

Raphanus sativus Linn. is found in temperate and warm countries; it is cultivated all over India up to 16,000 feet.

The radish has a hot, sharp, bitter taste; stomachic, binding, anthelmintic; nervine tonic; good in tumours, piles, and all inflammations; useful in diseases of the heart, amenorrhea, hiccough, prosy, cholera; the juice relieves earache. The flowers are bechic and cholagogue (Ayurveda).

The root is useful for urinary complaints and piles.. The seeds are sharp and bitter; laxative, tonic, emmenagogue, carminative; good for the spleen and in paralysis; produce alopecia; mixed with wine they counteract the effects of the bites of snakes and other poisonous animals (Yunani)

Radish seeds are peptic, expectorant, diuretic, laxative, carminative, and corrective. In the Punjab they are considered to be emmenagogue. They are much used in Indo-China by native practitioners.

The roots are used for urinary and syphilitic diseases; they are a reputed medicine for piles and gastrodynic pains.

The juice of the fresh leaves is also used as a diuretic and laxative.

The syrup of radishes is excellent for hoarseness, bronchial difficulty of breathing, whooping-cough, and other complaints of the chest.

These seeds are not an antidote to snake venom (Mhaskar and Caius).

Arabic: Bajrulkijal, Bokel, Boukir, Bou-toum, Fidgel, Figl, Fugel—; *Bengal*: Mula, Muli—; *Bombay*: Mula, Muro—; *Burma*: Monelahmungla, Moula—; *Canarese*: Bili, Mullangi, Mullangiyanne—; *Catalan*: Raba—; *Chinese*: Lai Fu, Tsai Fu Ken—; *Danish*: Redike—; *Deccan*: Mulli—; *Dutch*: Radijs—; *Egypt*: Fidjel—; *English*: Garden Radish, Rabone, Radish—; *French*: Navet rave, Radis, Petit radis, Radis rond, Radis rose, Petite rave, Rave rose, Tendrette—; *German*: Gartenrettich, Rettich, Retting, Sommerrettich, Winterrettich—; *Greek*: Raphanis—; *Gujerat*: Mala, Mura—; *Hausa*: Fijil—; *Hindi*: Fugil, Mula, Muli, Muro, Purabi sarisha, Sweti sarisha, Tora, Toorub—; *Hova*: Radisa—; *Hungarian*: Retek—; *Indo-China*: Cu cai, Lai bac, Lo bac, Lu bu—; *Iraq*: Fujl—; *Italian*: Rafano, Ramolaccio, Ravanello, Ravano—; *Japanese*: Daion—; *Kolami*: Morai—; *Konkani*: Mullo—; *Kurdish*: Tur—; *Languedoc*:

Rabasso, Rafe—; *Malaya* : Lobak, Loh fook—; *Malayalam* : Mullangi—; ***Malta*** : Radish, Radice, Rafano, Figel, Ravanall—; *Marathi* : Mula, Muri—; ***Mongolia*** : Laopang, Lobin, Turma—; *Mundari* : Morai, Moroe, Murai, Muroe—; *Mysore* : Magunigadde—; *Nepal* : Mulasinki—; *North-Western Provinces* : Muli—; *Persian* : Tukhmeturub, Turup—; *Philippines* : Rabano—; *Portuguese* : Rabanete, Rabano, Rabao, Rabao negro—; *Punjab* : Muli, Mungra, Taramira—; *Roumanian* : Ridiche—; *Russian* : Radis, Rediska—; *Sanskrit* : Bhumikakshara, Dirghakandaka, Dirghamulaka, Dirghapatraka, Hariparna, Hastidanta, Hastidantaka, Kandamula, Karukandaka, Ksharamula, Kunjar, Kahakanda, Mritakshara, Mulaka, Mulavha, Nilakantha, Rajaluka, Ruchira, Ruchishya, Shankhamula, Shimbiphala, Sita—; *Sind* : Muri—; *Sinhalese* : Rabu—; ***Spanish*** : Ràbano, Ràbano cultivado, Ràbano negro—; *Swedish* : Raettika—; ***Tamil*** : Mullangi, Mulinghie—; *Telugu* : Mullangi—; *Tibet* : La phug—; ***Turkish*** : Turp—; *Urdu* : Mulekebija—; *Uruguay* : Rabanito, Rabano—; ***Yemen*** : Bokel, Fidyl—.

SISYMBRIUM.

The genus includes 77 species, found chiefly in the temperate regions of both hemispheres.

The genus is diuretic and expectorant.

The following species are used medicinally in Europe—*S. altissimum* Linn., *S. austriacum* Jacq., *S. Irio* Linn., *S. Loeselii* Linn., *S. officinale* Scop., *S. strictissimum* Linn.—; in North America—*S. officinale* Scop., *S. pinnatum* Greene—.

A. Fruiting pedicles as broad as the pod ... ... 2. *S. altissimum.*

B. Fruiting pedicles slender
  1. Pods 3.8-5 cm., erect ... ... 1. *S. Irio.*
  2. Pods 2.5-3.8 cm., slightly curved ... ... 3. *S. Loeselii.*

1. **Sisymbrium Irio**, Linn. is found in Rajputana, Punjab, Peshawar, Baluchistan, and Kohat. It is distributed to Central Asia, Afghanistan, Arabia, the Mediterranean, and Eritrea.

The seed is expectorant and restorative, and used externally as a stimulating poultice.

It is also said to be febrifuge.

In Spain an infusion of the leaves is given in affections of the throat and of the chest.

Arabic : Khubah, silihg—; *Catalan* : Apagallums, Mata-llums—; ***Egypt*** : Figl-el-gemmal, Harra—; ***English*** : London Rocket—; ***Iraq*** : Huwairah—; *Malta* : London Rocket—; *Merwara* : Parjan—; ***Persian*** : Kharkasi—; ***Punjab*** : Janglisarson, Maktrusa—; ***Sind*** : Junglisurson—; ***Spanish*** : Matacandiles, Oruga leonina, Rabanillo amarillo—; *Urdu* : Khubakalan—.

2. **Sisymbrium altissium** Linn. occurs in Kashmir up to 8,000 feet and in Chitral. It is distributed to Persia, Western Asia, the Mediterranean, and Central Europe.

The leaves and flowers are considered astringent and antiscorbutic.

3. **Sisymbrium Loeselii** Linn. occurs in Kashmir up to 8,000 feet and in Peshawar. It is distributed to Central Asia, Siberia, Western Asia, Central Europe.

The leaves and flowers are given in scurvy and in scrofula.

Thlsapi.

The genus consists of 60 species inhabiting the Northern temperate regions.

Thlaspi arvense Linn. is found as a weed of cultivation throughout the temperate and subalpine Himalaya, ascending to 14,000 feet. It is spread over Europe and Asia, always in cultivated places.

The seeds and the shoots are used medicinally by the Chinese. In Spain the plant is considered astringent and the seeds stimulant.

Catalan: Traspic—; *Chinese*: Hsi Ming—; *English*: Mithridate Mustard, Penny Cress, Shepher's Furse—; *French*: Monnoyère, Tabouret des champs—; *Italian*: Erba storna—; *Spanish*: Mostaza salvaje, Talaspico, Tlaspios—.

MEDICINAL AND POISONOUS PLANTS OF INDIA:

CAPPARIDS, MIGNONETTES, VIOLETS, ROCKROSES, BIXADS.

BY

J. F. CAIUS, S.J., F.L.S.

I

The CAPPARIDACEÆ are herbs or shrubs, rarely trees, distributed nearly equally over the tropical and subtropical regions of both hemispheres. They abound in Africa and India; some are found in Europe and Canada; the frutescent species are mostly American. There are 40 genera with about 450 species.

The herbaceous members are acrid, and their therapeutic properties are similar to those of the Crucifers. These same properties persist in the roots and the leaves of the arboraceous members, the bark of which is usually bitter and laxative.

Quercetin has been isolated from the flowers of *Capparis spinosa* Linn.

The medicinal and poisonous Capparids of the world belong to 10 genera:—BOSCIA (tropical and South Africa); CADABA (tropics of the Old World); CAPPARIS (warm regions); CLEOME (tropical and subtropical regions); COURBONIA (tropical Africa); CRATAEVA (tropical regions); EUADENIA (tropical Africa); GYNANDROPSIS (tropical and subtropical regions); MAERUA (tropical Africa and Asia); RITCHIEA (western tropical Africa).

The medicinal and poisonous Capparids of India belong to 6 genera:—CADABA, CAPPARIS, CLEOME, CRATAEVA, GYNANDROPSIS, MAERUA.

A. Herbs. Fruit capsular
  I. Stamens sessile on the disk. Petals imbricate in bud ... ... CLEOME.
  II. Stamens on the gynophore. Petals open in bud ... GYNANDROPSIS.
B. Shrubs or trees. Fruit berried or capsular
  I. Calyx-tube lined by the disk, lobes valvate. Petals 4. Fruit moniliform ... ... MAERUA.
  II. Sepals 4, open in bud, disk hemispheric ... CRATAEVA.
  III. Sepals 4, biseriate, imbricate or 2 outer valvate
    a. Stamens 4-6, inserted high on the gynophore ... CADABA.
    b. Stamens 8-∞, inserted at the base of the gynophore ... ... CAPPARIS.

CADABA.

The genus consists of 20 species, natives of the Palaeotropics.

Leaves ovate or oblong-obtuse ... ... 1. *C. farinosa*.
Leaves palmately 3-foliate ... ... 2. *C. trifoliata*.

1. **Cadaba farinosa** Forsk. (= *C. indica* Lam.) is very common in India. It is found in the Punjab, Baluchistan, Sind, the Rajputana Desert, Central India, Gujerat, Konkan, the Southern Mahratta Country, the dry districts of the Northern Circars, the

Deccan and the Carnatic from Vizagapatam southwards to the Madura District. It extends to Arabia and tropical Africa.

In Sind the leaves and roots are considered deobstruent and anthelmintic, and are prescribed in uterine obstructions. In Pudukota the root and leaves are used in decoction as an anthelmintic.

The plant is used as a general antidote in Arabia Felix. The fresh young branches are either masticated or taken in powder form.

In Senegal a decoction or an infusion is used in pulmonary affections, dysentery, fever and rheumatism.

Arabic: Asal, El-bejadd, Gorrah, Gorreh, Korrah, Saerah, Sserah—; *Bombay*: Habab—; *Canarese*: Chegaviche, Maragache, Maragade, Shegurti—; *Cutch*: Budkiyal, Janglimirchi, Karopinjero—; *Gujerati*: Khordu—; *Hausa*: Bagayi—; *Hindi*: Kodhab—; *Jodhpur*: Dabi—; *Katagum*: Anza—; *Madras*: Viludi—; *Malayalam*: Kattakatti—; *Porebunder*: Kalokatkiyo, Kimiyatunjadvung, Thaniung—; *Sind*: Kodhab—; *Tamil*: Vili—; *Telugu*: Adamorinika, Chavukuttiyanku, Chemudu, Chikonadi, Chimurudu, Polumokinika—; *Tigré*: Astan—; *Tigrinia*: Sernak, Tueblischnai—; *Yemen*: Asal, El-bejad, Korreh, Serah, Toreh—.

2. **Cadaba trifoliata** Wight and Arn. occurs in the Carnatic and in Ceylon.

In India the roots and leaves are used by Ayurvedic practitioners. They are considered purgative, emmenagogue, anthelmintic and antiphlogistic; they are said to be useful in the treatment of amenorrhoea, and dysmenorrhoea, indigestion of children, and painful joints. The leaves are much employed in the preparation of medicated oils.

Sanskrit: Balaya—; *Tamil*: Manudukkurundu, Viluti—; *Telugu*: Chekonadi—.

CAPPARIS.

The genus includes 200 species, natives of tropical and warm regions, except North America.

The genus exhibits stimulant, antiscorbutic and antispasmodic properties. The following species are used medicinally in Europe—*C. spinosa* Linn.—; in Southern Africa—*C. citrifolia* Lam., *C. Gueinzii* Sond., *C. tomentosa* Lam.—; in Brazil *C. Yco* Mart.—; in Peru and Bolivia—*C. coriacea* Linn.—.

A. Flowers axillary, solitary or in fascicles of 2-3
- I. Prostrate shrubs. Leaves orbicular or ovate-orbicular. Thorns usually hooked. Lower sepal not very saccate ... ... 1. *C. spinosa.*
- II. Erect shrub. Leaves elliptic-lanceolate. Thorns minute or absent ... ... 2. *C. heyneana.*

B. Flowers corymbose (sometimes racemose in *C. grandis*)
- I. Mature branches leafless ... ... 3. *C. decidua.*
- II. Mature branches leafy. Leaves olive-green, pubescent when young ... ... 4. *C. grandis.*

C. Flowers in shortly peduncled or sessile umbels. Woody climber ... ... 5. *C. sepiaria.*

D. Flowers supraaxillary in a vertical line on the branches.
- I. Leaves $1\frac{1}{2}$-3 by $\frac{1}{2}$-$1\frac{1}{2}$ in. ... ... 6. *C. zeylanica.*
- II. Leaves 4-8 by 2-$3\frac{1}{2}$ in. ... ... 7. *C. micracantha.*

1. **Capparis spinosa** Linn. occurs in the plains between the Indus and the Jhelum, the Salt Range, the low inner valleys of the Himalaya, Chamba, Kumaon, Nepal; it is fairly common in the Bombay Presidency, being chiefly found in Sind, the Konkan, the Deccan and the Western Ghats; it is also met with in Waziristan and Baluchistan, whence it extends to Afghanistan, the Mediterranean region, North Africa, and Australia.

The root bark is a favourite Yunani remedy. It is said to be aperient, tonic, expectorant, anthelmintic, emmenagogue and analgesic. It is commonly prescribed for rheumatism, paralysis, toothache, enlarged spleen, and tubercular glands. The juice of the plant, more especially the fruit, is used to kill worms in the ear.

In Baluchistan the juice of the fresh berries—or, if they are dry, the liquid obtained by adding a very little water—is poured into the ear as a cure for earache. In Ormara and Las Bela a jelly prepared from the fruits is considered to be a cure for rheumatism and for snake bite.

In the Punjab the dried bark of the root is considered diuretic In Kangra, the macerated roots are applied to sores.

In Persia the root and root-bark are given for intermitten fever and rheumatism.

In Arabia the berries are given for toothache, and the bruise leaves applied to the head in cephalalgia.

In Europe the flower buds and fruits are considered aperient and diuretic. The young flower buds are the capers of the shops; they are esteemed antiscorbutic, stimulant, and aperient. The bark of the root passes for a diuretic. The leaves are bruised and used as poultices for gouty affections.

Afghanistan : Kabarra, Kabawa, Kabra— ; *Arabic* : Amdryan, Azuf, Kabar Kabbar, Kabur, Lassaf, Leysouf, Schok-el-homar, Uard-el-gebel— ; *Baluchi* Krap— ; *Bettani* : Kamarkuniae— ; *Bombay* : Kabar— ; *Canarese* : Mullu kattari— ; *Catalan* : Tapara, Taparera— ; *Diwana* : Pahinro— ; *Dutch* : Kapperboom— ; *Egypt* : Kabar, Lassaf, Shok-el-homar, Ward-el-gebel— ; *English* : Caper Plant— ; *French* : Caprier, Tapenier— ; *German* : Kapernbaum— ; *Greek* Kapparis, Rimoniaria— ; *Hamadan* : Risha kabar— ; *Hebrew* : Ezov— ; *Hindi* : Ber, Kabara, Kabra— ; *Iraq* : Kabar— ; *Italian* : Capparo, Cappero— ; *Jaunsar* : Bauri, Kiari— ; *Jhalawan* : Khafkhander, Khakandir— ; *Kumaon* : Bussar, Ultakanta— ; *Ladak* : Kabra— ; *Las Bela* : Kirap, Krap, Pahinro, Panetero— *Malta* : Caper-plant, Cappara, Capparo, Cappero— ; *Northern Baluchistan* : Khawarg— ; *Ormara* : Kirap, Krap— ; *Persian* : Kabar, Kebir, Kurak— ; *Portuguese* : Alcaparra— ; *Punjab* : Bandar, Barar, Barari, Bassar, Bauri, Ber, Kabarra, Kabra, Kakri, Kander, Kaur, Keri, Kiari, Taker— ; *Russian* : Kapersovyi kust— ; *Sind* : Kalvari— ; *Spanish* : Alcaparra, Alcaparro— ; *Syria* : Kabar— ; *Teheran* : Risha kavar— ; *Telugu* : Kokilakshamu— ; *Tibet* : Kabra— ; *Turkish* : Kabarish— ; *Urdu* : Kabar— ; *Waziri* : Katmavae— ; *Yemen* : Lasaf—.

var. *canescens* Coss. occurs in Baluchistan. The leaves are bitter; they are used by the Baluchis as a medicine for goats.

Baluchi : Khafkhandar, Krap— ; *Brahui* : Karap, Khafkhandar—.

var. *parviflora* Boiss. is found in Baluchistan, where the plant, boiled in water, is applied as a poultice for boils.

Baluchi : Krap— ; *Brahui* : Khafkhandar, Khalkandir—.

2. **Capparis heyneana** Wall. is found in the Western Ghats of South Konkan and North Kanara, spreading to the Tinnevelly District and Ceylon.

The leaves are used for rheumatic pains in the joints, and the flowers are made into a laxative drink.

Hindi : Chayruka—.

3. **Capparis decidua** Edgew. (=*C. aphylla* Roth.) is found in Sind, Baluchistan, Western Rajputana, Punjab, Central India, Gujerat, Deccan, and Tinnevelly. It extends to Socotra, Arabia, Egypt, and tropical Africa.

The bark and the fruit are Ayurvedic medicines of some repute. The whole plant is used by Yunani practitioners.

In the Punjab, the top shoots and young leaves are made into a powder and used as a blister; they are also applied to boils, eruptions and swellings, and used as an antidote to poison; they are prescribed in affections of the joints; when chewed, they are efficacious in relieving toothache.

The fresh young twigs (tips only) are crushed and soaked in water. The water is strained off. Sometimes this is done twice or thrice. The residuum is dried and allowed to solidify. A tiny piece of it is eaten with butter and gives relief from pain after a bruise or fall. Also makes a very strong plaster (Hotson).

Arabic : Hanbag, Margh, Sodab, Sodad, Tundub—; *Baluchistan* : Kaler, Kalir, Karar, Khirar—; *Berar* : Kari—; *Bombay* : Kari—; *Canarese* : Chippuri, Karira—; *Deccan* : Karyal—; *Egypt* : Hagbaq, Tundub—; *Gujerat* : Ker, Kera—; *Hindi* : Karel, Karer, Karu, Keril, Kurrel, Kurril, Lete, Satari—; *Jaisalmer* : Ban—; *Jodhpur* : Kair—; *Konkani* : Kiral—; *Marathi* : Karil, Ker, Kera, Nevati—; *North-Western Provinces* : Kair, Kari, Karil—; *Persian* : Bergesodab—; *Punjab* : Delha, Karel, Karia, Karil, Karis, Karril, Kerin, Kirra, Pinju, Tenti—; *Sanskrit* : Granthila, Gudhapatra, Kantaki, Karaka, Karira, Kataphala, Krakatha, Krakara, Krishashakha, Marubharuha, Mriduphala, Nishpatra, Nishpatrika, Shakapushpa, Shatakunta, Shonapushpa, Suhala, Tikshnakantaka, Tikshnasara, Ushnasundara, Vidahika, Vishvakpatra—; *Sind* : Dorakiram, Kirab, Kiral, Kirrur—; *Tamil* : Kulaladondai, Sengam, Sirakkali—; *Telugu* : Kariramu—; *Urdu* : Titali—; *Waziristan* : Karil, Kirra—.

4. **Capparis grandis** Linn. fil. is found in Mount Abu, Western Rajputana, Kanara, the Carnatic, the hill-forests of the Deccan, and on the eastern slopes of the Western Ghats from the Godaveri southwards.

An infusion of the bark and leaves is used internally for swellings and eruptions in Western Rajputana.

Bombay : Puchaonda, Ragota—; *Buldana* : Pacharan—; *Burma* : Hkawkwa—; *Canarese* : Revapi, Tarate, Totte, Tottulla—; *Deccan* : Pachaonda—; *Gujerati* : Dhuti—; *Konkani* : Ragot—; *Lambadi* : Raldero—; *Malayalam* : Waghutty—; *Marathi* : Kandel, Katarni, Kauntel, Pachenda, Puchunda, Ragot—; *Melghat* : Pacharan—; *Nimar* : Pachar—; *Porebunder* : Dhuti—; *Rajputana* : Antera—; *Tamil* : Nakkulinjan, Turattu, Vellaitturattu—; *Telugu* : Dridonda, Dudduppi, Guli, Nalluppi, Regutti—; *Yeotmal* : Pachunda—.

5. **Capparis sepiaria** Linn. is found in the dry parts of India and Ceylon. It extends to Malaya, Indo-China, Timor and Australia.

The plant possesses febrifugal properties. It is considered to be alterative and tonic; it has been found useful in skin diseases.

Bengal: Kaliakara, Kantagurkamai—; ***Canarese***: Kadukattari—; ***Gujerat***: Kantharo—; ***Hindi***: Hiun, Kanthari—; ***Marathi***: Kanthar, Kantharyel—; ***Merwara***: Katan, Kataran—; ***Porebunder***: Kalokantharo, Kantharo—; ***Punjab***: Hiungarna, Hius—; ***Sanskrit***: Ahimsra, Amlaphala, Dupravesha, Durdharsha, Guchchagulmika, Gridhranakhi, Hinsra, Jali, Kakadani, Kakatinduka, Kantha, Kanthari, Kantharika, Kanya, Kapalakulika, Krishnashriphalika, Kruragandha, Krurakarma, Tikshnagandha, Tikshnakantaka, Vakrakantaki, Vayastinduka—; ***Tamil***: Karindu, Karunjurai, Kattukkattiri, Kokkimullu, Sirukkattiri—; ***Telugu***: Nallapuyyi, Nallavuppi, Nalluppi, Puyyi, Uppi—; ***Uriya***: Hulubhi, Kantikapali, Koli, Kontoko, Nibido, Otaibe, Solorakoli—.

6. **Capparis zeylanica** Linn. (= *C. horrida* Linn. fil.) is to be found throughout the greater part of India; it extends to Java and the Philippine Islands.

Ayurvedists use the root-bark as a bitter and a cholagogue.

The root-bark is sedative, stomachic, and antihidrotic.

In Northern India, the leaves are used as a counter-irritant and as a cataplasm in boils, swellings and piles.

In Chota Nagpur, the bark, along with native spirit, is given in cholera. The Mundas use the thorns to prick the pustules of small pox. The root, well ground, and mixed with those of *Cassia occidentalis* and *Mimusops elengi*, is rubbed on the body in dropsy; also, even alone, on aching bodies, legs, or feet.

Ajmere: Gitoran—; ***Bengal***: Bagnai, Kalokera—; ***Bombay***: Anti, Taranti, Tarti, Wag, Wagatti—; ***Burma***: Nahmanitanget, Nahmanitanlyet, Nwamanithanlyet—; ***Canarese***: Mullukattari, Totte, Tottulla—; ***Ceylon***: Kattoddi, Kilaohedi, Vennachi—; ***Deccan***: Ardanda—; ***Dehra Dun***: Hins, His, Kalhins, Kalhis, Khalis—; ***Gond***: Katerni—; ***Hindi***: Ardanda, Jhiri—; ***Ilocano***: Talactac—; ***Kharwari***: Bagnahim—; ***Kolami***: Gaterna—; ***Konkani***: Govindphal, Wageti—; ***Kumaon***: Bipuwakanta, Ultakanta—; ***Marathi***: Govindi, Pachwa—; ***Matheran***: Taranti, Tarti—; ***Monghyr***: Bagnai—; ***Mundari***: Gaterna, Kularama, Marijanum, Ramakula—; ***Oudh***: Karrallura—; ***Punjab***: His, Hiungarna, Karvila—; ***Sanskrit***: Govindi, Granthila, Kantakalata, Karambha, Katukandari, Kinkani, Krishangi, Tapasapriya, Vartala, Vyaghraghanti, Vyaghrapada, Vyagranakhi—; ***Santali***: Buru asaria—; ***Sind***: Ardanda—; ***Sinhalese***: Welangiriya—; ***Tagalog***: Dauag—; ***Tamil***: Adondai, Igudi, Indu, Kaguturatti, Kottotti, Migupalattam, Tondai, Tulambikkiri, Viyanicham—; ***Telugu***: Adonda, Aridonda, Chittigara, Doddi, Palaki—; ***Uriya***: Asadua, Govindi, Lobhyotai, Osaro, Oserwa—; ***Visayan***: Laguino—.

7. **Capparis micracantha** DC. occurs on seashores and spots in Pegu, Tenasserim, the north of the Malay Peninsula, extending to Siam, Cambodia, Java, Timor and the Philippine Islands.

The roots and the wood are used medicinally in Cambodia. The roots are considered diuretic. The wood is given for bronchitis and for ulceration of the mucous membrane of the nose.

Cambodia: Kanhchoeu bai dak—; ***Tagalog***: Dáuag, Halugábat, Mararáyatcáhoy—; ***Visayan***: Alagúng-ung, Salimómo—.

CLEOME.

The genus consists of 70 species, inhabitants of tropical and subtropical regions.

The leaves are stimulant; the roots stimulant, antiscorbutic, and anthelmintic; the seeds rubefacient, vesicant, anthelmintic, and carminative.

The following are used medicinally:—in Indo-China—*C. Chelidonii* Linn., *C. viscosa* Linn.—; in the Philippine Islands—*C. viscosa* Linn.—; in North and Central America—*C. gigantea* Linn., *C. graveolens* Rafin., *C. heptaphylla* Linn., *C. serrata* Jacq., *C. viscosa* Linn.—; in the Gold Coast—*C. ciliata* Schum. & Thonn.—; in La Reunion—*C. viscosa* Linn.—; in Guiana—*C. frutescens* Aubl.—.

A. Leaves simple
  Stamens 6 or less
    Leaves oblong-lanceolate, penni-nerved; bracts petiolate ... ... 1. *C. monophylla.*
B. Leaves compound
  I. Stamens 6. Capsules sessile or subsessile
    Leaflets linear-oblong to obovate. Stem and leaves not asperous ... ... 2. *C. brachycarpa.*
  II. Stamens indefinite
    a. Flowers yellow. Capsule glandular pubescent ... 3. *C. viscosa.*
    b. Flowers purple or pink. Capsule glabrous, striate ... ... 4. *C. felina.*
    c. Flowers rosy. Capsule smooth ... ... 6. *C. Chelidonii.*
    d. Leaves 7-foliolate ... ... 5. *C. heptaphylla.*

1. **Cleome monophylla** Linn. is found in fields and waste places from Bihar and Orissa to Gujerat, the Deccan, the Konkan, the Southern Maratha country, and Ceylon. It also occurs in tropical Africa.

The pounded root is put on the lips by the Santals to restore consciousness when in a faint.

Kolami: Chamani—; *Mundari*: Carmani ara, Carmari ara—; *Sadani*: Hurhuriasag—; *Santali*: Harhara, Hurhura, Kedar Jhawar—.

2. **Cleome brachycarpa** Vahl is found in Sind, Baluchistan, the West Rajputana Desert and the Punjab Plain. It extends westwards to Arabia, Abyssinia and North Africa.

Mohammedan practitioners use the plant in the treatment of scabies, rheumatism and inflammations; they consider the leaves beneficial in leucoderma.

In Ormara the plant is said to be a useful medicine for persons suffering from heat. The Baluchis crush the leaves in oil and rub the paste on the arms and legs as an embrocation for high fever.

In Jodhpur and Jaisalmer the plant is used to cure worms in camels' noses.

The strongly aromatic leaves are often sold in the market of Aden. They are put between clothes in order to protect them against insects.

Arabic: Berberem, Chosam, Daf—; *Baluchi*: Miskok—; *Brahui*: Pawal—; *Egypt*: Berberan—; *Jhalawan*: Panwar—; *Ormara*: Shapako, Shawang—; *Rajputana*: Navli, Nodi—; *Sind*: Kasturi—; *Urdu*: Panwar—.

3. **Cleome viscosa** Linn. is found throughout the tropics of the world.

Both Hindu and Mohammedan practitioners credit this weed with great therapeutic value as an anthelmintic and carminative. The juice of the leaves, poured into the ears, is useful in deafness.

The juice of the leaves mixed with oil is a popular Indian remedy for purulent discharges from the ear.

The juice of the leaves is poured into the ear to relieve earache, and the bruised leaves are applied to the skin as a counter-irritant.

The seeds are used as anthelmintic and carminative; they also are given occasionally in fevers and diarrhoea.

In Indo-China the roots are considered to be stimulant and antiscorbutic; the whole plant, bruised, is used for counter-irritation and blistering.

In La Reunion the plant is said to be astringent and anti-spasmodic.

The leaves are given by the aboriginals of Australia to relieve headache.

In the United States, the roots are said to be used as a vermifuge.

In Ceylon the roots and the seeds are considered to be cardiac stimulants, and they are given internally in cases of snake bite (Roberts). But Mhaskar and Caius have shown experimentally that both the roots and the seeds are useless in the antidotal and symptomatic treatment of snake bite.

'The seeds of this plant are said to possess anthelmintic and rubefacient properties. Mohideen Sheriff recommends the pure dried seeds as substitute for santonin. The dose recommended is half to one drachm for adults and half the quantity to children to be given twice a day for two days and followed on the third day by a dose of castor oil. I tried the drug as per directions given above in an adult and in a child whose motions contained large number of round-worm ova, but the result was negative in both the patients' (Koman).

Arabic: Bantakalan, Oddar—; *Bengal*: Hurhuria—; *Biru*: Cirlinggid—; *Bombay*: Hurhuria, Kanphuti, Pivalatilavana—; *Canarese*: Huchasasavi, Nayibela—; *Deccan*: Choriajuwan, Churaiyajwani, Janglihulvul—; *English*: Sticky Cleome—; *French*: Brède puante, Herbe puante—; *Gangpur*: Cirliggid—; *Guam*: Mongos paloma—; *Gujerat*: Talvani, Tilwan, Tinmani—; *Hausa*: Namijan gasaya—; *Hindi*: Boogra, Hulhul, Hurhur, Hurhureh, Jangliharrar, Kamphutia—; *Indo-China*: Man man trang, Sa phac ron tien—; *Kolami*: Chamani—; *Malay*: Kutepeng—; *Malayalam*: Ariavila, Katkudagu—; *Marathi*: Harhuria, Kanphodi, Kanphuti, Pivalatilavana—; *Mundari*: Marang carmaniara, Marang carmari—; *Porebunder*: Pilitalvani—; *Portuguese*: Bredo mamma—; *Punjab*: Bugra, Hulhul—; *Sanskrit*: Adityabhakta, Arkabhakta, Arkakanta, Arikahita, Barbara, Bhaskareshta, Brahmasuvarchala, Karnasphota, Kukavaiminta, Kukavumitie, Mandukaparni, Manduki, Martandavallabha, Raviprita, Ravishta, Sauri, Satyanamni, Shunakabarbara, Surasambhava, Suryalata, Suryavarta, Suvarchala, Suteja, Svanabarbara, Tilparni, Varada, Vikranta—; *Santali*: Harhara—; *Sind*: Kattori—; *Sinhalese*: Ranmanissa, Walaba—; *Tagalog*: Apoyapoyan, Balabalanoyan, Silisian, Silisilihan—; *Tamil*: Nayikudagu, Nayivelai—; *Telugu*: Kukhavavulu, Kukhavominta—; *Urdu*: Hulhul—; *Visayan*: Hulayasangayan—.

4. **Cleome felina** Linn. fil. occurs in the Southern Maratha Country, the Northern Circars and the Deccan districts of the Madras Presidency.

The plant is reputed antiscorbutic.

The seeds are vesicant; they are given internally as a vermifuge.

The plant mixed with milk is applied externally to raise blisters.

Malayalam: Ariavila—.

5. **Cleome heptaphylla** Linn. is an American plant grown in Indian gardens.

It is used as a stomachic and as a vulnerary.

6. **Cleome Chelidonii** Linn. fil. is found in the Northern Circars and in most places throughout the Bombay Pesidency. It is also found in Indo-China and Java.

In Indo-China the roots are considered to be vermifuge. An infusion of the plant is commonly used in gingivitis and in the treatment of skin diseases.

Indo-China: Man man—; *Porebunder*: Ubhitalvani—; *Telugu*: Kar wominta—.

CRATAEVA.

This genus consists of 10 tropical species.

C. religiosa Forst. is used medicinally in Indo-China and the Philippine Islands, *C. tapia* Linn. in Brazil and Guiana, *C. Greveana* Baill. in Madagascar.

Crataeva nurvala Ham. (=*C. religiosa* Hook. fil. and Th.—non Forst.—) is found almost everywhere all over India and Burma, either wild or cultivated. Although commonly met with along streams, it may also be found in the dry deep boulder formations of the sub-Himalayan tract. A native of Malabar and Canara.

Its properties are described in Sanskrit and in Persian works. It is said to promote the appetite, increase the secretion of the bile, act as a laxative, and remove disorders of the urinary organs. In calculous affections it is used in a great variety of forms; thus a simple decoction of the bark may be given with the addition of treacle.

The bark is demulcent, antipyretic, sedative, alterative, and tonic; and the fresh leaves and root-bark are rubefacient.

The bark is a Pondicherry remedy for amenorrhoea. Boiled in oil it is considered good for rheumatism.

The bark is useful in some cases of urinary complaints and fever, and in some mild forms of skin diseases in which sarsaparilla is generally resorted to. It also relieves vomiting and other symptoms of gastric irritation. The fresh leaves and root-bark, particularly the former, are very efficacious in all the affections in which mustard poultice is indicated.

Bruised well with a little vinegar, lime-juice or hot water and applied to the skin in the form of a poultice or paste, the fresh leaves act as a rubefacient and vesicant. The bark and leaf pounded and tied in a cloth are used as a fomentation for rheumatism.

In Ceylon the leaves are used for gouty swellings. In Bombay they are a remedy for swelling of the feet, and a burning sensation in the soles of the feet. In the Konkan the juice is given in rheumatism. In caries of the bones of the nose, the leaf is smoked and the smoke exhaled through the nose. In China the leaves are used in drunkenness; an extract is given for diarrhoea.

A couple of buds pounded with salt are taken before meals to promote the appetite. In indigestion they must be given after meals.

In Ceylon the bruised roots, leaves and seeds are applied topically to snake bites and scorpion stings. But Caius and Mhaskar have shown experimentally that they are useless as an antidotal and symptomatic treatment.

Bengal: Barun, Tikoshak, Tiktoshak, Varuna—; *Bombay*: Bhatavarna, Hadavarna, Kawan, Kumla, Vayavarna, Waruna—; *Burma*: Kadat, Kadet, Katat—; *Canarese*: Bilpatri, Bitusi, Hoddelenage, Mavilinga, Narave, Neravambele, Neravele, Nervala, Nirvala, Tudemadirenge, Vitusi—; *Central Provinces*: Bel, Bela—; *Chinese*: Pa Yeh—; *Coorg*: Nerajane, Nerujani, Nirajani, Vittasi, Vitusi—; *English*: Holy Garlick Pear—; *French*: Tapier—; *Gujerati*: Varno, Vayavarno—; *Hansot*: Kagdakeri—; *Hindi*: Barna, Barua, Barun, Bila, Bilasi, Biliana, Brarna, Varuna, Varvunna—; *Konkani*: Nervol—; *Lao*: Mai fuk koom—; *Lepcha*: Purbong—; *Malay*: Cadat—; *Malayalam*: Kili, Niravila, Nirumaliyan, Varana, Vitusi—; *Marathi*: Haravarna, Karvan, Kumla, Nirvala, Ramola, Varun, Vayavarna—; *Mechi*: Bunboronda, Tailaslu—; *Punjab*: Barna, Barnahi—; *Rajputana*: Barna, Barnahi—; *Saharanpur*: Barna, Bilarsi, Brarna—; *Sanskrit*: Ajapa, Ashmarygna, Asmarighna, Barhapushpa, Kumara, Kumaraka, Mahakapittha, Marutapaha, Pasunadha, Sadhuvriksha, Setuka, Setuvriksha, Shikhimandala, Shvetadru, Shvetadruma, Shvetavriksha, Tamala, Tiktashaka, Urumana, Varana, Varuna, Vasaha—; *Sinhalese*: Lunuwarana—; *Tamil*: Adicharanam, Adimalam, Anjali, Inaivilai, Kattu mavilangai, Kuvilam, Maluram, Maralingam, Mavilangai, Miguttiyal, Narvala, Nilluvam, Nirumaliyam, Periamavilangai, Shuppigam, Shuvedan, Sinnamavilingam, Tiriburamerittan, Varanam, Villuvam—; *Telugu*: Bilvaram, Chinnavulumidi, Magalingam, Maredu, Peddamagalingam, Peddavulimidi, Tellavulimidi, Ulimidi, Urumatti, Urumudu, Usiki, Uskia—; *Urdu*: Barna—; *Uriya*: Barun, Boryno, Varuna—.

GYNANDROPSIS.

This genus includes 15 species spread over the tropical and subtropical regions of the world.

Gynandropsis gynandra Merrill (=*G. pentaphylla*) is used medicinally wherever it is found growing. It is a common weed in all tropical countries, and is abundant throughout the warmer parts of India.

It has long been known as a domestic remedy by the Hindus. The seeds, as well as the leaves, are administered in decoction in convulsive affections and typhus fever, to the quantity of half a teacup daily.

A decoction of the root is said to be a mild febrifuge.

The leaves are applied externally to boils to prevent the formation of pus. The bruised leaves are rubefacient and vesicant, producing a very copious exudation, affording in many cases the relief obtained from a blister without its inconveniences. The expressed juice is a popular remedy, in high repute as a local application in otalgia, both amongst the natives of India and the settlers in the West Indies, where the plant is also indigenous.

In Jodhpur and Jaisalmer the leaves are used against rheumatism; the green leaves applied to the skin and tied down form a good blister.

In Arabia the juice of the fresh leaves is made into an ointment with sesame oil, and used for prurigo.

In Yoruba the natives use the roasted leaves as a cure for earache; the jucie also is mixed with palm-kernel oil and squeezed into the ear.

In Gold Coast the juice of the leaves is commonly used for curing earache, and sometimes for curing headache. As it causes pain when applied freely to the eye or the ear it should be used with care.

The Shangaans of South Africa apply the pounded leaf as a counter-irritant in rheumatism, neuralgia, headache, and a stiff neck, taking care to withdraw the application before it produces a blister.

The seeds are anthelmintic and rubefacient, and are employed internally for the expulsion of round worms, and, externally as a counter-irritant. They are used as a substitute for mustard, and yield a good oil.

In Lakhimpur (Assam), a paste of the seeds is applied locally in headache.

In the Rajputana Desert the seeds infused in boiling water are used as a cure for coughs; bruised, they are applied as a poultice to sores that have maggots in them (Blatter). They are given to horses for stomach-ache.

The plant is used as a sudorific in Pondicherry and in the Nilgiris.

In Indo-China the plant is used as an antiscorbutic. In La Reunion it is mostly used as a diaphoretic, and as an emollient in the form of poultice.

A reputed cure for cobra bite in Ceylon; the bruised roots, leaves, and seeds are applied to the wounds (Roberts).

The plant is not an antidote to snake and scorpion venom; and it is useless as an external application in the treatment of snake bite or scorpion sting (Caius and Mhaskar).

Arabic: Gheifakan—; *Ashanti*: Tete—; *Awuna*: Sorbui—; *Bengal*: Ansarisha, Arkahuli, Hulhul, Hurhur, Hurhuria, Kanala, Kanalla, Karaila, Sadahurhuria—; *Bombay*: Mabli, Tilavana—; *Ceylon*: Tayirvalai—; *Chinese*: Pai Hua Ts'ai—; *Deccan*: Halhal—; *Egypt*: Abu-qarn, Arareg, Tamalika—; *Ewe*: Sorlwi—; *Ga*: Kete, Tete—; *Gujerati*: Adiyakarhan, Satitalvani, Tanmani—; *Hausa*: Gasaya—; *Hindi*: Churota, Hulhul, Hurhur, Karaila, Lalhulhul, Safedhulhul—; *Indo-China*: Man ma tia—; *Kano*: Gasaya—; *Kolami*: Chamani—; *Krobo*: Tete—; *La Reunion*: Pissat de chien—; *Loanda*: Mozambue, Mozembue—; *Malay*: Maman antu, Maman putih—; *Malayalam*: Karavela, Taivela, Vela—; *Marathi*: Kanphodi, Mabli, Motitilavan, Pandharitilavan, Tilavana—; *Mundari*: Carman, Ciarmari, Marangcarmani, Marangcarmari, Tundicarmani—; *New Caledonia*: Oua meti hakon—; *North-Western Provinces*: Kathalparhar—; *Philippines*: Cincocinco, Silisihan—; *Porebunder*: Dholitalvani,

Gandharitalvani, Satitalvani—; *Punjab*: Ghanduli, Hulhulsafed—; *Rajputana Desert*: Bagra, Pagra—; *Sanskrit*: Ajagandha, Arkapushpika, Avigandha, Barbaragandha, Bastagandha, Bodhayika, Bramhagrabha, Brahmi, Choraka, Hulhul, Kabari, Karnaspota, Kjarapushpa, Putimayurika, Sugandhika, Surjavarta, Tilaparni, Tungi, Ugragandha—; *Santal*: Setakataarak—; *Sind*: Kinro—; *Sinhalese*: Suriyavarta, Vela—; *Sokoto*: Yarungawa—; *Tagalog*: Apoyapayan, Balabalanoyan—; *Tamil*: Kadugu, Nayvelai, Velai—; *Telugu*: Vaminta, Velakura—; *Twi*: Tete—; *Visayan*: Hulaya—; *Yoruba*: Ekuya—.

MAERUA.

This genus includes 40 species, natives of tropical Africa and Asia.

M. angolensis DC. is said to be toxic, but is nevertheless used medicinally in Somaliland and Abyssinia, where *M. Denhardtiorum* Gilg. is similarly employed.

Maerua arenaria Hook. fil. and Th. is found in the Punjab, Sind, Gujarat, the Deccan, Central and South India, and Ceylon.

The root is used as an alterative, tonic, and stimulant.

Cutch: Dhoropinjero, Katkiyal—; *Gujerat*: Vika—; *Porebunder*: Dholokatkiyo, Dudhiyohemkand, Hemkand—; *Tamil*: Bhumichakkarai, Mulmurandai—; *Telugu*: Bhuchakramu, Makamettanitige, Menikatige, Morinika, Pattatige, Pattutige, Puttatige—.

II

The RESEDACEÆ are herbs, rarely shrubs, grouped into 6 genera including about 70 species. They grow in southern Europe, northern Africa, Syria, Asia Minor, and Persia; some reach the Indian frontier; a few inhabit central and northern Europe; three species belong to the Cape of Good Hope.

Medicinal Mignonettes belong to three genera: OCHRADENUS (southern Mediterranean); OLIGOMERIS (Africa, India, south-western North America); RESEDA (Mediterranean region, Europe).

1. Petals 4-7, lobed; ovary syncarpous ... RESEDA.
2. Petals 2; ovary syncarpous ... ... OLIGOMERIS.
3. Petals 0; ovary syncarpous; fruit a berry ... OCHRADENUS.

OCHRADENUS.

The genus consists of 5 species spreading from the Mediterranean to Sind.

Ochradenus baccatus Del. occurs in Sind and Baluchistan; extending to Persia, Arabia, Syria, Egypt, Nubia and Abyssinia.

In Baluchistan the twigs, leaves, and flowers are fried, ground to a powder, mixed with a little neshar and applied dry to wounds and sores to kill maggots, etc. (Hotson).

Arabic: Bliha, Gord, Gurdy, Gurssi, Gurzi, Qurdhah, Wuebeh—; *Baluchistan*: Kalirram, Kirmkush—; *Egypt*: Gurdy—.

OLIGOMERIS.

This genus includes 5 species distributed over Africa, Western Asia, India, and North America.

Oligomeris subulata (Del.) Boiss. is found in Sind, Baluchistan and the Punjab; it extends to Afghanistan, Persia, Arabia, the Mediterranean, and West North America.

In Kalat the plant is pounded, and the juice thus extracted is used by women to put on to their breasts to keep them soft.

Egypt: Deneban, Dhenebun—; *Kalat*: Shataki—.

RESEDA.

This genus numbers 55 species inhabiting the Mediterranean region, Europe, and Western Asia.

The root is considered sedative, aperient, diaphoretic, and diuretic; the seeds are used as a resolvent.

The following are used medicinally: in Europe—*R. alba* Linn., *R. glauca* Linn., *R. lutea* Linn., *R. luteola* Linn., *R. odorata* Linn., *R. phyteuma* Linn.—; in North America—*R. alba* Linn., *R. luteola* Linn., *R. odorata* Linn.—.

Reseda odorata Linn. is widely cultivated in Indian gardens.

In Spain the acrid root is used as a laxative, diaphoretic, and diuretic; the seeds are applied externally as a resolvent.

Arabic: Khuzam—; *Catalan*: Mardugi, Marduixi—; *Dutch*: Reseda—; *Egypt*: Tamr-el-hina frengi—; *English*: Mignonette—; *French*: Herbe d'amour, Mignonnette, Réséda, Réséda odorant—; *German*: Gartenreseda, Reseda—; *Roumanian*: Roseta—; *Russian*: Reseda—; *Spanish*: Miñoneta, Reseda, Reseda de olor, Resedal—.

III

The VIOLACEÆ are herbs or shrubs, natives of warm and temperate regions. The herbaceous species principally inhabit the northern hemisphere; they are rare in the temperate regions of the southern hemisphere and in the tropics; the woody species are chiefly natives of equatorial America; many species inhabit the intertropical region of both worlds, and especially of America. The family includes 21 genera with about 450 species.

The members of this Order are used for their sedative, diaphoretic, diuretic, emetic, purgative, and expectorant properties.

Among the products isolated from them may be mentioned:—(1) *essential oils*; (2) *glucosides*—iridin, violaquercitrin, violutoside—; (3) *a carbohydrate*—inulin—; (4) *an ester*—methyl salicylate—; (5) *a mustard oil*—phenylethyl—; and (6) *colouring matters*—luteolin, quercetin.

The medicinal and poisonous Violets of the world belong to 5 genera:—ALSODEIA (tropical and subtropical regions); ANCHIETEA (tropical South America); IONIDIUM (tropical and subtropical regions); SAUVAGESIA (tropics; Brazil); VIOLA (cosmopolitan).

The medicinal and poisonous violets of India belong to 3 genera:—ALSODEIA, IONIDIUM, VIOLA.

I. Corolla irregular. Lower petal dissimilar. Staminodes absent. Capsule loculicidal
  1. Sepals produced at the base ... ... VIOLA.
  2. Sepals not produced at the base ... ... IONIDIUM.
II. Corolla regular. Staminodes absent. Fruit a berry or loculicidal capsule ... ... ALSODEIA.

ALSODEIA.

The genus includes 80 tropical and subtropical species.

Alsodeia echinocarpa Korth. is common in woods from Singapore to Tringganu. It extends to Southern Siam, Laos, Cochin-China, Borneo, and Sumatra.

The Malays use the seeds as a purgative.

A decoction of the leaves is used in Annam in cases of myiasis of the nostrils in ruminants. The drug is administered per os. The same decoction, but in smaller doses, is given to pregnant women to procure abortion.

Malay: Aho lumut, Jangut baong, Jenta-jenta, Kayu rembut, Lalada, Lelada, Medang gombot, Sebilik, Segumpa betina—; *Penang*: Ardani—.

IONIDIUM.

The genus consists of 50 tropical and subtropical species.

The root is diaphoretic, diuretic, and in large doses emetic and cathartic.

The following species are used medicinally in China—*I. heterophyllum* Vent.—; in South America—*I. brevicaule* Mart., *I. circaeoides* H. B. & K., *I. glutinosum* Vent., *I. Ipecacuanha* Vent., *I. oppositifolium* Roem. & Schult., *I. Poaya* St. Hil., *I. polygalaefolium* Vent., *I strictum* Vent—.

Ionidium enneaspermum Vent. (=*I. suffruticosum* Ging.) is found in Bundelkhand, Agra, Bengal, almost throughout the Madras Presidency, Gujarat, Khandesh, the Carnatic, and Ceylon. It is common throughout tropical Asia, Africa and Australia.

The native physicians regard the plant as a tonic and diuretic, and prepare a confection of the whole herb.

The Santals employ the root in bowel complaints of children.

The leaves and tender stalks are demulcent, and are used in decoction and electuary; they are also employed in conjunction with some mild oil in preparing a cooling liniment for the head.

The plant is used in Southern India as a demulcent in gonorrhoea; and its demulcent properties are known in New South Wales, where the plant is common.

Arabic: Ssibbah—; *Bengal*: Nunbora—; *Bombay*: Ratanpuras—; *Deccan*: Ratanpuras—; *Hindi*: Ratanpurus—; *Malayalam*: Orelatamara—; *Marathi*: Ratanpuras—; *Naguri*: Pirikantara—; *Sanskrit*: Amburuha, Atichara, Avyatha, Charati, Chariti, Lakshmishreshta, Padma, Padmacharini, Padmavati, Padmavha, Pushkaranadi, Pushkarini, Ramya, Sarade, Sthalapadmini, Sthalaruha, Sugandhamula, Supushkara—; *Santal*: Birsurajmukhi, Tandisol—; *Tamil*: Orilaitamarai—; *Telugu*: Nilakobari, Purusharatnam, Suriyakanti—.

VIOLA.

The genus includes some 300 species distributed broadly in temperate regions, chiefly of the northern hemisphere.

The root is slightly emetic; the leaves are considered emollient and laxative; the flowers are used as an emollient, diaphoretic, antispasmodic, and stomachic.

The following species are used medicinally :—in Europe—*V. alba* Bess., *V. alpina* Jacq., *V. arenaria* DC., *V. calcarata* Linn., *V. campestris* Bieb., *V. canina* Linn., *V. cenisia* Linn., *V. collina* Bess., *V. biflora* Linn., *V. hirta* Linn., *V. mirabilis* Linn., *V. odorata* Linn., *V. palustris* Linn., *V. pedata* Linn., *V. pinnata* Linn., *V. sciaphila* Koch., *V. sepincola* Jord., *V. sylvestris* Lam., *V. tricolor* Linn., *V. uliginosa* Bess.—; in China—*V. Patrinii* DC.—; in Indo-China—*V. diffusa* Ging, *V. japonica* Lang., *V. mirabilis* Linn., *V. Patrinii* DC., *V. pinnata* Linn., *V. sylvestris* Lam., *V. vaginata* Max., *V. verecunda* A. Gray.—; in North America—*V. canadensis* Linn., *V. canina* Linn., *V. cucullata* Ait., *V. lanceolata* Linn., *V. palmata* Linn., *V. pedata* Linn., *V. pubescens* Ait., *V. scabriuscula* Schwein—; in Colombia : *V. odorata* Linn., *V. tricolor* Linn.—; in Madeira : *V. maderensis* Prim.—.

The toxic glucoside *iridin* has been found in *V. canina, V. odorata, V. sylvestris,* and *V. tricolor.*

A. Stigma obliquely 2-lobed, not beaked ; lobes short, spreading ... ... 7. *V. biflora.*
B. Stigma terminal, truncate, dilated, depressed, orbicular or lobed
 1. Flowers lilac or pale blue ... ... 5. *V. Patrinii.*
 2. Flowers pale purple ... ... 4. *V. diffusa.*
C. Stigma very oblique or quite lateral, often minute and perforated
 1. Flowers lilac or pale blue ... ... 1. *V. serpens.*
 2. Flowers bluish purple or white, more or less scented ... ... 2. *V. odorata.*
 3. Petals lilac, the lower with a few parallel simple veins, not reaching the margin ... ... 8. *V. sylvestris.*
 4. Flowers small, white with a violet blotch on the uppermost petal ... ... 3. *V. cinerea.*
 5. Flowers large, usually about 3 colours represented. 6. *V. tricolor.*

1. **Viola serpens** Wall. is found in hilly districts throughout India, Ceylon, Burma, and the Malay Peninsula. It extends to Java, Sumatra and China.

This species yields *Banafsha* of the bazaars, and is considered to have medicinal properties similar to those of *V. odorata.* In the Punjab, a medicinal oil, called *raughan-i-banafsha*, is prepared from it.

Hindi : Banafsha, Thungtu— ; *Kumaon* : Thungtu— ; *Punjab* : Banafsha— ; *Waziri* : Khojakai—.

2. **Viola odorata** Linn. occurs in Kashmir at 5,000-6,000 feet ; it is planted in most of the hill-stations. It is distributed to Northern and Western Asia, North Africa, and Europe.

A long account of its properties will be found in most Arabic and Persian works on Materia Medica ; it is generally considered cold and moist, and is especially valued as a diuretic and expectorant, and as a purgative in bilious affections. The diseases in which it is recommended are very numerous ; they are generally those in which a cooling treatment is thought to be indicated by the hakims.

Violets were used by the Athenians to moderate anger, to procure sleep and to comfort and strengthen the heart.

Pliny prescribes a liniment of violet root and vinegar for gout and disorder of the spleen, and states that a garland or chaplet of violets worn about the head will dispel the fumes of wine and prevent headache and dizziness.

The root is a powerful emetic, and is frequently used to adulterate ipecac. A dose of from forty to fifty grains of the powdered roots acts powerfully, inciting nausea and great vomiting and nervous affection.

The leaves are an old popular remedy for bruises.

Of late years, preparations of fresh violet leaves have been used both internally and externally in the treatment of cancer. It is stated that an infusion of the leaves, or a syrup of the petals and a liquid extract of the fresh leaves, have been used with benefit to allay the pain in cancerous growths, especially in the throat, and as a cure for cancer of the tongue. The fresh leaves are also prepared as a compress for local application.

For lubricating the throat, violet leaves are dried, powdered, and allowed to stand in olive oil for six hours in a water bath.

Violet leaves, stewed in lard for an hour till they are the consistency of cooked cabbage, are good as an application for superficial tubercles in the glands of the neck, violet leaves tea being drunk at the same time.

The leaves and petals are official in the pharmacopœia for Portugal.

The flowers are emollient and demulcent. They were used in olden times as remedies in many disorders, and were supposed to be especially serviceable in ague, sleeplessness, and inflammation of the eyes.

'The flowers', says Gerard, 'are good for all inflammations, especially of the sides and lungs; they take away the hoarsenesse of the chest, the ruggednesse of the winde-pipe and jawes, and take away thirst.'

He adds: 'There is likewise made of Violets and sugar certaine plates called Sugar violet, Violet tables, or Plate, which is most pleasant and wholesome, especially it comforteth the heart and the other inward parts.' Violet sugar was a favourite conserve in the days of Charles II; it was considered of excellent use in consumption, and was sold by all apothecaries. The flowers have undoubted expectorant qualities; they are officinal in France.

A syrup is made from the petals which is a favourite household remedy for infantile disorders. In France syrup of violets is a medicine for cough and hoarseness. In England violets are cultivated largely at Stratford-on-Avon for the purpose of making the syrup, which when mixed with almond oil is a capital laxative for children, and will help to soothe irritative coughs, or to relieve a sore throat.

The seeds are purgative and diuretic, and have been given in urinary complaints, and are considered a good corrective of gravel.

A modern homoeopathic medicinal tincture is made from the whole fresh plant, with proof spirit, and is considered useful for a spasmodic cough with hard breathing, and also for rheumatism of the wrists.

O'Shaughnessy experimented with the dry plant as a substitute for Ipecacuanha, but without success.

Mohideen Sheriff considers it antipyretic and diaphoretic, and very useful in relieving febrile symptoms and excitement in all forms of fever, particularly in combination with other drugs of the same class.

'This is the "banafshah" of the hakims who value it as a strong febrifuge in acute and chronic fevers when given in combination with other drugs. Mohideen Sheriff in his Materia Medica of Madras gives the ingredients of a compound decoction which he found useful in relieving pyrexia of obstinate and long standing cases of typhoid fever, after European medicines, generally used, had failed. The same decoction was administered to a case of chronic fever without any benefit at all. In the out-patients a simple infusion of *Viola odorata* was given to several cases of pyrexia and intermittent fevers without benefit' (Koman).

Arabic: Banafsaj, Banafshaj, Behussej, Benephig—; *Armenian*: Manischar—; *Bengal*: Banafsha, Banosa—; *Bombay*: Banafshah—; *Calabria*: Rovesciole—; *Catalan*: Viola, Viola boscana, Viola de Bosch, Viola d'olo, Viola d'olor, Viola vera—; *Czech*: Fialke—; *Danish*: Martzfioler—; *Deccan*: Banafsha—; *Dutch*: Tamme viol—; *Egypt*: Benefshig—; *English*: Appel-leaf, Bairnwort, Banwort, Bessy Banwood, Blaver, Blue Violet, March Violet, English Violet, Fine-leaf, Sweet Violet, Vilip, Violet—; *French*: Fleur de mars, Jacée de printemps, Violette, Violette cultivée, Violette de mars, Violette odorante, Violette des quatre saisons, Violier commun, Violier de mars—; *Genoa*: Viuleta, Viuretta—; *German*: Blauoesken, Blauvoegschen, Blauvoelken, Maertzveilchen, Veielotenblau, Veieloteykraut, Veilchen, Viole, Wohlriechende Veilchen—; *Greek*: Ion, Menexe, Menexes—; *Gujerati*: Banaphsa—; *Hindi*: Banafshah—; *Hungarian*: Ibolya—; *Italian*: Mammola, Mammoletta, Viola, Viola mammola, Viola marzia, Viola zoppa, Violetta—; *Languedoc*: Memog, Memoi—; *Lombardy*: Zoppina—; *Marathi*: Bagabanosa—; *Persian*: Banafshah—; *Polish*: Skopek—; *Portuguese*: Viola, Viola roxa, Violata—; *Roumanian*: Micsunea, Tamaioasa, Toporas—; *Russian*: Packutchaya fialka—; *San Remo*: Viureta—; *Sanskrit*: Jvarapaha, Nilapushpa, Sukshmapatra, Vanapsa—; *Sardinia*: Violedda, Viuletta—; *Spanish*: Violeta, Violeta de olor—; *Swedish*: Aekta fioler—; *Tuscany*: Mammoletta, Mammolina, Viola maura, Viola mammola—; *Urdu*: Banafshah—; *Verona*: Fior de San Bastian—.

3. **Viola cinerea** Boiss. is found in Sind, Baluchistan, Waziristan, Punjab, Western Rajputana, and Kathiawar. It extends to Afghanistan, Persia, and Arabia.

This plant is used medicinally in Sind as a substitute for *V. odorata*.

Porebunder: Jinkobanafsha—; *Punjab*: Banafsha—; *Sind*: Banafsha—.

4. **Viola diffusa** Ging. is found in the subtropical Himalaya from Nepal to Mishmi at 3,000-5,000 feet, and in the Khasia Hills. It also occurs in China and Indo-China.

In Indo-China the flowers are given in diseases of the chest as a pectoral and bechic.

Annam: Dia dinh—.

5. **Viola Patrinii** Ging. occurs in the temperate Himalaya, China and Japan.

In China, Indo-China, and Malaya the flowers are said to purify the blood and the plant is used as a pot-herb. The plants are also bruised and applied to ulcers, foul sores, and swellings. They are prescribed in syphilis, scrofula, and biliousness.

Annam : Tu hoa dia dinh—; *Chinese* : Tzu Hua Ti Ting—.

6. **Viola tricolor** Linn. is indigenous in Europe, North Asia and North America. It is cultivated in India.

The fresh plant is reputed emollient, slightly laxative, emetic, and pectoral; it has been used with success in the treatment of nephritic and cutaneous diseases.

The Wild Violet, or Pansy, is a household remedy in many parts of Europe. The plant is taken internally in infusion as a depurative in skin cruptions.

'For curing milk crust and scalled head in children, a small handful of the fresh plant, or half a drachm of the dried herb, boiled for two hours in milk, is to be taken each night and morning; also a bread poultice made with this decoction should be applied to the affected part. During the first eight days the eruption increases, and the urine, when the medicine succeeds, has a nauseous odour like that of the cat, which presently passes off; then, as the use of the plant is continued, the scabs disappear, and the skin recovers its natural clean condition.'

In Spain the plant is considered to be stimulant, and is used in rheumatism and in skin diseases.

A homoeopathic medicinal tincture is made from the entire plant with spirit of wine. Hahnemann found that the Pansy Violet, when taken experimentally by provers, served to induce cutaneous eruptions, or to aggravate them, and he reasoned out the curative action of the plant in small diluted doses for the cure of these symptoms, when occurring as disease.

The flowered plant is official in Austria, Germany, Portugal, Switzerland, and Turkey.

The leaves were once esteemed in the cure of cutaneous disorders, and they are still employed in Italy in *tinea capitis*.

The root is said to have similar properties to those of Ipecacuanha, and is often used beneficially as a substitute by country doctors in England. An infusion thereof is admirable for the dysentery of young children.

Catalan : Herba de la Trinitat, Pensaments, Trinitaria—; *Dutch* : Driekleurig viooltje—; *English* : Bird's Eye, Bouncing Bet, Bullweed, Call-me-to-you, Cuddle Me, Cuddle-me-to-you, Cull Me, Flame Flower, Flamy, Flower o' luce, Gardengate, Gentleman John, Godfathers and Godmothers, Heartsease, Herb Constancy, Herb Trinitatis, Jack-jump-up-and-kiss-me, John-of-my-Pink, Jump-up-and-kiss-me, Kiss-her-in-the-buttery, Kit-run-about, Kit-run-in-the-fields, Live-in-idleness Love-in-idleness, Love-lies-bleeding, Love Idol, Loving Idol, Meet-her-i-th-entry, Meet-me-in-the-entry, Pance, Pansy, Pansy Violet, Paunce, Pink-eyed-John, Pink-o'-the-Eye, Stepmother, Three-faces-under-a-hood, Tittle-my-fancy, Wild Pansy, Wild Violet—; *French* : Clavelée, Fleur de la Trinité, Herbe à la chevelée, Herbe clavelée, Herbe à la clavelée, Herbe de la Trinité, Jacée tricolore, Pensée sauvage, Petite jacée, Violette tricolore—; *German* : Abnehmkraut, Ackerveilchen, Blauoesken, Denkanmich, Denkbluemchen, Dreifaltigkeit, Dreifaltigkeitskraut, Faulschken, Fioelken, Freisamkraut, Fresern, Gedenkemein, Gengelkraut, Hundsveilchen, Maukraut, Schwiegerle, Schwiegermuetterchen, Stiefmuetterchen, Tagundnachtblume, Tausendschoen, Unnuetzesorgen, Dreifarbiges Veilchen, Veilchenkraut, Wergehundkommnichtwieder, Vergissmeinnicht—; *Hungarian* : Arvacska—; *Italian* :

Penserio, Pensieri, Viola tricolore—; *Languedoc*: Pensado—; *Malta*: Pansy, Viola di tre colori, Pensieri—; *North America*: Cupid's-delight, Garden Violet, Heart's-pansy, Johnny-jump-up, Pansy—; *Portuguese*: Amor perfeito, Herva seraphica, Violeta tricolor—; *Spanish*: Amor perfecto, Nueras, Pensamientos, Suegras, Trinitaria—; *Roumanian*: Catifeluta—; *Russian*: Anyutini glazki—; *Turkish*: Hercai menekse—.

7. **Viola biflora** Linn. occurs in the temperate Himalaya, and is found in the northern temperate regions of the world. In Kashmir it is common in damp woods and under big rocks at 8,000 feet at Gulmarg; it is found at Khelanmarg 10,000 feet, on steep rocky hill-side at Tosh Maidan 9,600 feet; about 10,000 feet at Basam Gali in Juniper tract.

This violet is much used medicinally in Spain. The root is given as an emetic; the flowers as an emollient, pectoral, diaphoretic, and antispasmodic; the leaves as an emollient and laxative.

English: Two-flowered Violet, Yellow Violet—.

8. **Viola sylvestris** Lam. is found in Kashmir and Kishtwar at 4,000-8,000 feet, extending to Northern Asia and westwards to the Atlantic.

The plant is used in chest troubles as a bechic and pectoral. The stems, leaves, and flowers are bruised and applied to wounds and foul sores.

IV

The COCHLOSPERMACEÆ are trees, shrubs or rhizomatous undershrubs with coloured juice. They are grouped under 3 genera with about 18 species.

COCHLOSPERMUM.

This genus consists of 12 tropical species.

C. tinctorium Perr. is used medicinally in Gambia and Guinea, *C. insigne* St. Hil. in Brazil.

Cochlospermum Gossypium DC. extends from Garhwal to Bundelkhand, Bihar, Orissa, Bengal, Burma, Central India, Deccan, the Western Peninsula, the Madras Presidency in dry forests, especially on stony hills, in all districts, but less common on the west coast.

In Punjab the dried leaves and flowers are given as stimulants.

The floss has been recommended as admirably suited for padding bandages and splints, and the gum has been proposed as a substitute for Tragacanth.

The gum, sold in the bazaars as *katira* or *kathira*, is used as a demulcent and astringent. It is given in cough and hoarse throat in the form of lozenges and mucilage. Mixed with curds or whey it is largely dispensed in cases of diarrhoea and dysentery.

The gum is a Yunani remedy prescribed for the treatment of gonorrhoea and syphilis, for eye troubles and trachoma.

Arabic: Katira—; *Bengal*: Golgol—; *Betul*: Galgal, Ganiar—; *Bhil*: Ganeri—; *Canarese*: Arasinaburaga, Arisinaburuga, Bettatvare, Buruga, Gagili,

Kaduburaga—; *Chota Nagpur*: Sisibaha, Udal—; *English*: White Silk-Cotton Tree, Yellow Silk-Cotton Tree—; *Gond*: Gangam, Ganiar—; *Gujerat*: Kada-chogund—; *Hasada*: Hupudaru—; *Hindi*: Gabdi, Galgal, Gangal, Ganiar, Gejra, Kumbi—; *Kolami*: Golgal, Hupu—; *Lambadi*: Hoghara—; *Malayalam*: Appakutakka, Chempanni, Chimappanni, Panninara, Parapanni—; *Marathi*: Galgal, Ganer, Ganeri, Ganglay, Gongal, Gulgul Gunglay, Kathalyagonda—; *Naguri*: Galgalolaru—; *Nimar*: Galgal, Ganiar—; *North-Western Provinces*: Gajra, Kumbi—; *Persian*: Gone, Katira-i-hindi, Kokamara—; *Punjab*: Kumbi—; *Saharanpur*: Arlu, Gejra—; *Santal*: Hopo—; *Saora*: Onkur—; *Sinhalese*: Elaimbul, Kinihiriya—; *Tamil*: Kannigaram, Kattilavu, Kattolaga, Katupanju, Kattuparutti, Kongilam, Kongu, Kumarai, Malaiparutti, Manjar-danakku, Nalal, Pachaigiluvai, Palini, Panjittanakku, Pinar, Sudinar, Tanakku, Turumorbalam—; *Telugu*: Adaviburaga, Akshotamu, Buraga, Gungu, Konda-gogu, Kongu, Parijatamu, Pratti—; *Urdu*: Katira—; *Uriya*: Beniyamrydami, Ganiari, Konokopolaso, Kontopolas, Pobosokoniari—.

V

The BIXACEÆ are shrubs or small trees with coloured juice. They comprise 3 genera with 6 species.

BIXA.

The genus consists of 2 species, natives of tropical America.

B. Orellana Linn. is used medicinally in Indo-China, the Philippine Islands, Brazil, Guiana, and Sierra Leone.

Bixa Orellana Linn. is largely cultivated throughout India.

The plant is used by Ayurvedists as an astringent and mild purgative, and is considered by them a good remedy for dysentery and kidney diseases.

The root-bark is antiperiodic and antipyretic, of great help in uncomplicated intermittent, remittent, and continued fevers.

The leaves are a popular febrifuge in Cambodia. In French Guiana they are considered detergent; an infusion is prescribed as a purgative in dysentery.

The seeds are cordial, astringent and febrifuge, and a good remedy for gonorrhoea. They possess the antiperiodic and anti-pyretic properties of the root-bark, but to a smaller extent.

The seed pulp, a well-known colouring matter, has the same astringent and febrifugal properties as the other parts of the plant. It is used by American Indians to paint their body all over for full dress, and this use of it is said also to prevent mosquito bites.

The root in combination with other drugs has been recom-mended for internal use in snake bite (Bapat); but it is not an antidote to snake venom (Mhaskar and Caius).

Angola: Quisafu—; *Antsianaka*: Vahinamalona—; *Ashanti*: Konin—; *Assam*: Jarat, Jolandhar—; *Awka*: Nkbul-ufia—; *Bengal*: Latkan, Latkhan, Watlnana; *Bombay*: Japhar, Kesari, Kesuri, Kisri, Sendri, Shendri—; *Brazil*: Urucu—; *Burma*: Thideng, Thidin—; *Cambodia*: Chompuh chralok—; *Canarese*: Arnattu, Bangarakayi, Kesari, Japhredu, Kuppamanhala, Ragna-malar, Rangamali, Rangumale, Sannajabbale, Sannajapali—; *Chittagong*: Powasi—; *Deccan*: Shalkepandu—; *English*: Annatto, Annotto, Arnatto, Roucou—; *Espiritu Santo*: Nepekara—; *Ewe*: Berniticu—; *French*: Achiote, Achit, Achote, Ananaie, Anate, Atole, Bichet, Cochehue, Rocouyer, Roucouyer—; *French Guiana*: Rocou, Roucou—; *Galibi*: Ururu—; *Guam*: Achiote, Ach-uete—; *Guarana*: Urucuy—; *Gujerat*: Sinduri—; *Hindi*: Latkan, Latkhan,

Rangphal, Vatkana—; ***Honduras*** : Anatto—; ***Hova*** : Sahy—; ***Ibo*** : Ula, Ula machuku—; ***Ibuzo*** : Ola, Ufie, Uhia nkum, Uhie aro—; ***Indo-China*** : Cham pou, Dieu nhuom, Som hou, Som phu, Xiem phung—; ***Konkani*** : Kesri—; ***Lambadi*** : Japrero—; ***La Reunion*** : Rocou, Roucou—; ***Limba*** : Kuburo—; ***Lokko*** : Mbundona—; ***Malayalam*** : Kuppamannal, Kurannamannal—; ***Manipur*** : Reipom—; ***Marathi*** : Kesari, Kesuri, Kisri, Sendri, Shendri—; ***Mende*** : Bundoi, Mbundoi—; ***Mexico*** : Achiotillo, Achiotl, Bija—; ***Mundari*** : Rangjodaru—; ***Onitsha*** : Ula, Ula machuku—; ***Pentecost*** : Waiang—; ***Philippines*** : Achiote, Achuete, Anate, Atola—; ***Portuguese*** : Arnotto—; ***Samoa*** : Loa—; ***Sanskrit*** : Karachchada, Raktabija, Raktapushpa, Shonapushpi, Sinduri, Sindurpushpi, Sunomala, Trivapushpi, Virpushpa—; ***Santal*** : Kougkuombi—; ***Sinaloa*** : Achiote—; ***Sinhalese*** : Kaha—; ***Spanish*** : Achote—; ***Susu*** : Kamunyi—; ***Tagalog*** : Achuete, Achute, Atsuiti—; ***Tamil*** : Amudadaram, Avam, Kungumam, Manjitti, Sappiravirai, Uragumanjal, Vennaivirai—; ***Telugu*** : Jabura—; ***Timne*** : Kam, Pu-bundoi—; ***Tschaudjo*** : Kirane—; ***Tupin*** : Urucu uva—; ***Twi*** : Brorfo agyama—; ***Uriya*** : Gulbas, Guliabha, Japhoran, Lotkons, Sakta—.

MEDICINAL AND POISONOUS PLANTS OF INDIA:

FLACOURTIADS, PITTOSPORADS, MILKWORTS, SEAHEATHS, PURSLANES, TAMARISKS.

BY

J. F. CAIUS, S.J., F.L.S.

I.

The FLACOURTIACEÆ are trees or shrubs inhabiting tropical and subtropical regions. They are arranged in 70 genera with about 500 species.

The seeds of some of the members yield fatty oils which are used with some success in the treatment of leprosy and other skin diseases. Narcotic and anthelmintic properties are also found in this Order.

The following have been isolated:—(1) *alcohols*—phytosterols—; (2) *acids*—hydrocyanic, linolic, palmitic, linolenic, isolinolenic, oleic, chaulmoogric, hydnocarpic—; (3) *glucosides*—gynocardin—; (4) *enzymes*—gynocardase—.

The medicinal and poisonous Flacourtiads of the world belong to 14 genera:—APHLOIA (South Africa, Madagascar, Mascarenes); ASTERIASTIGMA (India); CALONCOBA (tropical Africa); DIONCOPHYLLUM (tropical Africa); FLACOURTIA (tropical Asia and Africa); GYNOCARDIA (India); HOMALIUM (Tropics); HYDNOCARPUS (Indomalayan region); ONCOBA (tropical America and Africa; Madagascar); PANGIUM (Malay Archipelago); SCOLOPIA (warm regions); TARAKTOGENOS (Malaya); TRIMERIA (Southern and tropical Africa); XYLOSMA (Tropics).

The medicinal and poisonous Flacourtiads of India belong to 6 genera:—ASTERIASTIGMA, FLACOURTIA, GYNOCARDIA, HYDNOCARPUS, PANGIUM, TARAKTOGENOS.

A. Petals absent, disc present; fruit baccate ...	FLACOURTIA.
B. Petals present, with an adnate scale or basal appendage.	
1. Petals 5.	
a. Sepals connate; stamens very many ...	GYNOCARDIA.
b. Sepals free 5; stamens 5	HYDNOCARPUS.
2. Petals 5-6; sepals connate in a tube; stamens free 20 to 25	PANGIUM.
3. Petals 8; sepals free 4; stamens 20 to 30 ...	TARAKTOGENOS.
4. Petals 12-16; sepals imbricate 4; stamens 60-70 ...	ASTERIASTIGMA.

ASTERIASTIGMA.

The sole species, **Asteriastigma macrocarpa** Bedd., is a large tree found in the sholas or evergreen forests of Mysore, Tinnevelly and Travancore from about 1,000 to 4,000 feet.

The oil from the seeds is believed to be a valuable medicine.

Canarese: Doddasurante—; *English*: Cannonball Tree—; *Malayalam*: Aileyam, Vellananga—; *Tamil*: Vellainangu—.

FLACOURTIA.

This genus numbers 15 species distributed over tropical Asia and Africa.

The following are used medicinally in Indo-China—*F. cataphracta* Roxb.—; in La Reunion—*F. cataphracta* Roxb., *F. Ramontchi* L'Herit.—; in Madagascar—*F. Ramontchi* L'Herit., *F. sepiaria* Roxb.—; in Gold Coast—*F. flavescens* Willd.—.

A. Thorns not bearing flowers and fruit.

1. Drupe size of a plum, purple when ripe ... 1. *F. cataphracta.*
2. Drupe size of a pea. Stigmas 5-11 ... ... 2. *F. Ramontchi.*

B. Thorns bearing flowers and fruit ... ... 3. *F. sepiaria.*

1. **Flacourtia cataphracta** Roxb. is found in Kumaon, Orissa, Lower Bengal, Assam and Chittagong, whence it spreads to the Malay Peninsula and the Malay Archipelago.

Yunani practitioners often prescribe the bark and leaves, while Ayurveda doctors seem to favour the use of the fruit.

The bark, macerated in cold water, is given in Behar for the relief of hoarseness. A decoction is used in Lakhimpur for the treatment of biliousness; and it is much employed in La Reunion as an astringent and diuretic.

The leaves are said to have diaphoretic properties.

The leaves and young shoots taste like rhubarb, and are supposed to possess astringent and stomachic properties; they are prescribed in diarrhœa, weakness, and consumption.

The fruit is recommended as useful in bilious conditions and, like most acid fruits, it no doubt relieves the nausea and checks the purging.

Arabic: Talisfir, Zarnab—; *Bengal*: Beunch, Paniala—; *Bombay*: Jaggam, Jangama, Tambath—; *Burma*: Naydwed, Naywe—; *Canarese*: Chankali, Goraji, Kirinelli, Talisapatri—; *Dehra Dun*: Jamnua, Pachnala—; *French*: Prunier d'Inde—; *Gujerat*: Talispatra—; *Hindi*: Paniala, Panialah, Paniamalak, Paniaunvola, Talispatri—; *Indo-China*: Bo quan, Mu cuon, Mung quan—; *Kadir*: Saralanka—; *Konkani*: Jagomi—; *La Reunion*: Prune malgache—; *Malayalam*: Kanji, Talisam, Vayukattan—; *Marathi*: Tambat—; *Persian*: Talispatar—; *Portuguese*: Mamonga—; *Sanskrit*: Ahini, Kalameshi, Khadiraparni, Paniyamalaka, Prachinamalaka, Talishapatra, Varibadaram, Vidara—; *Saora*: Kuski, Kusus, Mullumanu—; *Sinhalese*: Ratangwassa, Rata uguressa—; *Tagalog*: Bitanhol—; *Tamil*: Saralu, Talisam, Talisapatri, Vayangarai—; *Telugu*: Kuragayi, Talisapatramu—; *Tulu*: Koraji—; *Uran*: Paniyara—; *Urdu*: Tulispatar—; *Uriya*: Baincha, Panionla—.

2. **Flacourtia Ramontchi** L'Herit. is common throughout India, wild and cultivated. It is found in the indaing and dry forests of Burma.

In Chota Nagpur the bark is applied to the body along with that of *Albizzia odoratissima*, at intervals of a day or so, during intermittent fever.

In La Reunion the bark is used as an astringent and diuretic.

In Madagascar the root is prescribed in nephritic colic; the fruit is considered diuretic.

The fruit is sweet, appetising and digestive. It is given in India for jaundice and enlarged spleen.

The gum is given, along with other ingredients, for cholera,

After child-birth, among natives in the Deccan, the seeds are ground to powder with turmeric, and rubbed all over the mother's body to prevent rheumatic pains from exposure to damp winds.

Akola: Gurgati, Parbekal, Parekal—; *Almora*: Kanel—; *Amraoti*: Kakai—; *Banswara*: Kanker—; *Bengal*: Binja, Kakai, Katai, Obir, Tambat—; *Betsimisaraka*: Valamoly, Voatronaka—; *Betul*: Kakai—; *Bombay*: Bhekal, Kaikun, Kakad, Kantaka, Pahar, Swadu, Tambat—; *Buldana*: Parbekal, Parekal—; *Burma*: Nayuwai—; *Canarese*: Gajabira, Gajale, Hettarimullu, Hunmunki, Muldindu, Mullutari, Nayibela—; *Central Provinces*: Bilati, Kakein, Kank, Kanki, Karumurukki—; *Coorg*: Gapra, Gupra—; *Deccan*: Bowchi, Bunj, Kanbowchi, Kundayi, Swadu—; *Dehra Dun*: Kango—; *English*: Ceylon Plum, Madagascar Plum, Mauritius Plum—; *Gond*: Armasuri, Katien—; *Gujerat*: Kankod—; *Hindi*: Bhanber, Bilangra, Bowchi, Bunj, Handi, Kakai, Kakein, Kandai, Kandei, Kanju, Katahi, Katar, Kukar, Kundayi—; *Khond*: Sapka—; *Kolami*: Mehrle, Mehrlo, Merlee, Sarlarka, Serali—; *Kumaon*: Bilangra, Bilangur—; *Kurku*: Gurgoti—; *Lambadi*: Lotpatar—; *La Reunion*: Prune malgache marronne—; *Madagascar*: Lamoty—; *Marathi*: Bhekal, Kaker, Paker, Tambat—; *Mauritius*: Prune malgache—; *Melghat*: Gurgati—; *Mhairwara*: Kaikun—; *Mundari*: Meriedaru—; *Nimar*: Parbekal, Parekal—; *Palamou*: Katail—; *Punjab*: Kakoa, Kandei, Kangu, Kukai, Kukoa—; *Sanskrit*: Swadukantaka—; *Santal*: Merlec—; *Saora*: Bejjuven—; *Seychelles*: Prunier—; *Sind*: Bavache, Bhutankas—; *Sinhalese*: Ugurassa, Uguressa—; *Tamil*: Maulkkarai, Sottaikala—; *Telugu*: Bontakandregu, Kanaregu, Kandregu, Nakkanaregu, Nakkaneredu, Nelli, Peddakanaregu, Putikatada—; *Uriya*: Balibhaincho, Boniso, Kontadhawra, Kontaikuli, Mamuri, Potnaboniso—; *Yeotmal*: Kakai—; *Zambesi*: Batoko Plum—.

3. **Flacourtia sepiaria** Roxb. occurs in the dry jungles throughout Bengal and the Western Peninsula; in the scrub forests of the Madras Presidency, especially on the Coromandel Coast and in the Deccan; in Bihar, Orissa, Upper Burma, the Andamans; Ceylon and Java.

In Madagascar an infusion of the leaves and roots is given in cases of snake-bite; the bark triturated in oil is used as a liniment in rheumatism; the ashes of the root are considered serviceable in kidney troubles.

Mhaskar and Caius have established experimentally that neither the leaves nor the roots are an antidote to snake venom.

Bombay: Atruna, Tambat—; *Canarese*: Mirde, Miridi—; *Central Provinces*: Bainch—; *Gujerat*: Lodri—; *Hindi*: Kondai, Kondari—; *Malayalam*: Kurumuli—; *Marathi*: Atran, Tambat—; *Porebunder*: Lodri—; *Punjab*: Dajkar, Jidkar, Khatai, Kingro, Sherawane, Zargal—; *Sakalave*: Lamoty—; *Tagalog*: Bitongol—; *Tamil*: Kodumundi, Sottaikala—; *Telugu*: Kanaregu, Kandregu, Pulivelaga—; *Uriya*: Baliboniso, Boniso, Botubo, Kankui, Sanubainchi—; *Visayan*: Sanasaua—.

GYNOCARDIA.

The only species, **Gynocardia odorata** R. Br., is an evergreen tree of moderate size to be found in Sikkim, the Khasia Hills, Chittagong, and the Martaban Hills.

The fruit is used as a fish poison.

The oil from the seeds is used in the treatment of leprosy and other forms of skin diseases.

In Indo-China the seeds are considered tonic, but they are used externally only; they are pounded, mixed with oil, and applied topically in skin diseases.

Bengal: Chaulmugra, Chaulmugri, Petarkura—; *Bombay*: Chaulmogra—; *Chinese*: Ta Foung Leao—; *Hindi*: Chaulmogra, Chhalmugra, Choulmungri—; *Indo-China*: Dai phong tu, Doc chat—; *Lepcha*: Tukkung—; *Magahi*: Tungpung—; *Nepal*: Kadu—; *Persian*: Brinjmogra—; *Sanskrit*: Alasakapaha, Kushthapa, Sagarodbhuta, Tuvaraka—; *Sinhalese*: Taliennoe—.

HYDNOCARPUS.

The genus consists of 25 Indo-Malayan species.

A medicinal oil is extracted from the seeds.

H. anthelmintica Pierre is used in China and Indo-China, *H. polyandra* Blanco in the Philippine Islands.

1. Petiole 6-9 mm. long. Flowers 9-12 mm. in diameter, white ... ... *H. Wightiana.*
2. Petiole 12 mm. long. Flowers 13 mm. in diameter. *H. venenata.*
3. Petiole 12-15 mm. long. Flowers rose, petals 15 mm. long ... ... *H. anthelmintica.*

1. **Hydnocarpus Wightiana** Blume occurs in the tropical evergreen forests of South-Western India along the Western Ghats, from the Konkan southwards, and below the Ghats in Kanara and Malabar in damp situations; it is common in Travancore, and is found up to an altitude of 3,000 feet.

The seeds have long been employed as a domestic remedy in cases of skin disease and ophthalmia, and as a dressing for wounds and ulcers. Their medicinal properties depend wholly on the oil they contain, which is much more useful and convenient as a drug than the seeds themselves.

The oil is employed as an external application to scabby eruptions, after being mixed with an equal portion of *Jatropha Curcas* oil, sulphur, camphor, and lime-juice. For scald-head equal parts of the oil and lime-water are used as a liniment. In Travancore half teaspoonful doses are given internally in leprous affections, and the oil, beaten up with the kernels and shells of castor-oil seeds, is used as a remedy for itch.

Bicol: Butungmanoc—; *Bombay*: Kadukavatha, Kauti, Kava, Kowti—; *Canarese*: Bhutahi, Garuduphala, Niradivittulu, Surante—; *Deccan*: Jangli-badam—; *Goa*: Kosto—; *Kadir*: Koti, Nirvetti, Vattai, Vetti—; *Konkani*: Konstel, Konxtti—; *Malayalam*: Koti, Maravetti, Maroti, Niralam, Nirvetti, Vetti—; *Marathi*: Kadukavata, Kantel, Kastel, Kowti—; *Sanskrit*: Garudaphala—; *Sinhalese*: Makulu—; *Tamil*: Maravattai, Maravetti, Niradimuttu—; *Telugu*: Adavibadamu, Niradi—; *Tulu*: Surante—.

2. **Hydnocarpus venenata** Gaertn. is found in Ceylon by the bank of rivers up to 2,000 feet, also in Malabar, Tinnevelly, and Travancore.

The seeds are employed to poison fish; but the fish thus killed are unfit for food.

The oil is used as an external application in certain cutaneous diseases, and has a special reputation in leprosy.

The oil from the seeds was administered to twenty cases of leprosy and was found very beneficial in early cases and slightly so in long standing cases (Koman).

Deccan: Jinglibadam—; *Hindi*: Janglibadam—; *Malayalam*: Moratti, Niredumuttu—; *Marathi*: Kauti—; *Sinhalese*: Makulu—; *Tamil*: Maravittai, Niradimattu, Niridumuttu—; *Telugu*: Niradi, Niruduvittulu—.

3. **Hydnocarpus anthelmintica** Pierre occurs in Siam and Indo-China up to elevations of 1,000 feet, and is in cultivation at Singapore. It is generally to be found on the banks of rivers and estuaries where the water table is high.

In Cambodia the bark is prescribed for incontinence of urine.

The seed is extensively used in Indo-China for the treatment of cutaneous diseases.

The oil from the seeds has for ages been employed by the Chinese in the treatment of parasitic pediculi, leprosy, and many skin affections.

Annam: Chum-bao lon—; *Cambodia*: Krabao phle thom—; *Chinese*: Feng Yu Tzu, Ta Feng Tzu, Ta Fung Tze—; *Indo-China*: Dai phong tu—; *Malaya*: Foong yau tze—; *Siam*: Lukrabao, Maikrabao—.

Pangium.

The sole species, **Pangium edule** Reinwdt., is common in Malaya by river banks and in villages. It is distributed to the Malay Islands.

Every part of the plant is used as an anthelmintic in the Philippine Islands.

The bark is used by the Malays and the Javanese as a poison to fish.

The poisonous properties of the plant are well known to the Malays. The seeds are the most toxic part, but apparently only when they are quite fresh and in the raw state. The fresh seeds are used by the Sakais in making dart poison, and the oil expressed from raw seeds is added to cakes by Malay criminals to cause death.

Oil expressed from sun-dried seeds is often used as an article of food, but is said to cause diarrhœa.

The seeds boiled, cut up and macerated in water are eaten, the oil extracted used in cookery. At Amboina the toxicity is destroyed by crushing the seed and macerating it for some time in pure water.

The toxic properties of the fresh seeds and oil are due to the presence of a poisonous glucoside breaking down and producing prussic acid.

Malay: Kepayang, Kepayung, Payung—; *Tagalog*: Pangi—; *Visayan*: Pangui—.

TARAKTOGENOS.

The genus numbers 12 species spread over Assam, Burma, Indo-China, Malaya, and the Dutch East Indies.

Taraktogenos Kurzii King occurs in Assam, Chittagong, the Minbu district of Upper Burma, the eastern and northern slopes of the Pegu Yoma, Tenasserim, and Martaban, where it is very common.

The seeds yield the Chaulmoogra Oil of commerce and medicine.

Assam : Lemtam—; *Burma* : Kalanzo, Kalaw, Kalawaso, Kalawni, Kalawso—; *Lepcha* : Tukakunga—.

II.

The PITTOSPORACEÆ are trees or shrubs, often climbing. They form 10 genera, which, except for *Pittosporum*, are confined to Australia.

PITTOSPORUM.

This genus numbers 70 species inhabiting the tropical and subtropical regions of the Old World.

The following species are used medicinally in Indo-China—*P. tobira* Ait.—; in the Philippine Islands—*P. brachysepalum* Turcz.—; in La Reunion—*P. senacia* Putterl.—; in South Africa—*P. viridiflorum* Sims.

Pittosporum napaulense Rehder and F. H. Wils. (=*P. floribundum* Wight and Arn.) is found in the subtropical Himalaya, from Sikkim to Garhwal, ascending to 5,000 feet on the hills, and at Mishmi; also in Ganjam, in the Western Peninsula from the Konkan to the Nilghiris, in the hills of South Arcot and Salem.

The bark is bitter and aromatic, and is said by the tribes of the Western Ghats to possess narcotic properties. It is used in doses of 5 to 10 grs. as a febrifuge, and in doses of 50 grs. is believed to be a specific for snake poisoning; 5 to 10 grains doses of the dried bark were given with benefit in chronic bronchitis. It is a good expectorant, but in one or two cases in which it was tried in Bombay, it gave rise to dysenteric diarrhœa.

Among the Mundas of Chota Nagpur the bark is ground with water and rubbed over inflammatory dropsical or rheumatical swellings, and on the limbs in all cases of rheumatism.

Mhaskar and Caius have shown experimentally that the bark is not an antidote to snake venom.

Bombay : Yekadi, Yekdi—; *Canarese* : Tammata—; *Kharwar* : Baghmuta—; *Kisan* : Baghmuta—; *Lepcha* : Bongzam, Prongzam—; *Marathi* : Vehkali, Vehyenti, Vikhari, Yekadi—; *Naguri* : Herekasmar, Umarsing—; *Nepal* : Tibiloti, Tibilti—; *Saora* : Pida, Rakamuki, Rongosani—; *Tamil* : Nanjundai, Tammata—; *Telugu* : Rakamuki—; *Uriya* : Debosundu—; *Visayan* : Balungcanayan—.

III.

The POLYGALACEÆ are herbs or undershrubs, rarely small trees, distributed over the tropical and warm southern temperate zones. They include 10 genera with about 700 species.

All the members are more or less tonic, and expectorant; some are bitter, and emetic; a few are acrid, and poisonous. The bark of genus *Krameria* is very astringent.

Saponins—senegin, polygalic acid—and *methyl salicylate* are amongst the few compounds isolated; also *glucosides*—gaultherin, polygalin—.

The medicinal and poisonous Milkworts of the world belong to 8 genera: BADIERA (West Indies), BREDEMEYERA (Australia, Tasmania, South America, West Indies); CARPOLOBIA (tropical West Africa); KRAMERIA (Mexico to Chili); MONNINA (Mexico to Chili); MUNDTIA (South Africa); POLYGALA (cosmopolitan; except New Zealand, Polynesia, and Arctic zone); SECURIDACA (tropics; except Australia).

All Milkworts used medicinally in India belong to the genus POLYGALA.

POLYGALA.

The genus numbers 475 species distributed over the whole world, except Tasmania.

The following species are used medicinally in Europe—*P. amara* Linn., *P. calcarea* F. W. Schultz., *P. major* Jacq., *P. monspeliaca* Linn., *P. nicaeenis* Risso, *P. rupestris* Pourr., *P. serpyllacea* Weihe, *P. vulgaris* Linn.—; in Japan and China—*P. sibirica* Linn.—; in Indo-China—*P. glomerata* Lour., *P. sibirica* Linn.—; in Malaya—*P. Reinii* Franch. & Sav., *P. sibirica* Linn.—; in North America—*P. boykini* Nutt., *P. polygama* Walt., *P. sanguinea* Linn., *P. Senega* Linn.—; in Mexico—*P. scoparia* H. B. & Kunt—; in Brazil—*P. angulata* DC., *P. fimbriata* A. W. Ben., *P. paniculata* Linn.—; in Guiana—*P. timoutou* Aubl.—; in West Africa—*P. arenaria* Willd., *P. butyracea* Heck., *P. guineensis* Willd., *P. micrantha* Guill & Perr.—; in Southern Africa—*P. amatymbica* E. & Z., *P. arenaria* Willd., *P. hottentota* Presl., *P. myrtifolia* Linn., *P. oppositifolia* Linn., *P. serpentaria* E. & Z., *P. tenuifolia* Link.—; in Madagascar—*P. macroptera* DC.—.

P. amara and *P. Senega* contain a toxic glucoside, polygalin.

Undershrubs, or herbs sometimes woody at the base. Calyx persistent. Keel crested.

A. Bracts persistent, at least till the flower expands.
  1. Undershrubs. Racemes axillary; wings petaloid, membranous; strophiole 2-appendiculate ... *P. crotalarioides.*
  2. Herbs. Wings herbaceous, strophiole 3-appendiculate.
     a. Flowers minute, 1/10 in. long, yellow. Racemes longer than the leaves ... ... *P. brachystachya.*
     b. Flowers 1/6 in. long. Racemes longer than the leaves ... ... *P. elongata.*
     c. Flowers 1/5-¼ in. long. Racemes much shorter than the leaves ... . *P. chinensis.*

B. Bracts caducous (before flowering), strophiole 3-appendiculate.
1. Capsule rather broadly winged, glabrous, not ciliate ... ... *P. sibirica.*
2. Capsule glabrous, margined, not ciliate ... *P. telephioides.*
3. Capsule narrowly winged, strongly ciliate ... *P. glomerata.*

1. **Polygala brachystachya** Bl. occurs in Malaya and is found in turf, roadsides, and grass plots.

The Malays use the plant as an antidote in cases of snake-bite.

2. **Polygala chinensis** Linn. is found throughout India, from the Punjab to Burma, South India, and Ceylon, up to 5,000 feet. It extends to tropical Asia and Australia.

In Chota Nagpur the root is given medicinally in cases of fever and dizziness.

English : Common Indian Milkwort—; *Gujerat* : Pilibhonyasana—; *Hasara* : Birminditasad—; *Hindi* : Meradu, Miragu—; *Marathi* : Negli—; *Nagpuri* : Danaminjo, Danaminju, Gurgur—; *Porebunder* : Pilibhoysana—; *Santali* : Gaighura—.

3. **Polygala crotalarioides** Ham. is found on the temperate Himalaya from Simla and Chamba to Sikkim, between 4,000 and 7,000 feet, also on the Khasia Hills.

The entire plant and root have a reputation as remedies for cough and pulmonary catarrh.

Among the Mundas of Chota Nagpur the root is chewed, or else ground and drunk with water, to expel phlegm from the throat; it provokes coughing.

Royle states that the plant was sent to him with the information that the root was employed as a cure for snake-bite by the hill people of the Himalaya.

The plant is not an antidote to snake venom (Mhaskar and Caius).

Gond : Bijnori—; *Hasada* : Birheremda—; *Mundari* : Heremola—; *Nagpuri* : Lilkanth, Nilkanth—; *Santali* : Gaighura, Lilkanthi, Lilkathi—.

4. **Polygala elongata** Klein occurs in the Western Peninsula from the Konkan southwards, and extends to Ceylon.

The plant is used in biliousness and constipation. It is said to be antidote to snake poison.

Malayalam : Periyananka—.

5. **Polygala glomerata** Lour. is found in Sikkim, Assam, the Khasia Hills, and Burma, whence it extends to the Malay Archipelago and China.

In Indo-China a decoction of the stems and leaves is given in inflammatory conditions.

Indo-China : Kim bat hoan—.

6. **Polygala sibirica** Linn. occurs in temperate and subtropical Himalaya at 1,000-6,000 feet; in Sikkim at 8,000 feet; from the Punjab and North-Western Frontier to Bhotan; on the Khasia

Hills at 4,000-6,000 feet; in the Western Ghats from the Nilgiris to Tinnevelly, chiefly above 6,000 feet; in Ceylon. It is distributed to China, Japan, and Siberia.

In Japan, China, and Malaya the roots are given as a substitute for senega in colds and coughs.

In Indo-China they are used as a diuretic; they are also given in bronchitis, amnesia, sexual impotency, and seminal losses.

Brahui: Kaj, Mashnakaj—; *Cantonese*: Yen Chi—; *Chinese*: Yuan Chih—; *English*: Japanese Senega—; *German*: Schlangenwurzel—; *Indo-China*: Nam vien chi, Re tieu thao, Vien chi—; *Japanese*: Hime hagi—; *Malaya*: Yoan chee—.

7. **Polygala telephioides** Willd. occurs in the Carnatic, Nellore, Chingleput, Travancore, Ceylon. It spreads to the Malay Archipelago and China.

The plant and the root are used medicinally in catarrhal affections.

IV.

The FRANKENIACEÆ are herbs or shrublets, which inhabit tropical and temperate maritime shores, and principally the Mediterranean and Atlantic, extending into central Asia and north-western India; they are comparatively rare in the tropics and southern latitudes. They form a small family of 4 genera with 60 species.

FRANKENIA.

This genus consists of 32 species, mostly sea-side plants inhabiting the temperate and subtropical regions of both hemispheres.

F. grandifolia Cham. and Schlecht. is used medicinally in California.

Frankenia pulverulenta Linn. occurs in Sind, Baluchistan, and Punjab. It is a wide-spread species on shores and in salt deserts, distributed over the Mediterranean regions, Arabia, Senegal, and the Cape.

Valued by native practitioners in the fresh state for its mucilaginous and aromatic properties; exhibited in the form of decoction in empyreuma (Murray).

Arabic: Garmal—; *Egypt*: Ghobeyra, Gurmeyl, Homra, Kheyet, Mulleygh—; *English*: Powdery Sea-Heath—; *Malta*: Mealy Sea-heath—; *Sind*: Khareeya—

V.

The PORTULACACEÆ are herbs, rarely undershrubs, usually glabrous and more or less succulent. The family includes 17 genera with 225 species, mostly natives of America and South Africa.

The medicinal Purslanes of the world belong to 3 genera: ANACAMPSEROS (South Africa); PORTULACA (tropical and subtropical regions); TALINUM (Africa, America, India).

The Purslanes used medicinally in India all belong to the same genus, PORTULACA.

PORTULACA.

The genus numbers 20 species distributed over the tropical and subtropical regions of the world.

Fleshy, insipid plants. They are considered cooling, diuretic, and recommended in scurvy and urinary affections.

The following are used medicinally in Europe—*P. oleracea* Linn.—; in China and Indo-China—*P. oleracea* Linn., *P. sativa* DC.—; in the Philippine Islands—*P. oleracea* Linn., *P. quadrifida* Linn.—; in the United States—*P. oleracea* Linn.—; in Brazil—*P. patens* Linn., *P. pilosa* Linn.—; in Guinea—*P. oleracea* Linn.—; in Gold Coast—*P. oleracea* Linn., *P. quadrifida* Linn.—; in Southern Africa and Portuguese East Africa—*P. quadrifida* Linn.—; in La Reunion and Madagascar—*P. oleracea* Linn.—.

A. Nodes without appendages. Flowers in clusters ... *P. oleracea.*
B. Nodes with a ring of hairs.
 1. Leaves opposite ... ... *P. quadrifida.*
 2. Leaves alternate, root tuberous ... ... *P. tuberosa.*

1. **Portulaca oleracea** Linn. occurs in all warm countries. It is found all over India, up to 5,000 feet in the Himalaya.

The plant has long been used as a domestic remedy in all the various countries where it is found growing. It is reputed vulnerary, antiscorbutic, refrigerant, and mildly diuretic; and it is said to be useful in catarrhal affections of the genito-urinary tract. Thus Culpeper: 'It is good to allay the heat of the liver, blood, veins, stomach, and hot agues: it stays hot and choleric fluxes of the belly, women's courses, the whites, and gonorrhœa, the distillation from the head, and pains therein proceeding from heat, want of sleep, or the frenzy.'

At the present day, the herb is chiefly valued as a refrigerant and alterative pot-herb, particularly useful as an article of diet in scurvy and liver disease.

In Colombia the plant is used as an emollient, and is applied to tumours and callosities. Similarly, in West Tropical Africa it is used for local application to swellings, bruises, whitlow, etc., crushed along with native natron and oil and applied by rubbing or as a poultice for abscesses and boils to bring them to a head.

In Siberia the herb is used as a gastric sedative; it is also given as a vegetable to a baby who has ceased to thrive; along with an aromatic Labiate plant it is a remedy for haematuria and other urinary troubles.

In Jamaica the herb is employed as a cooling and moistening remedy in burning fevers.

The herb, bruised, and applied to the forehead and temple, is said to allay excessive heat and pain, and applied to the eyes to remove inflammation. Quoth Culpeper: 'The herb if placed under the tongue assuayeth thirst. Applied to the gout, it easeth pains thereof, and helps the hardness of the sinews, if it come not of the cramp, or a cold cause.'

The expressed juice of the plant may be applied with advantage to prickly heat, as well as to the hands and feet when a burning sensation is felt. It is said to afford relief in cases of scorpion sting.

In Jamaica, the juice is of use in spitting of blood. In West Tropical Africa it is sometimes dropped in the ear for earache, and is used for toothache. With oil of roses, it was recommended in England for sore mouths and swollen gums, and also to fasten loose teeth.

The expressed juice of the herb is an ingredient in a Hausa prescription for syphilis.

In China the leaves are used for poulticing tumours, bad wounds and ulcers, œdematous swellings; also in blennorrhagia and leucorrhœa. In Nigeria they are applied topically to swellings.

In Indo-China a decoction of the leaves is given in dysentery.

In Gold Coast the leaves are ground, mixed with oil, and tied on boils to bring them to a head; they are eaten along with tiger nuts—*Cyperus esculentus*—as a remedy for skin diseases; also macerated in cold water and taken frequently they are considered to be a heart tonic and diuretic, used for palpitations.

In Indo-China the juice of the fresh leaves is applied to abscesses, and used as a collyrium. In North America it is considered a cooling diuretic.

In the Tamil country a thick paste obtained by mixing together the juice of the leaves, human milk, and sesamum oil is rubbed over the abdomen to help the expulsion of the retained placenta in cases of uterine inertia.

The bruised fresh leaves are prescribed by the Tamil practitioners as an external application in erysipelas; an infusion of them is also ordered as a diuretic in dysuria.

The seeds have always been credited with anthelmintic properties, and they are still given as a vermifuge in the Punjab. In England they were bruised and boiled in wine when used as a worm remedy for children.

In Cochin-China and in the West Indies the seed is frequently used as a stomachic and as a provocative of the menses, as well as an emollient and diuretic.

Caius and Mhaskar have shown experimentally that the plant is not an antidote to scorpion venom.

Afrikaans: Misbredic, Porselein, Postelein, Varkkos—; ***Akim***: Adwera—; ***Annam***: Rau sam, Rau tham—; ***Arabic***: Baglah, Baglatulhumga, Bogl, Buklut-ul-hakima, Buklut-ul-kukema, Dhoneb el farras, Khurfa, Kurfakara, Rigl—; ***Ashanti***: Adwera, Adwere, Asase-ne-aboo—; ***Australia***: Pigweed—; ***Baluchi***: Pechali, Pichali—; ***Bambara***: Missidi kumbaré—; ***Bengal***: Baraloniya, Chhota-lunia, Kulfi, Munya—; ***Betsileo***: Kalabotetraka—; ***Bicol***: Ausiman—; ***Bombay***: Gol, Kurfah, Motighol—; ***Brahui***: Mugher, Pechali, Pichali—; ***Burma***: Mayabyit—; ***Canarese***: Dudagorai—; ***Central Provinces***: Ghol, Gholu—; ***Ceylon***: Nilappachali, Pulichankirai—; ***Chinese***: Ma Ch'ih Hsien, Ma Tche

Han—; *Colombia*: Verdolaga—; *Dagomba*: Zinerigu—; *Danish*: Portulak—; *Deccan*: Mulfekibhaji—; *Dutch*: Porselein—; *Efik*: Efere makara—; *Egypt*: Bakle, Higl, Rigl—; *English*: Common Purslane, Garden Purslane, Golden Purslane, Pigweed, Purslane—; *Ewe*: Aflaa, Aflangtokpui, Devio-fe'ama, Dugba, Kugbe—; *French*: Porcelin, Porcellane, Pourcellaine, Pourcellane, Pourpier, Pourpier commun, Pourpier cultivé—; *Fulani*: Bechalhi, Endu nagge—; *Ga*: Awrorke—; *German*: Portulak—; *Greek*: Andrachni, Andrakla, Anuraklida, Antrakla, Glistrida—; *Gujerat*: Loni, Motiluni—; *Hausa*: Babba jibji, Da'burin saniya, Fasa k'aba, Fasa kumburi, Halshen rago, Halshen saniya, Halshen tunkiya, Rub da tukubya, Sarkin jibji—; *Hindi*: Baralaniya, Baralunia, Chhotalunia, Khurfa, Khursa, Kulfa, Kulfi, Kurfa, Kurfekasag, Lonia, Lunak, Lunia, Luniakulfah, Lunuk, Muncha, Munya, Nonkha, Nonkhalunuk, Sagkulpha—; *Hova*: Tsikobokobondanitra—; *Indo-China*: Ma si hien, Ma xi hieu, Rau sam—; *Iraq*: Barbin, Hamga—; *Italian*: Porcellana, Portulaca—; *Japan*: Osuberi-hiyu—; *Jhalawan*: Pichli, Shurdako—; *Kachhi*: Lunak, Pichlo—; *Kalmuk*: Assun obison—; *Katsina*: Fasa k'aba—; *Kohlu*: Lunak—; *Kolami*: Dali ara—; *Konkani*: Gol, Golchibagi—; *Krepi*: Aflaakupe, Aflaatokplei, Aflangtokpui—; *Krobo*: Nereyu, Nerufu—; *Kumaon*: Lunak, Luniyahkulfah—; *Kurdish*: Parpinah—; *Languedoc*: Poutoulaigo—; *La Reunion*: Pourpier, Pourpier rouge—; *Las Bela*: Manshuri—; *Malay*: Segan—; *Malayalam*: Koricchira—; *Malinke*: Mazabi—; *Malta*: Purslane, Porcellana, Sportellacchia, Burdliecka—; *Mano*: Toa p'lo—; *Marathi*: Bhuigholi, Ghole, Motighol—; *Mauritius*: Pourpier rouge—; *Mende*: Toge, Tondo wolii, Tooge—; *Mundari*: Dailara, Urialangara—; *Nasirabad*: Lunak—; *North America*: Common Purslane, Parsley, Pussly—; *North Queensland*: Thukouro—; *North-Western Provinces*: Desikulfah, Lunak, Luniya, Muniya—; *Nubia*: Segettemam—; *Persian*: Cholza, Kherefeh, Khurfah, Kurfah, Tirehkhurfeh, Turk, Turuk—; *Polish*: Kurza nega—; *Porebunder*: Mhotiluni—; *Portuguese*: Beldroega—; *Punjabi*: Sagkulfa—; *Pushtu*: Murlai, Tursbuk, Warkharai—; *Roumanian*: Iarba grass, Portulaca—; *Russian*: Portulak, Schrucha—; *Sadam*: Dailsag—; *Sakalave*: Fandrianomby—; *Sanskrit*: Brihalloni, Gholika, Lona, Loni, Lonika, Lunia—; *Santal*: Mota uric'alang—; *Shahrig*; Lunak—; *Sinaloa*: Verdolaga—; *Sind*: Lonk—; *Sinhalese*: Gendakola—; *Songhai*: San mafé—; *Spanish*: Verdolaga—; *Suto*: Selele—; *Swedish*: Portulak—; *Tagalog*: Colasiman, Golasiman, Olasiman, Sayiçan—; *Tamil*: Karikkirai, Kolhikirai, Paruppukkirai, Passalakkirai, Pulikkirai—; *Telugu*: Boddipavilikura, Gangapavilikura, Pappukura, Peddapavilikura—; *Toba*: Khulfa, Mirri—; *Tuareg*: Alokha, Alora, Aloza, Bender akech—; *Twi*: Adwera, Adwerair—; *Urdu*: Khurfah—; *Uriya*: Purunisag—; *Wolof*: Tang i mpiteurh, Tanguipeter—; *Yemen*: Brabra, Chamile, Denneb el farras, Rijlet el farras—; *Yoruba*: Papasan—.

2. **Portulaca quadrifida** Linn. is found throughout the warmer parts of India and Ceylon. It spreads to tropical Asia and Africa.

The plant is used as a substitute for *P. oleracea*.

The infusion is in Pondicherry a household remedy for cystitis in the last months of pregnancy. The Zulus use it as an emetic.

The natives of Portuguese East Africa take a decoction of the plant as an anthelminitic, and in the treatment of stomach complaints and gonorrhœa.

The plant is sometimes used as a pot-herb in Guam, either for food or as an antiscorbutic. In Gold Coast and in Lagos it is used to cure toothache.

The Arabs apply the bruised fresh leaves to the forehead in cephalalgia.

Arabic: Baglat-ul-arabbiyah, Baglat-ul-yamaniyah, Budelut-ul-mubarik, Kora-at errai, Rozzi, Sebib ed-dhan—; *Ashanti*: Asase-ne-aboo—; *Bengal*: Chhotaluniya, Muniya—; *Bettani*: Perkhadai—; *Bombay*: Barikaghola, Chavalkebhaji, Chowli, Gholkibhaji, Kota—; *Canarese*: Halibochcheli—; *Ceylon*: Chirupachalikkirai, Parupukirai—; *Deccan*: Chaunlayikibhaji, Chowli, Gholkibhaji—; *Ewe*: Dugba, Kugbe—; *Gold Coast*: Stone Crop—; *Gujerat*: Jhiniluni, Luni—;

Hindi: Baklatalhamka, Chaunlayi, Chotaluniya, Khatechawal, Loniya—; *Kolami*: Suniara—; *La Reunion*: Pourpier marron—; *Malinke*: Mazabi—; *Marathi*: Katechanval, Kathechawal, Ranghol—; *Porebunder*: Badhi, Jinkiluni—; *Portuguese East Africa*: Sanimarumbi—; *Pondicherry*: Baselle, Pourpier grimpant—; *Punjab*: Haksha, Lunak, Lunkibuti—; *Rajputana*: Lunki—; *Samoa*: Fiafiatuli—; *Sanskrit*: Kshudragholika, Laghughonika, Laghulonika, Upadyaki—; *Sinhalese*: Hingendakola—; *Spanish*: Yerba de pollo—; *Tagalog*: Sayican—; *Tamil*: Passalaikkirai, Passelikkirai, Sinnaparpukkirai, Siruppassaraikkirai—; *Telugu*: Batsalikura, Goddupavili, Kura, Pavili, Payalaku, Peddapavili, Saunapappu, Saunapavili, Sunapailkura—; *Waziri*: Zhinai—; *Yemen*: Mortah—.

3. **Portulaca tuberosa** Roxb. is found in Bihar, Sind, Gujarat, the dry districts of the Carnatic from South Arcot to Travancore, whence to Ceylon.

The fresh acid leaves are used medicinally; an external application is prescribed by native practitioners in erysipelas and an infusion in dysuria (Murray).

Marathi: Jangligajar—; *Sinhalese*: Urugenda—; *Sind*: Lunuk—; *Telugu*: Boddakura—.

VI.

The TAMARICACEÆ are mostly bushes or small trees, inhabiting the shores, steppes, and deserts of temperate and subtropical regions. They prefer sea-shores, the margins of brackish lakes, the banks of rivers and torrents, in sandy or clayey soils. There are 5 genera with about 100 species.

Medicinal properties are exhibited by members of 3 genera: MYRICARIA (northern temperate regions); REAUMURIA (eastern Mediterranean, Central Asia); TAMARIX (Europe, Mediterranean, Asia).

1. Seeds hairy all over, albuminous. Flowers solitary ... REAUMURIA.
2. Seeds hairy at the apex, exalbuminous. Flowers racemed or spiked.
   a. Stamens free, styles 3 ... ... TAMARIX.
   b. Stamens connate, stigma sessile ... ... MYRICARIA.

MYRICARIA.

The genus consists of 10 species distributed over the northern temperate regions of the Old World.

1. Bracts ovate, about twice as long as the pedicels ... *M. elegans*.
2. Bracts ovate-lanceolate, 3-4 times as long as the pedicels ... ... *M. germanica*.

1. **Myricaria elegans** Royle occurs in the Western Himalaya, Kunawar, Spiti, Lahul, Ladakh, Kumaon, up to 14,000 feet.

In Lahul the leaves are used as an application to bruises.

Garhwal: Wombu—; *Punjab*: Humbu, Umbu—.

2. **Myricaria germanica** Desv. is found in temperate and alpine Himalaya, from Sikkim to Kumaon, at altitudes of 10,000-14,000 feet. It spreads to Afghanistan, and westwards to Europe.

A decoction of the bark is used in Spain as an aperient, and in jaundice.

French: Petit tamarisc, Tamarisc d'Allemagne—.

REAUMURIA.

The genus numbers 15 species spreading from the Levant to Central Asia.

Reaumuria hypericoides Willd. is indigenous in Sind, Arabia, Syria, the Mediterranean coast, the Salt plains, and in the milder parts of Northern Asia.

It is used in Sind in the cure of porrigo and itch, the bruised leaves being applied externally, and a decoction of two-and-a-half to three ounces of the juice of the leaves diluted and administered internally.

Sind: Laneesah—.

TAMARIX.

The genus numbers 65 species, natives of Europe, the Mediterranean, and Asia.

The bark and the galls are astringent; the manna is detergent and aperient; the twigs and the leaves are vulnerary, carminative, and diuretic.

The following are used medicinally in Europe and North Africa—*T. gallica* Linn., *T. africana* Poir.—; in Arabia and Persia—*T. aphylla* Krst.—; in China—*T. chinensis* Lour.—; in Indo-China—*T. chinensis* Lour., *T. pallasii* Desv.—; in Malaya—*T. chinensis* Lour.—.

Stamens 5.

1. Flowers bisexual, in racemose panicles ... ... *T. Troupii.*
2. Flowers unisexual, in close cylindrical spikes ... *T. dioica.*
3. Flowers bisexual, in usually interrupted spikes ... *T. aphylla.*

1. **Tamarix aphylla** Lanza occurs in the Punjab plains, Sind, Cutch, and Baluchistan. It is distributed to Persia, Mesopotamia, Arabia, the Mediterranean, and South Africa.

The galls are employed as an astringent.

The bark is bitter, astringent; powdered, and in combination with oil and *Kamala*, it is used as an aphrodisiac. It is also employed as an application in eczema capitis, and other diseases.

Afrikaans: Abiekwas-gelhout, Dawee—; *Arabic*: Asl-ul-armar, Atl, Ethel, Oethel, Tarfa, Tarf-al-ahmar—; *Bengal*: Raktajhav—; *Deccan*: Laljhav—; *English*: Giant Tamarisk—; *Gujerat*: Laljhav—; *Helmand*: Kohragaz—; *Hindi*: •Laljhav—; *Iraq*: Athl—; *Kharan*: Siahgazz—; *Las Bela*: Kirri—; *Northern Baluchistan*: Kirri—; *Pahrod*: Gagaz, Shakargaz—; *Persian*: Farash, Gazesurkh, Khoragaz—; *Punjab*: Faraz, Farash, Farwa, Ghwa, Ghuz, Khagal, Kharlei, Narlei, Pharwan, Rukh, Ukhan—; *Pushtu*: Ghwa, Ghwaz—; *Sind*: Asrelei, Asri, Gaz, Gazlau—; *Southern Africa*: Tamarisk—; *Tamil*: Sivappattushavukku, Sivappukkottashavukku, Sivappusirushavukku—; *Telugu*: Ettashirisaru, Ettaverusaru—; *Tigre*: Obel, Owell—; *Tigrinia*: Obel, Owell—; *Tuareg*: Tabrakat—; *Waziristan*: Khugal—.

2. **Tamarix dioica** Roxb. is a gregarious shrub, found near rivers and on the sea-coast; throughout India from Sind to Burma. It is common along the Ganges, Hooghly, and forms extensive forests along the Indus in Sind. Also common in sandy river-beds and on the sea-coast of the Madras Presidency. It extends to Afghanistan.

The twigs and galls are used medicinally as an astringent.

Baluchi: Podhayaz—; *Bengal*: Jhao, Laljham—; *Betul*: Saran—; *Burma*: Byaungchedauk—; *Hindi*: Jau, Jhau, Lalijhau, Thar-thari—; *Kumaon*: Jhau—; *Ladak*: Reglta—; *Merwara*: Jhau, Kachlei, Pilchi—; *Oudh*: Jhau—; *Persian*: Ghuz—; *Punjab*: Faras, Farwan, Ghasiei, Harwan, Jhau, Kachlei Koan, Lai, Leh, Lei, Panj, Panjpilchi, Pilchi, Rukh—; *Pushtu*: Khwa—; *Sanskrit*: Pisula—; *Sind*: Gas, Jao, Lao, Lyi, Turunjabin—; *Tamil*: Attumari, Nirumari—; *Telugu*: Palivela, Palligi—.

3. **Tamarix Troupii** Hole occurs in the Punjab, the United Provinces, Sind, Baluchistan, and Mount Abu.

The galls are employed medicinally as an astringent, and administered internally in dysentery and diarrhœa.

The manna is considered to be detergent, aperient, and expectorant.

In Jaisalmer a sherbet made from the fruit is said to be very beneficial in cases of fever brought on from exposure to the hot wind. The shade of the tree is considered injurious to health; a heated traveller resting beneath its shadow is certain to suffer afterwards from fever or rheumatism.

Arabic: Asl, Taffa, Turfah—; *Baluchistan*: Gazkhera, Gazsurkh, Ghazlei—; *Bengal*: Jhau, Jhav—; *Catalan*: Tamarit, Tamariu—; *Central Provnces*: Jhau—; *Ceylon*: Kiri—; *Deccan*: Jhav—; *English*: Tamarisk—; *Gujerat*: Jhavnujhada—; *Hindi*: Jhau, Jhav—; *Italian*: Cipressina, Scopa marina—; *Konkani*: Jahu—; *Malayalam*: Jhavukam—; *Persian*: Gaz, Gazshakar, Shorgaz—; *Punjab*: Jhau, Koa, Lai, Lainya, Pilchi, Rukh—; *Pushtu*: Gazsurkh, Ghaz, Suragaz—; *Rajputana*: Imli—; *Sanskrit*: Aphala, Bahugranthika, Jhavu, Jhavuka, Pichula, Shavaka—; *Sind*: Jhaw, Lai, Lei—; *Spanish*: Taray—; *Tamil*: Attalari, Attuchavakku, Kodaichavukku, Sirusavukku, Sivappattuchavukku, Sivappukkodaichavukku, Sivappusirusavukku, Vannikkay—; *Telugu*: Erusaru, Etttasirusaru, Ettayerusaru, Pakke, Pakkepakki, Palligi, Prakke, Sirasaru—; *Tibet*: Regeeta, Telta—; *Urdu*: Jaheva—.

THE MEDICINAL AND POISONOUS COMPOSITES OF INDIA.

BY

J. F. CAIUS, S.J., F.L.S.

The COMPOSITÆ are found in all parts of the world, spread all over the warm and temperate regions, and extending even into the colder ones. They are herbs or shrubs, sometimes assuming arborescent forms in warm countries. This family far outstrips all others in the number of genera in the order, and the number of species in the genus. More than a tenth of all the species of flowering plants belong to this family, which includes about 900 genera and over 13,000 species.

Bitterness is the prevailing characteristic of the plants which compose the Order. Therapeutically these may be divided into two groups:—(1) the *Tubuliforæ* which are generally bitter, tonic, stimulant, and astringent, and in many instances diaphoretic, diuretic, and laxative; (2) the *Liguliforæ* which are milky, bitter, astringent, and sometimes narcotic.

The following are among the substances isolated:—(1) *hydrocarbons*—cadinene, dipentene, d-limonene, phellandrene, pinene—; (2) *alcohols*—cineol, borneol, mannitol, terpineol, thuyol—; (3) *phenolic ethers*—anethol—; (4) *acetones*—camphor, thuyone—; (5) *acids*—aconitic, alantic, angelic, caproic, carthamic, gallic, lactucic, nonylic, tiglic—; (6) *esters*—pyrethrins—; (7) *sugars*—glucose, laevulose—; (8) *starches*—inulin—; (9) *resinoids*—senecin, lactucopicrin—; (10) *glucosides*—absinthin, anthemic acid, arnicin, centaurein, cyanin, eupatorin, gaultherin, inulin, lactucenin, lactucin, moschatin—; (11) *alkaloids*—abrotine, achilleine, echinopseine, β-echinopsine, echinopsine, lobelanidine, lobelanine, isolobelanine, lobelidine, lobeline, moschatine, senecifolidine, senecifoline, senecine, senecionine—; (12) *miscellaneous drugs*—absinthiin, pyrethrin, santonin—.

The medicinal and poisonous Composites of the world belong to 196 genera:—ACANTHOSPERMUM (Hawaii, Singapore; Brazil); ACHILLEA (cosmopolitan; temperate northern regions); ACHYRACHAENA (Western North-America); ACTINOMERIS (North America); ADENOSTEMMA (cosmopolitan); AGERATUM (cosmopolitan, throughout the Tropics); AMBROSIA (Mediterranean region; America); AMPHIDOXA (southern and tropical Africa); ANACYCLUS (Mediterranean region); ANAPHALIS (cosmopolitan; Asia, America); ANTENNARIA (cosmopolitan, in mountains); ANTHEMIS (cosmopolitan); APLOPAPPUS (Chile); ARCTIUM (temperate Asia, Europe); ARNICA (Europe, Asia, North America); ARTEMISIA (cosmopolitan; northern temperate regions, South America); ASPILIA (tropical Africa; Madagascar; Mexico; South America, chiefly Brazil); ASTER (cosmopolitan, northern temperate regions); ATHANASIA (South Africa); ATHRIXIA

(Arabia, southern and tropical Africa, Madagascar, Australia); ATRACTYLIS (Mediterranean region; China, Japan); BACCHARIS (cosmopolitan; tropical America); BELLIS (cosmopolitan, northern hemisphere); BERKHEYA (South Africa); BIDENS (cosmopolitan, chiefly American); BIGELOWIA (northern and central America); BLEPHARIPAPPUS (western North-America); BLUMEA (tropical and subtropical Asia, Africa, Australia); BOJERIA (tropical Africa, Madagascar); BOLTONIA (northern subtropical Asia, North America); BRACHYCLADOS (Argentina, Patagonia, Chile); BRACHYLAENA (South Africa); BRICKELLIA (North America, Mexico, Brazil); CALEA (America, Jamaica, China); CALENDULA (cosmopolitan; northern Africa, southern Europe, western Asia); CALLILEPIS (South Africa); CARBENIA (Mediterranean region, Caucasus); CARDOPATIUM (Mediterranean region, North Africa); CARDUUS (cosmopolitan; temperate Europe, North Africa and Asia); CARLINA (Mediterranean region; Siberia, Canary Islands); CARPESIUM (Europe; temperate and subtropical Asia); CARTHAMUS (Mediterranean region, Central Europe, India, Abyssinia, Canary Islands); CENIA (South Africa); CENTAUREA (cosmopolitan; Europe, America, western Asia, Australia); CENTIPEDA (Asia, Australia, tropical South America); CENTRATHERUM (Asia, Australia, America); CHONDRILLA (Mediterranean to India; Siberia); CHRYSANTHEMUM (cosmopolitan, northern temperate regions); CHRYSOCOMA (Caucasus, Siberia; Egypt, Arabia; South Africa, South America); CICHORIUM (cosmopolitan, temperate regions); CINERARIA (Europe, South Africa); CLIBADIUM (Central and South America, West Indies); CNICUS (northern temperate regions); COLEOSANTHUS (Bolivia); CONYZA (tropical and subtropical regions); COTULA (tropical and temperate regions); CREPIS (cosmopolitan; northern temperate regions); CROSSOSTEPHIUM (Central Asia, China, Luzon); CYNARA (Mediterranean region, Canary Islands); DAHLIA (cosmopolitan; Mexico, Central America); DICHROCEPHALA (Asia, Africa, Madagascar); DICOMA (South Africa, India); DIMORPHOTHECA (South Africa); DORONICUM (temperate Europe and Asia); ECHINACEA (North America); ECHINOPS (cosmopolitan); ECLIPTA (cosmopolitan, tropical); ELEPHANTOPUS (cosmopolitan, chiefly tropical American); ELYTROPAPPUS (South Africa); EMILIA (India, tropical Africa, Madagascar); ENHYDRA (tropical and subtropical regions); EPALTES (tropical Asia, Africa, America); ERECHTITES (America, Australia, New Zealand); ERIGERON (cosmopolitan; northern temperate region); ERIOCEPHALUS (South Africa); ETHULIA (India, Sunda Archipelago, Java, eastern tropical Africa, Madagascar); EUPATORIUM (cosmopolitan, chiefly American); EURYOPS (South Africa); FAUJASIA (Mascarene Islands); FELICIA (southern and tropical Africa); FILAGO (Europe, North Africa, Argentina, Paraguay); FLAVERIA (Central and South America, Australia); GAMOLEPIS (South Africa); GARULEUM (South Africa); GAZANIA (South Africa); GEIGERIA (South Africa); GERBERA (cosmopolitan; temperate regions and mountains); GLOSSOCARDIA (India); GLOSSOGYNE (tropical Asia, Australia, New Zealand); GNAPHALIUM (cosmopolitan); GONGROTHAMNUS (tropical Africa); GRANGEA (tropical Asia and Africa); GRINDELIA (North and South America); GUIZOTIA (tropical

Africa); GYMNOSPERMA (Mexico); GYNURA (warm regions of Asia, Australia, and Africa); HAPLOCARPHA (South Africa); HELENIASTRUM (Chile); HELENIUM (North America); HELIANTHUS (America, mostly North America); HELICHRYSUM (cosmopolitan; temperate and subtropical regions); HEMIZONIA (California, Galapagos); HERDERIA (tropical Africa); HIERACIUM (cosmopolitan); HYPOCHOERIS (temperate regions and mountains; Europe, northern Asia, Mediterranean region, South America); HYSTERIONICA (Western America, Chile); INULA (cosmopolitan; Europe, Asia, Africa); JURINEA (Europe, Western and Central Asia); LACTUCA (cosmopolitan; northern temperate regions); LAGGERA (Asia, tropical Africa); LAMPRACHAENIUM (India); LASIOSPERMUM (South Africa); LAUNAEA (Europe, Canary Islands; South and North Africa to India; West Indies); LEONTONYX (South Africa); LEYSSERA (Mediterranean region, South Africa, North America); LIATRIS (North America); LYGODESMIA (western North America); MACROCLINIDIUM (Japan); MADIA (Chile, North America); MALACOTHRIX (North America, California, Mexico); MATRICARIA (cosmopolitan); MELANTHERA (Bahama, Yucatan); METALASIA (South Africa); MICROGLOSSA (Asia, Africa); MICROSERIS (South America, New Zealand); MIKANIA (cosmopolitan; America); MONOLOPIA (California); MONTAÑOA (cosmopolitan); NIDORELLA (southern and tropical Africa, Abyssinia, Cape Verde); NOLLETIA (Africa); NOTONIA (India, Burma, Abyssinia, tropical Africa); ONOPORDON (Mediterranean region, Europe); ONOSERIS (Mexico, South America); OSMITES (South Africa); OSMITOPSIS (South Africa); OSTEOSPERMUM (Africa); OTHONNA (cosmopolitan; South Africa); OTHONNOPSIS (Africa, Asia); PARTHENIUM (North America, Mexico); PECTIS (America, West Indies); PEGOLETTIA (Java; southern and tropical Africa); PENTZIA (South Africa); PEREZIA (Mexico, South America); PETASITES (northern temperate regions); PEUCEPHYLLUM (western North America); PICRIDIUM (Europe, Western Asia, North Africa); PICRIS (cosmopolitan; Mediterranean region, Europe, Abyssinia, temperate Asia); PIGUERIA (Mexico, Ecuador, Peru, Chile); PLEIOTAXIS (tropical Africa); PLUCHEA (tropical and subtropical regions); POLYMNIA (America); PRENANTHES (northern temperate regions); PRINTZIA (South Africa); PSIADIA (Arabia, tropical Africa, Mascarene Islands, Madagascar); PTERONIA (cosmopolitan); PULICARIA (cosmopolitan; Europe, Asia, Africa); RUDBECKIA (cosmopolitan; North America); SANTOLINA (cosmopolitan; Mediterranean region); SAUSSUREA (northern temperate regions, and mountains); SCHISTOSTEPHIUM (South Africa); SCLEROCARPUS (tropical Africa, North America, Mexico); SCOLYMUS (Mediterranean region, Nubia, Caucasus); SCORZONERA (Central and South Europe, North Africa, Asia); SENECIO (cosmopolitan; temperate climates, mountains of the tropics); SERRATULA (cosmopolitan; Europe, North Africa; western, central, and alpine Asia); SIEGESBECKIA (cosmopolitan; tropics and subtropics; Peru); SILPHIUM (North America); SILYBUM (Mediterranean region; Europe, North Africa, northern India); SOLIDAGO (North America; northern temperate regions); SONCHUS (cosmopolitan; northern temperate regions, Central Asia); SPARGANOPHORUS

Australia); SPILANTHES (tropical America); STENOCLINE (Madagascar, Brazil); STOEBE (South Africa); SYNEDRELLA (tropical America, Africa, Asia); TAGETES (cosmopolitan; Argentina to Arizonal); TANACETUM (north temperate regions); TARAXACUM (temperate and cold regions); TARCHONANTHUS (Africa); TRAGOPOGON (Mediterranean region; Europe, North Africa, temperate and subtropical Asia); TRICHOLEPIS (India, Afghanistan); TRILISA (North America); TRIPLOTAXIS (tropical Africa); TRIPTERIS (South Africa); TRIXIS (America); TUSSILAGO (cosmopolitan; northern temperate regions); URSINIA (cosmopolitan; South Africa); VENIDIUM (southern and tropical Africa); VERBESINA (tropical America); VERNONIA (cosmopolitan, chiefly tropical; mostly American); VOLUTARELLA (southern Europe, North Africa, western Asia, India); WEDELIA (tropical and subtropical regions); WYETHIA (western North America); XANTHIUM (cosmopolitan; Indo-Malaya, America); XERANTHEMUM (Mediterranean region to the East; South Africa).

The medicinal and poisonous Composites of India belong to the following 80 genera:—ACANTHOSPERMUM, ACHILLEA, ADENOSTEMMA, AGERATUM, ANAPHALIS, ANTHEMIS, ARCTIUM, ARTEMISIA, ASTER, BIDENS, BLUMEA, BOLTONIA, CALENDULA, CARDUUS, CARPESIUM, CARTHAMUS, CENTAUREA, CENTIPEDA, CENTRATHERUM, CHRYSANTHEMUM, CICHORIUM, CNICUS, COTULA, CREPIS, DICHROCEPHALA, DICOMA, DORONICUM, ECHINOPS, ECLIPTA, ELEPHANTOPUS, EMILIA, ENHYDRA, ERIGERON, ETHULIA, EUPATORIUM, GERBERA, GLOSSOCARDIA, GLOSSOGYNE, GNAPHALIUM, GRANGEA, GUIZOTIA, GYNURA, HELIANTHUS, HIERACIUM, HYPOCHOERIS, INULA, JURINEA, LACTUCA, LAGGERA, LAMPRACHAENIUM, LAUNAEA, MATRICARIA, MICROGLOSSA, MIKANIA, NOTONIA, OTHONNOPSIS, PEGOLETTIA, PICRIS, PLUCHEA, PULICARIA, SAUSSUREA, SENECIO, SIEGESBECKIA, SILYBUM, SOLIDAGO, SONCHUS, SPARGANOPHORUS, SPHAERANTHUS, SPILANTHES, SYNEDRELLA, TAGETES, TANACETUM, TARAXACUM, TRAGOPOGON, TRICHOLEPIS, TUSSILAGO, VERNONIA, VOLUTARELLA, WEDELIA, XANTHIUM.

A. Heads with the flowers all similar and tubular. Anthers cleft at the base. Leaves usually alternate. Flowers never yellow

I. Pappus absent.

a. Flower-heads 1 or more, glomerate and sessile in leaf axils ... ... SPARGANOPHORUS.

b. Flower-heads very numerous; peduncles short or long ... ... ETHULIA.

II. Pappus short, fugacious.

a. Achenes 8-10 ribbed ... ... CENTRATHERUM.

b. Achenes terete, shining ... ... LAMPRACHAENIUM.

III. Pappus chaffy. Heads few-flowered, crowded in dense bracteate clusters ... ... ELEPHANTOPUS.

IV. Pappus long, copious. Heads distinct, many-flowered ... ... VERNONIA.

B. Heads with the flowers all similar and tubular. Anthers subentire at the base. Leaves opposite or alternate. Corolla rarely orange, never yellow

I. Anther-tip truncate; pappus usually of 3 clavate hairs ... ... ADENOSTEMMA.

II. Anther-tip appendiculate.
- a. Pappus usually of 5 paleaceous scales ... AGERATUM.
- b. Pappus of slender hairs; involucral bracts ∞ ... EUPATORIUM.
- c. Pappus of slender hairs; involucral bracts 4 ... MIKANIA.

Heads with the flowers all similar or the outer ligulate. Anthers subentire (cells not tailed) at the base. Leaves usually alternate. Receptacle almost always naked. Disk-flowers yellow; ray-flowers yellow, white or purple

I. Flowers all yellow, those of the ray ligulate
Ligules few; heads in scorpioid panicles ... SOLIDAGO.

II. Flowers all yellow, ray absent. Pappus hardly any
- a. Receptacle conic or convex. Achenes with a terminal toothed or bristly ring ... ... GRANGEA.
- b. Receptacle usually elongate, top flat. Achenes obtuse, flat, with a thickened border ... DICHROCEPHALA.

III. Ray-flowers female, ligulate, never yellow. Disk-flowers yellow
- a. Pappus of short bristles and 2-4 long hairs ... BOLTONIA.
- b. Pappus hairs long, copious.
  - 1. Ligules long, uniseriate ... ... ASTER.
  - 2. Ligules 2-3-seriate ... ... ERIGERON.

IV. Ray-flowers very slender, tubular. Pappus-hairs copious, slender, 1-2-seriate, rather unequal ... MICROGLOSSA.

Heads with the flowers all similar or the outer ligulate. Leaves usually alternate. Disk- and ray-flowers usually both yellow

I. Heads androgynous. Receptacle naked. Style-arms of hermaphrodite flowers filiform, not truncate, or style of sterile flowers entire
- a. Heads corymbose or panicled.
  - 1. Herbs, often aromatic, with narrow bracts; pappus 1-seriate, caducous
    - i. Anther-bases tailed ... ... BLUMEA.
    - ii. Anther-bases not tailed ... ... LAGGERA.
  - 2. Shrubs or undershrubs with broad bracts; pappus 1-seriate, free or sometimes connate ... PLUCHEA.
- b. Heads in dense, globose or ovoid clusters. Herbs with decurrent leaves; bracts paleaceous, few or many. Pappus absent ... ... SPHAERANTHUS.

II. Heads androgynous or homogamous. Involucral bracts scarious, usually hyaline. Style-arms of hermaphrodite flowers truncate
- a. Hermaphrodite flowers all sterile with undivided or merely notched styles
  Heads corymbose. Pappus hairs quite free ... ANAPHALIS.
- b. Hermaphrodite flowers all or mostly fertile with divided styles
  Female flowers 2-∞-seriate. Pappus-hairs never barbellate ... ... GNAPHALIUM.

III. Heads heterogamous, radiate, rarely subdisciform or homogamous. Receptacle naked. Hermaphrodite flowers with linear style-arms, rounded or dilated at the top
- a. Heads rayed. Pappus-hairs few or many, sub-equal ... ... INULA.
- b. Heads rayed or discoid, heterogamous. Achenes ribbed. Outer pappus of scales, inner of hairs ... PULICARIA.
- c. Heads discoid, homogamous. Outer pappus of paleae, inner of hairs ... ... PEGOLETTIA.
- d. Heads discoid, heterogamous. Pappus absent ... CARPESIUM.

E. Heads usually radiate. Receptacle paleaceous. Anther-cells not produced into tails. Achenes 3-4-angled or terete or compressed

I. Heads heterogamous or unisexual. Hermaphrodite flowers sterile, with undivided styles; female flowers apetalous

Involucral bracts of male flowers free, of female forming a 2-flowered, 2-celled capsule armed with glochidiate spines ... ... XANTHIUM.

II. Heads hetero- or homogamous. Leaves usually opposite.

a. Hermaphrodite flowers sterile. Achenes spinous with 2 long horn-like spines at top ... ACANTHOSPERMUM.

b. Hermaphrodite flowers fertile.

1. Inner involucral bracts embracing the achenes.

i. Outer involucral bracts 5-glandular ... SIEGESBECKIA.

ii. Outer involucral bracts 4, broad, in opposite pairs ... ... ENHYDRA.

2. Inner involucral bracts not embracing the achenes. Outer involucral bracts many ... ... ECLIPTA.

3. Inner involucral bracts not embracing the achenes. Paleae of receptacle concave, embracing the achenes.

i. Ray-flowers with large yellow ligules. Achenes thick ... ... WEDELIA.

ii. Ray-flowers with white or yellow ligules. Achenes ciliate ... ... SPILANTHES.

III. Heads very large. Pappus of 1-3 deciduous bristles or scales or both. Ligule yellow ... ... HELIANTHUS.

IV. Heads hetero- or homogamous. Hermaphrodite flowers all fertile or neuter. Achenes dorsally compressed

a. Outer involucral bracts herbaceous, subequal. Inner shorter like the paleae of the receptacle

1. Achenes small, 4-angled, sheathed at the tip by the hairy corolla; pappus absent ... ... GUIZOTIA.

2. Achenes flat, margins lacerate and winged; pappus of 2 slender awns ... ... SYNEDRELLA.

b. Outer involucral bracts few, small; inner membranous, connate below

1. Style-arms with short appendages. Leaves alternate ... ... GLOSSOCARDIA

2. Style-arms with short appendages. Leaves opposite. Achenes usually 2-4-awned ... ... BIDENS.

3. Style-arms with long appendages. Achenes with 2 bristles ... ... GLOSSOGYNE.

F. Heads usually radiate and heterogamous. Bracts of involucre herbaceous. Receptacles without paleae. Pappus absent or of scales or bristles

Leaves opposite with oil-glands. Heads simple ... TAGETES.

G. Heads heterogamous, rayed or disciform or rayless and homogamous. Involucral bracts 2-∞-seriate. Receptacle naked or paleaceous. Style-arms truncate. Leaves usually alternate. Disk-flowers yellow, ray-flowers variously coloured

I. Receptacle paleaceous. Heads usually rayed

a. Achenes margined. Heads corymbose ... ACHILLEA.

b. Achenes 4-5-angled or many-ribbed. Heads peduncled ... ... ANTHEMIS.

II. Receptacle naked or with fimbriate pits
- a. Heads rayed, usually long-peduncled. Achenes 5-10-ribbed ... ... CHRYSANTHEMUM.
- b. Heads rayed. Achenes ventrally 3-5-ribbed, dorsally ∞-ribbed or plane ... ... MATRICARIA.
- c. Heads disciform, peduncled. Involucral bracts 1-2-seriate ... ... COTULA.
- d. Heads disciform, subsessile. Involucral bracts spreading in fruit ... ... CENTIPEDA.
- e. Heads disciform, corymbose. Involucral bracts ∞ seriate ... ... TANACETUM.
- f. Heads small, disciform, racemose or panicled ... ARTEMISIA.

H. Heads heterogamous, rayed or disciform. Involucral bracts usually 1-seriate and subequal. Style-arms of hermaphrodite flowers truncate or appendaged. Pappus of fine hairs. Leaves usually alternate. Disk-flowers yellow, rays usually also yellow

I. Involucral bracts 1-2-seriate, subequal, free nearly to the base. Style-arms of hermaphrodite flowers flattened or nearly clavate, papillose
Involucral bracts uniseriate. Heads rayed, bracteolate, solitary ... ... TUSSILAGO.

II. Involucral bracts 1-2-seriate, free nearly to the base, usually with a few smaller outer ones
- a. Involucral bracts herbaceous, broad, acuminate. Heads large, rayed, long-peduncled
  Receptacle hemispheric. Herbs, leaves alternate. DORONICUM.
- b. Involucral bracts narrow with a few much shorter outer ones, except *Emilia* and some *Senecio*
  1. Heads homogamous. Style-tips long, hairy ... GYNURA
  2. Heads homogamous, ebracteolate ... ... EMILIA.
  3. Heads homogamous, style-tips ovate. Fleshy shrubs ... ... NOTONIA.
  4. Heads rayed or disciform, homogamous. Style-tips truncate or short or absent. Leaves radical or alternate ... ... SENECIO.
  5. Heads heterogamous. Hermaphrodite flowers sterile. Glabrous undershrubs ... ... OTHONNOPSIS

I. Heads rayed. Involucral bracts 1-2-seriate, narrow, subequal. Style-arms of hermaphrodite flowers truncate or in sterile flowers undivided. Achenes without pappus
Achenes incurved. Herbs, with alternate leaves ... CALENDULA.

J. Flowers all tubular. Involucral bracts many-seriate, imbricate. Tips scarious, spinous or foliaceous. Achenes often hard. Pappus various, rarely absent. Leaves alternate, often spinous

I. Heads 1-flowered, crowded into involucrate globose balls ... ... ECHINOPS.

II. Heads ∞-flowered, separate. Achenes glabrous. Pappus-hairs distinct or connate
- a. Filaments free glabrous. Pappus hairs free, caducous. Involucral bracts with hooked bristles. Leaves unarmed ... ... ARCTIUM.
- b. Filaments papillose-hairy. Pappus-hairs connate into a deciduous ring
  1. Pappus-hairs simple, not feathery ... ... CARDUUS.
  2. Pappus-hairs feathery ... ... CNICUS.
- c. Filaments glabrous, connate
  Involucral bracts spinous. Receptacle bristly. Pappus simple ... ... SILYBUM.
- d. Filaments free, glabrous
  1. Pappus-hairs uniseriate, feathery, with a few outer simple ... ... SAUSSUREA.
  2. Pappus-hairs ∞-seriate, rigid, rough or feathery. JURINEA.

11

III. Heads separate, ∞-flowered. Achenes glabrous. Basal areole oblique or lateral

a. Involucre not bracteate at the base or rarely so

1. Filaments papillose. Anther-tails lacerate. Achenes smooth ... ... TRICHOLEPIS.

2. Achenes terete. 10-15-ribbed ... VOLUTARELLA.

3. Achenes compressed or 4-angled, 4-5-ribbed ... CENTAUREA.

b. Involucre bracteate at the base by spinous leaves Achenes compressed or angled. Pappus paleacous or absent ... ... CARTHAMUS.

K. Heads hetero- or homogamous, rayed or not. Involucral bracts ∞-seriate, unarmed. Corolla 2-lipped, ligulate, or deeply 5-fid. Style-arms rounded or truncate, not appendaged. Leaves rarely opposite.

a. Corolla tubular. Pappus hairs feathery. Rigid shrubs ... ... DICOMA.

b. Corolla 2-lipped or ligulate. Scapigerous herbs GERBERA.

L. Heads homogamous. Corollas all ligulate. Ligule truncate, tip 5-toothed. Style-arms slender, Herbs; juice usually milky. Leaves radical or alternate, never opposite

I. Achenes truncate. Pappus of scales with sometimes alternating hairs or absent

Rigid herbs. Flowers blue. Inner involucral bracts coriaceous, concave ... ... CICHORIUM.

II. Achenes usually contracted at both ends. Involucre calyculate, inner bracts often thickened. Usually leafy herbs, not woolly nor with stellate hairs.

a. Achenes ribbed, rugose; pappus hairs feathery ... PICRIS.

b. Achenes ribbed; pappus hairs simple, free, white. CREPIS.

III. Herbs, usually perennial, more or less stellately hairy. Achenes truncate. Pappus hairs simple.

Pappus hairs rigid, bristles usually rough, brown. HIERACIUM.

IV. Scapigerous herbs. Hairs various. Achenes contracted below. Pappus-hairs simple or feathery

a. Receptacle naked. Pappus-hairs simple ... TARAXACUM.

b. Receptacle paleaceous. Pappus hairs feathery ... HYPOCHOERIS.

V. Herbs, rarely shrubby below, glabrous or hispid, rarely scapigerous. Achenes usually narrowed below. Pappus-hairs many, simple

a. Achenes compressed, beaked, ribs smooth ... LACTUCA.

b. Achenes compressed, not beaked, many-ribbed. Ribs smooth or rough ... ... SONCHUS.

c. Achenes narrow, truncate at both ends, 4-5-ribbed. LAUNAEA.

VI. Herbs, glabrous, woolly or hispid. Achenes with a broad base, or curved hollow stipes, above narrowed or beaked ... ...

Involucral bracts 1-seriate. Achenes long-beaked. TRAGOPOGON.

ACANTHOSPERMUM.

This genus numbers three species known from tropical America and Hawaii.

In Gold Coast *A. hispidum* DC. is used in the treatment of leprosy.

One of the species, **A. Brasilium** Schrank, is now established in and about the town of Singapore. An infusion of the leaves—4 parts in 200 parts of boiling water—is used in Brazil as a diuretic and sudorific. The aromatic bitter leaves are also recommended in the treatment of blennorrhagia.

Brazil: Picao da praia—; *Uruguay*: Yerba de la oveja—.

ACHILLEA.

The genus includes 115 species, natives of temperate regions of the northern hemisphere.

The following species are used medicinally in Europe—*A. Ageratum* Linn., *A. atrata* Linn., *A. Millefolium* Linn., *A. moschata* Jacq., *A. nana* Linn., *A. nobilis* Linn., *A. Ptarmica* Linn.—; in China and Indo-China—*A. sibirica* Ledeb.—; in North America—*A. Millefolium* Linn.—; in India—*A. Millefolium* Linn., *A. Santolina* Linn.—.

1. Rays white, pink or purple ... ... *A. Millefolium.*
2. Rays yellow ... ... *A. Santolina.*

1. Achillea Millefolium Linn. is found in the western Himalaya, from Kashmir to Kumaon, at altitudes of 6,000 to 9,000 feet. It extends to Northern Asia, Europe, and North America. A very common plant on roadsides and pastures and banks.

Many healing virtues have been and are still ascribed to this herb which, whatever its virtues may be, is still regularly cultivated for medicinal purposes. The annual consumption in the British Isles alone is from 10 to 20 tons. The herb is officinal in Austria; and the leaves are officially recognized in Portugal, Russia, Sweden, and Switzerland.

The herb is diaphoretic, stimulant, tonic. It is most useful in colds, obstructed perspiration, and the commencement of fevers. It opens the pores freely and purifies the blood. It has also proved useful in hysteria, flatulence, heartburn, colic, and epilepsy.

It is much used in England as a vulnerary, and is given internally for the suppression of haemorrhages and of profuse mucous discharges. It is employed also in intermittents and as an antispasmodic in flatulent colic and nervous affections. Its hot infusion is used as an emmenagogue in France, and also in the suppression of the lochia; it is sometimes employed in low exanthematous fevers with difficult eruption. In the United States the infusion is occasionally used in acute suppression of the menses.

An ointment made by the Highlanders of Scotland of the fresh plant is good for piles, and is also considered good against the scab in sheep. Milfoil tea is held in much repute in the Orkneys for dispelling melancholy. A decoction of the whole herb is employed for bleeding piles, and is good for kidney disorders. It has the reputation also of being a preventive of baldness, if the head be washed with it.

A medicinal tincture is prepared from the whole plant with spirit of wine. This, when employed in a diluted form of the first or third decimal strength, and in small doses from five to ten drops in a tablespoonful of cold water, will act admirably in arresting bleeding from the lungs, the kidneys, or the nose, especially in florid hectic subjects. It has been found by healthy provers that stronger or larger doses of any preparation of the herb will induce or aggravate one or another of these bleedings.

To stimulate and promote the appetite the fresh juice of the plant may be had, a dessertspoonful three times in the day.

Italian peasants apply the fresh juice topically as an astringent to piles, varicose ulcers, and sore nipples. Among the Spanish-Californians the fresh plants are used for stanching the blood in recent wounds.

Linnaeus recommended the bruised herb, fresh, as an excellent vulnerary and styptic. It is employed in Norway for the cure of rheumatism, and the fresh leaves chewed are said to cure toothache.

In France the powdered dried leaves are used as a sternutatory. Or the leaf, being rolled up, is applied to the nostrils to cause bleeding and thus afford relief to headache.

In Scotland a warm decoction of the fresh leaves is regarded as a family specific against the colds and other ailments common to childhood.

Externally, a strong decoction of the leaves has been used as an injection into the nostrils to stay bleeding from the nose. It is similarly of service for piles, and for female floodings, because exerting a special local action on the organs within the middle trunk. It is further of benefit for sore nipples as a lotion, and for a relaxed sore throat as a gargle; also as a hair-wash.

An infusion was used by the Winnebag Indians of North America to bathe swellings. For earache a wad of the leaves, also the infusion was put into the ear.

In California the leaves steeped in hot water are considered very healing applications to cuts or bruises, and are used for poulticing skin rash.

The flowering plant or flowers are stimulant, aromatic, sudorific, tonic, astringent, diuretic, and vulnerary. An essential oil, extracted from the flowers, is taken in a dose of from three to five drops to stimulate and promote the appetite.

Mahommedan writers acknowledge the virtues of the plant collected in the wild state, when in flower.

In Sweden and in some parts of Africa the plant has been used in the manufacture of beer. Linnaeus considered beer thus brewed more intoxicating than when hops were used.

Achilleine, an amorphous bitter alkaloid of unknown composition, has been isolated.

Afghanistan: Buimaderan—; *Arabic*: Suila—; *Bombay*: Rojmari—; *California*: Milfoil, Old Man, Yarrow—; *Catalan*: Herba de tall, Marfull, Milfulles, Percala—; *Colombia*: Colchón de pobre—; *Cutch*: Biranjasif—; *Danish*: Roellike—; *Dutch*: Duizenblaad, Duizendblad—; *English*: Bad Man's Plaything, Bloodwort, Carpenter's Weed, Devil's Nettle, Devil's Plaything, Knight's Milfoil, Milfoil, Nosebleed, Old Man's Mustard, Old Man's Pepper, Sanguinary, Soldier's Woundwort, Staunch Grass, Staunchweed, Thousand Leaf, Thousand Weed, Yarrow, Yarroway—; *French*: Achilléine, Endove, Herbe à la coupure, Herbe au charpentier, Herbe aux charpentiers, Herbe aux cochers, Herbe aux coupures, Herbe aux militaires, Herbe aux voituriers, Herbe à saigner, Herbe de Saint Jean, Herbe militaire, Millefeuille, Saigne nez, Seignenez, Sourcils de Vénus—; *German*: Achilleskraut, Angerblume, Anserine, Bauchwehkraut, Edelgarbe, Fasankraut, Feldgarbe, Gachel, Gacht, Gaensezungen, Gaerb, Garwekraut, Gerbel, Gollenkraut, Gor, Gotteshand, Grillen, Grinsing, Gruensing, Gruettblume, Gruetzenkraut, Grundheil, Hasegerf, Hasengarbem, Heftkraut, Heidekraut, Jase, Judenkraut, Jungfraukraut, Kachelblume, Karweblume, Katzenfittich, Katzenschwanz, Kelkenkraut, Kerpen, Kraenzel, Krebselkraut, Loewenfusskraut, Margaretenkraut, Marienkranz, Reels, Reelse,

Relaka, Releke, Relik, Relkike, Roehlke, Roelken, Roellicke, Roelskraut, Rohlech, Ruels, Saegkraut, Schabab, Schabor, Schafgarbe, Schaffripchen, Schafrippe, Schafzunge, Schapschar, Schober, Sichelblume, Sichelschnitt, Siebengartenkraut, Tausendblatt, Weisses Garbenkraut, Zeiskraut, Zimmermannskraut—; *Greek*: Myriophyllon—; *Hindi*: Gandana—; *Hungarian*: Ezer-levelu-fa—; *Irish*: Athairthalmhna—; *Italian*: Millefoglie, Millefoglio—; *Kashmir*: Momadruchopandiga—; *Malta*: Common Milfoil, Yarrow, Haxixa tal morliti—, *Pacific Coast*: Dog-daisy, Green Arrow, Milfoil, Nose-bleed, Oldman's Pepper, Soldier's Woundwort, Yarrow—; *Persian*: Buimaderan—; *Polish*: Tysiacznik ziele—; *Portuguese*: Mil em rama, Milfolha, Milfolhada—; *Roumanian*: Coada soarecelului—; *Russian*: Tesyachelistnik—; *Saxon*: Gearwe—; *Spanish*: Filigrana, Manzanilla de los montes, Milefolio, Milenrama, Milhojas, Yerba de Aguiles, Yerba de San Juan—; *Swedish*: Roelleke, Roflleka—; *Urdu*: Biranjasifa—.

2. **Achillea Santolina** Linn. is found in Baluchistan, whence it extends to Afghanistan, Iran and the Mediterranean. It is widely distributed in the East and in Northern Africa.

The plant is in common use as a tonic and carminative in Iran and Sind. At Chaman, in Baluchistan, it is given to children to cure stomachache. The strong odour of the herb drives away fleas and other noxious insects.

Arabic: Rebey-el-ghebel—; *Egypt*: Ba'eytheran, Bishrin, Ghobeyra, Gesum—; *Iraq*: Gaisum, Qaisum—; *Kalat*: Bo-i-madaran—; *Mach*: Boemadran—; *Persian*: Biranjasib, Biranjasif, Bu-i-maderan—; *Pushtu*: Zawal—; *Teheran*: Bumadran—.

ADENOSTEMMA.

The genus consists of 5 or 6 American species, one of them cosmopolitan.

Adenostemma viscosum Forst. is found in all tropical countries. It occurs throughout India, Ceylon, and the Malay Peninsula. It is common in waste ground and in damp spots in forests, ascending to 4,000 feet in Malaya, to 5,000 feet on the Himalaya and other ranges, and to 6,000 feet in Ceylon.

The plant is used medicinally in La Reunion, the leaves are antispasmodic, and the fresh juice is a good stimulant and sternutatory.

Malay: Rumput babi, Rumput pasir, Sumbong gajah—.

AGERATUM.

The genus numbers 45 American species, one of them cosmopolitan.

Ageratum conyzoides Linn. is one of the commonest weeds of the Tropics. It is found throughout India, common everywhere in waste ground and on the outskirts of villages; it ascends the Himalayas to 5,000 feet.

The plant is applied externally in ague. Its juice is said to be a good remedy for prolapse of the anus. It is used in Togoland to cure fever. In Yoruba a decoction is given for 'craw-craw' externally and for fever internally. An infusion is prescribed in

Brazil and Guiana as a stimulant tonic in diarrhoea and flatulent colic.

The juice of the root is said to possess antidysenteric properties and, together with the leaves, is a common Indo-Chinese remedy for diarrhoea and dysentery. The juice of the root is moreover credited with the virtue of preventing the formation of stone or calculus in the bladder.

In Ceylon the leaves are commonly applied to wounds and sores; they act as a styptic and heal them quickly. In Gold Coast Colony the leaves are squeezed, and the juice is used as a lotion for the eyes. In Sierra Leone the chief use of the leaves is a remedy for craw-craw; they are also applied to chronic ulcers, and intravaginally for uterine troubles; crushed in water they are given as an emetic. In Southern Nigeria a decoction is both used as a lotion for craw-craw and taken internally for fever. In Siberia pneumonia in children is treated by rubbing an extract of the leaves on the chest. In South Cameroons the leaves pounded with *Ocimum* and macerated in water along with 'bush pepper' are prepared as a purgative enema. Some tribes in Portuguese Congo use it in the treatment of sleeping sickness.

The plant is a household medicine in Madagascar, Mauritius and La Reunion. As a fomentation the leaves and stems are used in skin diseases, more particularly leprosy; and they are prescribed as a bath to patients with ecchymoses. A poultice of the leaves is applied on boils; it is said to prevent tetanus if applied to a wound. A cold decoction of the roots is used as a lotion in purulent ophthalmia.

Ashanti: Gu-ekura, Guakuro—; *Australia*: Billy-goat weed—; *Bengal*: Dochunty, Uchunti—: *Benin*: Ebighoedore, Ehigodore—; *Betsimisaraka*: Fotsivony, Tinimbo—; *Bombay*: Osari, Sahadevi—; *Brazil*: Camara apeba, Fumo bravo, Mentrasto, Suacuaira—; *Ceylon*: Pumpillu—; *Efik*: Ikongifoiyen, Otitidahadaha—; *English*: Goatweed, White Weed—; *Ewe*: Mimang—; *Fanti*: Efungmormoe—; *French*: Herbe à Madame—; *French Guiana*: Raguet françois—; *Gold Coast*: Goat Weed—; *Gujerati*: Ajgandha, Gandharisedardi, Mankdamari—; *Ibo*: Ikbu odols—; *Ibuzo*: Ikbu odolo—; *Indo-China*: Bong thui, Bu xich, Cut lon, Thang hong ke—; *Karwar*: Ghayamari—; *Kathiawar*: Makadamari—; *Konkani*: Sahadevi—; *Lagos*: Imi-esu—; *La Reunion*: Herbe à bouc—; *Lepcha*: Nam-yu muk—; *Madagascar*: Hanitrinimpantsaka—; *Mano*: Dah vo—; *Malay*: Sianggit, Tahi ayam, Tombok jantan—; *Malayalam*: Appa, Muryanpacha—; *Marathi*: Ghanera osadi—; *Mauritius*: Herbe de bouc—; *Mende*: Ngugbe, Ngulugbei, Yanigbei—; *Mundari*: Purudumbu—; *Nepal*: Ilami—; *New Caledonia*: Adivijalakara—; *Oloke-Meji*: Imi-esu—; *Owerri*: Ahihia-nwa-oshi-naka, Akwukwo-moosinaka, Osu angweri ngwa—; *Sierra Leone*: Akan yunyun, Craw-craw plant—; *Sinhalese*: Hulantala—; *Timne*: Balkeyan, Ka-balkeyan, Keyan—; *Twi*: Guakuro, Gu airkuna, Gu-ekura—; *Umu Ahia*: Osu angrocri ngwa—; *Uriya*: Boksunga, Poksunga—; *Wassaw*: Ahaban kankan—; *Yoruba*: Akkaw-yungun, Imi-eshu—.

ANAPHALIS.

The genus consists of 50 species, chiefly temperate and mountain plants of Asia, Europe, and America.

A. margaritacea Benth. and Hook. is used medicinally in North America.

In the Nilghiris several species are used for cut wounds. The more commonly employed is **A. neelgerryana** DC., which occurs at an altitude of 7,000-8,250 feet. The fresh leaves are bruised and applied to the wound as a plaster.

Nilghiris: Katplaster—; *Sanskrit*: Raktaskandana, Vranapata—.

ANTHEMIS.

The genus includes 120 species, natives of Europe as far as Siberia—only one being native of Abyssinia—, and often naturalised elsewhere.

A. arvensis Linn., *A. Cotula* Linn., and *A. nobilis* Linn. are used medicinally in Europe and in North America; *A. Gayana* Boiss. and *A. odontostephana* Boiss. in Baluchistan; *A. Wiedemanniana* Fisch. and Mey. in Persia.

1. Achenes turbinate, tubercled, tip crenulate ... *A. Cotula.*
2. Achenes smooth, sulcate, slightly narrowed at the base, angular, truncate at the apex; margin acute, narrow ... ... *A. Gayana.*
3. Achenes cylindric, ribbed, tubercled; tip of outer toothed, of inner with a short auricle ... ... *A. odontostephana.*

1. **Anthemis Cotula** Linn. occurs in Baluchistan and Northern Asia. It is distributed westwards to Britain and the Canary Islands; and is found in Europe, North Africa, Siberia and West Asia. It has been introduced into North America.

The flowers and the leaves are tonic, antispasmodic, emmenagogue, and emetic. They have been used with success in sick headache, in convalescence from fevers, and in the absence or abnormal stoppage of the menses. The warm infusion of 1 ounce in 1 pint of boiling water is taken in wineglassful doses when required.

A strong decoction causes sweating and vomiting. It is said to be nearly as valuable as opium in dysentery. It has also been used in scrofula, dysmenorrhoea and flatulent gastritis. It is administered to induce sleep in asthma.

Applied to the skin fresh and bruised it is a safe vesicant. A poultice helpful in piles can be made from the herb boiled until soft, or it can be used as a bath or fomentation.

In hysteria the herb is used in Europe as an antispasmodic and emmenagogue. It is prescribed in North America for pulmonary and rheumatic affections.

The whole plant has a very foetid odour, which rubbing increases. It is full of an acrid juice which is liable to blister. Bees dislike it, and it is said to drive away flies and fleas.

Catalan: Camamilla pudenta—; *England*: Balder Brae, Baldeyebrow, Camomile, Camovyne, Cotula, Dog's Camomile, Dog-binder, Dog Daisy, Dog-fennel, Dog-finkle, Flowan, Hog's Fennel, Horse Daisy, Jayweed, Madder, Madenwede, Marse, Marg, Mathes, Mayweed, Morgan, Murg, Poison Daisy, Stinking Camomile, Wild Camomile—; *French*: Amarelle, Amouroche, Bouillot, Camomille fétide, Camomille puante, Chaillerie, Chamaran, Chamarou, Herboula, Maroule, Maroune, Maroute, Maroute puante, Oeil-de-vache, Queuneron—; *German*: Hundskamille, Hundskamillenkraut, Stinkende Kamille—; *Italian*: Camomilla puzzolente—; *Pacific Coast*: Bald-eyebrow, Chigger-weed, Dill-weed,

Camomile, Dog Daisy, Dog-fennel, Fetid Mayweed, Fieldwort, Hog-fennel, Horse Daisy, Madder, Mayweed, Poison Daisy, Stinking Camomile—; *Roumanian*: Mararui-cainalui, Mormorita, Romanita-puturoasa—; *Spanish*: Cotula fetida, Manzanilla hedionda—; *Uruguay*: Manzanilla—.

2. **Anthemis Gayana** Boiss. inhabits Baluchistan whence it extends to Persia.

In the Harboi Hills the leaves are eaten to cure pains in the chest.

Harboi Hills: Piunphuli—.

3. **Anthemis odontostephana** Boiss. is found at Peshawar, whence it spreads to Afghanistan, Baluchistan, and Persia.

A decoction of the flowers is used in Baluchistan as a febrifuge and carminative.

ARCTIUM.

The genus consists of 3 or 4 species spread over temperate Asia and Europe.

Arctium Lappa Linn. is found in Western Himalaya from Kashmir to Simla at altitudes of 6,000-8,000 ft., and also in Western Tibet where it ascends to 11,000-13,000 feet. It extends westwards towards the Atlantic and occurs in Syria, Persia, and Khorasan, as well as in Europe. It is common in North and Central China, and Manchuria, and is abundant in the United States.

The root is regarded throughout India as depurative and antiphlogistic. In Teheran the root, with that of sarsaparilla, is used as a remedy for syphilis. In the Pacific Coast States of America the dried first year root is considered alterative, aperient, diuretic, and diaphoretic.

The root, stem, and seeds are used medicinally in China. The seeds are alterative, depurative, diaphoretic, and diuretic.

The plant grows freely throughout Europe and the United States on waste ground and about old buildings, by roadsides and in fairly damp places. It had a great reputation in the past, and its fruits and roots still figure largely among the drugs commonly used by herbalists in Europe and America. It is alterative, diuretic, and diaphoretic; one of the finest blood purifiers in the herbal system. In all skin diseases it is a certain remedy and has effected a cure in many cases of eczema. The root is principally employed, but the leaves and fruits—erroneously called seeds—are equally valuable. Both root and seed may be taken as decoction of 1 ounce to a pint and a half of water, boiled down to 1 pint, in doses of a wineglassful three or four times a day. The antiscorbutic properties of the mucilaginous, demulcent root make the decoction very useful for boils, scurvy, and rheumatic affections; it has in addition been recommended for external use as a wash for ulcers and scaly skin disorders. It is officinal in Portugal and Spain.

An infusion of the leaves is useful to impart strength and tone to the stomach, for some forms of long-standing indigestion. When applied externally as a poultice, the leaves are highly resolvent for tumours and gouty swellings, and relieve bruises and inflamed surfaces generally. The bruised leaves have been applied by the peasantry in many countries as cataplasms to the feet and as a remedy for hysterical disorders.

From the seeds (or fruits), both a medicinal tincture and a fluid extract are prepared, of benefit in chronic skin diseases when taken, in doses of from ten to thirty drops with two tablespoonfuls of cold water three times a day, steadily for several weeks or months. Americans consider the seeds more efficacious and prompt in their action than the other parts of the plant. They are relaxant and demulcent, with a limited amount of tonic property Their influence upon the skin is due largely to their being of such an oily nature; they affect both the sebaceous and sudoriferous glands and restore to the skin that smoothness which is a sign of normal healthy action.

The infusion or decoction of the seeds is employed in dropsical complaints, more especially in cases where there is co-existing derangement of the nervous system, and is considered by many to be a specific for all affections of the kidneys, for which it may with advantage be taken several times a day, before meals.

The root contains inulin, mucilage, sugar, a bitter crystalline glucoside called lappin, a little resin, and tannic acid. The ashes of the plant, burnt when green, yield carbonate of potash abundantly, and also some nitre.

Anglo-Saxon: Fox's clote—; *Catalan*: Bardana, Llapassa, Repalassa—; *China*: E Shih, Niu P'ang, Niu Tzu, Ta Li, Wu Shih—; *Dutch*: Dokkebladen, Klis, Klisse, Klit—; *English*: Aireve, Airup, Beggar's Buttons, Burdock, Clithe, Clot-bur, Cockle-buttons, Cuckle-buttons, Cuckoo Button, Great Burdock, Happy Major, Hardock, Harebur, Herrif, Hurbur, Lappa, Love Leaves, Personata, Philanthropium, Turkey-bur, Thorny Bur—; *French*: Bardane, Bardane à grosses têtes, Bouillon noir, Coupeau, Dogue, Glouteron, Glotteron, Grande bardane, Gratteau, Grippe, Guippon, Herbe aux bardanes, Herbe aux teigneux, Lappe, Oreille de géant, Peignerolle, Picons, Poire de vallée—; *German*: Gemeine Klette, Grosse Klette, Klette—; *Greek*: Arkeion, Arktion, Prosopion, Prosopis—; *Italian*: Arsio, Bardana, Lappa maggiore, Lappola maggiore—; *Malaya*: Ngow chee—; *Pacific Coast*: Bazzies, Beggar's Buttons, Burdock, Cuckoo-button, Great Burdock, Harebur, Turkey-bur—; *Polish*: Lopian—; *Portuguese*: Bardana—; *Roumanian*: Brustur, Lipan—; *Russian*: Lapushnik, Rapeynik—; *Spanish*: Amores ruines, Bardana, Lampazo, Lapa—; *Teheran*: Bardane—.

ARTEMISIA.

This genus includes about 280 species, natives mostly of the northern hemisphere, a few from South Africa and South America.

The following species are used medicinally in Europe.—*A. Absinthium* Linn., *A. abrotanum* Linn., *A. arborescens* Linn., *A. caerulescens* Linn., *A. campestris* Linn., *A. Dracunculus* Linn., *A. gallica* Willd., *A. glacialis* Linn., *A. Herba-alba* Asso, *A. hispanica* Lam., *A. maritima* Linn., *A. Mutellina* Willd., *A. pontica* Linn., *A. scoparia* Waldst. and Kit., *A. spicata* Wulf., *A. vallesiana* Lam., *A. vulgaris* Linn.—; in Palestine and Syria—*A. Herba-alba* Asso, *A. judaica* Linn.—; in Arabia and Egypt—*A. judaica* Linn.—; in Persian and Turkistan—*A. Cina* Berg., *A. Vahliana*

Kostel—; in China *A. annua* Linn., *A. apiacea* Hance, *A. capillaris* Thunb., *A. japonica* Thunb., *A. Heiskeana* Miq., *A. stelleriana* Bess., *A. vulgaris* Linn.—; in Indo-China—*A. abrotanum* Linn., *A. annua* Linn., *A. apiacea* Hance, *A. capillaris* Thunb., *A. carvifolia* Roxb., *A. sina* Berg., *A. japonica* Thunb., *A. Heiskeana* Miq., *A. maritima* Linn., *A. vulgaris* Linn.—; in Malaya—*A. maritima* Linn., *A. vulgaris* Linn.—; in the Philippine Islands—*A. vulgaris* Linn.—; in North America—*A. Absinthium* Linn., *A. ludoviciana* Nutt., *A. Santonicum* Linn.—; in the Pacific Coast States—*A. Absinthium* Linn., *A. annua* Linn., *A. californica* Less., *A. dracunculoides* Pursh., *A. tridentata* Nutt., *A. vulgaris* Linn.—; in Mexico—*A. mexicana* Willd.—; in North Africa and the Canary Islands—*A. ramosa* C. Sm.—; in South Africa—*A. afra* Jacq.—.

A. Heads heterogamous, outer flowers female, disk-flowers hermaphrodite, sterile. Receptacle naked
- 1. Perennial ... ... 4. *A. Dracunculus.*
- 2. Annual. Very strong scented ... ... 2. *A. annua.*
- 3. Annual, or rootstock perennial. Faintly scented ... 8. *A. scoparia.*

B. Heads homogamous. Flowers all fertile. Receptacle naked ... ... 5. *A. maritima*

C. Heads heterogamous. Ray-flowers female, disk-flowers hermaphrodite, all fertile.
- 1. Receptacle puberulous. Perennial with rather large head ... ... 6. *A. persica.*
- 2. Receptacle naked.
  - a. Annual; quite glabrous ... ... 3. *A. carvifolia.*
  - b. Perennials with usually small heads.
    - i. Leaves large, ovate, lobed, laciniate or 1-2 pinnatipartite ... ...10. *A. vulgaris.*
    - ii. Leaves long-petioled, ovate- pinnatisect ... 7. *A. sacrorum.*

D. Heads heterogamous. Ray-flowers female, disk-flowers hermaphrodite, both fertile. Receptacle covered with long hairs
- 1. Perennial ... ... 1. *A. Absinthium.*
- 2. Annual or biennial ... ... 9. *A. Siversiana.*

1. **Artemisia Absinthium** Linn. inhabits Kashmir where it is found at altitudes of 5,000—7,000 ft. It is distributed over Northern Asia, Afghanistan, and westwards to the Atlantic. It is now naturalized and rather common in eastern Canada to Pennsylvania, growing along roadsides and waste places.

The herb has a strong fragrant odour which exercises a powerful influence over the nervous system. Its tendency to produce headache and other nervous disorders is well known by travellers in Kashmir and Ladak, who suffer severely when marching through the extensive tracts of country covered with this plant.

The whole plant is an aromatic tonic, and formerly enjoyed a high reputation in debility of the digestive organs. Wormwood tea, or the powdered herb in small doses, mixed in a little soup, will serve to relieve bilious melancholia, and will help to disperse the yellow hue of jaundice from the skin.

The juice of the larger leaves which grow from the root before the stalk appears has been used as a remedy for jaundice and dropsy, but it is intensely nauseous. A light infusion of the tops of the plant, used fresh, is excellent for all disorders of the stomach, creating an appetite, promoting digestion, and preventing sickness after meals, but it is said to produce the contrary effect if made too strong.

The flowers, dried and powdered, are said to be most effectual as a vermifuge, and used to be considered excellent in agues. Before the discovery of cinchona they were largely used in intermittents.

The plant yields by distillation a dark green or yellow oil which is used as worm-expeller. However, Caius and Mhaskar have shown experimentally that neither the herb nor the oil are effective as anthelmintics against hookworms. In large doses the volatile oil is an active narcotic poison. The main constituents of the oil are absinthol and absinthin, the former largely responsible for the characteristic odour of the plant and the latter for its intense bitter taste.

The herb is prescribed in the form of a poultice or fomentation as an antiseptic and discutient.

Arabic: Afsantin—; *Bologna*: Zicus—; *Catalan*: Donsell, Donzell, Donzell mascle—; *Como*: Medegh—; *Danish*: Malurt—; *Deccan*: Vilayatiafsantin—; *Dutch*: Alsem, Groote Alsem—; *English*: Absinth, Green Ginger, Madderwort, Mingwort, Mugwort, Warmot, Wermuth, Wormwood—; *French*: Absin menu, Absinthe, Absinthe amère, Absinthe commune, Absinthe vulgaire, Alliene, Aloïne, Alunine, Aluyne, Alvine, Alvuine, Aoussin, Armoise-absinthe, Armoise amère, Grande Absinthe, Herbe sainte, Herbe aux vers—; *Friuli*: Assinz, Sinz—; *Genoa*: Bonmegu, Erba medgu—; *German*: Aelsch, Allsam, Allsei, Alsam, Alsch, Alsen, Baermede, Bitteralsen, Bitterbeifuss, Bitterkraut, Elch, Else, Elzkraut, Ilsem, Kampferkraut, Kittelkraut, Knoopvanalsen, Magenkraut, Malnit, Meelrasch, Pardehan, Pardesan, Vraemte, Waermde, Waermkraut, Warmuede, Wermde, Weissrauch, Wermuth, Wiegenkraut, Woerken, Woermd, Woermke, Wormeth, Wraemte, Wroembk, Wuermken, Wurmei, Wurmet, Wurmtod—; *Greek*: Apsidia, Apsinthia, Apsinthion, Apsinthos, Apsithea, Apsithia—; *Hindi*: Vilayatiafsantin—; *Italian*: Assenzio—; *Lombardy*: Incens, Incens de bigat—; *Malayalam*: Shulabandha—; *Malta*: Wormwood, Assenzio, Assenziu—; *Norwegian*: Malurt—; *Pacific Coast*: Absinthium, Madderwort, Mingwort, Oldwoman, Wormwood—; *Padua*: Bun maja—; *Pavia*: Assenzi—; *Persian*: Afsantin—; *Piedmont*: Bonme, Bonmi, Dusang, Erba meja, Erba mia, Fort, Incenso, Insens, Insens grousser, Medighdt, Ourtmia, Ourtmiglia—; *Potenza*: Nascienzo—; *Reggio*: Aluina, Alvina, Asseinzi—; *Romagna*: Absanta, Absent, Absent rumen, Absent dal foj lerghi—; *Roumanian*: Pelin—; *Russian*: Polin—; *Sanskrit*: Damar—; *Spanish*: Ajenjó, Doncel—; *Treviso*: Bon maistro—; *Tuscany*: Assenzo—; *Verona*: Medego maistro—.

2. **Artemisia annua** Linn. occurs in the Punjab, from Peshawar to Waziristan, ascending up to 5,500 ft. It spreads to Afghanistan, Indo-China, China, Siberia, and westwards to Hungary.

In China and Malaya the dried leaves and stalks are used in skin diseases. The herb is a popular medicine in Indo-China, where it is considered a good stomachic and diuretic, and is prescribed in jaundice and in skin diseases.

Chinese: Huang Hua Hao, Ch'ing Hao—; *Indo-China*: Che noi, Hoang hoa cao, Nhan tran—; *Malaya*: Ching hoh—.

3. **Artemisia caruifolia** Roxb. is found in eastern Bengal, Assam, Nepal,. Burma, whence it extends over to China.

The toasted plant is much used medicinally in Annam. It is recommended for fevers, chronic diarrhoea, phthisis with emaciation, purulent scabies with ulcers, and intestinal troubles. Externally it is applied to wounds due to metallic instruments.

Annam: Cay thanh hao, Co thanh hao, Ran hao, Thanh cao, Thanh hao—.

4. **Artemisia Dracunculus** Linn. is found in Western Tibet and in the western Himalaya at altitudes of 14,000-16,000 ft. It is distributed to Afghanistan, Western Asia, South and Mid Russia.

The aromatic leaves are credited with aperient, stomachic, stimulant, and febrifuge properties.

Dutch: Dragon—; *English*: Little Dragon Mugwort, Tarragon—; *French*: Dragon, Dragonne, Estragon, Fargon, Herbe au dragon, Herbe dragonne—; *German*: Dragun, Kaisersalat—; *Italian*: Dragontea—; *Portuguese*: Estragao—; *Roumanian*: Tarhon—; *Russian*: Estragon—; *Spanish*: Dragoncillo, Estragon—.

5 **Artemisia maritima** Linn., in its many variations of forms, has an extremely wide distribution in the northern hemisphere of the Old World, occurring mostly in saltish soils. It is found in the salt plains of Western Tibet, where it grows in abundance and ascends up to 9,000-14,000 ft. It occurs at an altitude of 7,000-9,000 ft. in the Western Himalaya from Kashmir to Kumaon. It covers immense tracts in Northern Asia from Chinese Mongolia to Central Siberia, the region of the Caspian, and Southern Russia. It is found on saline soils in Hungary, on the coasts of the Baltic, of France and the Mediterranean, and in the salt marshes of the British Isles. In Britain it spreads as far as Wigton on the West and Aberdeen on the East; it also thrives in north-east Ireland and in the Channel Islands.

The medicinal properties of the plant are those of a slight tonic, inferior in efficacy to common wormwood; and it is not often employed except in discutient and antiseptic fomentations, for inflammations, tumours, and foul ulcers; where its topical application has been thought beneficial. It is still often made use of by country people for intermittent fever.

In Southern Afghanistan and Baluchistan the plant is much used as an antiperiodic. An infusion, and also decoction, of the fresh plant has been very successfully used in cases of ague, intermittent and remittent fever.

At Kirani, near Quetta, the plant is given to children for stomach-ache; at Sanjawi it is a cure for jaundice.

In Bombay, the Hakims prescribe it in doses of 2 to 3 drachms as an anthelmintic, and also deobstruent and stomachic tonic. In the form of a poultice, they use it to relieve the pain caused by the bites of scorpions and venomous reptiles. But Mhaskar and Caius have successfully demonstrated that, whether it be snake bite or scorpion sting, the treatment is useless.

The flower-heads are collected from the villages around Teheran and sold as a vermifuge. There is little doubt that they were used by the Greeks and Romans to expel intestinal worms, and it is a matter of certainty that Arabian and Persian physicians have always prescribed them as an anthelmintic. During the war both seeds and leaves were tried and found to be effective. Later Simonsen examined the leaves and flowering tops of *A. maritima* (*A. brevifolia* Wall.) growing in Chitral, Afghanistan, and Baluchistan, and obtained from 0 to 1 per cent of santonin, the amount being greater when the flowers are just appearing.

Arabic: Afsanthinulbarh, Sarifun, Sariqun, Shih, Shiharmani—; *Bhotia*: Sefedpurcha—; *Bombay*: Kiramaniova—; *Catalan*: Donsell mari—; *English*: Drooping Sea Wormwood, English Sea Wormwood, French Sea Wormwood, Levant Sea Wormwood, Roman Wormwood, Santonica, Sea Mugwort, Sea Wormwood, Wormseed—; *French*: Absinthe maritime, Armoise maritime—; *Gujerati*: Chhuvariajamoda, Kirmanidinechi—; *Hindi*: Ajavayana, Chhuari—, *Indo-China*: Chi nai—; *Kirani*: Zher—; *Marathi*: Kirmaniova, Surabandi—; *Persian*: Afsanthinulbarh, Darmaneh, Sariqun, Shih—; *Sanjawi*: Tarkhasperah—; *Sanskrit*: Chhara, Chauhara, Gandha, Jantunashana, Khurapushpika, Parasi, Yavani, Yavaniya—; *Spanish*: Ajengo maritimo—; *Teheran*: Dharmane—; *Urdu*: Darmanah—.

6. **Artemisia persica** Boiss. is found at 9,000-14,000 ft. in Western Tibet, whence it extends to Afghanistan, Southern Persia, and Kurdistan.

The plant is used as a tonic, febrifuge, and vermifuge.

Arabic: Afsantinulbarh, Sariqun, Shih—; *Gujerati*: Pardesidauno—; *Marathi*: Davana—; *Persian*: Afsantinulbarh, Sariqun, Shih—.

7. **Artemisia sacrorum** Ledeb. occurs in Western Tibet and Kunawar. It is found in Kumaon on the interior ranges bordering Tibet between 10,000 and 12,000 ft. It extends to Dahuria, Siberia, Central and South Russia.

The plant is given to horses in affections of the head.

Bhotia: Kalapurcha—; *Ladak*: Burmack, Tatwen—; *Punjab*: Burnak, Chumbar, Jau, Munya, Niurtsi, Tatwen, Zbur—.

8. **Artemisia scoparia** Waldst. and Kit. occurs in Western Tibet at altitudes of 7,000-12,000 ft., in the Western Himalaya from Kashmir to Lahul at 5,000-7,000 ft., in Sind, in the Punjab, in the Upper Gangetic Plain. Eastwards it extends to China and Japan, westwards to Afghanistan and Central Europe.

In the Punjab the infusion is given as a purgative, and the smoke is considered good for burns.

At Las Bela the plant is used as a cure for pain in the ear.

Baluchi: Gwatag—; *Bombay*: Churisaroj, Danti—; *Brahui*: Jir, Khisunjir, Luling—; *Las Bela*: Gajar—; *Punjab*: Biur, Dona, Durunga, Jhan, Lasaj, Marua, Pilajau—.

9. **Artemisia Siversiana** Willd. is found at 8,000-10,000 ft. in the Western Himalaya from Kashmir to Lahul, and at 12,000-14,000 ft. in Western Tibet. It extends to China whence it spreads westwards to South Russia.

The plant is much used in India where it is considered tonic, deobstruent, febrifuge and anthelmintic. It is applied externally as a discutient and antiseptic.

The hakims prescribe the drug freely.

The plant is also held in great esteem by the vaids and its place in Ayurveda is undisputed.

Arabic: Afsantin—; *Bengal*: Dana, Dona—; *Bombay*: Downa—; *Canarese*: Davana—; *Gujerati*: Damro—; *Hindi*: Dauna—; *Marathi*: Davana, Ranadavana—; *Persian*: Afsantin—; *Sanskrit*: Agnidamanaka, Bahukantaka, Brahmajata, Damana, Damanaka, Damani, Dandi, Danta, Devashekhara, Gandhotkata, Guchhaphala, Jatila, Kshudradussparsha, Kshudrakantakari, Kulapatra, Kulapatraka, Madanaka, Muni, Munipatra, Panduraga, Patri, Pavitraka, Pundarika, Pushpachamara, Tapasaptri, Tapaswipatra, Tapodhana, Vanadamanaka—; *Urdu*: Afsantin—.

10. **Artemisia vulgaris** Linn. is found throughout the mountainous districts of India, ascending up to 5,000-12,000 ft. in the Western Himalaya, and up to 5,000-8,000 feet in Sikkim, the Khasia, Ava and Martaban Mountains. It grows at Mount Abu, in Marwar, and on the Western Ghats, from the Konkan southwards to Ceylon. It extends to temperate Europe and Asia, to Siam and Java.

The herb has emmenagogue and antispasmodic properties; the root is tonic and antispasmodic.

The plant is considered to be a valuable stomachic, deobstruent, and antispasmodic; it is prescribed in infusion and electuary in cases of obstructed menses and hysteria. Externally it is used in fomentations given in skin diseases and foul ulcers as an alterative.

The expressed juice is used in diseases of children. It is applied to the head of young children for the prevention of convulsions.

The leaves and tops are administered in nervous and spasmodic affections connected with debility, in asthma and diseases of the brain. In Persia, in Afghanistan, and throughout India, a strong decoction is given as a vermifuge, and a weak one to children in measles. An infusion is given as a tonic.

The plant is much used medicinally in Indo-China, where the leaves and the flowers are considered as aromatic, emmenagogue, stomachic, antispasmodic, and anthelmintic. The boiled leaves are used as a poultice in headache; dried and cut into small fragments they are used to cauterize wounds.

In China, Japan and Indo-China inflammable cones or 'moxa' are obtained by grinding the leaves in a stone mortar with water, separating the coarser particles, and drying what remains. Moxibustion, or the method of cauterizing the skin by burning, is resorted to for a very large number of diseases, from itch to sterility.

In Annam the leaf is used in haemorrhage: epistaxis, blood vomiting, haematuria. It is also recommended for metrorrhagia, dysentery, intestinal and urinary troubles. It is highly praised as a vermicide. It is used in eczema, herpes, and purulent scabies. In Malaya the leaves are employed as a carminative and haemostatic.

The tonic and stomachic properties of the plant are well known in the Philippine Islands. An infusion of the leaves is commonly used as an emmenagogue.

The plant is prescribed by Sushruta in the treatment of snake bite and scorpion sting; but that the plant is not an antidote ɔ either snake or scorpion venom has been experimentally emonstrated by Caius and Mhaskar

Anglo-Saxon: Wyrmwyrt—; *Annam*: Bang dai, Benh thao, Cuu thao, La gai, La ngai cao, La ngai cuu, La thuoc cuu, Ngai diep, Thuoc cuu, Y hao—; *Arabic*: Isferaj—; *Bombay*: Nagdona—; *Canarese*: Davana, Manjiatri—; *Catalan*: Altimira—; *Chinese*: Ai—; *Deccan*: Dawan, Randawan—;)ehra Dun: Samri, Sarmi—; *Dutch*: Bijvoet, Sint Jans Gordel, Sint Jans Kruid—; *English*: Felon Herb, Felon Wort, Fleabane, Indian Wormwood, Maidenwort, Motherwort, Mugwort, St. John's Plant—; *French*: Armoise, Artémise, Ceinture de Saint Jean, Couronne de Saint Jean, Encens, Encens de mas, Fleur de Saint Jean, Herbe à cents gouts, Herbe de la Saint-Jean, Herbe de Saint Jean, Herbe Saint Jean, Remise—; *German*: Alsei, Beifuss, Beiposs, Besenkraut, Bibes, Biboth, Bibs, Biefoth, Biermersch, Buck, Budschen, Kamperkraut, Roterbock, Sant Johannis Guertel, Weiberkraut, Wiegenkraut, Wiesenkraut, Wildes Wurmkraut, Wurmtod—; *Garhwal*: Kunja—; *Greek*: Artemisia—; *Guam*: Yerba de Santa Maria—; *Hindi*: Dona, Gathivana, Majtari, Mastaru, Mastru, Nugduna—; *Indo-China*: Cuu ngai da, Lau cao, Ngai, Ngai cuu, Thuoc cau, Thuoc cuu, Yomogi—; *Italian*: Amarella, Artemisia, Canapaccia, Assenzio di siepe—; *Japanese*: Yomogi—; *Malaya*: Ai, Chi ai, Khee ngai, Ki ai, Ngai—; *Malayalam*: Appa, Damanakam, Kattuchatti, Makkippu, Mashipatri, Nilampala, Rirunittipacha—; *Marathi*: Gathona, Surband—; *Nepal*: Titapat—; *Pacific Coast*: Common Mugwort, Felon-herb, Green Ginger, Motherwort—; *Persian*: Absanthin, Absentin, Afsantin, Afsintin rumi—; *Philippines*: Santa Maria, Yerba de Santa maria—; *Punjab*: Afsuntin, Banjiru, Buimadaran, Chambra, Puujan, Tarkha, Tataur, Ubusha—; *Roumanian*: Pelin negru, Pelinita—; *Russian*: Cherdonilnik—; *Sanskrit*: Barha, Barhikusum, Barhipushpa, Granthika, Granthiparna, Granthiparnaka, Guchhaka, Gutthaka, Kakapushpa, Kukura, Nagadamani, Nilapushpa, Saraparni, Shuka, Shukabarha, Shukachhada, Shukapuccha, Sthauneya, Sugandha, Svaramaguchhaka, Tailaparnaka, Vnyadamanaka, Vishirnakhya—; *Saora*: Adavidhavanamu—; *Spanish*: Artemisa, Artemisia—; *Tagalog*: Camaria, Tinisas—; *Tamil*: Mashibattiri, Tirunama—; *Teheran*: Absint—; *Telugu*: Davanamu, Mashipatri—; *Uriya*: Doyona, Gonddhomaro, Nagodoyona—.

Aster

The genus includes 400 species inhabiting mostly the northern temperate regions of the world.

The following species are used medicinally in China—*A. fastigiatus* Fisch. and Max., *A. scaber* Thunb., *A. tataricus* Linn. f., *A. trinervius* Roxb.—; in Indo-China—*A. tataricus* Linn., *A. trinervius* Roxb.—; in South Africa—*A. erigeroides* Harv., *A. filifolius* Vent., *A. hispidus* Bkr., *A. serrulatus* Harv.—; in California—*A. puniceus* Linn.—

Aster trinervius Roxb. occurs in the Central and Western Himalaya, Nepal, Sikkim at 5,000—7,000 ft., the Mishmi Mountains, the Khasia Hills at 3,000—6,000 ft. Thence the plant spreads over to North China and Japan.

The Chinese use the root for coughs and pulmonary affections, and in the treatment of malaria and haemorrhages.

Cantonese: Tsz uen—; *Chinese*: Ma Lan—; *Indo-China*: Ma lan—; *Malaya*: Tsz yoon—.

Bidens.

The genus includes 150 cosmopolitan species, chiefly American.

B. bipinnata Linn. is used medicinally in West Africa; *B. pilosa* Linn. in China, Indo-China, North America, Brazil, Gold Coast,

Kenya, Tanganyika, and South Africa; *B. tripartita* Linn. also is used in China, and *B. graveolens* Mart. in Brazil.

Achenes cuneiform, tip truncate ... ... 2. *A. tripartita.*
Achenes long, slender, narrowed from the middle to the top. 1. *A. pilosa.*

1. **Bidens pilosa** Linn. is found throughout India, Ceylon, and most of the warm countries.

In Indo-China the dried flower buds ground and mixed with alcohol are used as a mouth wash in toothache. For sore eyes the pounded leaves are applied over the eyelids.

In Gold Coast and in Lagos the juice of the leaves is commonly squeezed into the eyes or the ears to cure eye complaints or ear complaints. In the latter case the leaves are first warmed in water with pepper. It is also used as a styptic to stop bleeding from a wound.

The Zulus chew the young shoots for the treatment of rheumatism. They also administer the powdered leaf in water as an enema for abdominal troubles and rub the burnt seeds into incisions on the sides for the relief of pain. The flower is a remedy for diarrhoea, and an infusion of the leaf and root a remedy for colic.

The leaves are used in Brazil as a styptic in stopping the flow of blood, and as a vulnerary. They are applied to foul ulcers and swollen glands. In Colombia the infusion is used as a sudorific.

Afrikaans : Duivelskerwel, Wevenaar—; *Antioquia* : Masiquia—; *Australia* : Black Fellows—; *Bogota* : Chipaca—; *Ceylon* : Ottupillu—; *Chinese* : Kuei Chen Ts'ao—; *English* : Beggar Ticks, Bur Marigold, Spanish Needles, Sweethearts—; *Ewe* : Dzani pipi—; *Gold Coast* : Black Jack—; *Gujerati* : Phusiun, Samara-kokadi—; *Iloilo* : Tubactubac—; *Indo-China* : Cuc ao, Quay cham thao, Tu to hoan—; *Kikuyu* : Michegi—; *Krepi* : Adzrorkpii—; *Krobo* : Dsetshi—; *Lagos* : Aberc oloko, Akesinmaso—; *Lepcha* : Mung-gu tuk-tsong, Tuk-tsong mung-ayap—; *Malay* : Rumput juala—; *Mano* : Zikilli wissi—; *Mauritius* : Ville-bague—; *Mende* : Tombolo, Tombo-makei—; *Meru* : Rathangi—; *Nandi* : Kipkoleit—; *South Africa* : Beggarsticks, Black Jack—; *Suto* : Moonyane—; *Timne* : Ebamp—; *Twi* : Ananse mpaane, Dwirantwi, Gyinantwi—; *Yoruba* : Abere-oloko—; *Zulu* : uQadolo—.

2. **Bidens tripartita** Linn. occurs in Western Tibet, and in the marshes of Central and Western Himalaya from Nepal to Kashmir at altitudes of 3,000-5,000 ft. It is distributed to Northern Asia and Japan, and westwards to North Africa, Western Europe, and North America.

The Chinese use the plant in chronic dysentery and for eczema.

Chinese : Lang Pa Ts'ao—; *English* : Bur Marigold, Water Agrimony—; *Greek* : Agria katephedia—.

BLUMEA.

The genus is eminently characteristic of India, and the species may be called the Groundsels of that country. The species number about 80; they are tropical and subtropical Asiatic, African, and Australian.

B. aurita DC. is used medicinally in Gold Coast.

A. Heads many, villous, small. Leaves thick, tomentose beneath ... ... 5. *B. lacera.*
B. Heads few, small, on long slender peduncles of dichotomous cymes ... ... 4. *B. eriantha.*

C. Heads very numerous, large or small. Leaves very large.
  1. Pappus white ... ... 2. *B. chinensis.*
  2. Pappus red
    a. Tall shrub, camphoraceous. Leaves thick usually silky above ... ... 1. *B. balsamifera.*
    b. Leaves 8-18 in. long, very woolly white beneath puberulous above ... ... 3. *B. densiflora.*
    c. Leaves 6-10 in. long, stiff, shining, quite glabrous ... ... 6. *B. myriocephala.*

1. **Blumea balsamifera** DC. is easily recognised by its strong smell of camphor. It is found in Tropical Himalaya, Nepal, Sikkim at 1,000-4,000 ft., Assam, the Khasia Hills, Chittagong, and Burma. It is very common in open places in the whole Malay Peninsula. It is also common in Java.

A warm infusion acts as a pleasant sudorific, and the decoction is a useful expectorant.

In Indo-China the leaves are considered to be stomachic, and antispasmodic, and they are used in leucorrhoea. In Cambodia they are used externally in scabies.

Taken internally the decoction of the leaves is an excellent diaphoretic in bronchitis. In Java and China it is given as an expectorant. It is stomachic, antispasmodic, emmenagogue, antiseptic. As a fumigation it is much used in the Philippine Islands for rheumatism and headache.

The leaves are officinal in Holland.

Burma: Ponmathein—; *Cambodia*: Baimat—; *China*: Ai Na Hsiang—; *Dutch India*: Semboeng octan—; *English*: Nagal Camphor—; *French*: Camphrée—; *Gujerati*: Kalahad—; *Hindi*: Kakaronda—; *Ilocano*: Sobsob—; *Indo-China*: Bai mat, Dai bi, Tu bi xanh—; *Malay*: Chapa, Chapu, Sembong—; *Marathi*: Bhangaruda—; *Pampangan*: Sambon—; *Tagalog*: Sambon, Sambong, Sambung—; *Visayan*: Alibhon, Alibun, Ayoban, Gabuen, Gintingintin, Guintinguintin, Guitinguitin, Hamlibon, Lacadbulan, Lacdanbulan, Lalacdan—.

2. **Blumea chinensis** DC. is found in the Eastern Himalaya, Sikkim and Bhutan at 2,000—4,000 ft., Assam, the Khasia Hills, and Penang. It is not rare in the woods of the Malay Peninsula, whence it spreads over to Java and Southern China.

The leaves or leafy stalks are used in Malaya as a stomachic, antispasmodic, and diaphoretic.

Chinese: Chin Li Ming—; *Malaya*: Kow lee meng, Tombak-tombak—.

3. **Blumea densiflora** DC. occurs in Tropical Himalaya, Sikkim, Assam, the Mishmi, Naga, and Khasia Hills at altitudes of 2,000—4,000 ft. It is found in Tavoy, and extends to the Malay and Fiji Islands.

The plant yields camphor. Its leaves are occasionally used as a sudorific.

Burma: Phummasin, Pungmatheing—.

4. **Blumea eriantha** DC. is found in Bundelkhand, Konkan, Deccan, the Western Ghats, the South Mahratta Country. It probably occurs on the West Coast of the Madras Presidency.

The juice of the plant is administered as a carminative, and the herb used along with the leaves of *Vitex Negundo* and *Careya arborea* for fomentations. A warm infusion is given as a sudorific in catarrhal affections, cold it is considered to be diuretic and emmenagogue.

Marathi: Nimurdi—.

5. **Blumea lacera** DC. occurs throughout the plains of India, from the north-west ascending to 2,000 ft. in the Himalaya, to Travancore and Singapore. It is a common roadside weed in Ceylon and Malaya. It is distributed to the Malay Islands, Australia, China, and tropical Africa.

The plant is described by Ayurvedists as hot, pungent, and bitter; antipyretic; good for bronchitis, diseases of the blood, fevers, thirst, and burning sensations. The root kept in the mouth is said to cure diseases of the mouth.

In the Konkan the plant is used to drive away fleas and other insects. It is prescribed as an antiscorbutic in West Africa.

The root mixed with black pepper is given in cholera.

The expressed juice of the leaves is used as an anthelmintic, febrifuge, astringent, and diuretic; mixed with black pepper, it is given in bleeding piles.

Arabic: Kamafitus—; *Bengal*: Burasuksung, Kukurmuta, Kukursunga—; *Bombay*: Nimurdi—; *Burma*: Maiyagan—; *Deccan*: Divarimulli, Janglikasni, Janglimulli—; *Golungo Alto*: Quitoco antiscorbutico—; *Gujerati*: Kalhar, Kokarunda, Pilo kapurio—; *Hindi*: Janglimuli, Kakronda, Kukkurbanda—; *Konkan*: Numurdi—; *La Reunion*: Lastron bâtard—; *Malay*: Lumai hitam—; *Marathi*: Bhamurda, Kukurbanda—; *Mundari*: Marangkuru—; *Porebunder*: Kapurio, Pilichanchadamari—; *Sanskrit*: Kukkuradru, Kukundara, Mriduchhada, Sukshmapatra, Tamrachuda—; *Tamil*: Kattumullangi, Narakkarandai—; *Telugu*: Advimulangi, Karupogaku—.

6. **Blumea myriocephala** DC. is found in the Sikkim Himalaya at an altitude of 2,000 ft., in Assam, Chittagong, Sylhet, and Burma. It occurs in the mountain forests of the Malay Peninsula.

The leaves are used as a sudorific in Indo-China; they are given in bronchitis and in aphthae.

Indo-China: Xang song. Xuong song—.

BOLTONIA.

The genus consists of 12 North American and North subtropical Asiatic species.

Boltonia indica Benth. is found in the Khakyen Hills of Upper Burma. It extends to China and the Malay Islands.

In Indo-China the plant is considered tonic, stomachic, and antipyretic.

Indo-China: Hai nhi cuc—.

CALENDULA.

The genus includes 15 species, natives of Central Europe and the Mediterranean region, as far as India, the Canaries; cultivated and naturalised elsewhere.

Ligules sulphur-coloured; achenes all curved, marginal dorsally echinate beaked, inner dorsally muricate ... 1. *C. arvensis.*
Ligules many, bright orange yellow; achenes all curved boat-shaped dorsally muricate not beaked, outer larger ventrally crested scarcely beaked ... ... 2. *C. officinalis.*

1. **Calendula arvensis** Linn. is a doubtful native of India; found in Kashmir; distributed over Western Asia and Southern Europe; naturalised in South America, Australia and Japan.

The plant is used medicinally in Spain; the leaves are considered sudorific; the flowers are reputed stimulant, antispasmodic, and emmenagogue.

Catalan: Galdirons, Llevagat, Llevamal, Llevamans—; *French*: Petit souci, Souci des champs, Souci des vignes—; *Malta*: Field Marygold, Calendula, Calta, Fior-rancio, Suffeira—; *Spanish*: Calendula silvestre, Flor de cada mes, Maravilla silvestre, Yerba del podador—.

2. **Calendula officinalis** Linn. is a native of the Mediterranean introduced in India and run wild. It is found in the fields of Punjab and Sind, extending to Afghanistan and westwards to South Europe.

The plant is signally valued for healing wounds, ulcers, burns, and other breaches of the skin surface; it is a precious vulnerary. The dried florets are vulnerary, antiemetic, bitter tonic, febrifuge, anthelmintic, and stimulant to wounds and ulcers.

The plant, especially its flowers, was used on a large scale by the American surgeons to treat wounds and injuries sustained during the last civil war, and obtained their warmest commendation. It quite prevented all exhausting suppurative discharges and drainings.

In Europe the herb is chiefly used as a local remedy. Its action is stimulant and diaphoretic. Given internally, it assists local action and prevents suppuration. The infusion of one ounce to a pint of boiling water is given internally, in doses of a tablespoonful, and externally as a local application. It is useful in chronic ulcer, varicose veins, etc. The plant was formerly considered to have much value as an aperient and detergent in visceral obstructions and jaundice.

The leaves, eaten as a salad, have been considered useful in the scrofula of children. The expressed juice has been given in cases of costiveness and proved very efficacious. Snuffed up the nose it excites sneezing and a discharge of mucus from the head. It has also been found useful as an extirpator of warts.

It has been asserted that a flower, rubbed on the affected part, is an admirable remedy for the pain and swelling caused by the sting of a wasp or bee. A lotion made from the flowers is most useful for sprains and wounds, and a water distilled from them is good for inflamed and sore eyes.

An infusion of the freshly-gathered flowers is employed in fevers, as it gently promotes perspiration and throws out any eruption—a decoction of the flowers is much in use among country people in England as a posset drink in measles and smallpox, a widely administered remedy and one of the few which everybody seems to know. Marigold flowers are in great demand for children's ailments.

Druggists now make a medicinal tincture which is advised as a sudorific stimulant in low fevers, and to relieve spasms. A saturated tincture of the flowers, when mixed with water, promotes the cure of contusions, wounds, and simple sores or ulcers; also the extract will allay chronic vomiting. One drop of the tincture with two grains of powdered borax when sprayed into the ear is very useful if a discharge has become established therefrom.

Anglo-Saxon: Merso-meargealla—; *Burma*: Htattaya—; *Catalan*: Clavellina de mort, Clavellines de mort, Gojato, Gojats, Graugets—; *Chinese*: Chin Chan Ts'ao—; *Danish*: Almindelige koeblomme—; *Dutch*: Afrikaantje goudsbloem, Goudsbloem, Tamme goudeloem—; *English*: African Marigold, Calendula, Common Marigold, Garden Marigold, Golds, Marigold, Mary Gowles, Pot Marigold, Ruddes—; *French*: Fleur de tous les mois, Souci, Souci cultivé, Souci des jardins—; *German*: Butterblume, Dannblaume, Dotterblume, Faerberblume, Gaehl, Gaehlgoelling, Gartringel, Gelken, Gilkenblume, Goeldeke, Goelling, Goldblume, Goldrose, Gugelkopf, Hauswirbel, Huehnernelke, Ingblume, Kohlblume, Kolblume, Morgenroete, Rinderblume, Ringelblume, Ringelken, Ringelrose, Sonnenblume, Sonnenrose, Studentenblume, Totenblume, Wagenblume, Warzenkraut, Wegroeslein, Zunenwirvel—; *Greek*: Chamobyoreta—; *Italian*: Calendula, Calendula ortense, Fiori d'ogni mese, Fiorrancio, Furrancio—; *Kurdish*: Hamaishabahar—; *Pacific Coast*: Garden Marigold, Goldbloom, Holligold, Marigold, Mary-bud—; *Polish*: Nogieter—; *Portuguese*: Calendula ortense, Maravilha bastarda—; *Punjab*: Aklelulmulk, Saldbargh, Zergul—; *Roumanian*: Galbinele, Hilimica—; *Russian*: Nogotki—; *Spanish*: Calendula, Calendula officinal, Flor del muerto, Flor de muerto, Maravilla, Mercaderes dorados, Mercaderes reales—; *Swedish*: Ringblomma—; *Turkish*: Qarah koz—; *Uruguay*: Flor de la virreina—; *Yemen*: Zobedje—.

CARDUUS.

The genus includes about 80 species, natives mostly of the Mediterranean region, but extending from Europe to Japan.

C. crispus Linn. is used medicinally in China, *C. nutans* Linn. in India.

Carduus nutans Linn. is found in Western Tibet at an altitude of 13,000 ft., in the Western Himalaya from Kashmir to Simla, and in the Punjab. It extends to Northern Asia and westwards through Baluchistan and Persia to North Africa and Western Europe.

The flowers are considered febrifugal in Sind and in the Punjab. In Kashmir they are used to purify the blood.

English: Musk Thistle, Nodding Thistle—; *Kashmir*: Gulibadaward—; *Punjab*: Badaward, Kanchhari, Tiso—; *Urdu*: Gulebadaward—; *Uruguay*: Cardo—.

CARPESIUM.

The genus consists of about 10 species in South Europe and in temperate and subtropical Asia.

Leaves subsessile, never truly petioled, lanceolate acuminate quite entire or serrate. Heads subsessile inserted along the whole length of the branches	1. *C. abrotanoides.*
Leaves shortly petioled elliptic-lanceolate obtuse sinuate-toothed. Heads terminal drooping with broad leafy bracts.	2. *C. cernuum.*

1. **Carpesium abrotanoides** Linn. is found in the Temperate Himalaya, from Kashmir at 5,000-1,000 ft. to Sikkim at 8,000-10,000 ft. It is distributed to North Persia and Austria, and to China and Japan.

The root, the leaf, and the seed are used medicinally in China. They are said to be diuretic and anthelmintic.

In Indo-China the seeds are considered laxative and bechic.

Chinese: Hao Shih, T'ien Ming Ching—; *Indo-China*: Cau nhi thai, Thien mong tong—.

2. **Carpesium cernuum** Linn. is one of the commonest and most variable Himalayan plants. It abounds in the Temperate Himalaya, the Khasia Hills, and the Nilghiri Mountains. It is distributed from the Caucasus to France, Java, and Japan.

The herb is used in China as an astringent, diuretic, and anthelmintic.

Chinese: Ho Shih—.

CARTHAMUS.

The genus includes about 25 species, natives of the Mediterranean region, Central Europe, India, Abyssinia and the Canaries.

A. Pappus made of scabrous toothed scales, the inner ones being thrice as long as the achene	1 .*C. lanatus.*
B. Pappus absent.	
1. Leaves entire and unarmed or spinulose-serrate ...	3. *C. tinctorius.*
2. Leaves oblong or oblong-lanceolate; lower shortly spinulose-toothed, upper half-amplexicaul, very spinous	2. *C. oxyacantha.*

1. **Carthamus lanatus** Linn. is found in Kashmir at 5,000—6,000 ft. It occurs in the Mediterranean region, Central Europe, Abyssinia, Madeira, the Canaries.

The plant is used medicinally in France as a sudorific, febrifuge, and anthelmintic.

English: Blessed Thistle, Distaff Thistle, Woolly Carthamus, Yellow Distaff Thistle—; *French*: Carthame laineux, Chardon bénit des Parisiens—; *Maltese*: Xeuk il far, Xeuc ta Cristu—.

2. **Carthamus oxyacantha** Bieb. occurs in the Punjab, whence, through Baluchistan and Afghanistan, it extends westwards to the Caucasus,

The oil extracted from the seeds is used medicinally in the Punjab as a dressing for bad ulcers and as a remedy for itch.

English: Wild Safflower—; *Hindi*: Kantiari, Karar, Kharara, Poli, Polian—; *Iraq*: Kassub asfar, Suffair, Summaina—.

3. **Carthamus tinctorius** Linn. is cultivated throughout a large part of India, Afghanistan, Persia, Syria, Egypt, and Southern Europe.

This plant is the Kusumbha of Sanskrit writers, a well known Hindu medicinal drug.

Mahomedan writers too ascribe many virtues to the plant.

The action of the flowers is laxative and diaphoretic. In domestic practice these flowers are used in children's and infants' complaints—measles, fevers, and eruptive skin complaints. An infusion is made of ½ oz. of the flowers to a pint of boiling water taken warm to produce perspiration.

In China and Indo-China the flowers are given in dysmenorrhoea and paralysis as a tonic and emmenagogue. In the Philippine Islands they are employed as a cure for jaundice.

The powdered seeds made into a poultice, are used to allay inflammation of the womb after childbirth. In Sind they are employed as a cooling medicine; they are sometimes boiled and made into a gruel. They are considered to be diuretic and tonic by the natives of the Punjab. In China and Indo-China they are reputed purgative.

Koman in Madras administered a decoction (1 in 20) of the powdered seeds to cases of constipation. The action was very mild, and in the majority of cases it did not produce the desired effect.

The oil from the seeds is considered a mild purgative in Sind. It is used as a dressing for bad ulcers, and as a liniment in rheumatism. In Bengal it is considered by the ryots as a valuable remedy for itch; a cure is said to be effected after three to six applications. The charred oil is used for healing sores and for rheumatism; as a veterinary medicine it occasionally finds use in healing sores on cattle.

The young green plant is said to be very efficacious in colds; it is believed to keep the system warm.

Safflower, in combination with other drugs, is prescribed for scorpion sting; but Caius and Mhaskar have experimentally shown that it is not an antidote to scorpion venom.

Arabic: Akhariza, Bazr-el-abris, Hab-ul-asfar, Hariz, Kirtum, Kurtum, 'Usfar, Za'faran—; *Bengal*: Kajirah, Kusamphul, Kusum, Kusumbha—; *Bombay*: Kardai, Karophi, Kusumba—; *Burma*: Heboo, Hshu, Su, Suban, Supan—; *Canarese*: Kossumba, Kusambe, Kusumba—; *Catalan*: Safrá bort, Safranó—; *Chinese*: Hong Hoa, Hong Lan Hoa, Hung Lan Hua—; *Cutch*: Kusumba—; *Dutch*: Basterd Saffraan, Wilde saffraan—; *Deccan*: Kusumb—; *Egypt*: Gartoom, Kurtim, Osfar, Qortom—; *English*: African Saffron, American Saffron, Bastard Saffron, Dyer's Saffron, Fake Saffron, Parrot Seed, Safflower, Wild Saffron—; *French*: Carthame, Cnique, Safran bâtard, Safran d'Allemagne, Safranon, Carthame des teinturiers, Safran faux—; *German*: Gartensafran, Falschesafran, Farberdistel, Safflor, Wildersafran—; *Greek*: Atractos, Atractylis, Knikos—; *Guilan*: Tokhme-Kafisheh, Tukm-i-kajrah, Tukm-i-kazirah—; *Gujerati*: Karada, Kusumbo—; *Hamadan*: Kaufsha, Qushon—; *Hindi*: Barre, Karrah, Kasumba, Kussum, Kusumba—; *Indo-China*: Daccam, Hong hoa,

Hong lam hoa, Rum—; ***Iraq*** : Qurtum—; ***Italian*** : Cartamo, Croco ortense, Zafferano saracinesco, Zaffrone—; ***Konkani*** : Kusbo—; ***Languedoc*** : Grano de perrouquet—; ***Malaya*** : Hong fah, Hsi hung hua, Hung hua, Hung lan hua, Sai hong fah—; ***Malayalam*** : Chendurakam—; ***Malta*** : Bastard Saffron, Zafferanone, Ghosfor—; ***Manipur*** : Galapmachu—; ***Marathi*** : Kadaya, Kararhi, Kardai, Kasdi, Kurdi, Sadhi—; ***Mauritius*** : Carthame, Safran bâtard—; ***North-Western Provinces*** : Barre, Kar—; ***Oceania*** : Kassoumbo—; ***Pampangan*** : Cachumba, Casubha, Castumba—; ***Persian*** : Gulekafshah, Gulemaskar, Kasakdanah, Kazhirah, Muasfir, Quortum—; ***Philippines*** : Azafran de la tierra—; ***Portuguese*** : Acafrao, Cartamo—; ***Punjab*** : Kar Karar, Kasumbha, Kurtam, Kusam, Kushumbha, Ma, Safir—; ***Rajputana*** : Bundi—; ***Roumanian*** : Brandusa de tvamna—; ***Russian*** : Saflor—; ***Sanskrit*** : Agnishikha, Gramyakunkuma, Kamalottara, Kamlottama, Kukkutashikha, Kusumbha, Lohita, Maharajana, Padmottara, Papaka, Pita, Rakta, Vanishikha, Vasraranjana—; ***Sind*** : Khoimbo, Quortum—; ***Spanish*** : Alazor, Azafran bastardo, Azafran romi, Azafranillo de Mejico, Azafranillo de papagayos—; ***Sudan*** : Essfar, Kurtum—; ***Swedish*** : Saffler—; ***Syria*** : Kashni—; ***Tagalog*** : Biri, Casabha, Casubha, Castumba, Catsumba, Lago—; ***Tamil*** : Chendurukam, Kusumba, Sendurgam—; ***Telugu*** : Agnisikha, Kushumba, Kusumbha—; ***Turkish*** : Kantawaras—; ***Urdu*** : Karha, Kusum—; ***Uruguay*** : Azafran bastardo, Azafran falso—; ***Visayan*** : Casabha—.

CENTAUREA.

The genus includes 600 species, mostly natives of the Mediterranean region, Central Europe and Western Asia, with a few North and South American and one Australian. The *Centaurea* are immigrants from the west into India and, except perhaps for *C. iberica* Stev. and *C. phyllocephala* Boiss. are visitors rather than even denizens.

C. calcitrapa Linn., *C. Centaurium* Linn., *C. cyanus* Linn., *C. jacea* Linn., *C. montana* Linn. are used medicinally in Europe; *C. Behen* Linn. in Persia; *C. alexandrina* Del., *C. praecox* Oliv. and Hiern., *C. rhizocephala* Oliv. and Hiern. in West Africa.

C. solstitialis Linn. is reputed poisonous to live stock when mature.

A. Flowers blue ... ... 2. *C. cyanus.*

B. Flowers purple or pink, very rarely white.
   1. Involucral bracts with narrow membranous margins terminating in a long strong spreading spine with short lateral ones at its base ... ... 1. *C. calcitrapa.*
   2. Involucral bracts ovate-rotund and oblong, herbaceous, cucullate with hyaline tomentose membrane ... ... 3. *C. picris.*

1. **Centaurea calcitrapa** Linn. is found in the Punjab and Kashmir, ascending to 3,500 ft. It also occurs in Mysore. It extends to Asia Minor, Central and South Europe, North and Tropical Africa.

The Arabs apply the bruised leaves to the head in cephalalgia.

In Europe the powdered root has long been considered a cure for fistula and gravel, and the seeds are made into powder and drunk in wine as a remedy for stone.

Arabic : Morrer, Shok—; ***Catalan*** : Cart estrellat, Floravia, Herba espitlera—; ***English*** : Star Thistle—; ***French*** : Centaurée chausse-trappe, Chardon étoilé, Chausse-trappe—; ***German*** : Flockenblume, Sterndistel, Sternflockenblume—; ***Italian*** : Calcatrepo, Calcatreppo, Calcatreppola, Calcatrippa, Ipofesto, Ippofresto—; ***Malta*** : Caltrops, Star-thistle, Calcatreppola, Ippofesto—; ***Spanish*** : Abrojos, Cardo estrellado, Trepacaballos encarnado—,

2. **Centaurea cyanus** Linn. is found in corn fields and cultivated places of North-West India. It is distributed to the Caucasus and westwards to the Atlantic.

The florets are mildly astringent. A water distilled from them was formerly in repute for weak eyes; it was famous in France under the name of *Eau de Casse-lunettes.*

Quoth Culpepper: 'The powder or dried leaves of the blue-bottle, or cornflower is given with good success to those that are bruised by a fall, or have broken a vein inwardly, and void much blood at the mouth'.

In modern herbal medicine the flowers are considered to have tonic, stimulant, and emmenagogue properties.

Catalan: Angelets, Blauets, Blauhet, Escombrera—; *Dutch*: Korenbloem—; *English*: Bachelor's Buttons, Blaver, Blaverole, Blawort, Blue Blawort, Blewball, Blewblow, Blueblaw, Blue-bonnets, Bluebottle, Bluebow, Blue-caps, Blue Poppy, Break-your-Spectacles, Brushes, Corn-binks, Corn Bluebottle, Corn-bottle, Corn Centaury, Cornflower, Cuckoo-hood, Hawdod, Hurt-sickle, Knobweed, Knotweed, Loggerheads, Thumble, Witch Bell, Witches' Thimble—; *French*: Aubefoin, Aubifoin, Aubiton, Aubitou, Barbeau, Barbot, Bavéolle, Blavelle, Blavéole, Blaverolle, Blavet, Blaveolo, Blavetta, Blavette, Bleuet, Bluet, Bluet des moissons, Boufa, Bouffa, Carconille, Casse-lunettes, Centaurée barbeau, Centaurée bleuet, Chevalon, Chevalot, Ciano, Cornaille, Cornillat, Cornille, Courcourille, Créconille, Fleur de Zacharie, Le Chevalier, Péréole, Pérole—; *German*: Blaue Kornblume, Kornblume, Roggenblume, Tremse, Ziegenbein—; *Italian*: Battisegola, Fiordaliso—; *Roumanian*: Albastrea, Dioc, Floarea griului, Ghioc, Sglavoc, Vinetea—; *Russian*: Vacilek, Vacilyok, Valoshky—; *Spanish*: Aciano, Azuleja, Azulejo, Escobilla, Flor celeste de sembrados, Flor del cielo español, Liebrecilla menor—.

3. **Centaurea picris** Pall. is found in Sind, Baluchistan, Afghanistan, the Levant, Central and Southern Russia, Siberia, and the Altai Mounts.

At Wad in Jhalawan the plant is pounded in water and used to cure worms. In Loralai it is a cure for wounds of sheep, used if wolves tear them.

Brahui: Talkh kah—; *Harboi Hills*: Tulkha—· *Loralai*: Kurakh—; *Wad*: Talkhakao—.

CENTIPEDA.

The genus numbers 5 species, inhabiting Chile, Madagascar, tropical Asia, and Australia.

Centipeda robicularis Lour. is found in moist places throughout the plains of India and Ceylon. It is distributed over Afghanistan, Eastern tropical Asia, Australia, and the Pacific Islands.

The plant, either whole or in parts, is used medicinally in India, China, the Philippine Islands, and New South Wales.

The powdered herb and the minute seeds are used as a sternutatory. The drug is administered in ozoena, headaches, and colds in the head. It is considered a hot and dry medicine, useful in paralysis, pains in the joints, and worms.

The Mundas of Chota Nagpur snuff the crushed plant in fevers and colds. In the Punjab the herb is boiled to a paste and applied to the cheeks in toothache.

Vyas and Sinha have shown that the herb contains an alkaloid, a glucoside, and traces of saponin; the watery extract increases the force of contraction of the frog's heart, prolongs the systole, and causes heart block in larger doses.

Arabic: Afkar, Makandash, Uffarkakudush—; *Bengal*: Chhikni, Hanchetagachha, Hanchuti, Mechitta, Nagdowana, Nakkchikni, Pachittie—; *Bombay*: Nagdowana, Nakkchikni, Pachittie—; *Chinese*: Shih Hu Sui—; *English*: Sneezeweed, Sneezewort—; *Gujerat*: Chhikani—; *Hindi*: Nagdowana, Nakchhikni, Nakkchikni, Pachittie—; *Indo-China*: Co the, Thach ho tuy—; *Malaya*: Chikkana, Chhikika, Chu tsao, E tai shih, O tai shih, Pe kong chau, Shih u sui, Yoo pak seek—; *Marathi*: Nakashikani, Narasinkani—; *Mundari*: Acuara, Acusing—; *New South Wales*: Sneezeweed—; *Persian*: Gawejahan—; *Sanskrit*: Chhikkani, Chhikkika, Ghranadukhada, Kruranasa, Kshavaka, Kshavakrita, Sanvedanapatu, Tikshna, Ugra, Ugragandha—; *Santal*: Bediachim—; *Sind*: Afkar—; *Tagalog*: Harangan—; *Urdu*: Nakachhikani—; *Victoria*: Gukwonderuk—; *Visayan*: Harangan, Pissic—.

CENTRATHERUM.

The genus consists of 15 tropical species, mostly Asiatic, with four American and one Australian.

Centratherum anthelminticum O. Ktze. (= *Vernonia anthelmintica* Willd.) is found throughout India and Ceylon, often cultivated.

The plant, roasted in a room, or powdered and thrown about the floor, is believed to expel fleas.

The juice of the leaves is given to cure phlegmatic discharges from the nostrils.

In Hindu medicine the seeds have long been esteemed as one of the principal remedies for white leprosy and other skin diseases. The drug is powdered with an equal quantity of black sesamum, and a drachm of the powder is taken in the morning with tepid water after perspiration has been induced by exercise or exposure to the sun; the diet should consist of milk and rice. In leucoderma a decoction of emblic myrobalans and catechu is given with the powdered drug. Externally the seeds are used in skin diseases in a variety of forms, such as powder, paste, oil, infusion, juice.

On the Malabar Coast an infusion of the seeds is given for coughs and against flatulency. In Travancore the bruised seeds, ground up to a paste with lime juice, are largely employed as a means of destroying lice.

In the Konkan the seeds enter into the composition of a popular antiperiodic powder. In the Punjab they are considered febrifuge; they are also given in anasarca, and used for plasters for abscesses. In Ceylon they are recommended for fever convulsions.

European doctors practising in India consider the seeds a valuable tonic, stomachic, and diuretic.

The seeds are considered as powerfully anthelmintic in South India, and are also an ingredient of a compound powder prescribed in snake bites.

The author of the *Makhzan-el-Adwiya* says that the seeds are given internally to remove phlegm and worms from the intestines, and that a poultice or plaster is used to disperse cold tumours. He concludes by stating that the drug is not often prescribed internally, as it is thought to have injurious effects, but that it is much used as a cattle medicine.

The seeds are used instead of quinine by the Mundas of Chota Nagpur. In paralysis of the legs the powdered seeds are applied externally. When the stomach of cattle swells, the powdered seeds are mixed in equal quantity with salt and soot from the fireplace. This is dissolved in water with the addition of two capsules of Spanish pepper, and given as a drink.

Wrench (1919), Caius and Mhaskar (1923) have reported unfavourably of the seeds in the treatment of hookworm infection.

There is conflict of opinions as to the value of the seeds in the expulsion of roundworms; not only in the past, but also in more recent times. Whilst Koman (1919-20) finds the seeds to possess considerable ascaricidal properties, Caius and Mhaskar (1923) report very unsatisfactory results.

The seeds contain a rather large quantity of resin. This was tried in a number of cases of helminthic infections at the Carmichael Hospital for Tropical Diseases, Calcutta. It appears to have very little effect on the ascaris. It is, however, distinctly effective in threadworms infections. In several children to whom the resin powder was administered, threadworms were expelled in the stools in large numbers and the symptoms which are often very troublesome, such as nocturnal enuresis and grinding of the teeth, were relieved.

Mhaskar and Caius have shown experimentally that the seeds are not an antidote to either snake or scorpion venom.

Arabic: Atar-i-lal, Itr-i-lal, Kamun-e-bari—; *Bengal*: Babchi, Bapchi, Bukshi, Hakuch, Kaliziri, Somraj—; *Bombay*: Kalenjiri, Kalijiri—; *Canarese*: Kadujirage, Kadujirige, Kalajirige, Sahadevi—; *Deccan*: Kalajira, Kalijiri, Karviziri—; *English*: Purple Fleabane, Wild Cumin—; *French*: Herbe aux mouches—; *Gujerati*: Kadvojiri, Kalijiri—; *Hindi*: Bakshi, Bukshi, Kalijhiri, Kaliziri, Somraj, Vapchi—; *Kumaon*: Kalijiri—; *Malayalam*: Kalajirakam, Kattasiragam, Kattujirakam, Puvankuruntala—; *Marathi*: Kalajira Kalenjiri, Kalijiri, Karalye, Ranachajire—; *Mundari*: Karigiri, Karijiri, Karijuri, Piritijudiring, Saoraj—; *Persian*: Atarilal, Itrilal—; *Porebunder*: Kalijiri—; *Punjab*: Bukoki, Kakshama, Kalazira, Kaliziri, Malwabakshi—; *Sanskrit*: Agnibija, Aranyajiraka, Avalguja, Atavijiraka, Brihanyali, Kana, Kananajiraka, Krishnaphala, Kshudrapatra, Putiphali, Sahadevi, Somaraji, Tiktajiraka, Vakushi, Vanajiraka—; *Sinhalese*: Sanninaegam, Sanninasang, Sanninavan—; *Tamil*: Katchiragam, Kattuchiragam, Neychitti, Nirnochi, Sittilai—; *Telugu*: Adavijilakatta, Garitikamma, Nelavavili, Vishakantakamulu—; *Tulu*: Kalajrdari—; *Urdu*: Janglijiri—; *Uriya*: Somraj—.

CHRYSANTHEMUM.

The genus includes 180 species, natives of the northern hemisphere.

The following are used medicinally in Europe—*C. Balsamita* Linn., *C. cinerariaefolium* Vis., *C. Leucanthemum* Linn., *C. Parenium* Bernh., *C. vulgare* Bernh.—; in Persia—*C. Marschallii* schers—; in China—*C. coronarium* Linn., *C. Decaisneanum* Max., *indicum* Linn., *C. sinense* Sab.—; in Indo-China—*C. indicum*

Linn., *C. sinense* Sab.—; in North America—*C. cinerariaefolium* Vis., *C. Leucanthemum* Linn.

Annual ... ... 1. *C. coronarium.*
Perennial ... ... 2. *C. indicum.*

1. **Chrysanthemum coronarium** Linn. is a native of the Mediterranean region, planted in Indian gardens.

The Yunanists consider the bark a purgative useful in syphilis; they apply the leaves topically to lessen inflammation.

The people of the Deccan administer the plant in conjunction with black pepper in gonorrhoea.

Arabic: Adharyun—; *Assam*: Pithogarkah—; *Bengal*: Guldaudi—; *Bombay*: Seoti—; *Canarese*: Hale—; *Chinese*: T'ung Hao—; *Deccan*: Gulchini—; *Egypt*: Mandiliye, Qehawan—; *Gujerati*: Guldaudi—; *Hindi*: Akurkurra, Gulchini, Guldaudi—; *Ladak*: Kalzang—; *Malta*: Crown Daisy, Bambagella, Fior d'oro, Lellux, Zigland—; *Marathi*: Gulesevati, Tursiphal—; *Persian*: Guledaudi—; *Punjab*: Bagaur, Zaenil—; *Sinhalese*: Lavulugas—; *Tamil*: Shamantippu—; *Telugu*: Chamanti—; *Urdu*: Gulechini—.

2. **Chrysanthemum indicum** Linn. is a native of China and Japan, grown in Indian gardens.

The natives of the Deccan administer the plant in conjunction with black pepper in gonorrhoea.

In China the flower heads are made into tonic and sedative preparations. Infusions are frequently applied as a collyrium in eye affections.

In Malaya the flowers are used for sore eyes and to promote longevity.

In Indo-China the leaves are used as a depurant; they are prescribed in migraine. The flowers are given for sore eyes and for inflammations of the abdomen.

The flowers in the form of an infusion are used by the natives of Guam as a remedy for intermittent fevers, and are valued by women as a remedy for hysteria and monthly irregularities.

Bengal: Chandramallika—; *Bombay*: Akurkura, Chevati—; *Chinese*: Yeh chu—; *English*: False Camomile, Indian Chrysanthemum—; *French*: Pyrèthre—; *Guam*: Manzanilla—; *Hindi*: Guldaudi—; *Indo-China*: Cuc rieng vang, Gia cuc, Kien cuc am dat, Kim cuc—; *Ladak*: Kalzang—; *Marathi*: Shevati—; *Philippines*: Manzanilla, Rosa de Japón—; *Punjab*: Bagaur, Gendi—; *Tagalog*: Dolontas—; *Tamil*: Akkarakkaram—; *Telugu*: Chamunti—; *Urdu*: Guledawoodi—.

CICHORIUM.

This genus includes 8 species scattered over the temperate regions of the Old World; some occur in America as an introduction.

A. Annual ... ... 3. *C. Noëanum.*
B. Perennial
 1. Stem leaves hastate at the base ... ... 1. *C. Endivia.*
 2. Stem leaves entire ... ... 2. *C. Intybus.*

1. **Cichorium Endivia** Linn. is a native of the Mediterranean region, cultivated in India.

It is much valued by the Hakims as a resolvent and cooling medicine, and is prescribed in bilious complaints.

The root is used in dyspepsia and fever as a tonic and demulcent; the fruit as a cooling remedy for fever, headache, and jaundice.

The root is considered warm, stimulating, and febrifuge; given in 'Munjus', the diluent taken preparatory to purging; the seed is used in sherbets.

The root is officinal in Portugal.

Bengal: Kassin—; *Bombay*: Kasini—; *Dutch*: Andijire—; [illegible] rukeyb, Endiwiya, Hendeba, Hindib, Shik[illegible]riya, Silis—; *En[illegible]*: Garden Endive—; *French*: Chicorée blanche Chic[illegible]ée endive, Chicorée frisée, Endive, Escarole, Scariole, Scarole—; *Germ[illegible]*: End[illegible]ie—; *Greek*: Radiki, Radikion—; *Hindi*: Kasini—; *Hova*: Saladingi[illegible]; *I[illegible]*: Hindiba—; *Italian*: Endivea, Indivia—; *Languedoc*: Enderio, En[illegible]io, Endive—; *Mundari*: Risasalad—; *Portuguese*: Chicorea, Endivia, Escarolla—; *Roumanian*: Laptuca—; *Russian*: Laktuk—; *Spanish*: Ekarola—; *Tamil*: Kashini—; *Telugu*: Koshi—; *Turkish*: Hiddiba—.

2. **Cichorium Intybus** Linn. is indigenous to Persia, and is cultivated in Europe and India. It is a common weed over all well-cultivated land in the vicinity of water-courses and wherever there is damp clay soil in Baluchistan, Waziristan, and North-West India up to 6,000 ft.

The Mahomedan writers recognize a cultivated sweet variety and a wild bitter variety.

In Loralai the plant is used as a cure for diarrhoea and bilious attacks.

In Persia, Baluchistan, and India the root is a resolvent and cooling medicine for bilious attacks. The seeds are also considered a cooling medicine.

In Europe the root is a bitter tonic, diuretic, and laxative. A decoction of 1 ounce of the root to 1 pint of boiling water is taken freely in jaundice, liver enlargements, gout, and rheumatic complaints. In France the leaf is considered as a very good substitute for the root; it is used in the form of an infusion, and as such is a very popular laxative.

The seeds are considered carminative and cordial. A decoction is used in obstructed menstruation and for checking bilious vomiting.

The root is officinal in Portugal; the root and the leaves are given official recognition in France.

Arabic: Hindubar, Indyba—; *Baluchistan*: Zral—; *California*: Chicory, Ragged Sailor, Saccory, Wild Bachelor's Buttons—; *Catalan*: Camaroja, Mastagueres, Xicoina, Xicoines, Xicoira, Xicoria amarga—; *Dutch*: Bitterste Cichory, Cichory, Wilde Cichory—; *Egypt*: Hendeb, Shikurie—; *English*: Chicory, Succory, Wild Endive, Wild Succory—; *French*: Barbe de capucin, Bois de corde, Cheveux de paysan, Chicorée amère, Chicorée sauvage, Ecoubette, Herbe à café, Herbe amère, Inthybe—; *German*: Blausamenwirbel, Cichorie, Hindeg, Verfluchte Jungfer, Weglunge, Wegwarte, Wegweiss, Wilde Endivie, Zichorie—; *Greek*: Kichora, Kikori, Kikorion, Korla, Seris Pikris—; *Gujerati*: Kasani—; *Hamadan*: Kashni—; *Hindi*: Hinduba, Kasni—; *Italian*: Cicorea, Cicoria, Radicchio—; *Loralai*: Kashin—; *Malta*: Chicory, Cicoria, Radicchio, Cicueira—; *Pacific Coast*: Blue Daisy, Blue Dandelion, Blue Sailors, Chicory, Succory—; *Persian*: Hinduba, Kasani, Kasni—; *Polish*: Podroznik—; *Portuguese*: Almeirao, Chicorea brava—; *Provence*: Cicoureio—; *Punjab*: Gul, Hand, Kasni, Suchal—; *Roumanian*: Cicoare—; *Russian*: Tsikorie—; *Spanish*:

Achicoria, Achicoria silvestre, Achicoria amarga, Chicoria—; *Swedish*: Waegwarda—; *Tamil*: Kashini—; *Teheran*: Kashni—; *Telugu*: Kasini—; *Urdu*: Kasani—; *Uruguay*: Achicoria—; *Yemen*: Hendibe—.

3. **Cichorium Noeanum** Boiss. is found in Baluchistan whence it spreads over to Mesopotamia.

In Baluchistan the flowers are soaked in water, and the water used for sore legs and also for the stomach derangement called 'dik'.

Brahui: Kashnen, Kashni, Talkə kah—.

CNICUS.

The genus numbers about 150 species found in the northern temperate regions.

C. japonicus Maxim., *C. sinensis* Gard. and Champ., *C. spicatus* Maxim. are used medicinally in China and Japan; *C. arvensis* Hoffm., *C. benedictus* Linn. in Europe; *C. arvensis* Hoffm. in the Pacific Coast States of North America.

Heads dioecious ... ... 1. *C. arvensis.*
Heads bisexual ... ... 2. *C. sinensis.*

1. **Cnicus arvensis** Hoffm. is found in Bengal and the Gangetic Plains extending from the Sunderbunds to the Punjab. It occurs in the Western Himalaya from Kashmir to Kumaon, and in Western Tibet at 11,000-13,000 ft. It is distributed to Northern Asia and westwards to the Atlantic.

The plant is emetic, tonic, diaphoretic, and reported to contain an alkaloid.

The root is astringent.

England: Creeping Field Thistle—; *Pacific Coast*: Canada Thistle—.

2. **Cnicus sinensis** Gard. and Champ. is a native of China extending to Burma and India. It occurs in marshy places on the Khasia Hills at 4,000-6,000 ft.

The stem and leaves are antiscorbutic. The root is given internally for flatulence; externally it is used for ulcers and abscesses.

Chinese: K'u Ao—.

COTULA.

The genus includes 50 cosmopolitan species found mostly in the southern hemisphere.

C. aurea Loefl. and *C. coronopifolia* Linn. are used medicinally in Europe; *C. anthemoides* Linn. and *C. villosa* DC. in South Africa.

Achenes ovate with thick narrow wings. Leaves bipinnatifid or bipinnatisect ... ... 1. *C. anthemoides.*
Achenes oblong hardly winged. Leaves 1-2-pinnatifid ... 2. *C. aurea.*

1. **Cotula anthemoides** Linn. is found in the Gangetic Plain from Rajmahal and Sikkim westwards to the Punjab. It is distributed to China and to Africa, Northern and Southern.

The plant heated with oil is applied externally in rheumatism. The infusion is used as a wash in most diseases of the eye.

A decoction is a Xosa remedy for head and chest colds. The nostrils are sometimes filled with the crushed leaf for colds.

The Sutos use a decoction of the leaf and root as a colic remedy.

Egypt : Ribbin—; *Hindi* : Babuna—; *Kashmir* : Tulobe—; *Punjab* : Babuna—; *Suto* : Hlapi-e-nyenyane—; *Urdu* : Babunah—; *Xosa* : umHlonyane—.

2. **Cotula aurea** Linn. is found in the Punjab. It is spread over Persia, Syria, Malta, Algeria and Spain.

In Spain the plant, chiefly the flowers, is used as a tonic, diaphoretic, anthelmintic, antipyretic, antihysteric, and for pain in the bowels. The flower heads are officinally recognized.

Catalan : Camamilla fina—; *Spanish* : Manzanilla fina—.

CREPIS.

The genus includes 170 species, natives mostly of Europe, Asia and North Africa.

Crepis acaulis Hook. fil. is common in the subtropical Himalaya from Jammu to Bhutan, in the Lower Gangetic Plain, Central India, Kanara and the Nilghiris. It is equally common in Burma.

Among the Mundas of Chota Nagpur the baked leaves, or the root ground and mixed with goat's milk, are taken to activate the secretion of milk in women. The root is also eaten raw in urinary complaints.

Mundari : Etete-ara, Pirieteke—.

DICHROCEPHALA.

The genus consists of 5 Asiatic and African species.

Dicrocephala latifolia DC. occurs on the Western Ghats from Bombay southwards, in the tropical and subtropical Himalaya from Simla to Sikkim at 8,000-9,000 ft., in the Khasia Hills, Cachar, Burma, and in the Malay Peninsula where it is rare. It is distributed over tropical and subtropical Asia and Africa.

The young shoots are used externally in Cambodia for the treatment of blennorrhagia in women, and for the bites and stings of insects.

Cambodia : Kbet choun thom, Kombet choun—; *Indo-China* : Phuc linh thai—.

DICOMA.

The genus numbers 30 species inhabiting Africa, Madagascar, and tropical Asia.

D. tomentosa Cass. is used medicinally in West Africa; *D. anomala* Sond., *D. capensis* Less., *D. speciosa* DC., *D. Zeyheri* Sond. are used in South Africa.

Dicoma tomentosa Cass. is spread over India from the North-West and the Punjab to Sind, Gujerat, the Deccan, the South Mahrata Country, the Carnatic, Mysore, Coimbatore, and the Nilghiris. It is fairly common in tropical Africa.

The herb is strongly bitter, and is used in the neighbourhood of Belgaum as a febrifuge, especially in the febrile attacks to which women are subject after childbirth.

In Hausa and Nigeria the herb is used as a local application to putrescent wounds.

Belgaum: Navananjichapala—; *Gujerati*: Gholoharnacharo—; *Hausa*: Dau da, Farin dayi, Kwardauda, Surandu—; *Katagum*: Dowda—; *Rajputana*: Vajradanti—.

DORONICUM.

The genus consists of 25 species, inhabitants of the northern temperate regions of the Old World.

D. austriacum Jacq., *D. Columnae* Tenore, *D. Pardalianches* Linn., *D. plantagineum* Linn. have been used medicinally in Europe.

A. Achenes all pappose. Heads 1-2 ... ... 2. *D. Hookeri.*

B. Achenes of the ray epappose, of the disk pappose.
 1. Herbaceous, 2-4 ft. high. Heads few or numerous ... 3. *D. Roylei.*
 2. Stout herb, 1-1½ ft. high. Heads 1-2 ... ... 1. *D. Falconeri.*

1. **Doronicum Falconeri** Clarke occurs in Kashmir at an altitude of 13,000 ft., and in the Karakoram Range at 14,000 ft.

The root is said to be useful in nervous depression.

2. **Doronicum Hookeri** Hook. fil. is found in the Sikkim Himalaya at Lachen and Tungu between 12,000 and 14,000 ft.

The root is an aromatic tonic.

3. **Doronicum Roylei** DC. is found in Kashmir and Garhwal at 10,000 ft.

The root is used to prevent giddiness on ascending heights.

ECHINOPS.

The genus includes 82 species, mostly Mediterranean distributed over Southern and Eastern Europe, tropical and North Africa, and Asia to Japan and Siberia.

E. dahuricus Fisch. is used medicinally in China; *E. longifolius* A. Rich. in Nigeria.

Turkish Manna is obtained from *E. persicus* Stev.

Three alkaloids—echinopseine, echinopsine, and B-echinopsine—have been insolated from the seeds of *E. ritro* Linn.

Echinops echinatus Roxb. occurs more or less throughout India and Afghanistan.

The plant is used by both Ayurvedists and Yunanists.

At Hesargai the roots are pounded and mixed with Acacia gum and applied to the hair to destroy lice; also the powdered roots are applied to wounds in cattle to destroy maggots.

Arabic: Ashtarkhar—, *Gujerati*: Shuliyo, Utkanto, Utkato—; *Hindi*: Gokhru, Utakanta, Utakatira—; *Marathi*: Kadechubak, Utanti, Utati, Utkatara—; *Ormara*: Gurgaj—; *Persian*: Astarkhar—; *Pushtu*: Chingamwali—; *Sanskrit*: Kantalu, Kantaphala, Karamadana, Mukhadantarujapaha, Raktapushpa, Shrigala, Shunakashana, Tikshnagra, Ushtrakanta, Utati, Utkantaka, Utkatotkata, Vrittaguchha—; *Urdu*: Untkatara—; *Yemen*: Jirdama—.

ECLIPTA.

The genus consists of 4 species, South American, Australian, and cosmopolitan.

Eclipta alba Hassk. is cosmopolitan in warm climates. It is found throughout India, ascending to 6,000 ft. in the Himalaya and other mountains.

The plant is an Ayurveda and Yunani medicine.

It is principally used as a tonic and deobstruent in hepatic and splenic enlargements, and in various chronic skin diseases. There is a popular opinion that the herb taken internally and applied externally will turn the hair black.

The fresh plant is applied with sesamum oil in elephantiasis, and the expressed juice is taken internally in affections of the liver and dropsy. When used in large doses, it acts as an emetic. It is also considered cooling. It is anodyne and absorbent, and relieves headache when applied with a little oil.

The plant is considered an astringent in China, and is used or checking haemorrhages and fluxes and strengthening the gums. The plant is rubbed on the gums for toothache, acting as a counter-rritant.

The plant is much used as a cure for asthma and bronchitis n Indo-China. In Ceylon it is used to purify the blood.

In La Reunion the plant is considered as pectoral and anti-asthmatic. The decoction is prescribed externally for skin diseases and elephantiasis.

In Bombay, the natives use the juice in combination with aromatics, as a tonic and deobstruent, and give two drops of it with eight drops of honey to new-born children, suffering from catarrh. In the Gujrat district of the Punjab, it is used externally or ulcers, and an antiseptic for wounds in cattle.

The root is given to relieve the scalding of urine. In Chota Nagpur it is applied in conjunctivitis and galled necks in cattle.

In Assam the leaves are reputed to cure sores when applied to them.

The juice of the leaves is generally given in one teaspoonful doses in jaundice and fevers. It is rubbed fresh on the shaven scalp for the purpose of promoting the growth of the hair. The pounded leaves are prescribed in haemorrhage.

In Gold Coast the leaves are ground and mixed with cold water; this mixture is then drunk to cure constipation.

The leaves are used in Brazil as a remedy for diarrhoea and as a black stain for the hair.

In scorpion sting the leaves are rubbed on the part affected as well as inhaled. In practice the leaves are first rubbed from above the inflamed part down to the sting; they are then made into a paste and applied as a poultice.

Koman, summing up his work with this plant, says: 'There are two varieties of the plant—the yellow flowered and the white flowered—the former variety has thicker leaves which are extensively used in catarrhal jaundice. The fresh leaves are well washed, ground with a few pepper corns, and a lump of the size of a lime is administered early in the morning in sour curd or buttermilk. I have found this drug very useful in curing the disease when so administered for five or six days. Occasionally a purgative may be required to aid the action of the drug. In its action, it resembles podophyllin and taraxacum. It may be administered in the form of a succus.'

Arabic: Kadim-el-bint, Tolak—; *Ashanti*: Ntum—; *Baluchi*: Murida—; *Bengal*: Kesari, Keshori, Kesuti, Keysuria—; *Brahui*: Bikgur—; *Canarese*: Garagadasappu, Kadiggagaraga—; *Ceylon*: Kaikechi, Kaivichi illai, Karichalankanni, Karippan—; *Chinese*: Han Lien Ts'ao, Li Ch'ang, Me Teou T'sao, Pa Ko Ts'ao—; *Egypt*: Sa'de—; *Gujerati*: Bhangra, Dodhak, Kalobhangro, Kaluganthi—; *Hasada*: Pirikesari—; *Hindi*: Babri, Bengraya, Bhangra, Mochkand, Mochrand—; *Ilocano*: Tintatinta—; *Indo-China*: Co muc, Lien tao, Nho noi, Phong trang, Phong truong—; *Iraq*: 'Arundis—; *Kolami*: Hatukesari—; *Lagos*: Abikolo—; *Mundari*: Benggaraj, Benggraj, Bhenggaraj, Bhenggraj, Huringsarsiranu, Sarsingranu—; *Rajputana*: Jal bangra—; *Sadani*: Bengaria—; *Sanskrit*: Ajagara, Angaraka, Bhekaraja, Bhringa, Bhringaraja, Bhringasodara, Bhringavha, Ekaraja, Karanjaka, Kesharaja, Kesharanjana, Keshya, Kuntalavardhana, Mahabhringa, Mahanila, Markara, Markava, Nagamara, Nilabhringaraja, Nilapushpa, Pankajata, Pararu, Patanga, Pitripriya, Rangaka, Shyamala, Sunilaka—; *Santal*: Lalkesari—; *Sind*: Tik—; *Sinhalese*: Kikirindi, Kikirindiha—; *Tagalog*: Higuismanoc—; *Tamil*: Kaikeshi, Kaivishiilai, Karishalanganni—; *Telugu*: Galagara, Guntagalijeru, Guntakalaagara—; *Urdu*: Bhangra—; *Uriya*: Kesarda—; *Uruguay*: Yerba de la oveja—; *Yoruba*: Abikolo, Arojoku—.

(*To be continued*).

THE MEDICINAL AND POISONOUS COMPOSITES OF INDIA.

BY

J. F. CAIUS, S.J., F.L.S.

COMPOSITAE

PART II

(*Continued from page 645 of this volume*).

ELEPHANTOPUS.

The genus consists of 20 tropical species. mostly American.

Elephantopus scaber Linn. is found throughout India. It is distributed over tropical Asia, Australia, and America.

The plant is an Ayurvedic and Yunani drug.

In Travancore a decoction of the roots and leaves is given in dysuria; the bruised leaves with rice are given internally for swellings or pains in the stomach.

In Chota Nagpur a preparation from the root is given for fever.

The Mundas use the roots as a remedy in diseases believed to be caused by a spell.

The plant is much used as a diuretic in Indo-China, as a diuretic and febrifuge in Madagascar, and as a vulnerary in Jamaica.

In the French Islands of the West Indies the herb is considered tonic, diaphoretic, and emmenagogue. The warm infusion is given in dyspepsia, intermittent fevers, loss of appetite, and menstrual derangements due to colds.

The herb is used in Brazil as an emollient and discutient in the form of a decoction or a poultice. A decoction of the root is prescribed in asthenic fevers.

Arabic: Kibnaite—; *Bengal*: Dadishaka, Gajialata, Gojialata, Samdulam, Shamdulum—; *Betsileo*: Tambakombako—; *Bombay*: Hastipata, Mhaka, Pathri—; *Brazil*: Fumo bravo—; *Burma*: Katoopin, Matoopin—; *Canarese*: Hakkarike—; *Ceylon*: Anichchovadi—; *Chanda*: Kaliagangawan—; *English*: Blue Elephant's-foot, Prickly-leaved Elephant's Foot—; *French*: Lastron marron, Pied d' éléphant—; *Gujerati*: Bhopathari, Kharsatbhopatri—; *Hindi*: Gobhi, Samdulam—; *Indo-China*: Chi thien—; *Jashpur*: Murgichundi, Talmuli—; *La Reunion*: Tabac marron—; *Malaya*: Tee tam tou, Ti tan, Tutup bumi—; *Malayalam*: Anashovadi—; *Marathi*: Gojibha, Pathari—; *Matheran*: Jirao, Junglijira, Lapoti—; *Mauritius*: Herbe la jouissance, Lastron marron, Libéralis—; *Mundari*: Maracuta—; *Persian*: Kalam-e-roomi—; *Porto Rico*: Lengua de vaca—; *Rio de Janeiro*: Erva do collegio—; *Sanskrit*: Adhomukha, Anadujivha, Darvi, Darvika, Darvipatrika, Gobhi, Gojivha, Gojivhika, Hdhapushpi, Kharapatri, Kurasa, Satamulika, Vatona—; *Santali*: Manjurjuti—; *Sinhalese*: Ataddeya, Etadi—; *Tagalog*: Tabatabacahan—; *Tamil*: Anashovadi—; *Telugu*: Eddumalikechettu, Enugabira, Hastikasaka—; *Tupi*: Suçuaya—; *Urdu*: Gobhi—.

EMILIA.

The genus consists of 5 palaeotropical species.

E. sonchifolia DC. is used medicinally in India, Indo-China, Gold Coast, La Reunion—; *E. flammea* Cass. in Lagos—; *E. sagittata* DC. in Lagos, Nigeria and Sierra Leone—; *E. adscendens* DC. and *E. graminea* DC. are used in Madagascar.

Involucral bracts much shorter than the scarlet flowers ... 1. *E. flammea.*
Involucral bracts about equalling the purple flowers ... 2. *E. sonchifolia.*

1. **Emilia flammea** Cass. is found in the Konkan. It occurs in the Philippine Islands, Lagos, Angola, Zanzibar.

It is used medicinally for children in Lagos.

Lagos: Aro agbe, Odundun odo—.

2. **Emilia sonchifolia** DC. is a common tropical weed found everywhere in India from the Punjab to Tenasserim and Ceylon, ascending to 4,000 ft. in the hills. It is common in waste ground and road sides over the whole Malay Peninsula, and over most of the tropical and subtropical regions.

In Malabar a decoction of the plant is said to be a febrifuge. Mixed with sugar it is given in bowel complaints.

In Travancore, pure juice of the leaves is poured drop by drop into the eyes in night-blindness. The natives consider the juice as cooling as rose-water and prescribe it in eye inflammations.

In Gold Coast the leaves mixed with Guinea grains and lime juice are a remedy for sore-throat.

In Indo-China a decoction of the leaves is prescribed as an antipyretic.

In La Reunion the plant is used as an astringent, antiasthmatic, and vulnerary.

Arabic: Oud el karah, Saela—; *Ashanti*: Guakuro—; *Bengal*: Sadimodi—; *Bombay*: Sadhimandhi—; *Canton*: Hong Pei Ts'ao—; *Cawnpore*: Hiranakhuri—; *Central India*: Hiranakhuri—; *Ceylon*: Kadupara—; *Indo-China*: Chua le, Chu le, Hen so, Hoa mat loi, La mat gioi, Mat gioi, Mat trang, Reang tuk, Tam toc, Tiet ga—; *La Reunion*: Goutte de sang, Lastron bâtard, Petit lastron—; *Malaya*: Katumbit jantan, Satumbak merah—; *Malayalam*: Mulshevi—; *Mende*: Gipoi, Kipoi—; *Mundari*: Kulaeara, Kulaebasu, Pirimaniare—; *Pampangan*: Tagulinao—; *Tagalog*: Tagulinao—; *Visayan*: Libun—.

ENHYDRA.

The genus consists of 9 tropical and subtropical species.

Enhydra fluctuans Lour. occurs in Eastern Bengal, Assam, Sylhet, and the Malay Peninsula. It extends to Sumatra, the Malay Islands, and China.

The leaves of this aquatic plant are regarded as laxative and useful in diseases of the skin and nervous system. The fresh juice of the leaves, in doses of about a tola, is prescribed by some kavirajas in Calcutta, as an adjunct to tonic metallic medicines, given in neuralgia and other nervous diseases.

The leaves are antibilious. The expressed juice of the leaves is used as demulcent in cases of gonorrhoea; it is taken mixed with milk, either of cow or goat. The leaves are pounded and made into a paste which is applied cold over the head as a cooling agent.

The Chinese prescribe the plant for torpidity of the liver.

Bengal: Hingcha, Hingeha—; *Hindi*: Harhuch, Haruch—; *Malaya*: Chinkro—; *Sanskrit*: Achari, Bramhi, Chakrangi, Helanchi, Hilamochi, Hilamochika, Jalabramhi, Mambi, Matsyakshi, Matsyangi, Mochi, Rochi, Shankhadhara, Trinittaparni, Vishaghni—.

ERIGERON.

The genus includes 180 cosmopolitan species, mostly North American.

The following species are used medicinally: in Indo-China—*E. aegyptiacus* Linn.—; in North America—*E. annuus* Pers., *E. canadensis* Linn., *E. philadelphicus* Linn.—; in La Reunion—*E. canadensis* Linn.—.

Heads $\frac{1}{4}$-$\frac{1}{2}$ in. diameter. Ligules bluish violet longer than the involucral bracts ... ... 1. *E. asteroides.*
Heads $\frac{1}{6}$-$\frac{1}{4}$ in. diameter. Ligules pale rose or purplish, scarcely exceeding the pappus ... ... 2. *E. canadensis.*

1. **Erigeron asteroides** Roxb. is found throughout India and Ceylon, up to 4,000 ft. in Sikkim.

The plant is a stimulant and diuretic, and is given in febrile conditions.

Bombay: Maredi, Sonsali—.

2. **Erigeron canadensis** Linn. is believed to be indigenous in North America. It occurs in all warm countries. In India it is found in the Upper Gangetic Plain, the Punjab Plain and the Western Himalaya where it ascends up to 3,000 ft.

According to the Ayurvedists the plant is styptic, diuretic, and astringent to the bowels.

The herb is used in America as a stimulant, astringent, and haemostatic. It has been found a useful remedy in the treatment of diarrhoea, dysentery, and dropsy. The oil is used in venereal diseases; it has been reported useful as a urinary and pulmonary haemostatic.

English: Cobbler's Pegs, Canada Fleabane, Squaw-weed—; *French*: Vergerette de Canada, Vergerolle—; *Iraq*: Dhail-as-saba'—; *Italian*: Impia—; *La Reunion*: Fausse camomille—; *Pacific Coast*: Fleabane, Horse-weed—; *Sanskrit*: Jarayupriya, Makshikavisha, Palita—.

ETHULIA.

The genus consists of 6 species spread over Asia, Africa, and South America.

Ethulia conyzoides Linn. is found in Eastern Bengal, Assam, and Sylhet. It extends to the Sunda Archipelago, Madagascar, and tropical Africa.

In Liberia the juice is squeezed into the eyes for headache; the root along with red pepper is given by enema for constipation; the leaves are eaten by pregnant women to prevent abortion; for pelvic pain a hot decoction of the leaves with those of *Baphia* is poured on the ground and covered with a mat on which the patient sits.

The Zulus use the plant as a remedy for intestinal parasites, for abdominal disorders, and for colic.

An infusion of the plant is used in Madagascar for dysentery, haemoptysis, and bruises. The pounded leaves are applied over sprains and fractures; the boiled leaves are used for wounds and traumatic haemorrhage.

In combination with other plants this herb is used to poultice phagedenic chancres and to wash purulent ulcers. The Betsileo use it in the treatment of scabies.

Betsileo: Tsimandra—; *Hova*: Homandra, Tangentsahona—; *Madagascar*: Kelihomandra—; *Mano*: Suo longo lah—; *Zulu*: umSokosoko—.

EUPATORIUM

The genus includes 450 species, mostly American; a few in Europe, Asia, and tropical Africa.

The following species are used medicinally in Europe—*E. cannabinum* Linn., *E. perfoliatum* Linn.—; in China—*E. chinense* Linn., *E. japonicum* Thunb.—; in Indo-China—*E. cannabinum* Linn., *E. stoechadosmum* Hance, *E. triplinerve* Vahl—; in Malaya —*E. japonicum* Thunb.—; in the Philippine Islands, La Reunion, Mauritius, the West Indies—*E. triplinerve* Vahl—; in Jamaica; *E. nervosum* Sw., *E. villosum* Sw.—; in North America—*E. ageratoides* Linn., *E. aromaticum* Linn., *E. incarnatum* Walt., *E. perfoliatum* Linn., *E. purpureum* Linn., *E. teucrifolium* Willd.—; in Mexico—*E. collinum* DC.—; in Central America and Guiana—*E. parviflorum* Sw.—; in Brazil, South America, Madagascar—*E. triplinerve* Vahl—; in Ecuador—*E. glutinosum* Lam.—; in Colombia—*E. aromatisans* DC., *E. scabrum* Linn.—.

Pubescent. Flowers nearly white ... ... 1. *E. cannabinum*.
Glabrous. Flowers slaty blue .. ... 2. *E. triplinerve*.

1. **Eupatorium cannabinum** Linn. is found in the temperate Himalaya at 3,000 ft. ascending up to 11,000 ft., in the Khasia Hills between 3,000 and 6,000 ft., and in Burma. It is spread over temperate Asia and Europe.

An infusion of a handful of the fresh herb acts as a strong purgative and emetic, and is excellent for fomenting ulcers and putrid sores.

The herb is a good wound herb, whether bruised or made into an ointment with lard. In Holland the peasants use it for jaundice with swollen feet, and also give it as an alterative or purifier of the blood in the spring and against scurvy.

Herbalists recognize its cathartic, diuretic, and antiscorbutic properties, and consider it a good remedy for purifying the blood, either used by itself, or in combination with other herbs. A homoeopathic tincture is prepared, given in frequent small well-diluted doses with water, for influenza or a similar feverish chill.

A tea made with boiling water poured on the dry leaves will give prompt relief if taken hot at the onset of a bilious catarrh or of influenza. The infusion is a valuable tonic.

The freshly gathered root, boiled in ale, purges briskly, but without producing any bad effects; it is to this day a popular remedy for dropsy.

The herb is used as an emmenagogue and a purgative in Indo-China.

Bombay: Bundar—; *Catalan*: Canabassa, Lladreca—; *Dutch*: Boelskenkruid, Koninginnekruid—; *English*: Hemp Agrimony, Holy Rope, St. John's Herb—; *French*: Chanvrin, Eupatoire, Eupatoire d'Avicenne, Eupatoire à feuilles de chanvre, Eupatoire chanvrine, Eupatoire commun, Eupatoire de Mésué, Herbe julienne, Herbe de sainte Cunégonde, Origan aquatique, Origan des marais, Pantagruelion—; *German*: Kunigundenkraut, Wasserdost, Wasserhanf—; *Indo-China*: Bach son, Thach lam—; *Italian*: Canapa acquatica, Eupatorio, Eupatorio di Avicenna—; *Malta*: Hemp Weed, Water Agrimony, Canapa acquatica—; *Roumanian*: Canipioala—; *Spanish*: Cannabina de agua, Cañamazo, Cañamo silvestre acuatico, Eupatorio, Eupatorio de los Arabes, Eupatorio de Avicenna, Yerba del angel—; *Swedish*: Floks—.

2. **Eupatorium triplinerve** Vahl is a native of America, planted in Indian gardens.

Ayapana may be compared with chamomile in its effects; it is stimulant and tonic in small doses, and laxative when taken in quantity; the hot infusion is emetic and diaphoretic, and may be given with advantage in the cold stage of ague and in the state of depression which precedes acute inflammatory affections.

An infusion of the whole plant is used as a stimulant, diaphoretic, and tonic in the Philippine Islands, Mauritius, La Reunion, and Madagascar.

The decoction of the plant and the juice of the leaves are considered detergent and applied to foul ulcers in the Philippines.

An infusion of the leaves is given in Indo-China and Guiana as an alexipharmic and a sudorific.

The herb is a snake remedy in Guiana, Brazil, the Philippine Islands, and India. The whole plant is given in the form of a decoction, and the juice of the leaves is given internally and applied externally to the part bitten. But Mhaskar and Caius have experimentally shown that the plant is useless in the symptomatic treatment of snake bite, and that the leaves are useless whether given internally or applied externally to the part bitten.

Annam: Cay bach son, La da yot, Trach lam, Trach lam diep—; *Bengal*: Ayapana—; *Betsimisaraka*: Ayapana—; *Brazil*: Ayapana—; *Ceylon*: Ayapeni—; *Cochin China*: La ba giot—; *English*: Ayapana Tea—; *French*: Ayapana vrai—; *Gujerati*: Alleppa—; *Indo-China*: Ba dot, Ca dot—; *Marathi*: Ayapana—; *Mauritius*: Ayapana—; *Sinhalese*: Ayiyapana—; *Spanish*: Ayapana—; *Tagalog*: Apana, Ayapana—; *Telugu*: Ayapani—.

GERBERA.

The genus numbers about 40 species, temperate or mountain herbs inhabiting Asia and Africa.

Gerbera piloselloides Cass. is found in the Central and Eastern Himalaya, ascending to 7,000 ft. in Sikkim and Nepal. It occurs in the Khasia Hills at 5,000 ft., and in the Karen Hills of Lower Burma. It extends to China and South Africa.

The Zulus apply an infusion of the root made with human urine, to the ear for earache.

The Sutos use a decoction of the root as a tonic and as ear-drops in earache, and a milk decoction or infusion for chest complaints. In addition they use the plant to fumigate the hut of a person suffering from a cold in the head.

Suto: Mothuntsetso, Tsebapelo, Tsebe-ya-pela—; *Zulu*: in Dlbe-yempithi—.

GLOSSOCARDIA.

The genus consists of 1 Indo-Malayan species.

Glossocardia Bosvallia DC. (=*G. linearifolia* Cass.) is found in the Upper Gangetic Plains, Central India, Konkan, the Deccan, the South Mahratta Country,. North Kanara, and all the plains districts of the Madras Presidency.

The plant is much used in complaints of women.

Bombay: Phatursuwa, Pithapra—; *Hindi*: Seri—; *Marathi*: Patharasuva—; *Naguri*: Pusikata—; *Poona*: Pittapapada—; *Sanskrit*: Pithari—; *Telugu*: Parapalanamu—.

GLOSSOGYNE.

The genus consists of 5 Indo-Malayan and Chinese species.

Glossogyne pinnatifida DC. occurs in the Punjab Plain, the Upper Gangetic Plain, Kumaon, Chota Nagpur, the South Mahratta Country, Western Bengal, the Northern Circars, and the Deccan of the Madras Presidency.

A preparation from the root is employed by the Santals as an application to snake-bite, and scorpion sting.

Mhaskar and Caius have shown that the root is useless in the antidotal and symptomatic treatments of snake-bite and scorpion-sting.

Gujerati: Pardeshibhangro—; *Santali*: Barangom bir barangom—.

GNAPHALIUM.

The genus includes 260 cosmopolitan species.

The following species are used medicinally in Europe—*G. luteo-album* Linn., *G. norvegicum* Gunner, *G. supinum* Linn., *G. sylvaticum* Linn., *G. uliginosum* Linn.—; in China—*G. luteo-album* Linn.—; in North America—*G. decurrens* Ives, *G. obtusifolium* Linn., *G. ramosissimum* Nutt., *G. uliginosum* Linn.—; in South Africa—*G. luteo-album* Linn., *G. undulatum* Linn.

Gnaphalium luteo-album Linn. is found throughout India. It occurs in most of the hot and warm temperate countries.

The leaves are used in the Punjab as an astringent and vulnerary.

The Sutos burn the plant in the sickroom to drive away sickness.

Afrikaans: Roerkruid—; *Arabic*: Ra'ra', Sabun'afrit—; *Burma*: Byaingchepiu—; *Egypt*: Luban, Ra'ra', Sabun'afrit—; *Hova*: Ahitrakohovavy, Tranomandrita—; *Igorrote*: Onanat—; *La Reunion*: Immortelle marronne—; *Punjab*: Balraksha—; *South Africa*: Cudweed—; *Suto*: Manku, Musuwane—.

GRANGEA.

The genus consists of 2 palaeotropical species.

Grangea maderaspatana Poir. is a common weed, growing flat on the ground, found throughout India, Baluchistan and Ceylon. It is distributed over tropical and subtropical Asia and Africa.

The herb is a Yunani drug.

The leaves are regarded as a valuable stomachic and to possess deobstruent and antispasmodic properties, and are prescribed in infusion and electuary in cases of obstructed menses and hysteria. They are also sometimes used in preparing antiseptic and anodyne fomentations.

The juice of the leaves is employed as an instillation for earache.

In Indo-China the leaves are considered as an excellent stomachic; they are also used as an antiseptic in fomentation. A decoction of the roasted leaves is given in cough and is used as an emmenagogue.

In Madagascar the leaves are used as a stomachic and antispasmodic.

In Brazil the plant is a popular substitute for chamomile.

Arabic: Afsantin, Khatareka—; *Bengal*: Namuti—; *Betsileo*: Angea—; *Brazil*: Marcella—; *Canarese*: Dovana—; *Gujerati*: Jhinkimundi, Nahanigorakhamundi—; *Hindi*: Mukhatari, Mustaru—: *Hova*: Motomaso, Voamontomosa—; *Indo-China*: Cai dong, Cuc rai, La chan vit—; *Loanda*: Macella—; *Malayalam*: Nelampala—; *Marathi*: Mashipatri—; *Persian*: Baranjasifkowhi, Murava—; *Santali*: Bhediachim—; *Sinhalese*: Velkolondu—; *Sudan*: Phunat—; *Tamil*: Mashipatri—; *Telugu*: Save—; *Urdu*: Afsantin—.

GUIZOTIA.

The genus consists of 8 African species.

Guizotia abyssinica Cass. is a native of tropical Africa, cultivated in various parts of India.

The oil from the seeds is employed as a substitute for ghee and sesamum oil.

Chemical data are available.

Bengal: Gunja, Ramtil, Sirguja—; *Bombay*: Kerani, Ramatila—; *Canarese*: Huchchellu, Ramtil—; *English*: Niger Oil Plant, Niger Seed Plant—; *Gujerati*: Kalatel—; *Hasada*: Jhatinggi—; *Hindi*: Kalatil, Ramtila, Surguja—; *Kolami*: Kallatil, Sarguja, Til—; *Marathi*: Kalitil—; *Naguri*: Maga, Magha, Surgunja—; *Portuguese*: Verbesina da India—; *Santali*: Surguja—; *Telugu*: Valesulu, Vulisi—.

GYNURA.

The genus includes 29 species, inhabiting the warm regions of Asia, Africa, and Australia.

G. auriculata Cass. and *G. pinnatifida* DC. are used medicinally in China; *G. cernua* Benth. in Madagascar; *G. amplexicaulis* Oliv. and Hiern, *G. cernua* Benth. *G. crepidioides* Benth., *G. sarmentosa* DC. in tropical West Africa.

Gynura Pseudo-China DC. is found in South India and Ceylon, in the Sikkim Himalaya at 2,000-4,000 ft, in Pegu and Martaban, extending to Java and Indo-China.

The plant is held in great repute in Indo-China as an emollient and resolvent; it is used whole as a poultice in erysipelas and for tumours on the breasts.

The powdered root, mixed with tea, is given as a drink to parturient women and also to regulate the menstrual flow.

The juice of the leaves is used as a gargle for inflammation of the throat.

Indo-China: Bau dat dai, Tam that gia—; *Mauritius*: Jacobée, Muel schévy, Shoudy mourdi—.

HELIANTHUS.

The genus includes 60 American species. *H. angustifolius* Linn., *H. annuus* Linn., *H. petiolaris* Nutt., and *H. tuberosus* Linn. are used medicinally in North America.

Helianthus annuus Linn., a native of America, is cultivated in India and in many other tropical and subtropical countries.

The growing plant is said to be a cure for malaria, and is used in malarial fever by the inhabitants of the Caucasus. The leaves are spread upon a bed covered with a cloth, moistened with warm milk, and then the patient is wrapped up in it. Perspiration is produced, and this process is repeated every day until the fever has ceased.

The boiled flower heads were used by American Indians for pulmonary affections.

A tincture of the flowers and leaves has been recommended in combination with balsamics in the treatment of bronchiectasis.

The seeds have diuretic and expectorant properties, and have been employed with success in the treatment of bronchial, laryngeal, and pulmonary affections, coughs and colds. If browned in the oven and then made into an infusion they are admirable for the relief of whooping cough.

In China the seeds are administered in dysentery.

A tincture prepared from the seed with rectified spirit of wine is useful for intermittent fevers and ague, instead of quinine. It has been employed thus in Turkey and Persia, where quinine and arsenic have failed, being free from the inconveniences which often arise from giving large quantities of the other drugs.

The oil from the seeds is officinal in Russia; its properties are the same as those of the seeds.

Arabic: Arzivana—; *Bengal*: Bhramoka, Shuriamukti, Suriamuki—; *Bombay*: Surajmaki, Suryakanta—; *Catalan*: Girasol, Mirasol—; *Chinese*: Toung K'oui Tzeu—; *Deccan*: Suriyakajhar—; *Dutch*; Zonnebolem—; *Egypt*: Habb-eshshems—; *English*: Lady Eleven o'Clock, Marigold of Peru, Sunflower—; *French*: Anthilion, Couronne de soleil, Girasol, Grand soleil, Grand soleil des jardins, Hélianthe, Soleil cultivé, Soleil rigide, Tournesol—; *German*: Sonnenblume—; *Greek*: Hilios, Hiliotropion—; *Gujerati*: Surajamukhi—; *Hindi*: Hurhuja, Surajmukhi—; *Iraq*: 'Abad-ash-shams, 'Ain-ash-shams, Shams-wa-gamar, Ward-ash-shams—; *Italian*: Clytia, Girasole, Harpalio—; *Konkani*: Jirasol, Suria-kamal—; *Kurdish*: Gulah barruzhah—; *Marathi*: Brahmoka, Surajmaka, Surya-phula—; *Pacific Coast*: Comb-flower, Common Sunflower, Garden Sunflower, Sunflower—; *Persian*: Aftabi, Guleaftabeparasta, Guliaftab—; *Portuguese*: Girasol, Heliotropio—; *Punjab*: Surajmukhi—; *Roumanian*: Floarea soarelui, Ruja soarelui—; *Russian*: Padsolnechnik—; *Sanskrit*: Adityabhakta, Surya-mukhi, Suryavarta, Suvarchala—; *Spanish*: Girasol, Mirasol—; *Telugu*: Adityabhaktichettu, Poddatringuddachettu, Poddutiruguduchettu—; *Turkish*: Qunah bagan—; *Urdu*: Surajamakkhi—.

HIERACIUM.

The genus includes about 400 species, mostly natives of Europe, but extending all over the Mediterranean region, Northern Asia, Japan, North America, the region of the Andes, and South Africa.

H. amplexicaule Linn., *H. murorum* Linn., *H. pilosella* Linn., *H. virosum* Pall. are used medicinally in Europe.

Hieracium virosum Pall. is found in Kashmir at a height of between 7,000 and 8,000 ft. It occurs in Northern Asia from Dahuria westwards to Europe.

The plant is considered aperient and vulnerary in Spain. It is said to be poisonous.

HYPOCHOERIS.

The genus includes about 45 species, natives of Europe, Northern Asia, the Mediterranean region, and South America. Two of them are widely diffused through cultivation.

Hypochoeris glabra Linn., a native of Europe, is found in the Nilghiris at an altitude of 7,000-8,000 ft.

The root is credited with diuretic, tonic, and slightly aperient properties.

The powdered leaves are a good astringent in haemorrhage. The green herb is vulnerary.

INULA.

The genus includes about 100 species, natives of the Old World.

The following species are used medicinally in Europe—*I. Conyza* DC., *I. crithmoides* Linn., *I. germanica* Linn., *I. graveolens* Desf., *I. Helenium* Linn., *I. viscosa* Ait.—; in China and Indo-China—*I. britannica* Linn.—; in Malaya—*I. japonica* Thunb.—; in North America—*I. Helenium* Linn.—.

The root of *I. Helenium* is officinal in Holland and Portugal.

A. Shrubby. Heads solitary ... ... 1. *I. grantioides.*
B. Herbs. Heads turbinate, terminating numerous slender branches .... 2. *I. graveolens.*
C. Tall, stout herbs. Heads large, solitary or racemose
  1. Heads racemed, $1\frac{1}{2}$-2 in. diameter ... ... 3. *I. racemosa.*
  2. Head solitary, 3-4 in. broad ... ... 4. *I. Royleana.*

1. **Inula grantioides** Boiss. is found in Cutch, Sind, Baluchistan, and Waziristan. It occurs also in South-eastern Arabia.

In Las Bela it is given steeped in water to patients suffering from asthma.

Baluchi: Kolmir, Kulmir, Pir—; *Brahui*: Kolbur, Kulmir—; *Las Bela*: Kolmur, Naro—.

2. **Inula graveolens** Desf. is found in North-Western India, extending westwards to Spain.

In the Punjab the plant is recommended as a diuretic, useful in calculous diseases.

At Basrah the plant is used as a specific for eye-diseases.

Basrah: Shuwasar—; *Cape Peninsula*: Khaki Bush, Khaki Weed—; *Malta*: Fleabane, Fulichi, Zghazigha—.

3. **Inula racemosa** Hook. fil. is found in the temperate and alpine Western Himalaya, ascending from 5,000 ft. to 14,000 ft.

The root and the seeds are used by Yunani practitioners.

The plant is commonly used as an expectorant, and as a resolvent in indurations.

In Kashmir the root is mixed with that of *Saussurea Lappa* as an adulterant.

Arabic: Rasan, Zanjabileshami—; *Kashmir*: Poshkar—; *Persian*: Gharsa, Pilgush, Rasan, Zanjabilishami—; *Urdu*: Rasan—.

4. **Inula Royleana** DC. is found in the temperate Western Himalaya between 7,000 and 11,000 ft.

The plant is considered poisonous.

In Kashmir the root is mixed with that of *Saussurea Lappa* as an adulterant.

Kashmir: Zahelniilkohee—.

JURINEA.

The genus includes 50 species, occurring in the Mediterranean region, Europe, and Asia.

Jurinea macrocephala Benth. is found in the Western Himalaya, from Kashmir to Kumaon at 11,000-14,000 ft.

The bruised root is applied to eruptions, and a decoction is given in colic. It is also considered cordial and administered in puerperal fever.

Kashmir: Dhup—; *Punjab*: Dhup, Dhupa, Gugal, Zhangar—.

LACTUCA.

The genus includes 100 species, inhabiting the Northern Hemisphere, chiefly the northern temperate regions of the Old World.

All lettuces possess a milky juice that flows freely from any wound. When dry, the juice hardens, turns brown, and is known as 'lactucarium'. It is a calmant, often used as a substitute for opium. It is officinal in Holland, Hungary, Portugal, and Spain.

The following species are used medicinally in Europe—*L. sagittata* Waldst. and Kit., *L. Scariola* Linn., *L. virosa* Linn.—; in China—*L. Scariola* Linn., *L. Thunbergii* Max.—; in Indo-China—*L. Fischeriana* DC., *L. indica* Linn.—; in North America—*L. canadensis* Linn.—; in Guiana—*L. quercina* Linn.—; in West Africa—*L. capensis* Thunb., *L. taraxacifolia* Schum. and Thonn.—.

A. Leaves runcinate, pinnatifid ... ... 2. *L. runcinata.*
B. Nearly entire (rarely sinuate-pinnatifid) smaller leaves ... 1. *L. remotiflora.*
C. Leaves suberect, sessile, runcinate-pinnatifid, spinulose-toothed; midrib prickly beneath ... ... 3. *L. Scariola.*

1. **Lactuca remotiflora** DC. is found in the Deccan, Banda, and Sind, extending to Arabia.

The whole plant is used as a remedy for chronic obstructions of the liver and bowels, and as a diuretic in calculous affections.

Goa: Taraxaco—; *Gujerati*: Pathardi—; *Marathi*: Undirachakan, Undirakani—.

2. **Lactuca runcinata** DC. spreads over the Punjab, the Upper Gangetic Plain, Sind, the Deccan, and most of the plains districts of the Madras Presidency.

The plant is diuretic, tonic and slightly aperient.

Goa: Taraxaco—; *Marathi*: Undirachakan—; *Pab*: Shamur—.

3. **Lactuca Scariola** Linn. occurs in Western Tibet at 9,000-12,000 ft., in the Western Himalaya from Murree to Kunawar at 6,000-11,000 ft. It extends to Siberia and westwards to the Atlantic.

The plant occupies an important place in the materia medica of the Yunanists.

The fresh plant is a mild sedative, anodyne, purgative, diuretic, diaphoretic, and antispasmodic. It has been found useful in the treatment of the coughs in phthisis, bronchitis, asthma, and pertussis.

A decoction of the seeds is commonly used as a demulcent. Boiled, and made into a confection, the seeds are given in cases of bronchitis, especially chronic ones.

Water distilled from lettuce is used in France as a mild sedative in doses of 2 to 4 oz., and the fresh leaves boiled in water are sometimes used as a cataplasm. Lettuce poultice is applied to burnings and scaldings, and to painful and irritable ulcers.

Arabic: Khas—; *Bengal*: Kahu, Salad—; *Catalan*: Ansiam llarch—; *Chinese*: Pai Chu, Wo Chu—; *Danish*: Laktuk—; *Dutch*: Latuw—; *English*: Garden Lettuce, Lettouce, Lettuce—; *French*: Laitue, Laitue cultivée, Laitue des jardins—; *German*: Gartenlattich, Gartensalat, Lattich, Salat, Sommerendivie—; *Greek*. Thridax—; *Hindi*: Kahu, Khas, Salad—; *Hungarian*: Kerti salata—; *Italian*: Guado, Lactuga, Lattuga—; *Languedoc*: Lacholebri, Lachougo, Lachuga—; *Malta*: Prickly Lettuce, Lettuce, Lattuga, Hass, Hassa salvagga—; *Pacific Coast*: Compass-plant, Horse Thistle, Prickly Lettuce, Wild Lettuce—; *Persian*: Kahu—; *Polish*: Loczyga, Loezya—; *Portuguese*: Alface brava, Alface menor, Leituga—; *Punjab*: Kahu—; *Roumanian*: Laptuc—; *Russian*: Laktuk—; *Sind*: Kahu—; *Sinhalese*: Salada—; *Spanish*: Lechuga, Lechuga larga, Lechuga romana—; *Swedish*: Laktuk—; *Tamil*: Salattu—; *Telugu*: Kavu—; *Urdu*: Kahu—; *Uruguay*: Lechuga silvestre—.

LAGGERA.

The genus consists of 10 species, tropical African and Indian.

Laggera alata Schultz-Bip. occurs in the tropical Himalaya between 1,000 and 5,000 ft. from Simla eastwards to Sikkim, and southwards in hilly districts ascending to 6,000 ft. in the Nilghiris and to 7,000 ft. in Ceylon. It is distributed over Java, the Philippine Islands, China, and tropical Africa.

The herb is used in Madagascar as a disinfectant.

Hova : Ahitandriana—; *Menabe* : Argodro—.

LAMPRACHAENIUM.

L. monocephalum Benth., the only species known, occurs at Mahableshwar, the Konkan, the South Mahratta Country, and Mysore.

The plant smells of chamomille, and is used as an aromatic bitter. The Ayurvedists recommend it for leucoderma and other skin diseases.

Bombay : Brahmadandi—; *Sanskrit* : Ajadandi, Brahmadandi—.

LAUNAEA.

The genus consists of 30 tropical and subtropical species.

A. Achenes winged ... ... 3. *L. glomerata.*
B. Achenes not winged
  I. Heads terminal on the paniculately branched flowering stems
    1. Radical leaves sinuate lobed or pinnatifid ... 1. *L. aspleniifolia.*
    2. Radical leaves rosulate, runcinate-pinnatifid ... 2. *L. chondrilloides.*
  II. Heads more or less racemose on the flowering stem or its branches ... ... 4. *L. nudicaulis.*
  III. Heads solitary or fascicled at the nodes of the flagelliform rooting flowering stems ... 5. *L. pinnatifida.*

1. **Launaea aspleniifolia** Hook. fil. is found throughout the plains of India, from the Punjab and the Upper Ganges to Bengal and Assam, the Sunderbunds, the Circars, Burma, Tenasserim, and the Andamans.

The root of this plant in combination with other drugs is used as a lactagogue by the Santals.

Bengal : Tikchana—; *Santali* : Birmalla—.

2. **Launaea chondrilloides** Hook. fil. is found in the Punjab, at Multan and in the low hills at Rawalpindi, in Sind, Baluchistan, Afghanistan, Persia, Arabia, and Egypt.

The plant is credited with the properties of a lactagogue.

Baluchi : Halaku, Mahari—.

3. **Launaea glomerata** Hook. fil. occurs in the Punjab, Baluchistan, Sind, Cutch, Kathiawar, and Mount Abu. It extends to Persia, Arabia, Egypt, and Algeria.

In Baluchistan the decoction, mixed with some wheat meal, is made into a poultice and applied to sore eyes.

Arabic: Halawat-el-ghazlan, Hufeifah, Huwwa, Huwwat-el-kilab—; *Baluchistan*: Aghut, Machi halako, Reki halaku—; *Egypt*: Abad, Helawan, Hudan Huveywa, Huwry, Huwweyt-el-kilab, Shegeret-el-libbeyne—; *Quetta*: Alko—.

4. **Launaea nudicaulis** Hook. fil. is found more or less throughout the plains of India, from Bengal and Bihar to the Punjab, ascending the Western Himalaya to 8,000 ft. in Kumaon. It is distributed over the Deccan, Sind, and Baluchistan, extending westwards to Afghanistan and the Atlantic.

In Las Bela the leaves are applied to the head of children suffering from fever.

Arabic: Hindibe, Hua—; *Baluchi*: Halaku, Kahoshonk, Khargoshkah, Machi malaku, Mahari halaku, Reki halaku—; *Brahui*: Halaco, Marubahi—; *Kharan*: Mahariatoko—; *Las Bela*: Bhattri—; *Ormara*: Shatrag—; *Pab*: Kharsagut—; *Punjab*: Batthal, Dudhlak, Spudukei, Tarizha—; *Rajputana*: Akria, Ban gobi, Jangli gobi—.

5. **Launaea pinnatifida** Cass. is found along the sandy coasts of India, from Bengal to Ceylon, Penang, Madras, and Malabar. It extends to the East African Coast, Egypt, and Mauritius.

In Bombay the plant is used as a lactagogue; in Goa it is given as a tonic, diuretic, and mild aperient.

The juice is applied in rheumatic affections, and used as a soporific for children.

Bombay: Pathri—; *Cutch*: Gewar—; *Goa*: Almirao—; *Gujerati*: Bhonpatri—; *Hindi*: Bankau—; *Indo-China*: Pissenlit maritime—; *Marathi*: Bhonpatri, Pathradi, Paththa—; *Porebunder*: Kormani—; *Sind*: Bankahu—.

MATRICARIA.

The genus includes about 50 species, mostly natives of Europe and the Mediterranean region as far as India, and of South Africa.

M. Chamomilla Linn. and *M. suaveolens* Pursh. are used medicinally in Europe; *M. glabrata* DC., *M. globifera* Fenzl., *M. multiflora* Fenzl., *M. nigellaefolia* DC. in South Africa; *M. Chamomilla* Linn. and *M. discoidea* DC. in North America.

Leaves bipinnatifid ... ...1. *M. Chamomilla.*
Leaves small in outline, narrowly oblong, cuneate, pinnatipartite ... ... 2. *M. lasiocarpa.*

1. **Matricaria Chamomilla** Linn. occurs in the Punjab and the Upper Gangetic Plain. It extends to Japan and Northern Asia, and westwards to the Atlantic.

The plant is a well known Yunani drug.

The flowers are considered stimulant, attenuant, and discutient. They are used in constitutional debility, hysteria, dyspepsia and intermittent fevers. The warm and strong infusion of the flowers is emetic, while a weak infusion acts as a tonic and febrifuge. In flatulence and colic, Chamomile oil is the most effectual of all remedies; in Gujerat it is used externally for rheumatism.

The flowers are considered an excellent remedy in children's ailments in many parts of Europe. They act as a nerve sedative and also as a tonic upon the gastro-intestinal canal. It is a useful remedy during dentition, in cases of earache, neuralgic pains, stomach disorders, and infantile convulsions.

The flowers are officinal in Austria, Denmark, Germany, Holland, Hungary, Italy, Japan, Norway, Portugal, Russia, Spain, Sweden, Switzerland, and Turkey.

In China the leaves are used as a depurant.

Arabic: Babunah, Shajrab-el-kaffoor, Tuffah-el-ard—; *Catalan*: Camamilla, Camamilla comú, Camamilla de Urgel—; *Chinese*: Houng Kiou—; *Danish*: Kamille—; *Egypt*: 'Ain-el-qutt, Babunngi, Babunnguy—; *English*: Bitter Chamomile, German Chamomile, Persian Chamomile—; *French*: Amarou, Astartife, Camomèle, Camomille d'Allemagne, Camomille commune, Camomille du Nord, Chamille, Malherbe, Mandiane—; *German*: Feldkamille, Gelbermaiblume, Haermelchen, Halmerl, Hameln, Harmel, Haugenblume, Helmerchen, Hermaennle, Hermchen, Hermeln, Herminchen, Hermligen, Hermuentzel, Hermunel, Kamilechte, Kamille, Karnille, Kummerblume, Laugenblume, Magdblume, Maidblume, Matronenkraut, Mutterkamille, Mutterkraut, Raimain, Ramerian, Romeien, Romer, Romerai, Romey, Saekfit, Stomeienblume, Thamillen, Zaeuwih—; *Greek*: Armen—; *Gujerati*: Babuna—; *Italian*: Camomilla comune, Capomilla—; *Malta*: Chamomille, Gowan, Camomilla, Camumilla—; *Naples*: Cambomilla—; *Norway*: Kamille—; *Parma*: Margariteine de pra—; *Pavia*: Crespola—; *Persian*: Babunah—; *Piedmont*: Amoreggiola, Camamia—; *Portuguese*: Camomilla dos Allemaes—; *Potenza*: Cambomilla—; *Punjab*: Babuna, Suteigul—; *Reggio*: Camamella—; *Romagna*: Camomella, Matricaria, Matricheria—; *Roumanian*: Musatel, Romanita—; *Russian*: Romashka aptetchnaya—; *Sardinia*: Carbomiglia, Capomilla, Crabumiglia—; *Sicily*: Calumidda—; *Spanish*: Manzanilla comun, Manzanilla de Urgel—; *Swedish*: Kamomill—; *Tuscany*: Camomila, Camumilla, Capomilla, Caprimilla, Capumilla—; *Urdu*: Babunah—; *Verona*: Erba Maria, Margarite, Samariza—.

2. **Matricaria lasiocarpa** Boiss. occurs in Baluchistan.

At Nushki a decoction of the flowers is used to cure fever.

Dak: Piunpulli—; *Nushki*: Painphuli—.

MICROGLOSSA.

The genus consists of 6 species, Asiatic and African.

M. Afzelii O. Hoffm. and *M. volubilis* DC. are used medicinally in tropical West Africa.

Microglossa volubilis DC. occurs in the Khasia Hills, Assam, the Naga Hills, Cachar, Burma from Pegu to Penang, the hill forests of Malaya and China. It is also found in some parts of tropical West Africa.

In Gold Coast the plant is given as an enema to cure fever in babies. In Liberia it is used as a vermifuge. In Cameroons Mountain area it is a remedy for severe cough.

A decoction of the root is used for colic in Malaya. The powdered root is used as a snuff to relieve colds in West Africa. The juice of the root is the part used as an eye medicine in West Tropical Africa, it is dropped through fine linen in the eye. The refuse from the crushed roots is mixed with shea butter, melted and smeared over the eyelids to reduce the slight swelling of the lids. This is applied at bedtime and is repeated weekly.

The juice of the warmed leaves is put in sore eyes, and applied also as a remedy for ringworm of the scalp. A tea-like infusion is taken for fever with headache, and also as a lotion or fumigation to cause perspiration, or by inhalation. A decoction is taken by women in labour. It is said to be a specific for yellow fever and for dropsy, and a reliable remedy in blackwater fever.

Bakwiri: Bendem-bende—; *Basa*: Gban-gbah—; *Fanti*: Esono-mbabae—; *Ibo*: Okbak kwu—; *Kibbi*: Asomerewa—; *Malay*: Poko ragin—; *Mende*: Gimbo yufi, Gimbui, Gimbu lufi, Kimbo yufi—; *Twi*: Asomerewa, Esono-mbabae—; *Ubuluku*: Okbak kwu—.

MIKANIA.

The genus includes about 60 species, all American, one cosmopolitan.

M. cordifolia Willd. and *M. officinalis* Mart. are used medicinally in Brazil.

Mikania scandens Willd. is found in Eastern Assam, in the Duphla Hills, Burma, and the Malay Peninsula from Tenasserim to Singapore. It is distributed over Siam, the Malay and the Philippine Islands.

The Thongas use the plant as a remedy for snake bite and scorpion sting.

Fiji: Wa batako—; *French Guinea*: Nore—; *Mexico*: Guaco—; *Thonga*: Kamele—; *Yoruba*: Iyawa—; *Zambesi*: Wedwedwi—.

NOTONIA.

The genus consists of 12 palaeotropical species.

Notonia grandiflora DC. occurs in the hilly districts of the Western Peninsula from the Konkan southwards: Konkan, Deccan and Western Ghats of the Bombay Presidency, South Deccan and Carnatic of the Madras Presidency. Also in the Batticaloa district of Ceylon.

The plant was brought forward in 1860, by Dr. A. Gibson, as a preventive of hydrophobia. The mode of administration is as follows:—About 4 ounces of the freshly-gathered stems, infused in a pint of cold water for a night, yield in the morning, when subjected to pressure, a quantity of viscid greenish juice, which, being mixed with water, is taken at a draught. In the evenings, a further quantity of the juice, made up into boluses with flour, is taken. These medicines are directed to be repeated for three successive days.

Dr. Waring says that from official documents placed at his disposal, it appears that the remedy has been tried in numerous cases; but as at the time of the infliction of the wound, caustic was applied locally in the majority of cases, it is difficult to determine how far the Notonia operated, if at all, as a prophylactic.

An extract of the herb was tried by the late Dr. Haines and myself on dogs, and afterwards at the European Hospital in Bombay (1864). In one drachm doses it had a feebly aperient action; no other effect was observed (Dymock).

Bombay: Gaidar—; *Marathi*: Wanderroti—; *Telugu*: Kundelucheviyaku—.

OTHONNOPSIS.

The genus consists of 8 species, inhabiting Africa and Western Asia.

Othonnopsis intermedia Boiss. is found in Waziristan at 6,000-7,000 ft. and in Baluchistan whence it spreads westwards to Persia.

In Baluchistan the plant is used as a poultice for the chest. The leaves are made into a paste with oil; they are then spread on a large chapati and laid on the head of a patient suffering from cold or fever.

In the Bolan Pass the plant is used as a cure for head aches; in Loralai for boils and pimples.

Baluchi: Manguli—; *Bolan Pass*: Mungli, Munglian—; *Brahui*: Manguli—; *Loralai*: Gango—; *Quetta*: Gangu, Manguli—; *Shahrig*: Gangu—; *Zhob*: Gangu—.

PEGOLETTIA.

The genus consists of 5 species, one in Java, the others in Africa.

Pegolettia senegalensis Cass., a native of tropical Africa, occurs in Baluchistan.

At Kharan, in Quetta, it is given with ghi to children to make them strong. It is boiled and the juice is applied to wounds of camels as a cure.

Kharan: Kuchitho—; *Kohlu*: Shin Bunt—

PICRIS.

The genus includes about 45 species, found in Europe, North Africa and Asia, one cosmopolitan.

Picris hieracioides Linn. occurs in the Temperate Himalaya from Murree to Bhutan at 6,000-8,000 ft., ascending to 10,000 ft. in Sikkim. It is also found in the Khasia Hills at 4,000-6,000 ft., and in the Nilghiris at 5,000-8,000 ft. It extends to North Asia and Japan, and to West Asia, Europe, and North Africa.

In Indo-China the bitter leaves are used by the Muong as a febrifuge.

Indo-China: Mao lien thai, Mat dat—.

PLUCHEA.

The genus includes 30 tropical and subtropical species.

The following species are used medicinally in Indo-China—*P. eupatorioides* Kurz., *P. indica* Less., *P. polygonata* Gagnep., *P. pteropoda* Hemsley—; in Brazil—*P. Quitoc* DC.—; in Somaliland—*P. ovalis* (Pers.) DC.—.

I. Leaves obovate or oblanceolate, subserrate, glabrous ... 1. *P. indica*.
II. Leaves sessile, very coriaceous, oblong or oblanceolate, hoary-pubescent ... ... 2. *P. lanceolata*.
III. Leaves pinnatifid ... ... 3. *P. pinnatifida*.

1. **Pluchea indica** Less. is found in salt marshes from the Sunderbuns to Malacca and Penang, the Malay Archipelago, and China.

The root and leaves are used in Patna as astringents and antipyretics.

In Indo-China the roots in decoction are prescribed in fevers as a diaphoretic, and an infusion of the leaves is given internally in lumbago.

Bengal: Kukronda, Munjurukha—; *Burma*: Kayu—; *Cambodia*: Pras anlok—; *Iloilo*: Tulolalaqui—; *Indo-China*: Cuc tan, Cuc tan o, La luc, Pros anloc, Tu bi—; *Malay*: Poko beluntas—; *Tagalog*: Lagundilati—.

2. **Pluchea lanceolata** C. B. Clarke is found in the Punjab and the Gangetic Plain as far as Cawnpore, in Sind, Afghanistan, westwards to North Africa.

The leaves are aperient.

Agra: Chotikalia—; *Aligarh*: Banserai—; *Cawnpore*: Sorahi—; *Gujerati*: Rasana, Rashana—; *Hindi*: Rasana, Rashana—; *Marathi*: Rashna, Rasna—; *Punjab*: Marinandai, Rasana, Reshae, Reshambuti, Reshami, Sarmei—; *Pushtu*: Marwande—; *Rajputana*: Chotakalia—; *Sanskrit*: Rasna—; *Sind*: Kurasanna—.

3. **Pluchea pinnatifida** Hook. fil. occurs in Baluchistan and Somaliland.

In Ormara the plant is rubbed on inflamed or wounded places.

The stems are employed by the Somalis for the cure of almost all diseases; but especially for pain in the bowels, lack or loss of the appetite for food, and debility after fevers.

Ormara: Majassar, Zika—.

PULICARIA.

The genus includes 30 species, European, Asiatic, and African.

P. dysenterica Gaertn. is used medicinally in Europe; *P. capensis* DC. and *P. scabra* Drme in South Africa; *P. crispa* Schultz-Bip. in West Tropical Africa.

I. Involucral bracts slender. Pappus double, outer of a crown of short scales, inner of slender hairs ... 2. *P. dysenterica.*

II. Involucral bracts very slender. Pappus-hairs bearded, thickened at the tips; outer scales connate with the hairs and deciduous with them ... ... 1. *P. crispa.*

III. Involucral bracts elliptic-lanceolate, acute. Pappus double; outer a short jagged crown, inner of flattened shining rigid scabrid hairs ... ... 3. *P. glaucescens*

1. **Pulicaria crispa** Schultz-Bip. occurs in the Punjab, the Upper Gangetic Plain, Behar, Bengal, Sind and Baluchistan; whence it spreads to Afghanistan, Persia, North and West Africa.

In the Salt Range the dried plant is applied as a vulnerary to bruises and sores of bullocks.

In Jaisalmer and Jodhpur the bruised leaves are applied to the head to relieve headache.

In Baluchistan the plant is boiled and the liquid given to children in whose stomach milk turns sour.

In Northern Nigeria it is applied locally for swellings and bruises, and rubbed on the temples for headache; a decoction is also taken for febrile conditions.

Arabic: Afrakh, Dehded, Desdas, Dithdath, Kanuf, Karwan, Qutqat, Ra'ra', Sabut, Tajar—; *Baluchi*: Bo-i-madaran, Boraku—; *Brahui*: Bo-i-madaran, Pihupulli, Rambo—; *Egypt*: Afrash, Dithdath, Gidiai, Khauf, Kutkat, Ra'ra', Sabat, Tagar—; *Fulani*: K'urar shanu—; *Hausa*: Bafuri, Balbela, Farar saura—; *Hindi*: Buhrna—; *Iraq*: Jifjaf, Juwaifah—; *Kano*: K'urar shanu—; *Katagum*: Bilbila—; *Pila-pila*: Paria zoupen—; *Punjab*: Bui, Gidi, Phatmer, Sutei—; *Rajputana*: Dhola lizru—; *Songhai*: Tarkunde dierfendu—.

2. **Pulicaria dysenterica** Gaertn. is found in Kashmir at 5,000-6,000 ft., extending westwards to Algeria and Britain.

The root is still a popular remedy for dysentery in many countries of Europe.

Arabic: Rara ejub, Rara jeub—; *English*: Cammock, Common Fleabane, Fleabane, Fleabane Millet, Herb Christopher, Middle Fleabane—; *French*: Aunée des prés, Aunée dysentérique, Conyse, Conyse des prés, Enule tonique, Herbe saint Roch, Inule conysière, Inule dysentérique, Pulicaise dysentérique—; *German*: Dumrians—; *Iraq*: Jifjaf, Juwaifah—; *Malta*: Common Fleabane, Menta selvatica, Mentastro—.

3. **Pulicaria glaucescens** Jaub. and Spach. is found in the Punjab, the Salt Range and Murree, Waziristan, Baluchistan, and Persia.

The plant is ravenously eaten by camels and is reputed to be strengthening to them; but it acts as a strong purgative

Baluchi: Kolbur, Kunchid, Kunchito—; *Brahui*: Kolbur, Ler Mazonk—*Makrani*: Kunchid—.

SAUSSUREA.

The genus includes 125 species, occurring in the northern temperate regions.

A. Heads sessile or shortly peduncled
- I. Pappus double, hairs all feathery. Stem 4-10 ft. ... 4. *S. Lappa.*
- II. Pappus brown, outer bristles scabrid or absent. Stem 1-18 in. ... ... 5. *S. obvallata.*

B. Heads long-peduncled
- I. Heads inclined, nodding ... ... 3. *S. hypoleuca.*
- II. Heads erect
  - 1. Inflorescence glabrous. Achenes smooth 10-ribbed. 1. *S. affinis.*
  - 2. Inflorescence cottony. Achenes 5-angled muricate. 2. *S. candicans.*

1. **Saussurea affinis** Spreng. is found in Bengal from Sylhet to the foot of the Nepal Hills. It also occurs in Burma, and it extends to China, Japan, and Eastern Australia.

In Assam the juice of the root is given with other medicines for diseases of women.

Assam: Gangamula—.

2. **Saussurea candicans** Clarke occurs from the Salt Range, Hazara and Kashmir to Bhutan at 2,000-7,000 ft. It is found in Baluchistan and Afghanistan.

The seeds are carminative. They are used for horses in the Punjab.

In tne Pab Hills in Jhalawan the seeds are considered as a cure for horse bite.

Pab Hills: Kareji—; *Punjab*: Batula, Kaliziri—.

3. **Saussurea ypoleuca** Spreng. occurs in the Temperate Himalaya from Kashmir to Sikkim at 7,000-13,000 ft. It extends to Indo-China.

The leaves are considered purgative and antisyphilitic in Indo-China.

Annam: Moc huong—; *Indo-China*: Mu hsiang—.

4. **Saussurea Lappa** Clarke is found in Kashmir between 8,000 and 12,000 ft.

The root is equally popular with Ayurveda and Yunani practitioners.

The root is prescribed as a stomachic and tonic, and in the advanced stage of typhous fever. In the Punjab, it is applied in powder to ulcers, for worms in wounds, and also in rheumatism; it is also considered depurative and aphrodisiac.

In Persia Costus root is prescribed externally and internally for various complaints, and is applied locally to ward off the effects of snake and animal bites.

For snake bite the root is given in Ceylon internally in powder form or in the form of a decoction.

Sanskrit writers on Materia Medica have prescribed the root and stem in snake bite and scorpion sting, but Mhaskar and Caius have shown experimentally that neither of these drugs is an antidote to either snake or scorpion venom.

As a medicine the root is considered carminative and stimulant in China, whereto it is exported in enormous quantities.

Chopra and De (*Ind. Journ. Med. Research*; October, 1929) are responsible for a very valuable investigation into the properties of the root. They thus summarise their work:—

(1) *Saussurea Lappa* or *kut* root grows on the moist slopes of the Northern Himalayas at a height of 8,000 to 13,000 ft. above the sea-level.

(2) The chief active constituents of the root are:—

(i) An essential oil 1.5 per cent, (ii) an alkaloid which has been named *saussurine* 0.05 per cent, (iii) Resin 6.0 per cent. Besides these there occur a fixed oil, traces of a bitter substance, small quantities of tannins, inulin, potassium nitrate, sugars, etc.

The leaves contain no essential oil but 0.025 per cent of the alkaloid *saussurine*.

(3) The essential oil has a strong aromatic, penetrating and fragrant odour. It has antiseptic and disinfectant properties; it relaxes the involuntary muscle tissue; it is a cardiac stimulant, a carminative, an expectorant and a diuretic.

(4) The alkaloid *saussurine* has a depressant action on the vagus centre in the medulla as well as on the involuntary muscle fibres of the bronchioles and gastro-intestinal tract. It produces a

slight but persistent rise of blood-pressure and increases the force of contraction and amplitude of the ventricles.

(5) The drug has a remarkable effect in controlling attacks of bronchial asthma, especially those of the vagotonic type. The paroxysms are cut short by the combined action of the essential oil and the alkaloid present in the root. The essential oil during its excretion in the lungs not only relaxes the bronchial muscle but has a marked expectorant action which relieves turgescence of the mucosa. It may be pointed out, however, that although the drug relieves asthmatic attacks, it does not produce permanent cure unless the casual factors are investigated and removed.

The drug is also useful in persistent hiccough.

(6) The drug has no anthelmintic action, nor has it any action against malaria, leprosy and rheumatism as has been claimed by writers of the indigenous systems in this country.

Arabic: Kust, Kustabeheri, Kustulhalu—; *Bengal*: Kur, Pachak—; *Bhote*: Rusta—; *Bombay*: Ouplate—; *Canton*: Muk heung—; *English*: Costus—; *Gujerati*: Kut, Upaleta—; *Hindi*: Kot, Kur, Kust, Kut, Pachak—; *Kashmir*: Postkhai—; *Malaya*: Mook heong, Mu hsiang—; *Malayalam*: Şepuddy—; *Persian*: Bughenagh, Khost, Koshnaha, Kust, Kutshirin, Kuttalkh—; *Punjab*: Kot, Kust, Kut, Kuth—; *Sanskrit*: Agada, Amaya, Apya, Bhasura, Dushta, Gada, Gadakhya, Gadavha, Gadavhaya, Haribhadraka, Jarana, Kadakhya, Kakala, Kashmirja, Kaubera, Kinjalka, Kushtha, Kuthika, Kutsita, Niruja, Padmaka, Pakala, Pakalam, Paribhadraka, Paribhavya, Pavana, Rama, Roga, Rogavhaya, Ruja, Ruk, Utpala, Vaniraja, Vyadhi, Vyapya—; *Sinhalese*: Godamahanel—; *Tamil*: Goshtam, Kostum, Putchuk—; *Teheran*: Butenak—; *Telugu*: Changala, Kustam—; *Urdu*: Kut—.

5. **Saussurea obvallata** Wall. is found in the Western Himalaya from Kashmir to Sikkim at 10,000-15,000 ft., spreading to the Altai Mountains.

The root is applied to cuts and bruises.

North-Western Himalaya: Bergandutongur—; *Punjab*: Birmkanwal, Kanwal—.

SENECIO.

The genus includes 1,450 cosmopolitan species, inhabiting chiefly the temperate regions and the mountains of the tropics.

The following species are used medicinally in Europe—*S Cineraria* DC., *S. Doria* Linn., *S. incanus* Linn., *S. Jacobaea* Linn., *S. viscosus* Linn., *S. vulgaris* Linn.—; in China—*S. campestris* DC. *S. nikoënsis* Miq., *S. palmatus* Pall., *S. scandens* Ham.—; in Indo-China—*S. Kaempferi* DC. *S. palmatus* Pall., *S. scandens* Ham.—; *in North America*—*S. aureus* Linn., *S. Cineraria* DC., *S. vulgaris* Linn.—; in Mexico—*S. cervariaefolius* Sch., *S. Grayanus* Hemsl.—; in Madagascar—*S. Ambavilla* Pers., *S. erechthitoides* Baker, *S. faugasioides* Baker, *S Hildebrantii* Baker, *S. microdontus* Baker—; in Mauritius and La Reunion—*S. Ambavilla* Pers.—; in South Africa—*S. albanensis* DC. var. *leiophyllus*, *S. asperulus* DC., *S. brachypodus* DC., *S. bupleuroides* DC., *S. concolor* DC., *S. deltoideus* Less., *S. Dregeanus* DC. var. *discoideus*, *S. erubescens* Ait., *S. Gerardi* Harv., *S. lasiorhizus* DC., *S. leontodontis* DC., *S. macrocephalus* DC. var. *hirsutissimus*, *S.*

orbicularis Sond., *S. othonnaeflorus* DC., *S. quinquelobus* DC., *S. rhyncholaenus* DC., *S. Serra* Sond., *S. serratuloides* DC., *S. speciosus* Willd., *S. subcoriaceus* Schltr., *S. tamoides* DC., *S. tanacetoides* Sond., *S. vulgaris* Linn.—; in Nigeria—*S. abyssinicus* Sch.-Bip., *S. baberka* Hutch., *S. Biafrae* Oliv. and Hiern—.

The genus SENECIO is of considerable toxicological importance in South Africa. Various species have from time to time been regarded as causing disease in man and animal.

Two alkaloids, senecifoline and senecifolidine, have been isolated from *S. barbellatus* DC.; one, senecionine, from *S. vulgaris* Linn.

A. Heads turbinate or obconic, all radiate, usually bracteolate. Involucral bracts 1-seriate ... ... 4. *S. tenuifolius.*

B. Perennial herbs with long-petioled, very broad orbicular, reniform or palmate radical leaves and cauline leaves with large sheaths
   Leaves palmately lobed. Heads corymbose 2. *S. Jacquemontianus.*

C. Heads few- or many-flowered. Flowers all pappose. Involucral bracts uniseriate
   1. Herbaceous, glabrous or sparsely pubescent. Leaves petioled, membranous, cordate or subreniform and 3-7-angled or palmately lobed ... ... 3. *S. quinquelobus.*
   2. Shrubby. Heads in axillary and terminal corymbs ... 1. *S. densiflorus.*

1. **Senecio densiflorus** Wall. is found in the Central and Eastern Himalaya from Nepal to Bhutan at 5,000-7,000 ft., in the Khasia Hills at 4,000-6,000 ft., and in Burma.

The leaves are applied to boils as an emollient and maturant.

Punjab: Chitawala—.

2. **Senecio Jacquemontianus** Benth. is apparently endemic in Kashmir between 8,000 and 13,000 ft.

The root is used in Kashmir as a nervine tonic.

Kashmir: Hatermul, Poshkar—.

3. **Senecio quinquelobus** Hook. fil. and Thunb. occurs in the Temperate Himalaya from Garhwal to Bhutan at 10,000-12,000 ft.

In Kanawar the seeds are given for colic.

Punjab: Morta—.

4. **Senecio tenuifolius** Burm. is found on the dry hills of the Western Ghats from the Konkan southwards. It also occurs in Java.

The leaves are used as an emollient and vulnerary.

Punjab: Mentog, Nimbar, Sanggye—.

SIEGESBECKIA.

The genus consists of 4 species, found in tropical and warm temperate regions.

Siegesbeckia orientalis Linn. occurs throughout India and Ceylon, ascending to 5,000 ft. in the Himalaya and other mountains. It is found in most of the tropical and subtropical regions of both hemispheres.

It has a high reputation as a valuable depurative, and also for its healing properties in gangrenous ulcers and sores. It is strongly recommended in diseases of the urethra.

Externally, a mixture of equal parts of the tincture and glycerine has been tried in Europe with good effect in ringworm and similar parasitic eruptions. Antiseptic properties have been ascribed to the fresh plant, applied to unhealthy ulcers.

In Indo-China the whole plant is prescribed as a cardiotonic.

In La Reunion the plant is used as a stimulant, diaphoretic, antiscorbutic, and sialagogue. It is considered anthelmintic in small doses.

In Tahiti the plant enters into the preparation of every cure for wounds, sprains, dislocations, contusions. It is also used to favour menstruation.

China: Hsi Hsien—; *French*: Guérit vite, Herbe de Flacq, Herbe de laque, Herbe divine, Herbe grasse, Herbe guérit vite—; *Garhwal*: Lichkura—; *Gujerati*: Pilibadkadi—; *Indo-China*: Cho de, Hy thiem thao, Kinh gioi dat, Luoi dong, Nu ao ria, Riem—; *La Reunion*: Colle-colle, Guérit-vite, Herbe divine, Herbe grasse, Herbe Saint Paul, Souveraine—; *Mauritius*: Herbe divine, Herbe de Flacq, Herbe grasse, Herbe guérit vite—; *Mundari*, Bindiramkata, Hatubirbiri, Ribribi—; *Sadani*: Latlati—.

SILYBUM.

The genus consists of 2 Mediterranean species.

Silybum Marianum (Linn.) Gaertn. occurs in the Punjab and the North-Western Himalaya at Peshawar and Hazara, and from Kashmir to Jamu at 6,000-8,000 ft. It extends westwards to North Africa and Europe.

The plant probably possesses antiscorbutic and lithotriptic properties. It is in popular use in Germany for curing jaundice and kindred biliary derangements. It also acts as a demulcent in catarrh and pleurisy. The decoction when applied externally is said to have proved beneficial in cases of cancer.

Disarmed of its prickles and boiled, it is worthy of esteem, and thought to be a great breeder of milk and proper diet for women who are nurses. The warm infusion scarcely ever fails to procure a proper supply of milk. It is considered one of the best medicines which can be used for the purpose.

The leaves are sudorific and aperient.

The seeds are pungent, antispasmodic, and demulcent; they have been found of special value in haemorrhages.

California: California Mustard—; *Catalan*: Card gallofer, Card mariá, Cardot, Cart gallofer, Cart de Maria—; *Dutch*: Mariendistel, Melkdistel, Vrouwendistel—; *Egypt*: Lekhlikh, Shok-el-ghazal—; *English*: Holy Thistle, Lady's Thistle, Marian Thistle, Milk Thistle—; *French*: Artichaut sauvage, Carthame maculé, Chardon argenté, Chardon lacté, Chardon Marie, Chardon Notre-Dame, Chardon taché, Epine blanche, Lait de Notre-Dame, Lait de sainte Marie—; *German*: Feedistel, Frauendistel, Froschdistel, Mariendistel, Silberdistel, Wehedistel, Wolldistel—; *Greek*: Silybon—; *Iraq*: Gul aghan—; *Italian*: Cardo del latte, Cardo di Maria, Cardo Mariano—; *Malta*: Blessed-Thistle, Milk Thistle, Cardo mariano, Xeuk baghli—; *Pacific Coast*: St. Mary's Blessed Thistle, St. Mary's Holy Thistle—; *Roumanian*: Armurariu—; *Russian*: Ostropestro—; *Spanish*: Borriquero, Cardo borriquero, Cardo de asno, Cardo de Maria, Cardo lechal, Cardo lechero—; *Swedish*: Sempertin—; *Uruguay*: Cardo asnal, Cardo lechero, Cardo maria, Cardo mariano—.

SOLIDAGO.

The genus includes 90 species, mostly North American, one of them cosmopolitan in the northern temperate regions.

The following are used medicinally in Europe—*S. Virgaurea* Linn.—; in Indo-China and Malaya—*S. Virgaurea* Linn.—; in North America—*S. canadensis* Linn., *S. californica* Nutt., *S. lepida* DC., *S. odora* Ait., *S. rigida* Linn., *S. sempervirens* Linn., *S. Virgaurea* Linn.—; in Brazil—*S. microglossa* DC.—.

Solidago Virgaurea Linn. is found in the temperate Himalaya from Kashmir eastwards at 5,000-9,000 ft., and in the Khasia Hills at 4,000-6,000 ft. It is distributed over Europe, temperate Asia and America, and Hongkong.

The plant is aromatic, stimulant, and carminative. It is astringent and diuretic, and efficacious for stone in the bladder. It allays sickness due to weak digestion.

In powder it is used for cicatrization of old ulcers. It has been recommended in many maladies, as it is a good diaphoretic in warm infusion, and is in this form also helpful in dysmenorrhoea and amenorrhoea.

As a spray and given internally, it is of great value in diphtheria.

Catalan: Vara d'or—; *Chinese*: Chi Nu, Lui Chi Mu—; *English*: Aaron's Rod, Banwort, Common Golden Rod, Golden Rod, Woundwort—; *French*: Herbe des Juifs, Verge d'or, Verge dorée—; *German*: Edelmindkraut, Goldenmundkraut, Goldenrautenkraut, Goldraute, Goldrute, Goldstengel, Gueldenwunderkraut, Gueldenwundkraut, Hainschwung, Heidnischwundkraut, Heilwundkraut, Klapperschlangenkraut, Nachtheil, Sankt Peterskraut, Sankt Petersstab, Stockschwungkraut, Stopfzu, Stoppkehrt, Stopploch, Wisselnkraut, Wunderkraut, Wundkraut—; *Indo-China*: Hoang Kam phuong, Nhat chi hoang hoa—; *Italian*: Erba pagana, Verga d'oro—; *Japanese*: Akino-kirinsô—; *Malaya*: Chec noo—; *Portuguese*: Vara d'oiro—; *Roumanian*: Splinuta de aur, Varga d'aur—; *Russian*: Solotoschnik—; *Spanish*: Vara de oro—; *Swedish*: Gulbris—.

SONCHUS.

The genus includes 45 species, natives of northern temperate regions and Central Asia, a few spread over the world with cultivation.

In Spain *S. arvensis* Linn., *S. asper* Hill., *S. crassifolius* Pourr., *S. maritimus* Linn., *S. oleraceus* Linn., *S. tenerrimus* Linn. are all used as emollients.

A. Annual
- 1. Glabrous or sparsely glandular above ... 2. *S. asper.*
- 2. Glabrous or sparsely glandular hispid ... ... 4. *S. oleraceus.*

B. Perennial
- 1. Leaves runcinate-pinnatifid, spinous-toothed ... 1. *S. arvensis.*
- 2. Leaves linear-oblong, quite entire or sparingly sinuate-toothed ... ... 3. *S. maritimus.*

1. **Sonchus arvensis** Linn. occurs throughout India, wild or cultivated, scarce in the plains, but more common on the hills where it ascends to 8,000 ft. in the Himalaya and the Khasia Hills. The

plant is found, wild or introduced, in all temperate and many tropical countries.

The plant is slightly bitter, diuretic; useful in chronic fevers, and in the treatment of the coughs in phthisis, bronchitis, asthma, and pertussis.

Among the Santals the root is given in jaundice.

Bengal: Banpalang—; *Ceylon*: Musalkatha, Poilaippillu—; *Hindi*: Sadhi, Sahadevibari—; *Punjab*: Bhangra, Kalabhangra—; *Santa*: Birbarangan, Birbarangom—; *Sinhalese*: Valdunkola—; *Telugu*: Nallatapata—; *Urdu*: Sahadevi—.

2. **Sonchus asper** Hill. is found throughout India, in fields and cultivated places, ascending to 12,000 ft. in the Himalaya. It occurs, wild or introduced, in all warm countries.

At Loralai the plant is pounded and applied to wounds or boils.

In Spain it is commonly used as an emollient.

La Reunion: Lastron piquant—; *Loralai*: Garwa, Machal—.

3. **Sonchus maritimus** Linn. is found in the Punjab and in Baluchistan, whence it extends westwards to the Atlantic.

In the Punjab it is used in the treatment of the coughs in phthisis, bronchitis, asthma, and pertussis.

Egypt: Libbeyn—.

4. **Sonchus oleraceus** Linn. is sparingly distributed throughout India and Ceylon; it ascends to 8,000 ft. in the Himalaya. It is found, wild or introduced, in all temperate and many tropical countries, in many varieties and forms.

An infusion of the root and leaves is used in Bengal as a tonic and febrifuge.

In Indo-China the stems are prescribed as a sedative and a tonic.

The plant is a popular laxative and diuretic in Italy.

The early Cape settlers applied the juice of the plant for cleansing and healing ulcers.

The brownish gum formed by evaporation of the common Sow-thistle, when taken internally in a dose of 2-4 grains, behaves as an 'intensely powerful hydragogue cathartic' and acts powerfully upon the liver, duodenum, and colon. In its general effects, it is said to most resemble elaterium, producing large and watery discharges so that it has proved a valuable therapeutic agent in ascites and hydrothorax. It requires, however, great care in its administration, since it has the disadvantage of griping like senna, and producing tenesmus like aloes.

The California Chinese use the juice as an anti-opium remedy.

Afrikaans: Melkdissel, Seidissel, Suidissel, Suigdissel—; *Antsianaka*: Beroberoka—; *Arabic*: Myrrejr—; *Betsileo*: Anamboraka—; *Bombay*: Mhatara—; *Brahui*: Aghud—; *Catalan*: Llacso, Lletso, Llicso, Llitso—; *Chagai*: Aghut—; *Chinese*: Ku Tsai—; *Egypt*: Besikh, Galail, Libbeyn, Qelawil, Tibsikh—; *English*: Du Tistel, Hare's Lettuce, Hare's Palace, Hare's Thistle, Milk Weed, Milk Thistle, Sow Thistle, Sprout Thistle, Turn Sole—; *French*: Labyrinthe étrille, Laceron, Laisseron, Lait d'âne, Laiteron, Laiteron commun, Laitron, Laitue de murailles, Lasseron, Liarge, Palais de lièvre—; *German*: Gaensedistel, Hasenkohl—; *Gippsland*: Thalaak—; *Hova*: Beroberokamboa—; *Indo-China*: Tuc doan—; *Iraq*: Murrair, Um-al-halib—; *Italian*: Cicerbita, Cicerchia,

Crespignia, Crespignoli, Crispignolo, Sonca, Sparaghella—; *Kharan*: Gokizaban—; *Kotra*: Gogru—; *Languedoc*: Engraisso por, Lachassou, Lachenca, Lacheyroun—; *La Reunion*: Lastron doux, Lastron tendre—; *Malta*: Sow-thistle, Cicerbita, Crespino, Tfief—; *Mastung*: Agut—; *Mexican*: Chichicaquihuitl—; *Pacific Coast*: Annual Sow-thistle, Common Sow-thistle, Hare's Lettuce, Hare's Thistle, St. Mary's Seed, Sow-thistle—; *Patna*: Titaliya—; *Punjab*: Dodak—; *Roumanian*: Sussiu—; *Russian*: Zayatchiya kapusta—; *Spanish*: Camaroja, Cerraja, Lechecilla, Lechugilla—; *Suto*: Bono-sa-lekxwaba, Lesabe—; *Telugu*: Ratrinta—; *Uruguay*: Cerraja, Lechon—; *Yemen*: Myrreja—.

SPARGANOPHORUS.

The genus consists of only one species, **S. Vaillantii** DC., a native of tropical America, now a weed in Africa and Asia. It is found on the edges of ponds, rice-fields, and ditches throughout the Malay Peninsula.

It is used medicinally for headache by the natives of East Akim and Gold Coast.

Sierra Leone: Ewaruda—; *Yoruba*: Ewuro odo—.

SPHAERANTHUS.

The genus consists of 25 palaeotropical species.

S. senegalensis DC. is used medicinally in Mauritania, Gambia and Nigeria.

Glabrous or pubescent; branches with wings entire ... 1. *S. africanus.*
Tomentose or villous; branches with wings toothed ... 2. *S. indicus.*

1. **Sphaeranthus africanus** Linn. is common in swamps, in damp waste ground, and in mud, throughout the plains of Bengal and Sylhet, and southwards to Ceylon and Malacca, extending to the Malay Islands, the Philippines and Australia, China, Persia, and Africa.

The herb is stomachic and diuretic.

In Indo-China the plant is used as an emollient and resolvent; it is applied as a poultice to any dolorous part of the body. The juice of the leaves is used as a gargle in inflammation of the throat.

Indo-China: Bo xet, Bo xit, Chung vit, Cuu vit—; *Malaya*: susu—.

2. **Sphaeranthus indicus** Linn. is common in the rice fields of India and Ceylon. In Tropical Himalaya it occurs from Kumaon to Sikkim, ascending to 5,000 ft. From Assam and Sylhet it spreads over the Malay Peninsula to Singapore, the Malay Archipelago, China, and Africa.

The plant is recognized by both Ayurved and Yunani systems of medicine.

The herb is a reputed general tonic, deobstruent, alterative and aphrodisiac.

The root and the seeds are considered as anthelmintic.

The powdered root is given as a tonic. The bark, ground and mixed with whey, is a valuable remedy for piles.

In the Punjab the flowers are highly esteemed as alterative, dipurative, cooling and tonic.

In Java the plant is considered as a useful diuretic.

This small plant was used in the form of a decoction as a diuretic in urethritis. The result was unsatisfactory (Koman).

The rind of the fruit, dried and pounded, is used as a fish poison.

Among the Mundas the bruised plant is thrown into rivers and ponds to kill fish. It is stuffed into the holes of crabs to kill those animals. The plant is pounded with a little water, and the juice is expressed and used as a styptic. The juice of the fresh leaves, similarly obtained, is boiled with a little milk and sugar-candy, and drunk for cough.

On the Gambia the plant is gathered and spread thickly on beds on which sick persons are laid, to relieve pain and to induce sleep.

Arabic: Kamazariyus, Kamdaryus—; *Bengal*: Chagulnadi, Ghorkmundi, Murmuriya—; *Bombay*: Gorakhmundi, Mundi—; *Cutch*: Munderi—; *Deccan*: Mundhri, Mundi—; *Gambia*: Lookidge—; *Gujerati*: Bodiokalara, Gorakha-mundi, Mundi—; *Hasada*: Najompuru, Paru—; *Hindi*: Gorakmundi, Mundi—; *Malayalam*: Adakkamanian, Attakkamanni, Mirangani—; *Marathi*: Barasa-vodi, Gorakhamundi—; *Mundari*: Kardanidudmuritasad, Kutunaara, Puru—; *Naguri*: Kardanidudumuritasad—; *Persian*: Kamaduriyus, Randarummitalkha—; *Porebunder*: Gorakhamundi—; *Punjab*: Gundhi, Gurukmundi, Khamadrus, Mundi, Mundibuti, Zakhmihaiyat—; *Sadani*: Khutlasag—; *Sanskrit*: Alam-busha, Aruna, Avyatha, Bhikshu, Bhukadambika, Bhukanda, Bhutaghni, Boda, Chhitragranthika, Kadambapushpa, Kadambapushpika, Krodachuda, Kumbhala, Lochani, Lotani, Mahamundi, Mahashvranika, Mata, Mundakhya, Mundi, Mundirika, Nadikadamba, Nilakadambika, Palankasha, Parivraji, Pravrajita, Shravana, Shravanashirshika, Shravani, Sthavira, Tapasvini, Tapodhana, Vika-cha, Vridha—; *Santal*: Belaunja—; *Sinhalese*: Etmadamahana, Mudamahana—; *Tagalog*: Sambonggala—; *Tamil*: Kottakkarandai—; *Telugu*: Bodasoram, Bodatarapu—; *Urdu*: Kamdaryus, Mundi—; *Uriya*: Buikadam, Murisa—.

SPILANTHES.

The genus includes 35 tropical species, mostly American.

Spilanthes Acmella Murr. occurs throughout India and Ceylon; it is found in all warm countries. It has been met with at an altitude of 5,000 ft. in the Himalaya and other mountains.

The whole plant is pungent to the taste, and the flower heads especially so, though when distilled with water these afford a distillate free from pungency.

The plant is used as a specific for toothache in Old Calabar. In South Africa the powdered leaf is placed in a carious tooth, and rubbed on the lips and gums for sore mouth in children.

The flower-heads are by far the most pungent part. They are chewed in India and Malaya to relieve toothache, which they do by producing redness of the gums and salivation. In Cameroons they are rubbed on the forehead for headache.

In Assam the plant is sometimes administered to women after childbirth. In Cameroons it is used in combination with others, chewed and swallowed, for snake bite, with local treatment of the wound.

Among the Mundas of Chota Nagpur the crushed plant is used as a fish poison. When in the rainy season, the children playing about in the water get their feet and legs covered with rash, they rub them with the leaves of this plant to soothe the itching. Poor people chew its seeds instead of pepper to provoke salivation when their mouth is dry.

In Indo-China the plant is boiled in water and the whole, liquid and solid, given in dysentery. In Ceylon the leaves and flowers are used for toothache and sore throat; they are also given to women at childbirth.

In the Philippine Islands a decoction of the root is given as a purgative. A decoction of the leaves is used as a bath in rheumatism, or as a lotion in scabies and psoriasis; the juice is considered vulnerary, as also the pounded leaves made into a poultice. Internally the decoction of the leaves is given as a diuretic and lithotriptic.

In Madagascar it is used as an antiscorbutic, diuretic, sialagogue, odontalgic, tonic, and digestive.

Dr. W. Farquhar has used and recommended a tincture of the flower-heads for toothache, in place of tincture of pyrethrum. He says it is a specific for inflammation of the periosteum of the jaws. A bit of lint, dipped in the tincture and laid on the gums, repeated three or four times a day, has a speedy effect in reducing the pain and swelling.

An active principle, *spilanthol*, with a local anaesthetic action, has been isolated from the flowers.

Ashanti: Nyamengen—; ***Assam***: Pirazha—; ***Awka***: Osana—; ***Bakwiri***: Sekke—; ***Betsileo***: Kimotodoha—; ***Betsimisaraka***: Anamafana—; ***Bombay***: Akarkara—; ***Brazil***: Agriao do Para, Mastruco, Pimenteira do Para—; ***Burma***: Henkala—; ***English***: Toothache Plant—; ***French***: Abéadaire, Abécédaire, Acmelle, Alcmelle, Cresson de l'Ile de France, Cresson des Indes, Herbe de Malacca—; ***French Guiana***: Cressonpara—; ***Hova***: Anamalaho, Anamalahokely, Anamalahombazalao, Anamalahoye—; ***Ibo***: Osana—; ***Indo-China***: Cuc ao, Ngo ao—; ***La Reunion***: Acmella—; ***Madagascar***: Cresson des Indes, Cresson du Para—; ***Malay***: Gutang—; ***Mauritius***: Acmella—; ***Mundari***: Barandu, Birbiri, Bocotupuri, Cirbiri, Dudmuritasad, Hatukesari, Hatukisari, Kesari, Khesari, Marceia—; ***Naguri***: Marcaia—; ***Oloke-Meji***: Awere-pepe—; ***Pampangan***: Palunag, Palunai—; ***Punjab***: Akarkarha, Pokarmul—; ***Russia***: Akmella—; ***Sakalave***: Anamafana—; ***Sinhalese***: Akmalla—; ***Tagalog***: Hagonog, Hagonoi—; ***Telugu***: Maratimogga, Maratitige—; ***Uganda***: Tonjatula—; ***Visayan***: Agonoi—; ***Yoruba***: Awerepepe—.

var. **oleracea** C. B. Clarke (=*S. oleracea* Jacq.) is cultivated in gardens.

The whole plant is very acrid, but the flower-heads are especially so, having a hot, burning taste, which causes profuse salivation. They are considered a powerful stimulant and sialogogue, and used in headaches, paralysis of the tongue, affections of the throat and gums, and for toothache. A popular remedy for children who stammer.

In Madagascar it is considered to have the same therapeutic properties as *S. acmella* proper.

Bengal: Roshaniya—, ***Betsileo***: Kimotodoha, Māngevitsa—; ***Betsimisaraka***: Anamafana—; ***Brazil***: Mastruco, Pimenteira do Para—; ***Catalan***: Creixans del Para—; ***English***: Brazil Cress, Para Cress—; ***French***: Brède malgache, Cresson du Brésil, Cresson du Para—; ***French Guiana***: Cresson-para—; ***German***: Parakresse—; ***Hova***: Anamalaho, Anamalahokely, Anamalahombazaho, Anamalahoye—; ***Russian***: Brazilskiy Kress, Tsarskiy Kress—; ***Sakalave***: Anamafana—; ***Spanish***: Berros del Para, Espilanto—.

SYNEDRELLA.

The genus consists of 2 tropical American species, one also African and Asiatic.

Synedrella nodiflora Gaertn. occurs in cultivated places in Assam, Martaban, and the Andaman Islands. It is found in tropical Africa and America.

The leaves are boiled and used as a laxative by the natives of Gold Coast.

Ashanti: Kwadupo, Nkwadupor, Tutummirika kohwe epo—; *Timne*: Bal-keyan-karuni—; *Yoruba*: Aluganbi, Zannaposo—.

TAGETES.

The genus includes 20 species, natives of warm America.

T. erecta Linn. is used medicinally in Mexico; *T. minuta* Linn. in Brazil; *T. pusilla* H. B. and K., and *T. Zypaquinensis* Humb. and Bonpl. are used in Colombia.

Tagetes erecta Linn. is a native of Mexico, grown in Indian gardens.

The leaves are used as an application to boils and carbuncles; their juice is given in earache.

The flowers are employed in diseases of the eyes and for unhealthy ulcers, internally they are said to purify the blood; their juice is given as a remedy for bleeding piles.

Arabic: Hajai, Hamahama—; *Bengal*: Genda, Gendha—; *Bombay*: Guljajari, Makhmal—; *Brazil*: Cravo de defunto—; *Ceylon*: African marigold—; *Egypt*: Gatife—; *English*: African Marigold, French Marigold—; *Gujerati*: Guljharo, Makhamala—; *Hindi*: Genda, Gendha, Gulatora, Kalaga, Lalamuraga, Makhamali—; *Hova*: Anantsinahimbazaha, Tsipolobazaha—; *Iraq*: Ja'fari—; *Kathiawar*: Gulgoto—; *Marathi*: Makhamala, Rojiachaphul, Zendu—; *Mexico*: Cempoalxochitt—; *Nasirabad*: Gulgaindo—; *Persian*: Kajekharusa, Sadabarg—; *Punjab*: Genda, Mentok, Sadbargi, Tangla—; *Sanskrit*: Sthulapushpa, Zandu, Zanduka—; *Telugu*: Banti—; *Urdu*: Genda—; *Uriya*: Gendu—; *Yemen*: Naufar, Ranjes—.

TANACETUM.

The genus includes 30 species, found in the north temperate regions.

T. vulgare Linn. is used medicinally in Europe, and its flower-heads are officinal in Belgium and Portugal.

Leaves $\frac{1}{4}$-$\frac{1}{2}$ in. 1-2 pinnatisect ... ... 1. *T. fruticulosum*.
Leaves $\frac{1}{2}$-1 in. few scattered palmately 2-pinnatisect ... 2. *T. gracile*.

1. **Tanacetum fruticulosum** Ledeb. is found at 12,000-15,000 ft. in the Western Himalaya. It is also found in Baluchistan and Afghanistan, and in the Altai Mountains.

The herb is used as a tonic and anthelmintic. Soaked in cold water it makes a decoction which Baluchis consider very good for colds.

Baluchi: Drana—; *Brahui*: Butau, Drani, Jir—; *Kalat*: Butau—; *Noshki*: Drani—.

2. **Tanacetum gracile** Hook. fil. and Th. is found at 13,500 ft. in the Alpine Western Himalaya, from Kashmir to north of Kumaon. It also occurs in Baluchistan.

At Pab in Jhalawan the plant is considered as a cure for fevers.

Pab : Drane—.

TARAXACUM.

The genus includes 25 species, inhabiting temperate and cold regions.

T. officinale Wigg. is used medicinally in Europe, China and La Reunion. The whole herb or some of its parts are officially recognized in several countries: the dried plant in Japan; the root in Great Britain, Hungary, Sweden, Switzerland; the root and the dried plant in Russia; the root and the leaf in Austria; the leaf in France.

Taraxacum officinale Weber in Wigg. is found throughout the Himalaya and Western Tibet from 1,000 ft. up to 18,000 ft. It also occurs in the Mishmi Hills. The plant is almost cosmopolitan; it is at home in all parts of the north temperate zone, in pastures, meadows and on waste ground, and is so plentiful as to be a troublesome weed.

The plant is largely cultivated in India where the root is a very popular remedy for liver complaints.

The root is diuretic, tonic, and slightly aperient. It is a general stimulant to the system, but especially to the urinary organs, and is chiefly used in kidney and liver disorders. It is, perhaps, one of the most generally prescribed remedies in Europe.

At Kironi in Baluchistan the leaves are used for fomentations.

In China the leaves are prescribed internally as a bitter depurant, externally they are used for poulticing after having been pounded.

In Derbyshire, the juice of the stalk is applied to remove warts.

Baluchi : Halako—; *Bogotá* : Diente de león—; *Brahui* : Aghud, Bahi, Halako, Pochaku, Pochku—; *Catalan* : Angelets, Apagallums, Caixals de vella, Dents de lleo, Llitso d'asa, Pixallit, Pixallits, Xicoina de burro—; *Chinese* : P'u Kung Ying—; *Cornwall* : Dashelflower—; *Deccan* : Pathri—; *Devonshire* : Dashelflower—; *Dutch* : Paardenbloem—; *English* : Blowball, Dandelion, Milk Gowan, Pissabed, Priest's Crown, Swinesnout, Time-Table—; *French* : Chopine, Cochet, Couronne de moine, Dent de lion, Florion d'or, Laitue de chien, Liondent, Pissenlit, Salade de taupe, Tête de moine—; *German* : Ackerzichorie, Apostemkraut, Augenmilch, Baerenzahn, Bettpisser, Bettseiger, Bissanliwurzel, Butterblume, Eierblume, Feldreis, Gaddeliese, Habichtskraut, Hundslattich, Hundszahn, Jungeblume, Kuhblume, Kuhlattich, Laternen blume, Loewenzahn, Luchten, Maistoeckel, Milchadistel, Milchroedel, Milchstoeckel, Mistfinke, Moenchsblume, Moenchskopf, Papankraut, Pfaffendistel, Pfaffenoerhlein, Pfaffenschnell, Pfaffenstiel, Pfefferoeslein, Pferdeblume, Saumelke, Saustock, Seherrkraut, Schweineroesl, Schwiblume, Sommerdorn, Sonnenwirbel, Teufelsrippen, Tiefstand, Weglattich, Wiesenlattich, Zunehmkraut—; *Irish* : Cais tsearbhan—; *Italian* : Dente di leone, Smirnio, Soffione, Tarassaco—; *Kironi* : Gulsagh—; *Ladak* : Rasuk, Yamaghika—; *Languedoc* : Lagaina, Pissolet—; *La Reunion* : Chicorée sauvage, Pissenlit—; *Lepcha* : Ta-klok bi—; *Malta* : Dandelion, Dente di leone, Pisciacane, Tarassacio, Cicueira salvagga—; *Pacific Coast* : Blow-ball, Dandelion, Irish Daisy, Lion's-tooth, Puff-ball, Yellow Gowan—; *Portuguese* :

Dente de leaõ, Taraxaco—; *Punjab*: Baran, Dudal, Dudhbattal, Dudli, Kanphul, Radam, Shamuke—; *Roumanian*: Papadie, Parasita gainelor—; *Russian*: Oduvanchik, Papovo gumentse—; *Sind*: Bathur, Buthur—; *Spanish*: Amargón, Cardillo, Dente de leon, Hocico de puerco, Tagarnina, Taraxacon—; *Tasmania*: English Dandelion—; *Uruguay*: Achicoria silvestre, Amargon, Diente de leon, Radicheta—.

TRAGOPOGON.

The genus includes about 24 species, natives of the Mediterranean region and Central Asia.

Peduncle much inflated under the head. Ligules purple ... 1. *T. porrifolium.*
Peduncle hardly thickened under the head. Ligules yellow ... ... 2. *T. pratense.*

1. **Tragopogon porrifolium** Linn. is found in cultivated places at Simla and in Western Tibet. It extends westwards to the Mediterranean.

Culpeper says of Purple Goat's Beard: 'The virtues of this are the same as the other, only less pleasant, therefore more bitter, astringent, detersive and medicinal. This, however, may be eaten in great quantities, and so will be useful in chronic complaints. The roots are particularly specific in obstructions of the gall and the jaundice; the best way to use them is stewed like chardoons.'

It ranks as one of the most salubrious of culinary vegetables, being antibilious, cooling, deobstruent, and slightly aperient.

The root is used in Spain as a diuretic, aperient, and pectoral.

English: Purple Goat's Beard, Salsafy, Salsify, Vegetable Oyster—; *French*: Barbebon, Salsifis blanc, Salsifis des jardins—; *Malta*: Salsify, Salsefica, Salsifi, Lehjet il botbot—; *Spanish*: Barba de cabra, Barbon, Salsifi blanca—.

2. **Tragopogon pratense** Linn. occurs in Western Tibet and the Western Himalaya, between 12,000 and 14,000 ft. It extends westwards to the Atlantic.

Culpeper tells us: 'A large double handful of the entire plant, roots, flowers and all, bruised and boiled and then strained with a little sweet oil, is an excellent clyster in most desperate cases of strangury or suppression of urine. A decoction of the roots is very good for the heart burn, loss of appetite, disorders of the breast and liver; expels sand and gravel, and even small stone. The roots dressed like parsnips with butter are good for cold, watery stomachs, boiled or cold, or eaten as a raw salad; they are grateful to the stomach, strengthen the lean and consumptive, or the weak after long sickness. The distilled water gives relief to pleurisy, stitches or pains in the side.'

The fresh juice of the young plant has been recommended as 'the best dissolvent of the bile, relieving the stomach without danger and without introducing into the blood an acrid, corrosive stimulant, as is frequently done by salts when employed for this purpose.'

In Italy the plant as a decoction is a home remedy for chest troubles and a popular sudorific.

In Spain the root is used as a diuretic, aperient, and pectoral.

Catalan: Barba de cabra, Barbata, Barbeta—; *Dutch*: Boksbaard—; *English*: Buck's Beard, Gait-berde, Goat's Beard, Go-to-bed-at-noon, Jack-by-the-hedge, John-go-to-bed-at-noon, Joseph's Flower, Nap-at-noon, Noon Flower, Noontide, Shepherd's Clock, Sleep-at-noon, Star of Jerusalem—; *French*: Barbe bleue, Barbe de bouc, Barbe de porc, Ratabout, Salsifis des prés, Salsifis sauvage—; *German*: Bocksbart, Ziegenbart—; *Italian*: Barba di becco—; *Roumanian*: Barba caprei, Barba popii, Barba tapului, Tita caprei—; *Spanish*: Barba cabruna, Barba de cabra, Barbajo, Barbon, Salsifi verdadera—.

TRICHOLEPIS.

The genus consists of 12 Indo-Malayan species.

Pappus absent ... ... 1. *T. angustifolia.*
Pappus short but conspicuous ... ... 2. *T. glaberrima.*

1. **Tricholepis angustifolia** DC. is found in the Western Peninsula, Canara, Mangalore, Travancore.

The plant is a bitter tonic and diuretic, used in cough.

Travancore: Utakatara—.

2. **Tricholepis glaberrima** DC. is common in western Rajputana, Central India, the Konkan, the Deccan, the South Maratha Country, the Western Ghats in the Bombay Presidency, South Canara, Coorg, and the hills of Mysore.

The plant is used in leucoderma and skin diseases, in inflammations. It is believed to be a nervine tonic and an aphrodisiac.

Bengal: Chhagaladandi, Vamanadandi—; *Bombay*: Motabor—; *Canarese*: Brahmadandi—; *Gujerati*: Brahmadandi, Phusiarun, Talakanto—; *Hindi*: Brahmadandi—; *Marathi*: Brahmadandi, Motachor—; *Sanskrit*: Ajadandi, Brahmadandi, Kantapatraphala—.

TUSSILAGO.

The genus consists of 1 species distributed over Asia, Europe, and North Africa.

Tussilago Farfara Linn. is found at 6,000-11,000 ft. in the Western Himalaya from Kashmir to Kumaon.

The Hindus consider that the leaves have the power of regulating the functions of the central and sympathetic nervous systems, and that they are valuable in the treatment of rheumatism; they also use the cotton-like down of the leaf as a styptic.

From the earliest times the plant has been esteemed useful in coughs and other pectoral affections. Hippocrates recommends the root with honey in ulcerations of the lungs. Dioscorides, Pliny, and Galen relate that the smoke of the leaves, received through a funnel or reed, is efficacious in coughs and dyspepsia.

Most of the Arabian and Persian medical writers describe the herb as bitter and carminative. The Yunanists use the roots and the leaves in chronic bronchitis, asthma, chest complaints, and inflammations; they are maturant and abortifacient.

Culpeper says: 'The fresh leaves, or juice, or syrup thereof, is good for a dry bad cough, or wheezing and shortness of breath. The dry leaves are best for those who have their rheums and distillations upon their lungs causing a cough: for which also the dried leaves taken as tobacco, or the root is very good. The distilled water hereof simply or with elder-flowers or nightshade is a singularly good remedy against all agues, to drink two oz. at a time and apply cloths wet therein to the head and stomach, which also does much good being applied to any hot swellings or inflammations. It helpeth St. Anthony's fire and burnings, and is singular good to take away wheals.'

Coltsfoot has been justly termed 'Nature's best herb for the lungs, and her most eminent thoracic'. The leaves are sometimes applied to wounds in Europe; and Sussex peasants esteem the white down of the leaves as a most valuable medicine. The dried leaves, when soaked out in warm water, will serve as an excellent emollient poultice.

In China the flowers are used as an expectorant in cough, asthma, apoplexy, and phthisis.

The leaves are officinal in Austria, Denmark, Germany, Norway, Portugal, Russia, Sweden, and Turkey. The flower heads are officinal in France.

Arabic: Ajanjiwun—; *Catalan*: Fills abans que el pare, Pota de caball, Pota de cavall, Pota d'euga—; *Chinese*: K'uan Tung Hua—; *Danish*: Foeldfod, Hestehov—; *Dutch*: Hoefblad, Klein hoefblad—; *English*: Asse's Foot, Ass's-foot, Bull-foot, Bull's Foot, Clatter-clogs, Clayt, Clayweed, Colt, Colt Herb, Colt's-foot, Coughwort, Coutfit, Cowheave, Dishalaga, Donnhove, Dove-dock, Dummy Weed, Fieldhove, Foalfoot, Foal's Foot, Foalswort, Foilefoot, Hallfoot, Hogweed, Hoofs, Horse-hoof, Horse-hove, Son-before-the-Father, Sow Foot, Tunhoof, Tushalan, Tushylucky Gowan—; *French*: Chou de vigne, Filius ante patrem, Herbe aux pattes, Herbe de saint Guérin, Herbe de saint Quirin, Pas d'âne, Pas de cheval, Pas de poulain, Procheton, Racine de peste, Taconnet, Tussilage, Tussilage commun—; *German*: Ackerlattich, Auflattig, Berglaetsche, Brandlattich, Brandlatsche, Brustlattich, Derrelatten, Eselfuss, Eselhuf, Fahlenfuss, Feldlattich, Foelfod, Fuelifuess, Haberlattich, Haferlattich, Hoaflotcher, Huflattich, Hundblume, Hustenkraut, Kuhblume, Labaschen, Latten, Lehmblatt, Logjehn, Loedke, Maergabluemli, Negenkraftkraut, Pappenmuentz, Pferdehuf, Querniskraut, Rosshuf, Rosslattich, Sandblatt, Sandkraut, Sommertuerle, Weisslabeschen, Werschlabeschen, Zeitroeslein—; *Greek*: Bichion—; *Hindi*: Watapana—; *Italian*: Farfarello, Farfaro, Farfugio, Fayfaro, Unghia cavallina—; *Languedoc*: Pepoulit—; *Malaya*: Kuan tung hua, Toong fah, Tung hua—; *Norway*: Hestehof—; *Persian*: Fanjiwun—; *Portuguese*: Farfara, Tussilagem—; *Punjab*: Watpan—; *Roumanian*: Podbeal, Podbol—; *Russian*: Bielakopitnik—; *Spanish*: Pie de caballo, Uña de caballo, Tusilago—; *Swedish*: Haesthov, Haetshof—; *Urdu*: Fanjiwun—; *Yorkshire*: Cleats—.

VERNONIA.

The genus includes 650 species, chiefly tropical, and mostly American.

The following species are used medicinally in Indo-China—*V. cinerea* Less., *V. teres* Wall.—; in La Reunion—*V. cinerea* Less.—; in Madagascar—*V. appendiculata* Less., *V. pectoralis* Baker, *V. scariosa* Baker—; in Nigeria—*V. amygdalina* Del., *V. Kotschyana* Schultz, *V. nigritiana* Oliv. and Hiern.—; in Gold Coast—*V. amygdalina* Del., *V. Biafrae* Oliv. and Hiern., *V. conferta* Benth., *V. nigritiana* Oliv. and Hiern., *V. senegalensis* Less.—; in Guinea—*V. nigritiana* Oliv. and Hiern., *V. senegalensis* Less., *V. Thomsoniana* Oliv. and Hiern.—; in Gambia—*V. colorata* Drake, *V. nigritiana* Oliv. and Hiern., *V. senegalensis* Less.—; in Lagos—*V. cinerea* Less.—; in Sierra Leone—*V. amygdalina* Del.,

V. nigritiana Oliv. and Hiern.—; in St. Thomas—*V. amygdalina* Del.—; in Angola—*V. senegalensis* Less.—.

A. Achenes glabrous. Pappus white, outer hairs few ... 2. *V. elaeagnifolia.*
B. Achenes slightly narrowed at the base, clothed with appressed white hairs. Pappus white or dirty white, outer hairs very short ... ... 1. *V. cinerea.*
C. Achenes silky. Pappus yellowish or reddish, outer hairs few short caducous ... ... 4. *V. teres.*
D. Achenes sparsely hairy. Pappus dirty white or reddish, outer hairs often absent ... ... 3. *V. Roxburghii.*

1. **Vernonia cinerea** Less. is one of the commonest Indian weeds, occurring throughout India ascending to 8,000 ft. in the Himalaya, and in the Khasia and Peninsular hills. It is spread over tropical Asia, Africa, and Australia.

In the Nighantas the plant is described as cold, sweet, tonic, stomachic, astringent, and correcting all the humors; useful in the treatment of asthma, bronchitis, and consumption. The flowers are reputed to cure fevers.

The plant is used in decoction to promote perspiration in febrile conditions. The expressed juice is given in piles. In Senegal and French Guinea an infusion of the plant is used to wash a newborn infant, and is given also to children with incontinence of urine. The bitter root is a vermifuge.

The seeds are employed in Patna as an alexipharmic and anthelmintic, and as a constituent of masalas for horses. In Chota Nagpur, the whole plant is given as a remedy for spasm of the bladder and strangury; the flowers are administered for conjunctivitis; the root is given for dropsy.

In Ceylon it is used for wounds and sores, and taken internally to promote perspiration.

This plant is considered to possess strong diaphoretic properties, and is generally given by vaidyans to produce perspiration in fever. By itself, it has no antiperiodic property, but when combined with a small dose of quinine, it appears to help the action of the latter in malarial fevers. At Cannanore I found that one of the vaidyans there was treating cases of malarial fever with *V. cinerea* and 5 grs. of quinine made into a bolus with lime juice administered every morning. I tried this method in several cases and found it to be a useful combination (Koman).

Sushruta recommends the plants for the treatment of scorpion-sting; but it is not an antidote to scorpion venom (Caius and Mhaskar).

Bengal: Kalajira, Kukshim, Kuksim—; *Brahui*: Aghud—; ***English***: Ash-coloured Fleabane—; *Ewe*: Osikonu—; *French*: Ayapana sauvage, Vernonia cendré—; *Gujerati*: Sadeori, Sadodi, Sedardi, Shedardi—; *Hindi*: Dandotpala, Sahadevi, Sadodi, Sadori—; *Indo-China*: Bac dau—; *Lagos*: Ewe jedijedi, Orungo—; *Malay*: Rumput sabagi, Rumput susor daun, Sembong hutan, Tahi babi, Tambak bukit, Tambak tambak—; *Malayalam*: Puvankuruntal—; *Malinke*: Kungueni—; *Marathi*: Sadodi, Sahadevi, Osari—; *Matheran*: Mothisardori, Sahadevi—; *Mauritius*: Ayapana sauvage—; *Merwara*: Kaliharr, Lalia—; *Mundari*: Toaara—; *Nasirabad*: Bhumbak—; *Oloke-Meji*: Elegbe oju—; *Porebunder*: Sadedi, Sadevi, Sadodi—; *Punjab*: Sahadevi—; *Sanskrit*: Dandotpala, Devasasha, Devi, Devika, Gandhavalli, Govandani, Saha, Sahadeva, Sahadevi, Vishamajvaranashini, Vishvadeva—; *Santal*: Bahututuri, Barangom, Birlopongarak, Duryaarak, Jhurjhuri—; *Sinhalese*: Monarakudumbia, Monerakudimbeya—; *Tamil*: Puvamkurundal, Sahadevi Sirashengalanir—; *Telugu*: Garitikamma, Gharitikamini—; *Timne*: Kubwo kerokani—; *Yoruba*: Bojure, Elegbe oju, Ewe jedijedi, Orungo—.

2. **Vernonia elaeagnifolia** DC. is found in Tenasserim and Pegu, extending to Siam and Indo-China.

In Cambodia the wood is used for troubles of the respiratory tract, more particularly of the nose. The wood is dried in the sun, finely scraped, and smoked with tobacco in the form of cigarettes. It is said to be very good for inflammation and ulceration of the mucous membrane.

Cambodia: Kanlak chak, Karleuk chak—.

3. **Vernonia Roxburghii** Less. is found in the Upper Gangetic Plain, Kumaon, Bengal, Burma, Central and Western India.

It is used as a substitute for *V. cinerea*.

Bonai: Banjatangi—; *Mundaṛi*: Kutunaba—; *Santali*: Dorabohok'—; *Uriya*: Agnijal—.

4. **Vernonia teres** Wall. occurs in the tropical Himalaya from Kumaon to Sikkim, Bihar, Central India, Pegu, Burma, Indo-China.

A popular medicine in Annam for luxations, ulcers and wounds. It is given for dysmenorrhoea and dropsy. The flower-heads are considered ascaricidal.

Annam: Cay man tuoi, Trach lan thao—; *Quang-tri*: Cay muoi tuoi—; *Tongking*: Man tuoi—; *Vinh*: Co xanh—.

VOLUTARELLA.

The genus consists of 5 species, occurring from the Mediterranean to India.

Volutarella divaricata C. B. Clarke is found in the Punjab Plain, Central and South India, Baluchistan and Afghanistan.

The plant is credited with tonic, aperient and deobstruent properties. It is slightly mucilaginous and used in coughs. It is also used as a febrifuge, and often prescribed in fevers and general debility.

Arabic: Shaukat-el-baida—; *Bombay*: Badaward—; *Gujerati*: Badaward, Bhonyadandi—; *Hindi*: Badaward—; *Marathi*: Sakayi, Sukayi—; *Persian*: Asfar-e-bari, Badavard, Kangar-e-sufid—; *Syria*: Sanskhurda—; *Turkish*: Lufiniki—; *Urdu*: Badavarda—.

WEDELIA.

The genus includes 65 tropical and subtropical species.

Leaves subsessile, linear-oblong or oblanceolate, acute or obtuse, entire or subcrenate ... ... 1. *W. calendulacea*.
Leaves petioled, ovate-acuminate, serrate ... ... 2. *W. Wallichii*.

1. **Wedelia calendulacea** Less. is found in wet places throughout Bengal, Assam, Sylhet and Burma, the Konkan, the plains

districts of the Madras Presidency, and Ceylon. It is distributed to the Malay Archipelago, China and Japan.

In decoction the plant is used as a deobstruent, and is given in uterine haemorrhage and menorrhagia.

The leaves are considered tonic, alterative, and useful in cough, cephalalgia, skin diseases and alopecia.

An infusion of the plant is given in Indo-China for the swelling of the abdomen. In Annam the plant is used to prevent the effects of bad waters in the hill tracts.

Bengal: Bangra, Bhimraj, Kesaraja, Kesraj, Kesuria—; *Bombay*: Pivala-bhangra—: *Chinese*: P'ang K'i Kiou—; *Deccan*: Pilabungra—; *Gujerati*: Bhangaro, Pilobhangro—; *Hindi*: Bhangra, Bhanra—; *Indo-China*: Hoa muc. Ngo nui, Tan sa—; *Marathi*: Pivalamaka—; *Sanskrit*: Bhringaraja, Devapriya, Haripriya, Harivesa, Kesaraja, Pavana, Pitabhringaraja, Pitabhringi, Svarna-bhringara, Vandaniya—; *Sinhalese*: Ranvankikirindi—; *Tagalog*: Hagonay—; *Tamil*: Patalaikaiantagerai—.

2. **Wedelia Wallichii** Less. is found in the tropical Himalaya up to 5,000 ft., from Kumaon to Bhutan, the Khasia and the Mishmo Hills, Burma and Java.

In Assam the plant is said to heal wounds when applied to them.

XANTHIUM.

The genus includes 25 species, probably all of American origin.

X. spinosum Linn. and *X. Strumarium* Linn. are used medicinally in Europe, North America, and Brazil; *X. canadense* Mill. is used in North America and Brazil; *X. Strumarium* Linn. in China, Indo-China, Malaya, and La Reunion.

Xanthium Strumarium Linn. is cosmopolitan in the warmer parts of the world. It is found throughout the hotter parts of India and Ceylon, usually near houses; ascending to 5,000 ft. in the Western Himalaya.

The root is a bitter tonic, useful in cancer and strumous diseases. The prickly fruit is considered cooling and demulcent and is given in small-pox. In Southern India, the prickly involucre is applied to the ear, or tied in bunch to the ear-ring, to cure hemicrania.

The whole plant is supposed to possess powerful diaphoretic and sedative properties. It is generally administered in the form of decoction, and is said to be very efficacious in long-standing cases of malarious fever.

In some parts of Germany the plant has a popular reputation as a remedy for ague, and in Russia it is considered to be a prophylactic in hydrophobia.

In China the burs are used as a tonic, diuretic, diaphoretic, and sedative.

The herb is used as an emollient and astringent in China and Indo-China. An extract of the root is applied to ulcers, boils, and abscesses.

In La Reunion the plant is considered bitter, astringent, antiscrofulous, and antiherpetic.

The leaves were formerly official in Europe and were administered internally in scrofula and in herpes.

In America and Australia, this plant has been observed to prove fatal to cattle and pigs.

Caius and Mhaskar have shown experimentally that the plant is not an antidote to either snake venom or scorpion venom.

Afrikaans: Boetebossie, Spitzklette—; *Assam*: Agara—; *Bengal*: Banokra—; *Bombay*: Shankeshvara—; *Brahui*: Bichudi—; *Burma*: Chosa, Koukpin—; *Catalan*: Bardana borda, Bardana menor, Escorpins, Llepassa borda, Llepassa menor, Rapalassa borda—; *Chinese*: Hsi Erh, Ts'ang Eul, Ts'ang N'i—; *Egypt*: Kharag-el-bahr, Shubhey—; *English*: Bur-weed, Clotbur, Cocklebur—; *French*: Glaiteron, Gleitron, Glèteron, Glouteron, Petit glouteron, Grappelles, Herbe aux écrouelles, Lambourde, Lampourde, Lampourde commune, Petite bardane—; *German*: Spitzklette—; *Gujerati*: Gadriyun—; *Hindi*: Banokra, Chhotagokhru, Chotadatura, Ghagra, Okura, Shankahuli—; *Indo-China*: Gi nhi, Ke, Ke dau ngha, Quyen nhi tu, Tuong nhi—; *Iraq*: Hasach, Lizzaij—; *Italian*: Lappola minore—; *Kashmir*: Lanetsuru, Tsur—; *La Reunion*: Grosse pagode—; *Malaya*: Buah anjang, Chong yee chee, Tsang ehr, Tsang yee—; *Marathi*: Dumundi, Dutundi, Sankeshwara—; *Pacific Coast*: Broad Cocklebur, Burweed, Button-bur, Bur Thistle, Common Cocklebur, Dike-bur, Lesser Burdock, Sea Burdock, Sheep-bur, Small Burdock—; *Punjab*: Chirru, Gokrakallan, Gudal, Jojre, Kuri, Sungtu, Wangantsuru—; *Pushtu*: Baggiari—; *Roumanian*: Scaietele-poppii—; *Sanskrit*: Arishta, Bhulagna, Chanda, Itara, Kambumalini, Kambupushpa, Kiriti, Malavinashini, Mangalyakusuma, Medhya, Pitapushpi, Raktapushpi, Sarpakshi, Shankhagalini, Shankhakusuma, Shankha puspi, Shankhavha, Shwetakusuma, Sukshmapatra, Supushpi, Vanamalini—; *Sind*: Gokhrukallan—; *South Africa*: Clotbur, Cocklebur, Noogoobur—; *Spanish*: Bardana menor—; *Tamil*: Marlumutta—; *Telugu*: Marulamatangi, Marulutige, Parsvapu, Veritallanappi, Talnoppi—; *Yemen*: Kavar el abid, Manj el ma—.

ADDENDUM.

E. IV. b.

4. Style-arms with long appendages. Achenes awnless ... *Chrysanthellum*.

CHRYSANTHELLUM

The genus consists of 4 tropical species.

Chrysanthellum indicum DC. is found in the Upper Gangetic Plains; Kumaon and Garhwal, at Hawalbagh and Naini-Tal, up to 4,000 feet; Nagpore. It is distributed to Madagascar and Tropical Africa.

The Mundas of Chota Nagpur bruise the leaves and rub them over aching heads; they miff the juice against headache. A piece of root is inserted into the hollow of a tooth to kill the maggot which is believed to live there and cause toothache.

Hasada: Peanggu—; *Naguri*: Piribaranggu—.

THE MEDICINAL AND POISONOUS FLAXWORTS OF INDIA.

BY

J. F. CAIUS, S.J., F L.S.

The LINACEÆ are herbs, shrubs or trees, broadly distributed, especially in the northern hemisphere. They consist of 9 genera with 150 species.

The medicinal and poisonous Flaxworts of the world belong to 5 genera: HUGONIA (Tropics); IXONANTHES (tropical Asia); LINUM (temperate and subtropical regions, especially Mediterranean); REINDWARTIA (North India, China); ROUCHERIA (Guiana).

Ixonanthes cochinchinensis Pierre, which is used medicinally in Annam, is not found in India.

A. Petals contorted, fugacious. Perfect stamens as many as the petals. Herbs rarely shrubs
   - I. Calyx glabrous or pubescent. Styles 5. Capsule 5-celled ... ... LINUM.
   - II. Calyx glabrous. Styles 3-4. Capsule 3-4-celled ... REINWARDTIA.

B. Petals contorted, fugacious. Perfect stamens 2-3-times as many as the petals. Fruit a drupe. Usually scandent shrubs
   - I. Sepals subacute, tomentose, ebracteolate ... HUGONIA.
   - II. Sepals obtuse, bracteolate ... ... ROUCHERIA.

HUGONIA.

The genus consists of 11 species, inhabiting the Tropics of the Old World.

H. Planchoni Hook. f. is used medicinally in Liberia and Sierra Leone, *H. serrata* Lam. in La Reunion.

Hugonia Mystax Linn. is found in the Konkan and North Kanara of the Bombay Presidency, throughout the dry forests of the Madras Presidency, and in Ceylon.

The bruised roots are employed externally in reducing inflammatory swellings, and as an antidote to snake-bites. In the form of a powder, it is administered internally as an anthelmintic and febrifuge. The bark of the root is also employed as an antidote to poisons.

Mhaskar and Caius have shown experimentally that the root is not an antidote to snake venom.

Canarese : Modirakkanni—; *English* : Climbing Flax—; *Malayalam* : Moderakkanni, Motirakkanni—; *Sinhalese* : Bugatteya, Mahagetiya—; *Tamil* : Agori, Kodivirai, Modirakkanni—; *Telugu* : Gatrinta, Kakibira, Penkebedali, Pisangi, Renangi, Tivvaputiki, Ungaralapidemu, Vendapa—; *Tulu* : Mullankola—; *Uriya* : Chulijinka—.

LINUM.

The genus numbers 90 species, inhabiting the temperate and subtropical regions of the world, but especially Mediterranean.

The seeds are oleaginous, emollient, and sometimes purgative.

Glucosides—linamarin, phaseolunatin—are among the products isolated. *Linin*, a peculiar drastic principle has been obtained from *L. catharticum* Linn.

The following are used medicinally in Europe—*L. alpinum* Linn., *L. angustifolium* Huds., *L. austriacum* Linn., *L. catharticum* Linn., *L. gallicum* Linn., *L. flavum* Linn., *L. hirsutum* Linn., *L. maritimum* Linn., *L. narbonense* Linn., *L. nodiflorum* Linn., *L. perenne* Linn., *L. strictum* Linn., *L. suffruticosum* Linn., *L. tenuifolium* Linn., *L. usitatissimum* Linn., *L. viscosum* Linn.—; in China—*L. perenne* Linn.—; in North America— *L. catharticum* Linn., *L. usitatissimum* Linn.—; in Brazil—*L. usitatissimum* Linn.— in Uruguay—*L. selaginoides* Lam.—; in Chili—*L. Chamissonis* Schiede, *L. Macraei* Benth., *L. selaginoides* Lam.—; in South Africa— *L. Thunbergii* Eckl. and Zeyh.—; in Mauritius— *L. usitatissimum* Linn.—.

A. Petals blue
   I. Capsule scarcely exceeding the narrowly white-margined sepals ... ... *L. usitatissimum.*
   II. Capsule much longer than the white-margined sepals ... ... *L. perenne.*

B. Petals yellow ... ... *L. strictum.*

1. **Linum perenne** Linn. is found in the North-Western Himalaya, extending westwards to the Canaries.

In Europe and in China the seeds are used as an emollient.

Country people, in England, boil the fresh herb and take it for rheumatic pains, colds, cough and dropsy.

Chinese : Ya Ma—; *English* : Perennial Flax—; *French* : Lin de Sibérie—.

2. **Linum strictum** Linn. occurs in the Punjab and the North-Western Himalaya, from Soongaria to the Mediterranean.

The seeds are used as an emollient in Spain.

Arabic : Farq, Kittan, Zir—; *Punjab* : Bab-basant, Basant—.

3. **Linum usitatissimum** Linn. is cultivated throughout India up to 6,000 feet.

The bark and the leaves are useful in gonorrhoea.

The bark when burnt and applied to wounds stops the bleeding and promotes the healing process.

The flowers are considered a nervine and cardiac tonic.

The roasted seeds are said to be astringent. Fumigation with the smoke is recommended for colds in the head and hysteria, and the tinder is used to staunch haemorrhages.

Linseed poultice is recommended for gouty and rheumatic swellings; as an emollient, the mucilage is dropped into the eye; with honey it is prescribed in coughs and colds.

The seeds are used internally for gonorrhoea and irritation of the genito-urinary system. The decoction is used extensively. Occasionally the seeds are powdered and taken with sugar.

In Iraq local druggists recommend linseed poultice for chest troubles.

An infusion of the seeds has long been given in Europe for soothing a sore chest or throat in severe catarrh or pulmonary complaints. As a domestic remedy for colds, coughs, and irritation of the urinary organs, linseed tea is most valuable. A little honey and lemon juice makes it very agreeable and more efficacious. It is taken in wineglassful doses, which may be repeated *ad libitum.*

The crushed seeds, or linseed meal, make a very useful poultice, either alone or with mustard. In ulceration and superficial or deep-seated inflammation a linseed poultice allays irritation and pain and promotes suppuration. It is commonly used for boils, abscesses and other local affections.

Tumours of a simple nature, and sprains, may be usefully rubbed with linseed oil. The oil forms, when mixed with lime water or with spirit of turpentine, a capital external application to recent burns or scalds.

Internally, the oil is sometimes given as a laxative; in cases of gravel and stone it is excellent, and has been administered in pleurisy with great success. It may also be used as an injection in constipation. Mixed with honey, linseed oil has been used as a cosmetic for removing spots from the face.

The oil enters into veterinary pharmacy as a purgative for sheep and horses, and a jelly formed by boiling the seeds is often given to calves.

Linseed has occasionally been employed as human food; but it affords little actual nourishment and is apparently unwholesome, being difficult of digestion and provoking flatulence. It has been mixed with corn for making bread, but it proved indigestible and hurtful to the stomach, and even was responsible for the death of many.

There is in England an Act of Parliament still in force which forbids the steeping of flax in rivers, or any waters which cattle are accustomed to drink, as it is found to communicate a poison destructive to cattle, and to the fish inhabiting such waters.

Cattle poisoning by linseed flowers is fairly common in Bengal and Bihar. The linseed flowers and the immature seeds contain the maximum amount of a cyanogenetic glucoside producing as much as 0.69 per cent of free hydrocyanic acid. About half a pound of flowers is sufficient to kill a bullock.

The seeds of *L. usitatissimum* Linn. and the oil from them are officinal in all pharmacopoeias, except that of Mexico.

Anam: Ho ma—; *Arabic*: Bazar-ul-katan, Bazr-ut-kattan, Kattan—; *Behar*: Chikna, Tisi—; *Bengal*: Masina, Mosina, Musni, Tisi—; *Bombay*: Alasi, Javas, Javasa—; *Brazil*: Linho—; *Bundelkhand*: Bijri—; *Canarese*: Agashi, Alashi, Alsi—; *Catalan*: Lli—; *Celtic*: Lin—; *Chinese*: Hou Ma—; *Danish*: Hoer—; *Deccan*: Alshi, Javas—; *Dutch*: Vlas—; *Egypt*: Kittan—; *English*: Common Flax, Flax, Linseed, Lint, Lyne—; *Finland*: Lin, Pellavan—; *Formosa*: Hu-ma—; *French*: Lin, Lin chaud, Lin commun, Lin cultivé—;

German : Flachs, Haarlinsen, Lein, Saatlein—; ***Greek*** : Linon—; ***Gujerat*** : Alshi, Alsi—; ***Hamadan*** : Basarak—; ***Hebrew*** : Bad—; ***Hindi*** : Alsi, Tisi—; ***Hungarian*** : Len—; ***Iraq*** : Kittan—; ***Italian*** : Lino—; ***Kashghar*** : Zighir—; ***Kashmir*** : Alish, Keun—; ***Kolami*** : Unshi—; ***Konkani*** : Sonnbium—; ***Kumaon*** : Alsi, Tisi—; ***Kurdish*** : Gosh—; ***Loralai*** : Alsi—; ***Madagascar*** : Rongonimbazaha —; ***Malayalam*** : Cheruchanavittintevilta—; ***Marathi*** : Alashi, Javas, Javasa—; ***North-Western Provinces*** : Bijri—; ***Norwegian*** : Lin—; ***Persian*** : Basarak katun, Bazarug, Bazr-ul-katan, Tukhm-e-katan, Zaghir, Zaghu—; ***Polish*** : Len—; ***Porebunder*** : Alasi, Javas—; ***Portuguese*** : Linhaca, Linho—; ***Punjab*** : Alish, Alsi, Tisi—; ***Roumanian*** : In—; ***Russian*** : Len, Lyon—; ***Sanskrit*** : Atasi, Auma, Chanaka, Devi, Haimwati, Kshauma, Kshaumi, Kshuma, Madagandha, Madotkata, Malina, Masina, Masrina, Masruna, Masuna, Matasi, Nilapushpi, Nilpushpika, Parvathi, Pichhila, Rudrapatni, San, Sunila, Suvarchala, Tailottama, Uma—; ***Spanish*** : Lino—; ***Swedish*** : Lin—; ***Tamil*** : Alshi—; ***Telugu*** : Atasi, Madanginjalu, Ullusulu—; ***Turki*** : Ziggar—; ***Turkish*** : Keten—; ***Urdu*** : Alasi—; ***Uriya*** : Pesu—; ***Yugoslavia*** : Lan—.

REINWARDTIA.

The genus consists of only 2 species, one Indian and the other Chinese.

Reinwardtia trigyna Planch. is found along the Himalaya from the Indus eastwards, the Salt Range, Simla, Kumaon to Sikkim up to 6,000 feet, Assam, Chittagong, Bihar, Mount Abu, Bombay Konkan and Ghats, the Western Ghats, the forests of South Kanara and Mysore. It is distributed to Siam, Tongking and China.

It is used as a medicine for 'founder' in cattle.

Deccan : Abai—; *Dehra Dun* : Basant—; *Jaunsar* : Pengun—; *Kumaon* : Piuli, Piunli—; *Matheran* : Abai—; *Punjab* : Balbasant, Basant, Gudbatal, Gulashruf, Karkun, Kaur—; *Saora* : Labodatar—.

ROUCHERIA.

The genus consists of 4 species, natives of Guiana.

Roucheria Griffithiana Planch. is found in Malaya; common in low country, Singapore, Malacca to Perak and Penang. It is distributed to Sumatra and Borneo.

The plant is used in the preparation of the dart-poison of the Sakais.

Malay : Akar biji, Akar kait putih, Akar musiang, Garam-garam, Ipoh akar putih—; *Sakai* : Bhoi—.

THE MEDICINAL AND POISONOUS LABIATES OF INDIA.

BY

J. F. CAIUS, S.J., F.L.S.

The LABIATAE are herbs or under-shrubs with stems usually quadrangular bearing the leaves on the flat side. They form a large family of 3,300 species grouped in 200 genera. They are distributed over both hemispheres, especially in temperate regions, being particularly abundant in the Mediterranean area.

This important Order has but few poisonous members, and includes a number of medicinal and sub-medicinal plants of great value. In the lists of drugs supplied to herbalists the Labiates are very strongly represented. They mostly exhibit aromatic or bitter-aromatic, stimulant, and astringent properties; and they are used as tonics, emmenagogues, diaphoretics, and antispasmodics.

Many members yield *essential oils* when distilled. *Alkaloids*—betonicine, stachydrine, turicine—, and *glucosides*—hyssopin—have been isolated.

The medicinal and poisonous Labiates of the world belong to 84 genera :—ACROCEPHALUS (Malay Archipelago to tropical Africa); ACROTOME (southern and tropical Africa); AEOLANTHUS (Africa); AGASTACHE (North America); AJUGA (palaeotemperate regions); AMARACUS (eastern Mediterranean); ANISOCHILUS (Asia, Africa); ANISOMELES (Indo-Malayan region); AUDIBERTIA (North America); BALLOTA (Europe, Mediterranean region; western Asia); BETONICA (cosmopolitan); BRUNELLA (cosmopolitan); CALAMINTHA (northern temperate regions; tropical mountains); COLEBROOKEA (India); COLEUS (palaeotropical regions); CUNILA (America); DRACOCEPHALUM (northern temperate regions); DYSOPHYLLA (eastern Asia, Australia); ELSHOLTZIA (Asia, Europe, Abyssinia); ENDOSTEMON (southern and tropical Africa); EREMOSTACHYS (central and western Asia); ERYTHROCHLAMYS (tropical Africa); GALEOPSIS (northern temperate regions); GENIOSPORUM (Africa, Madagascar, Indo-Malayan region); GLECHOMA (Old World); GLECHON (Brazil, Paraguay); GOMPHOSTEMMA (Indo-Malayan region, China); HEDEOMA (America); HOSLUNDIA (warm Africa); HYMENOCRATER (western Asia); HYPTIS (warm America); HYSSOPUS (Europe, Mediterranean region, Asia); IBOZA (southern and tropical Africa); LALLEMANTIA western Asia); LAMIUM (Europe, Asia, extratropical Africa); LASIOCORYS (Africa); LAVANDULA (Mediterranean region to India); LEONOTIS (tropical and southern Africa); LEONURUS (Europe, Asia; tropical regions); LEUCAS (tropical regions; Africa, Asia); LOPHANTHUS (central Asia; China); LYCOPUS (northern temperate regions); MARRUBIUM (Europe, North Africa, temperate Asia); MARSYPIANTHES (warm America); MELISSA (Europe, western Asia); MELITTIS (Europe); MENTHA (Old World); MERIANDRA (Himalaya; Abyssinia); MICROMERIA (cosmopolitan); MONARDA (North America); MONARDELLA

(Western North America); MOSCHOSMA (palaeotropical regions); MOSLA (Himalaya to Japan); NEPETA (northern hemisphere); OCIMUM (tropical and warm temperate regions); ORIGANUM (Europe; Mediterranean region); ORTHOSIPHON (Indo-Malayan region; tropical Africa); OTOSTEGIA (western Asia; Abyssinia); PELTODON (Brazil, Paraguay); PERILLA (India to Japan); PEROWSKIA (western Asia); PHLOMIS (northern palaeotemperate regions); PHYSOSTEGIA (North America); PLATYSTOMA (tropical Asia; Africa); PLECTRANTHUS (palaeotropical regions; eastern Asia); POGOGYNE (California); POGOSTEMON (Indo-Malayan region); PYCNANTHEMUM (North America); ROSMARINUS (Mediterranean region); ROYLEA (Himalaya); SALVIA (tropical and temperate regions); SATUREIA (warm regions); SCUTELLARIA (cosmopolitan, except South Africa); SOLENOSTEMON (western Africa); SPHACELE (warm America; Hawaiian Isles); STACHYS (cosmopolitan, except Australia); SYNCOLOSTEMON (southern Africa); TETRADENIA (tropical Asia; Australia); TEUCRIUM (cosmopolitan); THYMBRA (south-eastern Europe; western Asia); THYMUS (Old World); TRICHOSTEMA (North America); ZATARIA (Persia, Afghanistan); ZIZIPHORA (Mediterranean region; central Asia).

The medicinal and poisonous Labiates of India are included in the following 49 genera:—ACROCEPHALUS, AJUGA, ANISOCHILUS, ANISOMELES, BRUNELLA, CALAMINTHA, COLEBROOKEA, COLEUS, DRACOCEPHALUM, DYSOPHYLLA, ELSHOLTZIA, EREMOSTACHYS, GALEOPSIS, GENIOSPORUM, GOMPHOSTEMMA, HYMENOCRATER, HYPTIS, HYSSOPUS, LALLEMANTIA, LAMIUM, LAVANDULA, LEONOTIS, LEONURUS, LEUCAS, LYCOPUS, MARRUBIUM, MELISSA, MENTHA, MERIANDRA, MICROMERIA, MOSCHOSMA, NEPETA, OCIMUM, ORIGANUM, ORTHOSIPHON, OTOSTEGIA, PERILLA, PEROWSKIA, PLATYSTOMA, POGOSTEMON, ROYLEA, SALVIA, SATUREIA, SCUTELLARIA, STACHYS, TEUCRIUM, THYMUS, ZATARIA, ZIZIPHORA.

I. Perfect stamens 4, rarely 2, declinate. Anther-cells confluent. Ovary 4-partite. Nutlets dry. Basal scar small.

- A. Lower lip of corolla 1-lobed; stamens usually exserted; basal scar of nutlets not oblique.
  - 1. Lower lip of corolla flat or nearly so.
    - a. Calyx deflexed in fruit, upper lobe recurved, margin decurrent on the tube.
      - i. Corolla-tube not exceeding the calyx, stigma 2-fid ... ... OCIMUM.
      - ii. Corolla-tube exceeding the calyx, stigma entire ... ... ORTHOSIPHON.
    - b. Calyx sub-erect or declinate in fruit, upper lobe not recurved, margins not decurrent.
      - i. Flowers in globose heads or in simple spikes or racemes.
        - α. Calyx narrowly tubular in fruit.
          - * Calyx suberect, upper lip 1-lobed, lower lip entire or 4-toothed ... ACROCEPHALUS.
          - ** Calyx usually declinate, upper lip 1-lobed, with the lateral and lower teeth free, or the lateral teeth united with the upper lobe ... ... GENIOSPORUM.
        - β. Calyx widely campanulate in fruit ... PLATYSTOMA.
      - ii. Flowers in paniculate racemes ... ... MOSCHOSMA.

2. Lower lip of corolla conspicuously concave, or boat-shaped.
 a. Corolla distinctly 2-lipped, lower lip longer than the upper.
  i. Upper lip of calyx rounded, deflexed; or the calyx subequally 5-toothed ... COLEUS.
  ii. Upper lip of calyx beaked or truncate, curved downwards to close the mouth, or the calyx 1-lipped, the upper lip absent ... ANISOCHILUS.
 b. Corolla 5-lobed, the lowest lobe shorter or equalling the other lobes ... ... HYPTIS.

B. Lower lip of corolla 3-fid; stamens included; basal scar of nutlets oblique ... ... LAVANDULA.

II. Perfect stamens 4 with the upper pair longest or 2 straight diverging or ascending. Anthers 1- or 2-celled.

A. Stamens 2; anterior perfect, anther-cells parallel ... LYCOPUS.

B. Stamens 4.
 1. Stamens subequal not didynamous.
  a. Anthers 1-celled.
   i. Calyx 5-toothed, teeth neither feathery nor longer than the tube.
    α. Corolla sub-2-lipped, lower lip longer ... POGOSTEMON.
    β. Corolla equally 4-fid ... ... DYSOPHYLLA.
   ii. Calyx 5-partite, segments feathery, longer than the tube ... ... COLEBROOKEA.
  b. Anthers 2-celled.
   i. Fruiting calyx 2-lipped; corolla 5-lobed ... PERILLA.
   ii. Fruiting calyx 5-toothed; corolla 4-lobed ... MENTHA.
 2. Stamens didynamous.
  a. Anther-cells at length confluent, (2-celled when young ... ... ELSHOLTZIA.
  b. Anther-cells never confluent.
   i. Calyx 5-nerved ... ... ZATARIA.
   ii. Calyx 15-nerved
    α. Stamens exserted ... ... HYSSOPUS.
    β Stamens ascending under the hood ... HYMENOCRATER.
   iii. Calyx 10-13-nerved
    α. Corolla subequally 4-lobed, or obscurely 2-lipped; calyx-throat villous, mouth closed by the villi.
     * Whirls many-flowered, capitate; bracts conspicuous forming an involucre ... ... ORIGANUM.
     ** Whorls few-flowered, not capitate; bracts minute, not forming an involucre ... ... THYMUS.
    β. Corolla distinctly 2-lipped, calyx-throat glabrous or hairy, mouth not closed.
     * Corolla-tube straight.
      ¶ Calyx sub-equally 5-toothed, or obscurely 2-lipped; corolla-tube not exceeding the calyx.
       † Calyx usually 10-nerved (leaves linear-oblong in Indian species) ... ... SATUREIA.
       †† Calyx usually 13-nerved (leaves broadly ovate in Indian species) ... ... MICROMERIA.
      ¶¶ Calyx distinctly 2-lipped, corolla-tube exceeding the calyx ... CALAMINTHA.
     ** Corolla-tube recurved and ascending MELISSA.

III. Perfect stamens 2, ascending, parallel. Anther-cells linear, solitary or separated by a filiform connective.

A. Anther-cells equal, contiguous, pendulous from a small connective ... ... PEROWSKIA.

B. Anther-cells equal, distant, pendulous from a long connective ... ... MERIANDRA.

C. Anther-cells very unequal or solitary on a long connective ... ... SALVIA.

D. Anthers connate at the margin ... ... ZIZIPHORA.

IV. Perfect stamens 4 with the upper pair longest, rarely 2, ascending or diverging. Anthers 2-celled, cells at length diverging. Ovary 4-partite.

A. Calyx tubular, 5-toothed ... ... NEPETA.

B. Calyx 2-lipped, upper lip much the larger ... DRACOCEPHALUM.

C. Calyx 2-lipped, lateral lobes of upper lip on the face of the midlobe ... ... LALLEMANTIA.

V. Perfect stamens 4, ascending, lower pair longest. Calyx 5-10-nerved. Upper lip of corolla erect, lower spreading, trif.

A. Calyx deeply 2-lipped; mouth of fruiting calyx closed by the lips.

1. Lips of calyx entire, upper lip with a broad-plate ... ... SCUTELLARIA.

2. Upper lip of calyx 3-toothed, without any plate, lower lip 2-toothed ... ... BRUNELLA.

B. Calyx tubular or campanulate, 5-10-toothed, or obscurely 2-lipped, mouth not closed.

1. Upper lip of corolla flat, not hooded.

a. Stamens included in the corolla-tube. ... MARRUBIUM.

b. Stamens exserted from the corolla-tube ... ANISOMELES.

2. Upper lip of corolla hooded.

a. Calyx with a very widely expanded limb, with sinuate margin ... ... OTOSTEGIA.

b. Calyx limb not expanded, 5-lobed, or 5-10-toothed.

i. Anther-cells hairy.

α. Anther-cells transverse; nutlets compressed ... ... GALEOPSIS.

β. Anther-cells not transverse; nutlets triquetrous ... ... LAMIUM.

ii. Anther-cells glabrous.

α. Calyx-teeth much shorter than the tube.

*Calyx 8-10-toothed.

¶ Upper lip of corolla shorter than the lower ... ... LEUCAS.

¶¶ Upper lip of corolla longer than the lower ... ... LEONOTIS.

** Calyx 5-toothed, or lobed.

¶ Upper filaments not appendiculate at the base.

† Anther-cells not transverse, (leaves not sected) ... STACHYS.

†† Anther-cells transverse, (leaves much sected) ... ... LEONURUS.

¶¶ Upper filaments appendiculate at the base ... ... EREMOSTACHYS.

β. Calyx-teeth or -lobes as long as or longer than the tube ... ... ROYLEA.

V. Stamens straight or ascending. Nutlets with a small basal scar or areola, rugose with a thick and slightly fleshy pericarp ... ... GOMPHOSTEMMA.

VI. Stamens 4, ascending; calyx 10-nerved. Ovary 4-lobed. Nutlets with a large, oblique or lateral areola.

A. Corolla 1-lipped, upper lip absent, lower 5-lobed ... TEUCRIUM.

B. Corolla 2-lipped, upper entire or 2-fid, lower 3-lobed ... ... AJUGA[1]

ACROCEPHALUS.

The genus numbers 40 species, spreading from the Malay Archipelago to Africa.

A. lilacinus Oliv. is a Gold Coast remedy for headache.

Acrocephalus indicus O. Kunz. is found all over India and ascends to 5,000 ft. in the Himalayas.

The plant is used as an expectorant in Sind.

Sindi: Ustukudus—.

AJUGA.

The genus consists of 30 palaeotemperate species.

A. Iva Schreb. is used medicinally in Europe; *A. Chamaepitys* Schreb. in Europe and the M'Zab; *A. reptans* Linn. in Europe and North America; *A. ophrydis* Burch. in South Africa.

Ajuga bracteosa Wall. is a small herb inhabiting the North-West Frontier Province, Kashmir, Punjab, Kumaon, and Nepal. It extends to Afghanistan, Tibet, China, Japan, and Abyssinia.

In the Punjab the leaves are given in the treatment of fevers as a substitute for cinchona. The plant is considered a bitter astringent and an aromatic tonic, specially useful in ague.

On the Salt Range it is used to kill lice, and is regarded as depurative.

Jhelum: Kauriboti—; *Kumaon*: Ratpatha—; *Sutlej*: Karku, Nilkantihi—; *Trans-Indus*: Khurbauri—.

ANISOCHILUS.

The genus consists of 20 species, inhabiting the tropical areas of Asia and Africa.

Anisochilus carnosus Wall. is an erect annual herb found all over the country, from Western Himalaya to Burma and South India, and Ceylon.

The fresh juice of the leaves mixed with sugar-candy is given by the Tamil doctors in cynanche; and mixed with sugar and gingelly-oil, is used as a cooling liniment for the head.

---

[1] Dr. S. K. Mukerjee.—A Revision of the Labiatae of the Indian Empire. *Records Bot. Survey of India*; vol. xiv, no. 1; 1940.

The plant is a mild stimulant, expectorant, particularly useful in the cough of childhood. The juice of the leaves mixed with sugar and human milk is in Mysore a popular domestic remedy for coughs in children.

The juice of the leaves is commonly used in catarrh.

The plant yields a volatile oil which is credited with stimulant, diaphoretic, and expectorant properties.

Bombay: Choraonva, Kapurli—; *Canarese*: Doddapatri—; *Deccan*: Ajvan, Panjiri—; *Hindi*: Panjiri—; *Gujerati*: Ajama, Ajmamapatru, Ubhoratavelio—; *Malayalam*: Chomara, Kattukurkka, Kurkka, Patukurkka—; *Marathi*: Choraonva, Kapurli—; *Pondicherry*: Lavande—; *Sinhalese*: Galkapprawalliya—; *Tamil*: Karppuravalli—; *Telugu*: Karpuravalli, Omamu-aku, Rogachettu—.

ANISOMELES.

The genus consists of 6 Indo-Malayan species, spreading to Australia.

A. indica O. Ktze is used medicinally in Annam, the Philippine Islands, and La Reunion; *A. malabarica* R. Br. also is used in La Reunion.

Leaves broadly ovate. Nutlets black	*A. indica.*
Leaves oblong-lanceolate. Nutlets pale brown ...	*A. malabarica.*

1. **Anisomeles indica** O. Ktze is found throughout India, ascending to 6,000 feet in the Himalayas. It is distributed to Ceylon, the Malay Peninsula and Archipelago, the Philippine Islands, and China.

The plant has carminative, astringent, and tonic properties. It yields an essential oil which, in Ceylon, is used in uterine affections.

Bombay: Gopali—; *Cantonese*: Ts'in ts'o—; *Chinese*: Ch'ien Ts'ao—; *Malaya*: Hee chin choo—; *Sinhalese*: Yak-wanassa—; *Tagalog*: Talingharap—.

2. **Anisomeles malabarica** R. Br. is found in the Deccan peninsula, and is a well-known plant in Southern India, whence it spreads to Ceylon, Malaya, and Mauritius.

In Southern India, few plants are held in higher esteem, or are more frequently employed in native practice, than this. An infusion of the aromatic bitter leaves is in common use in affections of the stomach and bowels, catarrhal affections and intermittent fevers.

In addition to its internal use in the cure of fevers, patients are made to inhale the vapour of a hot infusion so as to induce copious diaphoresis. An infusion of the leaves is given to children in colic, dyspepsia and fever arising from teething. A decoction of the plant, or the essential oil distilled from the leaves, is used externally in rheumatism.

In La Reunion the plant is considered sudorific and antipyretic. It is given for rheumatism.

In Mauritius is credited with antispasmodic and emmenagogue properties. It is given for flatulence, and in hysteria.

An infusion of the leaves of this plant is given on the west coast to children in colic, dyspepsia and fever arising from teething.

The infusion acts as a diaphoretic. An infusion (1 in 10) was tried in fever accompanying teething and was found useful (Koman).

Bombay: Chodhara—; *Canarese*: Karitumbe—; *Deccan*: Mogbirekapatta—; *English*: Malabar Catmint—; *Gujerat*: Gholochodharo, Makhmalichodharo—; *Malayalam*: Karintumpa, Pemarutti, Peruntumpa—; *Marathi*: Chodhara, Pandrachodhara, Sundara, Sundraphal—; *Mauritius*: Menthe musquée des Malabars, Boutankoushum, Pémayretti—; *Sanskrit*: Oshthaphala, Vaikuntha—; *Tamil*: Irattaippeymarutti, Peymarutti, Peyverutti, Sadumbai—; *Telugu*: Chinnaranabheri, Magabira, Mogabheri—.

BRUNELLA.

This genus includes 5 more or less cosmopolitan species.

B. vulgaris Linn. is used medicinally in Europe, China, and Malaya.

Brunella vulgaris Linn. occurs in the temperate Himalaya from Kashmir to Bhutan from 4,000 to 11,000 feet, in the Khasia Hills at 4,000-6,000 feet, and on the hills of South India: Nilgiris, Pulneys, and Travancore mountains. It inhabits most of the temperate regions of the Northern Hemisphere, and is to be found in Australia.

The herb is aromatic and carminative; reputed useful in the treatment of haemorrhages and diarrhoea. It is esteemed by herbalists for relaxed throats.

'The decoction of Prunell', says Gerarde, 'made with wine and water, doth join together and make whole and sound all wounds, both inward and outward . . .' 'There is not a better Wound herbe in the world . . . ' The whole herb has astringent, styptic, and tonic properties.

Self-Heal is still in use in modern herbal treatment as a useful astringent for inward and outward use.

An infusion of the herb, made from 1 oz. to a pint of boiling water, and taken in doses of a wineglassful, is considered a general strengthener. Sweetened with honey, it is good for a sore and relaxed throat or ulcerated mouth, for both of which purposes it also makes a good gargle. For internal bleeding and for piles, the infusion is also used as an injection. It is a household remedy in Germany.

The Swiss peasants use the plant as a vulnerary.

Regarded by the Punjab Himalayan hill tribes as expectorant and antispasmodic (Stewart). The green leaves smeared with castor oil and warmed over the fire are applied externally to the anus in cases of painful piles.

The plant is used for fevers and coughs in China and Malaya, and is considered there anti-rheumatic, alterative, and tonic.

The Flambeau Ojibwe Indians of North America use the root to make a tea to drink before going hunting; it is supposed to sharpen their powers of observation. They also use the root mixed with others for a female remedy.

The whites of Wisconsin use the root as a pungent and bitter tonic and antispasmodic; they credit it with vermifuge properties, and regard it as slightly diuretic. The root has also been used for obstructions of the liver, cramps and fits.

Arabic: Anas-ul-rawah—; ***Chinese***: Hsia K'u Ts'ao—; ***English***: All-heal, Brown-wort, Brunel, Brunella, Bumble-bees, Carpenter-grass, Carpenter's Herb, Fly Flowers, Heart of the Earth, Herb Carpenter, Hercules' All-heal, Hercules' Wound-wort, Hook-heal, London Bottles, Pick Pocket, Pimpernel, Prince's Feather, Proud Carpenter, Prunella, Self-heal, Sickle-wort, Slough-heal—; ***French***: Bonnerette, Bonnette, Brunelle, Brunellier, Brunette, Charbonnière, Herbe au charpentier, Pâquerette, Petite consoude, Petite consyre, Prunelle—; ***German***: Braunelle, Braunheil—; ***Hindi***: Dharu—; ***Italian***: Brunella—; ***Malaya***: Lo han tsao, Look ham chow—; ***North America***: Blue Curls, Carpenter Weed, Heal-all, Heart of the Earth, Self-heal—; ***Ojibwe***: Basi' bûgûk—; ***Persian***: Ustekhadus—; ***Punjab***: Austakhadus—; ***Roumanian***: Busuioc de camp, Busuioc marunt—; ***Sind***: Ustukhudus—; ***Spanish***: Brunela, Consuelda menor—; ***Urdu***: Ustekhadus—; ***Yorkshire***: Black Man—.

CALAMINTHA.

The genus numbers 60 species, scattered over the northern temperate regions and the mountains of the Tropics.

C. Acinos DC., *C. alpina* Lam., *C. Clinopodium* Benth., *C. grandiflora* Moench, *C. Nepeta* Savi., *C. officinalis* Moench are used medicinally in Europe.

The flowered plant of *C. officinalis* Moench is official in France.

Calamintha Clinopodium Benth. is found in the western temperate Himalaya, from Kashmir to Kumaon, at 4,000-12,000 feet. It is distributed to northern and western Asia, Europe, North Africa, and Canada.

The plant is cephalic, astringent, carminative, and tonic to the heart.

Arabic: Asaba-el-tatiyat—; ***Catalan***: Alfábrega boscana—; ***French***: Clinopode—; ***Spanish***: Albahaca silvestre major, Angelotes, Clinopodio, Pie de cama, Perilla de cama—.

COLEBROOKEA.

A Himalayan genus, comprising only one species.

Colebrookea oppositifolia Sm. is one of the commonest and most abundant plants in the lower Himalaya. It occurs all over India, in the hilly parts, from Peshawar to Burma, and Simla to Travancore.

A preparation from the root is used by the Santals in epilepsy.

The leaves are applied to wounds and bruises.

The down on the stem and leaves is used by the Paharias of Sikkim to extract worms from bad sores on their legs.

Bombay: Bahmani, Bhamini, Dasai, Dasari, Dussarica—; ***Canarese***: Tuggigidda—; ***Dehra-Dun***: Binde, Bindu—; ***Garhwal***: Binda, Bindu—; ***Hindi***: Binda, Bindu, Pansra—; ***Jaunsar***: Bambher, Lulri—; ***Khond***: Darigopi, Merata—; ***Kumaon***: Binda, Bindu, Dulshat—; ***Matheran***: Bhaman—; ***Melghat***: Chotabhandara—; ***Nepal***: Dosul—; ***Punjab***: Barmera, Basuti, Briali, Dashane, Duss, Phisbekkar, Sampru, Shakardana, Suali—; ***Ramnagar***: Bhuriruderi—; ***Saharanpur***: Bintalakri, Kalabansa—; ***Santali***: Barsakapor, Bhainsa—; ***Saora***: Jolidi—; ***Trans-Indus***: Shakardana—; ***Uriya***: Bosiki, Darigopi—.

COLEUS.

This genus consists of 150 palaeotropical species.

C. Amboinicus Lour. is used medicinally in Cambodia, the Malay Archipelago, and the Philippine Islands; *C. atropurpureus* Benth. in the Philippine Islands; *C. Bojeri* Benth. in Madagascar; *C. dysentericus* Baker in Nyassaland, Nigeria, and Madagascar; *C. floribundus* Baker in Ubanghi-Shari.

Coleus Amboinicus Lour., a native of the Moluccas, is cultivated in gardens throughout India, and Ceylon. It occurs wild in Rajputana.

The leaves are said to have a specific action on the bladder and to be useful in urinary diseases, vaginal discharges, etc. The juice mixed with sugar is given in cases of colic in children, and acts as a powerful aromatic carminative.

In spite of its intoxicating properties the people of Bengal employ it in colic and dyspepsia.

The expressed juice of the leaves is considered an anodyne and astringent, and applied over and around the eyelids, in cases of conjunctivitis.

In Ceylon a decoction of the leaves is given for asthma, chronic coughs, etc.

In Cochin-China the juice of the leaves is considered carminative and is given to children suffering from wind colic. The decoction is given for asthma, chronic bronchitis, epilepsy, and convulsions.

The leaves are official in Holland.

Bengal: Paterchur—; *Bombay*: Owa, Pathorchur, Pathurchur—; *Cambodia*: Chi trasak damrey—; *Hindi*: Pathorchur—; *Malay Archipelago*: Daoen Koetjing, Djinten—; *Marathi*: Panacha ova, Patharchur—; *Philippines*: Oregano—; *Sanskrit*: Pashanabhedi—; *Sinhalese*: Kapprawalliya—; *Tagalog*: Suganda—.

DRACOCEPHALUM.

The genus consists of 40 species, inhabiting the temperate regions of the Northern Hemisphere.

Dracocephalum Moldavica Linn. is found in the western temperate Himalaya and Kashmir, at altitudes of 7,000 to 8,000 feet. It extends to northern Asia and Europe.

In Patna the seeds are used ground up in fevers, and as demulcent.

The seeds afford an opaque mucilage when soaked in water. The drug is esteemed in Persia as a carminative and tonic.

In Europe the plant is considered tonic, astringent, and vulnerary.

Arabic: Asaba-el-fatiyat, Baklat-el-utrujuya—; *Hindustani*: Tukhm- ferunjmishk—; *Persian*: Badrendj-bouyih, Badrendj-buya, Karanfalihostani—; *Spanish*: Melisa de Moldavia, Torongil de Turquia—; *Tabriz*: Badirash-bu—; *Teheran*: Badranj-buya—.

DYSOPHYLLA.

The genus consists of 20 species, inhabiting eastern Asia and Australia.

Dysophylla auricularia Blume is found in Sikkim, eastern Bengal, Assam, Burma, Singbhum, Poona, and South India. It is distributed to Ceylon, the Malay peninsula and islands, the Philippine Islands, and China.

The herb is a Malayan remedy for colic. The leaves are powdered with lime and rubbed on the abdomen.

Malay: Ekor kuching, Poko awi tana—.

ELSHOLTZIA.

The genus consists of 30 species distributed over Asia, Europe, and Abyssinia.

Elsholtzia cristata Willd. is to be found from Kashmir to Mishmi at 1,200-9,000 feet. It has been of late introduced in the Nilgiris. It spreads to Tibet, China, Japan, northern Asia, and Europe.

In China and Cochin-China the plant is used as a medicine, a pot herb, and a condiment. In Annam the flowering tops are given as a diuretic. In Japan the leaves are used for tea. In a general way, like other Labiates, the drug is regarded by the Chinese as carminative, stomachic, and astringent.

Cantonese: Heung ue—; *Chinese*: Hsiang Ju—; *Malaya*: Heong yee—.

EREMOSTACHYS.

The genus numbers 40 species, natives of Central and Western Asia.

Bracts and bracteoles spine-tipped, equalling calyx; bracteoles connate below in groups of 3 ... ... *E. acanthocalyx.*

Bracts and bracteoles not spiny, much shorter than calyx, all free ... ... *E. Vicaryi.*

1. **Eremostachys acanthocalyx** Boiss. is found in West Punjab and Baluchistan, whence it extends to Baluchistan.

The Baluchis consider the plant poisonous.

Kirani: Bishkhaf—.

2. **Eremostachys Vicaryi** Benth. occurs in Western Punjab and is common on the Salt Range, ascending to 2,500 feet. It is also found at Peshawar and in Baluchistan.

In the Punjab the seeds are given as a cooling medicine.

The plant is said to be used in the Eusufzai near Peshawar for poisoning fish.

Kila Saifulla: Khurzbin—; *Pab Hills*: Bischkhaf—; *Punjab*: Gurgunna, Khalatra, Rewandchini—.

GALEOPSIS.

A genus of 7 palaeotemperate species, inhabiting the Northern Hemisphere.

G. ochroleuca Lam. is official in the Pharmacopoeia of Austria.

Galeopsis Tetrahit Linn. occurs in Sikkim and Kashmir. It is distributed over Tibet, North and West Asia, Europe; and has been introduced into North America.

The plant is well spoken of as an expectorant, and in phthisical complaints. An infusion of the whole plant is used in the treatment of pulmonary troubles. It has also been employed as an anti-spasmodic resolvent, and a detergent.

English: Bastard Hemp, Bee-nettle, Blind Nettle, Dai-nettle, Day-nettle, Dea-Netle, De-Nettles, Deye-Nettle, Dog Nettle, Donnine-thell, Female Hems, Glidewort, Hemp Nettle, Holyrope, Nettle Hemp, Stinging Nettle, Sting Nettle—; *Forest Potawatomi*: Mena' kwûskûk—; *North America*: Common Hemp Nettle—.

GENIOSPORUM.

The genus includes 15 species scattered over Africa, Madagascar, and the Indo-Malayan region.

Geniosporum prostratum Benth. is found in sandy ground from the Konkan southwards, especially near the sea, and in the warmer parts of Ceylon.

In Pondicherry this plant is credited with febrifugal properties.

Pondicherry: Nazelnagai—.

GOMPHOSTEMMA.

A genus of 25 Indo-Malayan species spreading to China.

Gomphostemma crinitum Wall. occurs in Assam, Burma, and the Malay Peninsula.

var. *Griffithii* Prain, which is found in Tenasserim and Malaya, is administered in the form of decoction after confinements.

Malay: Munjulong bukit—.

HYMENOCRATER.

The genus consists of 9 species, natives of Western Asia.

Hymenocrater sessilifolius Benth. occurs in Baluchistan and Afghanistan.

In Baluchistan the leaves are left overnight soaking in water, and the infusion is given as a morning drink to children.

Baluchistan: Sursánda—

HYPTIS.

This large genus numbers 300 species, natives of warm America.

The following species are used medicinally in Madagascar and West Tropical Africa. *H. pectinata* Poit., *H. spicigera* Lam.—; in Guiana—*H. verticillata* Jacq.—; in Brazil—*H. canescens* Benth., *H. fasciculata* Benth., H. *fruticosa* Salzm., *H. graveolens* Salzm., *H. pectinata* Poit., *H. spicata* Poit., *H. suaveolens* Poit., *H. umbrosa* Salzm—.

Flowers not in globose heads.
    Calyx 3 mm. long in fruit ... ... *H. pectinata*
    Calyx 8-10 mm. long in fruit ... ... *H. suaveolens*

1. **Hyptis pectinata** Poit. is found in Bengal, Assam, and the Madras Presidency. It is distributed over tropical Africa and Asia.

In Ashanti and South Nigeria the plant is a medicine for young children, and used in connection with childbirth. Poultices of the leaves are applied for chest complaints.

In North Nigeria the leaf is used for fever. It is boiled for horses to inhale the vapour for diseases accompanied by mucous catarrh.

In Madagascar the plant is considered aromatic, tonic, anthelmintic, antispasmodic, emmenagogue, and odontalgic. It is mostly used to expel worms in children. A decoction or infusion of the flower heads is given in fever, and for chest troubles.

Adangme: Kadokeng—; *Angola*: Quibumbo, Quimbumba, Quinbumbo, Quinbungu—; *Ashanti*: Piaa—; *Awuna*: Awusa-kadi—; *Betsileo*: Afolava—; *Dagomba*: Baeba—; *Ga*: Suruwie—; *Hausa*: Kimba-kimba, Kimbar awaki, Kimbar dawaki—; *Hova*: Rombatsahona, Sangasanganandevolahy—; *Sakalave*: Sangasanganimarina—; *Twi*: Opea, Peaba—; *Yoruba*: Jogbo—.

2. **Hyptis suaveolens** Poit. occurs in the Deccan, in Chota Nagpur, Bengal, Assam, Burma, and the Andaman and Nicobar islands. It is distributed to the Malay Archipelago, the Philippine Islands, Formosa, Indo-China, Siam, and tropical Africa.

In Orissa the plant is pounded and applied to parasitical cutaneous diseases.

In West Tropical Africa an infusion of dried leaves is taken for fever. The plant is tied round the head for headache or applied to cure boils, and the juice of the pressed leaves along with lime juice is drunk for colic and stomach-ache.

In Brazil an infusion is used as a carminative and as a sudorific in catarrhal conditions.

Colombia: Botón morado, Chavito, Mastrantillo, Mastranto de perro, Mastranto de sabana, Yerba de las muelas, Yerba de la reuma—; *Malay*: Malbar hutan, Sapulut, Selasik hutan—; *Mal Paharia*: Purudo—; *Mandingo*: Furc-jambo—; *Mexico*: Chan—; *Panama*: Purgaperro, Salvia—; *Santali*: Gangatulsi—; *Tagalog*: Pansipansiyan, Soobcabayo—; *Timne*: Bupi-bupi—; *Uriya*: Gangatulsi, Purodo—; *Venezuela*: Mastranto—; *Visayan*: Locoloco—; *Yoruba*: Jogbo—.

Hyssopus.

Hyssopus officinalis Linn., the sole representative of the genus, is found in the Western Himalaya, from Kashmir to Kumaon, at 8,000-11,000 feet. It spreads to Western Asia, and grows wild in middle and southern Europe.

Hyssop is classed by the Arabians amongst their anthelmintics, stimulants, and deobstruents. In Europe it has long been known as a tonic and stimulant, and was at one time in great repute as a remedy for nervous diseases. *Pliny* also thought it useful in chest affections, and *Celsus* regarded it as an anthelmintic.

Hyssop tea is a grateful drink, well adapted to improve the tone of a feeble stomach, being brewed with the green tops of the herb. The same parts of the plant are sometimes boiled in soup to be given for asthma. To make Hyssop tea, one drachm of the dried herb should be infused in a pint of boiling water, and allowed to become cool. Then a wineglassful is to be given as a dose two or three times a day. It is of use in the ailments of women.

A distilled water of Hyssop is deemed a good pectoral medicine.

The essential oil, in doses of one to two drops, promotes expectoration in bronchial catarrh and asthma.

The green herb, bruised and applied, will heal cuts promptly. If it be steeped in boiling water and applied hot to the part, it will quickly remove the blackness consequent upon a bruise or blow, especially in the case of 'black' or blood-shot eyes. In America an infusion of the leaves is used externally for the relief of muscular rheumatism, as also for bruises and discoloured contusions.

The juice of the leaves made into a syrup with sugar and honey is used as a vermifuge for roundworms.

A decoction of the flowers of this plant from Persia was given in cases of asthma and chronic bronchitis and found to be a useful remedy in those complaints (Koman).

The flower-tops are official in France, Portugal, and Sweden; the leaves too are official in France.

Arabic: Zufah-e-Yabis, Zufah-ul-retih—; *Catalan*: Hisop—; *Danish*: Isop—; *Dutch*: Hysop—; *English*: Hyssop—; *French*: Herbe sacrée, Hysope—; *German*: Apothekerhysop, Eiserich, Gartenrispe, Hyssop, Isipo, Isop, Ispern, Issel, Josephskraut, Klosteruesopp, Lufftskraut, Ysop—; *Greek*: Hysopos—; *Hebrew*: Esof—; *Hindi*: Żufah yabis—; *Italian*: Isopo, Issopo—; *Languedoc*: Mariarmo—; *Persian*: Zufah-e-khuskka, Zufah-e-tar, Zufah-e-yabis—; *Polish*: Isopek—; *Portuguese*: Hissopo—; *Roumanian*: Cimbru cel brun, Isop—; *Russian*: Issop—; *Spanish*: Hisopo—; *Swedish*: Isop—; *Urdu*: Zufah yabis—.

LALLEMANTIA.

The genus consists of 4 species, inhabiting Western Asia.

Lallemantia Royleana Benth. occurs in the North West Province, the Punjab plains and hills, and Baluchistan; extending to Afghanistan, Persia, and Turkestan.

The plant is largely grown on account of its mucilaginous seeds, which are considered cooling and sedative, and are extensively employed in the preparation of a mucilaginous beverage. In Persia they are used for cough, and as a stimulant and aphrodisiac.

In the Punjab the seeds are used as cooling and sedative remedies.

Bombay: Tukhmibalangu—; *Hamadan*: Balingau—; *Harboi Hills*: Fuchkin, Yakhtali—; *Hindi*: Ghareikashmalu, Tukhmbalangu, Tukhmlealanga—; *Kashmir*: Tukhmibalunga—; *Punjab*: Ghareikashmalu, Tukhmbalangu, Tukhmmalanga, Tukmalanga—; *Persian*: Tukhmibalangu—; *Urdu*: Balanga—.

LAMIUM.

The genus includes 40 species distributed over Europe, North Africa, and temperate Asia.

Lamium album Linn. occurs in Kashmir, the Punjab, Kumaon, and Hazara. It is scattered over northern and western Asia, North Africa, and Europe.

The root is used medicinally in China. In Spain it is considered astringent and is used as a resolvent and vulnerary.

Catalan: Ortiga morta—; *Chinese*: Hsu Tuan—; *English*: Archangel, Bee Nettle, Blind Nettle, Day Nettle, Dead Nettle, Deaf Nettle, Dee Nettle, Dumb Nettle, Dummy Nettle, Dunny Nettle, Nettle, Snake Flower, Stingy Nettle, Suck-bottle, Suckie Sue, White Archangel, White Deadnettle, White Nettle—; *French*: Archangélique, Galéopsis, Lamier blanc, Lamion, Marachemin, Ortie blanche, Ortie morte, Pied de poule, Suçots blancs—; *German*: Bienenhuetel, Biemensang, Dahnnessel, Dangel, Daudelblume, Eddernessel, Eisblume, Gageneier. Heddernessel, Honigsugel, Huehnernessel, Hundnessel, Loeffelblume, Sengenessel, Suegede, Taubnessel, Taunessel, Urinblume, Wasserblume, Weisse Betonie, Weisser Ganzert, Weisser Kuckuck, Weisse Suegete, Wurmnessel—; *Italian*: Lamio bianco—; *Russian*: Sadovui dyagil—; *Spanish*: Ortiga blanca, Ortiga muerta, Ortiga muerta blanca—.

LAVANDULA.

The genus consists of 20 species spreading from the Mediterranean to India.

L. dentata Linn., *L. Spica* DC., *L. Stoechas* Linn., *L. vera* DC. are used medicinally in Europe.

Lavandula bipinnata O. Ktze. occurs in Chota Nagpur, Mount Abu, Konkan, Jubbulpore, Khandesh, and the Deccan.

It is reported that the villagers and shepherds of the Barda Hills in Kathiawar use the plant as a medicine.

The plant is supposed to act as an antidote against snake poison. The roots are rubbed with water and the solution or the paste is applied over the sting or the bite of poisonous animals. The powdered leaves are given for inhalation to the person who has been bitten by a serpent in order to prevent him from falling into sleep.

Gujerati: Aasmanigalgoto, Sarpnocharo—.

LEONOTIS.

The genus consists of 15 tropical and South African species.

L. dysophylla Benth., *L. Leonitis* R. Br., *L. Leonurus* R. Br., *L. microphylla* Skan., *L. mollis* Benth. are used medicinally in South Africa; *L. africana* Brig. is used in Angola and Northern Nigeria; *L. nepetaefolia* R. Br. in Northern Nigeria, Madagascar, and Brazil.

Leonotis nepetaefolia R. Br. is found, either cultivated or naturalized, in the Punjab and all the hotter provinces of northern and southern India, whence it spreads to Ceylon. It is distributed to tropical Asia, Africa, and America.

In Chota Nagpur the ashes of the flower-heads are applied to burns and scalds; in Bombay they are mixed with curds and applied to ringworm and other itchy diseases of the skin.

The plant is used medicinally by the Mundas of Chota Nagpur. The whorl, in flower or in seed, is pounded and fried in Koronj oil; this is rubbed on itch, also on head sores of small children. When a mother's breasts swell and milk does not pass through the nipples, the crushed root is rubbed on the breast (*Encyclopoedia Mundarica*).

In Loanda a decoction of the plant is used in diseases of the abdomen.

In Northern Nigeria the leaf pounded with natron is applied locally for swellings and ulcers supposed to be of syphilitic origin; the fresh leaves are applied for headache. A decoction is used to steam the head to relieve catarrh, fever, etc., and is taken also internally as a tonic and febrifuge, and for gastro-intestinal troubles. The plant is put amongst stored corn to keep away vermin.

In Madagascar the plant is considered emmenagogue, febrifuge, depurative, narcotic, bitter, and laxative; used in skin diseases, amenorrhoea, and fever.

In Porto Rico a decoction of the leaves is used as a tonic; the juice is also expressed and taken with limejuice and rum as a febrifuge.

The leaves are used in Brazil in the treatment of rheumatic affections.

Bengal: Hejurchei—; *Bombay*: Matijer, Matisul—; *Brazil*: Cordao do frade—; *Ceylon*: Kasitumpai—; *Efik*: Ubiom—; *Fulani*: Hore gujjo—; *Gujerati*: Matijer, Matisul—; *Hausa*: Chika saura, Jam'barawo, Kam mutum, Futar'yan sarki—; *Hindi*: Baraguma, Hejurchei—; *Krobo*: Nyeddo—; *Lagos*: Iku ekun—; *Loanda*: Maluvo, Maluvo-iamconeo, Maluvo-iamgilla, Maluvo-m'angilla—; *Mano*: Seto a yi—; *Marathi*: Dipmal, Ekri—; *Mauritius*: Dacca, Léonure—; *Mundari*: Agiajanum, Gharia, Hatusenggelsui, Senggelsui—; *Porto Rico*: Molonillo, Rascamoño—; *Sakalave*: Kilanjan-anahary—; *Santali*: Daredhompo, Janumdhompo, Jonumdhompo—; *Sinhalese*: Mahayakwanassa—; *Telugu*: Beri, Hanumantabira, Mulugolimidi, Ranabheri—; *Uriya*: Kontosidho—; *Uruguay*: Uña del diablo—; *Yoruba*: Iku ekun—.

LEONURUS.

The genus consists of 8 species inhabiting Europe, Asia, and tropical regions.

L. macranthus Maxim. and *L. sibiricus* Linn. are used medicinally in China, and *L. sibiricus* is also used in Malaya; *L. Cardiaca* Linn. is used in Europe.

Upper lip of corolla densely villous, midlobe of lower lip entire ... ... *L. Cardiaca*
Upper lip of corolla tomentose, midlobe of lower lip obcordate or 2-fid ... ... *L. sibiricus*.

1. **Leonurus Cardiaca** Linn. is found in Kumaon, Kashmir, the Punjab, Hazara, and the Kurrum Valley. It is distributed to western and northern Asia, and to Europe.

Says Culpeper: 'There is no better herb to take melancholy vapours from the heart, and to strengthen it. It may be kept in a syrup or conserve; it maketh mothers joyful, and settles the womb, therefore it is called Motherwort. It is of use for the trembling of the heart, fainting and swooning. The powder, to the quantity of a spoonful, drank in wine, helps women in sore travail, as also for the suffocating or rising of the mother. It provokes urine and women's courses, cleanses the chest of cold phlegm, kills the worms in the belly. It is of use to digest and disperse them that settle in the veins, joints, and sinews of the body, and to help cramps and convulsions.'

The herb is diaphoretic, aromatic, and stomachic.

Contact with the plant causes a dermatitis in susceptible individuals.

Catalan : Mà de Santa Maria— ; *English* : Motherwort— ; *French* : Agripaume, Cardiaire, Cardiale, Cardiaque, Cheneuse, Creneuse, Herbe aux tonneliers, Léonure, Patte de sorcier— ; *German* : Herzgespann, Loewenschwanz— ; *Italian* : Cardiaca— ; *Portuguese* : Cardiaca— ; *Roumanian* : Cione, Creasta cocosulin, Iarba flocoasa, Talpa gascei— ; *Russian* : Pustirnik— ; *Spanish* : Agripalma, Cardiaca, Cola de león— ; *United States* : Lion's Ear, Lion's Tail, Motherwort, Throwwort—.

2. **Leonurus sibiricus** Linn. is found in the plains of India from Bengal and Sylhet to Coorg. It is distributed to tropical Asia, Africa, and America.

The root and leaves are bitter and used as a febrifuge.

In Chinese medicine the seeds are considered to be constructive and aphrodisiac. The dried plant is prescribed as a tonic, alterative, vulnerary, and general remedy in puerperal and mentrual diseases.

Cantonese : Ch'ung wai— ; *Chinese* : Ch'ung Wei, I Mu, K'uen Ts'ao— ; *Malaya* : Choong wai, Kwan chor seranting, Tebung aga— ; *Patna* : Guma— ; *Tagalog* : Camariangsongsong—.

LEUCAS.

The genus includes about 100 Asiatic and African species.

L. aspera Spreng. is used medicinally in the Philippine Islands; *L. decurvata* Baker in Nyasaland; *R. deflexa* Hook. fil. in Gold Coast; *L. martinicensis* R. Br. in Northern Nigeria, South Africa, and Brazil.

A. Calyx-mouth oblique.
- I. Mouth of calyx much produced below ... *L. urticaefolia.*
- II. Mouth of calyx produced above
  Calyx-teeth very short
  - a. Calyx-mouth not villous within.
    - 1. Calyx smooth below, ribbed and hispid above ... ... *L. aspera.*
    - 2. Calyx smooth throughout or nearly so. Calyx-mouth very oblique ... *L. linifolia.*
  - b. Calyx-mouth villous within
    - 1. Leaves linear, less than 13 mm. broad, whorls less than 2.5 cm. diam. ; bracts linear, 6 mm. long ... ... *L. zeylanica.*
    - 2. Leaves ovate, more than 13 mm. broad,
      - i. Whorls 3.5 cm. diam. ; bracts lanceolate, 1.5-2.5 cm. long ... ... *L. Cephalotes.*
      - ii. Whorls 2.3 cm. diam. ; bracts linear-lanceolate or subulate, 6-9 mm. long ... *L. martinicensis*

B. Calyx-teeth stellately spreading in fruit, whorls terminal and axillary.
  Hairs on the stem erect or spreading. Calyx 5-10 mm. long ... ... *L. stelligera.*

1. **Leucas aspera** Spreng. occurs more or less throughout India, in the plains. It extends to Indo-China, the Philippine Islands, Java, and Mauritius.

The leaves are said to be useful in chronic rheumatism. The juice is applied in psoriasis and other chronic skin eruptions.

In North Bengal the flowers are given warmed in a little honey for coughs and colds to children. The juice of the leaves is applied to disperse painful swellings.

In Mauritius the plant is used as an emollient and pectoral.

Caius and Mhaskar have shown experimentally that the plant is not an antidote to snake venom.

Bengal: Chotahalkusa—; *Bombay*: Tamba—; *Deccan*: Thurduribaji—; *Hindi*: Chotahalkusa—; *La Reunion*: Herbe à mouches, Herbe Tombe, Tombe—; *Mauritius*: Madame Tombé, Marrube blanc, Couma, Halkasa, Poualla toumi—; *Mundari*: Gomaara—; *Sadani*: Gumhasag—; *Tagalog*: Carucansoli, Pansipansi, Solasolasian—; *Tamil*: Tumbai—; *Telugu*: Tummachettu—; *Visayan*: Paipaisi, Pansipansi, Paypaysi—.

2. **Leucas Cephalotes** Spreng. is found throughout India from Sind to Assam, and Kashmir to Deccan; scattered over all plains districts of the Madras Presidency. It extends to Afghanistan.

The plant is considered a mild stimulant and diaphoretic, and is considerably used in fevers and coughs.

The fresh juice is used in certain localities as an external application in scabies. The flowers are administered in the form of a syrup as a domestic remedy for coughs and colds.

The seeds yield an oil, which is used medicinally in Chota Nagpur.

The leaves, in combination with other drugs, are prescribed for scorpion sting (Vagbhata); but they are not an antidote to scorpion venom (Caius and Mhaskar).

Bengal: Barahalkasa, Ghalaghase—; *Gangpur*: Nakiara, Nakingara—; *Gujerati*: Doshinokubo, Khetraukubo, Kubi, Kubo—; *Hindi*: Deldóna, Dhurpisag, Goma, Guma, Motapati—; *Khandesh*: Kedari—; *Marathi*: Devkumbha, Kumbha, Shetvad, Tumba—; *Mundari*: Bananaki, Gomanaki—; *Punjab*: Chatra, Guldoda, Maldoda, Phuman, Sisalius—; *Sanskrit*: Chatraka, Chhatrani, Chitrakshupa, Chitrapatrika, Drona, Dronapushpi, Kaundinya, Kshavapatri, Kumbhayoni, Kumbhayonika, Kurumba, Kurumbika, Palindi, Phalepushpa, Shvasanaka, Supushpi, Vrikshasaraka—; *Santali*: Andiadhuruparak, Andiadura parak—; *Telugu*: Peddatumni, Tumni—.

3. **Leucas linifolia** Spreng. is met with in the plains, more or less throughout India. It is distributed to Malaya and Mauritius.

The natives of Central India believe that the leaves, when roasted and eaten with salt, have febrifugal properties.

The fresh juice is employed as a remedy against headaches and colds.

At Lakhimpur in Assam it is used for loss of appetite. The leaves are wrapped up in a plantain leaf, heated, and then eaten. The first effect of this treatment is that the appetite decreases to such an extent that the patient is unable to take any food at all, on the second day it passes off and he takes food with avidity (Carter).

In the North West, the leaves are bruised and a teaspoonful of juice given, which is snuffed up the nostrils as a remedy against snake-bite.

The leaves are equally useless in the treatment of snake-bite whether given internally, or used as an errhine or a collyrium, or applied locally to the part bitten (Mhaskar and Caius).

Assam: Dron—; *Bengal*: Halkasa, Halkussa—; *Deccan*: Goma—; *Gond*: Gumi, Kumbha—; *Gujerati*: Jhinanpannikubo—; *Hindi*: Guma, Halkusa, Kumbha—; *Lakhimpur*: Guma—; *La Reunion*: Herbe à mouches, Herbe Tombe, Tombe—; *Marathi*: Kuva—; *Philippines*: Parnipansi—; *Sanskrit*: Dronapushpi, Kumbhi, Rudrapushpa—; *Sundribuns*: Guma, Hal-kūsa—; *Telugu*: Pulátumni—; *Urdu*: Guma—.

4. **Leucas martinicensis** R. Br. is found all over South India. It is distributed to the Malay Peninsula, tropical Africa, and tropical America.

In Gambia the whole plant is made into an infusion and used as a wash in fevers.

In Northern Nigeria the plant is burnt for the purpose of expelling mosquitoes. An infusion is used for simple gastro-intestinal troubles and for colds, and as a wash or steam for fevers.

The infusion is used in Brazil as a bath for hysterical fits and for pain in the joints.

Betsileo: Kifilanjambola—; *Brazil*: Catinga da Mulata—; *Fulani*: Erisku, Hore gujjo, Risku—; *Gambia*: Wild Tea Bush—; *Hausa*: Kam'barawo—; *Hova*: Lanjananahary—; *Katsina*: Sarakuwar sauro—; *Sokoto*: Sarakuwar sauro—; *Zaria*: 'Dai'doyar gona—.

5. **Leucas stelligera** Wall. is found in the Konkan, Canara, Mysore, and the Nilgiris.

The plant has stimulant, carminative, and emmenagogue properties.

Gujerati: Dungaraukubo—; *Laos*: Jang nog gan—; *Marathi*: Barumbi, Guma—; *Matheran*: Borambi, Guma, Matasul—.

6. **Leucas urticaefolia** R. Br. occurs from Baluchistan and the Punjab to Behar, Sind, Gujerat, Rajputana, Central and South India, mostly in the plains. It is distributed to Arabia, Nubia, and Abyssinia.

At Gomawal in Baluchistan the plant is used as a cure for fever.

Gujerat: Kubo—.

7. **Leucas zeylanica** R. Br. occurs in Assam, Chittagong, and South India, whence it spreads to Ceylon. It extends to the Malay Peninsula and Archipelago, and to China.

In Malaya the herb is used as a cure for scabies.

In Ceylon the leaves are bruised and a teaspoonful of the juice given, which is sniffed up as a remedy in snake-bite. The juice is also employed in headache and colds.

The leaves are useless as an errhine in the treatment of snake-bite (Mhaskar and Caius).

Malay: Katumbit—; *Sinhalese*: Gattatumba, Getatumba—.

LYCOPUS.

This genus includes 10 species of the northern temperate regions.

L. sinuatus Ell. and *L. virginicus* Linn. are used medicinally in the United States of America.

Lycopus europaeus Linn. is found in Kashmir. It is distributed to Europe, West, North, and Central Asia.

The leaves are used externally as a poultice to cleanse foul wounds.

It is used in the Punjab as a cooling drug.

It has been employed in Europe as a substitute for quinine.

English : Gipsywort—; *French* : Marube aquatique—; *German* : Wolfsfuss—; *Kashmir* : Gandamgundu, Jalnim—; *Spanish* : Marrubio acuático, Pie de lobo—.

MARRUBIUM.

The genus includes 30 species, natives of Europe, North Africa, and temperate Asia.

M. Alysson Linn. is used medicinally in Spain; *M. vulgare* Linn. in Europe, California, and South Africa.

Marrubium vulgare Linn. occurs in Kashmir, the North West Frontier Province, and Baluchistan, whence it extends to Afghanistan, Europe, and North Africa.

The dried leaves and flowering tops are carminative, expectorant, laxative, aperient, diaphoretic, stimulant, and tonic.

The herb is a bitter tonic, expectorant, and diuretic. It is perhaps the most popular of herbal pectoral remedies in England. It is exceedingly valuable in coughs, cold, and pulmonary affections. In many parts it is brewed and sold as Horehound Ale, making an appetising and healthful beverage. Also a candy is prepared, and if properly made, is no doubt efficacious.

The infusion is much used in Europe as a domestic remedy for bronchitis with profuse expectoration. It is tonic, and in large doses, purgative.

The Europeans in South Africa use an infusion in febrile conditions, and especially in typhoid fever.

In America it is generally used in catarrhal states of the air passages.

The plant is especially esteemed by the Spanish-Californians as a remedy for colds and lung troubles.

In Mexico a preparation made from the leaves is used for rheumatism; it is added to mescal and applied as a liniment.

Afrikaans : Koorsbossie—; *Arabic* : Faracioūn, Farasiym, Hashishatelkalb, Sufelard—; *Brahui* : Borkaah—; *Catalan* : Malroig, Malrubé, Malrubi blanch, Malrubins, Marreus—; *Danish* : Hvidmarru, Hvidrubike, Rubike—; *Dutch* : Gemeene malrove—; *English* : Common Horehound, Hoarhound, Horehound, White Hoarhound—; *French* : Blanc roubi, Blanc rubi, Bonhomme, Bon riblet, Bon rubi, Grand bonhomme, Herbe vierge, Marchemin, Marinclin, Marrochemin, Marrochen, Marrube, Marrube blanc, Marrube bonhomme, Marrube commun, Mont blanc—; *German* : Andornkraut, Berghopfen, Chinakraut, Gotteshuelfe, Gottvergess, Gutverges, Kuckskraut, Lungenkraute, Maeuseoehrchen, Marienwurzelkraut, Markobell, Nagelkraut, Siebennagelspitzer, Wasserdorn, Weisse Leuchte,

Weisser Andorn, Weisser Anton, Weisser Daurand, Weisser Dorand, Weisser Gottvergess, Weisser Winderthon, Weissleuchterkraut—; ***Greek*** : Prasion—; ***Hindi*** : Paharigandana—; ***Italian*** : Erba apiola, Marrobio, Marrobio bianco—; ***Kurdish*** : Qutainah—; ***Languedoc*** : Maltraste, Maribbe—; ***Malta*** : White Horehound, Marrobio, Robbio, Erba apiola, Marrubija, Marrubjabajda—; ***Persian*** : Afnanesar—; ***Polish*** : Szanta biala—; ***Portuguese*** : Marroio, Marroio branco—; ***Provence*** : Bouan rible, Mentastre, Marroufe, Marrubi, Moun blanc—; ***Roumanian*** : Catusnica, Catusnica selbatica, Iarba flocoasa, Iarba mitei, Iarba vintului—; ***Russian*** : Marrub, Schandra—; ***South Africa*** : Horehound, Marvel, White Horehound—; ***Spanish*** : Marrubio, Marrubio blanco—; ***Swedish*** : Andorn—; ***Urdu*** : Farasiym—; ***Uruguay*** : Malva rubia, Marrubio, Yuyo del sapo—

MELISSA.

A genus of 4 species, natives of Europe and Western Asia.

M. officinalis Linn. is used medicinally in Europe. Its flowerheads and leaves are official in Austria, Belgium, Brazil, Denmark, France, Germany, Hungary, Italy, Norway, Portugal, Spain, Switzerland, Turkey, and Yugoslavia.

Melissa parviflora Benth. occurs in the temperate Himalaya, from Garhwal to Sikkim and Mishmi, and in the Khasia Hills. It is distributed to Java.

The plant is a good substitute for *M. officinalis.*

Arabic : Baklatelutrujiya, Mufarehulkalab—; ***Hindi*** : Bililotan—; ***Persian*** : Badrunjboya—; ***Urdu*** : Baranjboya—.

MENTHA.

The genus includes 25 Old World species.

The following species are used medicinally in Europe:—*M. aquatica* Linn., *M. arvensis* Linn., *M. cardiaca* J. G. Baker, *M. gentilis* Linn,, *M. piperita* Linn., *M. Pulegium* Linn., *M. rotundifolia* Huds., *M. sativa* Linn., *M. sylvestris* Linn., *M. viridis* Linn.—; in China, Indo-China, and Malaya—*M. arvensis* Linn.—; in Madagascar—*Mentha spp.*—; in South Africa—*M. aquatica* Linn., *M. capensis* Thunb., *M. crispa* Linn., *M. longifolia* Huds.—; in North America—*M. aquatica* Linn., *M. piperita* Linn., *M. sylvestris* Linn .—.

1. Leaves sessile, lanceolate to oblong, coarsely dentate, smooth above, glandular below ... ... *M. viridis.*
2. Leaves petioled, coarsely serrate, smooth above, rarely sparingly hairy on the nerves below ... ... *M. piperita.*
3. Leaves nearly sessile, sharply toothed, upper surface hoary pubescent, lower white tomentose ... ... *M. sylvestris.*
4. Leaves narrowed below, stalked, ovate, oblong, lanceolate, toothed ... ... *M. arvensis.*

1. **Mentha arvensis** Linn. is a herb of the Western Himalaya, found in Kashmir, the Punjab, Kumaon, and Garhwal. It is distributed to Afghanistan, West and North Asia, China, and Europe.

The dried plant is refrigerant, stomachic, diuretic, and stimulant. It possesses antispasmodic and emmenagogue properties. It is used in jaundice, and is frequently given to stop vomiting.

In China the leaves and stems are made into infusion, and used as carminative, sudorific, and antispasmodic.

In Annam the plant is considered an excellent diaphoretic. An infusion is given in fevers, indigestion, and cephalalgia. The juice of the leaves is applied to the sting or bite of poisonous animals. The leaves pounded with salt are applied to the whitlow.

It is said that the effect of this plant, when animals eat it, is to prevent coagulation of their milk, so that it can hardly be made to yield cheese.

Annam: Bac ha, Bac ha tay, Ke to, Kim tien bac ha, Nam ha, Thuy to, To lan—; *Arabic*: Fodanajihindi, Fotanajehindi, Habakjabuli, Habaqulhind, Naanaaehindi, Naanaaulhind—; *Bengal*: Podina—; *Bombay*: Pudinah—; *Burma*: Bhudina—; *Canarese*: Chetnimaragu, Maraga—; *Chinese*: Po Ho—; *English*: Chinese Mint, Corn Mint, Marsh Mint—; *French*: Baume des champs, Menthe des champs, Menthe du Japon, Pouliot thym—; *Guam*: Yerba buena—; *Gujerati*: Phudno, Pudina—; *Hindi*: Pudinah—; *Languedoc*: Fau pulegi—; *Malaya*: Pok ho—; *Malayalam*: Putiyina—; *Marathi*: Pudina—; *Persian*: Filfilmun, Pudinah—; *Sind*: Pfudnah—; *Sinhalese*: Odutalan—; *Spanish*: Yerba buena—; *Tamil*: Pudina, Yechakkirai—; *Telugu*: Igaenglikura, Pudina—; *Tongking*: Ba Kha—; *Urdu*: Pudinchkohi—.

2. **Mentha piperita** Linn. is cultivated in Indian gardens. It is found spontaneous and cultivated in most temperate regions of Europe, Asia, and North America.

The leaves and the tops of the flowering plant are carminative, stimulant, nervine, antispasmodic, analgesic, anodyne.

In Europe the herb is considered stimulant, stomachic, carminative; and is used for allaying nausea, flatulence, sickness, vomiting, and as an infants' cordial.

Its bruised fresh leaves, if applied, will relieve local pains and headache. A hot infusion, taken as tea, soothes stomach-ache, allays sickness, and stays colicky diarrhoea. This will also subdue menstrual colic.

This is one of the three plants used by the Menomini Indians of North America in the cure of pneumonia, the others being wild mint and catnep. The compound is drunk in the form of a tea and is also used as a poultice.

A volatile oil obtained from the plant is well-known in medicine for its antiseptic, stimulant, and carminative properties; and peppermint oil is the most extensively used of all the volatile oils, both medicinally and commercially.

The oil itself is often given on sugar and added to pills, also a spirit made from the oil, but the preparation in most general use is Peppermint Water, which is the oil and water distilled together. In flatulent colic, spirit of Peppermint in hot water is a good household remedy, also the oil given in doses of one or two drops on sugar.

The local anaesthetic action of Peppermint Oil is exceptionally strong. It is also powerfully antiseptic, the two properties making it valuable in the relief of toothache and in the treatment of cavities in the teeth.

Rats dislike Peppermint, a fact that is made use of by ratcatchers, who, when clearing a building of rats, will block up most of their holes with rags soaked in oil of Peppermint, and drive them by ferrets through the remaining holes into bags.

Arabic: Nāna—; *Berber*: Nemdar, Timedja, Timersitin—; *Bogotá*: Yerba buena—; *Brazil*: Hortela pimenta—; *Catalan*: Menta piperita—; *Chinese*: Po Ho—; *Danish*: Pebermynte—; *Dutch*: Pepermint—; *English*: Brandy Mint, Peppermint—; *French*: Menthe, Menthe anglaise, Menthe d' Angleterre, Menthe officinale, Menthe poivrée—; *German*: Edelmindkraut, Edelminze, Hausminze, Odernunze, Pfefferminze, Pfeffermuenze—; *Hungarian*: Borsos menta—; *Italian*: Menta piperita—; *Menomini*: Dakixkomik—; *Mexico*: Menta piperita—; *North America*: Brandy Mint, Lamb Mint, Lammint, Peppermint—; *Norwegian*: Peppermynte—; *Polish*: Pepparmynta—; *Portuguese*: Hortelana pimentosa, Hortelao apimentada, Hortelao pimenta—; *Russian*: Myata perechnaya—; *Spanish*: Menta inglesa, Menta pimentada, Menta piperita, Yerba buena de sabor de pimienta—; *Swedish*: Pepparmynt—; *Turkish*: Nana—; *Uruguay*: Menta—; *Yugoslavia*: Paprena metvica, Nana—.

3. **Mentha sylvestris** Linn. is met with in the temperate Western Himalaya and Western Tibet, at altitudes of 4,000 to 12,000 feet; in Kashmir, Kumaon, Garhwal, and the Punjab. It is distributed to Afghanistan, Western and Central Asia, and temperate Europe. It is frequently cultivated in gardens in the plains of India.

The dried leaves and flower heads are carminative, and stimulant.

The leaves and a decoction of the plant are considerably used by the natives of the Himalaya as a carminative. They are also employed as an astringent, and for rheumatic pains.

The leaves soaked in water give an infusion which is drunk as a cooling medicine in Shahrig. Trans-Indus, a decoction is used in fever and heat apoplexy.

In the Konkan the plant is much used as a domestic remedy on account of its mildly stimulant and carminative properties.

The herb is used as a carminative and stimulant in Europe. The dry herb yields about 1 per cent of essential oil, endowed with the same carminative and stimulant properties.

Arabic: Fudanajhabak, Fudanajnaanna—; *Baluchi*: Purchink—; *Bombay*: Pudina, Vartalau—; *Brahui*: Purchink—; *Burma*: Boodeema—; *Catalan*: Menta borda—; *Egypt*: Habag—; *English*: Horsemint—; *French*: Mentastre, Menthe sauvage—; *Hindi*: Podina—; *Khowas*: Wealni—; *North-Western Provinces*: Padina—; *Punjab*: Baburi, Belanne, Koshu, Pudnakushna, Vien, Yura—; *Pushtu*: Shamshabai—; *Sanskrit*: Ajirnahara, Pudina, Rochani, Ruchishya, Shakashobaana, Sugandhipatra, Vantihara, Vyanjana—; *Shahrig*: Shinshobae, Velanac, Vialangi—; *Spanish*: Mastranzo nevado, Menta silvestre—; *Urdu*: Pudina—; *Uruguay*: Menta—; *Zhob*: Shinshobae—.

4. **Mentha viridis** Linn. is commonly grown in native gardens all over the plains of India.

The seeds are mucilaginous. The leaves are given in fever and bronchitis, and a decoction is used as lotion in aphthae.

In Europe the herb is considered stimulant, carminative, and antispasmodic. It is added to many compounds on account of its carminative properties and its pleasant taste. For infantile troubles generally the sweetened infusion is an excellent remedy. A distilled water is made which will relieve hiccough and flatulence as well as the giddiness of indigestion.

The oil is less used than that of Peppermint,

A homoeopathic tincture prepared from the fresh plant in flower has been found serviceable in strangury, gravel, and as a local application in painful haemorrhoids. Its principal employment is for its febrifuge and diuretic properties.

Bengal: Pundia—; *Bombay*: Pahadipudina, Pudina—; *Brazil*: Hortela commum—; *Catalan*: Herba sana, Menta comuna—; *English*: Brown Mint, Fish Mint, Garden Mint, Green Mint, Lamb Mint, Mackerel Mint, Our Lady's Mint, Sage of Bethlehem, Spearmint, Spire Mint—; *French*: Baume vert, Menthe à épis, Menthe de Notre-Dame, Menthe romaine, Menthe verte—; *German*: Frauenmuenze, Gruene Muenze, Roemische Minze—; *Gujerati*: Phudino—; *Hindi*: Paharipudina, Podina—; *Italian*: Menta romana—; *Malta*: Spearmint, Menta comune, Naghnieh—; *Marathi*: Pudina—; *Mexico*: Herbabuena—; *North-Western Provinces*: Paharipudina—; *Persian*: Nagbo, Pudneh, Shahsufiam—; *Pishin*: Nana—; *Portuguese*: Hortelao, Ortelao vulgare—; *Punjab*: Paharipodina, Pudina, Pudinakuhi, *Pudna*—; Quetta: Nana—; *Roumanian*: Izma, Minta—; *Sind*: Phudina, Pudina—; *Sinhalese*: Meenchi—; *Spanish*: Costo, Erba Santa Maria, Menta romana, Yerba buena—; *Telugu*: Pudina—.

MERIANDRA.

This genus consists of 2 species, one Himalayan, and one Abyssinian.

Shrub finely tomentose or hoary. Spikes with distant clusters of flowers ... ... *M. bengalensis*.
Shrub flocculently woolly. Spikes uninterrupted ... *M. strobilifera*.

1. **Meriandra bengalensis** Benth., a native of Abyssinia, is cultivated in most of the provinces in India.

An infusion of the leaves is a useful application to aphthae and sore throats. It diminishes or arrests the secretion of milk.

Bombay: Kafurkapatta, Sesti—; *Deccan*: Kafurkapatta—; *English*: Bengal Sage—; *Hindi*: Kafurkapat—; *Tamil*: Sayayilai—; *Telugu*: Simakarpuramu—.

2. **Meriandra strobilifera** Benth. is found in the western temperate Himalaya, from Simla to Kumaon, at altitudes of 5,000 to 6,000 feet.

The decoction, when made strong, is a good lotion for ulcers and heals raw abrasions of the skin. It dries up the breast milk.

MICROMERIA.

The genus includes 130 cosmopolitan species.

M. Douglasii Benth. is used popularly as an emmenagogue along the Pacific Coast.

Micromeria capitellata Benth. occurs in Kumaon and Dehra Dun, the Upper Gangetic Plain, Chota Nagpur, Bihar, the Western Ghats, the Northern Circars, and the Nilgiris.

It is a fairly good substitute for *Mentha piperita*.

Moschosma.

The genus consists of 6 palaeotropical species.

Moschosma polystachium Benth. is found in Gujerat, Konkan, Deccan, Bengal, Bihar, and Burma. It extends to Ceylon, Java, the Philippine Islands, China, Australia, and tropical Africa.

In the Gold Coast Colony the juice of the plant is squeezed into the nostrils of children to cure headache.

Ashanti: Nwansing nwansing—; *Betsileo*: Karanjamboay, Karanjanamboay—*Hausa*: Kimbar rafi—; *Tamil*: Sanakki poondu—.

Nepeta.

The genus includes 150 species, natives of temperate Europe, North Africa, and Asia.

The following species are used medicinally in Europe—*N. Cataria* Linn., *N. glechoma* Benth., *N. italica* Linn.—; in China—*N. glechoma* Benth., *N. japonica* Maxim., *N. tenuifolia* Benth.—; in Annam and Malaya—*N. tenuifolia* Benth.—; in the United States of America—*N. glechoma* Benth.—.

I. Flowers in continuous spikes or heads; basal clusters rarely distinct ... ... *N. elliptica.*

II. Flowers in distinct clusters either axillary, or forming interrupted spikes or panicles.

A. Whorls all sessile.

1. Leaves as broad as long; hoary on both sides ... *N. glomerulosa.*
2. Leaves longer than broad, upper surface not hoary ... *N. ciliaris.*

B. Whorls peduncled, the lower conspicuously so.

1. Stem and branches stout; leaf-apex acute or acuminate; corolla dotted with purple ... *N. Cataria.*
2. Stem and branches slender; leaf-apex obtuse; corolla blue-purple ... ... *N. ruderalis.*

1. **Nepeta Cataria** Linn. is found in Kashmir, the North-Western Frontier Province, the Kurrum Valley, Baluchistan, and Afghanistan. It is distributed to Western Europe.

The dried leaves and flowering tops are carminative, tonic, diaphoretic, refrigerant and slightly emmenagogue, specially antispasmodic, and mildly stimulating.

Producing free perspiration, Catnep Tea is very useful in colds. It is a valuable drink in every case of fever, because of its action in inducing sleep, and producing perspiration without increasing the heat of the system. It is good in restlessness, colic, insanity, and nervousness, and is used as a mild nervine for children, one of its chief uses being, indeed, in the treatment of children's ailments.

The infusion of 1 oz. to a pint of boiling water may be taken by adults in doses of two tablespoonfuls, by children in two or three teaspoonfuls frequently, to relieve pain and flatulence. The tea may be drunk freely, but if taken in very large doses when warm, it frequently acts as an emetic. It has proved efficacious in nervous headaches and as an emmenagogue, though for the latter purpose, it is preferable to use Catnep, not as a warm tea, but to express the juice of the green herb and take it in tablespoonful doses, three times a day.

An injection of the tea also relieves headache, hysteria, and colicky pains.

Catmint is one of the ingredients in the Menomini Indian cure for pneumonia, the others being wild mint and peppermint. The compound is drunk in the form of a tea, and is also used as a poultice on the chest.

The Flambeau Ojibwe brew a tea of catmint leaves for a blood purifier. The mint water obtained by steeping the herb in lukewarm water is used to bathe a patient, to raise the body temperature.

Catalan : Herba gatera, Menta de gat, Nepta—; *Dutch* : Kattekruid, Nip—; *English* : Catmint, Catnep, Nep—; *Flambeau Ojibwe* : Tci'name'wûck—; *French* : Cataire, Chataire, Herbe aux chats, Menthe de chat—; *German* : Katzenminze, Katzennessel, Katzensterz, Neptenkraut, Steinminze, Steinnessel—; *Italian* : Cattaia, Cattaria, Erba gatta, Gattaria, Nepitella—; *Menomini* : Ka'saka muski'ki—; *North America* : Catmint, Catnep, Catnip—; *Roumanian* : Catusnic, Catusnica—; *Russian* : Kashachya myata—; *Spanish* : Menta gatera, Menta de gatos, Yerba gatera, Yerba de gatos—; *Swedish* : Kattmynta—.

2. **Nepeta ciliaris** Benth. is found in temperate western Himalaya, from Kashmir to Garhwal, at altitudes of 6,000 to 8,000 feet.

It is given in sherbet for fever and cough.

Punjab : Zufa yabis—; *Sind* : Jufa—.

3. **Nepeta elliptica** Royle occurs in western temperate Himalaya, from Kashmir to Kumaon, at altitudes of 5,000 to 8,000 feet.

An infusion of seeds in cold water is used in dysentery by the Punjabis.

Punjab : Tukhmmalanga—.

4. **Nepeta glomerulosa** Boiss. is found from the Punjab frontier to Baluchistan, whence it extends to Afghanistan and Persia.

It is a cure for pneumonia in Bolan, for itch in Sanjawi.

It is commonly used for indigestion.

Bolan : Simsok—; *Brahui* : Simsok—; *Chiltan* : Bhinjanbutai, Chanjanbutai—; *Quetta-Pishin* : Chinjanbutae—; *Sanjawi* : Chamjanbot—.

5. **Nepeta ruderalis** Hook. fil. occurs in the hilly parts of Bengal, Bihar, the Punjab, Kumaon, the North-West Frontier Province, Rajputana, Central India, Konkan, and Deccan. It is distributed to Afghanistan.

It is largely used in fevers, and as a cardiac tonic. The decoction is used as a gargle in sore throat.

In Nepal it is taken internally as a remedy for gonorrhoea.

Nepal : Niasbo—; *Punjab* : Badranj boya, Bebrang khatai, Billi lotan—.

Ocimum.

The genus numbers 60 species, inhabiting the tropical and warm temperate regions of the world.

The following species are used medicinally in Europe—*O. basilicum* Linn., *O. gratissimum* Linn.—; in China and Indo-China—*O. basilicum* Linn.—; in Japan and Malaya—*O. crispum* Thunb.—; in the Philippine Islands—*O. basilicum* Linn., *O. gratissimum* Linn., *O. sanctum* Linn.—; in Guiana— *O. micranthum* Willd.—; in Brazil—*O. canum* Sims., *O. gratissimum* Linn., *O. micranthum* Willd.—; in La Reunion—*O. basilicum* Linn., *O. gratissimum* Linn.—; in Ethiopia and Abyssinia—*O. gratissimum* Linn.—; in Guinea and the Gold Coast—*O. canum* Sims., *O. basilicum* Linn., *O. viride* Willd.—; in Sierra Leone and Liberia—*O. viride* Willd.—; in Ubangi-Shari : *O. canum* Sims.—.

The flower tops of *O. basilicum* Linn. are official in France.

A. Pedicels as long as or longer than the calyx ; 2 lower calyx-teeth longer than the upper lip ... ... *O. sanctum.*

B. Pedicels shorter than the calyx.
- I. Lower calyx-teeth longer than the upper lip.
  - a. Bracts stalked ... ... *O. canum.*
  - b. Bracts stalked. Fruiting calyx very shortly pedicelled *O. basilicum.*
- II. Lower calyx-teeth shorter than the upper lip ... *O. gratissimum.*

1. **Ocimum basilicum** Linn. grows throughout India. It is cultivated throughout the greater part of India, Ceylon, and Burma. It extends to the Malay Peninsula and Archipelago, Formosa, China, and Polynesia.

Diaphoretic and expectorant properties are ascribed to this plant, which is considered digestive and pectoral in La Reunion.

The roots are used for the bowel complaints of children.

The leaves are useful in the treatment of croup, for which the juice warmed with honey is given.

The expressed juice of the leaves forms the basis of a celebrated nostrum for the cure of ringworm; and the bruised leaves are applied to parts stung by scorpions to lessen the pain. The powdered dried leaves are said to be an effectual means of dislodging maggots.

The juice of the leaves or of the whole plant is dropped into the ears for the cure of headache and dullness of hearing. Mixed with ginger and black pepper it is given in the cold stage of intermittent fever.

The flowers possess stimulant, diuretic, and demulcent properties.

The seeds are largely employed, especially by the Mohammedans of Eastern Bengal, infused in water, to form a refreshing and cooling drink. When steeped in water they become immediately coated with a semi-transparent mucilage, and then form a mucilaginous jelly.

The seeds are much used medicinally in some parts of India, especially in the treatment of gonorrhoea and nephritic affections, and are regarded also as useful for dysentery and diarrhoea, especially in children for the diarrhoea of dentition. A cold infusion is said to relieve the after-pains of parturition.

The seeds washed and pounded are used in poultices for unhealthy sores and sinuses. They are also given internally with sherbet in cases of habitual constipation, internal piles, and fevers.

An infusion of the leaves is used in fevers by the natives of Gambia, where it is a popular common cooling drink.

In Guinea the decoction of the leaves and the stems is given in fevers, neuralgia, catarrh, and renal troubles; it is also used as a lotion for sore eyes.

In Annam an infusion of the plant is considered antiemetic and antidiarrhaeic. It is given for cephalalgia and gouty joints, and used as a gargle for foul breath. An infusion of the seeds is given in fevers. The seeds are chewed in cases of snake-bite, one portion is swallowed and the other portion applied to the bitten part.

The leaf is not an antidote to snake venom. In the treatment of snake-bite it is equally useless as a collyrium, an errhine, and a local application to the part bitten (Mhaskar and Caius).

Annam: Huong nhung, Nhat thong chi, Rau e lon trong, Rau e tia, Rau e trang, Rau que—; *Arabic*: Asaba-ul-feteyat, Badarwaj, Badruj, Baklut-ul-zub, Habaq, Habbok, Hebak, Hibag, Rihan, Schogor, Shahasfaram—; ***Baluchi***: Nazbu—; *Bengal*: Babuitulsi, Debunsha, Khubkalam, Pashanabeddie—; *Bombay*: Takmeria—; *Brahui*: Niazpu—; *Canarese*: Kamkusturi, Ramkasturi, Sajjebiya—; *Cantonese*: Hiang Hoa Tsao—; *Catalan*: Alfabrega de fulla ampla, Alfabrega de fulla petita—; *Chinese*: Hsun Ts'ao, Lo le—; *Culión*: Camange—; *Deccan*: Salzat, Subjah, Subze, Tirunitru—; *Dutch*: Basilienkruid, Basilikum—; *Egypt*: Rihan, Sa'atar hendy—; *English*: Basil, Common Basil, Common Sweet Basil, Garden Basil, Roman Basil, Sweet Basil—; *French*: Basilic, Basilic des cuisinières, Basilic cultivé, Basilic aux sauces, Grand basilic, Herbe royale—; *Fulani*: Soukora—; *Gambia*: Patmagi—; *German*: Basilienkraut, Basilikum, Koenigskraut—; *Greek*: Basilikos, Okimon—; *Guam*: Atbahakat—; *Gujerat*: Damaro, Damro, Nasabo, Sabje—; *Hamadan*: Tukhm-i-raihan—; *Hausa*: Deidoya, Dodoya, Doidoya—; *Hindi*: Babuitulsi, Babul, Bahari, Barbar, Kalitulsi, Niyakshbo, Rihan, Sabzah, Tukhmerihan—; *Ho*: Loba—; *Iraq*: Raihan—; *Italian*: Basilico, Bassilico, Ocimo, Ozzimo—; *Kathiawar*: Marvo—; *Kurdish*: Ruhan—; *Lagos*: Efirin wewe—; *Languedoc*: Embaimo—; *La Reunion*: Grand basilic—; *Las Bela*: Drar khatori—; *Malay*: Kemangi, Pokoh, Pokoh-pokoh hitam, Ruku, Ruku-ruku, Ruku-ruku itam, Selaseh hitam, Selaseh puteh, Selasih, Selasih antan—; *Malayalam*: Pachcha, Tirunitru—; *Malinke*: Sossoguena, Soughenfira—; *Marathi*: Marya, Sabja, Sabza, Tukhamariya—; *Mauritius*: Basilic à grandes feuilles—; *Peking*: Gai K'ang—; *Persian*: Dabanshah, Firanjmushk, Nazbu, Rehan-e-dash, Reyhane sibze, Tur-eh-korasani, Ungusht-kunizuckan—; *Polish*: Bazylik—; *Porebunder*: Marvi, Takmario—; *Porto Rico*: Albahaca—; *Portuguese*: Alfabaca, Mangericoo—; *Punjab*: Babri, Baburi, Purrunj mushk, Nigand, Niyazbo, Panr, Rehan, Tulsi—; *Roumanian*: Busuioc—; *Russian*: Bazilik—; *Sanskrit*: Ajaganothika, Apetarakhsasi, Asurasa, Barba, Barbara, Barbari, Karahi, Kharapushpa, Manjariki, Munjariki, Surabhi, Surasa, Tulasidvesha, Tungi, Varvara—; *Santali*: Bharbari, Dimbubaha, Malibuha—; *Serbian*: Bosilek, Bosiliak—; *Sind*: Nazbo, Sabajhi—; *Sinhalese*: Hintala, Sawandatala, Suwandutala—; *Spanish*: Albahaca, Albahaca fina, Albahaca de hoja ancha—; *Tagalog*: Solasi, Sulasi—; *Tamil*: Tirnutpatchi, Tirunitru ; *Telugu*: Bhutulasi, Rudrajada, Vepudupachcha, Yibudipatri—; *Tongking*: Hung gioi—; *Urdu*: Janglitulasi—; *Uriya*: Dhalatulasi—; *Visayan*: Bonac, Calooy, Canela—; *Yemen*: Hehak, Raihan—.

2. **Ocimum canum** Sims is found on the plains and lower hills of India, from Assam, Bengal, Bihar, and Central India, to the south Deccan and Ceylon. It is distributed to Java, Western Asia, tropical Africa, Madagascar; and is cultivated in America.

Among the Santals, during fever when the extremities are cold, the leaves made into a paste are applied to the finger—and toe-nails. The same preparation is used as a cure for parasitical diseases of the skin.

In Jodhpur the seeds are drunk in milk as a tonic; and a decoction of them with potash in water is used as a cooling drink in fever. A bunch of the plant hung in the corner of a room is said to attract mosquitoes and keep the rest of the room free from them.

In Persia the mucilaginous seeds are given for lung and chest complaints.

In French Guinea the herb is especially used by women before parturition, or as an emmenagogue, the leaves cooked with ground-nuts taken in small quantity.

In Ubangi Shari the decoction of the leaves is given to women after they have been delivered, as also to the newly born infant. The leaves ground in oil are applied to the temples for headache. Boiled with yam or maize flower they yield a pleasantly-flavoured soup which is eaten as a cure for cough, and bronchitis.

The plant is commonly grown in gardens in Tropical West Africa and used as 'tea leaf' for fevers, often given to children, or as a bath for febrile patients, the patient sitting and pouring it over the head, or with clothing opened out over a pot in which abundance of the plant is boiled. In Gold Coast the juice of the leaves is sometimes squeezed into the patient's eyes. At Accra a decoction is also given for dysentery, or as a mouth-wash to relieve tooth-ache.

For haemorrhage from the nose, the Sutos either inhale the smoke from burning the dried leaf or apply an ointment made with the powdered leaf.

In Madagascar the leaves and the seeds are considered aromatic, tonic, febrifuge, anticatarrhal, expectorant, sternutatory, anti-rheumatismal. For malaria the seeds are ground in an infusion of the leaves. The juice and the powder obtained by pounding together the seeds and the leaves are taken as an errhine in migraine.

Adangme : Danwe—; *Awuna* : Ahame, Defetsui—; *Bambara* : Chukula—; *Bengal* : Bharbari—: *Benin* : Ihiri—; *Bombay* : Ramtulsi—; *Canarese* : Nayi-tulasi, Ramatulasi—; *Efik* : Amana, Mfang—; *English* : American Basil, Hairy Basil, Hoary Basil, Hoary Tulsi, White Basil—; *Espiritu Santo* : Neruk—; *Ewe* : Ahame, Defetsui—; *French* : Basilic d'Amérique, Basilic commun—; *Fulani* : Katchukatchunga, Sukora, Urngol—; *Ga* : Kowe—; *Gbari* : Jemijemi—; *Gbaya* : Wélé—; *Golungo Alto* : Machericao, N-xilica—; *Guam* : Atbahakat—; *Hamadan* : Tukhm-i-sherbati—; *Hausa* : 'Da'ddoya, 'Dai'doya, 'Dai'doyar masar, 'Do'dowa, 'Do 'doya, 'Dwai'dwaiya, Sarakuwar sauro, 'Yar kan masallachi—; *Ho* : Loba—; *Jodhpur* : Bapji—; *Jukun* : Afyinu, Shinu—; *Kanuri* : Kabur—; *Kaolak* : Guguniece—; *Koranko* : Sogoi—; *Krobo* : Danwe—; *Lagos* : Efirin nla, Efirin oshu—; *Limba* : Fufurufuru—; *Madagascar* : Kiranjay—; *Malay* : Kemangi—; *Malayalam* : Katturamatulasi—; *Malinke* : Su-kola—; *Mandingo* : Sisejambo—; *Manja* : Sâhagna—; *Mbi* : Gouroulou—; *Moorish* : Gumuguei—; *Old Calabar* : Iyino—; *Oloke Meji* : Efinrin marag-bosanyan—; *Persian* : Badrudge ibieze, Reyhane kuhi, Tukhm-chirbati—; *Porto Rico* : Abahaca cimarrona—; *Sadani* : Garaighasi—: *Sango* : Tété—, *Sanskrit* : Ajaka, Arjaka, Gambhira, Gandhapanirjjaka, Jambira, Kathinjara, Kshudraparna, Kshudratulasi, Kuthera, Mukharjaka, Ugagandha—; *Santali* : Bharbhari—; *Sierra Leone* : Patnage—; *Sinhalese* : Hintalla—; ***Spanish*** :

Albahaca—; *Suto*: Mmavwatwane—; *Tamil*: Ganjamkorai, Kanjankorai, Naitulasi—; *Telugu*: Kukkatulasi—; *Timne*: An-soro, E-soro, Koe suru—; *Tivi*: Kungulaku—; *Togbo*: Biroulou—; *Twi*: Emeng—; *Woloff*: Ngumgume—; *Yoruba*: Aruntantan, Efinrin aja, Efinrin ata, Efinrin marugboshanyan, Efinrin wewe—.

3. **Ocimum gratissimum** Linn. is found throughout India and Ceylon, often cultivated. It is distributed to Java, tropical Africa and America.

Aromatic baths or fumigations prepared with the plant are advised in the treatment of rheumatism and paralysis. In the aphthae of children a strong decoction has been found effectual.

A decoction of the leaves is of value in cases of seminal weakness, and is an esteemed remedy in gonorrhoea.

The seeds are given in headaches and neuralgia.

In China the leaves and flowers are used as a sudorific in diseases of the lungs.

Considered digestive and pectoral in La Reunion. On the Gold Coast the leaves are mashed and used as an enema by newly delivered women. It is also used for young infants.

A very popular remedy in Madagascar. It is considered aromatic, digestive, tonic, pectoral, antiemetic, antispasmodic, antineuralgic, The Betsileo chew the leaves for toothache, and sniff the juice of the leaves or the powdered seeds in headache.

Arabic: Furanjmishk—; *Bengal*: Ramtulshi, Ramtulsi—; *Betsileo*: Romba—; *Bombay*: Ramatulasa, Rantulsi, Tulsi—; *Chinese*: Tzeu Sou—; *Deccan*: Ramtulsi—; *English*: Large Basil, Lemon Basil, Shrubby Basil—; *Ewe*: Daiblorsi—; *French*: Basilic *crépu*—; *Gujerati*: Avachibavachi, Ramtulasi—; *Hindi*: Bantulsi, Malatulsi, Ramtulsi—; *La Reunion*: Baumier, Gros braume—; *Malay*: Ruku-ruku hitam, Selaseh besar—; *Malayalam*: Kattutrittavu, Ramatulasi—; *Marathi*: Ramatulasi, Ranatulasu—; *Mauritius*: Basilic, Toulashi, Toulsi—; *Mundari*: Dimbubaha—; *Persian*: Palangmishk—; *Porebunder*: Ramtulasi—; *Punjab*: Banjere—; *Sadani*: Dimbu—; *Sanskrit*: Ajaka, Doshakleshi, Nidralu, Ramatulasi, Shophahari, Sugandhi, Sukshmapatraka, Sumukha, Suprassanaka, Suvakra, Vanabarbarika, Vishaghna, Vriddhatulasi—; *Sinhalese*: Gastala, Kiritala, Otala—; *Tamil*: Elumichantulasi, Peruntulasi, Ramtulasi—; *Tankay*: Romba—; *Telugu*: Nimmatulasi, Ramatulasi—; *Urdu*: Ramtulasi—; *Uriya*: Ramotulosi, Sondabhogohulono—; *Visayan*: Coloncogon—; *Yemen*: Hobokbok, Shajaret eszir, Vusab—

4. **Ocimum sanctum** Linn. is found throughout India, Burma, and Ceylon, and distributed to the Malay Archipelago, Australia, the islands of the Pacific, Western Asia, and Arabia.

The root is given in decoction as a diaphoretic in malarial fevers.

The leaves have expectorant properties, and their juice is used in catarrh and bronchitis. This preparation also is applied to the skin in ring-worm and other cutaneous diseases. An infusion of the leaves is used as a stomachic in the gastric disorders of children, and in hepatic affections. The dried leaves are powdered and employed as a snuff in ozaena. They are also an effectual means of dislodging maggots. The juice dropped into the ear is said to be a good remedy for earache.

The leaves are a South Indian substitute for tea. In the Tamil country the flower-tops are ground with sesamum oil and given internally to help the expulsion of the foetus.

In Arabia the leaves are given together with pepper in tertian and quartan fevers.

The seeds are mucilaginous and demulcent, and are given in disorders of the genito-urinary system.

In Ceylon the herb is used in decoctions for cough and catarrh, sometimes chewed as a substitute for betel.

The fresh roots are ground with water and applied to the stings of wasps and bees and the bites of worms and leeches. The bruised fresh roots, stems, and leaves are applied to the bites of mosquitoes (Roberts).

Every part of the plant finds its application in the treatment of snake-bite and scorpion sting; but Caius and Mhaskar have shown experimentally that every part of it is equally useless in the antidotal and symptomatic treatment of snake-bite and scorpion sting.

The juice of the leaves or decoction of the same is considered to possess diaphoretic, antiperiodic and stimulating expectorant properties. A compound decoction of *O. sanctum*, *Tinospora cordifolia*, and *Evolvulus alsinoides* was given in cases of malarial fever with no benefit (Koman).

Arabic: Dohsch, Schadjant eszirr, Vusab—; ***Badaga***: Kapputulasi—; *Bengal*: Kalatulsi, Kural, Tulshi, Tulsi—; ***Bombay***: Tulas, Tulasa—; ***Burma***: Lun—; *Canarese*: Kalatulasi, Karitulasi, Sritulasi, Tulasi—; ***Deccan***: Tulsi—; ***English***: Holy Basil, Monk's Basil, Rough Basil, Sacred Basil—; ***Gujerati***: Talasi—; ***Hindi***: Baranda, Kalatulsi, Krishnatulsi, Tulsi, Varanda—; ***Ilocano***: Biday—; ***Konkani***: Tulsi—; *Malay*: Oku, Ruku-ruku merah, Selaseh hitam—; *Malayalam*: Krishnatulasi, Kunnakam, Nallatrittavu, Punya, Sivatulasi, Surasam, Trittavu, Tulasi—; *Marathi*: Tulasa, Tulasichajadha—; ***Pampangan***: Locoloco—; ***Philippines***: Albahaca—; *Portuguese*: Mangericao—; ***Punjab***: Bantulsi, Tulsi—; ***Queensland***: Bulla-bulla, Mooda—; ***Sanskrit***: Ajaka, Arjaka, Amrita, Apetarakshasi, Bahupatri, Bharati, Bhutaghni, Bhutaka, Bhutapatri, Brinda, Devadundubhi, Divya, Gandhaharini, Gauri, Gramya, Haripriya, Kathinjara, Kayastha, Krishnamula, Krishnatulasi, Kutheraka, Laxmi, Madhavi, Malashreshtha, Manjari, Papaghni, Parnasa, Patrapushpa, Pavani, Pavitra, Pretarakshasi, Punya, Sarasa, Shrikrishnavallabha, Shyama, Subhaga, Sugandha, Sulabha, Surabhi, Suradundubhi, Surasa, Suravallari, Suravalli, Surejya, Suvaha, Tivra, Tridashamanjari, Tulasi, Vaishnavi, Vishnupatni, Vishnuvallabha, Vrinda—; *Sinhalese*: Madurutala, Mudurutulla—; ***Spanish***: Albahaca morada—; *Tagalog*: Balanoi, Locoloco—; *Tamil*: Alangai, Karuttulasi, Kulimittan, Kullai, Kumuli, Malgodai, Malmurugu, Mudi, Nediyon, Pirundam, Sirttulay, Surasa, Surasam, Suriyagarandai, Savadugundi, Suvi, Tulasi, Tulavam, Tulay—; *Telugu*: Brynda, Gaggera, Krishnatulasi, Kukkatulasi, Nallagaggera, Nalla-tulasi, Tellatulasi, Tulasi—.

ORIGANUM.

The genus consists of 7 species, natives of the Mediterranean region.

O. dictamnus Linn., *O. hirtum* Link., *O. Majorana* Linn., *O. virens* Hoff. and Link, *O. vulgare* Linn. are used medicinally in Europe; *O. vulgare* Linn. is also used in China and Malaya.

1. The 2 longest stamens and sometimes all 4 projecting beyond the corolla ... ... *O. Majorana*.
2. Stamens 4 in unequal pairs slightly protruding ... *O. vulgare*.

1. **Origanum Majorana** Linn., a native of Europe, North Africa and Asia Minor, is extensively cultivated in India.

The herb is carminative and stomachic; useful in measles.

The leaves and seeds are considered astringent and a remedy for colic. The essential oil from the leaves is used for hot fomentations in acute diarrhoea.

In Europe an infusion made from the fresh plant is given to relieve nervous headaches; and externally the herb is applied in bags as a hot fomentation to painful swellings and rheumatism, as likewise for colic. The volatile oil is considered an excellent external application for sprains, bruises, etc.

Arabic: Mardakusch, Merdkouche, Mizunjush—; ***Bogotá***: Mejorana—; ***Bengal***: Murru—; ***Catalan***: Moraduix, Moradux—; ***Deccan***: Murwa—; ***Dutch***: Marjolein—; ***Egypt***: Bardaquesh, Mardaqush—; ***English***: Sweet-knotted Marjoram, Sweet Marjoram—; ***French***: Marjolaine à coquille, Marjolaine des jardins, Marjolaine d'Orient, Origan marjolaine—; ***German***: Badkraut, Blaudoste, Blauer Tarant, Blaudunst, Costenzkraut, Doschte, Dost, Felddoste, Frauendosten, Kostenskraut, Kunerle, Kuttelkraut, Majoran, Maraun, Margrankraut, Marieleine, Masaran, Maseran, Masoran, Meiran, Meyran, Muellerkraut, Ohrkraut, Schusterkraut, Sommermajoram, Thorant, Walddosten, Wohlgemut, Wurstkraut—; ***Greek***: Amarakon, Masuran—; ***Hindi***: Murwa—; ***Italian***: Maggiorana, Marjorana, Persa—; ***Kumaon***: Bantulsi—; ***Languedoc***: Majourana, Majourena, Majurena, Mayran—; ***Malta***: Sweet Marjoram, Maggiorana, Persia, Mertkux—; ***North America***: Knotted Marjoram, Marjoram, Sweet Marjoram—; ***Roumanian***: Maghiran—; ***Russian***: Mayoran—; ***Sanskrit***: Ajanmasurabhipatra, Bahuvirya, Gandhapatra, Jambira, Kharapatra, Kulasaurabha, Maricha, Maru, Marubaka, Maruta, Marutaka, Phani, Phanijjaka, Prasthakusuma, Prasthapushpa, Samirana, Shitalaka, Suravha—; ***Sind***: Murwo—; ***Spanish***: Almoradux, Mejorana—; ***Tamil***: Marru—; ***Urdu***: Marvakhusha—; ***Yemen***: Mardakush—.

2. **Origanum vulgare** Linn. occurs plentifully in the Himalaya, from Kashmir to Sikkim, between altitudes of 7,000 and 12,000 feet. It is distributed to Europe, North Africa, western and northern Asia.

The herb is carminative, anodyne, stomachic, and emmenagogue. In China it is considered an excellent refrigerant.

Externally, the dried leaves and flower-tops may be applied in bags as a hot fomentation to painful swellings and rheumatism, as well as for colic. An infusion made from the fresh plant will relieve nervous headache.

The whole herb is medicinal and contains a volatile oil, which is separated by distillation. Perspiration may be produced by a warm infusion, and this is useful in the commencement of measles to bring out the eruption; it is also taken to promote the menstrual flow, when suppressed by cold; it is also valuable in spasms, colic, and to give relief from pain in dyspeptic complaints.

The oil is stimulant and rubefacient, and often used as a liniment; it is given as a stimulant and tonic in colic, diarrhoea, and hysteria. It is also applied in chronic rheumatism, tooth-ache, and ear-ache.

Arabic: Buklutulgezal, Mirzanjosha, Sutur—; ***Bohemian***: Dobramyssl—; ***Catalan***: Orenga—; ***Chinese***: Ching Chieh, Yin Ch'en—; ***Danish***: Tost, Vild merian—; ***Dutch***: Orego—; ***English***: Common Marjoram, Organ, Organy, Origany, Wild Marjoram—; ***French***: Grande marjolaine, Grande marjolaine bâtarde, Grande marjolaine sauvage, Grand origan, Marjolaine d'Angleterre, Marjolaine bâtarde, Marjolaine sauvage, Pied de lit, Origan, Origan commun, Origan vulgaire—; ***German***: Blauer Orant, Brauner Dosten, Dosten, Wilder Majoran—; ***Greek***: Origanos—; ***Hindi***: Mirzanjosh, Sathra—; ***Italian***: Origano,

Regano—; *Malaya*: Yan chan—; *Malta*: Common Marjoram, Pot Marjoram, Regamo, Acciughero, Riegnu—; *Marathi*: Marvā, Mayarona—; *North America*: Common Marjoram, Origanum, Pot Marjoram, Wild Marjoram, Winter Marjoram, Winter-sweet—; *Norwegian*: Vild merican—; *Persian*: Mizangosch, Mirzanjosha, Oushneh—; *Polish*: Lebiotka—; *Portuguese*: Ouregao, Ouregos—; *Punjab*: Mirzanjosh—; *Russian*: Dushitsa—; *Spanish*: Oregano—; *Swedish*: Dosta—; *Telugu*: Mridumaruvamu—; *Urdu*: Mirzanjosha—.

ORTHOSIPHON.

The genus includes 50 species, natives of Indo-Malaya and tropical Africa.

Orthosiphon stamineus Benth. occurs in Assam, Burma, the Nicobar Islands, and South India. It is distributed to the Malay Archipelago, the Philippine Islands, and Australia.

In Java the leaves are made into a tea, and used in the treatment of diseases of the kidneys and bladder. They are official in Holland.

English: Cat's Whiskers, Java Tea—; *French*: Barbiflore, Thé de Java—; *Malay Archipelago*: Koemis Koetjing, Remock djoeng—.

OTOSTEGIA.

The genus consists of 10 species, inhabitants of Western Asia and Abyssinia.

1. Leaves spine-tipped; calyx naked within ... *O. Aucheri*,
2. Leaves not spine-tipped; calyx bearded within ... *O. limbata*.

1. **Otostegia Aucheri** Boiss. occurs in Baluchistan, whence it spreads to southern Persia.

At Kohlu in Baluchistan the drug is administered in suppressed small-pox if the pox do not appear, after which the pox appear (Hughes-Buller).

Baluchi: Samar, Shinisg—; *Brahui*: Sadikh, Samar—; *Kohlu*: Suraghzai—.

2. **Otostegia limbata** Hook. fil. is found on the lower hills of the Punjab, west of the Jhelum to the Salt Range.

The juice of the leaves is applied to children's gums, and to ophthalmia in man and beast (Stewart).

Hazara: Chitiboi—; *Punjab*: Agzhan, Awanibuti, Bui, Jandi, Kandiari, Lana, Phutkanda—.

PERILLA.

The genus consists of 3 species, spreading from India to Japan.

Perilla ocimoides Linn. is found from Kashmir to Bhotan at altitudes of 1,000 to 10,000 feet; also in the Khasia Mountains from 3,000 to 6,000 feet; it is cultivated in Chittagong; from Champaran it extends to Burma, and is distributed to Indo-China, China, and Japan.

In China and Indo-China the leaves, the stems and the seeds are considered resolvent, diaphoretic, and cephalic. They are prescribed in cephalalgia, hypochondriasis and mania. The leaves are eaten raw or cooked, or administered as an infusion; the seeds are either toasted or dispensed in the form of a decoction.

The plant is a popular remedy in Annam, where it is used as a sedative, an antispasmodic, and an antidote. At Vinh the herb, without the roots, is made into a decoction and taken on an empty stomach for sunstroke. At Nghe-an the decoction is used as a uterine stimulant. At Nam-o it is combined with an equal amount of *Ocimum sp.*, and used as a fumigant.

Annam: Bach to, Rau hung, Thai am mau, Tu to, Xich to—; *Chinese*: Sou Yeh—; *Hindi*: Bhanjira—; *Kumaon*: Bhangara, Jhutela—; *Naga*: Kenia—; *Vinh*: Tia to, Tu to—.

PEROWSKIA.

The genus consists of 4 species, natives of Central Asia.

1. Leaves linear-oblong, incised or pinnatisect. Calyx closed with long cottony wool ... ... *P. abrotanoides*.

2. Leaves oblong-ovate or lanceolate, crenate-serrate or incised. Calyx hispid ... ... *P. atriplicifolia*.

1. **Perowskia abrotanoides** Karel. is met with in the Western Himalaya at altitudes of 8,000 to 13,000 feet, and in Baluchistan; it is distributed to Afghanistan, Persia, and Turkestan.

At Ziarat the plant is used as a cooling medicine. The flowers are soaked in water; and this is very cooling when applied to the body of a patient suffering from fever.

Brahui: Gwari drani—; *Pushtu*: Shanshohai—; *Seistan*: Maur—.

2. **Perowskia atriplicifolia** Benth. occurs in Kashmir at 7,500-10,000 feet, and in Baluchistan, whence it spreads to Afghanistan.

In Baluchistan the plant is used as a cooling medicine.

Baluchistan: Gwaree dumoo, Gwaridrane, Tirk—; *Waziristan*: Sansobe—.

PLATYSTOMA.

The genus consists of 4 species, natives of tropical Asia and Africa.

Platystoma africanum Beauv. occurs in Bombay, Dharwar, North Kanara, and Mysore. It inhabits tropical Africa.

In Northern Nigeria it is used for fever, feverish chills or rheumatic symptoms. In Gold Coast the leaves and seeds are a remedy for children's coughs, and are also chewed with salt to cure sore throat. The juice is squeezed into the eyes to cure headache and fever. The leaves are used in Southern Nigeria as a local haemostatic.

Ashanti: Asiresidie, Saman meng—; *Benin*: Eborukhu—; *Ewe*: Sesebli—; *Hausa*: Kimbar-rafi—; *Nzima*: Siresireke—; *Twi*: Asirisiri—.

Pogostemon.

The genus consists of 36 Indomalayan species.

A. Bracts ovate, imbricating, more or less concealing the calyx.

1. Stems and leaves nearly glabrous. Flowers purple ... *P. parviflorus.*
2. Stems and leaves below hoary-puberalous. Flowers pink or white ... ... *P. plectranthoides.*

B. Bracts narrow, lanceolate, not imbricating, not concealing the calyx.

1. Spikes more or less continuous, about 1 cm. broad ... *P. purpurascens.*
2. Spikes much interrupted, 5-6 mm. broad ... ... *P. Heyneanus.*

1. **Pogostemon Heyneanus** Benth. (=*P. Patchouli* Hook. fil.) occurs in Kanara, the Western Ghats, and the Nilgiris. It extends to Ceylon, and is distributed to Java and the Philippine Islands.

The plant is diuretic, carminative, and insecticide. It yields an essential oil largely employed in perfumery.

Bengal : Pachapat— ; *Bombay* : Patchpan— ; *Dutch* : Patchoeli— ; *English* : Patchouli— ; *French* : Patchouli— ; *Gujerati* : Patchpanadi— ; *Hindi* : Pacholi, Peholi— ; *Malaya* : Bur kalif, Poko nilam— ; *Marathi* : Mali, Patcha, Patchpan— ; *Sind* : Panel— ; *Sinhalese* : Gang-kolang-kola— ; *Spanish* : Pachuli— ; *Straits Settlements* : Tilam wangi— ; *Visayan* : Çablan—.

2. **Pogostemon parviflorus** Benth. occurs more or less throughout India.

The fresh leaves, bruised, are applied as a cataplasm in order to clean wounds and promote healthy granulation. In Satara the juice is given in colic and fever.

The root is a reputed remedy for haemorrhage, and has been given successfully in uterine haemorrhage.

The roots are used in the Ratnagiri District as an antidote for the poison of *Echis carinata,* a common snake in that district. The plant, in combination with other drugs, is prescribed as an antidote to snake and scorpion venoms; but Caius and Mhaskar have shown experimentally that no part of the plant is an antidote to either snake or scorpion venom.

Bombay : Pangla, Phang, Phangla— ; *Konkani* : Pangla— ; *Marathi* : Pangli— ; *Matheran* : Pangli—.

3. **Pogostemon plectranthoides** Desf. is found in the Western Himalaya, Lower Bengal and Upper Burma, Bihar, Gujerat, Konkan, and South India.

The properties are said to be the same as those of *P. parviflorus.*

Bengal : Jin— ; *Deccan* : Pangla— ; *Garhwal* : Lujra— ; *Khond* : Ishwarjata— ; *Konkani* : Pangla— ; *Haldwani* : Kalabasinga— ; *Ramnagar* : Rudera— ; *Telugu* : Kusurijang— ; *Uriya* : Dumobadotoko, Gondripulu, Poksunga—.

4. **Pogostemon purpurascens** Dalz. is found in South India, Manipur, and Burma.

It is used as a substitute for *P. parviflorus.*

ROYLEA.

Roylea elegans Wall., the only species, is found in the sub-tropical Western Himalaya from Kashmir to Kumaon, at altitudes between 2,000 and 5,000 feet.

An infusion of the leaves is drunk for contusions produced by blows, and about Kumaon the same preparation is used as a bitter tonic and febrifuge.

Garhwal: Karui, Titpati—; *Hindi*: Patkarru—; *Jaunsar*: Karanoi—; *Kumaon*: Kauri, Titpatti—; *Punjab*: Kaur, Kauri—.

SALVIA.

The genus numbers 550 species, inhabiting the tropical and temperate regions of the world.

The following species are used medicinally in Europe—*S. aethiopis* Linn., *S. canariensis* Linn., *S. glutinosa* Linn., *S. Grahami* Benth., *S. hispanica* Linn., *S. Horminum* Linn., *S. officinalis* Linn., *S. pratensis* Linn., *S. Sclarea* Linn., *S. triloba* Linn. fil., *S. Verbenaca* Linn., *S. verticillata* Linn., *S. viridis* Linn.—; in China—*S. japonica* Thunb., *S. miltiorhiza* Bunge., *S. plebeja* R. Br.—; in Malaya—*S. miltiorhiza* Bunge—; in North America—*S. lanceolata* Bruce, *S. officinalis* Linn., *S. Verbenaca* Linn.—; Mexico—*S. polystachya* Ort., *S. hispanica* Linn.—; in Colombo—*S. palaefolia* H. B. K.—; in Madagascar—*S. leucodermis* Bak.—; in La Reunion—*S. gerardiana* Benth.—; in Mauritius—*S. coccinea* Juss.—; in South Africa—*S. africana* Linn., *S. aurea* Linn., *S. coccinea* Juss., *S. paniculata* Linn., *S. repens* Burch., *S. rugosa* Thunb., *S. runcinata* Linn. fil., *S. scabra* Thunb., *S. sisymbrifolia* Skan., *S. stenophylla* Burch., *S. triangularis*. Thunb.—.

I. Small shrubs or undershrubs; leaf-blade usually .5-2 cm. long
- 1. Corolla 15-25 mm. long ... ... *S. cabulica.*
- 2. Corolla 5 mm. long ... ... *S. aegyptiaca.*

II. Herbs; leaf-blades usually much exceeding 2 cm.
- A. Leaves more or less thick, rugose or rugulose; calyx teeth more or less spinous.
  - 1. All leaves more or less sessile, oblanceolate ... *S. lanata.*
  - 2. All or lower leaves petiolate, ovate to oblong.
    - a. Calyx 20-25 mm. long, in fruit.
      - i. Leaves subentire or serrate; bracts usually whitish ... ... *S. macrosiphon.*
      - ii. Leaves coarsely dentate or lobed; bracts usually green ... ... *S. spinosa.*
    - b. Calyx 10-15 mm. in fruit. Perennial ... *S. Moorcroftiana.*
- B. Leaves membranous, neither rugose nor rugulose; calyx teeth not spinous.
  - 1. Corolla 5 mm. long, white or lilac. Annual ... *S. plebeia.*
  - 2. Corolla 25-30 mm. long, yellow. ... ... *S. glutinosa.*

1. **Salvia aegyptiaca** Linn. occurs in the Punjab plains, Sind, and Baluchistan, whence it spreads westwards to Afghanistan. It is distributed to Persia and North Africa.

At Pab, in Jhalawan, the plant is used as a cure for eye diseases.

In Sind the seeds are used as demulcent in diarrhoea, gonorrhoea, and haemorrhoids.

Arabic: Ra 'al, Raale, Sadjaret el ghasal, Shajaret-el-ghazal—; *Baluchi*: Kohi maur—; *Bolan*: Chamimar, Chammamaor—; *Egypt*: Ra 'al, Shegeret-el-ghazal—; *Pab*: Maur—; *Punjab*: Tukhmmalanga—; *Tank*: Paska, Rangboti—.

2. **Salvia cabulica** Benth. extends from the Punjab frontier to Baluchistan and Afghanistan.

In Baluchistan the plant is considered a cure for fever, also for colds and lung diseases.

Baluchistan: Mateto—.

3. **Salvia glutinosa** Linn. occurs in the North-West Frontier Province, Kashmir. the Punjab, Kumaon, Gharwal, Sikkim, and Bhootan. It is distributed to Tibet, Western Asia, and Europe.

In some parts of Europe an infusion of the leaves and flower tops is used as a diaphoretic, stimulant, and stomachic. It is an excellent lotion for ulcers, and to heal raw abrasions of the skin.

Spanish: Cetro de Júpiter—.

4. **Salvia lanata** Roxb. is found in the Western Himalaya, from Murree to Kumaon, at altitudes of 5,000 to 8,000 feet.

'It is a good substitute for *S. Moorcroftiana.*

5. **Salvia macrosiphon** Boiss. extends from the Punjab frontier to Baluchistan, Afghanistan and Persia.

At Pre Ghal, in Waziristan, the plant is used as a poultice for gangrene.

6. **Salvia Moorcroftiana** Wall. is found in the Western Himalaya, from Kashmir to Kumaon, at altitudes of 6,000 to 9,000 feet.

The root is given in cough, and the seeds are used as an emetic. The leaves are a medicine for guinea-worm and itch, and in the form of poultice applied to wounds. At Lahore, the seeds are given in colic and dysentery, and are applied to boils. The seeds are given for haemorrhoids.

Punjab: Gurgumma, Halu, Kallijarri, Laphra, Papra, Shobri, Thut—.

7. **Salvia plebeia** R. Br. is found throughout India. It is distributed to China, the Malay Islands and Australia.

The seeds are used as demulcent in gonorrhoea, menorrhagia, diarrhoea, and haemorrhoids.

The herb is used as a diuretic, astringent, and anthelmintic in China.

Bengal: Bhutulsi, Khokaburadi—; *Bombay*: Kammarkas—; *Chinese*: Ching Chieh, Ho Shih—; *Punjab*: Samundarsok, Sathi—; *Sind*: Kiuro, Summundursok—.

8. **Salvia spinosa** Linn. occurs in Baluchistan, and is distributed to Mesopotamia, Syria, and Arabia.

At Burj, in Toba Achakzai, the seed is powdered and applied to the teeth to cure tooth-ache.

Egypt: Na'eyme, Shadjeret-el-gemal, Ta'elbe, Tha'alaba—; *Mosul*: Hamham—; *Pushtu*: Ganacha—; *Quetta-Pishin*: Ganacha—.

SATUREIA.

The genus numbers 130 species dispersed through the warm regions of the world.

S. hortensis Linn., *S. montana* Linn., and *S. Thymbra* Linn. are used medicinally in Europe.

Satureia hortensis Linn. occurs in Kashmir. It is distributed to Afghanistan, Western Asia, Europe, America, and South Africa.

The herb is aromatic and carminative. The leaves and flower-tops are used as a stimulant.

Catalan: Sajulida, Sajurida—; *Dutch*: Boonenkruid—; *English*: Summer Savory—; *French*: Herbe de Saint Julien, Sadrée, Sarriète, Sarriette, Sarriette des jardins, Sauriette, Savorée, Savourée—; *German*: Bohnenkraut—; *Italian*: Santoreggia, Satureja—; *Portuguese*: Segurelha—; *Roumanian*: Galbinare—; *Russian*: Chaber—; *Spanish*: Ajedrea, Tomillo real—.

SCUTELLARIA.

The genus includes 200 cosmopolitan species.

S. canescens Nutt., *S. cordifolia* Muehl., *S. galericulata* Linn., *S. integrifolia* Linn., *S. lateriflora* have, at some time or other, been used medicinally and also as a source of 'scutelarin'—; *S. baicalensis* Georg. is used in China.

Scutellaria galericulata Linn. is found in Kashmir. It inhabits Central and Northern Asia, Europe, North Africa, and North America.

If a decoction of the plant is made with 2 ounces of the herb to 8 ounces of water, and is taken for some weeks continuously in recent epilepsy, or when the disease has only functional causes, it will often prove very beneficial. Likewise, this decoction, in common with an extract of the herb, has been given curatively for intermittent fever and ague, as well as for some depressed and disordered states of the nervous system.

A homoeopathic tincture has been successfully given in cases of the epileptiform 'petit mal'.

The crushed plant is applied fresh to old ulcers.

The Flambeau Ojibwe, of North America, use the plant for medicine having something to do with heart trouble.

Dutch: Glidkruid—; *English*: Blue Scullcap, Greater Scullcap, Helmet Flower, Hooded Willow Herb, Hoodwort—; *Flambeau Ojibwe*: Tcatcabonû 'ksik—; *French*: Grande toque, Scutellaire en toque, Toque—; *German*: Helmkraut, Stumpfhelmkraut—; *North America*: Hooded Scullcap, Marsh Skullcap—; *Spanish*: Escutelaria—.

STACHYS.

The genus consists of 200 cosmopolitan species, absent from Australia.

The following species are used medicinally in Europe—*S. annua* Linn., *S. arvensis* Linn., *S. Betonica* Linn., *S. germanica* Linn., *S. hirta* Linn., *S. maritima* Linn., *S. palustris* Linn., *S. recta* Linn., *S. sylvatica* Linn.—; in China—*S. aspera* Michx., *S. Sieboldi* Miq.—; in Colombia—*S. bogotensis* H. B. K.—.

I. Stems and leaves beneath densely white woolly ... *S. parviflora*.

II. Stems and leaves densely or sparingly hairy, but not woolly.
1. Leaves petioled; corolla-tube exserted.
a. Petiole 2 cm. or shorter, margin crenate ... *S. Sieboldi*.
b. Petiole 4 cm. or longer, margin serrate ... *S. sylvatica*.
2. Leaves sessile; corolla-tube included ... ... *S. palustris*.

1. **Stachys palustris** Linn. grows in Kashmir. It is found in Northern and Western Asia, Europe, and North America.

The plant has always had in Europe a great reputation as a vulnerary.

In modern herbal medicine the herb, collected when just coming into flower and dried, is employed for its antiseptic and antispasmodic properties. It relieves gout, cramp and pains in the joints, and vertigo. The bruised leaves when applied to a wound will stop bleeding and heal the wound. The fresh juice is made into a syrup and taken internally to stop haemorrhages, diarrhoea, and dysentery.

In North America the plant is used as a vulnerary, antispasmodic, emetic, and emmenagogue.

English: All-Heal, Clown's Woundwort, Downy Woundwort, Husbandman's Woundwort, March Woundwort, Marsh Woundwort, Opopanewort, Panaij—; *French*: Crapaudine, Epiaire des marais, Ortie morte, Ortie rouge, Stachys des marais—; *North America*: Clown's All-heal, Clown's Woundwort, Dead-nettle, Hedge-nettle, Marsh Woundwort, Rough-weed—.

2. **Stachys parviflora** Benth. is found in the Punjab plains and hills; it extends to Afghanistan.

In the Salt Range the bruised stems are applied to the guinea-worm.

Punjab: Baggibuti, Kirimar—; *Pushtu*: Speraghunai—.

3. **Stachys Sieboldi** Mig. occurs in Upper Burma; it extends to China.

The plant is used medicinally in China as a febrifuge, astringent, and vulnerary.

Chinese: Ts'ao Shih Ts'an—.

4. **Stachys sylvatica** Linn. is found in Kashmir; it extends to Northern Asia and Europe.

An old authority has it that this herb 'stamped with vinegar and applied in manner of a pultis, taketh away wens and hard swellings, and inflammation of the kernels under the eares and jawes', and also that the distilled water of the flowers 'is used to make the heart merry, to make a good colour in the face, and to make the vital spirits more fresh and lively'.

The plant is credited in France with tonic, emmenagogue, and diuretic properties.

English : Hedge Woundwort—; *French* : Epiaire des bois, Ortie à crapaud, Ortie morte des bois, Ortie puante, Stachys des bois—.

TEUCRIUM.

The genus includes 100 cosmopolitan species.

The following species are used medicinally in Europe—*T. Botrys* Linn., *T. Chamaedrys* Linn., *T. fruticans* Linn., *T. majorana* Pers., *T. marum* Linn., *T. montanum* Linn., *T. Polium* Linn., *T. pyrenaicum* Linn., *T. scordium* Linn., *T. scorodonia* Linn.—; in South Africa—*T. capense* Thunb., *T. riparium* Hochst.—.

1. Leaves small, subsessile, ovate or obovate, cuneate-attenuated, obtuse ... ... *T. Stocksianum.*
2. Leaves small, sessile, oblong, obtuse ... ... *T. scordium.*

1. **Teucrium scordium** Linn. is found in Kashmir; it extends to Afghanistan, Northern and Western Asia, Europe, and North Africa.

The herb is considered in Europe antiseptic, diaphoretic, and stimulant. An infusion gives excellent results in all inflammatory diseases.

In Spain the flower tops and the leaves are considered aromatic, bitter, astringent, and are mostly used as diaphoretics and vermifuges.

Arabic : Skordeon—; *Bohemian* : Wodnj czessner—; *Catalan* : Escordi—; *Danish* : Skordium—; *Dutch* : Waterlook—; *English* : Water Germander—; *French* : Germandrée aquatique, Germandrée d'eau, Scordium—; *German* : Batengel, Knoblauch gamander, Lachenknoblauch, Lachinsknopfloch, Laeusekraut, Moosknoblauch, Peterskraut, Rosslauch, Schlagkraut, Schorseken, Scordienkraut, Sonnenschiit, Wasserbathengel, Wasserknoblauch—; *Greek* : Scordion, Scordochorto—; *Hungarian* : Vizi foghadyma—; *Italian* : Scordio—; *Lithuanian* : Embutti—; *Polish* : Czosnkowc ziele—; *Portuguese* : Escordio—; *Russian* : Dikiy tchesnok, Scordiya, Zayatchiy tchesnok—; *Spanish* : Escordio, Escordio oficinal—.

2. **Teucrium Stocksianum** Boiss. occurs in the Western Punjab, the North-Western Frontier Province, Baluchistan, and Afghanistan.

In the Harboi Hills the herb is given in cases of pain at the heart. At Kirani, near Quetta, it is used as a cure for fever.

In Persian Baluchistan the plant is boiled and allowed to stand all night; the water is then drunk as a remedy for colds.

Baluchistan : Kalpora, Kalporag ; *Waziristan* . Kastoral—.

THYMUS.

The genus consists of 33 species, inhabiting the temperate regions of the world.

T. mastichina Linn., *T. serpyllum* Linn., and *T. vulgaris* are used medicinally in Europe.

Thymus serpyllum Linn. is found in the Western temperate Himalaya, from Kashmir to Kumaon, at altitudes between 5,000 and 13,000 feet. It is distributed to Northern and Western Asia, Europe, and North Africa.

In the Punjab the herb is given in weak vision, complaints of the stomach and the liver, suppression of urine and menstruation (Honigberger). On the Chenab the seeds are given as a vermifuge (Stewart).

The oil is sometimes applied as a remedy in toothache.

In Europe the herb is considered antispasmodic, carminative, and tonic. An infusion is given in convulsive coughs, whooping coughs, catarrh, and sore throat; it is good for nervous or hysterical headaches, for flatulence, and the headache which follows inebriation. The infusion is also profitably applied for healing skin eruptions of various characters.

In Persia the leaves are considered carminative. In Teheran they are given for 'too much water in the stomach'.

Arabic : Saatar—; ***Catalan*** : Farigola de montanya, Serpoll—; ***Danish*** : Vildtimian—; ***Dutch*** : Quendel, Wilde Thym—; ***English*** : Bank Thyme, Brotherwort, Creeping Thyme, Hill-wort, Horse Thyme, Mother of Thyme, Pella Mountain, Penny Mountain, Piliol, Puliali Mountain, Running Thyme, Serpell, Shepherd's Wild Thyme, Wild English Thyme, Wild Thyme—; ***French*** : Mignotise des Génevoix, Pillolet, Piolet, Poleur, Pote, Pouilleux, Pouliet, Pouliot bâtard, Serpolet, Serpoule, Thym bâtard, Thym rouge, Thym sauvage, Thym serpolet—; ***German*** : Ameisenkraut, Bienenkraut, Choelm, Citronenquendel, Demutkraut, Fallboll, Feldkoehm, Feldkuemmel, Feldquendel, Feldthymian, Gundelkraut, Gundling, Gunnerle, Heidequendel, Heublume, Huehnerklee, Huehnerkull, Immenkraut, Jungfernzucht, Kandelkraut, Karwendel, Kindelkraut, Kingle, Koschmes, Kostenskraut, Kudelkraut, Kuechenpolei, Kuekenkuemmel, Kuenlein, Kuhkraut, Kundelkraut, Kunerle, Kutelkraut Magaro, Marienbettstroh, Mattenkolen, Neugelenk, Quandel, Quangelchen, Quendel, Quengel, Raenderpolei, Rainkuemmel, Rainpolei, Schmergel, Simio, Steinquendel, Thymchen, Thyrmann, Tuemchen, Unsererlieben-frauenbettstroh, Zymis—; ***Greek*** : Erpylos—; ***Hindi*** : Banajwain—; ***Italian*** : Pepolino, Sermollino, Serpillo, Serpolino, Serpollo—; ***Languedoc*** : Faligouleto, Ferigouleto—; ***North-Western Provinces*** : Banajwain—; ***Persia*** : Djusha, Seetere, Zatar—; ***Polish*** : Mauerzanka—; ***Portuguese*** : Serpao, Serpil, Serpilhi, Serpol—; ***Punjab*** : Kalandarzatar, Marizha, Masho, Rangsbur, Shakei—; ***Roumanian*** : Cimbru de camp, Cimbrusor, Serpun, Serpunel—; ***Russian*** : Bogorodskaya trava, Chaber, Chabietz—; ***Spanish*** : Serpol—; ***Swedish*** : Backtimjan, Bracktimian—; ***Teheran*** : Joshan shirazi—; ***Turkish*** : Sateriberri—; ***Urdu*** : Hasha—; ***Yemen*** : Saatar—.

ZATARIA.

Zataria multiflora Boiss. is found in Baluchistan, whence it spreads to Afghanistan and Persia.

In Jhalawan the plant is used as a cure for stomach-ache.

Baluchistan : Isghand, Izgun—; ***Hamadan*** : Zatar—; ***Jhalawan*** : Izghand— ***Urdu*** : Saatar—

ZIZIPHORA.

The genus includes 12 species, natives of the Mediterranean region and Central Asia.

1. Much-branched, at the base more or less suffruticose ... *Z. clinopodioides.*
2. An annual herb ... ... *Z. tenuior.*

1. **Ziziphora clinopodioides** M. Bieb. is found in Baluchistan. It is distributed to Afghanistan, Persia, the Caucasus, Armenia, and Syria.

At Jelga in Baluchistan the whole of the plant is dried and kept, and a decoction is used to cure typhus fever; also leaves are soaked in water at night and the infusion is drunk in the morning in cases of heat. In the Harboi Hills the juice is used as a tonic after recovery from fever.

Baluchi : Purchink— ; *Harboi Hills* : Purchink— ; *Persian* : Pudina— ; ***Pushtu*** : Maurai— ; *Ziarat Hills* : Maurai—.

2. **Ziziphora tenuior** Linn. occurs in the North-West Frontier Province, and Baluchistan. It is distributed to Afghanistan, Persia, Asia Minor, Turkestan, Soongaria, and Siberia.

The herb is used in Baluchistan to allay fever. In Teheran it is employed as a cordial and stomachic.

The seeds powdered and mixed with butter milk are used to cure dysentery in Kila Abdulla. They are used as a cure for fevers in Kharan.

Arabic : Mishkatar-el-mashih— ; *Hindubagh* : Maurai— ; *Iraq* : Za 'tar— ; *Kalat* : Mashnapurchink— ; *Kharan* : Kohipurchink— ; *Khawas* : Tukhammelanga, Tukhumimalayan— ; *Kila Abdulla* : Kahkuti, Tukhammalangai— ; ***Kurdish*** : Jata, Pung— ; *Quetta-Pishin* : Kahkuti— ; *Shiriz* : Rang— ; *Tank* : Paparboti— ; *Teheran* : Kakuti—.

THE MEDICINAL AND POISONOUS PLANTS OF INDIA.

Dammers, Guttifers, Silk-cottons, Teas, Tutsans, Water-peppers.

BY

J. F. Caius, S.J., F.L.S.

I

The Dipterocarpaceae are all natives of the hot damp forests of India and Malaya, except a few African species. Many are among the biggest trees in the forests and are valued for their magnificent timber. Many also produce useful gums, and some yield a very good quality of camphor. The seeds are generally oleaginous. There are 25 genera with 350 species.

The medicinal and poisonous Dammers belong to 5 genera, all represented within our limits: Dipterocarpus (Indo-Malaya); Dryobalanops (Borneo, Malaya); Hopea (Indo-Malaya); Shorea (Ceylon to Philippine Islands); Vateria (Seychelles, South India).

A. Calyx scarcely enlarged in fruit, segments reflexed ... Vateria

B. Calyx much enlarged in fruit, segments erect.

1. Fruiting calyx with a distinct tube covering the fruit. Fruit free.
   a. Sepals developed into long wings, 4 ... ... Dryobalanops
   b. Sepals developed into wings 2, the others very short ... ... Dipterocarpus
2. Fruiting calyx with an obscure tube; aestivation of the calyx imbricate.
   a. Wings 3 long, 2 shorter ... ... Shorea
   b. Wings 2 large, 3 small ... ... Hopea

Dipterocarpus.

The genus consists of 70 Indo-Malayan species.

Many species produce a valuable oleo-resin, known as wood oil, which is used in medicine.

The following are used in Dutch India—*D. gracilis* Blum., *D. littoralis* Blum., *D. retusus* Blum., *D. Spanoghei* Blum., *D. trinervis* Blum.—; in the Philippine Islands—*D. affinis* Brandis, *D. grandiflorus* Blanco, *D. pilosus* Roxb., *D. polyspermus* Blanco, *D. trinervis* Blum., *D. vernicifluus* Blanco—; in Indo-China—

D. alatus Roxb., *D. artocarpifolius* Pierre, *D. intricatus* Dyer, *D. Jourdainii* Pierre, *D. tuberculatus* Roxb., *D. turbinatus* Gaertn.—.

A. Calyx-tube in fruit without ribs or wings.
  1. Young branches glabrous or canescent ... *D. turbinatus.*
  2. Young branches more or less hairy ... ... *D. pilosus.*

B. Calyx-tube in fruit 5-angled or winged.
  1. Angles confined to the upper portion of the calyx-tube ... ... *D. tuberculatus.*
  2. Angles or wings prolonged to the base of the calyx-tube or nearly so.
     Angles winged ... ... *D. alatus.*

1. **Dipterocarpus alatus** Roxb. occurs in Chittagong, Burma, Tenasserim, the Andamans, Penang; extending to Siam and Indo-China.

In Cambodia the bark is considered tonic and depurative and prescribed in rheumatism. The bark of the young plant is used externally in rheumatism and in liver troubles. The oil is applied to ulcerated wounds.

In general the oleo-resin is applied externally in gonorrhoea.

Andamans: Gurjun—; *Bengal*: Battisal, Garjan, Shwetagarjan—; *Burma*: Kanyin, Kanyinbyu, Kanyinni, Kanyinwettaung—; *Cambodia*: Chhoeuteal, Trach—; *Chittagong*: Duliagurjan—; *Indo-China*: Chhoeu teal thom, Chhoeu teal tuc, Chor tuc, Dau con rai trang—; *Sinhalese*: Horagaha—.

2. **Dipterocarpus pilosus** Roxb. occurs in Sylhet, Chittagong, South Tipperah, Burma, Martaban, Mergui, the Andamans, the Malay Peninsula, and Siam.

The balsam is used in the treatment of gonorrhoea, gleet, and similar affections of the urinary organs.

Assam: Hollong—; *Tagalog*: Hagachac—.

3. **Dipterocarpus tuberculatus** Roxb. is distributed over Burma, Siam, and Cochin-China.

The oleo-resin is used with asafoetida and cocoanut oil as an application for large ulcers.

In Cambodia the roots are used in the treatment of hepatic troubles.

Burma: Eng, In Inbo, Kanyingok—; *Cambodia*: Khlong—; *Laos*: Mai bao, Mai tung—; *Siamese*: Mai rang—; *Taleing*: Sooahn—.

4. **Dipterocarpus turbinatus** Gaertn. f. occurs in Assam, Chittagong, Burma, the Andamans; extending to Siam and Cochin-China.

The oleo-resin is applied externally to ulcers, ring-worm, and other cutaneous affections. It is stimulant of mucous surfaces, particularly that of the genito-urinary system, and also diuretic. In gonorrhoea and other affections in which copaiba is generally employed, it has proved an effectual remedy.

Assam: Kuralsal, Kuroil, Tiliagurjan—; *Bengal*: Gurjan, Shwetagurjan, Tihyagurjan—; *Burma*: Inbo, Inkanyin, Kanyinbyu, Kanyingyi, Kanyinni, Kanyinnin, Kanyinwettaung, Kanyinwettaw, Kanyinywetgyi, Kanyinywetthe,

Maihow—; *Canarese*: Challane, Guge, Valivara—; *Chittagong*: Kaligurjan, Teliagurjan, Tiliagurjan—; *English*: Wood Oil Tree—; *Gujerat*: Gurjun—; *Magahi*: Kanyoung—; *Tagalog*: Mayapis—.

DRYOBALANOPS.

The genus consists of 7 species, natives of the Malay Peninsula and Islands.

Dryobalanops aromatica Gaertn. inhabits the forests and lowlands on low hills of the Malay Peninsula and Archipelago.

The tree produces camphor oil and camphor, much used as a tonic and sudorific. The camphor is chiefly exported to China, where it is employed as a tonic and aphrodisiac; in Borneo itself it is used as a diuretic and in nephritic affections, and as a popular remedy for rheumatism.

China: Lung Nao Hsiang—; *Indo-China*: Camphrier de Baroum, Camphrier de Barros, Camphrier de Bornéo, Long nao thu, Ping pien—; *Malay*: Kapur barus—.

HOPEA.

This genus numbers 50 Indo-Malayan species.

Hopea odorata Roxb. occurs from Pegu and Tenasserim to Cochin-China. It is also found in the Andamans.

In Cambodia the bark is used as an astringent in gingivitis.

Among the Burmese the resin, reduced to powder, forms a popular styptic.

Andamans: Rimda—; *Burma*: Thengan—; *Cambodia*: Koki—; *Lao*: Mai takien—.

SHOREA.

This genus numbers 90 species spread from Ceylon to the Philippine Islands.

All the species abound in various kinds of copalline resins.

The following are used medicinally:—in the Philippine Islands—*S. Guiso* Blum., *S. malaanonan* Blum., *S. mangachapuy* F. Vill.—; in Cambodia—*S. cambodiana* Pierre, *S. Harmandii* Pierre.

The oil from the seeds of *S. stenoptera* Buck is officinal in Holland. The resin from S. Wiesneri Stapf. and various other species are officially recognized in Austria, Russia, and Spain.

Stamens 20-∞.

1. Leaves 15-25 by 10-15 cm. Petiole 2-2.5 cm. ... *S. robusta.*
2. Leaves 6.3-20 by 3.2-11.5 cm. Petiole 2.5-5 cm. *S. Tumbuggaia.*

1. **Shorea robusta** Gaertn. f. occurs in the Kangra district of the Punjab, from the Kalesar forest in the Ambala district along the sub-Himalayan tract to the Darrang district of Assam,

sometimes in the outer Himalayan valleys up to 5,000 feet; in the Garo Hills, Kamrup, the Khasia Hills, the Jaintia Hills, from the Santal Parganas through Chota Nagpur and Orissa to Ganjam, Jeypore, the Central Provinces, Vizagapatam.

Ayurveda practitioners freely prescribe the bark, leaves, fruit, and resin. Among Yunani doctors the resin and the oil seem to be more especially favoured.

The resin is regarded as astringent and detergent. It is used in dysentery, and for plasters and fumigations. It is commonly given for weak digestion, gonorrhoea, and as an aphrodisiac.

Arabic : Kaikahr—; *Bengal* : Sakher, Sakhu, Sakhua, Sakoh; Sal, Sala, Salwa, Sarj, Shal—; *Bhumij* : Sargi—; *Bombay* : Sal—; *Burma* : Enkhyen—; *Canarese* : Asina, Asu, Ashvakarna, Guggula, Kabbu, Sarja, Vamsa—; *Central Provinces* : Rinjal, Sal, Sarai—; *Deccan* : Ral—; *English* : Common Sal, Indian Dammer, Sal Tree—; *Garhwal* : Kandar—; *Garo* : Bolsal—; *Gujerati* : Ral—; *Hindi* : Sakher, Sakhu, Sakhua, Sakoh, Sal, Sala, Salwa, Shal—; *Kharwar* : Sakwa—; *Khond* : Jargi—; *Kolami* : Sarjum, Sekura—; *Kumaon* : Sal—; *Lepcha* : Takral, Teturl—; *Malayalam* : Maramaram, Mulappumarutu—; *Marathi* : Guggilu, Rala, Sajara—; *Nepal* : Sakwa—; *North-Western Provinces* : Kandar, Koron, Sakhu, Sal—; *Oudh* : Koroh—; *Persian* : Lalemoabbari, Lalemohari—; *Punjab* : Sal, Seral—; *Sanskrit* : Agnivallabha, Ashvakarna, Ashvakarnika, Chiraparna, Dhanya, Dirghaparna, Dirghashakha, Divyasara, Jaladashara, Jaranadruma, Kala, Kalalajodhbhava, Karshya, Kashayi, Kaushika, Kaushikahva, Kushika, Lalana, Latashankha, Latataru, Rala, Ralakarya, Sala, Salaniryas, Salaveshta, Sarja, Sarjakarya, Sarjarasa, Sarjjaka, Sasyasambara, Sayasamvera, Shankataru, Shankurriksha, Shasyasambara, Shura, Sidhaka, Sureshtaka, Tarkshyaprasava, Vallivriksha, Vansha, Vastakarna, Yakshadhupa—; *Santal* : Sarjom—; *Sinhalese* : Dammala—; *Tamil* : Attam, Kungiliyam, Shalam—; *Telugu* : Gugal, Guggilamu, Saluva, Sarjakamu, Sarjamu—; *Tharu* : Sakwa—; *Uraon* : Sekwa—; *Urdu* : Ral—; *Uriya* : Rengal, Sagua, Salo, Salwa, Shalua, Sodingi, Soringhi, Sorjjo—.

2. **Shorea Tumbuggaia** Roxb. is found in the forests of the Cudappah, North Arcot, and Chingleput hills, up to 3,000 feet.

The resin is used as an external stimulant.

English : Green Dammer—; *Malayalam* : Tampakam—; *Tamil* : Karundanbai, Karuppudamar, Tambagam, Tambai, Tambugai—; *Telugu* : Guggilamu, Jalari, Nalladammara, Nallaguggilamu—.

VATERIA.

The genus consists of 3 species confined to South India and the Seychelles.

Vateria indica Linn. is found from North Kanara to Travancore up to 3,500 or 4,000 feet, chiefly in evergreen forests, but occasionally along rivers in deciduous forests. In Coorg both in the Ghat forests and east of the Ghats up to 3,500 feet, in the latter locality always in evergreen forest.

The fruit yields a solid fatty oil, which has obtained considerable repute as a local application in chronic rheumatism and some other painful affections.

Under the influence of gentle heat, the resin combines with wax and oil and forms an excellent resinous ointment; it is a good substitute for officinal resin. Fine shavings are administered internally to check diarrhoea.

Arabic: Mukilijraka—; *Bengal*: Chundrus—; *Bombay*: Ral—; *Canarese*: Bilidhupa, Biliguggula, Dhupa, Dhupada, Dupa, Guggula, Maddidhupa, Mandadupa, Rala, Shandike—; *Ceylon*: Pinai—; *Coorg*: Bilidupa—; *Deccan*: Sufed-damar—; *English*: Indian Copal Tree, Piney Varnish Tree, White Dammer Tree—; *Greek*: Sandaraki—; *Hindi*: Kahruba, Sageddamar, Sandras—; *Malayalam*: Kunturukkam, Pantam, Payani, Payin, Perumpayani, Telli, Vellakkunturukkam—; *Persian*: Buejhudan—; *Sanskrit*: Ajakarna, Dhupa, Kundura, Mandadhupa, Marichapatraka, Pitaphada, Sarjaka, Shala—; *Sinhalese*: Hal, Haldumula—; *Tamil*: Attam, Kukkil, Kukkulu, Kundurukkam, Kungiliyam, Kungulu, Sadagulai, Tubam, Vellaikkundurukkam, Vellaikkungiliyam—; *Telugu*: Dupadamaru, Telladamaru, Tellaguggilamu—; *Tulu*: Lobhana, Paini, Tandoligeda—; *Urdu*: Guggul—.

II

The GUTTIFERAE are trees or shrubs with a resinous yellow or greenish juice. Except a few natives of the warm regions of North America, they are all intertropical; they are more numerous in America than Asia, and are comparatively rare in Africa. There are 40 genera with about 630 species.

The members are the source of gum resins endowed with emetic and cathartic properties. The seeds are mostly oleaginous and the oils and fats are used medicinally. Some of the barks are diuretic. The rind of the fruit may be astringent.

Among the gum-resins gamboge may be mentioned as containing α—, β—, and γ— garcinolic acids.

An essential oil was found to consist of terpene and camphor.

Cambogin, a toxic resin, has been obtained from *Garcinia cambogia* Desr.

The medicinal and poisonous Guttifers of the world belong to 11 genera: ALLANBLACKIA (tropical Africa); CALOPHYLLUM (tropics, chiefly Old World); CLUSIA (warm America); GARCINIA (tropical Asia, Africa, and Polynesia); MAMMEA (West Indies); MESUA (tropical Asia); MORONOBEA (Guiana, North Brazil); OCHROCARPUS (palaeotropical); PENTADESMA (Western tropical Africa); RHEEDIA (tropical America, Madagascar); SYMPHONIA (Madagascar, tropical Africa and America).

The medicinal Guttifers of India belong to 4 genera: CALOPHYLLUM, GARCINIA, MESUA, OCHROCARPUS.

A. Ovary-cells 1-ovuled; stigma sessile or subsessile, peltate, entire or with radiating lobes; berry indehiscent.
   1. Calyx of 4 or 5 sepals ... ... GARCINIA.
   2. Calyx closed in bud, bursting in 2 valves ... OCHROCARPUS.

B. Ovary with 1, 2, or 4 erect ovules; styles slender (rarely styles 2); stigma peltate or 4-fid or acute; fruit fleshy, rarely dehiscent.
   1. Ovary 1-celled, 1-ovuled, style 1, stigma peltate CALOPHYLLUM.
   2. Ovary 2-celled, 4-ovuled, style 1, stigma peltate MESUA.

CALOPHYLLUM.

The genus consists of 80 tropical species, mostly inhabiting the Old World.

The bark is diuretic; the resin is emetic and cathartic.

The following species are used medicinally: in China, Indo-China, and the Islands of the Indian Ocean—*C. inophyllum* Linn.—; in Madagascar— *C. laxiflorum* Drake, *C. parviflorum* Bojer, *C. Tacamahaca* Willd.—; in the West Indies—*C. Calaba* Jacq.—; in Brazil—*C. brasiliense* Camb.—; in Colombia—*C. Calaba* Jacq., *C. Mariae* Pl. and Tr.—.

1. Leaves 10-20 by 7.5-10 cm., petiole 3.8-3.2 cm. ... *C. inophyllum.*
2. Leaves 7.5-12.5 by 3.2-5 cm.; petiole 13-20 mm. ... *C. elatum.*
3. Leaves 5-10 by 3.2-5 cm.; petiole 4 mm. ... *C. apetalum.*

1. **Calophyllum apetalum** Willd. occurs in the Western Ghats of the Bombay Presidency, and from Mysore to Travancore up to 1,000 feet; on the banks of rivers and backwaters on the West Coast from North Kanara southwards.

The resin acts as a vulnerary resolvent, and anodyne.

The oil obtained from the seeds is used as medicine in leprosy and cutaneous affections, and in infusion, mixed with honey, in scabies and rheumatism.

Bombay: Cherupinnai, Sarapuna—; *Canarese*: Babbe, Bobbe, Bobbi, Irai, Iria, Kalhonne, Kalpun, Kirihonne, Kullponne—; *Marathi*: Bobbi, Irai—; *Tamil*: Sirubinnai—; *Travancore*: Attupunna, Kattapunna, Mannapunna, Purapunna, Serupunna—; *Tulu*: Kaiponne, Kalponne, Seruponne, Siriponne—.

2. **Calophyllum elatum** Bedd. is found in the evergreen forests of the Western Ghats, and adjoining hills, from South Kanara to the Pulneys.

The gum is feebly astringent.

Bombay: Pun, Sirpon—; *Canarese*: Bobbi, Holehonne, Kuve, Siripunehuve, Siripunekuve, Siruponne, Srihonna, Surponnebobbi—; *English*: Malabar Poon, Poon Tree, Sirpoon Tree—; *Kadir*: Karanguttiyan, Viri—; *Malayalam*: Kattupunna, Malampunna, Pun, Pungu, Punna, Punnappai—; *Marathi*: Nagani—; *Sinhalese*: Kina—; *Tamil*: Kattupinnai, Pinnai, Pongu, Pungu—; *Tulu*: Pune—.

3. **Calophyllum inophyllum** Linn. is essentially a littoral species to be found along the East and West coasts of the Peninsula, Burma, the Andamans, Malaya, and Ceylon. It is distributed to the sea-shores and the islands of East Africa, Australia, and Polynesia.

The bark is astringent and useful in internal haemorrhages. The juice is used as a purgative, and is said to be very powerful in its action.

In Java the tree is supposed to possess diuretic properties.

In Cambodia the leaves are prescribed as an inhalation in migraine and vertigo. The oil from the kernels of the seeds is used in scabies.

The leaves, soaked in water, are employed as an application to inflamed eyes, in the Malay Archipelago.

In Madagascar the leaves are applied to sore eyes; the pounded bark is used topically in orchitis; the gum resin is considered vulnerary, resolvent, and anodyne; the oil from the seeds is a reputed antisoric, and is much used in the treatment of rheumatism; a decoction of the root is used in dressing ulcers.

The tears which distil from the tree and its fruit are emetic and purgative (Rheede).

The gum which flows from the wounded branches, mixed with strips of the bark and leaves, is steeped in water, and the oil which rises to the surface is used as an application to sore-eyes. It is said to be a useful remedy for indolent ulcers.

In New Caledonia the gum resin is applied to ulcerous wounds.

The fixed oil, expressed from the kernels of the seeds, is said to cure scabies. It exercises a great beneficial influence over the mucous membrane of the genito-urinary organs, and is therefore highly useful in the treatment of gonorrhoea and gleet. Externally, it is a good and useful embrocation in rheumatism and gout. The watery paste of the kernel of the seeds, applied to the painful joints and dried by the heat of fire, often affords a great relief in the same diseases, and may be resorted to in the absence of the oil.

Bengal: Punnag, Sultanachampa, Sultanachampaca—; *Betsileo*: Foraha—; *Betsimisaraka*: Forho—; *Bicol*: Dancalan—; *Bombay*, Udi, Undi—; *Burma*: Pengnyet, Phounniya, Phungnyet, Ponnyet, Pungnyet—; *Cambodia*: Kchyuong—; *Canarese*: Honne, Huhonne, Nameru, Pinnaikai, Ponne, Surahonne, Vuma, Wuma—; *Ceylon*: Domba, Dommakkottai, Punnai—; *Chinese*: Hu T' ung Lei—; *Cutch*: Udi—; *Deccan*: Surfan, Surpanda, Undi—; *English*: Alexandra Laurel, Alexandrian Laurel—; *Fiji*: Dilo—; *Guam*: Daog, Daok—; *Hawaii*: Ramani, Kamanu—; *Hindi*: Sultanachampa, Surpan, Surpunika, Undi—; *Hova*: Tsindelo—; *Ilocano*: Bitaog, Bitog, Pamitaoguen, Pamitlaten—; *Indo-China*: Ho dong, Mu u—; *Konkani*: Unddi—; *Madagascar*: Foraha, Vintanina—; *Malay*: Betan, Bintangor bunga, Penaga laut, Pudik—; *Malayalam*: Pouna, Punna—; *Marathi*: Nagchampa, Puniag, Surangi, Undag, Undela, Undi, Wundi—; *Mauritius*: Bois Tacamaca—; *New Caledonia*: Pit, Tamanu, Vara ; *Oceania*: Ndilo—; *Pampangan*: Bitaog, Palomaria—; *Philippines*: Birog, Bitanhol, Bitoc, Dingcalin, Palomaria de playa, Tamanian, Vitam—; *Rarotonga*: Tamanu—; *Samoa*: Fetau—; *Sanskrit*: Nagachampa, Nameru, Punnaga, Purosakeshara, Surangi, Tungakeshara—; *Seychelles*: Takamaka—; *Sind*: Duggerphul, Dugurphort, Purraya, Purreya, Surangi, Undi—; *Sinhalese*: Domba, Dombagaha, Dombatel, Teldomba—; *Spanish*: Palo Maria, Palo de Santa Maria—; *Tagalog*: Bancalan, Bitanhol, Dancalan, Dincalin, Palomaria, Tamauian—; *Tahiti*: Tamanu—; *Taimoro*: Voakotry—; *Tamil*: Nagam, Nameru, Pinnai, Punnagam, Punnai—; *Telugu*: Nameru, Pouna, Punnaga—; *Tongking*: Cay mu hu—; *Tulu*: Ponne—; *Uriya*: Polang, Punang, Punnango—; *Visayan*: Bitaog, Dancalan—; *Zambales*: Bitaoi—.

GARCINIA.

The genus numbers 200 species scattered over the tropical regions of the Old World.

Astringent properties are met in the bark and in the rind of the fruit of some species. The gum resins are powerful, drastic cathartics.

The following are used medicinally:—in China—*G. Mangostana* Linn., *G. Morella* Desc.—; in Indo-China—*G. Hanburyi* Hook, f., *G. Mangostana* Linn.—; in Cambodia—*G. Harmandii* Pierre, *G. Lanessanii* Pierre, *G. Mangostana* Linn., *G. Olivieri* Pierre, *G. Vilersiana* Pierre—; in Malaya—*G. dulcis* Kurz., *G. Mangostana* Linn.—; in the Philippine Islands—*G. Mangostana* Linn., *G. spicata*

Hook. f., *G. venulosa* Chois.—; in New Caledonia— *G. corallina* Vieil.—; in West Tropical Africa—*G. Conrauana* Engl., *G. Elliotii* Engl., *G. gnetoides* Hutch. and J. M. Dalz., *G. guineensis* Willd., *G. Kola* Heckel, *G. Mannii* Oliv., *G. ovalifolia* Oliv., *G. polyantha* Oliv.—; in Southern Africa— *G. Livingstonii* T. And.

A. Sepals and petals 4 each.	
1. Stigma entire. Stamens of male flowers in a globose central mass	*G. atroviridis.*
2. Stigma divided into rays, or deeply 4-lobed.	
I. Stamens of male flowers in 4 masses or in a 4-lobed mass surrounding the rudimentary ovary; anthers oblong, dehiscing vertically.	
a. Leaves 15-25 cm.	*G. Mangostana.*
b. Leaves 10-15 cm.	*G. cornea.*
II. Stamens of male flower in a central shortly-stalked 4-angled or columnar mass; anthers quadrate, dehiscing vertically; rudimentary ovary usually absent.	
Male flower in 3-∞-flowered, terminal and axillary fascicles; fruit subglobose or ovoid, tip mamillar.	
a. Leaves 6.3-9 cm.	*G. indica.*
b. Leaves 7.5-12.5 cm.	*G. Cowa.*
III. Stamens of the male flowers in a subglobose mass; anthers adnate, orbicular, dehiscence circumsciss, rudimentary ovary absent.	
a. Leaves 10-15 by 3.8-7.5 cm.	*G. Morella.*
b. Leaves 15-20 by 7.5-10 cm.	*G. heterandra.*
B. Sepals and petals 5, very rarely 4; filaments connate in 5, rarely 4 erect distant pedicelled spathulate bodies, antheriferous at the top, free portions very short.	
I. Leaves 23-45 by 5-10 cm. long	*G. xanthochymus.*
II. Leaves 12.5-25 cm. long	*G. dulcis.*

1. **Garcinia atroviridis** Griff. is found in Upper Assam, at Tabong, in Singapore, Johor, Malacca, Perak, and Penang.

The fruit is used medicinally by the Malays.

Malay: Asam gelugur—.

2. **Garcinia cornea** Linn. is found from East Bengal to Tenasserim, whence it extends to the Malay Archipelago.

The tree yields an inferior kind of gamboge used medicinally in Burma.

3. **Garcinia Cowa** Roxb. occurs in Eastern Bengal, Assam, the Eastern Peninsula, and the Andaman Islands.

The gum-resin is used medicinally by Burman practitioners.

Bengal: Kowa—; *Burma*: Madow, Taungthale, Toungdalai—; *Hindi*: Cowa—; *Ho*: Soroa—; *Uriya*: Sarbana—.

4. **Garcinia dulcis** Kurz is found cultivated in the Malay Peninsula. It is distributed to the Malay Islands.

The oily seeds are sold in the drug shops of Malaya as a remedy for dysentery and chronic diarrhoea.

Malay : Bijimundu, Mundu—.

5. **Garcinia heterandra** Wall. is found in Pegu and Tenasserim up to 4,000 feet.

The gum-resin is occasionally, though not extensively, employed as a medicine by Burman native practitioners.

Burma : Thanatdau, Thanattau—.

6. **Garcinia indica** Chois. is often cultivated; it is found in Konkan, North Kanara, Goa, the Western Ghats of Bombay, South Kanara, Coorg, and Wynaad.

The fruit, either green or ripe, is much used by Ayurveda practitioners.

The Apothecaries of Goa prepare trom the juice of the fruit a very fine purple syrup, which is used in bilious affections. The bark is astringent, and the young leaves, after having been tied up in a plantain leaf and stewed in hot ashes, are rubbed with cold milk and given as a remedy for dysentery.

The oil of the seeds is much used for the preparation of ointments, suppositories and other pharmaceutical purposes. It has been used as a local application to ulcerations, fissures of the lips, hands, etc., by partly melting it and rubbing on the affected part.

Bombay : Kokam—; *Canarese* : Ambsol, Dhupadamara, Murgala, Murginahali, Tittidika—; *Deccan* : Kokamb, Ratambi—; *English* : Wild Mangosteen—; *Goa* : Brindao, Brindoeiro—; *Gujerati* : Kokam—; *Hindi* : Kokam—; *Konkani* : Birondd, Birondi, Ratambi—; *Malayalam* : Punampuli—; *Marathi* : Amsole, Bhirand, Chirand, Katambi, Kokam, Kokamb, Ratambe—; *Matheran* : Kokam, Ratamba—; *Sanskrit* : Amlabija, Amlapura, Amlashaka, Amlavriksha, Amlavrikshaka, Atyamla, Bijamla, Chudamla, Chukra, Chukramla, Chukraphala, Phalamlaka, Puramla, Raktapuraka, Rasamla, Shakamla, Shreshthamla, Tintidika, Tittidiphala, Vrikshamla—; *Tamil* : Murgal—; *Tulu* : Puranapuli—.

7. **Garcinia Mangostana** Linn. is cultivated on the Western Coast of the Madras Presidency, the Nilgiris, in Goa, but very rarely in the Bombay Presidency. It is also cultivated all over the Malay Peninsula.

According to Rumphius, the bark and young leaves are employed by the Macassars in diarrhoea, dysentery and affections of the genito-urinary tracts, and also as a wash for aphthae of the mouth.

In Cambodia the bark of the plant and the rind of the fruit are used as astringents in dysentery and diarrhoea.

The rind of the fruit is a well-known astringent useful in the treatment of diarrhoea and dysentery. Waring and others have strongly recommended its use in chronic diarrhoea of children. It has also been employed successfully as a febrifuge. A very strong decoction is also recommended as an external astringent application.

The powdered rind of the dried fruit in daily doses of 60-120 grains dispensed in 3-4 portions gave satisfactory results in 63.8 per cent of amoebic—36 treated—and 66.6 per cent cases of non-amoebic—15 treated—dysentery, and in 72.0 per cent cases of diarrhoea other than dysentery—45 treated—. Mangostin was extracted from the dried rind and tried clinically. It was found to be very inferior to the powdered rind as an antidiarrhoeal agent (Caius and Mhaskar).

Bengal : Mangustan—; *Bombay* : Mangostin, Mangustan—; *Burma* : Mangkob, Mengkop, Mengut, Mimbu, Mingut, Youngzalai—; *Cambodia* : Mongkhut—; *Chinese* : Shan Chu Kuo, Tu Nien Tzu—; *English* : Mangosteen—; *French* : Mangoustan cultivé—; *Hindi* : Mangustan—; *Indo-China* : Mang cut, Mung khut—; *Jolo* : Manguis—; *Malaya* : Manggis, San chook hok—; *Malayalam* : Manggusta, Sulampuli—; *Marathi* : Mangastin—; *Portuguese* : Mangosta—; *Sinhalese* : Mangus—; *Spanish* : Mangostan, Mangostan de la India—; *Tamil* : Sulambuli—.

8. **Garcinia Morella** Desrous. occurs in Eastern Bengal and the Khasia Hills; in the evergreen forests of North Kanara and the Western Ghats, from South Kanara and Mysore to Travancore, up to 3,000 feet; in Ceylon, Malacca, Singapore, and Siam.

The gamboge is considered a valuable hydragogue cathartic. It also possesses anthelmintic properties. It is used in dropsical affections, amenorrhoea, obstinate constipation, and as a vermifuge.

The stem rubbed with water is a household remedy as a local application to rising pimples and boils, and often cuts them short.

Bengal : Tamal—; *Burma* : Thamengut—; *Canarese* : Aradala, Arasinagurse, Devanahuli, Jarize, Jarigehuli, Kankutake, Kankutgal, Punarapuli—; *Ceylon* : Korakkappuli, Makki—; *Chinese* : T'eng Huang—; *English* : Gamboge Tree—; *Hindi* : Tamal—; *Kadir* : Sikiri—; *Malayalam* : Daramba, Karukkampuli, Makki, Pinnarpuli, Pulinjakka, Valakkanna, Valogam—; *Marathi* : Tamil—; *Sanskrit* : Amritadruma, Kalaskandha, Kalatala, Lokaskandha, Mahabala, Niladvaja, Nilatala, Tama, Tamala, Tapichcha, Tapinja, Tapitha—; *Sinhalese* : Gokatu, Kanagoraka, Kanogoraka, Kokatiya—; *Tamil* : Irevalsinni, Makki, Solaipuli—; *Telugu* : Pasupuvarne, Revalchinni—; *Tulu* : Jarige, Jarigepuli, Kanakotekay—.

9. **Garcinia xanthochymus** Hook. f. occurs in Eastern Bengal and Eastern Himalaya, Burma, Penang, the Andamans, the Northern Circars, Ganjam, the Western Peninsula on the Ghats from Bombay to North Kanara, Mysore, Coorg, the Nilgiris and North Travancore up to 3,500 feet.

The fruit, which is yellow and of the size of a small apple and very acid, sweetish when ripe, edible, is used for the same purposes as that of *G. indica;* it is dried and made into a kind of Amsul. A sherbet made with about 1 oz. of the Amsul, with a little rock-salt, pepper, ginger, cumin and sugar, is administered in bilious conditions.

Assam : Tepor, Tezpur, Tihur—; *Bengal* : Chalate, Dampel, Tamal—; *Bombay* : Dampel, Onth, Osth—; *Burma* : Madau, Matau—; *Canarese* : Devangi, Deavkai, Devagarige, Devajarige, Divarige, Gansargi, Gurse, Hirekanigu, Janagi, Javangi, Neralemavu—; *Ceylon* : Egg-tree, Simaigoraka—; *Coorg* : Divarige, Nelamavu, Vate—; *Garo* : Manhola—; *Gujerat* : Karamala,

Ota—; *Hindi*: Dampel, Ota, Tamal—; *Kadir*: Anavaya—; *Konkani*: Dharambe—; *Lao*: Mai dah—; *Marathi*: Jharambi, Ota—; *Sanskrit*: Avika, Bhavana, Bhavishya, Bhavya, Kalakhanda, Kusumodar, Lamphala, Pichchalabija, Samputanga, Tamala, Tapinjha, Vakrashodana—; *Saora*: Lollorimanu—; *Sinhalese*: Cochingoraka, Ratagoraka—; *Tamil*: Kulavi, Malaippachai, Malaippuli, Pachilai, Pachumbadi, Tabinjam, Tamalam—; *Telugu*: Ivarumamidi, Sikatimramu, Sitakamraku, Tamalamu—; *Tulu*: Jarige—; *Uriya*: Cheoro, Chiuri, Sitambu—.

MESUA.

The genus consists of 3 species, natives of tropical Asia.

Mesua ferrea Linn. occurs in the mountains of Eastern Himalaya and Eastern Bengal, Assam, Tenasserim, Burma, the Andamans; in the evergreen rain-forests of North Kanara and South Konkan; in the forests of the Western Ghats from South Kanara to Travancore, up to 5,000 feet; and in Ceylon.

The bark and roots are considered to be, in infusion, an excellent tonic bitter.

The bark is mildly astringent and feebly aromatic. Combined with ginger it is given as a sudorific.

The flowers are astringent and stomachic. In many localities they are used for cough, especially when attended with much expectoration. A paste made of the flowers with butter and sugar is used in bleeding piles and burning of the feet.

In North Kanara and Bengal the oil of the seeds is used as an embrocation in rheumatism, and found useful in the treatment of itch. It is also employed as an antidote for snake poison.

The unripe fruits of this plant are aromatic and sudorific. The flower buds are used in dysentery. A syrup of flower buds 1 in 10 was used in acute cases of dysentery in the out-patient department; mild cases were cured by its use, but in very acute and severe cases, it was found to be inefficacious (Koman).

The leaf and the flower are among the best snake remedies of Ayurveda, and they are still very much used in Bengal as an antidote to snake and scorpion venoms. But Caius and Mhaskar have shown experimentally that no part of the plant is an antidote to either snake or scorpion venom.

Assam: Nahor—; *Behar*: Nagkeshur—; *Bengal*: Nagesar, Nagkesar—; *Bombay*: Nagchampa, Thorlachampa—; *Burma*: Gangau, Kangau—; *Canarese*: Kanchana, Kasara, Nagachampa, Nagakesara, Nagasampage, Nagasampige—; *Ceylon*: Naka—; *Coorg*: Atta, Iruppumara—; *Deccan*: Nagchampa—; *English*: Ceylon Iron-wood, Ironwood of Assam—; *French*: Arbre de fer, Bois d'anis, Bois de fer—; *Hindi*: Naghas, Nagkesar—; *Indo-China*: Thiet luc moc, Vap—; *Kadir*: Peri, Suruli—; *Konkani*: Nagchampa—; *Magahi*: Kainggo—; *Malay*: Matopus, Penaga kunyet, Penaga lilin, Penaga patih, Penaga sabat, Penaga suga—; *Malayalam*: Nagachempakam, Nanga, Peri, Vainavu, Veluttachempakam, Veluttapala—; *Marathi*: Nagachampa, Nagchampa, Nagchampha, Nagchapha—; *Michi*: Nahshor—; *Persian*: Narmishka—; *Pulaiya*: Atuponnai—; *Punjab*: Nagkesar—; *Sanskrit*: Bhujangakhya, Champeya, Hema, Hemakinjalka, Ibhakhya, Kanchana, Kanchanavhaya, Kanakavhaya, Kesara, Keshara, Kinjalka, Mahaushadha, Naga, Nagakeshara, Nagakhya, Nagakinjalka, Nagapushpanaga, Nagaya, Phalaka, Phanikeshara, Pinjara, Sunnagakeshara, Pushparachana, Bajapushpa, Rukma, Shatapadapriya, Suvarna, Suvarnakhya,

Svaraghatana—; *Sind*: Nakesuru—; *Sinhalese*: Deyana, Na, Nagaha—; *Tagalog*: Malabocboc, Malabucbuc—; *Tamil*: Irul, Karunangu, Malainangu, Mannainangu, Naganchambagam, Nagappu, Nagesuram, Nangu, Nirnangu, Patai, Pudangoli, Sirunagappu, Tadinangu—; *Telugu*: Gajapushpamu, Kesaramu, Kinjalkamu, Nagachampakamu, Nagakesaramu, Sikatimanu, Suvarnamu—; *Tinnevelly*: Nang—; *Tulu*: Kesara, Nagasampai—; *Urdu*: Narmishka—; *Uriya*: Nageshvar, Nageshvoro, Nagokesoro—.

OCHROCARPUS.

The 10 species of this genus inhabit the tropical regions of the Old World.

O. pentapetalus Blanco is used medicinally in the Philippine Islands, *O. Harmandii* Pierre in Indo-China, and *O. africanus* Oliv. in Tropical West Africa.

Ochrocarpus longifolius Benth. and Hook. fil. occurs in the Western Ghats of the Konkan, North Kanara, Malabar and Coimbatore; it is cultivated in the Northern Circars.

The flower-buds possess astringent and aromatic properties. They are used as a tonic in Persia.

The flowers are stimulant and carminative, useful in some forms of dyspepsia and in haemorrhoids.

Bengal: Nagakesar Punnangachcha, Rajachampaka—; *Bombay*: Suringi, Tambranagkesar—; *Canarese*: Gadhavunate, Phatapale, Punay, Puniye, Punnaga, Suragi, Surgi, Surungi, Unate, Wundy—; *Deccan*: Gardundi—; *Gujerat*: Punnaga, Ratinagkesar, Surapunnaga—; *Hamadan*: Normush—; *Hindi*: Nagkesar, Pulaga, Punnaga, Sultanachampaka—; *Konkan*: Ramundi, Suringi, Surong, Surongi—; *Malayalam*: Surampurna—; *Marathi*: Godiyundina; Punnag, Suringi, Undali—; *Matheran*: Godundipuwag, Harkia, Satwin, Surangi—; *Persian*: Nagkeshur, Tambra—; *Sanskrit*: Aruma, Devavallabha, Kamboge, Kesari, Keshara, Keshava, Kumbhika, Nagakeshara, Nagapushpa, Pandunaga, Pataladruma, Punnaga, Purusha, Purushakhya, Raktakeshara, Raktapushpa, Raktarenu, Raktavriksha, Surangi, Tunga— *Tamil*: Surabunnagam, Surabunnai, Valai—; *Telugu*: Suraponna, Surapunnagamu—; *Uriya*: Churiana, Surongo—.

III

The BOMBACACEAE number about 150 species grouped in 21 genera. They are all arborescent, and principally tropical, and include some of the largest trees.

Medicinal properties are exhibited by members of 4 genera: ADANSONIA, BOMBAX, CEIBA, PACHIRA. The last named is not represented in India.

A. Calyx 5-cleft ... ... ADANSONIA.

B. Calyx truncate or irregularly 3-5-lobed.

- I. Branches of the staminal tube 1-antheriferous ... BOMBAX.
- II. Branches of the staminal tube 2-3 antheriferous ... CEIBA.

ADANSONIA.

This genus consists of 10 species inhabiting the tropics of the Old World.

A. digitata is used medicinally in whatever country it is found growing. *A. Grandidieri* Baill. and *A. Za* Baill. are similarly used in Madagascar, *A. madagascariensis* Baill. in Madagascar and South Africa, and *A. Gregorii* F. Muell. in North Australia.

Adansonia digitata Linn., indigenous in tropical Africa, is grown in many places in India. It is pretty common about Madras, but is chiefly met with in Bombay, being plentiful on the coast.

The fruit is composed largely of a dry acid pulp which, in Bombay, is mixed with butter-milk and used as an astringent in diarrhoea and dysentery. In the Konkan, the pulp with figs is given in asthma; with the addition of cumin and sugar it is made into a sherbet, and administered in bilious dyspepsia.

According to the Bombay Gazetteer, the wood possesses antiseptic properties, and the bark is antiperiodic and a useful substitute for quinine in low fever.

The bark was at one time exported to Europe and used as a febrifuge in lieu of cinchona bark. It is still highly commended as an antiperiodic in the West Indies and in some parts of South Africa. In Cameroons it is sometimes used medicinally like the leaves.

In Gold Coast the bark is used instead of quinine for curing fever. The cream of tartar surrounding the seeds is made into a cooling drink in cases of fever. The pulp of the fruit is considered a specific in putrid pestilential fevers.

The native medicinal use of the pulp as a remedy or palliative and diaphoretic for fever and dysentery is widespread in tropical Africa, and the article was for a time exported to Europe from the East Sudan for this purpose. In Senegal, dysentery is treated by giving rice-water in which rust has been boiled and the baobab pulp added. Smallpox is similarly treated, and also a thick emulsion of the pulp is put in the eyes of the patient several times a day.

In Gambia the root is used as a febrifuge.

The dried leaves in powder form are credited with medicinal properties, promoting perspiration and preventing kidney and bladder troubles; they have been proved serviceable in diarrhoea, fevers, and other maladies. In Senegal an infusion of the powder is used internally or locally applied for a variety of inflammatory conditions, as a preventive of fever, for dysentery, internally or as a hip-bath, and similarly for genito-urinary conditions; also as a lotion for earache, ophthalmia, etc.

In Nigeria the leaves coarsely pounded are boiled with bran and salt or plant ashes, and administered to horses in bolus form. As a food, given in large quantity, this preparation keeps a horse in good condition without fattening, or maintains strength on a journey; as a medicine, given in small bulk, it is a tonic and blood maker, and is also given for conditions with subcutaneous swellings due to insect bites.

In Sierra Leone the leaves are used as a prophylactic against fever in the rains, to check excessive perspiration, and as an astringent. An infusion of both flowers and leaves is used for respiratory and digestive disorders and for eye inflammations. Locally the leaves are applied to reduce inflammatory swellings, or boiled to make a hot lotion, or burnt as a fumigation.

In Hausa a cooling drink is prepared by breaking off a piece of the shell, pouring in water, loosening and mixing the pulp by stirring, and then boiling the emulsion; the beverage is taken cold. The pulp burns with an irritating smoke, and has been used as a fumigant to keep biting insects at a distance from domestic animals.

In Senegal the inner fibrous part of the shell is made into a decoction and used as an emmenagogue. The seeds, roasted and pulverised, as well as the latex, are applied for toothache and inflammation of the gums. A decoction of the young plant from which the bark has been removed is used as an eye remedy. The gummy fluid from the bark of the tree and a powder scraped from the outside of the fruit are applied to cleanse foul sores, stimulating and promoting granulation.

In Madagascar, La Reunion, and Mauritius the leaves are given as an emollient in dysentery and inflammatory fevers. The pulp of the fruit is used in haemoptysis, dysentery, and diarrhoea.

Afrikaans: Krimmetatboom—; *Ajmere*: Kalbriskh, Kalpbriskh—; *Akwapim*: Dadee, Odadee—; *Angola*: Imbondeiro, Nbondo—; *Arabic*: El omarah, Gongoleis, Homar, Hujed, Humar, Oufa, Tabaldi—; *Awuna*: Alagba—; *Bafo*: Njobwih—; *Bakossi*: Njobwele—; *Bakundu*: Ngubwele—; *Balondo*: Ngubwele—; *Balong*: Njobwih—; *Bambara*: Molodo, Sira, Tedum—; *Basari*: Niturr—; *Batanga*: Ngubwele—; *Bemoba*: Toreg—; *Benin*: Ushi, Usi—; *Bombay*: Choyarichinch, Gorakhaamli, Gorakhchincha, Gorokhchintz—; *Brong*: Ala—; *Canarese*: Brahmamlika, Magimavu—; *Chiswina*: Muwugu—; *Dagomba*: Tua—; *Dagarti*: Tuo—; *Deccan*: Hathikhattyan—; *Delhi*: Kalbriskh, Kalpbriskh—; *Dutch*: Apenbroodbloom—; *English*: African Calabosh, Baobab, Cork Tree, Cream of Tartar Tree, Ethiopian Sour Gourd, Maputa, Sour Gourd—; *Ewe*: Adido, Alagba—; *Falor*: Ba—; *Fanti*: Efuo-bodedwo, Efuo-brodidwa—; *French*: Arbre de mille ans, Baobab, Baobab d' Afrique, Boabab, Calebassier du Sénégal, Grosmapou, Pain de singe—; *Fufulde*: Bokki—; *Fulani*: Boha, Boki, Bokki—; *Ga*: Shadzo—; *Gbari*: Kwahi—; *German*: Adansonie, Boabab—; *Gold Coast*: Cream of Tartar Tree—; *Guam*: Toto—; *Gujerat*: Bukha, Gorakhaamli, Moramlreli, Rukhdo—; *Hausa*: Kuka—; *Hindi*: Gorakamali, Gorakhamli, Goramlichora—; *Italian*: Baobab—; *Kabure*: Taelu, Telu—; *Kamba*: Mwamba—; *Kanuri*: Kuka—; *Khartoum*: Gongalasu—; *Konkomba*: Nitule—; *Kontagora*: Kuka—; *Koranko*: Sire, Sirele—; *Kratchi*: Kellai, Kelle—; *Krepi*: Dindo, Dodo—; *Krobo*: Saletcho, Salo—; *Lagos*: Ushe—; *Limba*: Kutidi—; *Lokko*: Hokbawai, Sakwi mbawi—; *Losso*: Telo—; *Makalanga*: Mguya—; *Malinke*: Boki—; *Mandingo*: Sito—; *Marathi*: Gorakhchinch—; *Masai*: Ol imisora—; *Matabele*: Umkomo—; *Mauritius*: Baobab, Gros mapou, Anai pouliya, Bumni umli—; *Mbonge*: Ngubwele—; *Mende*: Bo-wulni—; *Morar*: Vilaytiyimbi—; *Mossi*: Toyega—; *None*: Boh—; *Nupe*: Muchi—; *Porebunder*: Gorakhambali—; *Portuguese Africa*: Embondeiro—; *Portuguese Guinea*: Calabaceira—; *Russian*: Adansonia, Baobab—; *Sakalave*: Sefo—; *Sanskrit*: Chitrala, Choramli, Dirghadandi, Gandabahula, Gopali, Gorakshi, Panchaparnika, Sarpadandi, Sudandika—; *Serere*: Bak—; *Shuwa Arabic*: Hamar, Hamaraya—; *South Africa*: Baobab, Cream-of-Tartar Tree, Lemonade Tree, Monkey Bread Tree—; *Spanish*: Baobab—; *Susu*: Kiri—; *Swahili*: Mbuyu—; *Tamil*: Anaippuli, Papparappuli, Perukku, Puri—; *Telugu*: Barhmamlika, Maggimavu, Simachinta—; *Tigre*: Hoemmer, Hoemret—; *Tigrinia*: Dima, Dumma—; *Timne*: Diaroebai—; *Tschaudjo*: Taelu, Telu—; *Tukulor*: Boki—; *Twi*: Adade, Ototaa, Ototowa—; *Walo*: Alu—; *Wolof*: Gui—; *Yoruba*: Ose, Oshe—.

BOMBAX.

This genus includes 60 tropical species.

B. Ceiba Linn. is used medicinally in China, Cambodia, Malaya, and the Philippine Islands; *B. brevicuspe* Sprague in Liberia; *B. buonopozense* P. Beauv. in tropical West Africa.

Bombax Ceiba Linn. (= *B. malabaricum* DC.) occurs throughout the hotter parts of India, Burma, and Ceylon. It is abundant on the eastern side of India, ascending to 4,000 feet in altitude. It is distributed to Malaya, Sumatra and Java.

The root has stimulant and tonic properties. The bark and the root are emetic. The young roots, dried in the shade and powdered, form the chief ingredient in the *muslasemul*, a medicine highly thought of as an aphrodisiac; it is also given in impotence.

The roots of saplings upto about three years old are known as 'Semarkanda' in the Central Provinces and are used as a nerve tonic and as an astringent.

The gum or dried juice, *mocha-ras*, which the tree yields, is used as an aphrodisiac. It contains a large proportion of tannic and gallic acids, and may be successfully employed in cases requiring astringents. The gum has also tonic and alternative properties, and is used in diarrhoea, dysentery, and menorrhagia.

The dried flowers, with poppy seeds, goats' milk, and sugar, are boiled and inspissated, and of this conserve two drachms are given three times a day in haemorrhoids.

The flowers and fruit in combination with other drugs are recommended for the treatment of snake-bite and scorpion sting; but Caius and Mhaskar have shown experimentally that neither the flower nor the fruit have any antidotal value against snake or scorpion venoms.

In Cambodia the bark is used as a styptic in abnormal uterine bleeding, the root is considered diuretic, and the gum is occasionally administered in water for gonorrhoea.

In China the flowers are applied externally to boils, sores, and itch.

Basim: Khatsawar—; *Bengal*: Roktosimul, Simul—; *Bhil*: Khatseori—; *Bombay*: Katesaveri, Saer, Saur, Semul, Shembal, Somr—; *Burma*: Didu, La-i, Lepanbin, Letpan—; *Cambodia*: Roka—; *Canarese*: Apurani, Buraga, Burga, Burla, Dudi, Elava, Hatti, Kempuburaga, Kempuburga, Mullelava, Mulluburaga, Pishphele, Sauri—; *Central Provinces*: Semar, Semur—; *Ceylon*: Parutti—; *Chinese*: Mu Mien—; *Deccan*: Kantonkakhatyan, Kantonkasemul, Lalkhatyan—; *English*: Cotton Tree, Red Cotton Tree, Red Silk-cotton Tree, Silk-cotton Tree—; *Formosa*: Moc-main, Pun chi—; *French*: Bombax de Malabar, Cotonnier Mapou, Kapokier du Tonkin—; *Garhwal*: Shimal—; *Garo*: Bolchu, Panchu—; *Gond*: Vallaiki—; *Gujerat*: Ratoshemalo, Sauvor, Sawar, Shemalo, Shimar, Shimlo, Shimul—; *Hazara*: Simbal—; *Hindi*: Kantisembal, Pagun, Ragatsemal, Ragatsembrel, Raktasimul, Semal, Semul, Semur, Shimbal, Simal, Somr—; *Indo-China*: Gao, Sich moc mien thu—; *Khond*: Kamba—; *Kolami*: Del, Edel, Idel—; *Konkani*: Sanvor, Sauvor—; *Kumaon*: Shimbo—; *Lambadi*: Chamblero—; *Lepcha*: Sunglu, Tung-glu—; *Magahi*: Lapaing—; *Malaya*: Mook min, Simur—; *Malayalam*: Ilavu, Mocha, Mullilava, Pichila, Pula, Purani, Unnamuriku—; *Mal Pahari*: Simue—; *Marathi*: Kantasair, Kanterisamar, Kantesavar, Khatsawar, Sair, Sairi, Samar, Savara, Savari, Sayar, Semal, Shevari, Simlo, Tamari—; *Matheran*: Sarvar, Tambdi savar—; *Melghat*: Saori—; *Mund'ari*: Edelsanga—; *Palkonda*: Wuraga—; *Persian*:

Simbal—; *Portuguese*: Algodoeiro do matto, Arvore de panha, Panheira sumauma—; *Punjab*: Sum—; *Sanskrit*: Apurani, Bahuvirya, Chirayu, Chirjivi, Dirghadruma, Dirghapadapa, Dirghayu, Duraroha, Kadala, Kalpavriksha, Kantakadruma, Kantakaria, Kantakashtha, Kukkutavandaka, Kukkutti, Mahavriksha, Mocha, Mochani, Nirgandhapushpi, Nissara, Pnachaparni, Pichhala, Purani, Raktapushpa, Raktotpala, Ramyapushpa, Salmili, Shalmali, Shalmalini, Shimulu, Sthirayu, Sthulaphala, Tulavriksha, Tulini, Tuliphala, Yamadruma—; *Santal*: Edel—; *Saora*: Buroh—; *Sinhalese*: Kattuimbul—; *Sutlej*: Shirlan—; *Tagalog*: Bobuygubat, Buboygubat, Malabulac—; *Tamil*: Agigi, Ilavam, Ilavu, Kongu, Mullilavu, Parutti, Pongar, Pulai, Purami, Sallagi, Samani, Sanmali, Selavagu, Sitten, Surabu—; *Telugu*: Buraga, Kondaburaga, Mundlaburaga, Pinnaburaga, Salmali—; *Tulu*: Ala, Mullala—; *Uriya*: Buro, Mochoroso, Salmali, Simuli—; *Visayan*: Quesero, Salay, Talutu—.

CEIBA.

This genus consists of 10 tropical species, mostly American.

Ceiba pentandra (Linn.) Gaertn. (=*Eriodendron anfractuosum* DC.) occurs in the forests throughout the hotter parts of India and Ceylon; it is distributed to South America, the West Indies, and tropical Africa.

The juice obtained from the roots is considered a most valuable cure for diabetes.

The tree yields a gum which is astringent and used as a remedy for bowel complaints. It is a popular remedy for diarrhoea and dysentery.

The unripe fruit is regarded as demulcent and astringent.

In Cambodia the bark is used as a diuretic, astringent, and febrifuge; the very young plant with its leaves and bark is used externally in fevers as a lotion or a bath, and is given internally as an emetic to cure drunkenness; the fruit is prescribed in migraine and vertigo.

In Annam the bark is considered emetic; the flowers are given in lochiorrhoea, and in plague; the oil from the seeds is used as an emollient.

In Ubangi-Shari a decoction of the root is used internally as a drink and externally as a lotion in cases of emaciation and general debility.

In Guinea the young leaves are used as an emollient, and a decoction or infusion of the flowers is given as a laxative.

In La Reunion the bark is used as an emetic.

The leaves and fruits have emollient properties and are used in various parts of tropical West Africa to wash the head and face in fever, headache, etc. A decoction of the flowers is used in French Guinea for constipation. In parts of Cameroons the bark is pounded and macerated in cold water to apply to a swollen finger. In Liberia an infusion of the bark is used as a mouth-wash.

Abé: Gbi—; *Abuan*: Umum—; *Agni*: Enya—; *Akposso*: Ju, Juna—; *Anago*: Ogufé—; *Anang*: Ukum—; *Angola*: Mafuma, Mafumeira, Suma-uma—; *Annam*: Cay gao, Cay gon, Cay moc khoang, Moc mien—; *Aowin*: Enya—; *Aro-Chuku*: Agpu-ugu—; *Ashanti*: Ongina, Onyang, Onyina—; *Atakpame*: Huti, Wuti—; *Attié*: Muong, Nguéhié—; *Awuna*: Fuleng, Futi, Vu, Vuti—; *Bakwiri*: Buma, Wuma—; *Balundu*: Bum—; *Bambara*: Bana, Banan—; *Banda*: Kopou—; *Bariba*: Gouma—; *Baulé*: Nyé—: *Basari*: Bubumbu, Bufu—;

Bemoba: Gbang—; *Bengal*: Shwetsimul—; *Benin*: Okha—; *Bicol*: Cayo—; *Bobo-Dioulasso*: Pi—; *Boki*: Bokum—; *Bondoukou*: Ton'go, Ton'ko—; *Brignan*: Etchoui—; *Brong*: Ekile—; *Burma*: Thinbawle—; *Bussanke*: Gbée—; *Cambodia*: Kor—; *Canarese*: Apurani, Bilibarlu, Biliburaga, Biliburga, Buraga, Burga, Dudi, Elava—; *Cebu*: Bulacdondol—; *Central America*: Ceiba—; *Ceylon*: Elavam, Illaku, Imbul, Pulunimbul—; *Cuba*: Ceiba—; *Dagarti*: Goni—; *Dagomba*: Gunga—; *Dahomey*: Dehon, Gué dehonsu, Hunti—; *Dakpwa*: Kapou—; *Deccan*: Khatyan, Sufedkhatyan—; *Dendi*: Bantan—; *Diola*: Busaira, Busana—; *Ebrié*: Agué—; *Efik*: Ukem, Ukum—; *Engenni*: Akawu—; *English*: Cotton Tree, Kapok, Kapok Floss, White Cotton Tree, White Silk-cotton Tree—; *Ewe*: Ewu, Futi, Vu, Vuti, Wu, Wudesé—; *Fanti*: Nyahene, Nyina—; *Fong*: Huti, Wuti—; *French*: Bois épineux, Fromager commun, Kapokier, Ouatier—; *Fulani*: Bantahi, Bantignei, Bentegnievi, Linihi, Rinihi—; *Ga*: Onyai-tsho—; *Gbari*: Gehi, Gyehi—; *Gbaya*: Guela, Guila—; *Gouin*: Belon—; *Gouro*: Ngoué—; *Grand Popo*: Hunti—; *Guam*: Algodon de Manila—; *Guene*: Bentan habu—; *Gujerat*: Doloshamlo—; *Gurunshi*: Gung—; *Hausa*: Rimi, Rini—; *Hindi*: Hattian, Katan,, Safedsemal—; *Hova*: Landihazobe—; *Ibibio*: Ukem akabi, Uman ukem—; *Ibo*: Akbo, Akbu, Akpu—; *Ijaw*: Asisaga, Kpasukaro, Shakka, Shakra, Sishara—; *Ilocano*: Capas, Capassanglay, Dondol—; *Indo-China*: Gon, Kok niou, Kor—; *Java*: Kapok—; *Jekri*: Egungun—; *Jolo*: Capoc—; *Kabure*: Komu—; *Kalabari*: Afalafasi—; *Kanuri*: Tom—; *Khandesh*: Katsavan—; *Konkomba*: Bufu-sogbum, Kpugbum—; *Konno*: Banda—; *Kontagora*: Rimi—; *Kpandu*: Loe—; *Kratchi*: Keshafu—; *Krepi*: Ofwho, Ovua, Wuti—; *Krobo*: Leno—; *Kukuruku*: Okho—; *Kwale*: Akpo—; *Lagos*: Eggun—; *Lahu*: Egna—; *Langwasi*: Kepou—; *Laos*: Kokuiyu—; *La Reunion*: Ouatier—; *Losso*: Bahun, Ubonbe—; *Lower Amazon*: Sumaumeira—; *Malayalam*: Ilavu, Mullillapappula, Nakuli, Panni, Panniyala, Pula—; *Malinke*: Bana, Banan, Busana—; *Mamprussi*: Gunga—; *Mandingo*: Bantang, Bantango, Bentafore, Bentang, Bentango, Bintaforo—; *Mano*: Geh—; *Marathi*: Pandhari, Pandhrasaur, Salmali, Safetasavara, Samoli, Shamieula—; *Mende*: Nguwe—; *Nago*: Ogufe—; *Nankani*: Gonga—; *None*: Len—; *Nzima*: Eguina, Eniémé, Enyan'goua—; *Nupe*: Kuchi—; *Onele*: Tiou—; *Onitsha*: Akpu-ogwu, Ok-akpu—; *Onitsha Olona*: Kpakpa—; *Owerri*: Akpu-udele, Mbom—; *Pampangan*: Bulac castila—; *Portuguese Guinea*: Pelon, Poilao—; *Sakalave*: Moraingy, Pamba, Pemba—; *Salaga*: Kakre—; *Sango*: Ndoulou—; *Sanskrit*: Chiragu, Kutashalmali, Kutsitashalmali, Mocha, Rochana, Salmali, Shvetasalmali, Sthiraya—; *Sarracole*: Batiugne—; *Sassandra*: Go—; *Savalou*: Gué dehunsu—; *Sefwi*: Enya—; *Serere*: Bouday, Mboudaye—; *Shuwa Arabic*: Rum—; *Sinhalese*: Imbul, Kottapulung, Pulung—; *Sobo*: Okahen—; *Spanish*: Algodon de Manila—; *Susu*: Konole—; *Tagalog*: Boboy ,Bubuy, Bulac, Bulac sino—; *Tagouana*: Sérigné—; *Tamil*: Ilavam, Karukkanam, Panji—; *Telugu*: Kadami, Tellaburaga—; *Timne*: Am-polong, A-pullo, Ke-polong, Ma-pullo—; *Tivi*: Vambe, Yambe—; *Tschaudjo*: Komu—; *Tulu*: Ala—; *Turca*: Blo—; *Twi*: Ongina, Onyang, Onyina—; *Umu Ahia*: Akpo—; *Urdu*: Sambal—; *Visayan*: Daldol—; *Wassaw*: Enyina—; *Wele*: Tiou—; *West Indies*: Silk-cotton Tree—; *Wolof*: Bantang, Bantango, Bentang, Bentango, Bentegnié, Benten, Betenbi—; *Yoruba*: Araba—.

IV

The TERNSTROEMIACEAE are trees or shrubs of tropical Asia and America. A few are to be found in Africa.

Stimulant, astringent, and antidysenteric properties. Some barks are powerful vesicants.

The stimulants owe their action to the alkaloids caffeine and theophylline. Methyl salicylate and quercitrin have been isolated from some of them.

Medicinal properties are to be found in 9 genera: ACTINIDIA (Eastern Asia); ANNESLEA (Indo-Malayan); CAMELLIA (India, China, Japan); GORDONIA (Indo-Malaya, China, North America); KIELMEYERA (South Brazil); NESOGORDONIA (Madagascar); SAURAUJA (tropical Asia, America); SCHIMA (Eastern Indo-Malaya); TERNSTROEMIA (South America, Asia).

The medicinal species of India belong to 6 genera: ANNESLEA, CAMELLIA, GORDONIA, SAURAUJA, SCHIMA, TERNSTROEMIA.

A. Peduncles many-flowered ... ... SAURAUJA.

B. Peduncles one-flowered.

1. Fruit half inferior, drupaceous ... ... ANNESLEA.
2. Fruit superior.
   a. Anthers basifixed ... ... TERNSTROEMIA.
   b. Anthers versatile.
      i. Seeds wingless ... ... CAMELLIA.
      ii. Seeds winged.
         α Radicle inferior. Fruit globose ... SCHIMA.
         β Radicle superior. Fruit oblong ... GORDONIA.

ANNESLEA.

The genus consists of 2 species inhabiting Burma, Camodia, and the Malay Peninsula.

Anneslea fragrans Wall. occurs in the Eastern Peninsula, Moulmein and Martaban.

The bark and the flowers are used medicinally in Cambodia. The former is considered antidysenteric and anthelmintic; the latter are said to be antiperiodic.

Cambodia: Sauphi—.

CAMELLIA.

The genus consists of about 10 species spreading from India to Japan.

Root and bark astringent and antidysenteric; leaves astringent, antidysenteric, digestive, tonic, and diaphoretic.

C. japonica Linn. is used medicinally in China, *C. Thea* Link in China and Indo China.

Flowers erect, sepals deciduous ... ... *C. drupifera.*
Flowers nodding, sepals persistent ... ... *C. Thea.*

1. **Camellia drupifera** Lour. occurs in Eastern Himalaya, at 3,000-7,000 feet, from Nepal to Bhutan; in Assam and the Khasia Mountains, at 5,000-8,000 feet; in Tenasserim and the Andaman Islands.

The oil-cake from the seeds is used in Tongking to stupefy fish.

Indo-China: Cha mai, Dan-che, so—.

2. **Camellia Thea** Link (= *Thea sinensis* Linn.) occurs in Assam and the hilly country to the East of it, and in Upper Burma. It has been cultivated for ages in China and Japan. It is now extensively cultivated in Assam, Cachar, Sikkim, North-Western Himalaya, the Nilgiris, and Ceylon.

Tea is astringent and gently excitant, and exerts a decided influence over the nervous system.

Arabic: Chha—; *Assam*: Hilkat—; *Burma*: Letpet—; *Cachar*: Dullicham—; *Canarese*: Cha, Chaha, Theyale—; *Chinese*: Ming—; *Danish*: The—; *Dutch*:

Thee—; *English*: Assam Tea, China Tea, Indian Tea, Tea—; *French*: Thé, Thé de Chine, Théier—; *German*: Thee—; *Gujerati*: Chah—; *Hindi*: Cha, Chha—; *Hungarian*: Tea—; *Indo-China*: Cao lo, Cha, Cha ginh, Che, Te ve, Tra, Tra hue—; *Italian*: Te—; *Japanese*: Teh, Tsja—; *Malaya*: Cha, Te—; *Malayalam*: Chaya, Teyila—; *Marathi*: Chaha—; *Mundari*: Cadaru—; *Muttack*: Khlap, Misaphlap, Phlap—; *Persian*: Ca, Chha, Chaikathai—; *Polish*: Te—; *Portuguese*: Cha, Cha da India—; *Punjabi*: Cha—; *Roumanian*: Ceaiu—; *Russian*: Tshay—; *Sanskrit*: Chaha, Chavika—; *Scotch*: Te—; *Sinhalese*: Thaygas—; *Spanish*: Te, Te de China—; *Swedish*: Thee—; *Tamil*; Karupputteyilai, Pachaitteyilai, Teyilai—; *Telugu*: Nallateyaku, Teyaku, Tiyaku—; *Tulu*: Cha—; *Turkish*: Chai—; *Urdu*: Chai—; *Uriya*: Cha—.

GORDONIA.

The genus numbers 20 species, found in the Indo-Malayan region, China, and North America.

Gordonia obtusa Wall. occurs in the Konkan, the Western Ghats of the Madras Presidency chiefly of the Eastern side, usually from 5,000 to 7,000 feet, lower in Travancore.

In the Nilgiris an infusion of the leaves is given as a stomachic, stimulant, and appetiser.

Badaga: Nagatte—; *Kadir*: Attangi, Ola—; *Nilgiris*: Nagetta—; *Tamil*: Miyilai—.

SAURAUJA.

The genus numbers 60 species distributed over tropical Asia and America.

Saurauja napaulensis DC. occurs in Temperate Himalaya, from Bhutan and Sikkim—at 5,000-7,000 feet—to Garhwal—at 2,400-5,000 feet; in the Khasia Mountains, at 5,000 feet; in the Mishmi Hills.

In Tongking the bark is used as a poultice to help the extraction of splinter imbedded in the flesh.

Hindi: Gogina—; *Jaunsar*: Ratendu—; *Kumaon*: Gogin, Gogna, Gugna—.

SCHIMA.

The genus consists of 10 Indo-Malayan species.

The bark is vesicant.

S. Noronhae Reinw. is used medicinally in Indo-China.

Flowers in a loose terminal corymb, peduncles smooth slender ... ... *S. crenata.*

Flowers in a short terminal raceme, peduncles with minute white warts ... ... *S. Wallichii.*

1. **Schima crenata** Korth. occurs in the Eastern Peninsula from Tenasserim to Penang, and in Burma. It is distributed to Borneo and Sumatra.

The stem and the sap are used medicinally in Cambodia. The stem is given in nausea; its bark is vesicant. The extremities of the young shoots are gently warmed and the sap that exudes is dropped in the ear for otitis.

Cambodia: Trathok—; *Indo-China*: Rma, Sang soc, sat hat—.

2. **Schima Wallichii** Choisy occurs in Nepal, Sikkim up to 5,000 feet, Bhutan, Assam, the Khasia Hills, Manipur, Chittagong, and Upper Burma.

The bark is nearly black externally, with deep clefts; the liber is made up of an abundance of white, needle-shaped cells, which are readily detached and act as cowage, in producing painful irritation, when brought into contact with the skin.

Anthelmintic and rubefacient.

Assam: Chilauni, Makria, Makusal, Mukriasal—; *Bhutia*: Samching—; *Burma*: Ananpho, Laukya, Theetya—; *Cachar*: Jam—; Duars: Chilauni—; *Garo*: Boldak—; *Goalpara*: Gugera—; *Hindi*: Chilauni, Makriya, Makriyachilauni, Makusal—; *Khasia*: Dingan—; *Lepcha*: Sambrong—; Sung-brong kung, Sung-sung kung—; *Nepal*: Chilauni, Goechassi—; *Sema*: Michi-sii—; *Sikkim*: Chilauni—.

TERNSTROEMIA.

The genus numbers 35 species inhabiting tropical Asia and America.

Ternstroemia japonica Thunb. occurs in Eastern Bengal and the Eastern Peninsula, from the Khasia Mountains at 4,000-5,000 feet to Moulmein; in the Western Peninsula, in the Nilgiris; and in Ceylon. It is distributed to Sumatra, China, Japan, and the Loochoo Islands.

The bark and the root are astringent. They are used in Japan as an antidysenteric.

Indo-China: Giang nui, Hoa bi huong, Son cha hoa—.

V

The HYPERICACEAE are herbs, shrubs or trees, spread over the temperate and hot regions of the globe, and especially in the northern hemisphere. There are 8 genera with about 210 species.

The medicinal Tutsans of the world belong to 5 genera: CRATOXYLON (Indo-Malayan); HARONGA (tropical Africa, Madagascar, Mauritius); HYPERICUM (cosmopolitan, temperate regions); PSOROSPERMUM (tropical Africa, Madagascar); VISMIA (tropical America).

Two of the above, CRATOXYLON and HYPERICUM, are represented in India.

1. Capsule dehiscing loculicidally. Seeds winged ... CRATOXYLON.
2. Capsule dehiscing septicidally or at the placentas. Seeds not winged ... ... HYPERICUM.

CRATOXYLON.

The genus consists of 12 species inhabiting tropical Asia.

C. Hornschuchii Blume is used medicinally in Java, *C. neriifolium* Kurtz in Indo-China.

Cratoxylon neriifolium Kurtz is found in Chittagong, and in Burma: Teiyet, Pegu, Prome hills.

In Tongking, an infusion of the leaves is considered a very powerful digestant.

Indo-China: Nganh nganh, Thank nganh—

HYPERICUM.

The genus numbers 220 species, spread over the whole world, chiefly in temperate regions.

Vulnerary, astringent, and anthelmintic. The seeds may be diuretic and antispasmodic.

The following are used medicinally:—in Europe—*H. Androsaemum* Linn., *H. barbatum* Jacq., *H. Coris* Linn., *H. elegans* Steph., *H. hircinum* Linn., *H. hirsutum* Linn., *H. humifusum* Linn., *H. montanum* Linn., *H. perfoliatum* Linn., *H. perforatum* Linn., *H. pulchrum* Linn., *H. quadrangulum* Linn., *H. Richeri* Vill., *H. tetrapterum* Fries.—; in Indo-China—*H. ascyron* Linn., *H. chinense* Linn., *H. erectum* Thunb., *H. japonicum* Thunb. *H. patulum* Thunb., *H. Sampsoni* Hance—; in Malaya—*H. japonicum* Thun.—; in Southern Africa—*H. aethiopicum* Linn., *H. lalandii* Chois.—; in La Reunion—*H. angustifolium* Lam.—; in Madagascar— *H. japonicum* Thunb.—; in North America—*H. Ascyron* Linn., *H. maculatum* Walt., *H. mutilum* Linn., *H. perforatum* Linn.—; in Colombia—*H. Brathys* Lam., *H. Chamaemyrtus* Tr. and Pl., *H. Humboldtianum* Steuder, *H. mexicanum* Linn., *H. platyphyllum* Gleason, *H. thesiifolium* H. B. K.—; in Brazil—*H. connatum* Lam., *H. laxiusculum* St. Hil.—.

A. Sepals 5, unequal; petals deciduous; stamens 5-adelphous at the base; ovary 5-celled.
  1. A glabrous shrub, 30-90 cm. high ... ... *H. patulum.*
  2. Stem none but branches innumerable ... ... *H. chinense.*

B. Sepals 5, connate at the base, equal or unequal; petals persistent; stamens 3-adelphous at the base; ovary 3-celled.
  1. Styles twice the length of the ovary, equalling the stamens ... ... *H. perforatum.*
  2. Styles very short ... ... *H. Sampsoni.*
  3. Styles half the length of the ovary ... *H. humifusum.*

C. Sepals 5; petals persistent; stamens connate at the base; ovary 1-celled ... ... *H. japonicum.*

1. **Hypericum chinense** Linn., a native of China, is cultivated in many Indian gardens.

The plant is astringent and alternative. In Indo-China the leaves and the green stems are made into a paste and applied to bites from dogs and stings from bees.

Indo-China: Kim ty dao—.

2. **Hypericum humifusum** Linn. grows in the Nilgiris. It is distributed over Europe, the Atlantic Isles, and South Africa.

In Europe the flowers are infused in olive oil or in alcohol and used as a vulnerary, chiefly for old sores and eczema.

English: Trailing St. John's Wort—.

3. **Hypericum japonicum** Thunb. is found in temperate and subtropical Himalaya, the Khasia Hills, Assam, Burma, Eastern and Western Peninsula, and Ceylon. From Japan it spreads to China and Indo-China, the Philippine Islands, Java, Australia, New Zealand, and Madagascar.

In China and Indo-China the plant is credited with astringent and alterative action, and externally it is used as vulnerary.

In Madagascar the plant is used as a vulnerary, styptic, anti-asthmatic, and antidysenteric.

Betsimisaraka: Manitsorohina—; *Cantonese*: Thin Kee Wang—; *Chinese*: T'ien Chi Wang—; *Hova*: Anangoaika, Anantatara, Tsikotrakotra, Voantrotroka—; *Imerina*: Tsikotrokotroka—; *Indo-China*: Ban—.

4. **Hypericum patulum** Thum. occurs in the Khasia Hills 5,000-6,000 feet; throughout the temperate Himalaya (except Sikkim) at 3,000-7,000 feet, from Bhutan to the Ravi. It is also found in China, Formosa, and Japan.

The scented seeds are employed as an aromatic stimulant in Patna, where they are imported from Nepal (Irvine).

In Indo-China they are used both externally and internally as a remedy for the bite of dogs and the sting of bees.

Behari: Tumbhul—; *Indo-China*: Kim ty mai—.

5. **Hypericum perforatum** Linn. occurs in the temperate Western Himalaya at 6,000-9,000 feet; in Kashmir and Simla, apparently not in Kumaon. It is distributed to Northern temperate Asia, Europe, and North Africa.

The plant has been, and is still, recognized in Europe as aromatic bitter, astringent, resolvent, expectorant, and nervine. It is used in all pulmonary complaints, bladder troubles, in suppression of urine, dysentery, worms, diarrhoea, hysteria and nervous depression, hoemoptysis and other haemorrhages, and jaundice. For children troubled with incontinence of urine at night an infusion or tea given before retiring will be found effectual; it is also useful in pulmonary consumption, chronic catarrh of the lungs, bowels, or urinary passages. Externally for fomentations to dispel hard tumours, caked breasts, ecchymosis, etc.

It is recommended in Arabian medicine as a vermifuge; and is also used to cure piles, prolapsus uteri and ani.

The red juice is esteemed in Europe and China as one of the most popular and most curative applications for excoriations, wounds, and bruises.

The juice gives a red colour to the spirit of wine with which it is mixed, and to expressed oils. The oil of St. John's Wort is made from the flowers infused in olive oil. This oil is highly useful for healing bed sores, and is commended as excellent for ulcers.

The flowering tops are bitter, terebinthinate, acrid, and vulnerary. A salve compounded from the flowers, and known as St. John's Wort Salve, is still much used and valued in English villages.

Homoeopathic practitioners prepare a medicinal tincture with spirit of wine from the entire fresh plant, collected when flowering, or in seed, and this proves of capital service for remedying injuries to the spinal cord, both by being given internally and used externally. It has been employed in like manner with benefit for lock-jaw.

Yunani practitioners consider the leaves a good application for scorpion sting. However, Caius and Mhaskar have shown experimentally that the leaves are not an antidote to scorpion venom.

The weed is poisonous to horses.

Arabic: Lioufarikoun, Khashkhash-el-asoued, Mesmoune—; ***Catalan***: Herba de cop, Herba foradada, Herba de Sant Joán, Hypericon, Pericó groc—; *Chinese*: Ia Iou—; *Danish*: Jordhumbe, St. Hans urt—; ***Dutch***: St. Jans Kruid—; *English*: Amber, Devil's Scourge, Grace of God, Hard Hay, Hundred Holes, Lord God's Wonder Plant, St John's Grass, St John's Wort, Terrestrial Sun, Witch's Herb—; *Finland*: Werdu heino—; ***French***: Chasse diable, Herbe à millepertuis, Herbe à mille trous, Herbe aux piqûres, Herbe de la Saint-Jean, Herbe saint Jean, Millepertuis, Millepertuis officinal, Trascalon perforé, Trescalan, Trucheron jaune, Truscalan—; *German*: Christiankraut, Christignadenkraut, Christikreuzblume, Christikreuzblut, Christiwundkraut, Christwundkraut, Conradskraut, Durchwachs, Elfenbeutkraut, Frauvonwurde, Gartheu, Harthau, Hartenan, Hartheu, Hasenkraut, Hexenkraut, Jagemichel, Jageteufelkraut, Johannesblume, Johanniskraut, Hohannesblut, Hohanneskraut, Hohannsinkrut, Kannsblut, Sankt Johanniskraut, Scharnokel, Schernekel, Tausendloch, Teufelsflucht, Teufelsraub, Unsereliebenfrauenbettstroh, Urldgartheil, Wolfkraut—; *Hindi*: Bassant, Dendhu—; *Italian*: Cacciadiavoli, Erba di San Giovanni, Iperico, Ipericone, Perforata, Pilatro—; *Languedoc*: Trescalan, Trescoulaou—; *Malta*: Pitted St. John's Wort, Cacciadiavoli, Erba di San Giovanni, Iperico—; *North America*: Balm-of-warrior's-wound, Cammock, Common St John's Wort, Penny-John, Rosin-rose, Tipton-weed—; ***Polish***: Dziurawice—; *Portuguese*: Herva de San Joao, Hypericao, Milfurada—; ***Provence***: Herbo de l'oli rouge, Herbo de la San-Jean—; ***Punjab***: Bassant, Dendlu—; ***Roumanian***: Iarba lui Sfant Ioan, Iarba sfantului Ioan, Pojarnita, Sunatoare—; ***Russian***: Zweroboi—; ***Spanish***: Corazoncillo, Hierba de las heridas, Hypericon, Pericon, Yerba de San Juan—; *Swedish*: Johannisoert—; ***Urdu***: Balsana—.

6. **Hypericum Sampsoni** Hance occurs in the Khasia Mountains. It is distributed to South China and Formosa.

In Tongking the plant is used as a vulnerary.

Indo-China: Nguyen bao thao—.

VI

The Elatinaceae are herbs or undershrubs, widely dispersed, especially in the Old World. They inhabit ditches and the submerged shores of ponds and rivers. There are 30 species grouped in 2 genera.

The plants credited with medicinal properties all belong to the genus Bergia.

Bergia.

The genus consists of 25 species, inhabiting tropical and temperate regions.

B. decumbens Planch. is used medicinally in South Africa, *B. guineensis* Hutch. and J. M. Dalz. in Northern Nigeria.

Bergia odorata Edg. is found in Sind, Western Rajputana, and Gujerat. It is distributed over Persia, Egypt, and tropical Africa.

Used for cleaning teeth and, in Jodhpur, applied to broken bones. The leaves rubbed down in water are used as a poultice for sores (Macadam).

Hadeija: Babargiwa—; *Porebunder*: Gangharun, Lavadiyun, Okharal—; *Rajputana*: Karbuja, Kakria, Rohwan—; *Sokoto*: Bushi, Dushiya, Jisshiya—

THE MEDICINAL MALLOWWORTS OF INDIA.

BY

J. F. CAIUS, S.J., F.L.S.

The MALVACEAE include 35 genera, and over 700 species distributed chiefly in the warmer regions of both hemispheres. They are essentially tropical, diminishing rapidly as they recede from the equator, and they are more numerous in the northern tropics and in America than in the Old World. All the members agree in containing mucilage freely, and in possessing no unwholesome properties. Some contain free acids, and are employed as refreshing drinks; others yield volatile oils, and are classed among stimulants. The seeds contain a fixed oil, and their testa is often woolly; the bark of many is very tenacious.

Among the products obtained by analysis may be mentioned:—(1) *mucilage;* (2) *furfurol;* (3) *glucosidal pigments*—althaein, gossypetin, malvin—; (4) *glucosides*-gossypitrin, isoquercitrin, quercimeritrin—; (5) *nitrogenous substances*—B-asparagin, betaine, cnoline—; (6) *alkaloids*—ephedrine, pseudophedrine—.

The medicinal Mallowworts of the world belong to 17 genera: ABUTILON (tropical and subtropical regions); ALTHAEA (temperate regions); CIENFUEGOSIA (America, Africa, Australia); GOSSYPIUM (tropical and subtropical regions); HIBISCUS (tropical and subtropical regions); KYDIA (India); LAVATERA (Mediterranean, Australia, mid-Asia); MALACHRA (warm America, West Indies); MALVA (northern temperate regions); MALVASTRUM (America, South Africa); MALVAVISCUS (warm America); NAPAEA (North America); PAVONIA (tropical and subtropical regions); SIDA (cosmopolitan); SPHAERALCEA (Cape Colony, America); THESPESIA (warm regions); URENA (tropical and subtropical regions).

The medicinal Mallowworts of India are included in 11 genera: ABUTILON, ALTHAEA, GOSSYPIUM, HIBISCUS, KYDIA, MALACHRA, MALVA, MALVASTRUM, PAVONIA, SIDA, THESPESIA, URENA.

A. Herbs or shrubs. Ripe carpels separating from the axis. Styles as many as the carpels.
  I. Ovule solitary, ascending.
    a. Stigmas linear.
      1. Bracteoles 6-9 ... ... ALTHAEA.
      2. Bracteoles 3 ... ... MALVA.
    b. Stigmas capitate ... ... MALVASTRUM.
  II. Ovule solitary, pendulous.
      Carpels convergent at the points or beaked. SIDA.
  III. Ovules 2 or more.
      Carpels with no false partition ... ABUTILON.

B. Styles or stigmatic branches twice as many as the carpels,
  I. Carpels opposite the petals.
    a. Carpels unarmed ... ... MALACHRA.
    b. Carpels beset with spines ... ... URENA.
  II. Carpels opposite the sepals ... ... PAVONIA.

C. Herbs or shrubs. Fruit capsular. Sepals leafy. Staminal tube truncate or 5-toothed at the apex.

I. Stigmas spreading. Seeds reniform. Stamens numerous. Bracteoles 5 or more ... ... HIBISCUS.

II. Stigmas coherent in a club-shaped mass.

a. Bracteoles 3-5, small ... ... THESPESIA.

b. Bracteoles 3, large, cordate ... ... GOSSYPIUM.

D. Trees. Sepals leathery. Styles connate or free. Fruit capsular or indehiscent. Leaves simple or lobed. Bracteoles 4-5 ... ... KYDIA.

ABUTILON.

The genus numbers 120 species, natives chiefly of the tropical and subtropical regions of both hemispheres.

The leaves of all the species contain a large quantity of mucilage. The roots are regarded as cooling, astringent, and tonic. The seeds are considered antidysentric.

The following species are used medicinally in China-*A. Avicennae* Gaertn.—; in Indo-China—*A. Avicennae* Gaertn., *A. indicum* G. Don—; in Malaya and the Philippine Islands—*A. indicum* G. Don—; in Gold Coast —*A. asiaticum* G. Don, *A. mauritianum* Sweet—; in La Reunion—*A. exstipulare* G. Don, *A. glaucum* Sw., *A. hirtum* G. Don, *A. indicum* G. Don—; in Somaliland—*A. glaucum* Sw.—.

Carpels more than 10, usually 15-20.

A. Carpels pointed or with a distinct mucro.

Carpels 8-13 mm. long.

a. Carpels hairy, ultimately glabrate, shining ... *A. indicum*.

b. Carpels densely hairy, ultimately shaggy ... *A. asiaticum*.

B. Carpels obtuse without a mucro.

I. Corolla 5 cm. diam.

a. Fruit globose, densely silky villous ... *A. glaucum*.

b. Fruit roughly hairy ... ... *A. hirtum*.

II. Petals hardly exceeding the sepals ... ... *A. Theophrasti*.

1. **Abutilon asiaticum** G. Don occurs in the Western Peninsula and in Ceylon. It is scattered over the tropics of both hemispheres.

In Gold Coast an infusion of the leaves macerated in water along with spices is drunk of venereal troubles.

In Southern Nigeria the juice of the plant is applied as an emollient to relieve soreness of the nates in young children.

Arabic: Hambok, Hanbuk—; *Bengal*: Petari—; *Ewe*: Kusiti—; *Fanti*: Nwarha—; *Ga*: Fufuba-tsho—; *Hindi*: Jhampi, Kangahi, Kanghi—; *Krepi*: Didinglome—; *Kuka*: Mulu ; *Marathi*: Chakrabhenda, Kangori, Petari—; *Sanskrit*: Balbija, Kalikanghi—; *Sinhalese*: Anoda—; *Songhai*: Talca-ouragna—; *Tamil*: Perundutti, Tutti—; *Telugu*: Botlabenda, Nugubenda, Peddabenda, Tutturubenda—; *Twi*: Apongo, Mmofra forowa—; *Uriya*: Jhonkapedi—.

2. **Abutilon glaucum** Sw. is fairly common in India and Ceylon It extends to Afghanistan, the Mediterranean region, Arabia, Egypt, tropical Africa, the Cape, and Australia.

At Nasirabad the mucilaginous leaves are considered as a cure for piles. In La Reunion they are used as a pectoral.

Kachhi: Gidarwar, Tutare—; *Las Bela:* Bur—; *Nasirabad:* Giddarwal—; *Somali:* Balambal morodi—; *Swahilli* Balambal morodi—.

3. **Abutilon hirtum** G. Don occurs in the United Provinces and the Central Provinces, whence it spreads to South India and Ceylon; in Sind and Baluchistan, extending to Arabia and tropical Africa. It is distributed to Malaya and Australia.

The roots, leaves, and seeds are medicinal, and good substitutes for those of *A. indicum.*

Bengal: Barkhangi—; *Cawnpore:* Barabanghi, Barkhangi,—; *Hindi:* Barkhangi—; *Sunderbuns:* Bara kanghi, Bar potari—; *Tamil:* Vadattutti—; *Telugu;* Belabenda—; *Uriya:* Karpuripotro—.

4. **Abutilon indicum** Sw. is found throughout the tropics.

A decoction of the bark, leaves, and seeds together has been long used by the Hindus on account of its mucilaginous and diuretic properties.

Boiled milk, whisked with the fibrous twigs, coagulates; the fluid obtained by decantation is regarded by hakims as efficacious in hoemorrhoids when given internally.

The infusion of the root is prescribed in fevers as a cooling medicine, and is considered useful in strangury, haematuria, as also in leprosy.

The bark is valued as a diuretic.

The leaves are cooked and eaten in bleeding piles. A decoction is used in bronchitis, in catarrhal bilious diarrhoea, in gonorrhoea and inflammation of the bladder, and in fevers; it is prescribed as a mouth-wash in cases of tooth-ache and tender gums.

The root and leaves boiled with raisins and strained, make a pleasant diluent and demulcent.

The seeds are reckoned aphrodisiac, and are used as a laxative in piles, and in the treatment of coughs. They are burned on charcoal and the rectum of children affected with thread-worms is exposed to the smoke. Infused in hot water they form a cooling drink.

According to the Chinese in Hong-Kong, the seeds are employed as an emollient and demulcent; the root is used as a diuretic and pulmonary sedative, and the flowers and leaves as a local application to boils and ulcers. Porter Smith states that the seeds and the entire plant are used as 'demulcent, lenitive, diuretic, laxative and discutient remedies. Puerperal diseases, urinary disorders, chronic dysentery and fevers are treated with the seeds.

The slightly bitter bark is considered diuretic in the Philippines. The root, leaves, and flowers are used as an emollient.

Arabic: Deishar, Masht-ul-ghola, Masht-ul-ghoul—; *Bengal:* Potari—; *Bombay:* Chakrabenda, Etari, Kangoi, Kangori, Pamaolni—; *Brahui:* Baibaro, Gogharo—; *Burma:* Bonkhoe, Bonkhoye, Thanurchok—; *Canarese:* Gidutingi, Hettukisu, Hettutti, Kisangi, Srimudre, Srimudrigida, Tutti—; *Chinese:* Kuan Sha Yuan—; *Cutch:* Balbij—; *Deccan:* Chakrabenda, Etari, Kangoi, Kangori, Pamudni—; *Goa:* Petari, Tupkadi ; *Guam:* Malbas, Malva, Matbas—; *Gujerati:* Dabali, Kantaki-; *Hindi:* Jhampi, Kandhi, Kanghani, Kanghi, Potari Tepari, Tutri—; *Hongkong:* Tung K'uci—; *Ilcano:* Lulupao—; *Indo-China:* Coi xay, Dok tok lai—; *Kharwar:* Kakhi—; *Konkani:* Voddlipettari—; *Malaya:*

Kwan sa yin —;*Malayalam*: Katturam, Katturan, Pitik-kapattu, Tutti, Tuvatti, Uram, Velluram—; *Marathi*: Akakai, Kansuli, Karandi, Madmi, Mudra, Mudrika, Pidari, Vikankati —; *Mauritius* Mauve du pays—; *Mundari*: Pusikatadaru—; *Niger*: Abeokuta—; *Persian*: Darakhteshanah, Darakhteshane—; *Philippines*: Malvas, Malvas de Castilla, Tabing—; *Porebunder*: Bapat—; *Portuguese*: Fruta gargontilha, Malwa—; *Punjab*: Pilibuti, Sanbal—; *Rajputana*: Dabi, Jhili, Tarakanchi-; *Sanskrit*: Atibala, Balika, Balya, Bhuribala, Ghanta, Kankati, Rishiprokta, Shita, Shitapushpa, Vikankata, Vatyapushpika, Vrishyagandha, Vrishyagandhika—; *Santal*: Mirubaha—; *Siamese*: Klorb tabart—; *Sind*: Khapato, Pilibuti, Sunbul—; *Sinhalese*—; Anodagaha—; *Sunderbuns*: Kanghi, Potari—; *Tagalog*: Cuacuacohan, Giliggiligan, Guilingguilingan, Melbas—; *Tamil*: Nallatutti, Paniyarattutti, Perundutti, Tutti—; *Telugu*: Adavibenda, Botlabenda, Dudi, Muttavashirubenda, Nugubenda, Peddabenda, Tutirichettu, Tutti, Tutturubenda—; *Tulu*: Urki—; *Twi*: Mmofraforowa—; *Urdu*: Kanghi—; *Uriya*: Nakochono—; *Visayan*: Dulupang, Malis, Pilis, Taratacopes, Yampong—;

5. **Abutilon Theophrasti** Medic. (=*A. Avicennae* Gaertn.) occurs in North-Western India, Sind, Kashimir and Bengal. It extends to Arabia, Egypt, the Mediterranean, and South-Eastern Europe. It is naturalized in many parts of Asia, Africa and America.

The leaves, seeds, and roots are put to the same uses as those of *A. indicum*.

In Indo-China a decoction of the seed is given in dysentery, fistulae, and eye sores.

Chinese: Ching Ma—: *English* American Jute, Indian Mallow—; *French*: Jute de Chine—; *Indo China*: Manh ma—; *Meskwaki* Menakwusk—,

ALTHAEA.

The genus consists of about 30 species, natives of the temperate regions of the Old World, and rarely found in the tropics.

This genus is well-known for the emollient and resolvent properties of its members.

The following are commonly used medicinally in Europe—*A. cannabina* Linn., *A. hirsuta* Linn., *A. officinalis* Linn., *A. pallida* Waldst. u. Kit., *A. rosea* Cav.—in China and Indo-China—*A. rosea* Cav.—.

1. Stem 15-30 cm. high. Flowers 13-20 mm. diam ... *A. Ludwigii*.
2. Stem 60-90 cm. Flowers 2.5-5 cm. diam. ... ... *A. officinalis*.
3. Stem often exceeds 1.8 m. in height. Corolla often exceeds 10 cm. across. ... ... *A. rosea*.

1. **Althaea Ludwigii** Linn. occurs in Agra, Bundelkhand, the Punjab, Sind. Baluchistan, and the Deccan. It extends to the Mediterranean, and South Africa.

In Wad the plant is used as an aperient, being for the purpose pounded and mixed with sugar and water, and strained.

Arabic: Ketmi, Khutmi—; *Baluchi* Baharo—: *Brahui* Utper—; *Kech*: Gwaragpad—; *Kohlu*: Gardaibuti, Girdaebuate—; *Panjgur*: Nampacho, Pochako—: *Wad* Pachko—; *Waziristan*: Nagamboti—:

2. **Althaea officinalis** Linn. is found in Kashmir. It is distributed over the Mediterranean region and Central Europe as far as Siberia.

The great demulcent and emollient properties of Marsh Mallow make it useful in inflammation and irritation of the alimentary canal, and of the urinary and respiratory organs.

Decoctions of the plant, especially of the root, are very useful where the natural mucus has been abraded from the coats of the intestines. They are excellent in painful complaints of the urinary organs, exerting a relaxing effect upon the passages, as well as acting curatively. They are also effective in curing bruises, sprains or any ache in the muscles or sinews.

The powdered root boiled in milk is useful in haemorrhage from the urinary organs, and in dysentery. The powdered or crushed fresh roots make a good poultice that will remove the most obstinate inflammation and prevent mortification; they have proved more serviceable than the ointment albeit its popular reputation.

Boiled in wine or milk, Marsh Mallow will relieve diseases of the chest, constituting a popular remedy for coughs, bronchitis, whooping-cough, etc., generally in combination with other remedies. It is frequently given in the form of a syrup, which is best adapted to infants and children.

The fresh leaves, steeped in hot water and applied to the affected parts as poultices also reduce inflammation, and bruised and rubbed upon any place stung by wasps or bees take away the pain, inflammation and swelling.

In France, the young tops and tender leaves are added to salads, as stimulating the kidneys healthily.

The flowers, boiled in oil and water, with a little honey and alum, have proved good as a gargle for sore throats.

Arabic: Bazarulkhatme, Khetmia, Kobbeiza. Medja-el-Abiod, Moudjdjir, Ouerd-ez-zoual—; *Berber:* Amejjir, Binesar, Tebencert—; *Bombay:* Gulkhairo, Khaira, Khairakajhor, Khitmi, Khitmikajhor—; *Catalan:* Malvi, Malvins—; *Chinese:* Fou Ioung—; *Cutch:* Gulkhair—; *Danish:* Altaea, Ibisk—; *Deccan:* Gulkhairo, Khaira, Khairakajhor, Khitmikajar—; *Devonshire:* Drunkards. Meshmellish—; *Dutch*: Heemst—; *English:* Bread and Cheese, Bulls-eyes, Marsh Mallow, White Mallow, Wymote—; *French:* Althee, Althee officinale, Guimauve, Mauve blanche—; *Friuli:* Altee—; *Genoa:* Artea, Arteja, Marvaiscu, Marvavestc-; *German:* Adewurzel, Attigkraut, Eibisch, Fliesskrautwurzel, Flusskrautblume, Gilfwurz, Gimorwurzel, Guimauvewurzel, Heiligkraut, Heilwurz, Heilwurzblume, Heinisch, Hemisch, Hemstwurzel, Henest, Hibisch Hilfskraut, Hilfswurzel, Hustewurzel, Ibisch Ibischpappel, Ibsche, Ivisch, Kinderbettee, Wilde Malvenwurzel, Weisse Pappel, Sammtpappel, Schleimwurzel, Stockwurzkraut, Weisse Suessholzwurzel, Wollkraut—; *Greek*: Altea, Ibiscos—; *Gujerat:* Gulkhair, Khaira—; *Hindi:* Gulkhairo, Khaira, Khairakajhor, Khitmikajhar—; *Hungarian:* Fejer Malval, Ziliz—; *Italian:* Altea, Avisch, Bismalva, Benefischi, Buonvischio, Davisch., Malvaccioni, Malvavischio, Malvischio, Marvaiscu—; *Languedoc:* Maoula blanca—; *Madagasear:* Fiandrilavenombazaha—; *Malta:* Marsh Mallow, Altea, Malvaccioni—; *Marathi:* Khaira, Gulkhair—; *Norway:* Altaea-; *Persian:* Gulkhairo, Jukhamekhatme, Khaira, Khairakajhor, Khitmi, Khitmikajhar—; *Portuguese:* Malvaisco—; *Reggio:* Bonaves'c—; *Romagna:* Maibon—; *Roumanian:* Nalba mare, Zamosita—; *Russian:* Altei, Dikaya roja, Podswonok, Proskurniak, Prosvirniak—; *Sardinia:* Narbaonia, Pramacisca, Rosa d' Ispagna—; *Somerset:* Bull Flower, Pool Flower—; *Spanish:* Altea, Hierba canamera, Malvavisco—; *Swedish :* Altea—; *Tamil:* Simaitutti—; *Turkish :* Hatmi—; *Tuscany* : Erba que fa pisciar i buoi, Ibisco—; *Urdu:* Khatme, Khitmi—; *Venice*: Malvaviscio—.

3. **Althaea roesa** Cav., a native of Crete and Greece, is found planted in Indian gardens.

The seeds of this plant are demulcent, diuretic and febrifuge. The flowers have cooling and diuretic properties. The roots are supposed to be astringent and demulcent, and are much used in France to form demulcent drinks.

In the Punjab, the flowers are given in rheumatism, and the root in dysentery (Stewart).

The leaves and roots are also used for the same purposes as those of *A. officinais*.

Bogota : Malvarrosa—; *Catalan* : Malva doble, Malva vera—; *Chinese* : Shu K'uei— *Dutch* : Stokroos—; *English* : Althea Rose, Hock Herb, Hollyhock, Round Dock—; *French* ; Alcee, Alcee rose, Althee, Althee rose, Baton de Saint-Jacques, Bourdon de Saint-Jacques, Guimauve rose-tremiere, Mauve arboree, Mauve des jardins, Mauve rose, Passe-rose, Rose alcee, Rose a baton, Rose de mer, Rose d'outre-mer. Rose papale, Rose tremiere, Rose treniere—; *German* : Augenpappel, Baummalve, Baumrose, Brandrose, Braunrose, Erdntrose, Ehrenrose, Felriss, Feuerbluete, Gartenmalve, Glockenpappel, Glockrose, Halsrose, Herbstrose, Herzleuchte, Hochleuchte, Kohlrose, Roemische Malve, Mundrose, Nackrose, Rosenpappel, Schwarzmalve, Siegmarsblume, Stangenrose, Stockmalve, Stockpappelrose, Stock-rose, Ungerblume, Weinrose, Wetterrose, Winterrose—; *Greek* : Altaia, Malachi rodoides—; *Hindi* : Gulkhairu, Gulkhirri—; *Indo-China* : Thuc guy—; *Iraq* : Ward-al-khathmi—; *Italian* : Malvarosa, Malvoni—; *Kurdish* : Gul hailu—; *Malta* : Hollyhock, Malvarosa, Malvoni, Rosoni, Bastun ta San Guisepp— ; *Punjabi* : Gul khaira—; *Roumanian* : Nalba de gradina; *Russian* : Chernaya rosa, Shtok-rosa—; *Spanish* : Malva arborea, Malva Doncella, Malva Isabela, Malva de los jardines, Malva loca, Malva del principe, Malva real, Malva de la reina, Malvarosa—.

GOSSYPIUM

The genus consists of 12 tropical and subtropical species.

The following are used medicinally in the Philippine Islands— *G. arboreum* Linn., *G. barbadense* Linn., *G. herbaceum* Linn., *G. perenne* Linn.—; in Cochinchina and China — *G. herbaceum* Linn.—; in North America — *G. barbadense* Linn., *G. herbaceum* Linn., *G. hirsutum* Linn.—; in Porto Rico—*G. racemosum* Poir.—; in Peru—*G. peruvianum* Cav.—; in Brazil—*G. barbadense* Linn.—; in Guinea *G. barbadense* Linn., *G. herbaceum* Linn., in Guinea— *G. hirsutum* Linn.—.

A. Flowers wholly yellow or yellow with purple base. Involucral bracts laciniate.
   - I. Cotton adherent to the seeds.
     - a. Seeds with underlying down ... ... *G. herbaceum.*
     - b. Seeds with firmly adherent downy down underlying cotton of the same colour or white ... *G. hirsutum.*
   - II. Cotton easily seperable from the seeds ... *G. barbadense.*

B. Flowers wholly purple or yellow with purple base. Involucral bracts subentire or toothed, not laciniate *G. arboreum*

I. **Gossypium arboreum** Linn. is grown in gardens and about temples.

In bombay, the root is used in the treatment of fever.

In the Konkan, the root, rubbed to a paste with the juice of patchouli leaves, has a reputation as a promoter of granulation in wounds; and the juice of the leaves, made into a paste with the seeds of *Vernonia anthelmintica*, is applied to eruptions of the skin following fever. In Pudukota, the leaves ground and mixed with milk, are given for strangury.

The petals squeezed and soaked in human or cow's milk, are used as a soothing and effective application for conjunctivitis of infants.

The cotton is a very useful external remedy in burns, scalds, and some other surgical diseases. The seeds exercise some good influence over gonorrhoea, gleet, chronic cystitis, consumption and some catarrhal affections. The fresh young capsules and shoots have been observed to produce good effects in some cases of dysentery and gonorrhoea. The control of the seeds over gonorrhoea and gleet is more manifest when combined with some other drugs.

In Southern India an emulsion of the seeds is used as an antidote for opium poisoning, and in general for poisoning with narcotics; a decoction of the root is said to be emollient and diuretic

In Tropical West Africa the leaves, flowers and seeds are used medicinally. A cold infusion of the leaves in water with lime juice has been found to give relief in dysentery. The root is believed to be emmenagogue with an action like that of ergot. As the use of the root to procure abortion is known amongst American negroes, but not in the Orient, the knowledge of this property appears to be indigenous to Africa. The active principle resides in the root-bark. The leaves and crushed seed-kernels are applied to sores or as a poultice to bruises and swellings, and the lint as a dressing to wounds. A paste of the seeds is applied to the fore-head and temples for headache.

Barranguilla: Algodon pajarito—; *Bombay*: Deokapas—; *Bundelkhand:* Boojali, Nurma—; *Burma*: Nuwa—; *Canarese*: Anji, Hanji, Kari arale, Kari-hatti—; *Central Provinces*: Deo, Mannua—; *Dacca*: Borailly—; *Egypt*: Gotn—; *English*: Tree Cotton—; *French*: Cotonnier arborescent—; *Guam*: Algodon—; *Hausa:* Abduga, Auduga—; *Hindi:* Deokapas, Nurma—; *Malayalam:* Chemparutti—; *Marathi:* Devakapusa—; *Mysore:* Deokurpas—; *North-Western Provinces:* Manua, Nurma, Radhia—; *Pondicherry:* Cotonnier—; *Punjab:* Kapas—; *Sanskrit:* Karpasamu—; *Santal:* Bhogakuskom, Budikaskom -; *Sokoto*: Kada—; *Spanish:* Algodonero, Algodonero arboreo—; *Tagalog:* Bulacnabundoc, Bulacnatotoo—; *Tamil:* Sembarutti—; *Telugu:* Patti—; *Visayon:* Bulacngabisaya—; *Yemen:* Odjaz, Oth, Zejt.

2. **Gossypium barbadense** Linn. is cultivated in India.

The seeds, in the form of an emulsion, are given in dysentery and are supposed to be pectoral. They yield by expression an oil which is much used to clear the skin of spots and freckles.

A tea made of the young leaves is recommended in lax habits, and for preparing a vapour bath for the anus in cases of tenesmus.

Arabic: Lekid—; *Betsimisaraka:* Hasina, Landihazo—; *Canarese:* Kari-hatti, Vilayatihatti—; *Egypt:* Ashmuny, Gotn, Gotn-esh-sheger—; *English:* Sea-island Cotton—; *Gambia:* Duloboro—; *Guam:* Algodon—; *La Reunion:* Gros coton—; *Malayalam:* Chemparutti—; *Philippines:* Bulac Fernambuco, Bulac Pernambuco, Canton, Pernambuco—; *Sanskrit:* Maghani, Purvam—; *Tamil:* Arattam. Mayiliyam, Sembanja, Sembarutti, Simaipparutti—; *Tahiti:* Vavai—; *Telugu:* Ettappatti, Paidipatti, Pamidipatti—; *Uriya:* Rongokopa—; *Uruguay:* Algodonero—.

3. **Gossypium herbaceum** Linn., probably indigenous in Northern Arabia and Asia Minor, is found cultivated in the North-Western

Frontier Province and Baluchistan. It is also met with in Afghanista, Persia, Mesopotamia, Syria, Egypt, the Mediterranean region, and the United States of America.

A hip bath of the young leaves and roots is recommended in uterine colic. The Tamil doctors prescribe a decoction of the root in strangury and gravel.

The juice of the leaves is considered a good remedy in dysentery, and the leaves with oil are applied as a plaster to gouty joints.

A syrup of the flowers is prescribed in cases of morbid depression on account of its stimulating and exhilarant effect; a poultice of them is applied to burns and scalds.

The seeds are used as a laxative, expectorant, and aphrodisiac. Pounded cotton seed, mixed with ginger and water, is applied in orchitis.

Cotton is used as a moxa. Burnt cotton is applied to sores and wounds to promote healthy granulation.

In Cambodia every part of the plant in used medicinally. The whole plant is considered febrifuge; the flowers and leaves are said to be pectoral and sudorific.

In Annam the flowers are given in amenorrhoea and dysmenorrhoea; the oil from the seeds is applied to wounds, and used in scabies and herpes.

In North America the root bark is used in large doses as an abortifacient. The seeds are considered antidysenteric and galactagogue. The juice of the leaves is administered as an emollient in diarrhoea and in mild forms of dysentery.

In South America cotton-seeds in the form of decoction are employed in the treatment of intermittent fevers.

In India the juice of the fresh leaves is considered very efficacious in the treatment of snake-bite, and the root and leaves are recommended for the treatment of scorpion-sting. But Caius and Mhaskar have shown experimentally that the leaves are not an antidote to snake venom, and that neither the root nor the leaves have any effect in the symptomatic treatment of scorpion-sting.

Annam: Bong se, Bong tau, Cat boi, Cay bong, Cay bong lam vai, Cay vai bong, Thao Khoang—; *Arabic:* Fitan, Kotonefezzani, Kurtamussul—; *Bengal:* Kapas, Tula—; *Bombay:* Kapas, Rui—; *Bulandshahr:* Binaula—; *Burma:* Wa, Wah—; *Cambodia:* Krabas—; *Canarese:* Ambara, Arale, Arali. Badari, Dudi, Hatti, Kariharivale, Karpasa, Tula—; *Central Provinces*: Kapas—; *Deccan:* Kapas—; *Egypt:* Gotn. Gotn-el-sadiar—; *English:* Common Cotton, Indian Cotton—; *French:* Cotonnier, Cotonnier herbace, Cotonnier de Malte—; *Greek:* Bombaki—; *Gujerati:* Kapas, Ru—; *Hindi:* Deokapas, Kupas Narma, Rui—; *Indo-China:* Bong, Fai, Krabas. Mien boa, Thao mien—; *Iraq:* Ou'n.-iraqi—; *Konkani:* Kapsini, Kapus—; *Malayalam:* Badara, Karppas. Karppasi, Karuparutti, Kuruparatti, Pangni, Parittti, Parutti, Pishu, Tulam. Tundikeri—; *Malta:* Common Cotton, Indian Cotton, Short stapled Cotton, Cotone, Koton Malti, Koton ta Malta—; *Marathi:* Kapus, Ru—; *Mundari:* Kadsom, Kaksom, Karsom, Kaskom, Kasom, Katom—; *Pampangan:* Balac Castila— *Persian:* Pambah—; *Portuguese:* Algoda, Algodoeiro—; *Punjab:* Broach Kapa, Rui—; *Raiputana:* Kapas—; *Roumanian:* Bumbac—; *Russian:* Khlopchatnik—; *Sakalave:* Tsiahilika—; *Sanskrit:* Anagnika, Badara, Chavya, Chhadana Guda, Kalakanta, Karpasasarini, Karpasi, Maghani, Marudbhava, Patada. Patahena, Pichu, Samudranta, Sutrapushpa, Tula, Tundakerika, Tundikeri, Vadara—; *Sind:* Vaum—; *Sinhalese:* Kapus—; *Spanish:* Algodonero—; *Tagalog:* Bulac, Bulag—; *Tamil:* Hadambarutti, Karbasam, Karppasam, Panji, Pari, Parutti, Samuttrandam, Taiparutti, Tulam. Tulavam, Uppambarutti, Uttiri Vanapparutti-; *Telugu*: Badari, Badarika, Edudi, Karpasamu, Patti, Pinja, Pishulamu, Pishuvu, Pratti—; *Tuareg*: Tabdouk -; *Tulu*: Parti—; *Uraon*: Rabji—; *Uriya*: Karpaso, Kopa, Korpaso—; *Urdu*: Rui— *Visayan*: Cadaba, Candaba, Gapas—.

4. **Gossypium hirsutum** Linn. is found cultivated in India.

In Guinea the seeds and the leaves are considered emollient, the roots emmenagogue.

Fulani: Hotollo—; *Malinke*: Koroni—; *Malta*: Koton ta Gallipoli—; *Punjabi*: Safed narma—; *Sinaloa*: Algodon— ; *Soussou*: Guessefonte—; *Zambesi*: Tonje-manga—

HIBISCUS.

The genus numbers 160 tropical and subtropical species.

The roots are much used as demulcent. The seeds are considered stimulant and antispasmodic.

The following are used medicinally in Europe—*H. esculentus* Linn., *H. rosa-sinensis* Linn., *H. syriacus* Linn., *H. Trionum* Linn.—; in Egypt—*H. Abelmoschus* Linn., *H. syriacus* Linn.—; in Tropical West Africa—*H. Abelmoschus* Linn., *H. asper* Hook. f., *H. cannabinus* Linn., *H. esculentus* Linn., *H. rostellatus* Guill & Perr., *H. Sabdariffa* Linn., *H. tiliaceus* Linn., *H. vitifolius* Linn—; in South Africa—*H. aethopicus* Linn., *H. leiospermus* Haw., *H. malacospermus* E. Mey., *H. pusillus* Thunb., *H. Trionum* Linn., *H. surattensis* Linn.—; in East Africa—*H. Sabdariffa* Liin.—;in La Reunion—*H. esulentus* Linn., *H. liliiflorus* Cav., *H. rosa-sinensis* Linn., *H. Sabdariffa* Linn. –; in Madagascar—*H. diversifolius* Jacq., *H. phanerandrus* Baker, *H. tiliaceus* Linn.—; in Mauritius—*H. esculentus* Linn., *H. liliiflorus* Cav., *H. mutabilis* Linn., *H. rosa-sinensis* Linn., *H. tiliaceus* Linn.—; in the Philippine Islands—*H. Abelmoschus* Linn., *H. grewiaefolius* Hassk., *H. Lampas* Cav., *H. mutabilis* Linn., *H. rosa-sinensis* Linn., *H. surattensis* Linn., *H. tiliaceus* Linn.—; in Malaya—*H. mutabilis* Linn., *H. rosa-sinensis* Linn., *H. syriacus* Linn., *H. tiliaceus* Linn., *H. Trionum* Linn.—; in Indo-China—*H. Abelmoschus* Linn., *H. Manihot* Linn., *H. mutabilis* Linn., *H. rosa-sinensis* Linn., *H. syriacus* Linn., *H. tiliaceus* Linn., *H. Trionum* Linn.—; in China—*H. Manihot* Linn., *H. mutabilis* Linn., *H. rosa-sinensis* Linn., *H. syriacus* Linn.—; in the West Indies and South America—*H. Abelmoschus* Linn.—; in Guiana—*H. Abelmoschus* Linn., *H. digitiformis* DC., *H. esculentus* Linn., *H. mutabilis* Linn., *H. Sabdariffa* Linn.. *H. tiliaceus* Linn.—; in Tahiti—*H. rosa-sinensis* Linn.—; in North America—*H. californicus* Kell., *H. esculentns* Linn., *H. rosa-sinensis* Linn.

A. Style distinctly lobed. Involucre not caducous. Calyx terminated by 5 distinct lobes. Capsule woody, dehiscent, many-seedep ... ... *H. Lampas.*

B. Capsule 5-celled

- I. Calyx membranous, inflated ... ... *H. Trionum.*
- II. Calyx not inflated
    - a. Involucral bracts distinct, 8-12, forked or provided with a leafy appendage
        - 1. Stipules semicordate, auricled ... ... *H. surattensis.*
        - 2. Stipules lanceolate ... ... *H. furcatus.*

b. Involucral bracts without appendages. Seeds cottony.
Leaves broad, ovate, not lobed ... ... *H. micranthus.*

c. Involucral bracts without appendages, sometimes adnate to the calyx-tube. Seeds smooth or hairy, not cottony.
1. Capsule 5-winged. Seeds tubercled ... ... *H. vitifolius.*
2. Capsule not winged. Seeds nearly glabrous or smooth.
α. Sepals with a gland at the back of each ... *H. cannabinus.*
β. Sepals not glandular ... ... *H. diversifolius*

d. Involucral bracts adnate to the calyx-tube, accrescent, thick, fleshy, purple ... ... *H. Sabdariffa.*

e. Involucral bracts 4, distinct ... ... *H. Manihot.*

f. Involucral bracts more than 5, linear.
1. Involucral bracts shorter than the calyx ... *H. Abelmoschus.*
2. Involucral bracts equalling the calyx ... *H. esculentus.*

C. Capsule with false dissepiments, spuriously 10-celled. Involucral bracts connate at the base. A tree ... *H. tiliaceus.*

D. Cultivated species.
I. Corolla 7.5 cm. diam. red ... ... *H. rosa-sinensis.*
II. Corolla 7.5-10 cm. diam. white or pink ... *H. mutabilis.*

1. **Hibiscus Abelmoschus** Linn. is found in cultivation throughout the hotter parts of India. It is met with in most other tropical countries.

Ayurveda and Yunani parctitioners recognize the seeds as cold, dry, tonic and stomachic. The latter recommend a mucilage made from the root and leaves in gonorrhoea.

The natives of Western Africa use the leaves as an emollient in inflammations of the eyes, and the pounded seeds as a stomachic and digestive tonic.

Antsianaka: Sondraranjaza—; *Arabic:* Habb-ul-mishk, Habb-ul-mushk—; *Bambura:* Suma-diala—; *Bengal:* Mushakdana—; *Betsimisaraka:* Mana—; *Bombay:* Mishkdana, Mushkdana—; *Brazil:* Guicombo,Guigombo do cheiro—; *Burma:* Baluwa—; *Canarese:* Kasturibende—; *Ceylon:* Vattilaikasturi—; *Colombia:* Abelmosco, Almizclillo, Majaguito de playa—; *Deccan:* Kasturibenda, Kasturubenda, Mushkbenda—; *English:* Musk Mallow—; *French:* Abelmosc, Ambrette,Gombo musque, Graine de musc,Guimauve veloutee, Ketnie odorante—; *French Guiana:* Ambrette, Calalou musque—; *German:* Ambre, Ambrette, Bisorn—; *Gold Coast:* Osamankroma, Saman nkuruma—; *Gujerati:* Mushakdana—; *Hindi:* Kalakasturi, Mushkaana—; *Indo-China:* Bong trang nui, Bong vang, Nhan sam, Vong vang—; *Italian:* Ambretta—; *Malay:* Kapas hantu, Kapas hutan—; *Malayalam:* Kasturiventa, Kattukasturi-; *Marathi:* Kasturibhenda—; *Mende:* Feo-bonde, Grabondo, Musukui—; *New Caledonia:* Adlivigona-gahako—; *Pampangan:* Castocastolian—; *Persian:* Mushkdana—; *Roumanian:* Pesma—; *Russian:* Dushishtayan ketmia, Mukushnaya trava—; *Sakalave:* Tsindraranjaza—; *Sanskrit:* Latakasturika-; *Sinhalese:* Kapukimissa, Kapuyinessa-; *Southern Nigeria:* Okoloo n-yelogolo—; *Spanish:* Abelmosco, Ambarina—; *Susu:* Baminkame, Soumari—; *Tagalog:* Castio, Castiocastiogan, Sastiogan, Castoli, Castuli, Dalupan, Putucan—; *Tamil:* Kasturivendia, Kattukasturi—; *Telugu:* Karpurabenda, Rasturibenda, Nelabenda—; *Tukulor:* Kundialana—; *Urbu:* Mushkadanah—; *Visayan:* Ducum, Marapoto, Maricum, Maropoto, Marucum—.

2. **Hibiscus cannabinus** Linn. is generally cultivated in most tropical countries.

The seeds are used as an external application to pains and bruises, and are said to be aphrodisiac and fattening.

One tola of the juice of the flowers, with sugar and black pepper is a popular remedy for biliousness with acidity.

The leaves are purgative. In Northern Nigeria they are powdered and used as a local application for Guinea-worm sores.

Arabic: Til—; *Australia* : Okra, Surm—; *Bakundu* : Belokaloke—; *Balondo*: Bolokoloko—; *Bambara* : Da dian, Da fou, Da ian, Da wulu—; *Basari* : Inangae—; *Basrah* : Gunnab—; *Bauchi*: Jirin da rani—; *Behar* : Kudrum—; *Bengal* : Ambari, Chandana, Mestapat, Nalita, Nalki, Patsan, Pulu, Pulua—; *Bombay* : Ambari—; *Bozo* : Fo—; *Canarese* : Holadapundrike, Pundi—; *Central Provinces*: Ambari—; *Chanda*: Ambari—; *Chota Nagpur*: Kudrum—; *Dagomba* : Balaga—; *Delhi* : Tukhmibhang—; *Djenne* : Bargin—; *Egypt:* Til, Tylbeledy—; *English*: Ambari Hemp, Ambasi Hemp, Bastard Jute, Bimlipatam Jute, Bombay Hemp, Deccan Hemp, Hemp Bendy, Hemp-leaved Hibiscus, Indian Hemp—; *Ewe* : Abema, Egbepetri—; *French* : Chanvre de Guinee—; *Fulani* : Follere pa'bi, Baji, Gabai, Polle, Polli—; *Gambia* : Wild Saur—; *Gbari* : Sawung—; *Godauari* : Gaynara—; *Guierati* : Bhindiyamboi, Sheria—; *Hausa* : Farar rama, Jar rama, Rama—; *Hindi*: Ambari, Kudrung, Mestapat, Nalita, Nalki, Patsan, Pulu, Pulua, San, Sankarkra, Sankukra—; *Ibo* : Odu agu, Udo ocha—; *Iraq* : Jiljil—; *Jhelum* : Shan—; *jukun* : Azhi—; *Kano* : Dirin da rani—; *Kanuri*: Ngabbai—; *Katisina*: Karamapnowa, Koka rani—; *Kolami* : Ji, Kotle—; *Konkomba*: Ditoanje, Ditotono—; *Kratchi*: Niaripari, Riaripari—; *La Reunion*: Chanvre de Gombo—; *Malayalam*: Kanjaru—; *Malinke*: Da le—; *Mandingo*: Da julo, Da sitto—; *Marathi* : Ambada, Ambadi, Khadbind, Khar bhendi—; *Martiniqe* : Chanvre de Gombo, Gombo—; *Mbonge* : Belokaloka—; *Mundari*: Kotole, Kotoleara, Kotoleipilara—; *North-Western Provinces*: Patsan, Pitrva, Rattiasan—; *Persian* : Kanabe, Kanaff, Kanaspe. Sujjado—; *Punjab* : Patsan, Sankokla, Sankokra, Sankukri, Sinjubara—; *Rajputana*: Ambari—; *Sanskrit*: Amla, Ambalika, Ambashtha, Ambika, Balika, Bhurimalli, Chhinnapatri, Chitrapushpi, Dridhvalka, Gandhapatri, Garnikura, Keshi, Machika, Maryurika, Mayurvidala, Mukhavachika, Nali, Phalamla, Prashthika, Rajjudambashtha, Sahasravatamulika, Shathamba, Shreyasi—; *Santal*: Darekudrum, Kudrung—; *Sind* : Sujjado—; *Sokoto* : Kare aiki. Karama mowa—; *Songhai* : Koro guisouma—; *South Africa*: Wild Stock Rose—; *Susu* : Forto bamingui—; *Tamil* : Kachurai, Palungu, Pulichai, Pulicharkirai, Pulimaniai—; *Telugu* : Gogu, Gongura, Gonkura, Gulungu, Pundikura-; *Tschaudio*: Rama—; *Tukulor*: Pollengagina—; *Twi* : Ishoho, Ishoho vesen—; *Uran* : Kudrundora, Tape—; *Uriya*: Bhanga, Kanuriva, Kornniya—; *Wolof* : Pondori—; *Yoruba* : Ida orisha, Oja ikoko, Yemoro, Yewuru—.

3. **Hibiscus diversifolius** Jacq. occurs in Burma. It is distributed over tropical Africa and Australia.

Every part of the plant is used as an emollient in Madagascar.

The native physicians of Fiji use the juice of the leaves to procure abortion.

Betsileo: Tsotsona—; *Imerina:* Roibevavy—; *Madagascar:* Roibe—; *North Queensland:* Cooreenyan—.

4. **Hibiscus esculentus** Linn., is found cultivated throghout India and in all tropical countries.

The plant is a household medicine, and highly esteemed by both Hindu and Mahommedan practitioners as an emollient and demulcent.

The leaves are used to form emollient poultices.

The mucilage from the fruits and seeds is useful in gonorrhoea and irritation of genito-urinary system.

The immature capsules are employed in the form of a decoction as an emollient, demulcent, and diuretic in catarrhal affections, ardor urinae, dysuria, and gonorrhoea.

In Guinea the fruit and the leaves are used as emollient. In some parts of West Africa mild aphrodisiac properties are attributed to the seeds.

In Guiana the plant is very much used as a cooling mucilage; every part ofthe plant is considered emollient and employed as such; the fruit boiled in milk is given for cough.

Arabic: Baeledi, Bamir, Bamiya, Guenaouia, Uaeki; *Ashanti*: Nkuruma—; *Bafo*: Ndando—; *Bakossi*: Ndando—; *Bakundu*: Mabune—; *Bakwiri*: Ndando— *Balondo*: Mawuni—; *Balong*: Ndando—; *Bambara*: Gan, Guan, Guaniala—; *Banda*: M'Veke—; *Basari*: Imo—; *Batanga*: Mabune—; *Bengal*: Dhenras, Dhenrus, Dheras, Ramtorai—; *Benin*: Ikhiavbo—; *Bombay*: Bhenda, Chendi-; *Bota*: Ndando—; *Brazil*: Guiabo, Guimgombo ; *Burma*: Younpadisi—; *Canarese*: Bendakainaru, Bendekai, Bhende—; *Central Porvinces*: Bhendi—; *Ceylon*: Bandakkai—; *Cuba*: Quimbombo—; *Dagomba*: Mana, Manvale—; *Deccan*: Bhendi—; *Efik*: Etighi—; *Egypt*: Bamia, Belledi, Ueki—; *English*: Edible Hibiscus, Gobba, Gumbo, Lady's Fingers, Ochro, Okra—; *Ewe*: Fetri—; *Fanti*: Nkuruma—; *Fra-Fra*: Marna—; *French*: Bamia, Gombault, Gombeau, Gombo, Ketmie comestible, Mauve comestible, Okra-; *French Guiana*: Calalou Calou—; *Fufulde*: Baskoje, Takkei—; *Fulani*: Candie, Takaere, Takeyi, Taku-; *Ga*: Engmomi—; *Gbari*: Okmi—; *Gujerati*: Bhinda, Bhindu, Binda—; Gurunshi: Paula—; *Hausa*: Guro, Ku 'baiwa, Ku 'bewa, Yau 'di—; *Hindi*: Bhendi, Bhindi, Bindi, Katavandai, Ramturai, Ranturi—; *Ibibio*: Etikhe—; *Ibo*: Okolo, Okuro, Okworo, Okwulu, Okwuru—; *Iraq*: Bamia—; *Kabura*: Mana—; *Kanuri*: Gebalgo, Gobasko, Gubaltu, Njita kime—; *Kissi*: Gbasa—; *Kolami*: Mindijinga—; *Konkani*: Bendo—; *Konkomba*: Emoi—; *Konno*: Bondoi, Gbondui—; *Koranko:* Bonde-; *Krepi*: Alokoe. Atise, Bodro, Mesedi-; *Krobo*: Muomi, Pingpesi Pui—; *La Reunion*: Lalo—; *Lepcha:* Hryok kun-tsu-; *Louisiana*: Gobbo, Gumbo—; *Losso:* Mana, Meni- ; *Malayalam:* Vanta—; *Malinka*: Gavu, Gau, Guan—; *Mandingo*: Kainjo, Kanjo, Nah—; *Manjia:* Gona, M' Beyi, Yoga—; *Mano*: Zambe—; *Marathi*: Bhenda, Bhendi—; *Mauritius*: Bhindi, Lalo, Ramtouraye, Vendekaye—; *Mbonge*: Mabuna—; *Mende*: Bonde, Bondebondogbamai, Bonde nande, Bonde valiangoi, Bondo—; *Mexico*: Chinbombo, Gombo, Quingombo—; *Mundari*: Mindidiring, Ramjhingga—; *Nupe*: Kpamfine, Kpanmi—; *Panama:* Naju—; *Peking*: Yang Chiao T'ou—; *Persian:* Bamiyah—; *Porto Rico*: Guingambo—; *Portuguese*: Quiabo— *Portuguese Guinea*: Quingombo—; *Punjab*: Bhindatori, Bhindi, Bhinditori, Ramturai—; *Sanskrit*: Asrapatraka, Bhenda, Bhinda, Bhindatika, Chatupunda Chatushpada, Darivka,Gandhamula, Karaparna, Kshatrasambhava, Pichchila, Sushaka, Tindisa, Vrittabija—; *Santal*: Ramjinga—; *Sarracole*: Diakatame—; *Shanghai*: Yang Koh Deu—; *Serere*: Kandia—; *Sherbro*: Lonto-le—; *Shuwa Arabic*: Bamiya, Daraba gabesco, Daraba umm gurun—; *Sind*: Bhendi, Ramturai—; *Sinhalese*: Bandakai, Bandakka, Okra—; *Sobo:* Ishavbo—*Sokoto*: Guro—; *Songhai*: Gombo, Karas, Lahoil—; *Spanish*: Guimbombo—; *Susu:* Sulegni, Sulinyi, Sulunyi—; *Tamil*: Vendai, Vendi—; *Telugu*: Benda, Penda, Venda—; *Timne*: A-lontho, E-lunto—; *Tivi*: Ityulugh—; *Tschaudjo*: Gmana—; *Twi*: Nkuruma—; *Vai*: Gbongbong—; *West Africa*: Bendi kai, Ochro, Okra—; *West Indies:* Bendi kai, Common Okra, Gombeau,Gombo,Guiabo, Long Green Okra, Ochro, Okra—; *Wolof*: Kandia—; *Yemen*: Bamia—; *Yoruba*: Ila, Ilasa, Ilasado, Ilasha—.

5. **Hibiscus furcatus** Roxb. occurs in the hotter parts of India and Ceylon. It is distributed over the tropics of the Old World.

An infusion of the roots in water is a good cooling drink for the hot weather.

Canarese: Huligowri—; *Malayalam*: Naranampuli, Pachapuli, Suriyamani—; *Sinhalese*: H nnapiritta, Napiritta—; *Telugu*: Kondagogu, Kondagongura—; *Uriya*: Piripirika—.

6. **Hibiscus Lampas** Cav. occurs in the Himalaya upto 4,000 feet Bengal, Burma,Konkan,Kanara, the Western Ghats up to 3,000 feet,

the Northern Circars, Deccan and Ceylon. It extends to Java and Eastern tropical Africa.

The root and fruit are used in Chota Nagpur as a remedy in gonorrhoea and syphilis.

Assam : Bonkapash—; *Bengal*: Bankapas, Bankapsi, Bankapus—; *Canarese:* Adavibende, Turuve—; *Central Provinces* : Jangli bhendi—; *Dehra Dun* ; Ban kapasi—; *Gujerati*: Paruspiplo—; *Hindi* : Bankapas, Bankapasi, Kakhi, Ksyapal—; *Ho* : Reke—; *Kolami* : Birkatsom—; *Lepcha* : Ka-fal-muk—; *Malayalam* : Daraba, Kattuparatti—; *Marathi* : Janglibhenda, Ranbhendy—; *Matheran* : Lahanbhendi, Ranbhendi—; *Mundari* : Birkadsom, Birkaksom, Birkarsom, Birkaskom, Birkasom, Birkatsom—; *Pangasinan* : Banaro—; *Porebunder* : Adbauporushpiplo, Jangliparushpiplo—; *Ramnagar:* Bankapasi—; *Santal* : Birkatsom, Bonkapsi—; *Tagalog* : Banagapula, Bannago—; *Telugu* : Adavibende, Adavipratti, Bharadvaji, Kondapatti, Pagadipatti, Pattinga—; *Uriya:* Bonokopa—; *Visayan* : Bannago, Bulacan—.

7. **Hibiscus Manihot** Linn. is found in Bengal, Mount Abu, Gujerat, Konkan, the Western Ghats, and along the West Coast from South Kanara to Travancore.

In Indo-China the bark is considered emmenagogue. It is used in the form of mucilage.

Chinese : Huang Shu K'uei, Shu Kuei—; *Ga* : Laga laga—; *Indo-China* : Hoang thuc quy—.

8. **Hibiscus micranthus** Linn. f. is found in the hotter parts of India from North-Western India eastwards and southwards to Ceylon. It spreads to Arabia and tropical Africa.

The plant is considered a valuable febrifuge in Ceylon.

Acrur : Ligat atolo, Rigaget elo—; *Asmara* : Chercurantel—; *Baluchi:* Zwangir—; *Ceylon* : Perumaddi—; *Cutch* : Darianujhad, Kurudwel—; *Ga* : Lagalaga—; *Gujerati* : Chanakbhindo—; *Mensa* : Konatal—; *Porebunder* : Abaubuporio, Darianujhad—; *Tamil* : Sittamutti—; *Telugu* : Chalabharata, Tuturubenda—.

9. **Hibiscus mutabilis** Linn., indigenous in China, is cultivated in India.

In Malaya and China the flowers are an established remedy for pectoral and pulmonary complaints; they are prescribed as a stimulant. The leaves are applied to swellings.

In Guiana the plant is used as an emollient.

America : Confederate Rose, Cotton Rose—; *Bengal* : Sphalpadma, Thulpadma—; *Canarese:* Nelaolavare, Suryakanti—; *Chinese:* Fu Jung, Mu Fu Jung—; *Colombia* : Amistad del dia, Rosa de engano Variedad—; *English* : Changeable Rose—; *French Guiana* : Caractere des dames, Rose changeante—; *Guam:* Mapola—; *Hindi:* Gul-i-ajaib, Shalapara, Sthalkamal—; *Indo-China* : Moc phu dung Phu dung—; *La Reunion*: Passe-rose—; *Malaya*: Foo yoon—; *Malayalam*: Chinapparatti, Hinaparutti—; *Mauritius*: Passe rose, Satoula padma—; *Mexico*: Amistad—; *Philippines*: Mapola—; *Porto Rico* : Maravilla— *Portuguese:* Amor inconstante, Rosa mudavel—; *Punjab*: Gul i ajaib—; *Sanskrit*: Padmacharini, Sthalpadma—; *Tamil*: Irattaivellaichembarattam,Sembarattai—; *Uriya*: Sthalopidmo, Tholopodmo—; *Uruguay*: Farolito japones, Flor de la vida, Rosa de mayo—.

10. **Hibiscus rosa-sinensis** Linn. is found cultivated in gardens throughout India.

The flowers are considered emollient,and an infusion of the petáls is given as a demulcent. Fried in ghee they are administered for checking excessive menstruation. In La Reunion they are reputed to be emmenagogue.

The flowers are a household remedy in the Philippine Islands. Externally they are used in all kinds of inflammation; internally they are prescribed in the form of decoction in bronchial catarrh as a bechic and sudorific.

The root is mucilaginous and demulcent, valuable in coughs. In the Konkan the powdered roots is used in menorrhagia, and the juice of the fresh root is given for gonorrhoea

The bark is said to act as an emmenagogue.

The levaes are emollient and aperient. In China they are used in paralysis and dysmenorrhea.

Arabic : Angara-e-hindi—; *Bengal* : Jaba, Japa, Jiwa, Joba, Juwa, Oru, Patkili—; *Bombay* : Jasavanda—; *Burma* : Kaungyan, Koungyan—; *Canarese* : Dasanihu, Dasavala Dasavana, Kempupundrika, Nadeya—; *Cantonese* : Ch 'uen Kan—; *Chinese* : Ch 'uan Chin, Fu Sang, Houng Hoa—; *Colombia* : Alsaciana, Astromelia, Cayena, Escandalosa, Islena, Resucitado, Roja-; *Deccan*: Gudel, Jasum, Jasum, Jasut, Kudhal—; *English* : Scarlet Rose-mallow, Shoe Flower—; *French* : Ketmie de Cochinchine, Rose de Chine—; *Guam* : Gumamela—; *Gujerati* : Jasuva—; *Hindi* : Jasum, Jasun, Jasut—; *Ilocano* : Cayanga—; *Indo-China* : Dam but, Dok mai, Hong can, Phu tang—; *Konkani* : Doxini—; *Malaya* : Choon kin—; *Malayalam* : Ayamparutti, Jampa, Japa, Shemparatti—; *Marathi*: Desindachaphula, Jasavanda, Jasavandi—; *Mauritius*: Djassoun, Foulsapate, Sapatoucheddi—; *Nasirabad* : Badshapashand—; *North America*: China Rose, Chinese Hibiscus, Chinese Rose—; *Pampangan*: Cayanga, Gomamila, Tapolanga, Tapuranga, Tarocanga—; *Persian* : Angara-e-hindi—; *Philippines:* Cayaga, Gumamela, Tapulanga, Tapuranga—; *Portuguese*: Flor de sapato—; *Rarotonga*: Kaute—; *Samoa* : Aute—; *Sanskrit* : Arkapriya, Aruna, Harivallabha, Japapushpa, Java, Joba, Odhrapushpa, Ondrakhya, Pratiko, Raktapushpi, Rogapushpi, Rudrapushpa,Trisandhya—; *Sinhalese*: Sapatthumal, Wadamal—; *Spanish* : Rosa de China—; *Tagalog* : Antolangan, Aroganan, Cayanga, Gomamila, Tapolanga, Tarocanga—; *Tamil* : Arattam, Irattaichegappuchembarattam, Irattaimanjajembarattam, Mandaram, Sapattuppu, Sembarattai, Sevarattai. Sivandavesai—; *Telugu* : Dasanamu, Dasani, Japapushpamu—; *Tulu*: Dasanapu—; *Uriya*: Mondaro, Odophulo, Onghribollika—; *Uruguay*: Flor de obispo, Rosa de China—; *Visayan* : Antolangan, Aroganam, Cayanga, Gomamila, Tapolanga, Tapuranga, Tarocanga—.

11. **Hibiscus Sabdariffa** Linn. is generally cultivated in the hotter parts of India and Ceylon. It is distributed through the tropics of the Old World.

The leaves are emollient. They are much used in Guinea as a diuretic, sedative, and refrigerant.

The leaves, seeds, and ripe calyces possess diuretic and antiscorbutic properties.

The succulent calyx, after the flower fades, is used for the preparation of jelly, chutney, curries, etc.; also as a refreshing drink or made into a wine or syrup. In bilious conditions, a diet drink is prepared by boiling it with water and adding a little salt, pepper, asafoetida and molasses.

East of Chad an infusion of the calyces, "Sudan Tea", is taxen to relieve symptoms of plethora, for bronchitis and cough, and is also used as a vehicle for various native medicines. Externally the leaves, flamed to cause the sap to exude, are applied to assist the healing of sand-cracks of the feet, and a lotion is used for sores.

In Hausa the oil from the seeds is used as a substitute for castor oil and applied to camel sores.

Antsianaka: Voamahombazaha—; *Arabic:* Karkade, Karkangi—; *Basari:* Injangbam—; *Bengal:* Lalmista, Mesta, Patwa—; *Bombay:* Lalambari, Patwa—*Burma:* Chinbaung, Chinpoungni; *Canarese:* Kempupundrike, Pulachakiri, Pundibija—; *Ceylon:* Pulinchakira ; *Dagomba:* Dibemre, Digbemre—; *Deccan:* Lalambari, Patwa—; *Egypt:* Kerkadeb—; *English:* Indian Sorrel, Jamaica Sorrel, Red Sorrel, Rosella, Rozelle, Rozelle Hemp—; *Ewe:* Abema—; *Fra-Fra:* Bito—; *French:* *Guiana* Oseille-de-Guinee rouge—; *Fufulde:* Fulle—; *Fulani:* Follere, Follere ba'di, Follere boleyo, Hologo, Hoyoro—; *Ga:* Sakpa—; *Gbari:* Amma—; *Guinea:* Oseille de Guinee—; *Gujerati:* Lalsheria—; *Gurunshi:* Nangana—; *Hausa:* Sure, Yakuwa—; *Hindi:* Lalambari, Patwa—; *Ho:* Arhaipila—; *Honduras:* Sorrel—; *Hova:* Divay—; *Ibo:* Ojo, Okworo-ozo—; *Konkomba:* Tingyanbam—; *Konno:* Sandoi—; *Kratchi:* Riaripari—; *La Reunion:* Groseille—; *Limba:* Bu-santor—; *Malayalam:* Polechi—; *Malinke:* Da—; *Mandingo:* Kucha—; *Marathi:* Lal ambadi, Patva, Tambadi ambadi—; *Mende:* Satoi, Satui—; *Mundari:* Arharjora, Jengaipillara, Jojoara, Telengaipilara—; *Nupe:* Emagi—; *Portuguese:* Rosela—; *Santal:* Arakkudrum, Togotarak—; *Sierra Leone:* Crincrin, Sour-sour-; *Sind:* Lalambari ; *Sinhalese:* Ratabilincha—; *Sokoto:* Sure—; *Susu:* Santon-belli, Santui—; *Tamil:* Simaikkasuru, Sivappukkasuru—; *Telugu:* Ettagomgura, Ettagongaka, Ettagonguru, Shimogonguru—; *Timne;* A-santor, Ka-santor, Koe-santoor—; *Tivi:* Aishwe, Ashwe-; *Tschaudjo:* Dibemre, Digbemre—; *Uran:* Kudrung—; *West Indies:* Red Sorrel, Rozelle—; *Wolof:* Bissab—; *Yoruba:* Amukai, Isepa —.

12. **Hibiscus surattensis** Linn. occurs in the hotter parts of India, from Bengal to Penang, and Ceylon. It extends to tropical Asia, Africa, and Australia.

The mucilaginous flowers are much used as an emollient and pectoral in La Reunion.

The Zulus use a lotion of the leaves and stem for the treatment of penile irritation of any sort, including venereal sores and urethritis. It is sometimes applied as an ointment for the same purposes. An infusion is also used as an injection into the urethra and vagina for gonorrhoea and other inflammations.

Benin: Akenye *Betsimisaraka:* Sirangabalala—; *Bombay:* Ranbhendy—; *Burma:* Welmachinpoung—; *Ibo:* Ile ago—; *Lagos:* Wongo—; *La Reunion:* Oseille melabare—; *Malay:* Asam susor—; *Mende:* Koli-nei—; *Popo:* Pode—; *Sinhalese:* Hinnapiritta, Naapiritta—; *Tamil:* Kashlikirai—; *Telugu:* Mulugogu—; *Timne:* Da-mirakanka, Damirasip, Ramirasip—; *Visayan:* Labog—; *Yoruba:* Awon-ekun, Wonjo—; *Zulu:* in Cathucathu—.

13. **Hibiscus tiliaceus** Linn. occurs in all tropical regions, especially near the coast. It is particularly plentiful in the Sundarbans and on the river banks in Burma.

The root is said to be febrifuge, and employed in the preparation of embrocations (Irvine).

The bark yields a dark hairy fibre which is used as an absorbent in Malaya.

In the Philippines the powdered bark is given as an emetic; an infusion of the leaves is employed to wash ulcers and wounds; the flowers, boiled in milk, are used in the treatment of earache.

In Gold Coast the yellow juice from the young fruits rubbed on the skin to cure weakness.

In Indo-China the leaves are considered laxative and resolvent.

In Madagascar the plant is used as an emollient.

Australia: Cotton Tree, Talwalpin—; *Bengal*: Bala, Bhola, Bola, Chelwa—; *Betsimisaraka* : Baro—; *Bombay* : Bellipata—; *Burma* : Thengben, Thinban—; *English* : Corkwood, Lime-tree-leaved Hibiscus—; *Fiji* : Fau—; *French* : Bois de flot, Bois de liege, Grand Nahot, Varo—; *French Guiana* : Grand Mahe, Maho—; *Ga* : Fef—; *Guam* : Pago—; *Hawaii* : Hau—; *Hindi* : Bola—; *Honduras* : Mahoe—; *Hova* : Varo—; *Indo-China* : Huu nap, Tua lam chieu—; *Java* : Waru—; *Konkani* : Bellipata—; *La Reunion* : Foulsapate—; *Luabo* : Milola—; *Malay* : Ambaru, Baru, Baru laut, Dedap laut—; *Malayalam* : Nir-parutti, Talipparutti—; *Mandingo* : Burfando—; *Mauritius* : Vaur—; *Mende* Bome-kowei ; *Mexico*: Mahagua, Masagua, Masahua-; *Mortlocks*: Gili-fau—; *New Caledonia* : Borao, Borao de marais, Borao rouge, Bourao, Eemi, Eimi, He, Ven—; *Nzima* : Nwofwea—; *Pampangan* : Balibago, Raguindi—; *Panama* : Majagua—; *Philippines*: Malabayo—; *Ponape* : Kalahau, Kalau—; *Porto Rico* : Emajagua—; *Rarotonga* : Au—; *Sakalave* : Masaizano—; *Samoa* : Fau—; *Sanskrit*: Bala—; *Seychelles*: Bois var—; *Sherbro* : Papam—; *Sinhalese* : Bele-patta, Beligobel, Belipatta , Bellipatta -; *Swahili* : Mtakawa—; *Tagalog* : Balibago—; *Tahiti*: Fau—; *Tamil* : Nirpparutti—; *Telugu* : Ettagogu—; *Uriya* : Baniah, Baniya, Bariya, Kurubeli—; *Visayan* : Balabago, Malabago—; *West Indies*: Mahagua, Mahoe—; *Yap* : Kal—; *Zulu* : umLolo, umLolwa—.

14. **Hibiscus Trionum** Linn. occurs in the Western Himalaya, Kashmir, Simla, Bengal, Sind, Konkan. It is distributed over Southern Europe and the tropics of the Old World.

In China and Malaya an infusion of the flowers is taken for itching and painful skin diseases, and as a diuretic. The dried leaves are held to be stomachic.

Brahui : Lasura, Pihupulli—; *Cantonese* : Wo Sheung T'au—; *Chinese*: Ho Shang T'ou—; *Egypt* : Aimbaggara, Shebbet, Til-shitami—; *Iraq* : Jiljil—; *Malaya*: Woh seong tow—; *Spanish*: Aurora, Malva vejigosa, Malva vesicaria—.

15. **Hibiscus vitifolius** Linn. is a common, herbaceous bush, in the jungles and brushwoods of the hotter parts of India, from the North-West Provinces to Ceylon. It extends to tropical Africa and Australia.

The plant is mucilaginous. The roots provide a preparation used by Gold Coast women to kill head-lice.

Bengal: Bankapas—; *Fanti* : Ekyi, Nkyetan—; *Sanskrit* : Bharadvaji, Vana karpasa—; *Telugu* : Karupatti—; *Yoruba* : Ofo odan, Okun agutan—.

(To be continued).

MEDICINAL AND POISONOUS MALLOWWORTS

BY

J. F. CAIUS, S. J., F.L.S.

PART II

(Continued from page 241 of this volume).

KYDIA

The genus consists of 2 species, both Indian.

Kydia calycina Roxb. is common in the forests of the sub-tropical regions of India and Burma, except the arid tracts.

Among the Santals, the leaves are pounded and made into a paste and applied to the body for pains. They are also chewed, when there is a deficiency of saliva (Campbell).

Almora : Pata—; *Amraoti* : Bhoti—; *Assam* : Boldabak—; *Bhil* : Bothi—; *Bihar* : Bitagonyer—; *Bijnor* : Palao, Pattra—; *Bombay* : Motipotari, Varanga, Varangada, Varung—; *Buldana*: Bhoti—; *Burma*: Bokemaiza, Dwabote, Dwalok Myethlwa, Tabo-; *Canarese*: Belagu, Bellaka, Bende, Bendenaru, Bendi, Billulhendi, Kadubende, Kolibnde, Nayibende-; *Central Provinces*: Baranga, Bargha, Bhotti—; *Garhwal*: Pillu, Pulao—; *Garo*: Boldobak-; *Gond*: Bosha, Burkapa Buruk, Kunji— *Gujerati*: Mhotihirwani Nihotiliwani—; *Hasada* : Bitabororo- *Hindi*: Baranga Choupultea, Pasha, Pola, Potari, Pula, Puli, Pulipasha, Pulu—; *Kharwar* : Derki—; *Khond* : Wala—; *Kolami* : Bitagoinr, Bittia, Gonyer, Pata-dhamin—; *Konkani* : Varang—; *Lambadi* : Charpili—; *Lepcha* : Sedangtaglar, Ta-gla kung hlosa—; *Malayalam*: Velukku, Venta—; *Marathi* : Bhendi, Bhoti, Iliya, Potari, Varung—; *Matheran* : Bhoti, Potasi, Warung—; *Mechi* : Mahow, Moshungon—; *Melghat*: Bhoti—; *Nepal* : Kubinde—; *Nimar* : Safed dhaman; *North-Western Provinces* : Puta, Puttiya—; *Oudh* : Kakahi—; *Porebunder* : Mhotihirwani—; *Punjab* : Pola, Pula, Pulli—; *Ramnagar* : Pula—; *Sadani* : Baranga, Bicra, Jhari—; *Santal* : Poshkaolat, Poskaolat—; *Saora* : Erukutada, Pulan—; *Shan* : Dwabok—; *Sinhalese* : Pule—*Tamil* : Vendai—; *Telugu* : Kondapotari, Pachabotuka, Pandiki, Peddakunji, Paddapotari, Potari—; *Tharu*: Patar—; *Uriya* : Bankopasia, Bharimo, Khopashya—.

MALACHRA

The genus consists of 6 species, natives of the warmer regions of America, and of the West Indies. Two of them are naturalized in Asia and Africa.

Malachra capitata Linn., a native of tropical America, is found throughout the hotter parts of India where it has become naturalized.

In La Reunion the plant is used as an emollient and pectoral. It is a popular cough remedy in Mauritius.

Bombay : Ranbhendi—; *Marathi*: Ranbhendi, Vilayatibhendi—; *Mauritius* Mauve a fleurs jaunes—; : *Porebunder* : Pardeshibhindo—; *Sunderbuns* : Ban bhindi, Bandheras—.

MALVA

The genus numbers 30 species inhabiting temperate Europe and Asia, and North Africa.

All species are remarkable for their mucilaginous and cooling properties.

The following are used medicinally in Europe—*M. Alcea* Linn., *M. althaeoides* Cav , *M. hispanica* Linn., *M. moschata* Linn., *M. nicaeensis* All., *M. parviflora* Linn , *M. rotundifolia* Linn., *M. sylvestris* Linn., *M. Tournefortiana* Linn—; in China—*M. verticillata* Linn—; in Indo-China—; *M. sylvestris* Linn., *M. verticillata* Linn.—; in North America and the West Indies—*M. moschata* Linn., *M. parviflora* Linn., *M. rotundiofoia* Linn , *M. sylvestris* Linn.—; in Colombia—*M. sylvestris* Linn.—; in Madagascar—*M. crispa* Linn.—; in La Reunion—*M. parviflora* Linn., *M. sylvestris* Linn.—; in Southern Africa—*M. parviflora* Linn., *M. rotundifolia* Linn.—.

1. Carpels 10-12 enclosed within the accrescent calyx, prominently ribbed at the back, ... *M. verticillata*
2. Carpels 13-16 not reticulated on the back, the margins of the back rounded. ... ... *M. rotundifolia*
3. Carpels about 10-12, 2-keeled on the back ... *M. sylvestris*
4. Carpels 10, reticulated on the hack, the 2 margins of the back keeled. ... ... *M. parviflora*

1. Malva parviflora Linn. occurs in Bengal, the United Provinces, Kashmir, the Punjab. Baluchistan, Sind, the Bombay Deccan, Mysore and Madura. It is distributed through Afghanistan, Persia, the Mediterranean region, Arabia, Nubia.

The seeds are used as a demulcent in coughs, and ulcers in the bladder.

In La Reunion the plant is used as an emollient and pectoral.

In South Africa Europeans drink an infusion of the leaf as a nerve tonic, and apply the leaf as a hot poultice to wounds and swellings. The Xosas have a similar use to the latter, and often use a decoction as a lotion.

The Sutos use the decoction of the leaf as a remedy for tapeworm and for profuse menstruation. They also make a lotion for bruised limbs from the plant.

Afrikaans: Kasieblaar, Kiesieblaar—; *Arabic:* Bakoul—; *Egypt:* Esh-sheytaniyeh, Khobbeyze, Khobbeyzeh—; *Iraq* : Khubbaz—; *Kurdish* : Tulakah—; *Kuwait:* Khubaiz—; *Makran:* Guragpad—; *Mexico:* Malva—; *North America* : Mallow—; *Punjab:* Gogisag, Nanna, Narr, Panirak, Sonchal, Supra—; *Sinaloa:* Malva castilla—; *South Africa* : Mallow—; *Spanish* : Chica malva de flor, Pequena malva de flor—; *Suto*: Mosalasuping, Qena, Thibapitsa, Tikamotse—; *Turkish* : Tulah—; *Xosa* : a Nomolwana—; *Waziri* : Ngankai, Tikalai—.

2. Malva rotundifolia Linn. occurs in Sind, Baluchistan, Waziristan, Kumaon, ascending to 10,000 feet, and in the plains of Northern India. It is distributed through the Oriental regions, Europe, and North Africa.

The leaves are mucilaginous and emollient, employed externally in scurvy, and reckoned useful in piles (Honigberger).

The seeds possess demulcent properties; they are prescribed in bronchitis, cough, inflammation of the bladder, and haemorrhoids; the seeds are also externally applied in skin diseases.

At Hindubagh, in Baluchistan, the plant is used as a cooling drug. At Kalat the plant is ground up, soaked in water, mixed with sweetmeat, and drunk to relieve retention of urine.

In the Transvaal, Europeans apply a poultice of the leaf in inflammations of the breast.

The leaf has been used in Europe and America, in the from of an infusion or a decoction, for catarrh, dysenteries, and nephritis. As decoction, fomentation, or poultice, it is given in sore throat and ophthalmia, or for maturing abscesses.

Afrikaans: Kiesieblaar-; Arabic : Gobhaiz, Honzebey, Khobeiz, Khubbaizi, Roboueis—; *Baluchi* : Gwaragpad, Pochako—; *Brahui* : Utper—; *Canarese*: Kadukadalegida—; *Catalan* : Malva de fulla petita, Malva petitia de fulla rodona—; *Danish* : Lille Katost—; *Dutch* : Rundbladige maluwe—; *English* : Cheese Cake Flower, Dwarf Mallow—; *French* : Fromageon, Fromagere, Herbe de Simon, Mauve a feuilles rondes, Mauve ronde, Menthe de cimetiere, Petite mauve—; *German* : Kaesepappel—; *Harboi Hills* : Fochako—; *Hindi* : Khubasi, Sonchala—; *Hindubagh*: Sarkukar—; *Italian*: Malva comune, Malva minore—: *Kalat* : Uthpar—; *Kharan* : Pachko—; *Languedoc*: Malbre—; *Portuguese* : Malva—; *Pushtu* : Kukerai—; *Raumanian* : Casui popei, Nalba marunta, Nalba mica—; *Sind*: Chandiri, Khabazi—; *South Africa* : Dwarf Mallow, Mallow—; *Spanish*: Malva de hoja redonda—; *Telugu*: Trikalamalli—; *Waziri*: Naghankai, Nahankai, Wahankai—.

3. **Malva sylvestris** Linn. is found in the West temperate Himalaya from the Punjab to Kumaon, at an altitude of 2, 000 — 8, 000 feet. It is met with as a weed of cultivation in Bombay, Mysore and Madras. It is distributed to Siberia, the Caucasus, Europe and North Africa.

The plant is popular demulcent and emollient with Muhammadan physicians. It is used for coughs and colds, and in affections of the urinary bladder.

In Teheran the drug — flowers with the small rotate carpels of the fruit in an immature state—is used for whooping cough.

Afghanistan : Gulikhadmi—; *Arabic* Bekoul, Khitimi, Khubaji Kobbeiza, Moudjdjir, Querd-ez-zoual—; *Bombay*. Khubasi—; *Canarese* : Sannabindigegida— ; *Catalan* :Malva de cementiri, Malva comuna, Malva major—; *Colombia*; Malva, Malva azul—; *Danish*: Katost—; *Deccan* : Vilayati kangoie—; *Dutch* ; Kaasjeskruid, maluwe Malve—: *Egypt* : Khobbeyze frengiye—: *English*; Blue Mallow Bread-and-Cheese, Cheese-cake, Cheese Log, Cheese flower, Chockcheese, Chucky cheese, Common Mallow, Country Mallow, Custard Cheeses, Dock, Frogcheese,Ground Dock, High Mallow, Loaves-of-Bread, Mallow,Marsh Mallow,Maul, Maws,Pancake Plant,Pick Cheese-; *French*: Beuret, Beurrat, Fouassier, Fromagelet, Fromageon, Fromageot, Herbe a fromage, Herbe a lacher, Herbe Saint-Simon,Maule,Mauve, Grande mauve,Mauve officinale, Mauve sauvage, Mauve verte, Mauve de ville, Meule—;*German*: Baerwinde, Blaugloeckel, Blaumalve, Feldmalve,Feldpappel, Feldriss. Feldwinde, Fuenfaderkraut, Gaenselaetsche, Gaensepappel, Hanfpappel, Hanfwurzel, Kaselskraut. Katzenkaese, Katzenlee, Kesselkraut- Kornwinde, Malve, Milzkraut, Papelle, Papellraut. Pappelkaese, Pferdepappel, Rossmalve, Rosspappel, Schwelkraut Schwollkraut, Siegmarskraut, Toeppel, Waldmalve, Wewinne, Blaue Winde, Ziegerli, Zwischenkraut—; *Greek*; Malachi—:*Hindi*: Gulkhair, Kunzi, Vilayatikangai—; *Hungarian* ; Malyva, Papsajt— ; *Indo-China* : Cam quy—; *Italian*: Malva, Malvone—; *La Reunion*:Mauve-; *Malta*: Common Mallow, Malva, Hobbeiza—; *Mauritius*: Mauve be France—; *North America* : Common Mallow, Doll-cheeses, Dutch-cheese. Fairy-cheeses, Low Mallow, Mallow—; *North Western Provinces*: Kanji,

Tilchuni—; *Patna* : Khatmi— ; *Persian* : Khib-bazi, Khitmiikuchak, Khubazi, Nanakillagah, Nanikulagh, Penirek, Towdrie-; *Polish* : Slaz ziele-; *Portuguese*: Malva, Malva ordinaria—; *Roumanian* : Nalba—; *Russian* : Lyesnaia malva, Prosvirka—; *Sind* : Khabaji—; *Spanish* : Malva, Malva comun, Malva lisa, Malva silvestre, Malva yedra —; *Sussex* : Maller—; *Swedish*: Katost—; *Teheran*: Pasurak —. *Turki* : Hamam komandji—; *Turkish* : Ebegumeci—; *Urdu* : *Khubaji*—; *Waziristan* : Chota gul kharru—.

4. **Malva verticillata** Linn. occurs in the temperate Himalaya up to 12,000 feet. It is distributed to Northern Asia, Northern Africa, and Europe.

In Indo-China the root is used to produce vomiting in whooping-cough. The leaves and young stems are considered digestive and they are given to women in the advanced stage of pregnancy. The ash of the dried leaves is employed in the preparation of a drink which is given in scabies.

Assam : Laffa—; *Chinese* : K'uei—; *Indo-China* : Dong quy—.

MALVASTRUM

The genus consists of about 60 American and South African species, two of them cosmopolitan in the tropics.

M. coromandelianum Garcke is used medicinally in the West Indies, *M. spicatum* A. Gray in French Guiana, *M. coccineum* A. Gray in North America, *M. peruvianum* A. Gray in Colombia.

Malvastrum coromandelianum Garcke is found in the Madras and Bombay Presidencies, the Punjab and the United Provinces, Orissa and Bengal.

In the West Indies the plant is considered emollient, resolvent, and bechic. The leaves are applied to inflamed sores and wounds as a cooling and healing salve. The flowers are given as a pectoral and diaphoretic.

French Guiana: Ouade-ouade-; *Mauritius*: Herbe a balais, Mauve sauvage-.

PAVONIA

The genus numbers 70 species inhabiting tropical and subtropical regions.

The following species are used medicinally in Gambia—*P. zeylanica* Cav.—; in La Reunion—*P. columella* Cav., *P. urens* Cav.—; in Madagascar—*P. Bojeri* Baker, *P macrotis* Baker—; in Brazil—*P. diuretica* St. Hil.—; in West Tropical Africa— *P. hirsuta* Guill and Perr.—; in Somaliland—*P. propinqua* Garcke—.

Involucral bracts 8-14, subulate.

1. Ripe carpels unarmed wingless. Leaves slightly lobed ... ... *P. odorata.*
2. Ripe carpels glabrous, narrowly winged, Leaves deeply lobed ... ... *P. zeylanica.*
3. Ripe carpels with 2 short prickless. Leaves oblong creneate ... ... *P. propinqua.*

1. **Pavonia odorata** Willd. is found in North-West India, Bundelkhand, Sind, Baluchistan, West Rajputana, Bengal, Konkan, Southern Maratha Country, North Circars, Deccan, Carnatic, Ceylon. It extends to East tropical Africa.

The fragrant root is esteemed by Ayurveda practitioners, who consider it aromatic, cooling, and stomachic, It is much used, in combination with other drugs, as a remedy for fever, inflammation, and haemorrhage from internal organs. It is a well-knawn South-Indian febrifuge. In Bombay it is prescribed as an astringent and tonic in cases of dysentery.

In Las Bela the plant is used as a cure for rheumatism.

Bombay : Kalavala—; *Canarese* : Balarakkasigida, Mudivala—; *Gujerati* : Kalowalo—; *Hindi* : Sugandhabala—; *Jodhpur* : Chirikenahl—; *Las Bela* : Zordar—; *Malayalam* : Kuruntotti—; *Marathi* : Kalavala, Sughandabala—; *Sanskrit* : Ambunamaka, Bala, Barhishtha, Haribera, Hrivela, Kachamoda, Keshanama, Keshanamaka, Keshya, Kuntala, Kuntaloshira, Lalanapriya, Toya, Udichya, Vajra, Vala, Valaka, Varapinga, Vari, Varida, Varinamaka—; *Sinhalese* : Pushpa-bevila—; *Tamil* : Avibattam, Peramutti, Suvesagam—; *Telugu*: Chittibenda, Ettakuti, Muttavapulagamu, Tigebenda—.

2. **Pavonia propinqua** Garcke is found in Sind and Baluchistan; it is distributed to Abyssinia and Somaliland.

In Somaliland a decoction of the leaves and branchlets is used as a wash, in the from of either a bath or a douche, to prevent abortion in syphilitic women.

Somali : Brumo—.

3. **Pavonia zeylanica** Cav. is found in Northern India, West Rajputana, Sind, the Western Peninsula, the Circars, Carnatic, and Ceylon. It is distributed to tropical Africa, and Mauritius.

The plant is a South Indian febrifuge.

In Gambia the plant is used as a vermifuge and a purgative by the Mandingo natives.

Canarese : Antutogari, Balarakshasi, Chittamutti—; *Mandingo* : Ratach—; *Sinhalese* : Gasbeuila—; *Tamil*: Kurundotti, Mammatti, Sevagan, Sittamutti—; *Telugu* : Chinnamutavapulagamu, Chinnamutennu, Karubenda, Peramutti—.

SIDA.

This genus includes 120 cosmopolitan species.

The roots of the different species are regarded as cooling, astringent, and tonic.

The following are used medicinally in the Canary Islands—*S. rhombifolia* Linn. ; in West Tropical Africa—*S. acuta* Burm., *S. cordifolia* Linn., *S. linifolia* Cav., *S. rhombifolia* Linn.—; in La Reunion—*S. acuta* Burm., *S. cordifolia* Linn., *S. rhombifolia* Linn.—; in Madagascar—*S. rhombifolia* Linn.—; in Southern Africa—*S. cordifolia* Linn., *S. longipes* E. Mey.—; in Indo-China—*S acuta* Burm.—; in the Philippine Islands—*S. acuta* Burm., *S. rhombifolia* Linn.—; in Australia—*S. rhombifolia* Linn.—; in Central and South America—*S. paniculata* Linn., *S. viscosa* Linn —; in Brazil—*S. acuta* Burm., *S. multiflora* Cav., *S. rhombifolia*

Linn.—; in North America—*S. obliqua* Nutt., *S. paniculata* Linn., *S. rhombifolia* Linn.

A. Prostrate trailing herb ... ... *S. veronicaefolia*

B. Erect or suberect, herbs or undershrubs. Stellately hairy

1. Petiole spiny at the base ... ... *S. spinosa.*
2. Petiole not spiny at the base
   - *a.* Awns exceeding the calyx ... ... *S. cordifolia.*
   - *b.* Awns short.
     - i. Leaves linear lanceolate, glabrous or nearly so *S. acuta.*
     - ii. Leaves rhomboid-lanceolate, or obovate, hoary beneath ... ... *S. rhombifolia.*
     - iii. Leaves oblong-ovate obtuse, downy on both surfaces ... ... *S. grewioides.*

Sida acuta Burm. is generally distributed throughout the hotter parts of India. It is found all over the tropics.

The root is regarded as cooling, astringent, tonic and useful in nervous and urinary diseases, and also in disorders of the blood and bile. It is intensely bitter, and is prescribed in infusion, and in conjunction with ginger, in cases of intermittent fever. It is considered by the Hindu practitioners as a valuable stomachic and useful remedy in chronic bowel complaints; the dose, a small tea-cupful, twice daily.

The authors of the Bengal Dispensatory, after a trial of the roots, were unable to satisfy themselves as to its febrifuge action, but it was found to promote perspiration, to increase the appetite, and to act as a useful bitter tonic. In Goa, the Portuguese value it as a diuretic, especially in rheumatic affections. They also use it as a demulcent in gonorrhoea, and Muhammadans believe this to have aphrodisiac properties.

In the Konkan, the root is applied with sparrow's dung to mature boils.

In Southern India the leaves, made warm and moistened with a little gingelly oil, are employed to hasten suppuration. In the Konkan they are applied with other cooling leaves in ophthalmia. In Bengal, the expressed juice is used in the form of an electuary in the treatment of intestinal worms.

In West Tropical Africa the leaves are sometimes an ingredient in prescriptions for intestinal worms in horses. They are used frequently by native witch doctors to procure abortion.

A cold infusion of the leaves is a common Yoruba remedy for gonorrhoea, and they are sometimes taken medicinally cooked with rice or other food. Rubbed up with water, they are used to scrub the body and as a wash for dogs. goats, etc., for scab and parasitic skin diseases. The bitter bark in infusion is taken as an appetiser and stomachic tonic.

In Gold Coast the plant is used to cure venereal disease. The leaves when bruised are slimy and are put on the hands of midwives when they are about to remove dead children from the womb; they are mashed in water and the liquid used as an enema for paralysed children to help them to walk; they are frequently used to cause abortion.

Sanskrit writers recommend the plant in combination with other drugs for the treatment of snake-bite and scorpion-sting. But Caius and Mhaskar have proved experimentally that the plant is not an antidote to either snake or scorpion venom.

Andamans: Sirivadibdibila—; *Ashanti*: Sowa—; *Awnna*: Xa - gbe—; *Bengal*: Bonmeshi, Pilabarelashikar, Shvetberelakoreta—; *Bombay*: Bala, jang limeshl—; *Brazil*: Malva, Vassoura—; *Burma*: Katsayani, Pyendangnalin—; *Canarese*: Bhimankaddi, Vishakaddi—; *Ceylon*: Malaitanki, Vaddatiruppi, Vishapati—; *Deccan*: Isarbadi, Isbadi—; *Eure*: Ademe, Ademe, Afidemii—; *Ga*: Shwuoblo—; *Goa*: Cha de India—; *Gujerati*; bala, Jungumenthi—; *mundi*; Bariara, Kareta, Kharenti, Paharibarara—; *Indo-China*: Bat choi, Vai coni—; *Kolami* : Ipirpijon—; *Krepi*: Didinglome—; *La Reunion:* Grosse herbe dure, Herbe a balais—; *Malay*: Kelulut puch, Ketumbar hutan, Poko lidah ular: Sada turi, Telor belangkas—; *Malayalam*, Malatanni, Sharuparuva—; *Murarhi*: Chikana, Pata, Tukan, Tupkaria—; *Mauritias*: Herbe panner, Herbe a paner a fleurs roses, Mauve du pays—; *Mende*: Helui —; *Mundari*: Ipirpichig Marci baear, Marcijono—; *New Hebrides*: Herbe a babais—; *Pampangar*; Higotbolato Pamalis—; *Porbunder*: Bala, Dungrabal—; *Sanskarit*: bala Bribnunagabala, duma—; *Tagalog*: Escobagnaba, Higotblat, Pamalis—; *Tamil*; Arivalmanaip pundu, Arivahmukan, Kayappundu, Malaikkurumrlali, Mayirmanik kam, Ponmusutti—; Vattatiruppi—; *Tanna*: Teranmatln—; *Telugu*: Chittimu, Gayapaku, Muttaveapulagamu, Nelabenda, Sahadevi, Visaboddi—; *Tinne*: Ekboentomer, Ka-tanta, Koe-tanta, Makboentonber—; *Tupin*; Tupitcha—; *Twi*: Apono, Obranetuata Petekuku, Siwiabiri, Sowa—; *Uriya* Ancharna, Siobola, Sunakhodika—; *Visayan*: Higotbalato, Pamalis, Silhigon—; *Yoruba*; Oshepotu, Oshepotufua—.

2. **Sida cordifolia** Linn. occurs in the tropical and subtropical regions of both hemispheres. It is generally distributed in moist places.

A decoction of the root with ginger is given by Hindu physicians in intermittent fever. It is also administered in fever accompanied by shivering fits and strong heat of body. The powder of the root-bark is given with milk and sugar for the relief of frequent micturition and leucorrhoea. In diseases of the nervous system the root is used alone, or in combination with other medicines.

The bark of the root with sesamum oil and milk is very efficacious in curing cases of facial paralysis and sciatica when caused by the inflammation of the nerves concerned (Koman).

In Cambodia the root is considered diuretic and depurative; it is given in the treatment of gonorrhoea and ringworm.

The Thongas of Portuguese East Africa use the plant as a children's medicine.

The leaves, mixed with rice, are given to alleviate the blooby flux.

In the Konkan, the leaves, with other cooling leaves are applied in ophthalmia; the root - juice is used to promote the healing of wounds, and the juice of the whole plant pounded with a little water is given in ¼ seer doses for spermatorrhoea.

The seeds are reckoned aphrodisiac, and are administered in gonorrhoea. They are also given for colic and tenesmus.

Bengal: Bala, Barila, Brela, Svetberela—; *Cambodia*: Kantrang bai sa—; *Canarese*: Chittuharalu, Hettutti—; *Ewe*; Didinglome—; *Fanti*; Kumpa—; *Gujerati*: Baladana, Khareti—; *Hausa*: Farar hankufa, Frain grammi, Garmani, K'ardfi; Mai-k'alo—; *Hindi*; Barial, Bariala, Baria: Kharenti, Khoroeti, Kungyi—; *Katogum*: Kardafi—; *Konkan*: Kobirsirbhaji, Muttava—; *La Re.*

Union : Mauve—; *Malay* : Kelulut putih—; *Malayalum* : Katturam—; *Marathi*: Chikana, Khiranti—; *Mauritius* : Herbe dure, Herbe panier, Herbe a 'panier Mauve a teuiles veloutees, Arivamouc, Barjala, Barriar—; *Mundari*; Huring-mindilata—; *Porebunder* : Bal, Bala, Baldana, Balncchotvo—: *Punjab*: Kharenti—; *Sanskrit* : Badiyalaka, Bala. Baladhya, Balini, Bhadra, Bhadrabala Bhadrodani, Brela, Jayanti, Kalyanini Kanaka 'Kathorayashtika, Kharakakash tika, Kharayashtika Krura, Motapati, Nilaya, Odanavha, Odani, Odanika, Phani jivaka, Prahasa, Raktatandula, Samanga, Samansha, Shitapaki, Suvarna, Sveibe-rela, Variga, Vataghni Vatyalaka, Vatyali, Vatyapushpi, Vilala—; Sind : Bur-rayra—; *Sokoto* Garamani, Faringaramani—; *Tamil* Arivahnanaippundu. Nila-tutti—; *Telugu* Antisa, Chirubenda, Muttavapulagamu, Suvarnamu, Tellagorra, Telluntisa—; *Thonga* : Sitjhesinyana sanIlhaba—; *Timme* : Makboentamoer—; *Uriya* : Badiananla, Bisookopari—; *Yemen* : Ren—.

3. **Sida grewioides** Guill. and Perr. occurs in the North Wets-Provinces and Sind. It is distributed to Arabia and tropical Africa

In Jodhpur and Jaisalmer the seeds, ground and mixed with goor, are used as a cure for lumbago.

Rajputana : Ball, Dabi—.

4. **Sida rhombifolia** Linn. is found as a weed of waste places, throughout the tropics of both hemispheres.

The root is held in great repute in the treatment of rheumatism.

The Mundas apply the pounded leaves on swellings.

In Lakhimpur (Assam) the roots are taken internally to help childbirth. The herb is also tied round the abdomen for the same purpose (Darter).

The stems abound in mucilage, and are employed as demulcents and emolients both for external and internal use.

In Europe the plant has been regarded as a valuable remedy in pulmonary tuberculosis and rheumatism. Conflicting reports of it sefficacy as a tuberculosis remedy have been published.

In Madagascar the plant is mostly used as an emollient; an infusion of the root is given in dysentery; the leaves are pounded and applied to tumours, or chewed and applied to boils.

Sanskrit authors recommend the use of the plant in combination with other drugs for the treatment of snake-bite and scorpion-sting. But Caius and Mhaskar have shown experimentally that no part of the plant is an antidote to either snake or scorpion venom.

Afrikaans : Pretoria-bossie, Taaiman—; *Australia* ; Common Sida Weed, Jelly Leaf, Native Lucerne, Queensland Hemp—; *Bambara* : Balan-balan—; *Bengal* Lal berela, Pitabala, Pithala, Svetbarela—; *Benin*: Aramabi—; *Betsileo*: Kisin-dahorina—; *Canarese*: Bennegaragu, Gubethadegida, Kallangadale—; *Central America*: Escoba—; *Ceylon*; Atipaiacheddu, Chittamadi—; *Cuba*: Malva de cochino—;*English*: Paddy's Lucerne, Queensland Hemp, Sida Hemp— ;*Espiritu Santo*:Niveding irin irin—;*French*: Herbe a balais, Fausse guimauve, Guimauve des Indes ; *Fulani*: Gogi—; *Guam*: Escobilla—; *Gujerati*; Baladana—; *Hausa*: Faskara saiwo, Miyar tsanya, Namijim hankufa—; *Hawaii* : Ilima—; *Hindi* : Bariara, Bhiunli, Kharenti, Pitabala, Sahadebi, Sahadeva, Swetbarela—; *Hova*: Tsindahoro—; *Ibo*: Azu uzo mbo, Udo agheregha—; *Indo-China*: Choi duc, Ke dong tien, Kc hoa vang—;*Katsina*: Miyatsanya—: *Kolami*: Ipirpijon—; *Korambo*: Sose, Susi; *La Reunion*: Faux the, Herbe dure—; *Malay*: Bunga padang, Poko lidah ular, Seliguri panang, Senda guri padang—; *Malayalum*: Anakkuruntotti, Totti, Valangoruntotti, Vatturam—; *Malinke*: Mali- sobo—; *Marathi*: Chikna, Sadeda, Sahadevi—; *Mauritius*: Herbe balais, Herbe a balasi, The batard, Kareti Vattatirippi—; *Mundari*: Ipiripiung, Pipirpiung—; *Mende*: Helui—; *New South Wales*: Paddy Lucerne—; *Nimar*: Bariari—; *North: America*: Canary-island Tea-plant, Jelly Leaf, Paddy's Lucerne, Queensland

Hemp, Sida Hemp, Tea Plant—; *Nyasaland*: Denji—; *Owerri*: Azu uzo mbo, Udo agheregho—; *Panama*: Escobilla—; *Pampangan*: Escobanghabo—; *Porebunder*: Betraubal, Betraubaldana—, *Samoa*: Mantofu—; *Sanskrit*: Ahikhanda Atibala, Bala, Barela, Brihadagala, Devaarha, Devabala, Devasaha, Gandhavallari, Gandhavalli, Jyeshthabala, Karambhara, Kesarika, Keshawardhini, Keshruha, Lalbarila, Mahabala, Mahagalarthaprasadini, Mahagandha, Mriga, Mrigadini, Mrigarasa, Pitapushpa, Pitapushpi, Prasadini, Sahadeva, Sahadevi, Samanga, Sarini, Varshapushpa, Varshpushpi, Vataghni, Vatya, Vatyayani—; *Sema*: Thopfu-ga-bo—; *Smaloa*: Malva colorada—: *Sinhalese*: Kotikanbevila—*Sokoto*: Miyatsanya—; *South Africa*: Queensland Hemp—; *Tagalog*: Escobanghaba—; *Tahiti*: Burume—; *Tamil*. Anaikurundotti, Kurundetti, Tenacham—; *Telugu*: Atibala, Gubatada, Mayilumanikyamu, Muttavapulagamu—; *Tigrinia*: Deche-dabro—; *Tulu*: Kadiru—; *Uriya*: Dholabadeanla, Nalobadianla—; *Uruguay*: Afata, Malvavisco, Mata alfalfa, Tipicha—; *Venezuela*: Escoba—; *Visayan*: Basengbaseng, Escobanghaba—; *Wolf*: Ndetalmnor, Ndetinor—; *Yoruba*: Ewe ifin—; *Zulu*: iVavana—.

5. **Sida spinosa Linn.** occurs throughout the hotter parts of India from the North-West to Ceylon. It is distributed over the tropical and subtropical regions of both hemispheres.

The leaves are demulcent and refrigerant, and are useful in cases of gonorrhoea, gleet and scalding urine. They are bruised in water, strained through cloth, and administered in the form of a draught.

The decoction of the root-bark and root is used as a demulcent in irritability of the bladder and in gonorrhoea.

The root acts as a gentle tonic and diaphoretic, and is employed in mild cases of debility and fever. They are useful also in the treatment of some forms of cattle disease.

Arabic: Kulbahebarri—; *Bengal*: Bonmethi, Gorakchaulia, Pilabarela—; *Canarese*: Kadumenthya—; *Deccan*: Janglimethi—; *Egypt*: Agul, Melukhiyet iblis—; *Gujerati*: Kantalobal—; *Hindi*: Bariar, Bariara, Gangeran, Gulsakari, Janglimethi, Khareti—; *La Reunion*: Herbe dure—; *Malayalam*: Kattaventiyam Mayirmanikkam-; *Marathi*: Gandedhaman, Gangeti, Kanteritukati-; *Mundari*: Mindilat, Mindilate-; *Persian*: Shambalidebarri, Shamlithedashti-; *Porebunder*: Balnansavenan, Kantalobal—; *Sanskrit*: Avishta, Bhadrandani, Chatupala, Devadanda, Gangeruki, Ghanta, Gorakshatandula, Hrisvagavedhuka, Jhasha, Kharagandha, Kharagandhini, Kharavallika, Kharyashthika, Mahagandha, Mahapatra, Mahaphala, Mahashakha, Mahodaya, Nagabala, Pila, Pitberela, Vishvadeva, Vishvadevi-; *Sinhalese*: Girivedi-bevila, Kotikambabila, Manmanikam-; *Tamil*: Arivalmanaippundu, Mayirmaniakkm-; *Telugu*: Chinnamuttamu, Chinnamuttavapulagmu, Mayilumanikyam. Muttavapulagamu, Ternallabenda, Tirinelabenda—; *Uruguay*: Malvavisco—.

6. **Sida veronicaefolia Lam.** is distributed throughout the hotter parts of India, and throughout the tropical and subtropical regions of the world.

The flowers and unripe fruit are given together with sugar for burning sensation in micturition.

Among the Santals the leaves are pounded and used as a local application to cuts and bruises. They are also given in the diarrhoea of pregnancy. In the Coimbatore District, they are ground up with cummin seeds, onions, and the succulent portion of aloe leaves mixed with buffalo butter-milk, and given to cattle suffering from rinderpest.

Bengal: Junka—; *Ceylon*: Palampadu—; *Gujerat*: Bhoyabala—; *Hindi*: Bananiyar, Bhiunli, Kharenti—; *Kolami*: Bariar—; *Marathi*: Bhoybal, Bhui-chikna—; *Mundari*: Jangki, Mindilat, Mindilatanari—; *Sanskrit*: Bhumibala—; *Santal*: Bariar, Bir, J.,khasakan, Tandi– ; *Sinhalese*: Bevila—; *Tamil* : Palampasi—; *Telugu*: Gayapuwaku, Nellabenda—.

THESPESIA

The genus consists of 5 species inhabiting warm regions.

T. populnea Corr. is used medicinally in Indo-China, the Malay Archipelago, the Society Islands, Mauritius, Mozambique, and Upper Guinea; *T. campylosiphon* Rolf. is used in the Philippine Islands.

Thespesia populnea Soland. occurs in the coast forests of India and Burma. It is largely grown as a roadside tree in tropical regions

In the Central Provinces, the root is taken as a tonic.

In the Konkan, the flowers are employed in the cure of itch: and the leaves are employed as local application to inflamed and swollen joints.

The fruit yields a yellow, viscid juice, which forms a valuable local application in scabies and other cutaneous diseases. The affected parts of the body are daily washed with a decoction of the bark.

A decoction of the bark is given internally as an alternative.

The bark is astringent and is prescribed in the Philippines for the treatment of dysentery in the form of a decoction. The fruit, leaves, and root are applied externally in scabies and other skin affections.

In Tahiti, the fresh capsules, bruised and applied to the forehead are said to cure migraine; the yellow sap exuding from the peduncles is considered a cure for the bites of insects. especially of the centipede; it is also useful in sprains, bruises, and all cutaneous affections. In Mauritius, the bark is described as depurative, and used in dysentery, haemorrhoids; the juice of the fruits being applied to warts.

In Madagascar a decoction of the bark is commonly used in chronic dysentery and cutaneous diseases; the sap is applied externally for herpes.

Rumphius speaks highly of the value of the heartwood as a remedy for bilious attacks and colic, and in a kind of pleurodynia from which the Malays often suffer.

Waring tried the bark in scabies and other cutaneous diseases; in some cases, it exercised a favourable influence, but in the majority it was productive of little or no benefit.

The leaves are ground into a paste and applied externally in children's eczema; so also an oil prepared by boiling the ground bark in cocoanut oil is applied externally in psoriasis and scabies. The contents of the fruit which is a capsule are applied externally to ringworm. A decoction of the bark is given internally in skin diseases. The ground leaves and contents of the capsule applied externally in eczema and ringworm respectively were found to be useful. A compound oil of the bark and capsule was given in cases of urethritis and gonorrhoea with beneficial results (Koman).

Ahanta: Tamsi—; *Bengal*: Dumbla, Gajashundi, Palaspipal, Parash, Pares, Paresh, Parespipal, Porash Prash—; *Bombay* : Bhendi, Bhindi, Palaspiplo, Parsipu—; *Bougainville Straits* : Kaikaia—; *Canarese* : Arasi, Asha, Bangali, Bugari, Gandarali, Hurvashi, Huvarasi, Jogiyarale, Kandasola—; *Central Provin* : Ranbhendi—; *Ceylon* : Karavachu, Suriyagas—; *Cuba* : Majagua de Florida—; *Deccan* : Paraspippal, Paris—; *English* : Bhendi Tree, Portia Tree, Tulip Tree, Umbrella Tree—; *Eve.*: Bobosenya—; *Fanti*: Adcmba, Fref, Frefi-; *Fiji* : Mulo—; *French* : Porcher—; *Ga*: Adengkra, Fetsho, Fez, Foz—; *Gilbert; Islands* : Bengibeng—; *Guam* : Kilulu—; Quilulu—; *Gujerat* : Bendi, Bhindi, Parasapiplo—; *Hawaii*: Milo—; *Hindi*: Bhendi, Gajadanda, Gajhand, Parashajhad, Paraspipal, Parispipul, Parsipu, Pipal, Pippul, Porush—; *Honduras*: Cork Tree—; *Hova* : Valo—; *Indo-China* : Chrey sramol, Tra bo de, Tra bua, Tra lam vo—; *Konkani* : Benddy, Maner—; *La Reunion* : Porcher—; *Malay* : Baru, Buah keras laut—; *Malayalam*: Chandamaram, Chilanti, Kallal, Pupparutti, Puvarasu ; Puvvarasha—; *Marathi* : Bendi, Bhenda, Bhendi, Paraspipar, Parsachajjhada. Ranbheudi—; *Mauritius* : Porche, Porcher, Valou, Poursong have marom, Paraspipal—; *New Caledonia*: Bois de rose, Daleni, Kabaoui—; *Nzima:* Eijan, Tamsi—; *Ponape* : Pana, Pena, Pona—; *Porebunder* : Paraspiplo—; *Porto Rico* : Palo de jagueca—; *Portuguese*: Pau rosa, Pau de rosa—; *Punjab*: Paharipipal, Paraspipal—; *Rarotonga* : Miro—; *Sakalave*: Valomena—; *Samo* : Milo—; *Sanskrit*: Gardabhanda, Kamandalu, Kandarala, Kapichuta. Kapitana, Kuberaksha, Kund: h Nandi, Parisha, Phalisha. Suparshvaka—; *Sefwi*: Eijan—; *Sinhalese*: Gansurigaha, Suriya, Suriyagha—; *Sunderbunds*: Dumbla, Paras, Paraspipal—; *Tagalog*: Baboigubat, Babuy, Banalo, Boboigubat, Bubuygubat, Malasantol, Malibago—; *Tahiti* : Milo—; *Tamil* : Kallal, Piram, Pupparutti, Puvarasu—; *Telugu*: Gangaravi, Gangareni, Munigangaravi—; *Tonga*: Milo—; *Tulu*: Jogi, Jogiyattasa—; *Twi* : Ayedru, Benorsenye—; *Uriya* : Gunjausto. Habali, Porosopippoli—; *Visayan*: Bulacan—; *Yap*: Bonabeng—.

URENA.

The genus consists of 3 species inhabiting the warmer regions of both hemispheres.

The root is considered cooling and is used as a resolvent; the leaves are emollient; the flowers mucilaginous and bechic.

The following are used medicinally in the Philippine Islands and Indo-China—*U. sinuata* Linn.—; in South America—*U. lobata* Linn., *U. sinuata* Linn.—; in West Africa and La Reunion—*U. lobata* Linn.—.

A. Carpels armed with bristles.
  I. Leaves not divided below the middle ... ... *U. lobata.*
  II. Leaves divided below the middle ... ... *U. sinuata.*
B. Carpels smooth, unarmed ... ... *U. repanda.*

1. **Urena lobata** Linn. is a weed of waste places, forest-clearing and roadsides, occurring in all tropical countries. It is very common in the bamboo and mango clumps of Bengal.

The root is a very popular diuretic in Lakhimpur. In Chota Nagpur it is employed as an external remedy for rheumatism.

In West Africa the plant is used medicinally as a mucilage.

In La Reunion and Mauritius the leaves and roots are made in poultices and used as an emollient; the flowers are considered pectoral.

In Guiana an infusion of the flowers is used as a gargle for aphthae and sore throat.

A decoction of the root and stem is used in Brazil as a remedy in windy colic; the flowers as an expectorant in dry and inveterate coughs.

Akim : Sowa—; *Akwapim* : Petekuku—; *Bakwiri* : Sofo—; *Bambara*: So-so—; *Basari* : Mangemboleba, Tingmile—; *Bengal* : Benochra—; *Benin*: Orònhon—; *Betsileo* : Paka, Pampaho, Pampana, Pàmpano—; *Brazil* : Aramına, Carrapicho, Guaxima, Malvaisco, Malvisco, Uaixyma—; *Burma* ; Katsaenai, Whetkhya-panai—; *Canarese* : Otte—; *Dehra Dun* : Unga—; *Espiritu Santo* : Neapsara—; *Fanti* : Akyeng, Finina—; *Florida* : Caesar Weed—; *Ga* : Asofotoro—; *Gambia* : Bubobubo, Toja—; *Hausa*: Bak'in garmani, Igiyar rafi, Jan tsu, Ka-fi-rama, Ramaniya, Ramarama, Ramo-ramo, Uwarmaganni—; *Hindi* : Bachata, Bachit, Bachita, Banokra, Brachta—; *Ibo* : Udo agheregha, Udo azuro—; *Kobure* : Kaluka—; *Katsina* : Ramàniya—; *Konkan*: Villiah—; *Koranko*: Sose, Susi—; *Lagos*: Toja—; *La Reunion*: Herisson rouge—; *Losso*: Mitti, Mondondo, Tangmuid--; *Malay*: Pepulut, Perpulut, Poko kelulut, Pulut-pulut—; *Malayalam* : Udiram, Uram, Uran, Vatto—; *Malinke* : Susu—; *Mandingo*: Bubo bubo, Da ba julo, Da julo—; *Mano* : Gbo su—; *Marathi* : Rantupkada, Rantupkuda—; *Mauritius*: Herbe panier, Herbe panier a feuilles incisees—; *Mende* : Sogbei, Subwe, Tosbwe Tsogbei ; *Mundari* : Mindilat, Mindilata—; *North Western Provinces* : Bachita—; *Nzima* : Meteku—; *Porebunder* : Wagdaubhindi —; *Sakalave* : Kirijy, Kiriza, Kisilenjo, Tsikilenza—; *Sanskrit*: Vanabhenda—; *Santal*: Bhidijanetet—; *Sierra Leone* : Horse Whip—; *Sinhalese*: Patta apele, Pattu epala, Valta epala—; *Sokoto*: Ramaniya—; *Susu*: Nore, Nure—; *Tamil*: Ottatti, Ottuttutri—; *Telugu*: Peddabenda—; *Timne*: Defen, Defenfen, Ekboentomoer, Ka-fenfen, Ma-denfen, Nafenfen—; *Tivi*: Ishoho—; *Tschaudjo*: Kadjangbea—; *Tupin*: Guaxima—; *Twi*: Petekuku—; *Umu Ahia*: Udo agheregha, Udo azuzo—; *Uriya*: Bilokopasia, Jotyaholo—; *Wassaw*: Nsa-ne-nsa—; *Wolof*: Paddat, Padet, Pundere—; *Yoruba*: Ake-iri, Akeri, Ake-riri, Bolo-bolo, Ilasaagbonrin, Ilasa-omode, Ilasa-oyibo—.

2 **Urena repanda** Roxb. is found in the Punjab, Dehra Dun, the Central Provinces, the Northern Circars, the sal forests of Ganjam, and Ava.

The root and bark are believed by the Santals to be a cure for hydrophobia.

Santal : Sikuar—; *Uriya*: Jotofotia, Sikhini—.

3 **Urena sinuata** Linn. is found throughout the hotter parts of India. It is distributed over the tropics of both hemispheres.

In Chota Nagpur the root is used as an external application for lumbago.

In the Philippines the root is considered emollient, refrigerant, and maturant; the leaves are prescribed in inflammation of the intestines and the bladder.

The plant is considered emollient in Brazil and a decoction is given in colic; an infusion of the flowers is used in bronchitis.

Bengal : Kunjia—; *Bombay*: Tapgote—; *Brazil*: Carapicho, Carapicu, Urucurano—; *Chota Nagpur*: Berilat—; *Fulani*: Uwarbaganeji—; *Guam*: Dadangsi, Dadanse—; *Hausa*: Ramaniya, Uwar maganni—; *Hindi*: Kunguya, Kunjuya, Lotloti—; *Indo-China*: Bay, Ke dau nguo Phu thien hoa, Vai—; *Japan*: Bondenkiva—; *Malayalam*: Uram—; *Marathi*: Lichi, Ramkapasi—; *Matheran*: Lichi, Rankapshi—; *Mundari*: Mindilat, Mindilata—; *Nupe*: Rama—; *Pampangan*: Colotan, Colotcolotan, Dalupan, Molopolo—; *Philippines*; Culutan—; *Porebunder*: Wagdaubhindo—; *Proto Rico*: Cadillo pata-de-perro—; *Samoa*: Mautofa—; *Santal*: Mothabedijanetet—; *Sinhalese*: Hinappele. Hinepala—; *Tagalog*: Coloton, Colotcolotan, Dalupan, Dalupang, Molopolo—; *Tamil*: Ottatti Ottuttutti—; *Telugu*: Nallabenda. Padanikada, Piliyamankena—; *Visayan*: Colotan, Colotcolotan, Culuculutan,Dalupan, Dalupang, Molopolo—

THE MEDICINAL AND POISONOUS LINDENBLOOMS OF INDIA.

BY

J. F. CAIUS, S. J., F.L.S.

The TILIACEAE are trees or shrubs, rarely herbs. Most of them are tropical; a few inhabit the temperate northern hemisphere, and some are found behind the tropic of Capricorn; but none extend into the Arctic Circle, or ascend to great mountain elevations. The 380 species are distributed into 35 genera.

In general the leaves are mucilaginous and emollient; the bracts and flowers are aromatic, antipasmodic, and slightly sudorific. The bark may be bitter and astringent.

Among the products obtained may be mentioned:—(1) *a volatile oil*; (2) laevo-rotatory *phytosterols*; (3) *vanillin*; and (4) *glucosides*—capsularin, corchorin, tiliacin

The medicinal and poisonous Lindenblooms of the world belong to 12 genera: ANCISTROCARPUS (tropical West Africa); CISTANTHERA (tropical Central and West Africa); CORCHOROPSIS (Japan, China); CORCHORUS (warm regions); ELEOCARPUS (tropical); GLYPHAEA (tropical Africa); GREWIA (Asia, Africa, Australia; mostly tropical); LUEHEA (tropical America, West Indies); SLOANEA (tropical); SPARMANNIA (tropical and South Africa); TILIA (northern temperate regions); TRIUMFETTA (tropical).

The medicinal and poisonous Lindenblooms of India belong to 4 genera: CORCHORUS, ELAEOCARPUS, GREWIA, TRIUMFETTA.

A. Anthers opening by slits
- I. Petals usually foveolate or glandular at the base ; stamens springing from the apex of a raised torus.
  - a. Fruit without prickles ... ... *Grewia*
  - b. Fruit prickly.
    - Herbs or undershrubs. Fruit small ... *Triumfetta.*
- II. Petals not foveolate or glandular at the base; stamens springing from a contracted torus ... ... *Corchorus.*

B. Anthers opening by a terminal pore.
Petals sepaloid ... ... *Elaeocarpus.*

CORCHORUS.

The genus consists of 40 species widely dispersed throughout the warm countries of the world.

The following are used medicinally in the Philippine Islands—*C. acutangulus* Lam., *C. capsularis* Linn.—; in Indo-China—*C. capsularis* Linn. ; in Egypt *C. olitorius* Linn. ; in Gold Coast *C. acutangulus* Lam.—; in Southern Africa—*C. asplenifolius* Burch., *C. serraefolius* Burch.—.

A. Capsule globose, not beaked ... ... *C. capsularis.*

B. Capsule elongated.
I. Beak 3-fid spreading ... ... *C. acutangulus.*

II. Beak entire.

1. Capsule glabrous, cylindric, 10-ribbed. Beak long, erect ... ... *C. olitorius*

2. Capsule scabrous or aculeate, 3-angled. Beak short, erect ... ... *C. trilocularis.*

C. Capsule short, beak entire.

I. Annual, Capsule pubescent, 3-valved ... *C. fascicularis.*

II. Perennial. Capsule glabrous, 4-valved ... *C. depressus.*

1. **Corchorus acutangulus** Lam. is found throughout the hotter parts of India and Ceylon. It is distributed to Australia, tropical Africa, and the West Indies.

The bitter seeds are given by the Mundas of Chota Nagpur in pneumonia and stomach ache.

In Gold Coast the whole plant, including the root, is used in the form of an injection for urethral discharges.

Akim : Survabiri— ; *Awuna* : Lolui-tsu— ; *Bengali* : Titapat— ; *Ewe* : Have— ; *Ibo* : Aheheara, Ahihira— ; *Krepi* : Ademademaa, Adomadoma ; *Krobo* : Togatoya—; *Mende* : Ndogbo-ngengele—; *Mundari*: Karigiri, Karijiri, Karijuri, Piritijudiring— ; *Tubu* : Koebulu— ; *Twi* : Padeedee, Siwabiri—; *Yoruba* : Abojaga, Abo jaja—.

2. **Corchorus capsularis** Linn. occurs throughout the hotter parts of India. It is cultivated in most tropical countries.

A decoction of the dried root and unripe fruit is given in diarrhoea.

In cases of dysentery the dried leaves are eaten at breakfast time with rice. The cold infusion is also administered as a tonic in dysenteric complaints, fever, and dyspepsia.

It is a common practice in the jute-growing districts of Bengal to keep a small stock of the dried jute leaves in the house. whereof an infusion—a so-called tea—is made and taken by those suffering from any disorder of the liver. From the reports received by the Indigenous Drugs Committee it appears that the infusion is used as a popular domestic medicine for disorder of the lives and is of great value when there is a trouble with burning sensation in hands and feet. It is also used as a bitter tonic, stomachic, laxative, carminative, stimulant to increase appetite and flow of saliva and gastric juice; also as an antiperiodic, anthelmintic, astringent and intestinal antiseptic. It has been reported as efficacious in fever, bilious troubles, worms of children, dysentery, hepatic and intestinal colic, gastralgia and gastric catarrh, skin diseases especially itches, atonic dyspepsia, slight jaundice, and in the disorder of the digestive system.

In Indo-China the flowers are given in epistaxis; the fruits are applied to swelling and abscesses, and prescribed in diseases of the bladder.

Assam, Marasag, Tithmara—; *Bengal*: Ghinalitapat, Koshta, Narcha Titthapat—; *Bombay* : Chouchen, Ghinaltapat, Narcha—; *English* : Jute—; *Formosa* : Ma-p'i—; *French* : Chanvre de Calcutta, Jute—; *Hindi* : Chinalita, Koshta, Narcha, Pat—; *Indo-China* : Bo day, Floang ma, Ta ma—; *Marathi* Chaunchan—; *Mundari* : Hatularita- ; *Philippines* : Patas, Ponglopongloan—, *Porebunder* : Borachhunchh, Chhunchh—; *Portuguese* : Juta—; *Sadani* : Larita— ; *Sanskrit* : Kalasaka, Nadikapatta— ; *Santali* Kaskomrau—;

Shahjahanpur : Harrawa—; *Sinhalese* : Jaladara—; *Tagalog* : Pasaonabilog—; *Tamil* : Pirattikirai—; *Visayan* : Lamhay—.

3. **Corchorus depressus** Stocks is found in the Punjab, Sind, Baluchistan, Cutch, Gujerat, Deccan. It is distributed to Afghanistan, Arabia, North Africa, Cape Verde Islands, and tropical Africa.

The plant has tonic properties. It is given as a cooling medicine in fever.

The mucilage is used in Sind for gonorrhoea.

The leaves are emollient. The Baluchis make a cooling drink out of them. In Jaisalmer and Jodhpur they are applied to wounds, and a decoction is said to be efficacious in cases of skin eruption.

The seeds in decoction, with milk and sugar, are given as a tonic.

Arabic : Ueki—; *Baluchistan* : Mandira, Munderi, Mundheri—; *English* : Shrubby Jute—; *Gujerati* : Bahuphali, Bethibahuphali, Chhikni, Chhunchh—; *Hindi* : Baphuli—; *Marathi* : Bahuphali—; *Punjab* : Babuna, Bahuphalli, Bophalli, Kurana—; *Rajputana* : Hadekhaket—; *Sanskrit* : Bhedani, Chanchu, Katuka, Kshudra, Ksudrachanchu, Patupatrika, Shunakuchanchuka, Tvakasara—; *Sind* : Mudhiri, Munderi—; *Uriya* : Bojoromuli—.

4. **Corchorus fascicularis** Lam. occurs throughout the hotter parts of India: It is distributed to tropical Africa and Australia.

It is very mucilaginous and somewhat astringent, and is valued as a restorative.

In Bombay, a watery extract, mixed with sugar-candy, is taken as a nutritive tonic.

Bengal : Banpat, Bilnalita, Janglipat—; *Bombay* : Bhauphali, Hirankhori, Mothibahuphali—, *Gujerat* : Chhunchhadi, Ubhibahuphali—; *Hindi* : Bankostoa, Janylipat, Khetapat—; *Marathi* : Hirankuri, Motibuhuphali—; *Poona* : Mayarmithi—; *Porebunder* : Chhunchhadi, Ubhibahuphali—; *Sanshkrit* : Bhirupatrika, Chanchu, Chanchuputra, Chanchura, Chhunchhu, Chincha, Diaghpatri, Kalabhi, Kshestrachhunchhu, Kshestrasambhava, Sushaka, Vijala—.

5. **Corchorus olitorius** Linn is generally distributed by cultivation in all tropical countries. The chief seat of its Indian cultivation is near Kulna in the Burdwan district of Bengal.

The dried plant, roasted and powdered, is used in visceral obstruction. In South India it is used as a demulcent.

The leaves are demulcent, tonic and diuretic, useful in some cases of chronic cystitis, gonorrhoea and dysuria.

The leaves and tender shoots are eaten, and in the dried state, known as *nalita* ; they are used in infusion by the natives as a domestic medicine, being tonic and slightly febrifuge, and hence used as a fever drink.

The dried leaves are sold in the market. A cold infusion is used as a bitter tonic, and is devoid of any stimulating property. It can be safely given to patients recovering from acute dysentery to restore the appetite, and improve the strength.

The seeds are purgative. In Egypt they are put into the eye to cause inflammation by way of malingering.

Abeokuta : Eyo—; *Arabic* : Melochia, Meloukhia, Meluhhije, Molochia, Molukhyia, Muluhhije—; *Awuna* : Ademe—; *Bengal* : Banpat, Bhungipat, Phunjipat, Koshta, Lalitapat, Pat—; *Berar* : Chooich ; *Bombay* : Chhunchh, Motichunchh, Tankla—; *Efik* : Etinyon—; *Egypt* : Melukhiye—; *Ekoi* : Etinyung—; *English* : Jew's Mallow, Jute, Long-fruited Jute—; *Ewe* : Sigli,

Singili, Singli—; *Fanti* : Otoro, Oturo—; *French* : Corete, Corette, Corette potagere, Guimauve pot-gere, Melochie—; *Fulani* : Lolo, Lolo memel—; *Ga* : Otoro, Oturo—; *Gambia* : Alo—; *German* : Gemuese Corchorus—; *Greek* : Corchoros—; *Gujerati* : Chhunchho—; *Hausa* : Lalo, Malafiya, Marafiya, Tungurnuwa—; *Hindi* : Banpat, Koshta Pata, Sanpat, Singinjanasha—; *Ho* : Hatempa—; *Ibo Owerri* : Ahu ·hara—; *Ijebu-Ode* : Ayo—; *Ilesha* : Yoyo—; *Iraq* : Malukhiyah—; *Kanuri* : Darraba, Gamgaino—; *Katsina* : Turgunnuwa—; *Kissi* : Yo-ngengeyo—; *Konno* : Kokoe —; *Krutchi* : Aposse—; *Lagos*: Ewedu—; *Mandingo*: Kiringere—; *Marathi* : Chunch, Motichunch—; *Mende* : Gingere, Ngenge, Ngengele, Yengei, Yeingei—; *Mundari* : Larita—; *Nimar* : Rajan—; *North-Western Provinces* : Banphal—; *Oloke Meji* : Eyo-ganbi—; *Porebunder*: Chhunchho, Chunchhdo, Motichhunchh—; *Punjab* : Banpat, Banphal—; *Sadani*: Larita—; *Sanskrit*: Brihatchanchu, Dirghapatri, Divyagandha. Kalasa, Mahachanchu, Nadika, Patta, Singgika, Sthulachanchu, Suchanchuka Vishari—; *Santali* : Birnarcha—; *Senoufo* : Sobo—; *Serere* : a Koud—; *Sherbro* : Krinkrin-de—; *Sierra Leone*: Bush Okra, Crincra, Crincrin, Ingle—; *Sind* : Banpat—: *Sinhalese* : Wanuk—; *Sobo*: Oyoyo—; *Susu* : Suri—; *Tamil* : Peratti, Perumpunnakhuppoondu, Punaku—; *Telugu* : Parinta, Parintakura—; *Timne* : Kenkerin, Kirinkirin—; *Umu Ahia*: Agheregha, Aheheara, Ahihira—: *Uriya*: Jhoto, Joto, Kaunria—; *Wolf* : Mbali—; *Yemen* : Melochia—: *Yoruba* : Ewedu-ga-nbe, Eyo, Eyo-ga-nbe. Oyo—.

6. **Corchorus trilocularis** Linn. is found in Bihar, the Deccan and Carnatic of the Madras and Bombay Presidencies, Khandesh, Gujerat, Cutch, Sind, Baluchistan; whence it extends to Afganistan, Arabia, tropical and southern Africa.

The plant, macerated for a few hours in water, yields a mucilage, prescribed as a demulcent.

The seeds are bitter and administered in doses of about 80 grains in fever and obstruction of the abdominal viscera.

Arabic : Melochia, Molukhiya—; *Arago* : Emakikoho—; *Bombay* : Kaduchhunchh, Kuruchuntz—; *Canarese* : Tanolassir—; *Egypt* : Melukhiye—: *Fulani* : Lalo—; *Gujerati* : Kadvi-chhunchhdi—; *Hausa* Lalo, Turgunuwa—: *Hindi* : Kadukosta, Kadvapat—; *Malay* : Ramput baya roaza—; *Marathi* : Kaduchunch—; *Nasirabad* : Datrab—; *Porebunder* : Lambichhunchh—: *Rajputana* : Hardikeket, Kaglekitamaku, Karak—; *Sanskrit* : Dirghachanchu, Kaunti, Nadika—; *Sinhalese* : Jahedara, Wanuk—; *Songhai* : al Muluguia—; *Tamil* : Peratti, Punnakkuppoondu, Talakkaippoondu—.

ELAEOCARPUS

The genus numbers 90 tropical species, most abundant in the hotter part of India and the Indian Archipelago; a few are found in some of the South Sea Islands, New Zealand, and Australia.

E. madepolatus Pierre is used medicinally in Cambodia, *E. photiniaefolia* Hook. and Arn. in Indo-China.

A. Anthers not terminated by an awn.
- I. Drupe 5-celled ... ... *E. Ganitrus*
- II. Drupe 1-3-celled.
  - a. Anthers bearded, Filaments short, straight ... *E. serratus*
  - b. Anthers rarely bearded. Filaments long, twice bent ... ... *E. oblongus*

B. Anthers terminated by a long awn.
Leaves obovate. Stone much tubercled ... ... *E. tuberculatus*

1. **Elaeocarpus Ganitrus** Roxb. is found in the Western Ghats and the Konkan of the Bombay Presidency, Nepal, Bengal, Burma, Siam, the Malay Peninsula and Archipelago.

The fruit is used by Hindu practioners in diseases of the head and epileptic fits.

Bengal : Rudrakya—; *Canarese* : Rudraksha—; *English* : Utrasum Bead Tree—; *Hindi* : Rudrak—; *Malayalam* : Rudraksham—; *Marathi* : Rudraksh—; *Sanskrit* : Amara, Bhutanashana, Harksha, Nilakanthaksha, Pavana, Pushpachanamara, Rudraksha, Sharvaksha, Shivaksha, Shivapriya, Trinameru—; *Tamil*: Akkam, Irattaiyuruttirasham, Kaurichangamani, Uruttiradcham, Uruttirakkam—; *Telugu* : Rudrachallu—; *Tulu* : Rudraksha—; *Uriya* : Rudrakhyo—.

2. **Elaeocarpus oblongus** Gaertn. occurs in the Western Peninsula, and in Malaya.

The fruit is used as an emetic

Badaga : Bikki—; *Canarese* : Bikki, Hanaltadi, Hanillatade, Hennalatade—; *Malayalam* : Kattakara, Malankara—; *Tamil* : Kattukkarai—; *Visayan* : Cabalte, Cabilte—.

3. **Elaeocarpus serratus** Linn. is found in the Western Peninsula, Ceylon, and Malaya.

The leaves are used in rheumatism, and as an antidote to poison.

The fruits are prescribed in dysentery and diarrhoea.

Bengal : Jalpai—; *Canarese* : Bigada, Guddarenje, Perinkara—; *English* : Wild Olive—; *Hindi* : Jalpi—; *Malayalam* : Avil, Karamavu, Nallakara, Perunkara, Valiyakara—; *Sanskrit* : Chiribilva—; *Sinhalese* : Weralu—; *Tamil* : Ulangarai, Uruttiracham—; *Tulu* : Rahubija—; *Uriya* : Jolopari—.

4. **Elaeocarpus tuberculatus** Roxb. is found in the Western Peninsula, and in Malaya.

A decoction of the bark is given in vomiting of blood, and in biliousness.

The nuts are used in rheumatism, typhoid fever, and epilepsy.

Badaga : Rudraksha—: *Canarese* : Bhutali, Dandele, Dandla, Rudrak, Rudraksha, Rudrakshi—; *Hindi* : Rudrak—; *Kadir* : Navati, Pagumbal, Pillahi, Pulandi—; *Malayalam* : Ammakkaram, Kotuvasi, Maggara, Nakara, Navati, Pillahi, Pulanthi—; *Mysore* : Danala—; *Tamil* : Pagumbal, Rudraksham, Uruttracham—.

GREWIA

The genus numbers 150 species spread over Asia, Africa and Australia, mostly tropical.

Mucilaginous and demulcent, useful in the treatment of diarrhoea and dysentery.

The following are used medicinally in Indo China-*G. paniculata* Roxb.—; in the Philippine Islands- *G. paniculata* Roxb., *G umbellata* Roxb. —; in Australia—*G. hirsuta* Vahl—; in Tropical Africa—*G. carpinifolia* Juss., *G. mollis* Juss., *G. populifolia* Vahl, *G. villosa* Willd.—; in South Afric -*G. occidentalis* Linn. —.

A. Inflorescence terminal and axillary (sometimes extra-axillary). Flowers in umbellate cymes

I. Scandent shrubs. Torus long. Drupes obscurely lobed, purple ... ... *G. umbellata.*

II. Shrubby. Torus short. Drupes yellowish, gener-a y 2-lobed, lobes the size of a small pea, orange- ... ... *C. populifolia.*

B. Inflorescence leaf-opposed and axillary. Leaves glabrous or nearly so ... ... *G. tenax*

C. Inflorescence axillary (rarely extra-axillary).

I. Leaves usually hoary, at least beneath. Leaves 5-6-nerved

1. Stipules leafy, auricled ... ... *G. tiliaefolia*
2. Stipules linear-lanceolate ... ... *G. asiatica*

II. Leaves not hoary beneath, orbicular. Anthers glabrous

a. Drupes with a crustaceous rind

1. Levaes 3-6 by 4 in ... ... *G. sclerophylla.*
2. Leaves 3 by 1½ in. ; petiole ¼ in. ... *G. ca r inifolia*
3. Leaves 1-3 in.; petiole ½-1 in. ... *G. villosa.*

b. Drupes fleshy. Flowers polygamous. Branched shrub. Stamens more then 40 ... *C. hirsuta.*

D. Inflorescence terminal, in panicled cymes. Florwers involucrate.

I. Petals oblong, entire or shortly emarginate. Thrice shorter than sepals ... ... *G. Microcos*

II. Petals entire, much shorter then sepals ... *G. paniculata.*

1. **Grewia asiatica** Linn. is cultivated throughout India.

The fruit is a well-known Ayurveda medicine credited with astringent, cooling, and stomachic properties.

An infusion of the bark is used as a demulcent. The Santals use the root-bark for rheumatism.

The leaf is employed as an application to pustular eruptions, and the bud is also prescribed by native practitioners.

Ajmere : Dhamni—; *Arabic* : Phalasah—; *Bengal* : Phalsa, Shukri—: *Burma* : Pintayaw, Tagaw—; *Bombay* : Phalasi—: *Canarese* : Buttiyudippe, Jana, Tadasala—; *Central Provinces*: Dhamru, Dhamun—; *Deccan* : Phulsha, Pulsha—; *Gujerat*: Phalsa—; *Hindi* : Dhamin, Karra, Parusha, Phalsa, Pharvah, Pharsa, Phulsa, Shukri—; *Kohlu* : Pistawan—; *Kolami* : Gonyia, Singhindamin—; *Konkani* : Phalsi—; *Kotra* : Pharwan—; *Malayalam* : Chadicha—; *Marathi* : Phalsi—; *Nepal* : Sialposra—; *Newari* : Fussi—; *North - Western Provinces* : Dhaman, Phalsa, Pharsiya—; *Parbut* : Fulsa, Fursu—; *Persian* : Falseh, Palasah—; *Punjab* : Phalsa, Phalso, Pharoah, Phalue—; *Pushtu* : Pastaoni, Shikarimaiwah—; *Sanskrit* : Alpasthi, Dhanvanchhada, Giripilu, Mriduphala, Nagadalapam, Nilacharma, Nilamandala, Parapara, Paravata, Parusha, Parushaka, Porusha, Purusha, Roshana—; *Santal* : Jangolat—; *Sind* : Phalsa, Phalso, Pharaho, Pharoah—; *Sinhalese* : Dowaniya, Hin-damaniya —; *Tamil* : Palisa. Tadachi, Unnu—; *Telugu* : Jana, Nallajana, Peddajana, Phutiki, Phutiki—; *Urdu* : Phalasah—; *Uriya* : Pharosakoli—.

2. **Grewia carpinifolia** Juss. is found in the Western Peninsula and is distributed to Tropical Africa.

In West Tropical Africa the plant is used medicinally or to combine with other medicines. Women use it in washing the hair to remove or prevent lice, and put the roots in soup when approach ing childbirth.

In East Tropical Africa the broth obtained by boiling the head of a kid in a decoction of the root is taken in cases of diuresis and polyuria.

Ashanti : Ntabanu—; *Ewe* : Gayalige, Tupoetupoe—; *Fanti* : Nkukumbe, Ntaanta—; *Ga* : Asegetete-ntaata, Asegremi-tsho—; *Gosha* : Donoi—; *Krobo* : Akpe-tsho—; *Somali* : Debi—; *Yoruba* : Itakun okere—.

3. **Grewia hirsuta** Vahl occurs in the sub-Himalayan tract up to 4 500 feet from the Indus eastwards, the Salt Range, Bihar, Orissa, Burma, the southern Provinces, and Ceylon. It is distributed to North Australia.

The fruit is employed as a medicine by the Santals, in diarrhoea and dysentery. The root pounded is also prescribed for the same diseases, and powdered in water is applied externally to hasten suppuration. and as a dressing for wounds. The paste dries and forms a hard coating, thus effectually excluding air from the raw surface.

This plant is used by the aborigines of North-Western Australia as a remedy for dysentery.

Arabic : Kamafetusa—; *Australia* : Plain Currant—; *Betul* : Kharota—; *Bombay* : Gowali, Gowli—; *Burma* : Kyettayaw—; *Canarese* : Chikkudippe, Jana, Udippe—; *Hasada* : Setalele—: *Hindi* : Gursukri, Kakarundah, Kukurbicha—; *Kharwar* : Gursukri—; *Kolami*; Setaandir, Setabeli, Setakata—; *Marathi* ; Govli—; *Merwara* : Phrongli—; *Naguri* : Seta-bili, Seta-kata—; *Persian* : Karafasrumi—; *Porebunder* : Khoddhramani—; *Queensland*: Karoom, Kooline, Outaie—; *Santal* ; Setaandir, Setabeli, Setakata—; *Tamil* : Tavidu—; *Telugu* : Bidarachipuru, Budda, Chmnachipuru, Chipuru, Chittijana, Jibilike, Muvalchu, Tellajana—; *Urdu* : Kakarundehrumi—; *Uriya* : Kulo, Sonaranga—

4. **Grewia Microcos** Linn. occurs in Eastern Bengal, Assam, Burma, the Western Peninsula, Mysore, and Ceylon. It is distributed to Java and China.

The plant is much used medicnally; it is given for indigestion, eczema and itch, smallpox, typhoid fever, dysentery, syphilitic ulceration of the mouth.

Bombay : Ansale, Shiral—; *Burma* : Myaza—; *Canarese* : Abhhrangu, Biliyabhhrangu, Majjigesoppu—; *Malayalam* : Kottakka—; *Sinhalese* : Keliya—; *Tamil* : Kadambu, Visalam—; *Tulu*: Abroni—.

5. **Grewia paniculata** Roxb. is common in the Malay Peninsula and in Indo-China, in open country.

In the southern parts of Indo-China a decoction of the roots is given as a cure for cough.

Indo-China : Co ke, Poplear thom—; *Malay* : Chenderai hutan—; *Visayan*: Bangalad—.

6. **Grewia populifolia** Vahl extends from the arid tracts of the Punjab, Sind, Rajputana, and Western India, down to the Nilgiri Hills. It is distributed to Afghanistan, Arabia, tropical Africa, Mauritius.

A mucilage of the bark is used by the women of West Tropical Africa to cleanse the hair of vermin.

Somali women take a decoction of the roots to incite the foetus in the womb and thus ascertain that it is alive.

Falor: Sonne—; *Gosha*: Moris—; *Kohat*: Shikari mewa—; *Kolami*: Bursa-; *None* : Sonne—; *Punjab*: Ganger, Gangi, Kanger, Khircha, Inzarre—; *Pushtu* : Khircha, Madzar—; *Rajputana* : Ganegam, Gangan, Gangerun—; *Serere* : Nghel, Nguel—; *Sind*: Gangi, Gango—; *Somali*, Morio-; *Telugu*: Gundukadira, Kadadari, Kaladi, Kattekolupu—; *Tuareg* : Tarkat—; *Tukulor* : Keli, Kelli—; *Wolof* : Kel—.

7. **Grewia sclerophylla** Roxb. occurs throughout the sub-Himalayan tract and outer hills of Kumaon, up to 3,500 feet, Sikkim, Assam, Chittagong, Ava.

The root is prescribed in coughs and irritable conditions of the intestines and bladder. The decoction is also used as an emollient enema.

Bengal : Phalsa—; *Burma* : Petshat—; *Canarese* : Darsuk, Kadukadele—; *Cutch* : Susaka—; *Dehro Dun* : Gurbheli—; *Gujerati* : Padekado—; *Haldwani* : Phalsa—; *Hindi* : Pharsia—; *Kumaon* : Pharsia—; *Marathi* : Khatkhati, Pandharidhaman—; *Mundari* : Gaphni—; *Tamil* : Kattukkadali, Punaippidukkan—; *Telugu* ; Bankajana—; *Tharu* : Dapher—.

8. **Grewia tenax** Fiori is found in the Punjab, the desert of Western Rajputana, Sind, Baluchistan, Cutch, Southern Maratha Country, the Deccan and Carnatic of the Madras Presidency, and Ceylon. It is distributed to Afghanistan, Persia, Arabia, tropical Africa, and Mauritius. It is most probably identical with *G. populifolia* Vahl.

A Baluchi medicine for all sorts of diseases and very highly esteemed. The berries are given for colds.

In Jhalawan a decoction of the wood is given as a cure for coughs and pains in the side.

Arabic : Chodar, Moucken, Nabba—; *Baluchi* : Gwangir, Buzi putrunk—; *Brahui* : Gwangir—; *Jhalawan* : Gwangi, Kango—; *Kalat* : Rango—; *Mandi* : Putrunk—; *Rajputana* : Ganneran, Gangerun, Gangi, Gango, Kankeran—; *Sinhalese* : Katuperatti—; *Tamil* : Achu—; *Telugu* : Gundukadira, Kadadari, Kaladi Kattekolupu—.

9. **Grewia tiliaefolia** Vahl occurs in the sub-Himalayan region from the Jumna to Nepal up to 4, 000 feet, Central India, all the districts of the Madras Presidency, Bihar, Orissa, Burma, and Ceylon. It is distributed to East Tropical Africa.

In the Konkan the bark, after removal of the tuber, is rubbed down with water, and the thick mucilage strained from it and given in 5-tola doses, with 2 tolas of the flour of *Panicum miliaceum* (warri) as a remedy for dysentery.

The bark is also employed externally to remove the irritation from cow-itch.

The wood reduced to a powder acts as an emetic, and is employed as an antidote to opium poisoning.

Bengal : Dhamani, Pharsa—; *Bhil* : Dhamnak—; *Bombay*: Damana, Karakana—; *Canarese* : Batale, Butale, Buttele, Dadsal, Jana, Tadagana, Tadasali, Tadasalu, Thadasal, Thadsal-; *Central Provinces*: Damun, Dhaman, Dhamni—; *Ceylon*: Chadachi—; *English*: Dhaman—; *Gond*: Ainlubaranda, Kasul, Khesla—; *Gujerat* : Dalmon, Dhamana—; *Hindi* : Dhaman, Dhamani, Dhamin, Pharsa—; *Kadir* : Chatachi, Una—; *Kharwar* : Dhaman—; *Khond* : Kahal, Karkana—; *Kolami* : Ahsing, Dhamin, Gonyer—; *Konkani* : Butale, Dadsale, Damoni—; *Kumaon* : Pharsia—; *Kurku* : Dhamni—; *Malayalam* : Satchi, Una—; *Marathi* : Daman, Damni, Dhaman, Karavarani, Karakarani—; *Matheran* : Dhaman—; *Mundari* : Asin, Asindaru, Bengadaru—; *Mayurbhanj* : Kultho—; *Porebunder* : Dhaman, Dharman—; *Sanskrit*: Dhamsi, Dlanuvriksha, Dharmana, Dharwana, Manabala, Pichhilaka, Pichhilatvaka, Raktakusuma, Rujasaha, Ruksha, Swaduphala—; *Santali* : Janoolat, Olat—; *Saora* : Inuputada—; *Sinhalese* : Daminiya, Daminne—; *Tamil* : Sadachi, Tada, Tarra, Una, Unnu—; *Telugu* : Charachi, Ettatada, Jana, Nulijana, Tada, Tadajana, Udupai—; *Tulu* : Kanapadi—; *Uriya* : Bhangia, Dhaman, Dhamono, Dhamuro—.

10. **Grewia umbellata** Roxb. is found in the Malay Peninsula, Siam, and Borneo.

The leaves are applied to cuts and wounds.

Malay : Akar chenderai, Akar kapialu, Akar sekapu, Akar sempelas lida kuching—; *Philippines* : Danloy—.

11. **Grewia villosa** Willd. occurs in the Trans-Indus, the Punjab, Rajputana, Sind, Cutch, Kathiawar, the Deccan and Carnatic of the Madras Presidency from the Kistna southwards. It is distributed to the Cape Verde Islands and tropical Africa.

The juice of the fresh bark is used with sugar and water for gonorrhoea and urinary complaints attended with irritability of the bladder.

The root is employed for diarrhoea in Chota Nagpur.

East of Chad, in tropical Africa, the root is an ingredient in native prescriptions for syphilis and smallpox.

The roots macerated and triturated with milk are a Somali cure for belly ache in babies; the adults apply a poultice to the abdomen, the roots are ground with as little water as possible.

Ajmere : Dhohan—; *Baluchi* : Pachini putrunk—; *Canarese* : Buttigaragale, Garakele, Sannudippe—; *Cutch*: Luskano—; *Ewe*: Adzadze—; *Falor*: Lekit—; *Gujerati* : Padekhado, Parekhado—; *Hausa* : Gwiwar rak umi—; *Kamba* : Mugu—; *Kolami* : Gaphni—; *Mangu* : Yumbu—; *Marathi* : Kharmati—; *Merwara* : Dhokelan—; *None* : Ngomen—; *Punjab* : Jalidar, Kaskusri, Thamther—; *Pushtu* : Inzarra, Pastuwanne—; *Rajputana* : Lonkas—; *Santali* : Tarsekolap—; *Tamil* : Kullai—; *Telugu* : Benta, Chenula—; *Tigre* : Hafule. Khafule—; *Wolof* : Horom sap—.

TRIUMFETTA

The genus consists of 75 tropical species.

All the species are mucilaginous, demulcent, bechic, and dieuretic; but those more commonly used are *T. rhomboidea* Jacq. and *T. semitriloba* Linn. generally, and *T. Lappula* Linn. in Guiana.

Fruit the size of a small pea ; spines hooked, glabrous or ciliated ... ... *T. rhomboidea*
Fruit the size of a large pea ; spines bristly or shaggy ... *T. semitriloba.*

1. **Triumfetta rhomboidea** Jacq. occurs in tropical and subtropical India, Ceylon, the Malay Peninsula and Archipelago, from which it extends to China. It is distributed to Africa and America.

The root is bitter and acrid, and used as a diuretic.

The bark and fresh leaves are given in diarrhoea and dysentery.

The leaves, flowers and fruits are mucilaginous and astringent. They are given in gonorrhoea. The fruit is believed to promote parturition.

Zulu women take a hot infusion of the root to facilitate childbirth or to hasten the inception of parturition when it is delayed.

In a Reunion and Mauritius the mucilaginous roots and flowers are used for their demulcent, bechic, and diuretic properties.

The pounded root is applied to sore eyes and to boils in Madagascar; the leaves and the stems bruised together are used for poulticing all sorts of tumours. In the district of Menabe the plant, associated with *Cynodon dactylon,* is applied to burns and sore eyes.

In Brazil a decoction of the herb or of the fruit is used as an injection in inveterate gonorrhoea.

In Northern Nigeria the leaves, along with natron and bran, are given to horses for internal troubles. Similarly in Sierra Leone the seeds are mixed with Guinea-corn and given to horses for worms, and constipation.

Agolo : Abali oicha—; *Aguku* : Abali ilenza—; *Akim* : Betekuo—; *Antsianaka* : Besofimbavy, Besofinantanana—; *Bakwiri* : Toli—; *Bengal* : Bunokra—; *Betsileo* : Besofina, Kihasinkasina—; *Bombay* : Nichardi—; *Brazil* : Carapicho, Garapicho da Calcada—; *Ceylon*: Aaiyooldi—; *English* ; Paroqueet Burr—; *Ewe* : Bobui—; *Goa* : Tupsado—; *Gujerati* : Jhipato—; *Hansot* : Bharvado—; *Hausa* : 'Danka' dafi, Suren fadama—; *Hindi* : Banokra, Chikti :, Chiriyari—; *Hova* : Tsindailay, Tsitiamoty—; *Ibo* : Udo—; *Ijebu* : Epafo—; *La Reunion* : Herisson blane—; *Madagascar* : Besofina, Tsindailay—; *Marathi* : Jhinjhira, Jhinjudi, Kutrevandare, Nichardi—; *Matheran* : Nechardа—; *Mauritius* : Herbe a panier, Herisson blanc, Aadaye otti—; *Mende* : Chobwe, Chohobwe, Sogbe, Sogbei—; *Mundari* : Mindiat, Mindilata—; *Onitsha* : Udo mberi—; *Owerri* : Azuze, Azu uzo, Oke udo Udo mberi—; *Sakalave* : Kisalenjy, Tsirrijy—; *Sanskrit* : Jhinjharita, Jhinjhirita, Jhinjhurdi, Jhirpata, Kantaphali—; *Sinhalese*: Epala—; *Tamil* · Ottuppullu, Puramutti—; *Telugu* : Chirusitrika, Tutturubenda—; *Timne* : Agbint-marabana, Egbuntemoer, Ekboentemoer—; *Twi* : Petekuku—; *Umu Ahia* : Ngbo—; *Uriya* : Bojoromuli, Jotojoti—; *Wolof* : Doupereit—; *Yoruba*: Ako-bolobolo, Bokopupa, H sa-omde—; *Zulu* : in Dolaencane, iNbrthwane—.

2. **Triumfetta semitriloba** Linn. is a tropical weed.

The leaves and fruit are mucilaginous and astringent.

The leaves are used in parts of Sierra Leone as a medicine for dysentery. In Cameroons a cold infusion of the leaves along with those of *Microdesmis puberula* is tyken as a remedy for diarrhoea.

Ia Brazil a decoction of the herb or the fruit is used as an injection in inveterate gonorrhoea.

Aguku : Agamevumawo--; *Ashanti* : Mfo--; *Bakwiti* : Toli--; *Brazil* : Carapicho Carapicho da Calcada—; *Cameroons* : Okon, Okung—; *Duala*: Wonge-; *English* : Burweed—; *Ewe* : Bobui—; *Ibo* : Udo, Udo ji—; *Koranko* : Segbe—; *Limba* : Ubama-wuyeremi—; *Mende* : Bongewuli, Bonhui, Chobwe, Kpolohui, Kponhui, Pohui, Sogbe—; *Nzima*: Egurie—; *Onitsha*: Udo umuaka—; *Owerri*: Udo umuaka—; *Tagalog* : Calotang-bilok, Colotan—; *Timne* ; E-boloni, Kaboloni, Ka-bononi, Raka—; *Visayan* : Daracot—; *Yoruba*: Esura—.

THE MEDICINAL AND POISONOUS STERCULIADS OF INDIA.

By

J. F. Caius, s.j., f.l.s. (Deceased)

The STERCULIACEÆ are herbs, shrubs or trees, sometimes lianes, which belong to tropical and subtropical regions. There are 660 species, distributed into 48 genera.

They contain an abundant mucilage, combined in the old bark of the woody species with a bitter astringent extractable matter, and exhibit stimulant and emetic properties. Many yield gums which in their appearance resemble tragacanth, and are reputed tonic. The fruit is often astringent and antidysenteric. The seeds may be diuretic and have a stimulating effect on the central nervous system.

Among the products isolated we may mention :—(1) *glucosides*—kolatin—; (2) *tannins*—kolatannin—; (3) *gums*; (4) *alkaloids*—caffeine, theobromine—; (5) *acids*—tartaric—; (6) *amino-acids*—asparagin—; (7) *fats* and *carbohydrates*.

The medicinal and poisonous Sterculiads of the world belong to 23 genera :—Abroma (tropical Asia to Australia); Buettneria (tropical); Cola (Africa); Dombeya (Africa, Madagascar); Eriolaena (Indomalayan); Fremontia (California); Guazuma (tropical America); Helicteres (tropics, except Africa); Hermannia (tropical and subtropical, chiefly Africa); Kleinhovia (tropical Asia); Leptonychia (tropical Africa and Asia); Mahernia (Africa); Melochia (tropical); Myrodia (tropical); Pentapetes (Indomalayan); Pterocymbium (Burma, Malaya); Pterospermum (tropical Asia); Pterygota (tropical); Scaphium (Burma, Malaya, Borneo); Sterculia (tropical); Tarrietia (Eastern Asia, Australia); Theobroma (tropical America); Waltheria (tropical America).

The medicinal and poisonous Sterculiads of India belong to 16 genera :—Abroma, Buettneria, Eriolaena, Guazuma, Helicteres, Kleinhovia, Melochia, Pentapetes, Pterocymbium, Pterospermum, Pterygota, Scaphium, Sterculia, Tarrietia, Theobroma, Waltheria.

A. Flowers unisexual; petals none; stamen-column with anthers on the apex, clustered or in a ring.

I. Ovarian cells 1-ovuled; fruit of thick samaras with foliate wings ... ... Tarrietia.

II. Ovary with 2 or more ovules in each cell; fruit dehiscent

1. Radicle remote from the hilum. Follicles coriaceous or woody ... ... Sterculia.

2. Radicle next the hilum.

a. Follicles woody. Seeds many, winged at the apex, albuminous .. ... Pterygota.

b. Follicles membranous, opening long before maturity.
  i. Calyx small, lobes lanceolate; follicles very large, green, one seed at base of each ... ... SCAPHIUM.
  ii. Calyx campanulate; 2 seeds in each follicle ... ... PTEROCYMBIUM.

B. Flowers bisexual.
I. Petals deciduous; stamen-column bearing a cup edged with anthers and staminodes, 5 to 15.
  1. Shrub; flowers pink, capsule membranous inflated ... ... KLEINHOVIA.
  2. Shrub; flowers purple or white; capsule woody not inflated; seed not winged ... HELICTERES.
  3. Trees; flowers large, white; capsule woody, large; seed winged ... ... PTEROSPERMUM.

II. Petals deciduous; stamen-column tubular, conical, antheriferous for nearly its whole length; staminodes none. Capsule ovoid, woody or coriaceous ... ... ERIOLAENA.

III. Petals persistent; anthers 10 to 20 on the tip of a cup on the column. Bracteoles caducous. Herb; petals red ... ... PENTAPETES.

IV. Petals marcescent, flat. Stamen column tubular at the base only; stamens 5; staminodes none.
  1. Shrubby herb or shrub; capsule 5-celled; flowers pinkish ... ... MELOCHIA.
  2. Shrubby; capsule 1-celled, 1-seeded; flowers yellow ... ... WALTHERIA.

V. Petals concave at the base; stamen-column tubular; anthers marginal, solitary or in groups between the staminodes.
  1. Fertile stamens 5 opposite the sepals ... THEOBROMA.
  2. Stamens uniseriate.
    a. Anthers solitary between each staminode. Petals clawed with 2 lateral lobes and a long appendage. Climber ... ... BUETTNERIA.
    b. Anthers 2-4 between each staminode.
      i. Petals with a stipitate ovate blade. Capsule turbinate, membranous ... ABROMA.
      ii. Petals with a linear bifid blade. Capsule oblong, woody, tubercled. Flowers yellow ... ... GUAZUMA.

ABROMA.

The genus consists of 10 species spread over tropical Asia and distributed to Australia.

A. augusta Linn. *fil.* is used medicinally in Indo-China and in the Philippine Islands. In the latter country *A. alata* Blanco is also credited with therapeutic properties.

Abroma augusta Linn. *fil.* is found indigenous or cultivated throughout the hotter parts of India, Java, the Philippine Islands, and China.

In Lakhimpur, Assam, the bark is used for sores. A part of the root is used internally and externally by the Santals to cure abscess.

The root bark has long been used in Bengal as an emmenagogue. As far back as 1873, Thornton wrote in the *American Journal* of *Medical Science* (p. 276): '*The fresh viscid juice* of the root-bark is useful in the congestive and neuralgic varieties of dysmenorrhoea; it regulates the menstrual flow and acts as an uterine tonic.'

In the *Indian Medical Gazette* of May 1900, Sircar reported as follows:—

'Forty years ago I first came to know the medicinal properties of this indigenous plant as a good emmenagogue in menstrual disorders. . . . The official part of the plant is the fresh viscid sap, which abounds in the thick easily separable bark of the root and is insoluble in water. . . . Attempts have been made to administer the drug in the more acceptable forms of tincture, pill or powder, *but none prove so efficacious as the fresh viscid sap in substance*, in which form I have used it with wonderful results.'

On analysis, Chopra, Ghosh and Chatterjee obtained from the root-bark a mixed oil, resins, an alkaloid—in minute quantities—, and water-soluble bases. The alkaloid and the various other fractions were then 'passed through pharmacological tests, but no remarkable activity was manifested on the gastro-intestinal tract, circulation, respiration, etc., nor was there any marked effect on the uterus whether virgin or pregnant, isolated or *in situ*. In the absence of any signs of physiological activity, clinical trials were not carried out.'

It was left to K. C. Bose to solve the problem satisfactorily. 'The root-bark is an emmenagogue and uterine tonic. The action of the dried root as well as the sap of the fresh root, has been studied in my laboratory. It showed a tonic contractile action on the uterus and its use is, therefore, indicated before the menses, to help the uterine contractions to bring about a proper flow of menstrual blood. The active principle of the drug is totally destroyed if mixed with alcohol or any other preservative; either the fresh root-bark or dried root-bark should be used.'

Assam: Gunakhiakarai—; *Bengal*: Olatkambol, Ulatkambal, Ulutkambal—; *Bombay*: Olatkambol—; *Canarese*: Melpundigida; *Ceylon*: Ulatkambal—; *English*: Devil's Cotton—; *French*: Abrome—; *German*: Abrome—; *Hindi*: Kumal, Ulatkambal, Sanukapashi—; *Indo-China*: Bom vang, Tai meo—; *Philippines*: Anabo—; *Tagalog*: Anibog, Anibong—; *Tamil*: Sivapputtutti—; *Uriya*: Pisachogonjai—.

BUETTNERIA.

The genus consists of 60 species inhabiting the tropics; mostly American, a few Asiatic and African.

Buettneria herbacea Roxb. occurs in the Western Peninsula, Orissa, the Northern Circars, the Deccan, the Carnatic, and the Konkan.

The rootstock is ground and rubbed on swellings of the legs by the Kols. It is used in combination with Bael fruit, hesel gum, and Banyan root in cholera and diarrhoea. It is given in the female complaint known in Santali as 'prodhol' (Campbell).

Kolami: Idel sanga—; *Porebunder*: Adbaubal, Vagdaubal—; *Santali*: Deku sindur—.

ERIOLAENA.

The genus consists of 8 Indo-Malayan species.

Eriolaena quinquelocularis Wight is found in the Bombay Presidency: Konkan, Western Ghats, Deccan, Southern Maharatta Country; in the Madras Presidency: Deccan, the Sandur Hills of Bellary, the hills of Coimbatore. the Western Ghats from Mysore to Travancore at 2,000-4,000 feet.

A poultice made of the roots is said to heal wounds.

Bombay: Budjari-dha-mun;— *Chota Nagpur*: Bhawat—; *Canarese*: Gomaj-jige, Kattale—; *Konkan*: Bujaridamu—; *Malayalam*: Vetinar—; *Mundari*: Bunduddaru—; *Tamil*: Malamtutti, Nayunnu, Vattanunnu—.

GUAZUMA.

The genus consists of 5 species, natives of tropical America.

G. ulmifolia Lam. is used medicinally in Colombia, Brazil, and La Reunion.

Guazuma tomentosa H. B. and K., a native tropical America, is found frequently cultivated in India.

The bark is tonic and demulcent, and is used with benefit in some of those cases in which calumba and gentian are indicated (Moodeen Sheriff).

The inner bark is esteemed as a remedy for elephantiasis in the West Indies; the infusion of the old bark is considered sudorific, and useful in cutaneous diseases and diseases of the chest.

In Mauritius the fruit is used as a pectoral in cases of bronchitis.

Bengali: Nipaltunth—; *Deccan*: Bandog—; *Canarese*: Bucha, Rudrakshi—; *Ceylon*: Patti-parutti—; *English*: Bastard Cedar, Honey Fruit Tree, Musket Tree—; *Honduras*: Bay Cedar—; *Malayalam*: Rudraksham—; *Mauritius*: Caca poule, Orme pyramidale, Thain pachai—; *Sinaloa*: Guasima—; *Tamil*: Rudrasam, Tenbachai, Tengai, Tubakki, Uruttiracham—; *Telugu*: Rudraksha, Udrikpatta—; *Tulu*: Rudrakshi—; *Uriya*: Debodaru—.

HELICTERES.

The genus consists of 40 species, distributed over the tropics of both worlds.

The following species are used medicinally in Indo-China and the Philippine Islands—*H. Isora* Linn.—; in Brazil—*H. brevispira* A. Juss., *H. corylifolia* Nees and Mart., *H. ovata* Lam., *H. Sacarolha* A. Juss., *H. Vuarame* Mart.

Helicteres Isora Linn. is rarely found in open country; it occurs in the dry forests throughout Central and Western India, and from the Punjab and Bengal to Ceylon and Burma. It is distributed to Malaya, Australia, and the West Indies.

The juice of the root is said to have a beneficial effect in empyema and stomach affections. In the Konkan it is used in diabetes, and is a favourite cure for snake-bite

The bark is used in diarrhoea and dysentery.

The fruits are made into a liniment for sores of the ear, and they are administered internally for colic. They are demulcent.

mildly astringent, and useful in the griping of bowels and flatulence of children.

The Santals credit the root, the bark, and the fruit with the same therapeutic properties: demulcent, expectorant, astringent to the bowels, a cure for scabies when applied topically.

The fruit, under the name of Chabei Pintal, is sold as a drug in the shops of Malaya.

A decoction of the leaves is used as an enema in Jamaica.

Mhaskar and Caius have shown experimentally that neither the bark nor the root is an antidote to snake venom.

Arabic: Altwa-al-latu—; *Banwara*: Anteri—; *Bengal*: Antamora, Antomura, Atmora—; *Berar*: Marorphal, Muradsheng—; *Betul*: Aitha, Aithi—; *Bombay*: Kawun, Kevana, Kewan, Khiran—; *Burma*: Khungiche, Thuguaykhyae, Thungeche—; *Canarese*: Bhutakarulu, Edamuri, Kadukalnaru, Kaiyuri, Kavargi, Murudi, Narukolu, Thunshulla, Yadamuri, Yedamuri—; *Central Provinces*: Boltuka—; *Ceylon*: Kawa, Vullum-puri—; *Deccan*: Dhameenee, Dhamni, Kevan, Kewanne, Maradsing, Marorikephalli—; *Godaveri*: Itah—; *Gond*: Aita—; *Gujerati*: Murdasing—; *Haldwani*: Benwa—; *Hindi*: Bhendu, Jonkaphal, Kapasi, Maraphal, Maraphali, Marorephalli, Marorphal, Marorphali, Marosi, Merosi, Murad, Muradsing—; *Indo-China*; Cay do tron—; *Kharwar*; Aita, Aitem, Aiteni—; *Khond*: Kavali, Korajbothi, Pitabaranda—; *Kolami*: Goinr, Potoporla, Renta, Sakomsang, Sinkari—; *Konkani*: Kivani—; *Khmaon*: Marorphal—; *Kurku*: Korajbothi, Koributa—; *Lambadi*: Moldaphaliro—; *Malay*: Mori—; *Malayalam*: Ishvaramuri, Kaivalanara, Kaiyuna, Valampiri—; *Marathi*: Kewan, Muradsing—; *Matheran*; Dhamni, Kewan, Muradsing—; *Monghyr*: Ainthiadhamin—; *Mundari*: Rantadara, Rentadaru—; *Nimar*: Aitha, Aithi—; *North-Western Provinces*: Bhendu, Marorphal—; *Persian*: Kishtburkisht, Pechaka—; *Porebunder*: Anted, Antedi—; *Punjab*; Kupasi, Marorphali—; *Saharanpur*: Kapasi—; *Sanskrit*: Avartani, Mrigashinga—; *Santal*: Petcamra, Petchamra—; *Sind*: Vurkatee—; *Sinhalese*: Liniya, Liniyagass, Zimiagaha—; *Tamil*: Vadampiri, Valamburi, Valampuri, Valumberi—; *Teheran*: Pachman-i-puh—; *Telugu*: Adasamanti, Adasyamali, Gubadarra, Gubalada, Kavanchi, Nulitada, Nuliti, Peddasamanti, Sadala, Samanti, Syamali, Tada, Uttrasi, Valambiri—; *Tulu*: Kaiycli, Muriga—; *Urdu*: Marorphali—: *Uriya*: Murmuria Orola, Kaval, Modimodi, Murimuri.

KLEINHOVIA.

K. Hospita Linn., the sole representative of the genus, inhabits the tropics of both hemispheres. It is found in the Eastern and Western Peninsulas of India, and distributed to Malacca, Singapore, Ceylon, Java, the Philippine Islands, and East Tropical Africa.

A decoction of the leaves is used in the Philippine Islands as a remedy for scabies, and as a lotion for cleaning skin eruptions.

Ilocano: Bignon, Bitnog, Bitnong—; *Indo-China*: Tra—; *Pampangan*: Panampat—; *Philippines*: Bitang, Bitnon, Pampar—; *Sunderbuns*: Bhola—; *Tagalog*: Tan-ag—; *Visaan*: Hamitanaga, Tan-ag—.

MELOCHIA.

The genus consists of 65 tropical species.

M. borbonica Cav. is used medicinally in La Reunion, *M. pyramidata* Linn. in Mauritius.

Melochia corchorifolia Linn. is found in the hotter parts of India from Kumaon to Sikkim and the Malacca Peninsula; in Cutch

Gujerat, the Konkan, and North Kanara; in most of the districts of the Madras Presidency, and in Ceylon.

The whole of this plant, with the exception of the root, boiled in oil, is supposed on the Malabar Coast to be an efficacious remedy for preventing bad consequences from the bite of a water snake (Ainslie).

Water snakes are harmless (Caius).

Bengal: Tikiokra, Tikiopra—; *French Sudan*: Togoyo—; *Fulani*: Tukurra—; *Hausa*: Tukurra—; *Hindi*: Bilpat—; *Malay*: Lumah ketam; *Malayalam*: Seruvuram—; *Mende*: Ndopa-yengei, Ngengele—; *Mundari*: Dela ara, Delka ara—; *Philippines*: Calingan—; *Santali*: Thuiakarak—; *Sinhalese*: Hingalkara, Kingalkara—; *Sokoto*: Tukurra—; *Susu*: Suri—; *Tamil*: Chiraparam, Pinnakkuppundu, Punnakkukkirai—; *Telugu*: Ganugapindikura, Sittantakura, Tutturubenda—; *Timne*: Kirkinkirin-Kobanabana—; *Uriya* Chyeron, Dasokerotan, Nolita—; *Wolof*: Pag hou gor—.

PENTAPETES.

The only representative, *P. phoenicea* Linn., is indigenous in North-Western India, Gujerat, and Bengal. It is distributed to Malaya, Java, the Philippine Islands, and China.

The fruit is mucilaginous. The root is employed as a medicine by the Santals. The plant is used as an emollient in Annam and in China.

The Chinese use the plant in indigestion and in atony of the stomach.

Bengal: Bandhuli, Doopahuria, Katlala—; *Chinese*: Ou Che Hoa, Ou Ts'ai Hoa—; *Gujerati*: Duporio, Sowbhagyasundari—; *Hindi*: Dopaharia, Daparia, Dopohoria—; *Indo-China*: Da lac kim tien, Hoa ti ngo, Ngo thi tra—; *Marathi*: Tambridupari—; *Punjab*: Guldupaharia—; *Sanskrit*: Arkavallabha, Bandhujiva, Bandhujivaka, Bandhuka, Bandhula, Bandhura, Haripriya, Jvaraghna, Madhyadina, Madhyanika, Oshthapushpa, Pushparakta, Rakta, Raktaka, Raktapushpa, Saratapushpa, Supushpa, Suryabhaka—; *Santal*: Barebaha—; *Tamil*: Nagappu—.

PTEROCYMBIUM.

The genus consists of 2 species, natives of Burma, the Nicobars, and Malaya.

Pterocymbium javanicum R. Br. is found in the open forests of Malaya by creeks or open woodland. It is distributed to Burma, Nicobars, Java, and Philippines.

The tree yields a gum which is a fair substitute for tragacanth.

Burma: Tshaw—; *Malay*: Kluet, Kulunot—.

PTEROSPERMUM.

The genus consists of 20 species, natives of tropical Asia.

A. Bracteoles linear, entire, very caducous. Leaves 5-15 by 2.5-5 cm. ... ... *P. suberifolium*.

B. Bracteoles laciniate or palmately divided.

1. Leaves 25-35 by 15-30 cm. ... ... *P. acerifolium*
2. Leaves 10-15 by 5-9 cm. ... ... *P. heyneanum*

1. **Pterospermum acerifolium** Willd. is found in the sub-Himalayan tract and outer Himalayan valleys and hills up to 4,000 feet, from the Jumna easterwards to Bengal, Chittagong, and Burma. It is extensively planted in the Bombay Presidency.

The down on the leaves is used by the hill-people in Sikkim to stop bleeding in wounds.

In the Konkan the flowers and bark, charred and mixed with kamala, are applied in suppurating small-pox.

The flowers are used as a general tonic. They are employed by Bengalis as a disinfectant and to keep insects away from bed clothes. They are also said to render water gelatinous.

Bengal: Kanakchampa, Muchokunda, Mus—; *Bombay*: Kanakchampa, Karnikara—; *Burma*: Tha-ma jam wai-soke, Toungpetwun—; *Canarese*: Kanakachampaka, Rajataru—; *Hindi*: Kanakchampa, Kanier, Kathachampa—; *Jaunsar*: Mayeng—; *Kachin*: Magwi napa—; *Kolami*: Muchukundi—; *Konkani*: Kanokchampo—; *Lepcha*: Numbung—; *Magahi*: Gaik—; *Michi*: Laider—; *Mundari*: Makcunddaru—; *Nepal*: Hattipaila—; *Philippines*: Bayoc—; *Sanskrit*: Karnikara, Mushukunda, Padapotpala, Parivyadha—; *Santali*: Machkunda—; *Tamil*: Vennangu—; *Telugu*: Matsakanda—; *Tharu*: Machkan—; *Uriya*: Konokochompa, Mushukundo.

2. **Pterospermum heyneanum** Wall. occurs in the Western Peninsula.

The leaves are used in leucorrhoea; they are smoked like tobacco.

Khond: Baili—; *Tamil*: Polavu—; *Telugu*: Duddika, Loluga, Machchakanda, Nolika—; *Uriya*: Bailo, Giringa, Machkunda—.

3. **Pterospermum suberifolium** Lam. is found in the Northern Circars, and Carnatic. It is sometimes planted in the Bombay Presidency.

The flower rubbed into paste with rice vinegar is an ancient and well known application for hemicrania. It is said to render water gelatinous.

Bengal: Muchkand, Muchukunda, Muskunda—; *Bombay*: Muchunda, Muchukunda—; *Burma*: Naji—; *Canarese*: Muchukunda—; *Ceylon*: Tada, Toddi, Vinanku, Vinnaku—; *Hindi*: Muchkand—; *Konkani*: Munchund—; *Marathi*: Muchkund—; *Sanskrit*: Arghyarhalakshanaka, Bahupatra, Chitraka, Dirghapushpa, Harivallabha, Kshatravrikha, Muchukunda, Mundivrikshamukaraka, Prativishnuka, Raktaprasava, Sudala, Supushpa—; *Sinhalese*: Velanga, Velenge, Venangu—; *Tamil*: Sembolavu, Sittilaipolavu, Tadai, Vennangu—; *Telugu*: Lolagu, Narudu, Tada—; *Uriya*: Baelo, Giringa, Guputi, Gureno—.

PTERYGOTA.

The genus consists of 4 palaetropical species.

Pterygota alata R.Br. (=*Sterculia alata* Roxb.) occurs in the Western Peninsula, Sylhet, Chittagong, Pegu, and Martaban down to Tenasserim. It is also found on the Andamans.

In Sylhet the seeds are used as a substitute for opium.

Assam: Tula—; *Burma*: Letkope—; *Canarese*: Bekaro, Jaynkatala—; *Chittagong*: Buddhanarikella—; *Malayalam*: Anattonti, Porila, Porutonti—; *Tamil*: Anaittondi, Kodaittondi—.

SCAPHIUM.

The genus consists of 4 species found in Martaban, the Malay Peninsula, and Borneo :

Leaves 15-20 by 10-12.5 cm. ... ... 1. *S. affine.*
Leaves 30-35 by 15 cm. ... ... 2. *S. Wallichii.*

1. **Scaphium affine.** Ridley is fairly common in the forests in the south of the Malay Peninsula.

The fruit acts as a demulcent.

The seed, immersed in water for a night, exudes a gum containing bassorin, to such an extent as to fill a small cup. This, mixed with sugar-candy, is eaten as a stomachic, and is very pleasant and wholesome.

Malay : Kembang semangkok, S'ilayer—.

2. **Scaphium Wallichii** Schott and Endl. (=*Sterculia scaphigera* Wall.) is common in the tropical forests along the eastern and central slopes of the Pegu Yomah and Martaban. It is also found in Chittagong, Tenasserim, and Malacca.

The fruit is used in China as a remedy for piles and dysentery, and for diseases of the lungs and the kidneys.

In Cambodia, Siam, and Malaya the fruit is macerated in water, when the outer shell or pericarp increases enormously in volume forming a large gelatinuous mass. The jelly is sweetened and eaten as a delicacy, and is said to be a specific in diarrhoea and dysentery.

Chinese : Ta Hai—; *Malaya* : Kembang semangkok, Tai hoi—; *Siam* : Bungtalai—.

STERCULIA.

The genus consists of 100 tropical species.

Many of the species yield gums in considerable quantity, most of which resemble tragacanth in their appearance and properties.

The following are used medicinally in Japan—*S. javanica* R.Br.—; in the Philippine Islands—*S. Balanghas* Linn., *S. foetida* Linn., *S. oblongata* R.Br., *S. rubiginosa* Vent, *S. stipularis* R.Br., *S. urens* Roxb.—; in Indo-China—*S. hypochroa* Pierre, *S. lychnophora* Hance, *S. platanifolia* Linn., *S. foetida* Linn., *S. Thorelii* Pierre—; in China—*S. platanifolia* Linn.—; in West Africa—*S. tomentosa* Guil. and Perr., *S. Tragacantha* Lindl.—; in Australia—*S. acerifolia* A. Cun., *S. diversifolia* G. Don., *S. rupestris* Benth.—; in Brazil—*S. chicha* St. Hil., *S. striata* A.St.Hil. and Naud.—.

A. Leaves digitate ... ... *S. foetida.*
B. Leaves palmately lobed ... ... *S. urens.*
C. Leaves simple, not lobed, 1-nerved.
Calyx-lobes narrow, incurved, frequently cohering at the tips, gaping at the sides.
1. Leaves pubescent beneath ... ... *S. Balanghas.*
2. Leaves with the nerves rusty-hispid beneath. *S. rubiginosa*

1. **Sterculia Balanghat** Linn. is found throughout the hotter parts of India, on the coasts of Tenasserim, in Ceylon, and the Andaman Islands.

In the Philippines the fruit is considered cooling and laxative.

French: Bois de cavalan—; *Malayalam*: Kavalam—; *Philippines*: Balanghas—; *Sinhalese*: Nawa—.

2. **Sterculia foetida** Linn. occurs in the forests of Western and South India at low elevations. It is also found in Burma, Ceylon, and the Malay Peninsula. It is distributed to East Tropical Africa, the Philippine Islands, and North Australia.

The oil from the seeds is a mild laxative and is used as a carminative. The seeds when swallowed incautiously bring on nausea and vertigo.

The leaves are repellent and aperient.

A decoction of the fruit is mucilaginous and astringent.

In Java the fruit is employed in gonorrhoea.

Bombay: Junglibadam, Pun—; *Burma*: Hlyampyu, Letkok, Shawbyu, Showbju—; *Canarese*: Bhatala, Bhutaili, Gotinakayi, Jaynkatala, Jenukayitaili, Kuduregotu, Penari—; *Ceylon*: Kaditeni, Pinari—; *French*: Arbre puant, Bois caca, Bis Bois de cavalam, Bois de merde, Bois puant—; *Goa*: Junglibaddam, Kuomhad, Virhoi—; *Hindi*: Jangalbadam, Janglibadam—; *Ilocano*: Bangar—; *Indo-China*: Chim chim rung, Trom—; *Jolo*: Caumpang—; *Konkani*: Cunvemruc, Viroi—; *Malay*: Kelompang—; *Malayalam*: Pinari, Pottakavalam—; *Marathi*: Goldaru, Janglibaddam, Nagalkuda—; *Muduva*: Kaludaivitte—; *Pampangan*: Calumpan—; *Portuguese*: Puna, Puna bastarda—; *Pulaiya*: Kesalamaram—; *Sinhalese*: Telambu—; *Tagalog*: Calompan, Calumpag, Calumpang—; *Tamil*: Arali, Attirappidukkan, Kudiraippidukku, Malaittengai, Pinari, Pudagarappanbattai—; *Telugu*: Guttapubadamu, Manjiponaku, Piyyattiponaku—; *Tulu*: Pinari—; *Visayan*: Bobag—.

3. **Sterculia rubiginosa** Vent. is common in woods in open country in Burma and the whole of the Malay Peninsula from Singapore to Penang. It is distributed to Cochin-China, Java, and Sumatra.

The fruit is used as a mild laxative.

Malay: Dundanak, Saburu, Sakelat, Unting-unting besar—; *Tagalog*: Nato—.

4. **Sterculia urens** Roxb. occurs in North-Western India, Assam, Behar, the Eastern and Western Peninsulas, and Ceylon.

The leaves and tender branches steeped in water yield a mucilaginous extract, useful in pleutro-pneumonia in cattle.

The gum, known as *karai-gond*, is used as a substitute for tragacanth in Bombay.

The Santals consider the gum a useful medicine in throat affections.

In Philippines the root bark is pounded, made into a poultice, and applied to wounds, and fractures; it is also used in orchitis.

Ajmere: Kalru, Katila—; *Assam*: Hatchanda, Odla—; *Banda*: Kulu—; *Berar*: Guru, Kairu, Karai, Kulu, Taklej—; *Betul*: Guru, Karai, Kulu—; *Bombay*: Gular, Gulu, Gwira, Kadai, Kandai, Kando, Kandol, Kandula, Kawalee, Kullin, Kulu, Pandruka, Sardol, Sardora—; *Canarese*: Bhutali, Kempudaili, Kempudala, Kempudale, Pinari, Punike—; *Central Provinces*: Gulu, Gurlu, Karhar, Kulu—; *Ceylon*: Kavali—; *Deccan*: Kurdu—; *French*: Sterculia démangeant, Touroutier—; *Gond*: Hittum, Pinoh—; *Gujerati*: Kada, Kagdol, Karai—; *Hindi*: Bali, Gular, Gulu, Kabru, Karrai, Katira, Kuli, Kulu, Tabsi, Tanuku—; *Jeypore*: Kaddu, Karr—; *Khandesh*: Kud—; *Khawar*:

Kaunji—; *Khond*: Gauduli, Gendule—; *Kolami*: Feley, Kaunji, Teley—; *Konkani*: Pandruk—; *Kurku*: Takli—; *Malayalam*: Tonti—; *Manbhum* Keonge—; *Marathi*: Gwira, Kandol, Kandula, Karai, Kavali, Pandruka, Soldawar—; *Matheran*: Bandruk, Kaoli, Kuari, Sardol—; *Melghat*: Teklej—; *Merwara*: Kor—; *Monghyr*: Kanaunji, Karaunji, Mogul—; *Mundari*: Burkunda, Gurkaranj, Makchund—; *Nimar*: Guru, Karai, Kulu—; *North-Western Provinces*; Gulli, Kuli—; *Panch Mahals*: Kalauri—; *Porebunder*: Kadai, Kadayo—; *Santali*: Telhec'—; *Tagalog*: Banilad—; *Tamil*: Kavalam, Puttali, Sendalai, Sendanakku, Sigapputtanakku, Tanakku, Vellaipputtali—; *Telugu*: Ettaponaku, Kavali, Kavili, Ponaku, Tanuku. Tapasi—; *Uran*: Keunji—; *Uriya*: Girungilia, Gudalo, Kavili—; *Visayan*: Banilad—.

TARRIETIA.

The genus consists of 7 species inhabiting Malaya, Cochin-China, and Australia.

Tarrietia unifoliolata Ridl. occurs in Malacca. The plant is used medicinally by the Malays.

Malay: Kamuning hutan—.

THEOBROMA.

The genus consists of 20 species, natives of tropical America.

The seeds are analeptic. They yield a fixed oil known as 'oil of Theobroma' or 'Cacao Butter', and they are extensively used for the extraction of the alkaloid 'theobromine'.

The species used are all originally South American: *T. bicolor* H. and B., *T. Cacao* Linn., *T. glauca* Krst., *T. martiana* D. Dietr., *T. macrantha* Bernoulli. *T. microcarpum* Mart., *T. syvestris* Aubl.—

Theobroma Cacao Linn. has long been introduced into India, and is cultivated in the Southern Presidency and Ceylon.

A light yellowish, opaque, solid oil, known as 'Cacao butter', is prepared for use in pharmacy, by pressing the warmed seeds. It is officinal in the pharmacopœcias of Austria, Belgium, Brazil, Denmark, Finland, France, Germany, Great Britain, Holland, Hungary, Italy, Japan, Mexico, Norway, Portugal, Russia, Spain, Sweden, Switzerland, Turkey, United States, Yugoslavia.

The seeds contain from 1.5 to 2.4 per cent theobromine.

Ceylon: Chocolathgas, Cocomaram—; *Dutch*: Cacaoboom, Kakauboom—; *English*: Cacao, Chocolate Tree, Cocoa—; *French*: Cabasse, Cabosse, Cacao, Cacaotier, Cacaoyer—; *German*: Kakaobaum—; *Guam*: Kakao—; *Hungarian*: Kakao, Kokuszfa—; *Italian*: Albero del Cacao, Cacao—; *Mexican*: Cacacoatl—; *Portuguese*: Cacao—; *Roumanian*: Cacaotier—; *Russian*: Kakaovoi derevo—; *Spanish*: Arbol del cacao, Cacao, Haba de Mejico—; *Swedish*: Kakao—; *Turkish*: Kakao—.

WALTHERIA.

The genus consists of 30 species, mostly American.

W. Douradinha A. St. Hil. is used medicinally in Brazil.

Waltheria indica Linn. is a widely diffused tropical weed, common in all the hotter parts of India, from Kumaon at 4,000 feet to Malacca and Ceylon.

In Mauritius the plant is credited with emollient and bechic properties, and is used as a cough medicine.

In West Tropical Africa the plant is a common medicine for infants. In Northern Nigeria the infant is given to drink or to sniff or inhale a decoction of the root boiled along with native natron. In Yoruba and in Togo the plant is a constituent of various infusions, both to drink or as wash for a few weeks to strengthen the child's resistance against fevers and other troubles.

Amongst the Hausas the root is considered to be purgative; also a decoction of the plant is believed to be a preventive of syphilis, and, if taken frequently, to afford immunity. It is also drunk by agriculturists as a restorative during the labours of harvesting.

In Togo a spoonful of the pulverised plant with hot water is taken morning and evening as a cough medicine.

In Gold Coast the plant is used to cause abortion.

The Thangaan women take a decoction of the root for barrenness.

The Sutos either chew the raw root or drink a decoction of it for internal haemorrhages.

Bambara: Dabada—; *Fulani*: Kafaffi, Kapappi—; *Ga*: Fufuñoba, Nansuru—; *Gbari*: Nyiknapmiyaya—; *Hausa*: Hankufa, Yankufa—; *Malinke*: Bourai—; *Mauritius*: Guimauve, Guimauve créole, Toutti—; *Mende*: Ngala-yuyui—; *Shangaan*: Simbongana—; *Suto*: Lexutasela—; *Telugu*: Nallabenda—; *Togo*: Adu weti, Bati—; *Twi*: Apem—; *Uriya*: Bojoromuligna—; *Wolof*: Bourai—*Yoruba*: Korikodi, Opa-emere—.

INDEX TO BOTANICAL NAMES OF PLANTS

Indian plants are in **bold**, their synonyms in *italics* and others in roman. Family names are in CAPITALS.

Abroma 483, 484
augusta 484
Abuta 166
ALutilon 445, 446
asiaticum 446
avicennae 446, 448
exstipulare 446
glaucum 446
hirtum 446, **447**
indicum 446, **447,** 448
mauritianum 446
theophrasti 446, **448**
Acalypha 182, 183, 184, 185
ciliata 185
fruticosa 185, **186**
hispida 185, **186**
indica 185, **186**, 187
integrifolia 185
ornata 185
paniculata 185, 186, **187**
peduncularis 185
petiolaris 185
siamensis 185, 186, **187**
spiciflora 185
Acampe 100, 101, 102
wightiana 102
Achthopoenix 1
nobilis 2
rubra 3
Acanthospermum 302, 305
brasilium 309
hispidum 309
Achillea 302, 305, 307, 310
ageratum 310
atrata 310
millefolium 310
moschata 310
nana 310
nobilis 310
ptarmica 310
santolina 310, **312**
Achlys 177
Achyrachaena 302
Achyranthes aspera 23
Aconitum 137, 138
anthora 138
balfourii 138, 139, **141**
chasmanthum 138, 139, **140**
cammarum 138
deinorrhizum 138, 139, **141**
elwesii 140, **142**
falconeri 140, **142**
ferox 140, **142**
fischeri 138
heterophyllum 139, **141**
kusnezoffii 138
laciniatum 140, **142**
lethale 140, **142**
luridum 139, **140**
lycoctonum 138, 139
napellus 138
palmatum 139, **141**
spicatum 140, **142**
uncinatum 138
variegatum 138
violaceum 139, **140**
Acorus 109, 110
calamus 110
gramineus 110, **112**
Acrocephalus 381, 385
indicus 385
lilacinus 385
Acrostichum 34
Acrotome 381
Actaea 137, 138, 143
alba 143
eburnea 143
spicata 143
Actiniopteris 34, 35, 36
dichotoma 36

Actinomeris 302
Adansonia 433
  digitata 434
  grandidieri 434
  gregorii 434
  madagascariensis 434
  za 434
Adenostemma 302, 305, 312
  viscosum 312
Adiantum 34, 35, 36
  aethiopicum 37, **40**
  capillus-veneris 37 **39**
  caudatum 37, **38**
  cuneatum 37, **41**
  emarginatum 40
  flabellulatum 41
  lunulatum 37
  monochlamys 37
  pedatum 37, **41**
  radiatum 37
  subcordatum 37
  tenerum 37, **39**
  venustum 37, **38**
Adonis 137, 138, 144
  aestivalis 144
  autumnalis 144
  microcarpa 144
  vernalis 144
Aeolanthus 381
Agastache 381
Ageratum 302, 305, 306, 312
  conyzoides 312
Aglaonema 109, 110, 113
  angustifolium
    var. **undulatum 113**
Agropyron 55, 56, 58, 59
  repens 59
Agrostema 129
Ajuga 381, 385
  bracteosa 385
  chamaepitys 385
  iva 385
  ophrydis 385
  reptans 385
Albizzia
    odoratissima 288
Alchornea 182, 183, 184, 187
  cordifolia 187
  hirtella 187
  floribunda 187
  rugosa 187
Aleurites 182, 183, 184, 188
  cordata 188
  moluccana 188
Allanblackia 426
Alliaria 247, 248, 249
  officinalis 249
Allium macleanii 102
Alocasia 109, 110, 113
  denudata 113
  indica 113
  macrorrhiza 113, **114**
  montana 113, **114**
  singaporensis 113
Alphonsea 159, 160
  ventricosa 160
Alsodeia 278, 279
  echinocarpa 279
Alsophila 34
Althaea 445, 448
  cannabina 448
  hirsuta 448
  ludwigii 448
  officinalis 448
  pallida 448
  rosea 448, **449**
Alyssum 247, 248, 249
  maritimum 249
Amaracus 381
Ambrosia 302
Amorphophallus 109, 110, 114
  campanulatus 114
  dracontioides 114
  flavovirens 114
  johnsonii 114
  prainii 114, **115**
  rivieri 114
Amphidixa 302
Amphilophis 55, 56, 57, 60
  odorata 60
Anacampseros 296
Anacamptis 100
Anacyclus 302

Anadendrum 109, 110, 115
montanum
var. **longirostris** 115
Anamirta 166, 167
paniculata 167
Anaphalis 302, 305, 306, 313
neelgerryana 314
margaritacea 313
Anaxagorea 159, 161
scortechinii 161
Anatherum 55
Ancistrocarpus 473
Anchietea 278
Anchomanes 109
Andrachne 182, 183, 188
cordifolia 188
ovalis 188
Andropogon 55
citratus 69
contortus 76
halepensis 92
jwarancusa 69
laniger 69
muricatus 97
nardus 69
odoratus 60
schoenanthus 69
sorghum 93
squarrosus 97
Aneilema
beninense 124
conspicuum 124, **125**
lineolatum 124, **125**
nudiflorum 124, **125**
ovato-oblongum 124
scapiflorum 124, **125**
Anemone 137, 138, 144
alpina 144
acutiloba 144
altaica 145
caffra 145
caronaria 144
cernua 145
cylindrica 145
dichotoma 145
hepatica 144, 145
japonica 145
multifida 145
narcissiflora 144, **145**
nemorosa 144, 145
nikoensis 145
obtusiloba 145
parviflora 145
patens 145
pratensis 144, 145
pulsatilla 144, 145
ranunculoides 144, **145**
reflexa 145
stellata 144
sylvestris 144
thalictroides 145
trifolia 144
transvaalensis 145
vernalis 144
virginiana 145
Angraecum 100
Anisochilus 381, 383, 385
carnosus 385
Anisomeles 381, 383, 385
indica 386
malabarica 386
Anneslea 438, 439
fragrans 439
Anona 160, 161
ambotay 161
cherimolia 161
marogravii 161
muricata 161, **162**
palustris 161
pisonis 161
reticulata 161, **162**
senegalensis 161
spinescens 161
squamosa 161
Anomospermum 161
ANONACEAE 159
Ansellia 100
humilis 102
Antennaria 302
Anthemis 302, 305, 307, 314
arvensis 314
cotula 314
gayana 314
nobilis 314

odontostephana 314
wiedemanniana 314
Anthephora 55
Anthoxanthum 55, 56, 60
odoratum 60
Anthurium 109
Antidesma 182, 183, 184, 189
bunius 189
ghaesembilla 189
madagascariense 189
membranaceum 189
venosum 189
zeylanicum 189
Aphloia 287
Aplopappus 302
Aplostellis 100
Aporosa 182, 183, 184, 190
lindleyana 190
Aquilegia 137, 138, 145
alpina 145
canadensis 145
formosa 145
kitaibelii 145
parviflora 145
pyrenaica 145
vulgaris 145
Arabidopsis 247, 248, 249, 250
thaliana 250
Arabis 247
ARACEAE 109
Arctium 302, 305, 308, 315
lappa 315
Areca 1, 3
catechu 3
concinna 3, **7**
dicksonii 24
nagensis 3, **7**
rubra 3
Arenaria 130
serpyllifolia 130
Arenga 1, 2, **7**
engleri 7
obtusifolia 7
saccharifera 8
westerhoutii **7**
Argemone 237
alba 237
mexicana 237
Arisaema 109, 115
japonicum 109, 115
leschenaultii 115
ringens 115
speciosum 116
thunbergii 115
tortuosum 115, **116**
triphyllum 115
Arisarum 109
Aristida 55, 56, 61
adscenscionis 61.
Ârnica 302
Artabotrys 160, 163
odoratissimus 163
suaveolens 163
Artemisia 302, 305, 308, 316
abrotanum
absinthium 316, **317**
afra 317
annua 317, **318**
apiacea 317
arborescens 316
brevifolia 320
caerulescens 316
californica 317
campestris 316
capillaris 317
carvifolia 317, **318**
cina 316
dracunculoides 317
dracunculus 316, 317, **319**
gallica 316
glacialis 316
heiskeana 317
herba-alba 316
hispanica 316
japonica 316
judoica 316
ludviciana 317
maritima 316, 317, **319**, 320
mexicana 317
mutellina 316
persica 317, **320**
pontica 316
ramosa 317

sacrorum 317, **320**
santonicum 317
scoparia 316, 317 **320**
sina 317
siversiana 317, **320**
spicata 316
stelleriana 317
tridentata 317
vahliana 316
vallesiana 316
vulgaris 316, 317, **321**
Arthraxon 55, 56, 57, 61
ciliaris 61
Arum 109
Arundinella 55, 56, 57, 61
brasiliensis 61
Arundo 55, 56, 58, 62
donax 62
Asimina 160
Aspidium 34, 35, 41
aculeatum var. pungens 41
athamanticum 41
filix-mas 48
falcatum 41
fragile 41
marginale 41
polymorphum 41, **42**
roeticum 41
spinulosum 41
trifoliolatum 41
wildeanum 48
Aspilia 302
Asplenium 34, 35, 42
adiantum-nigrum 42, **43**
ceterach 49
cuneatum 42
falcatum 42, **43**
furcatum 42, **44**
monanthemum 42
regulare 42
ruta-muraria 42, **43**
sulcatum 42
trichomanes 42
Aster 302, 305, 306
erigeroides 322
fastigiatus 322
filifolius 322
hispidus 322
puniceus 322
scaber 323
serrulatus 322
tataricus 322
trinervius 322
Asteriastigma 287
macrocarpa 287
Astrocaryum 1
Athanasia 302
Athrixia 302
Athyrium 34, 35, 44
dichotoma 36
filix-foemina **44**
Atractylis 308
Attalea 1, 2, 3
speciosa 8
Audibertia 381
Avena 55, 56, 58, 62
fatua 62
sativa **62**
var. **orientalis 63**
Baccharis 303
Badiera 293
Balantium 34
Baliospermum 182, 183, 185,
axillare 190
montanum 190
Ballota 381
Bambusa 55, 56, 59, 63
arundinacea 64
blumeana 64, **66**
vulgaris 64
Barbarea 247, 248, 250
vulgaris 250
Beckmannia 55
Bellis 303
Bentonica 381
BERBERIDACEAE 177
Berberis 177
aquifolium 177
aristata 177, 178, **179**
asiatica 178, **180**
aetnensis 178
buxifolia 178

glauca 178
japonica 178
lycium 178, **180**
nervosa 178
pinnata 178
sieboldii 178
trifoliatus 178
vulgaris 177, **178**
Bergia 444
decumbens 444
guineensis
odorara 444
Berkheya 303
Berteroa
Biarum 109
Bidens 303, 305, 307, 322
bipinnata 322
graveolens 323
pilosa 322 **323**
tripartita 323
Bigelowia 303
Bischofia 182, 183, 184, 190
javanica 190
Bixa 285
orellana 285
BIXACEAE 285
Blechnum 34, 44
orientale 44
Blepharipappus 303
Blumea 303, 305, 306, 323
aurita 323
balsamifera 324
chinensis 324
densiflora 324
eriantha 323, **324**
lacera 323, **324**
myriocephala 324, **325**
Bletia 100
Bocagea 160
Bocconia 237
Bojeria 303
Boltonia 303, 305, 306, 325
indica 325
BOMBACACEAE 433
Bombax 433, 436
brevircuspe 436
buonopozense 436
ceiba 436
malabaricum 436
Bongardia 177, 180
rauwolfii 180
Borassus 1, 2, 9
flabellifer 9
var. aethiopum 9
gomutus 8
Boscia 267
Botrychium 34, 36
lunaria 45
ternatum 45
virginianum 45
Brachiaria 55
Brachyclados 303
Brachylaena 303
Brassica 247, 248, 250
campestris 250, 251, **252**
var. **rapa 252**
cernua 250 251, **253**
intcgrifolia 250, 251, **253**
napus 250, **251**
var. **chinensis 252**
nigra 250, 251, **253**
oleracea 250, **251**
rapa 250
Brasenia 232
schreberi 233
Bredemeyera 292
Breynia 182, 183, 184, 191
patens 191
rhamnoides 191
Brickellia 303
Bridelia 182, 183, 191
atroviridis 191
cambodiana 191
ferruginea 191
micrantha 191
montana 191, **192**
retusa 191, **192**
schlechteri
Bromus 55, 56, 58, 66
catharticus 66
mollis 66
purgans 66
Brunella 381, 384, 387

vulgaris 387
Buettneria 483, 485
herbacea 485
Bufonia 129
Bunias 247
Burasaia 166
Buxus 182, 183, 192
sempervirens 192
Cadaba 267
farinosa 267
indica 267
trifoliata 267, **268**
Caesalpinia sappan 196
Cakile 247
Caladium 109
Calamintha 381, 383, 388
acinos 388
alpina 388
clinopodium 388
grandiflora 388
nepeta 388
officinalis 388
Calamus 1, 2, 11
draco 11
margaritae 11
rheedei 11, **12**
rotang 11
travancoricus 11, **12**
verus 11
Calea 303
Calendula 303, 305, 307, 326
arvensis 326
officinalis 326
Calla 109
Callilepis 303
Caloncoba 287
Calophyllum 426
apetalum 427
brasiliense 427
calaba 427
elatum 427
inophyllum 428
laxiflorum 427
mariae 427
parviflorum 427
tacamahaca 427
Calotropis procera 227
Caltha 137, 138, 146
palustris 146
Camelina 247
Camellia 438, 439
drupifera 439
japonica 439
thea 439
Canangium 160, 164
latifolium 164
odoratum 164
CAPPARIDACEAE 267
Capparis
aphylla 270
citrifolia 268
coriaceae 268
decidua 268, **270**
grandis 268, **270**
gueinzii 268
heyneana 268, **270**
horrida 271
micracantha 268, **271**
sepiaria 268, **270**
spinosa 267 268, **269**,
var. canescens 269
tomentosa 268
yeo 268
zeylanica 268, **270**
Capsella 247. 248, 249, 255
bursa-pastoris 255
Carbenia 303
Cardamine 247, 248, 254
alpina 254
amara 254
asarifolia 254
gelida 254
hirsuta 254
impatiens 254, **255**
parviflora 254
pratensis 255
resedifolia 254
sylvatica 254
trifolia 254
Cardopatium 303
Carduus 303, 305, 308 327
crispus 327
nutans 327
Carex 26
arenaria 26

cernua 27
distícha 26
hirta 26
macrocephala 26
siderosticta 26
Carissa carandas 194
Carlina 303
Carpesium 303, 305, 306, 328
abrotanoides 328
cernuum 328
Carpolobia 293
Carthamus 303, 305, 309, 328
lanatus 328
oxyacantha 328
tinctorius 329
CARYOPHYLLACEAE 129
Caryota 1, 2, 12
mitis 12, **13**, 82
urens 12
Cassia occidentalis 271
Caulophyllum 177
Ceiba 433, 437
pentandra 437
Cenchrus 55, 56, 57, 66
catharticus 66
Cenia 303
Centaurea 303, 305, 309, 330
alexandrina 330
behen 330
calcitrapa 330
centaurium 330
cyanus 330, **331**
iberica 330
jacea 330
montana 330
phyllocephala 330
praecox 330
picris 330, **331**
rhizocephala 330
solstitialis 330
Centipeda 303, 305, 308, 331
robicularis 331
Centratherum 303, 305, 332
anthelminticum 332
Cerastium 129, 130
vulgatum 130
Ceratopteris 34, 35, 46
thalictroides 46
Cephalostachyum 55, 56, 59, 66
capitatum 66
Chamaerops 1, 2, 13
excelsa 25
humilis 13
Cheilanthes 34, 35, 46
hirta 46
tenuifolia 46
Cheiranthus 247, 248, 256
cheiri 256
Chelidonium 237
Chloris 55, 56, 58, 67
petraea 67
virgata 67
Chondrilla 303
Chrozophora 182, 183, 184, 193
plicata 193
prostrata 193, **194**
rottleri 193, **194**
senegalensis 193
Chrysanthellum 376
indicum 376
Chrysanthemum 303, 305, 308, 333
balsamita 333
cinerariaefolium 333, 334
Coronarium 333, **334**
decaisneanum 333
indicum 333, **334**
leucanthemum 333, **334**
marschallii 333
parthenium 333
sinense 333, 334
vulgare 333
Chrysocoma 303
Chrysopogon 55, 56, 57, 67
serrulatus 67
Cibotium 34, 35, 46
barometz 46
glaucum 46
Cicca 182, 183, 184, 194
disticha 194
Cichorium 333, 305, 309, 334
endivia 334
intybus 334, **335**
noeanum 334, **336**
Cienfuegosia 443
Cimicifuga 137, 138, 147
foetida 147
racemosa 144, 147

Cineraria 303
Cissampelos 166, 167
angustifolia 168
caapeba 168
capensis 168
mauritiana 168
mucronata 168
ovalifolia 168
owariensis 168
pariera 168, **169**, 173
torulosa 168
Cistanthera 473
Cleistanthus 182, 183, 194
collinus 194
Clematis 137, 138, 147
alpina 147
apiifolia 147
bojeri 147
brachiata 147
chinensis 147
crispa 147
dioica 147
flammula 147
florida 147
grandiflora 147
gouriana 147, **148**
graveolens 147, **148**
integrifolia 147
lingusticifolia 147
mauritiana 147
napaulensis 147, **148**
paniculata 147
patens 147
recta 147
saxicola 147
smilacifolia 147, **148**
stanleyi 147
stewartiae 147
triloba 147, 148
thunbergii 147
vitalba 147
viticella 147
virginiana 147
viorna 147
Clematopsis 137
Cleome 267, 271
brachycarpa 272
chelidonii 272, **274**
ciliata 272
felina 272, **274**
frutescens 272
gigantea 272
graveolens 272
heptaphylla 272, **274**
monophylla 272
serrata 272
viscosa 272, **273**
Clibadium 303
Clusia 426
Clypeola 247
Cnesmone 182
Cnicus 303, 305, 308, 336
arvensis 336
benedictus 336
joponicus 336
sinensis 336
spicatus 336
Cocculus 166, 167, 170
hirsutus 170
laurifolius 170, **171**
macrocarpus 170, **171**
pendulus 170, **171**
thunbergii 170
trilobus 170
Cochlearia 247, 248, 257
armoracia 257
flava 257
officinalis 257
COCHLOSPERMACEAE 284
Cochlospermum 284
gossypium 284
insigne 284
tinctorium 284
Cocos 1, 2, 13
coronata 13
flexuosa 13
nucifera 13
schizophylla 13, **16**
yatai 13
Coix 55, 56 67
lachryma-jobi 67
Cola 483
Colebrookea 381, 383, 388
oppositifolia 388
Coleosanthus 303
Coleus 381; 389
amboinicus 389
atropurpureus 389
bojeri 389

dysentericus 389
floribundus 389
Colocasia 109, 110, **116**
antiquorum 116
esculenta 116
Commelina 124, 125
africana 126
barbata 126
benghalensis 126
communis 125
deficiens 126
madagascarica 126
nudiflora 126
obliqua 126, **127**
pallida 125
salicifolia 126, **127**
suffruticosa 127
tuberosa 126
COMMELINACEAE 124
COMPOSITAE 302
Conyza 303
Copernicia 1, 2, 17
cerifera 17
Coptis 137, 138, 148
anemonaefolia 148
teeta 148, **149**
trifolia 148
Corchorus 473
acutangulus 473, **474**
asplenifolius 473
capsularis 473, **474**
depressus 474, **475**
fascicularis 474, **475**
olitorius 473, 474, **475**
serraefolius 473
trilocularis 474, **476**
Corchoropsis 473
Corydalis 244
ambigua 244
bulbosa 244
claviculata 244
fabacea 244
govaniana 244
incisa 244
lutea 244
ochroleuca 244
ramosa 244, **245**
Corypha 1, 2, 17
pilcaria 18
umbraculifera 17
Coscinium 166, 167, 171
fenestratum 171
Cotula 303, 305, 308, 336
aurea 336, **337**
anthemoides 336
coronopifolia 336
villosa 336
Courbonia 267
Crambe 247, 248, 249, 257
cordifolia 257
hispanica 257
maritima 257
Crataeva 267, 274
greveana 274
nurvala 274
religiosa 274
tapia 274
Cratexylon 441
hornchuchii 441
neriifolium 441
Cremastra 100
Crepis 303, 305, 309, 337
acaulis 337
Crossostephium 303
Crossotropis 55
Croton 182, 183, 184, 195
amabilis 195
antisiphiliticus 195
argyratus 195
californicus 195
campestris 195
cascarilla 195
caudatus 195, **196**
cumingii 195, **196**
dioicus 195
draco 195
elliotianus 195
eluteria 195
flavens 195
fulvus 195
glabellus 195
gratissimus 195
gubouga 195
leptostachyus 195
lobatus 195
lucidus 195

macrobuxus 195
maieuticus 195
malasubo 195
niveus 195
oblongifolium 195, **196**
origanifolius 195
penduliflorus 195
polianei 195
reticulatus 195, **196**
suberosus 195
sylvaticus 195
tiglium 195, **197**
tonkinensis 195
vulpinus 195

CRUCIFERAE 247

Cryptocoryne 109, 110, 117
spiralis 117
Cryptolepis 204
buchanani 204
Cucubalus 129, 131
bacciferus 131
Culcasia 109
Cunila 381, 383
Curotella 158
Cyanotis 124, 127
axillaris 127, **128**
tuberosa 128, **128**
Cyclea 166
Cyclophorus 34
Cymbidium 100, 102
aloifolium 103
Cymbopogon 68
citratus 68, 69, **70**
excavatus 68
jwarancusa 68, **69**
marginatus 68
nardus 68, 69, **70**
schoenanthus 68, **69**
validus 68
winterianus 68
Cynaria 303
Cynodon 55, 56, 58, 71
bradleyi 71
dactylon 71, 482
Cynosurus 55
Cyperus 26, 27
aequalis 27
alboviridis 27
articulatus 27, **28**
elegans 27
esculentus 27, 30, 33, 297
fastigiatus 27
iria 27
longus 27, **28**
platyphyllus 31
rotundus 27, **29**
stoloniferus 30
sexangularis 27
scariosus 27, **28**
Cypripedium 100
Cyrtopodium 100
Cystopteris 34, 35, 47
fragilis 47
Dactyloctenium 55, 56, 58, 73
aegyptium 73
Dahlia 303
Dalechampia 182
Daphniphyllum 182
Davallia 34
tenuifolia 54
Davilla 158
Delphinium 137, 138, 149
ajacis 149
brunonianum 149, **150**
caeruleum 149, **150**
consolida 149
denudatum 149
elatum 149, **150**
grandiflorum 149
hybridum 149
laxiflorum 149
montanum 149
occidentale 149
orientale **149**
palmatifidum 149
peregrinum 149
pubescens 149
Staphisagria 149
triste 149
Dendrobium 100, 103
ceraia 103
chlorops 103
macraci 103
moniliforme 103
nobile 103,
ovatum 103

pumilum 103
reptans 103
Dendrocalamus 55, 56, 59, 73
strictus 73
Dentaria 247
Descurainia 247, 248, 257
sophia 257
Desmos
chinensis 164
cochinchinensis 164
Desmostachya 55, 56, 58, 74
bipinnata 74
Desmotrichum 100, 103
fimbriatum 103
Dianthus 129, 131
anatolicus 131
armeria 131
barbatus 131
carthusianorum 131
caryophyllus 131
chinensis 131
crenatus 131
plumarius 131
scaber 131
superbus 131
sylvestris 131
Diarrhena 55
Dicentra 244
Dichostemma 182
Dichrocephala 303, 305, 306, 337
latifolia 337
Dicksonia 34
Dicoma 303, 305, 309, 337
anomala 337
capensis 337
speciosa 337
tomentosa 337, **338**
zeyheri 337
Dieffenbachia 109
Digitaria 55
DILLENIACEAE 158
Dillenia 158
baillonii 158
indica 158
ovata 158, **159**
Dimorphotheca 303
Dioncophyllum 287
Diphylleia 177
Diplotaxis 247
Diplothemium 1
DIPTEROCARPACEAE 422
Dipterocarpus 422
affinis 422
alatus 423
artocarpifolius 423
gracilis 422
grandiflorus 422
intricatus 423
jourdainii 423
littoralis 422
pilosus 422
polyspermus 422
retusus 422
spanoghei 422
trinervis 422
tuberculatus 423
turbinatus 423
vernicifluus 422
Domleya 483
Doronicum 303, 305, 308, 338
austriacum 338
columnae 338
falconeri 338
hookerii 338
pardalianches 338
plantagineum 338
roylei 338
Draba 247, 248, 258
muralis 258
Dracocephalum 381, 389
moldavica 389
Dracontium 109
Dracunculus 109
Drimys 155
Drymaria 129, 130, 132
cordata 132
glandulosa 132
Drymoglossum 34, 36, 47
carnosum 47
Drynaria 34, 47
quercifolia 47
Dryobalanops 422, 424
aromatica 424
Dryopteris 34, 48
athamontica 48

crassi-rhizoma 48
filix-mas 48
inaequalis 48
marginalis 48
rigida 48
sophoroides 48
Drypetes 182
Dysophylla 389, 381, 383
auricularia 390
Echinacea 303
Echinochloa 55, 56, 57, 74
colona 74
var. **frumentacea 74**
crus-galli 74, **75**
Echinops 303, 305, 308, 338
dahuricus 338
echinatus 338
longifolius 338
persicus 338
ritro 338
Eciipta 303, 305, 307, 339
alba 339
Elaeis 1, 2, 17
guineensis 17
melanococca 17
Elaeocarpus 473, 476
ganitrus 476
madapolatus 476
oblongus 476, **477**
photiniaefolia 476
serratus 476, **477**
tuberculatus 477
Elaecphorbia 182, 187
ELATINACEAE 444
Elephantopus 303, 305, 341
scaber 341
Eleusine 56, 58, 75
aegyptiaca 73
coracana 75
indica 75, **76**
Elionuchloa 55, 56, 57, 74
Elionurus 55
Elsholtzia 381, 383, 390
cristata 390
Elymus 55
Elytropappus 303
Emilia 303, 305, 308, 342
adscendens 342
flammea 342
graminea 342
sagittata 342
sonchifolia 342
Enantia 160
Endostemon 381
Enhydra 303, 305, 307, 342
fluctuans 342
Epaltes 303
Epidendrum 100
Epimedium 177
Epipremnum 109, 100, 117
giganteum 117
mirabile 117
Eragrostis 55
cynosuroides 74
Eranthis 137
Erechtites 303
Eremocarpus 183
Eremostachys 381, 384, **390**
acanthocalyx 390
vicaryi 390
Erigeron 303, 305, 306, 343
aegyptiacus 343
annuus 343
asteroides 343
canadensis 343
philadelphicus 343
Eriodendron anfractuosum 437
Eriolaena 486
quinquelocularis 486
Eriocephalus 303
Eriophorum 26
Erophila 247, 248, 258
vulgaris 258
Eruca 247, 248, 258
sativa 251, **259**
Erythrochlamys 381
Erythrocucca 183
Erysimum 247, 248, 259
repandum 259
Eschscholzia 237
Ethulia 303, 305, 343
conyzoides 343
Euadenia 267
Eugenia jambolana 234
Eulophia 100, 104
arenaria 104

campestris 104
flaccida 104
herbacea 104
hians 104
nuda 104, **105**
pratensis 104, **105**
robusta 104

Eupatorium 303, 305, 306
ageratoides 344
aromaticum 344
aromatisans 344
cannabinum 344
chinense 344
collinum 344
glutinosum 344
incarnatum 344
japonicum 344
nervosum 344
parviflorum 344
perfoliatum 344
purpureum 344
scabrum 344
stoechadosmum 344
teucritolium 344
triplinerve 344, **345**
villosum 344

Euphorbia 183, 198
abyssinica 199
adenochlora 198
aegyptiaca 198
albomarginata 198
aleppica 198
amygdaloides 198
anagalloides 198
antiquorum 198, **200**
apios 198
atoto 198, 199, **201**
balsamifera 198
basutica 199
bojeri 198
brasiliensis 198
bupleurifolia 199
buxifolia 198
caecorum 198
canariensis 198, 199
candelobrum 199
cerebrina 198
chamaesya 198
chamaesyce 198
characias 198
clavarioides 199
continoides 198
convolvuloides 198
cotinifolia 198
coralloides 198
corollata 198
cyparissias 198
dendroides 198
dissimilis 198
dracunculoides 200, **201**
drummondii 198
edulis 198
emirnensis 198
erythroxyloides 198
esula 198
gerardiana 198
glyptosperma 198
goliana 198
granulata 199, **201**
helioscopia 198, 199, 200, **201**
heterodoxa 198
heyneana 198
hirta 198, 199, **202**
humifusa 198
humistrata 198
hybernia 198
hypericifolia 198, 199, **204**
ipecacuanha 198
irigens 199
kamerunica 198
lateriflora 198
lathyris 198
laurifolia 198
leonensis 198
linearis 198
longifolia 200, **204**
maculata 198
marginata 198
mauritanica 199
microphylla 199, **204**
myrsinites 198
neriiflora 198
neriifolia 199, 204
nivulia 200, **205**
ocellata 198
officinarum 198
palustris 198
papillosa 198

parviflora 198
pekinensis 198
peplus 198, 199
phosphorea 198
pilulifera 20
poissonii 198
polycnemoides 198
portulacoides 198
preslii 198
procumbens 199
prostrata 198, 199
pulcherrima 198
punicea 198
resinifera 198, 199
restituta 199
rosea 198, 199, **206**
royleana 206
sanguinea 199, **206**
sapini 199
serpyllifolia 198
serrata 198
sieboldiana 198
spinosa 198
splendens 198
striata 199
thomsoniana 200, **206**
thymifolia 198, 199, **206**
tirucalli 198, 199, **207**
trichophylla 198
trigona 200, **208**
turcomanica 199, **208**
truncata 199
virosa 199

EUPHORBIACEAE 182

Euptelea 155

Euryale 232, 233
ferox 233
Euryops 303
Euterpe 1
Evolvulus alsinoides 410

Excoecaria 183, 185, 208
acerifolia 208
agallocha 208

Farsetia 247, 248, 259
aegyptiaca 259, **260**
hamiltonii 259, **260**
jacquemontii 259
ramosissima
var. macrocarpa 259
Faujasia 303
Felicia 303
Festuca 55

Fibraurea 167, 172
chloroleuca 172
Filago 303
Filices 34
Fimbristylis 26, 31
junciformis 31
Flacourtia 287, 288
cataphracta 288
flavescens 288
ramontchi 288
sepiaria 288, **289**

FLACOURTIACEAE 287

Flaveria 303
Floscopa 124, 128
scandens 128
Flueggea 183, 184, 209
leucopyrus 209
microcarpa 210
virosa 210

Frankenia 295
grandifolia 295
pulverulenta 295

FRANKENIACEAE 295

Fremontia 483

Fumaria
capreolata 245
densiflora 245
indica 245
media 245
officinalis 245
parviflora 245
schleicheri 245
spicata 245
vaillantii 245

FUMARIACEAE 244

Gahnia 26
Galeandra 100

Galeopsis 381, 384, 390
tetrahit 391
ochroleuca 390

Gamolepis 303
Garcinia 426, 428
- **atroviridis** 429
- conrauna 429
- corallina 429
- **cornea 429**
- **cowa 429**
- **dulcis** 428, 429, **430**
- elliotii 429
- gnetoides 429
- guineensis 429
- hanburyi 428
- harmandii 428
- **heterandra 429**
- **indica** 429, **430**
- kola 429
- lanessanii 428
- livingstonii 429
- **mangostana** 428, 429, **430**
- mannii 429
- **morella** 428, 429, **430**
- olivieri 428
- ovalifolia 429
- polyantha 429
- spicata 428
- venulosa 429
- vilersiana 428
- **xanthochymus 429**

Garuleum 303
Gastrodia 100
Gazania 303
Geigeria 303
Gelonium 183, 185, 210
- **multiflorum** 210

Gerbera 303, 305, 309, 345
- **piloselloides 345**

Geniosporum 381, 391
- **prostratum 391**

Glaucium 237
Glechoma 381
Glechon 381
Gleichenia 34, 35, 49
- **dichotoma 49**

Glochidion 183, 184
- **hohenackeri 211**
- **zeylanicum 211**

Glossocardia 303, 305, 307, **346**
- **bosvallia 346**
- *linearifolia* 346

Glossogyne 303, 305, 307, 346
- **pinnatifida 346**

Glycerina 55
Gnaphalium 303, 305, 306, 346
- decurrens 346
- **luteo-album 346**
- norvegicum 346
- obtusifolium 346
- ramosissimum 346
- supinum 346
- sylvaticum 346
- uliginosum 346
- undulatum 346

Gomphostemma 381, 385, 391
- **crinitum 391**
  - var griffithii 391

Gongrothamnus 303
Goniothalamus 160, 164
- **macrophyllus** 165

Goodyera 100
Gordonia 438, 439, 440
- **obtusa 440**

Gossypium 445, 446, 450
- **arboreum 450**
- **barbadense** 450, **451**
- **herbaceum** 450, **451**
- **hirsutum** 450, **453**
- perenne 450
- peruvianum 450
- racemosum 450

Govenia 100
GRAMINEAE 55
Grangea 303, 305, 306, 347
- **maderaspatana 347**

Grewia 473, 477
- **asiatica 478**
- **carpinifolia** 477, **478**
- **hirsuta** 477, 478, **479**
- **microcos** 478, **479**
- mollis 477
- occidentalis 477
- **paniculata** 477, 478, **479**
- **populifolia** 477, **479**
- **sclerophylla** 478, **480**
- **tenax** 478, **480**
- **tiliaefolia** 478, **480**
- **umbellata** 477, **481**
- **villosa** 477, 478, **481**

Grindelia 303
Guazuma 483, 486
tomentosa 486
Guizotia 303, 305, 307, 347
abyssinica 347
GUTTIFERAE 426
Gymnadenia 100
Gymnogramme 34
Gymnosperma 304
Gynandropsis 267, 275
gynandra 275
pentaphylla 275
Gynerium 55
Gynocardia 287, 289
odorata 289
Gynura 304, 305, 308, 347
amplexicaulis 348
auriculata 348
cernua 148
cerpidioides 348
pinnatifida 348
pseudo-china 348
sarmentosa 348
Gypsophila 129
Habenaria 100, 101, 105
commelinifolia 105
foliosa 105
Habzelia 160
Haplocarpha 304
Haronga 441
Hedeoma 381
Heleniastrum 304
Helenium 304
Helianthus 304, 305, 307, 348
angustifolius 348
annuus 348
petiolaris 348
tuberosus 348
Helichrysum 304
Helicteres 484, 486
brevispira 486
corylifolia 486
isora 486
ovata 486
sacarolha 486
vuarame 486
Heliophila 247
Helleborus 137
Helminthostachys 34, 36, 49
Hemidictyum 34, 49
ceterach 49
zeylanica 49
Hemizonia 304
Herderia 304
Hermannia 483
Herniaria 129
Hesperis 247
Hetaeria 100, 101, 105
obliqua 105
Heteropogon 55, 56, 57, 76
contortus 76
Hevea 183
Hexalobus 161
Hibiscus 445 446, 453
abelmoschus 453, **454**
aethopicus 453
asper 453
californicus 453
cannabinus 453, **454**
digitiformis 453
diversifolius 453, 454, **455**
esculentus 453, 454, **455**
furcatus 453, **456**
grewiaefolius 453
lampas 453, **456**
leiospermus 453
liliflorus 453
malacospermus 453
manihot 453, 454, **457**
micranthus 454, **457**
mutabilis 453, 454, **457**
phanarandrus 453
pusillus 453
rosa-sinensis 453, 454, **457**
rostellatus 453
sabdariffa 453, 454, **458**
surattensis 453, **459**
syriacus 453
tiliaceus 453, 454, **459**
trionum 453, **460**
vitifolius 453, 454, **460**
Hieracium 304, 305, 309, 349
amplexicaule 349

murorum 349
pilosella 349
virosum 349
Hifrochloa 55
Hippomane 183, 185, 211
mancinella 211
Holosteum 129, 130, 132
umbellatum 132
Homelomena 109, 110, 118
aromatica 118
ovata 118
rubescens 118
Homonoia 183, 185, 211
riparia 211
Hopea 422, 424
odorata 424
Hordeum 55, 56, 58, 77
vulgare 77
Hoslundia 381
Hugonia 377
mystax 377
planchoni 377
serrata 377
Hura 183, 212
crepitans 212
Hydnocarpus 287, 290
anthelmintica 290, **291**
polyandra 290
venenata 290
wightiana 290
Hydrastis 137
Hygrorhiza 55, 56, 58, 78
aristata 78
Hymenocardia 183
Hymenocrater 381, 383, 391
sessilifolius 391
Hypecoum 237, 239
pendulum 239, **240**
procumbens 239, **240**
HYPERICACEAE 441
Hypericum 441, 442
aethiopicum 442
androsaemum 442
angustifolium 442
ascyron 442
barbatum 442
brathys 442
chamaemyrtus 442
chinense 442
coris 442
elegans 442
erectum 442
hircinum 442
hirsutum 442
humboldtianum 442
humifusum 442
japonicum 442
lalandii 442
laxiusculum 442
maculatum 442
mexicanum 442
montanum 442
mutilum 442
patulum 442, **443**
perfoliatum 442
perforatum 442, **443**
platyphyllum 442
pulchrum 442
quadrangulum 442
richeri 442
sampsoni 442, **444**
tetrapterum 442
thesiifolium 442
Hyphaene 1, 2, 18
thebaica 18
Hypochoeris 309, 349
glabra 349
Hyptis 381, 383, 391
canescens 391
fasciculata 391
fruticosa 391
graveolens 391
pectinata 391, **392**
spicata 391
spicigera 391
suaveolens 391, **392**
umbrosa 391
verticillata 391
Hysterionica 304
Hyssopus 381, 383, 392
officinalis 392
Iberis 247
Iboza 381
Ichthyothere 183

Illicium 155
anisatum 155
floridanum 155
griffithii 155, **156**
parviflorum 155
verum 155
Imperata 55, 56, 79
arundinacea 79
cylindrica 79
Inula 304, 305, 306, 349
britannica 349
conyza 349
crithmoides 349
germanica 349
grantioides 349, **350**
graveolens 349, **350**
helenium 349
japonica 349
racemosa 349, **350**
royleana 349, **350**
Iodanthus 247
Ionidium 278, 279
brevicaule 279
circaeoides 279
enneaspermum 279
glutinosum 279
heterophyllum 279
ipecacuanha 279
oppositifolium 279
poaya 279
polygalaefolieum 279
strictum 279
suffruticosum 279
Isatis 247, 248, 249, 260
tinctoria 260
Isupyrum 137
Ixonanthes 377
cochinchinensis **377**
Jatropha 183, 184, 212
capensis 212
curcas 212
glandulifera 212, **215**
gossypifolia 212, **215**
hirsuta 212
macrorhiza 212
multifida 212, **215**
nana 212, **216**
zeyheri 212
Jatrorrhiza 167, 172
palmata 172
strigosa 172
Johannesia 183
Julocroton 183
Juncellus 26, 31
inundatus 31
Jurinea 304, 305, 308, 350
macrocephala 350
Kadsura 155, 156
japonica 156
scandens 156
Kielmeyera 439
Kleinhovia 483, 484
hospita 487
Knowltonia 137
Krameria 293
Kydia 445, 446, 461
calycina 461
Kyllinga 26, 31
brevifolia 31, **32**
monocephala 31
odorata 31
triceps 31, **32**
LABIATAE 381
Lactuca 304, 305, 309, 350
canadensis 351
capensis 351
fischeriana 351
indica 351
quercinata 351
runcinata 351
remotiflora 351
sagittata 351
scariola 351
taraxacifolia 351
thunbergii 351
virosa 351
Lagenandra 109, 118
ovata 118
toxicaria 118
Laggera 304, 305, 306, 352
alata 352
Lallemantia 381, 384, 393
royleana 393

Lamium 381, 384, 393
album 394
Lamprachaenium 304, 305, 352
monocephalum 352
Lasia 109, 110, 118
aculeata 118
heterophylla 118
Lasiocorys 381
Lasiopermum 304
Lastrea filix-mas 48
Launaea 304, 305, 309, 352
aspleniifolia 352
chondrilloides 352
glomerata 352
nudicaulis 352, **353**
pinnatifida 352, **353**
Lavandula 381, 383, 394
bipinnata 394
dentata 394
spica 394
stoechas 394
vera 394
Lavatera 445
Leonotis 381
africana 394
dysophylla 394
leonotis 394
leonurus 394
microphylla 394
nepetaefolia 394
Leontice 177
Leontonyx 304
Leonurus 381, 384, 395
cardiaca 395
macranthus 395
sibiricus 395, **396**
Lepidium 247, 248, 249, 260
campestre 260
capensis 260
crassifolium 260, **261**
graminifolium 260
iberis 260
draba 260, **261**
latifoliatum 260, **262**
medium 260
perfolium 260, **262**
ruderale 260, **262**
sativum 260
sohinzii 260
virginicum 260
Leptonychia 483
Leucas 381, 384, 396
aspera 396, **397**
cepholotes 396, **397**
decurvata 396
linifolia 396, **397**
martinicensis 396, **398**
stelligera 396, **398**
urticaefolia 396, **398**
zeylanica 396, **398**
Leyssera 304
Liatris 304
Licuala 1, 2, 18
spinosa 18
Limodorum 100
LINACEAE 377
Linum 377, 378
alpinum 378
angustifolium 378
austriacum 378
catharticum 378
flavum 378
gallicum 378
hirsutum 378
maritimum 378
narbonense 378
nodiflorum 378
perenne 378
strictum 378
suffruticosum 378
tenuifolium 378
usitatissimum 378, 379
viscosum 378
Liriodendron 155
Lissochilus 100
Lodoicea 1, 2, 18
seychellarum 18
Lolium 55, 56, 58, 79
temulentum 79
Lomaria 34
Lophanthus 381
Lophatherum 55, 56, 58, 80
gracile 80
Lophodium filix-mas 48
Loxococcus 1, 19
rupicola 19
Luehea 473

Luisia 100, 101, 105
tenuifolia 105
Lunaria 247
Lychnis 129, 132
alba 132,
chalcedonica 132
coronaria 132
dioica 132
flos-cuculi 132
fulgens 132
githago 132
grandiflora 132
senno 132
Lycopus 381, 383, 399
europaeus 399
sinuatus 399
virginicus 399
Lygodesmia 304
Lygodium 34, 36, 50
flexuosum 50
japonicum 50, **51**
pinnatifidum 50
Mabea 183
Macaranga 183, 185, 216
huraefolia
indica 216
peltata 216
Macroclinidium 304
Madia 304
Maerua 267, 277
angolensis 277
arenaria 277
denhardtiorum 277
Maesobotrya 183
Magnolia 155
MAGNOLIACEAE 155
Mahernia 483
Mahonia 177, 180
fascicularis 180
napaulensis 180
Malachra 445
Malacothrix 304
Mallotus 183, 185 216
apelta 217
cochinchinensis 317
furctianus 217
japonicus 217
oppositifolius 217
philippinensis 217
Malva 445, 462
alcea 462
althaeoides 462
crispa 462
hispanica 462
moschata 462
nicaeensis 462
parviflora 462
rotundifolia 462
sylvestris 462 **463**
tournefourtiana 462
verticillata 462, **464**
MALVACEAE
Malvaviscus 445
Malvastrum 445, 464
coccineum 464
coromandelianum 464
peruvianum 464
spicatum 464
Mammea 426
Manisuris 55, 56, 80
granularis 80
Manihot 183, 184, 218
loureirii 218
ultilissima 218
Manniophyton 183
Maprounea 183
Mareyo 183
Marrubium 381, 384, 399
alysson 399
vulgare 399
Marsilia 34
Marsypianthes 381
Matricaria 308, 353
chamomilla 353
discoidea 353
glabrata 353
globifera 353
lasiocarpa 353, **354**
multiflora 353
nigellaefolia 353
suaveolens 353
Matthiola 247, 248, 262
incana 262
tristis 262
varia 262

Mauritia 1
Maximiliana 1
Meconopsis 237, 240
aculeata 240
napaulensis 240
Melandrium 122
Melanthera 304
Melica 55
Melinis 55
Melissa 381, 383, 400
officinalis 400
parviflora 400
Melittis 381
Melochia 483, 488
borbonica 487
corchorifolia 487
pyramidata 487
Melodorum 160, 165
kingii 165
MENISPERMACEAE 166
Menispermum 167
Mentha 381, 383, 400
aquatica 400
arvensis 400
capensis 400
cardiaca 400
crispa 400
gentilis 400
longifolia 400
piperita 400, **401**, 403
pulegium 400
rotundifolia 400
saliva 400
sylvestris 400, **402**
viridis 400 **402**
Meraurialis 183
Meriandra 381, 384, 403
bengalensis 403
strobilifera 403
Mertensia 34
Mesua 426, 432
ferrea 432
Metalasia 304
Metroxylon 1, 2, 19
sagus 19
rumphii 19, **20**
Michelia 155
champaca 156
nilagirica 156, **157**
montana 156, **158**
Microdesmis 183, 184, 219
caseariaefolia 219
puberula 219, 482
zenkeri 219
Microglossa 304, 305, 306, 354
afzelii 354
volubilis 354
Micromeria 381, 383, 403
douglasii 403
capitellata 403
Microseris 304
Microstylis 100
Mikania 304, 305, 306, 355
cordifolia 355
officinalis 355
scandens 355
Miscanthidium 55
Mohria 34
Mollugo nudicaulis 221
Monadenium 100
lugardae 102
Monarda 381
Monardella 381
Monnina 293
Monodora 160
Monolopia 304
Monstera 109
Montanoa 304
Montrichardia 109
Moricandia 247, 248, 263
tortuosa 263
Moronobea 426
Moschosma 404
polystachium 404
Mundita 293
Myosurus 292
Myricaria 137
elegans 299
germanica 299
Myrodia 483
Nandina 177
Nannorhops 1, 2, 20
ritchieana 20
Nanthosoma 109
Napaea 445

Nasturtium 247, 248, 263
barbareaefolium 263
curvisiliqua 263
fontanum 263
indicum 263
montanum 262, **264**
palustre 263
sylvestre 263
Nastus 55
Necepsia 183
Nelumbium speciosum 233
Nelumbo
nucifera 233
Neoaypsis 1
Neottia 100
Nepeta 384, 404
cataria 404
ciliaris 404, **405**
elliptica 404, **405**
glechoma 404
glomerulosa 404, **405**
italica 404
japonica 404
ruderalis 404, **405**
tenuifolia 404
Nephrodium 34
filix-mas **48**
Nephrolepis 34
Nidorella 304
Nigella 137, 138, 150
arvensis 150
damascena 150
hispanica 150
orientalis 150
sativa 150
Nipa 1, 2, 20
fruticans 20
Nolletia 304
Nothoelaena 34
Notonia 304, 305, 308 355
grandiflora 355
NYMPHAEACEAE 232
Nymphaea 232, 233
alba 235
caerulea 235
lotus 255
odorata 235
polysepala 235
pubescens 235, **236**
rubra 235
stellata 235, **236**
tetragona 235
Oberonia 100, 106
anceps 106
Ochradenus 277
baccatus 277
Ochrocarpus 426, 433
africanus 433
harmandii 433
longifolius 433
pentapetalus 433
Ocimum 406
basilicum 406
canum 406, **407**
crispum 406
gratissimum 406, **409**
micranthum 406
sanctum 406, **409**, 410
viride 406
Oenocarpus 1
Oldfielda 183
Oligomeris 277
subulata 278
Olyra 55
Omphalea 183
Oncoba 287
Onoceris 304
Onoclea 34
Onopardon 304
Ophioglossum 34, 36, 51
vulgatum 51
Ophrys 100
Opthalmoblapton 183
Orania 1
ORCHIDACEAE 100
Orchis 100, 101, 106
bifolia 106
fusca 106
hircina 106
latifolia 106
laxiflora 106
maculata 106
mascula 106
militaris 106
morio 106
pyramidalis 106

sambucina 106
simica 106
Origanum 383, 410
dictamnus 410
hirtum 410
majorana 410
virens 410
vulgare 410, **411**
Orontium 109
Orophea 160, 165
hirsuta 165
setosa 165
Orthosiphon 412
stamineus 412
Oryza 55, 56, 58, 81
sativa 81
Oryzopsis 55
Osmites 304
Osmitopsis 304
Osmunda 34, 36, 51
claytoniana 51, **52**
regalis 51
Osteospermum 3C4
Othonna 304
Othonnopsis 304, 305, 308, 356
intermedia 356
Otostegia 384, 412
aucheri 412
limbata 412
Oxymitra 160
Oxytenanthera 55, 56, 59, 82
sinuata 82
Pachygone 167, 173
ovata 173
Paeonia 137, 138, 151
albiflora 151
brownii 151
emodi 151
corallina 151
moutan 151
officinalis 151
peregrina 151
tenuifolia
PALMAE 1
Pangium 287, 291
edule 291
Panicum 56, 57, 82
antidotale 83, **84**
crus-corvi 75
crus-galli 75
frumentaceum 74
miliaceum 83, 480
miliare 83
sarmentosum 88, **84**
Papaver 237, 240
alpinum 240
argemone 240, 241, **242**
dubium 240, **242**
hybridum 240, **242**
nudicaule 240, **244**
orientale 241, **244**
rhoeas 240, **241**
Somniferum 240, 241, **243**
PAPAVERACEAE
Paris 183
Parthenium 304
Paspalum 55, 56, 57, 84
conjugatum 84, **85**
sanguinale 84, **85**
scrobiculatum 84, **85**
Patisola 124
Pavonia 445, 464
bojeri 464
columella 464
diuretica 464
hirsuta 464
macrotis 464
odorata 464, **465**
propinqua 464, **465**
urens 464
zeylanica 464, **465**
Pectis 304
Pedilanthus 183
Pegolettia 304, 305, 306, 356
sanegalensis 356
Pellaea 34, 35, 52
calomelanos 52, 53
consobrina 52
hastata 62
involuta 52
Penianthus 167
Pennisetum 56, 57, 86
compressum 86
japonicum 86
spicatum 86
typhoideum 86

Pentadesma 426
Pentapetes 483, 484, 488
  phoenicea 488
Pentzia 304
Perezia 304
Pericampylus 167, 173,
  glaucus 173
Perilla 383, 412
  acimoides 412
Perowskia 384, 413
  abrotanoides 413
  atriplicifolia 413
Petalostigma 183
Petasites 304
Peucephyllum 304
Phalaris 56, 86
  arundinacea 86
Philodendron 109
Phoenix 1, 2, 21
  acaulis 23
  dactylifera 21, **22**
  paludosa 21
  pusilla 21
  sylvestris 21
Phragmites 56, 58, 87
  communis 87
  karka 87
  maxima 87
  roxburghii 81
Phyllanthus 183, 184, 219
  amarus 220
  casticum 220
  conami 219
  discoideus 220
  distichus 194
  elegans 219, **220**
  emblica 219, **220**
  epiphyllanthus 219, 220
  floribundus 220
  guyanensis 220
  ichthyomethius 219
  lathyroides 219
  madagascariensis 220
  maderaspatensis 220, **221**
  niruri 219, 220, **222**
  pentandrus 220
  phillyreaefolius 220
  reticulatus 219, 220, **223**
  salviaefolius 219
  simplex 220, **224**
  urinaria 219, 220, 222, **224**
  wildemannii 220
Phyllostachys 56, 58, 87
  bambusoides 87
Picridium 304
Picris 304, 305, 309, 356
  hieracioides 356
Pigueria 304
Pinanga 1, 24
  dicksonii 24
Pincellia 109
Pistia 109, 110, 119
  stratiotes 119
PITTOSPORACEAE 292
Pittosporum 292
  brachysepalum 292
  floribundum 292
  nepaulense 292
  senacia 292
  tobira 292
  viridiflorum 292
Platanthera 100
Platystoma 413
  africanum 413
Pleiotaxis 304
Pleopeltis 34, 36, 53
  lanceolata 53
Plesmonium 109, 110, 120
  margaritiferum 120
Pluchea 304, 305, 306, 356
  eupatorioides 356
  indica 356, **357**
  lanceolata 356, **357**
  ovalis 356
  quitoc 356
  pinnatifida 356, **357**
  polygonota 356
  pteropoda 356
Poa 56
Podophyllum 177, 181
  emodi 181
  peltatum 181
Pogonarthria 56
Pogostemon 383, 414
  heyneanus 414
  parviflorus 414
  patchouli 414
  plectranthoides 414
  purpurascens 414

Pollia 124
Pollichia 129
Polyalthia 160, 165
  longifolia 165
  oliveri 165
  simiarum 165, **166**
Polycarpaea 129, 130, 133,
  arenaria 133
  corymbosa 133
  linearifolia 133
Polycarpon 129, 130, 133
  indicum 133
Polygala 293
  amara 293
  angulata 293
  arenaria 293
  amatymbica 293
  boykini 293
  brachystachya 293, **294**
  butyracea 293
  calcarea 293
  chinensis 293, **294**
  crotalarioides 293, **294**
  elongata 293, **294**
  fimbriata 293
  glomerata 293, **294**
  guineensis 293
  hottentota 293
  macroptera 293
  major 293
  micrantha 293
  monspeliaca 293
  myrtifolia 293
  nicaeenis 293
  oppositifolia 293
  paniculata 293
  polygama 293
  reini 293
  rupestris 293
  sanguinea 293
  scoparia 293
  senega 293
  serpyllacea 293
  serpentaria 293
  sibirica 293, **294**
  telephoides 293, **294**
  tenuifolia 293
  timoutou 293
  vulgaris 293
POLYGALACEAE 293
Polymnia 304
Polypodium 34
  filix-mas 48
  nemorale 48
  quercifolium 47
Polystichum 34
  abbreviatum 48
  durum 48
  filix-mas 48
  induratum 48
Polytoca 86, 87
  barbata 88
Portulaca 296
  oleracea 296
  patens 296
  pilosa 296
  quadrifida 296, **298**
  sativa 296
  tuberosa 296, **299**
PORTULACACEAE 295
Pothos 109, 110, 120
  cathcartii 120
  scandens 120
    var. **cognatus 120**
Prenanthes 304
Printzia 304
Psorospermum 441
Psiadia 304
Pseudolachnostylis 183
Pteridium 34
Pteris 34, 35, 53
  aquilina 53
  buchanani 53
  leptophylla 53
  multifida 53
  palmata 53
  pedata 53
Pterocymbium 483, 488
  javanicum 488
Pteronia 304
Pterospermum 483, 484, 488
  acerifolium 489
  heyneanum 489
  suberifolium 489
Pterygota 413, 489
  alata 489

Pulicaria 304, 305, 306, 359
capensis 357
crispa 357
dysenterica 357, **358**
glaucescens 357, **358**
scabra 357
Pulsatilla 137
Putranjiva 183, 184, 225
roxburghii 225
Pycnocoma 183
Pycreus 26
RANUNCULACEAE 137
Ranunculus 137, 138, 152
abortivus 152
acris 152
arvensis 152, **153**
asiaticus 152
bulbosus 152
capensis 152
falcatus 152, **154**
Ficaria 152
flammula 152
glacialis 152
gramineus 152
hederaceus 152
langsdorfii 152
lingua 152
monspeliacus 152
multifidus 152
muricatus 152, **154**
parviflorus 154
pedatus 152
pensylvanicus 152, **153**
pubescens 152
repens 152
sceleratus 152, **153**
ternatus 152
thora 152
trichophyllus 152
trilobus 152
Raphanus 247, 248, 249, 264
sativus 264
Raphia 1, 2, 24
pedunculata 24
ruffia 24
vinifera 24
Rapistrum 247
Reaumuria 299, 300
hypericoides 300
Reinwardtia 377, 380
trigyna 380
Remirea 26, 32
maritima 32
Remusatia 109, 110, 120
vivipara 121
Reseda 277, 278
odorata 278
RESEDACEAE 277
Rhaphidophora 109, 110, 120, 121
pertusa 121
Rheedia 426
Rhynchostylis 100, 101, 106
retusa 106
Richardia 109
Richeria 183
Ricinodendron 183
Ricinus 183, 185, 225, 227
communis 225
Ritchiea 267
Rollinia 160
Romneya 237
Roripa 247
Roucheria 377, 380
griffithiana 380
Roylea 384, 415
elegans 415
Rudbeckia 304
Saccharum 56, 88
arundinaceum 88, **90**
ciliare 90
munja 90
officinarum 89
spontaneum 88
Saccolabium 100, 101, 106
papillosum 106
praemorsum 102
wightianum 102
Sageraea 160, **166**
laurifolia 166
Sagina 129
Saguerus rumphii 8
Sagus laeris 19
rumphii 20

Salvia 384, 415
- **aegyptiaca 415**
- aethiopis 415
- africana 415
- aurea 415
- **cabulica** 415, **416**
- canariensis 415
- coccinea 415
- gerardiana 415
- **glutinosa** 415, 416
- grahami 415
- hispanica 415
- horminum 415
- japonica 415
- **lanata** 415, **416**
- lanceolata 415
- leucodermis 415
- **macrosiphon** 415, **416**
- miltiorhiza 415
- **moorcroftiana** 415, **416**
- officinalis 415
- palaefolia 415
- paniculata 415
- **plebeia** 415, **416**
- plebeta 415
- polystachya 415
- pratensis 415
- repens 415
- rugosa 415
- runcinata 415
- scabra 415
- sclarea 415
- sisymbrifolia 415
- **spinosa** 415, **416**
- stenophylla 415
- triangularis 415
- triloba 415
- verbenaca 415
- verticillata 415
- viridis 415

Sanguinaria 237

Santolina 304

Sapium 183, 185, 229
- ellipticum 229
- grahamii 229
- **indicum 229**
- **insigne 229**
- **sebiferum 229**

Saponaria 129, 133
- ocymoides 133
- officinalis 133
- **vaccaria 133**

Sasa 56

Satureia 383, 417
- **hortensis 417**
- montana 417
- thymbra 417

Saurauja 438, 439, **440**
- **napaulensis 440**

Sauromatum 109, 121
- **guttatum 121**

Sauropus 183, 184, 229
- **quadrangularis 229**

Saussurea 304, 305, 308, 358
- **affinis 358**
- **candicans 358**
- **hypoleuca** 358, **359**
- **lappa** 358, **359**
- **obvallata** 358, **360**

Sauvagesia 278

Scaphium 483, 490
- **affine 490**
- **wallichii 490**

Schistostephium 304

Schima 439, 440
- **crenata 440**
- noronhae 440
- **wallichii** 440, **441**

Schizandra 155

Scimirea 26

Scindapsus 109, 110, 121
- **hederacea 121**
- **officinalis 121**
- *pertusus* 121

Scirpus 32
- **articulatus** 32, **33**
- cernuus 32
- **corymbosus** 32, **33**
- **grossus** 32, **33**
- **lacustris** 32, **33**
- **maritimus** 32, **33**
- paludicola 32

Scleria 26

Sclerocarpus 304

Scolopendrium 34

Scolopia 287
Scolymus 304
Scorzonera 304
Scutellaria 384, 417
baicalensis 417
canescens 417
cordifolia 417
galariculata 417
integrfioila 417
lateriflora 417
Sebastiana 183, 184, 229, 230
chamaelea 230
Secale 56
Securidaca 293
Senebiera 247
Senecio 304, 305, 308, 360
abyssinicus **361**
albanensis 360
var. leiophyllus 360
ambavilla **360**
asperulus 360
aureus 360
baterka 361
barbellatus 361
biafrae 361
brachypodus 360
bupleuroides 360
campestris 360
cervariaefolius 360
cineraria 360
concolor 360
deltoideus 360
densiflorus 360, **361**
doria 360
dregeanus 360
var. dicoideus 360
erechthitoides 360
erubescens 360
faugasioides 360
gerardi 360
grayanus 360
hildebrantii 360
incanus 360
jacobaea 360
jacquemontianus 360, **361**
kaempferi 360
lasiorhizus 360
macrocephalus 360
var. hirsutissimus 360
microdontus **360**
nikoensis 360
orbicularis **360**
othonnaeflorus 360
palnotus 360
quinquelobus 360, **361**
rhyncholaenus 361
scandens 360
serra 361
serratuloides **361**
speciosus **361**
subcoriaceus **361**
tamoides **361**
tanacetoides **361**
tenuifolius 360, **361**
viscosus 360
vulgaris 360, **361**
Serenoa 1
Serratula 304
Seseli 247
Setaria 56, 57, 91
italica 91
plicata 91
sulcata 91
viridis 92
Shorea 422, 424
cambodiana 424
guiso 424
harmandii 424
malaanonan 424
mangachapuy 424
robusta 424
stenoptera 424
tumbuggaia 425
Shybum 308
Sida 445, 465
acuta 465, **466**
cordifolia 465, 466, **467**
grewioides 466, **468**
linifolia 465
longipes 465
multiflora 465
obliqua 466
paniculata 465, 466
rhombifolia 465, 466, **468**
spinosa 466, **469**

veronicaefolia 466, **469**
viscosa 465
Siegesbeckia 304, 305, 307, 361
orientalis 361
Silene 129, 133
apetala 134, **135**
burchellii 134
capensis 134
cerastoides 134
conoidea 134, **135**
cucubalus 134
gallica 134, **135**
hirsuta 134
inaperta 134
italica 134
laciniata 134
macrosolen 134
nicaeensis 134
noctiflora 134
nocturna 134
nutans 134
otites 134
rupestris 134
saxifraga 134
stellata 134
virginica 134
viscosa 134
Silphium 304
Silybum 304, 305, 362
marianum 362
Simodorum giganteum 102
Sisymbrium 247, 248, 265
altissimum 265
austriacum 265
irio 265
loeselii 265
officinale 265
pinnatum 265
sophia 257
strictissimum 265
Sloanea 473
Solidago 304, 305, 306, 363
californica 363
canadensis 363
lepida 363
microglossa 363
odora 363
rigida 363
sempervirens 363
virgaurea 363
Sonchus 304, 305, 309, 363
arvensis 363
asper 363, **364**
crassifolius 363
maritimus 363, **364**
oleraceus 363, **364**
tenerrimus 363
Sorghum 56, 57, 92
guineense 92
halepense 92
vulgare 92, **93**
Spadix 10
Sparganophorus 304, 305, 365
vaillantii 365
Sparmannia 473
Spathe 10
Spergula 129, 130, 135
arvensis 135
Spergularia 129
Sphaeralcea 445
Sphaeranthus 305, 306, 365
africanus 365
indicus 365
senegalensis 365
Sphenocentrum 167
Spilanthes 305, 307, 366
acmella 366
var. **oleracea** 367
oleracea 367
Spirospermum 157
Spondianthus 183
Sporobolus 56
Stachys 384, 418
annua 418
arvensis 418
aspera 418
bogotensis 418
germanica 418
hirta 418
maritima 418
palustris 418
parviflora 418
recta 418
sieboldi 418
sylvatica 418
Stellaria 129, 135
aquatica 135, **136**

media 135
uliginosa 135, **136**
Stenocline 305
Stenoloma 34, 35, 54
chinensis 54
Stenotaphrum 56, 57, 93
glabrum 94
Stephania 167, 173
abyssinica 173
dinklagei 173
glabra 173, **174**
hernandifolia 173
Sterculia 483, 490
acerifolia 490
alata 489
balanghas 490, **491**
chicha 490
diversifolia 490
foetida 490, **491**
hypochroa 490
javanica 490
lychnophora 490
oblongata 490
platanifolia 490
rubiginosa 490, **491**
rupestris 490
striata 490
thcrelli 490
tomentosa 490
tragacantha 490
scaphigera 490
stipularis 490
urens 490, **491**
STERCULIACEAE 483
Stillinga 183
Stipa 56, 94
capillata 94
inebrians 94
sibirica 94
vaseyi 94
viridula 94
Stoebe 305
Stylochiton 109
Symphonia 426, 440
napaulensis 440
Symplocarpus 109
Synadenium 183
Synantherias 109, 110, **122**
sylvatica 122
Synedrella 305, 307, **368**
nodiflora 368
Tagetes 305, 307, 368
erecta 368
minuta 368
pusilla 368
sypaquinensis **368**
Talauma 155
Talinum 296
TAMARICACEAE **299**
Tamarix 299, 300
africana 300
aphylla 300
chinensis 300
dioica 300, **301**
gallica 300
pallasii 300
troupii 300, **301**
Tanacetum 305, 308, 368
fruticosum 368
gracile 368, **369**
vulgare 368
Taraktogenos 287, 292
kurzii 292
Taraxacum 305, 309, 369
officinale 369
Tarchonanthus 305
Tarrietia 483, 492
unifoliolata 492
Tectarea filix-mas 48
Ternstroemia 439, 441
japonica 441
TERNSTROEMIACEAE 438
Terra japonica 5
Tetracarpidium 183
Tetracera 158, 159
alnifolia 159
aspera 159
assa 159
indica 159
laevis 159
macrophylla 159
oblongata 159
ovalifolia 159
potatoria 159
Tetrorchidium 183
Teucrium 385, 419
botrys 419

capense 419
chamaedrys 419
fruticans 419
majorana 419
marum 419
montanum 419
polium 419
pyrenaicum 419
scordium 419
scorodonia 419
stoksianum 419
Thalictrum 137, 138, 154
angustifolium 154
aquilegifolium 154
flavum 154
foliolosum 154
minus 154
polycarpum 154
Thea sinensis 439
Thelepogon 56, 57, 94
elegans 94
Themeda 56, 57, 94
triandra 94
Theobroma 483, 484, 492
bicolor 492
cacao 492
glauca 492
martiana 492
macrantha 492
microcarpum 492
sylvestris 492
Thespesia 445, 446, 470
campilosiphon 470
populnea 470
Thlaspi 248. 249, 266
arvense 266
Thymus 383, 419
mastichina 419
serpyllum 419, **420**
vulgaris 419
Thysanocarpus 248
Thysanolaena 56, 57, 95
acarifera 95
agrostis 95
procera 95
Tilia 473
TILIACEAE 473
Tiliacora 167, 174
acuminata 174
Tinantia 124
Tinomiscium 167, 174
petiolare 174
Tinospora 167, 174
bakis 174
cordifolia 174, **175**, 410
crispa 174, **175**
malabarica 174, **175**
rumphii 174
Tissa 129
Toxicodendrum 183
Trachycarpus 1, 2, 25
excelsa 25
Tradescantia 124
Tragia 183, 185, 230
angustifolia 230
cordifolia 230
involucrata 230
volubilis 230
Tragopogon 305, 309, 370
porrifolium 370
pratense 370
Trewia 183, 185, 231
nudiflora 231
Tricholepis 305, 309, 371
angustifolia 371
glaberrima 371
Trigonostemon 183
Trilisa 305
Trimeria 287
Triplotaxis 305
Tripteris 305
Tritaxis 183
Triticum 56, 58, 95
aestivum 95
amyleum 95, **97**
durum 95, **97**
repens 59
sativum 95
spelta 95, **97**
Triumfetta 473, 481
lappula 481
rhomboidea 481
semitriloba 481, **482**
Trixis 305
Trollius 137

Tussilago 305, 308, 371
farfara 371

Typhonium 109, 122
trilobatum 122

Typhonodorum 109

Unona 160

Upaca 183

Urena 445, 471
lobata 471
repanda 471, **472**
sinuata 471, **472**

Ursinia 305

Uvaria 160, 166
afzelii 166
calamistrata 166
chamae 166
dulcis 166
narum 166

Vanda 100, 101, 106
roxburghii 107
spathulata 102, 106. **107**
tessellata 106, **107**

Vanilla 100, 101, 107
griffithii 102, **107**
planifolia 102, 107

Vateria 422, 425
indica 425

Venidium 305

Verbesina 305

Vernonia 305, 372
amygdalina 372, 373
anthelmintica 332, 450
appendiculata 372
biafrae 372
cinerea 372, 373
conferta 372
colorata 372
elaeagnifolia 373, 374
nigritiana 372
pectoralis 372
scariosa 372
senegalensis 372
teres 372, 373, 374
thomsoniana 372
roxburghii 373, 374

Vesicaria 248

Vetiveria 56, 57, 97
zizanoides 97

VIOLACEAE 278

Vismia 441

Viola 278, 279
alba 280
alpina 280
arenaria 280
biflora 280, **284**
calcarata 280
campestris 280
canadensis 280
canina 280
cenisia 280
collina 280
cinerea 280, **282**
cucullata 280
diffusa 280, **282**
hirta 280
japonica 280
lanceolata 280
maderensis 280
mirabilis 280
odorata 280
palmata 280
palustris 280
patrinii 280. **282**
pedata 280
pinnata 280
pubescens 280
scabriuscula 280
sciaphila 280
sepincola 280
serpens 280
sylvestris 280,**284**
tricolor 280, **283**
uliginosa 280
vaginata 280
verecunda 280

Volutarella 305, 309, 374
divaricata 374

Wallichia 1, 2, 25
disticha 25

Waltheria 483, 492
douradinha 492
indica 492

Wasabia 248
Wedelia 305, 307, 374
  calendulacea 374
  wallichii 374, **375**
Woodwardia 34
Wyethia 305
Xanthium 305, 307, 375
  canadense 375
  spinosum 375
  strumarium 375
Xeranthemum 305
Xylopia 160
Xylosma 287
Zanthorhiza 137
Zataria 383, 420
  multiflora 420
Zea 56, 98
  mays 98
Zeuxine 100, 101, 108
  stateumatica 108
  sulcata 108
Zizania 56
Ziziphora 384, 420
  clinopodioides 420, **421**
  tenuior 420, **421**